4th Asia Pacific Conference on Manufacturing Systems and 3rd International Manufacturing Engineering Conference (APCOMS-iMEC 2017)

IOP Conference Series: Materials Science and Engineering
Volume 319

Yogyakarta, Indonesia
7 – 8 December 2017

ISBN: 978-1-5108-6037-7
ISSN: 1757-8981

Printed from e-media with permission by:

Curran Associates, Inc.
57 Morehouse Lane
Red Hook, NY 12571

CURRAN ASSOCIATES INC. proceedings.com

Some format issues inherent in the e-media version may also appear in this print version.

Printed by Curran Associates, Inc. (2018)

For permission requests, please contact the Institute of Physics
at the address below.

Institute of Physics
Dirac House, Temple Back
Bristol BS1 6BE UK

Phone: 44 1 17 929 7481
Fax: 44 1 17 920 0979

techtracking@iop.org

Additional copies of this publication are available from:

Curran Associates, Inc.
57 Morehouse Lane
Red Hook, NY 12571 USA
Phone: 845-758-0400
Fax: 845-758-2633
Email: curran@proceedings.com
Web: www.proceedings.com

Table of contents

Volume 319

The 4th Asia Pacific Conference on Manufacturing Systems and the 3rd International Manufacturing Engineering Conference

7–8 December 2017, Yogyakarta, Indonesia

Accepted papers received: 16 February 2018
Published online: 21 March 2018

Preface

Papers

IOP Conf. Series: Materials Science and Engineering **319** (2018) 011001 doi:10.1088/1757-899X/319/1/011001

FOREWORDS

It is a great honor for Bandung Intitute of Technology (ITB) in collaboration with Universiti Malaya Pahang to host the second joint conference of The **The 4th Asia Pacific Conference on Manufacturing Systems** and **The 3rd International Manufacturing Engineering Conference** (APCOMS-iMEC 2017) at Yogyakarta. The special region that is very important to history of our country, Indonesia. It is also a great honor for ITB to work along with Sebelas Maret University in organizing APCOMS-iMEC 2017 locally in Yogyakarta.

We sincerely express our gratitude to Vice Chancellor of UMP; Dean of Sebelas Maret University; Professor Zahari Taha from the Fakulti Kejuruteraan Pembuatan Universiti Malaysia Pahang; The Head Secretary of The Creative Economy Agency of Indonesia (BEKRAF), Doctor Mesdin Simarmata; Professor Shih-Ming Wang, from Mechanical Engineering Department of Chung-Yuan Christian University, and; All parties, which have been very supportive to holding this conference in Indonesia and being part of our important milestone.

To date, we are witnessing another revolution of industrial development. The internet of things era has lead the change on consumer behavior, innovate production system technology, transform global manufacturing network. All of these may reshape our future. Our conference theme is about "**Taking The Factories to The Next Level**". All technologies which may bring the seamless integration (vertically-horizontally) within or between manufacturers are the hot topics to realize the concept of Industry 4.0. The current requirement analysis and future plan for a country within Asia Pacific will be needed accordingly.

On behalf of Bandung Institute of Technology, I encourage all participants of APCOMS-iMEC to work together and contribute to the future development of Manufacturing Technologies to strengthening and harmonizing as well the competitiveness of Asia Pacific countries and for our better society.

We wish all participants to have an excellent conference, expand our academic network, and enjoy the historical city of Yogyakarta.

Bandung, December 2017

The APCOMS-iMEC 2017 Organizing Committee
apcoms@itb.ac.id

IOP Conf. Series: Materials Science and Engineering **319** (2018) 011001 doi:10.1088/1757-899X/319/1/011001

COMMITTEE

Chair: Dr. Mohammad Mi`radj Isnaini

Co-Chairs:

Prof. Abdul Hakim Halim
Prof. Dradjad Irianto
Assoc. Professor Dr. Ahmad Razlan Yusoff
Dr. Muhammed Nafis Osman Zahid

Steering and Organizing Committee:

Dr. Anas Ma`ruf
Prof. Bermawi Priyatna Iskandar
Mr. Fariz Muharram Hasby
Dr. Iwan Inrawan Wiratmadja
Mr. Muhammad Akbar
Dr. Rachmawati Wangsaputra
Dr. Sukoyo
Dr. TMA. Ari Samadhi
Mr. Wildan Trusaji
Dr. Wisnu Aribowo
Dr Ahmad Shahrizan Abdul Ghani
Prof. Dato' Dr. Daing Nasir Ibrahim
Assoc. Prof. Dr. Ismed Iskandar
Dr. Mohd Azmir Mohd Azhari
Mr. Mohd Khairulnazri bin Saidi
Dr. Nafrizuan Mat Yahya
Prof. Ir. Dr. Wan Azhar Wan Yusoff
Dr. Zamzuri Hamedon
Mr. Zulkifli Md. Yusof
Ms. Zuryaty binti Zol

Local-Organizing Committee from Sebelas Maret University:

Dr. Cucuk Nur Rosyidi
Mr. Pringgo Widyo Laksono
Mr. Wakhid Ahmad Jauhari
Ms. Aditya Isnaini Setyargo P
Mr. Aldy Fajrianto
Ms. Anida Norma Cahyani
Ms. Endah Budiningsih
Ms. Fathin Kusumo Pramesti
Ms. Fitri Nur Muqodimah
Ms. Intan Dewi Melinda
Mr. Iqbal Wahyu Saputra
Mr. Kuncoro Sakti Pambudi
Ms. Kusumaningtyas Tika
Mr. Nurrudin Baidowi
Ms. Octavia Riskadayanti
Mr. Seamus Tadeo Marpaung
Ms. Shanella Nestri Hapsari
Ms. Stephani Diesya Floratiara
Ms. Viny Sartika
Ms. Yuka Sato

IOP Conf. Series: Materials Science and Engineering **319** (2018) 011001 doi:10.1088/1757-899X/319/1/011001

INTERNATIONAL SCIENTIFIC COMMITTEE

IOP Conf. Series: Materials Science and Engineering **319** (2018) 011002 doi:10.1088/1757-899X/319/1/011002

Peer review statement

All papers published in this volume of *IOP Conference Series: Materials Science and Engineering* have been peer reviewed through processes administered by the proceedings Editors. Reviews were conducted by expert referees to the professional and scientific standards expected of a proceedings journal published by IOP Publishing.

e-Learning Content Design for Corrective Maintenance of Toshiba BMC 80.5 based on Knowledge Conversion using SECI Method: A Case Study in Aerospace Company

Ayu Permata Shabrina[1], Rayinda Pramuditya Soesanto[1], Amelia Kurniawati[1,2], and Mochamad Teguh Kurniawan [1], Luciana Andrawina[1]
[1]Industrial Engineering Faculty, Telkom University, Bandung 40257, Indonesia
[2]Industrial Engineering Department, Bandung Institute of Technology, Bandung 40132, Indonesia
E-mail: ayupermatashabrina@gmail.com, raysoesanto@telkomuniversity.ac.id, amelia.kurniawati@gmail.com, teguhkurniawan@telkomuniversity.ac.id, luciana@telkomuniversity.ac.id

Abstract. Knowledge is a combination of experience, value, and information that is based on the intuition that allows an organization to evaluate and combine new information. In an organization, knowledge is not only attached to document but also in routine value creating activities, therefore knowledge is an important asset for the organization. X Corp is a company that focused on manufacturing aerospace components. In carrying out the production process, the company is supported by various machines, one of the machines is Toshiba BMC 80.5. The machine is used occasionally and therefore maintenance activity is needed, especially corrective maintenance. Corrective maintenance is done to make a breakdown machine back to work. Corrective maintenance is done by maintenance operator whose retirement year is close. The long term experience of the maintenance operator needs to be captured by the organization and shared across maintenance division. E-learning is one type of media that can support and assist knowledge sharing. This research purpose is to create the e-learning content for best practice of corrective maintenance activity for Toshiba BMC 80.5 by extracting the knowledge and experience from the operator based on knowledge conversion using SECI method. The knowledge source in this research is a maintenance supervisor and a senior maintenance engineer. From the evaluation of the e-learning content, it is known that the average test score of the respondents who use the e-learning increases from 77.5 to 87.5.

1. Introduction

Knowledge is a combination of experience, value, and information that is based on the intuition that allows an organization to evaluate and combine new information. In an organization, knowledge is not only attached to document but also in routine value creating activities, therefore knowledge is an important asset for the organization [1]. To gain benefit from the knowledge, organizations must explicitly manage their knowledge, the problem occurs when managing the knowledge because knowledge is difficult to imitate, intangible and also embedded in organization members and from past experience of the organization member [2]. An organization can obtain great profit if investing in knowledge asset, this can be done because knowledge is one of the un-depreciated assets, on the contrary knowledge, can be increased as long the knowledge is being used by another [3]. X company is a manufacturing company that focused on manufacturing part of aircraft. In order to maintain the productivity in optimum level, there is maintenance department which focused on maintain the machine in the company. Table 1 shows the machine that is used in Company X.

IOP Conf. Series: Materials Science and Engineering **319** (2018) 012001 doi:10.1088/1757-899X/319/1/012001

Table 1. Machine List in Company X

Machine Group	Machine Name	Quantity
Sheet Metal	Cyril bath	1
	Sheridan	1
	ABB	1
Gantry	Cincinnati	9
5 Axis	HAAS 5 Axis	1
	Toshiba	3
	Droop Rein	2
	Deckel Maho	4
4 Axis	6H	4
	HAAS EC 500	5
	YAASDA/JB	2
	Toshiba	4
3 Axis	SHW	4
	HAAS VF 6	8
	Quasser	5

Toshiba is one of the machine that is used in Company X. Toshiba BMC 80.5 is a machining center machine in the production process in Company X. The machine is used to make parts from roughing process until finishing process. The part that is manufactured in Toshiba BMC 80.5 is varied depending on the machine programs. Figure 1 shows the Toshiba BMC 80.5 Machine.

Figure 1. Toshiba BMC 80.5

All machine that is in the production process is considered important therefore maintenance process is needed. Maintenance department in company x is a division that specifically handles the maintenance operation of the machine for production. This division has 36 employees that consist of 25 permanent employees, 3 outsourcing, and 8 laborers. The activity of maintenance department is divided into two big group which is preventive maintenance and corrective maintenance. In general, corrective maintenance is not scheduled activity, because the purpose of the activity is to repair the machine right after the machine is breakdown/failure [4]. Corrective maintenance needs to be done in order to fulfill the production schedule, the problem occurs when only specific maintenance operator that can handle the specific machine, therefore when the operator is retired there is no one in the department that can replace the operator. The knowledge from such operator is within the mind of a person, therefore it is hard to articulate and convert into tangible form. According to Davenport and Prusak [5], knowledge is a combination of experience, values and contextual information that provides a framework for the evaluation of information and new experiences. According to Frost [6], knowledge is divided into two type, which is tacit knowledge and explicit knowledge. Tacit knowledge is a knowledge that is hard to communicate in form of written or oral and hard to transfer into people minds where explicit knowledge is a knowledge that is written, archive, and can be used as a reference for others.

To prevent the loss of knowledge in the department, it is needed knowledge management. Knowledge conversion with SECI (Socialization, Externalization, Combination, Internalization) method by Nonaka and Takeuchi [7] can be used as a solution to externalized the knowledge from the experienced operator. The long term experience of the maintenance operator needs to be captured by the organization and shared across maintenance division. E-learning is one type of media that can support and assist knowledge sharing. This research purpose is to create the e-learning content for best practice of corrective maintenance activity for Toshiba BMC 80.5 by extracting the knowledge and experience from the operator based on knowledge conversion using SECI method.

2. Theoretical Background

2.1. Knowledge

Knowledge is a main essential for any companies because, in the knowledge based era, the knowledge itself became the source of competitive advantage [8]. According to Drucker [9] knowledge is an information that changes something or someone.

2.2. SECI Model

SECI model first mentioned in the research conducted by Nonaka and Takeuchi [7]. The model describes four phase of converting knowledge which is socialization, externalization, combination, and internalization. Socialization is defined as sharing process and creation of tacit knowledge through interaction and direct observation. Externalization is defined as an articulation of tacit knowledge into explicit knowledge through dialogue and reflexion. The combination is defined as conversion process of explicit knowledge into new explicit knowledge through systemization and information. Internalization is defined as a learning process and knowledge acquisition by organization member towards explicit knowledge that spread into the organization through self-experience so it can be tacit knowledge of organization member. Figure 2 shows the SECI method by Nonaka and Takeuchi [7].

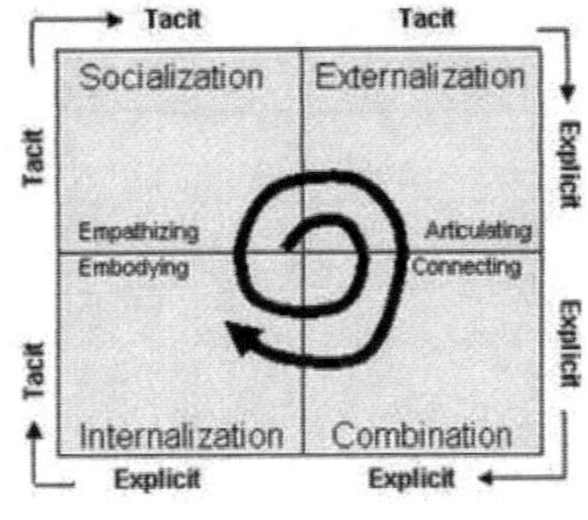

Figure 2. SECI Method

3. Research Methodologies

The purpose of this research is to create the e-learning content for best practice of corrective maintenance activity for Toshiba BMC 80.5 by extracting the knowledge and experience from the operator based on knowledge conversion using SECI method. The focus of this research is the corrective maintenance activity. SECI method is used in this research for managing knowledge that is held by senior employees regarding the experience of handling Toshiba BMC 80.5 machine. The SECI method converts the knowledge of a person's tacit knowledge into an explicit knowledge so that the knowledge can be used by others. The knowledge that is gathered from SECI method is converted into best practice for e-learning content. In socialization phase, tacit knowledge is transformed into another tacit knowledge, in this case, tacit knowledge will be converted into knowledge about corrective maintenance activity of Toshiba BMC 80.5. The data is gathered by interviewing to the maintenance operator. The corrective maintenance activity is repairing the bearing in the machine spindle. In externalization phase, the previous knowledge about the corrective maintenance activity is articulated into explicit knowledge. In combination phase, explicit knowledge is converted into new explicit knowledge through a combination of knowledge. The combination process is done by conducting a brainstorming session with two senior operators. The discussion focused on what is the best action about how to maintain the machine. In internalization phase, the best practice is shared into all maintenance department employee. The process is done by giving insight about new best practice through informal communication.

4. Result and Discussion

4.1. Socialization

In this phase, socialization process is done by interviewing the respondent. The result of this phase is a tacit knowledge about how to maintain the bearing in the machine spindle of Toshiba BMC 80.5. From the interview result, it is known that there is two alternative way to repair the bearing.

4.2. Externalization

In this phase, tacit knowledge from the previous phase is articulated into explicit knowledge. The explicit knowledge is the detailed diagram about how to repair the bearing. Figure 3 shows the externalized corrective maintenance procedure.

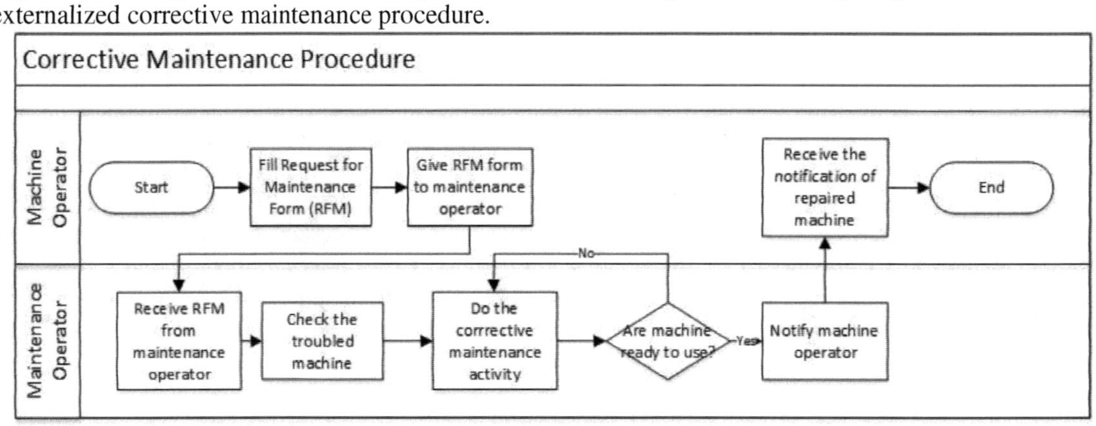

Figure 3. Externalized Corrective Maintenance Procedure

After the procedure is defined, the next step is to externalized the tools that are used to repair the machine. Table 2 shows the externalized result from corrective maintenance tools requirement.

Table 2. Externalized Corrective Maintenance Tools Requirement

Items	Quantity	Notes
Ring Key no. 8-24	1 Set	To open the cover and spindle mechanic
L Key	1 Set	To open the cover and spindle mechanic
Dial Indicator	1	To measure the run out of the spindle
Mandrel BT60	1	To help calibrate the machine
Screw	1	-
Majun Cloth	1	To clean parts of the machine
Solar fluid		To clean up the spindle
Grease		To lubricate the bearing in the spindle
Sand Stone	1	To clean rust in the spindle
Sand Paper CC 1000 CW	1	To finishing the rust cleaning process in the spindle
Crane	1	To lift the spindle from the machine
Paintbrush	1	To lubricate the solar
Pallet Jack	1	To handle the spindle

4.3. Combination

In combination phase, the externalized knowledge from the previous step is combined into one best practice, the externalized activity from the two operators will be merged into one best practice, this step is done by conducting a brainstorming session. Figure 4 shows the combined result of the corrective maintenance process.

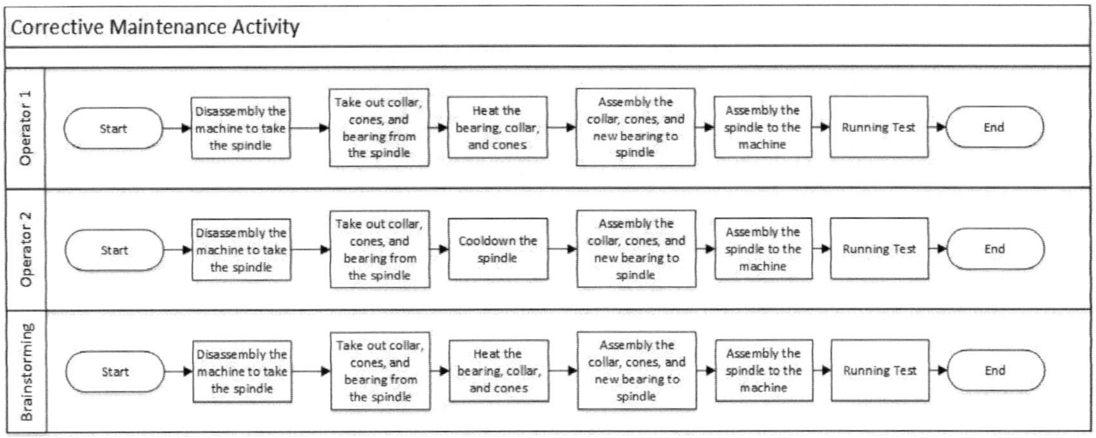

Figure 4. Combined Result of Corrective Maintenance Process

From figure 4, it is known that the best practice that is used after brainstorming session is from the operator 1. The differences between the two operators are that in the third activity. Operator 1 heats the collar, cones, and bearing first using the heater to put the parts into the spindle. Operator 2 cool down

IOP Conf. Series: Materials Science and Engineering **319** (2018) 012001 doi:10.1088/1757-899X/319/1/012001

the spindle using dry ice first before taking the bearing. From brainstorming session, it is known that the effective way to put back the parts is by heating the parts first. Using dry ice is not recommended because after the part is assembly, there is a chance that the residue of the dry ice creates rust in the machine.

After the information of the best practice is gathered, the next step is designing the content of e-learning about the corrective maintenance activity. The proposed e-learning focused on the topic of corrective maintenance activity, in this case, the corrective maintenance process of replacing the bearing in Toshiba BMC 80.5 machine. The content is designed for the initial phase of replacing the bearing until doing run test for the machine. Figure 5 shows the example of the e-learning content for replacing the bearing.

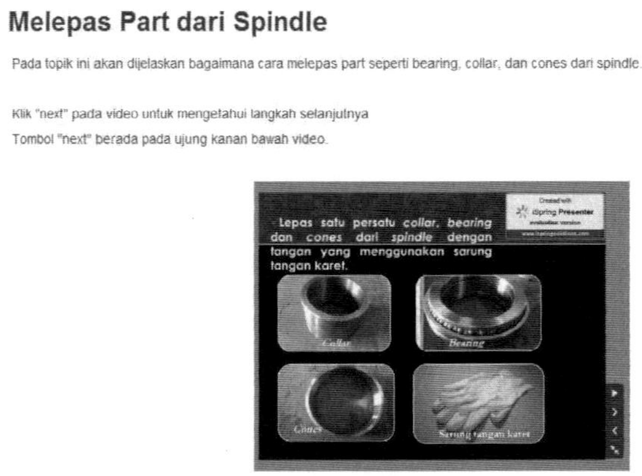

Figure 5. Example of Corrective Maintenance Content in e-learning

In the end of the section, there is a quiz to evaluate the result of the learning. The evaluation is done by answering the multiple choice question in the e-learning system. Figure 6 shows the example of the evaluation in e-learning.

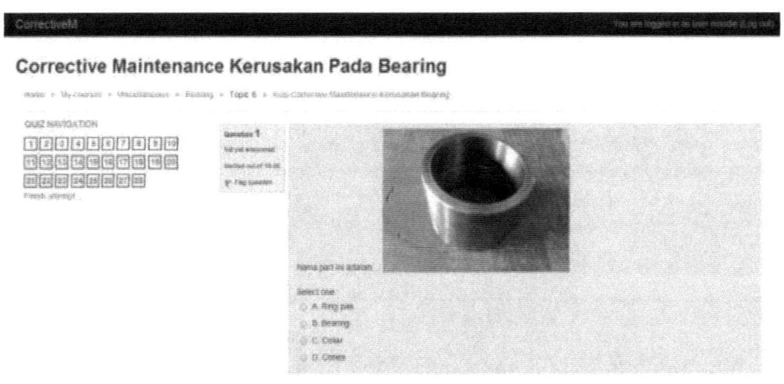

Figure 6. Example of Corrective Maintenance Evaluation in e-learning

4.4. Internalization

The best practice is shared into all maintenance department employee. The process is done by giving insight about new best practice through informal communication. The internalization phase is done to verbally communicate the detailed step about replacing the bearing from the spindle in Toshiba BMC 80.5 machine. In internalization process, the e-learning is tested, from the result, it is known that the maintenance operator understands the topics and the e-learning are considered as the effective way to understand the maintenance process because the operator just only sees the e-learning. The pre-preliminary test is conducted for the operator to know the initial score before using the e-learning. After taking the e-learning, it is known that the average test score of the respondents who use the e-learning increases from 77.5 to 87.5.

5. Conclusions

The purpose of this research is to create the e-learning content for best practice of corrective maintenance activity for Toshiba BMC 80.5 by extracting the knowledge and experience from the operator based on knowledge conversion using SECI method. The focus of this research is the corrective maintenance activity. SECI method is used in this research for managing knowledge that is held by senior employees regarding the experience of handling Toshiba BMC 80.5 machine. From the result, it is known that the average test score of the respondents who use the e-learning increases from 77.5 to 87.5. From the interview, the maintenance operator understand the topics and the e-learning is considered as the effective way to understand the maintenance process

References

[1] T. Davenport and L. Prusak, Working Knowledge, Cambridge: Harvard Business School Press, 1998.

[2] M. Zack, "Developing a knowledge strategy," *California Management Review,* pp. 125-145, 1999.

[3] A. Kurniawati, T. A. Samadhi, I. I. Wiratmadja and R. P. Soesanto, "The Impact of Source, Recipient, and Tacit Knowledge Characteristics on Tacit Knowledge Transfer Effectiveness," Taipei, 2016.

[4] M. Ben-Daya, U. Kumar and D. P. Murthy, Introduction to Maintenance Engineering: Modeling, Optimization, and Management, Wiley, 2016.

[5] T. H. Davenport and L. Prusak, Working Knowledge: How Organizations Manage What They Know, Boston: Harvard Bussiness School, 2000.

[6] A. Frost, "Knowledge Management Tools," 2013. [Online]. Available: http://www.knowledge-management-tools.net/knowledge-management-systems.html. [Accessed 22 November 2014].

[7] I. Nonaka and H. Takeuchi, The Knowledge-Creating Company, New York: Oxford University Press, 1995.

[8] R. P. Soesanto, L. Andrawina, I. R. Pertiwi and A. Kurniawati, "Relationship of Knowledge Management Process and the Performance from Human, Customer, and Organizational Perspective," in *KMICe*, Chiang Mai, 2016.

[9] P. F. Drucker, Harvard Business Review on Knowledge Management, Massachusetts: Harvard Business School Press, 1998.

IOP Conf. Series: Materials Science and Engineering **319** (2018) 012002 doi:10.1088/1757-899X/319/1/012002

Effect of laser parameters on surface roughness of laser modified tool steel after thermal cyclic loading

Annie Lau Sheng[1], Izwan Ismail[1, 2], Syarifah Nur Aqida[2, 3]

[1] Faculty of Manufacturing Engineering, University Malaysia Pahang, Malaysia
[2] Automotive Engineering Centre, University Malaysia Pahang, Malaysia
[3] Faculty of Mechanical Engineering, University Malaysia Pahang, Malaysia

izwanismail@ump.edu.my

Abstract. This study presents the effects of laser parameters on the surface roughness of laser modified tool steel after thermal cyclic loading. Pulse mode Nd:YAG laser was used to perform the laser surface modification process on AISI H13 tool steel samples. Samples were then treated with thermal cyclic loading experiments which involved alternate immersion in molten aluminium (800°C) and water (27°C) for 553 cycles. A full factorial design of experiment (DOE) was developed to perform the investigation. Factors for the DOE are the laser parameter namely overlap rate (η), pulse repetition frequency (f_{PRF}) and peak power (P_{peak}) while the response is the surface roughness after thermal cyclic loading. Results indicate the surface roughness of the laser modified surface after thermal cyclic loading is significantly affected by laser parameter settings.

1. Introduction

Die casting is a process that can produce geometrically complex parts with closer dimensional tolerances by using permanent metallic moulds. This process generally has a high production rate that produce as fast as 200 parts per hour and 300,000 parts per batch. Molten aluminium that temperature of 670-710°C is injected into the mould at velocities of 30-100 $m{\cdot}s^{-1}$ and injection pressure of 50-80 MPa [1]. Extreme working conditions in form of thermal cyclic loading lead to surface failure of the die casting mould. The mould usually affected by corrosion, soldering, erosion, oxidation and thermal fatigue which occur on the die surface [2]. To prolong the tool life, die repair process is performed on the failed surface. The repair process involves grinding, surface recovering by welding and machining [3]. This process is unreliable as chances of repeated failure is high. Alternatively, improving the surface properties by laser surface modification technique was proven to tackle this problem [4]. The surface is irradiated by high power laser beam that generate austenitizing temperature and rapidly self-quenched to produce hard surface of the tool steel. The self-quenching process was due to high-speed heating and cooling which is also known as thermal cyclic loading [4]. Surface roughness of the laser modified tool steel was tailored for specific purpose. However, the surface roughness might be altered when the modified surface slides with molten aluminium in cyclic manner. This paper presents the experimental study that try to explain the effect of laser parameters on surface roughness of laser modified tool steel after thermal cyclic loading.

IOP Conf. Series: Materials Science and Engineering **319** (2018) 012002 doi:10.1088/1757-899X/319/1/012002

2. Methodology

Properties of the AISI H13 tool steel used in this study are shown in Table 1. Test specimen of the AISI H13 tool steel was cut into desired dimension and shape by using milling machine. Figure 1 shows dimension of the samples with four side of test surface. Four samples were prepared to be lasered. The surface the test specimens were then treat with laser surface modification process by using Nd:YAG laser with pulse mode.

Table 1. Typical chemical compositions of the AISI H13 tool steel used in the experiment. [4]

Element	C	Mn	Si	Cr	Mo	V	P	S	Fe
wt%	0.36	0.38	0.96	4.82	1.19	0.86	0.017	0.004	Bal.

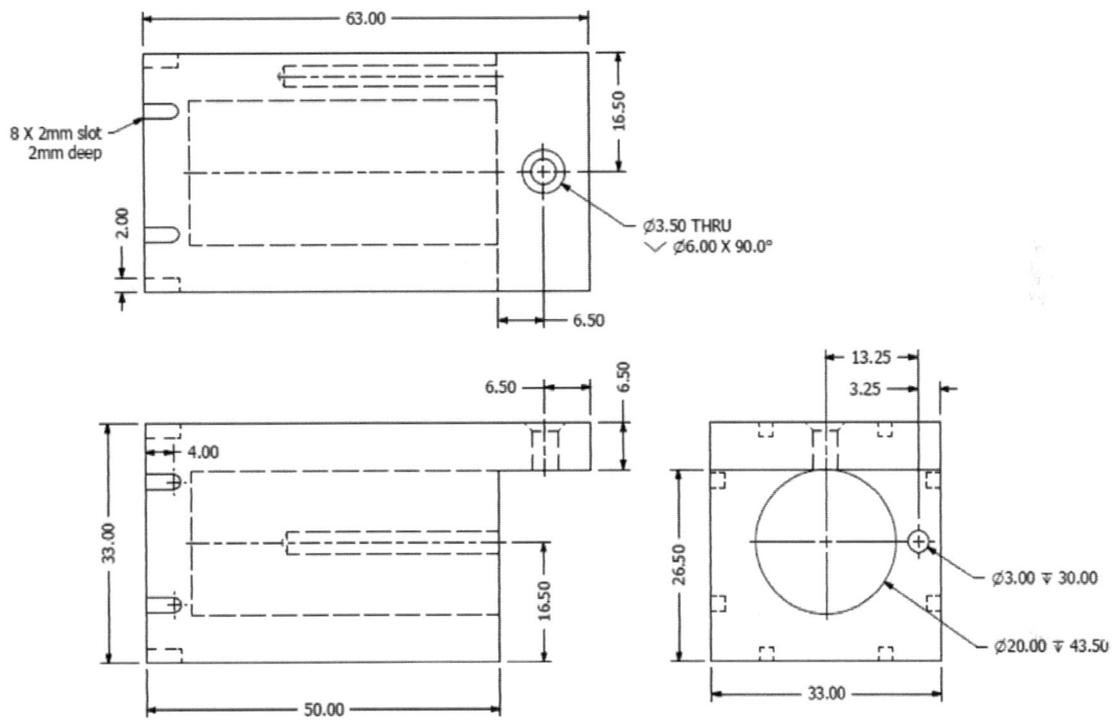

Figure 1. Dimension of the AISI H13 tool steel test specimen.

To produce small number of test specimen block, each surface of test specimen was lasered with two different laser parameters setting. Therefore, eight sample surfaces were prepared for each test specimen block. The laser parameter settings were designed using 3^3 full factorial design of experiment (DOE). The factorial design produced 28 different parameter setting. Three parameters of the DOE are (a) overlap rate, (b) pulse repetition frequency and (c) peak power while the response is the surface roughness after cyclic loading experiment. The specification of the laser processing is shown in Table 2. The overlap rate is set between 30% to 70%. The PRF is set between 50 Hz and 70 Hz. Meanwhile, the peak power is set between 1700 W and 2500 W.

Table 2. Summary of specification of laser surface modification process.

Laser specification	Value	
	Low	High
Peak power	1700 W	2500 W
PRF	50 Hz	70 Hz
Overlap rate	30 %	70 %
Pulse width	0.6-1.2 ms	
Energy	1.4-2.0 J	
Spot size	0.7 mm	

The lasered samples were treated with thermal cyclic loading by conducting alternate immersion of test specimens in molten aluminium and water bath. The cyclic immersion test was used to imitate the process of aluminium die casting [1]. Temperature of the molten aluminium and the water bath was controlled at 800°C and 27°C respectively. Cycle time was set constant at 31 s with 11.2 immersion time in both molten aluminium and water. The average surface roughness (RA) was measured using contact surface profilometer. The evaluation length and measurement speed were set to 4.0 mm and 0.6 mm/sec respectively.

3. Results and Discussion

Design of experiment analysis indicates variation in the parameter effect on the surface roughness of the lasered surface after thermal cyclic loading. Figure 2 (a) and (b) depicted the effect of peak power and PRF on the surface roughness at 30% and 70% beam overlap percentage respectively. It can be seen that, the surface roughness relationship with peak power and overlap are inversed at low and high PRF. At low PRF (50 Hz), the surface roughness after thermal cyclic loading is high at combination of high peak power and large overlap. This finding was opposite at high PRF (70 Hz) where the surface roughness is increase with decreasing of both PRF and overlap. The inverse trend at low and high PRF indicates contrast contribution of the factor. Higher PRF will increase surface energy density. This behaviour allows surface hardening to be occurred ad low peak power and overlap.

Figure 2 (c) and (d) show the effect of overlap and PRF on surface roughness at peak power of 1700 and 2500 W respectively. Similar inverse trend also depicted here. At low peak power, the surface roughness is increase with increasing PRF and decreasing of overlap. While at high peak power the relationship is inversed. It was notice that, there are no significant changes of Ra at any value of overlap when the PRF and peak power at lowest magnitude. This behaviour was due to domination of peak power and PRF as contributed factor to material properties of laser modified sample. By providing low frequency and low peak power, the energy intensity bombarded on the sample surface is reduced. Low energy intensity makes the surface temperature reduced which the rapid cooling process from austenitic temperature is not happened. The consequence of this behaviour is the hard surface is difficult to achieve and erosion due to cyclic thermal cyclic loading is prone to be happened.

Effects of peak power and PRF on surface roughness are shown in Figure 2 (e) and (f). When overlap is at 50%, combination of low peak power and low PRF failed to show significant changes of Ra. While at larger overlap (70%) the behaviour changed. The combination of low peak power and high PRF is failed to show significant surface roughness.

Analysis of variant (ANOVA) of the data shows the model of two factor interaction (2FI) is significant with the model F-value of 2.69. The chance of noise to influence the response is 4.24% which is relatively low. The terms are significant with prob>F is less than 0.05. The result of ANOVA test also indicates acceptable value of lack of fit 3.46 which is not significant, there for the model is fit.

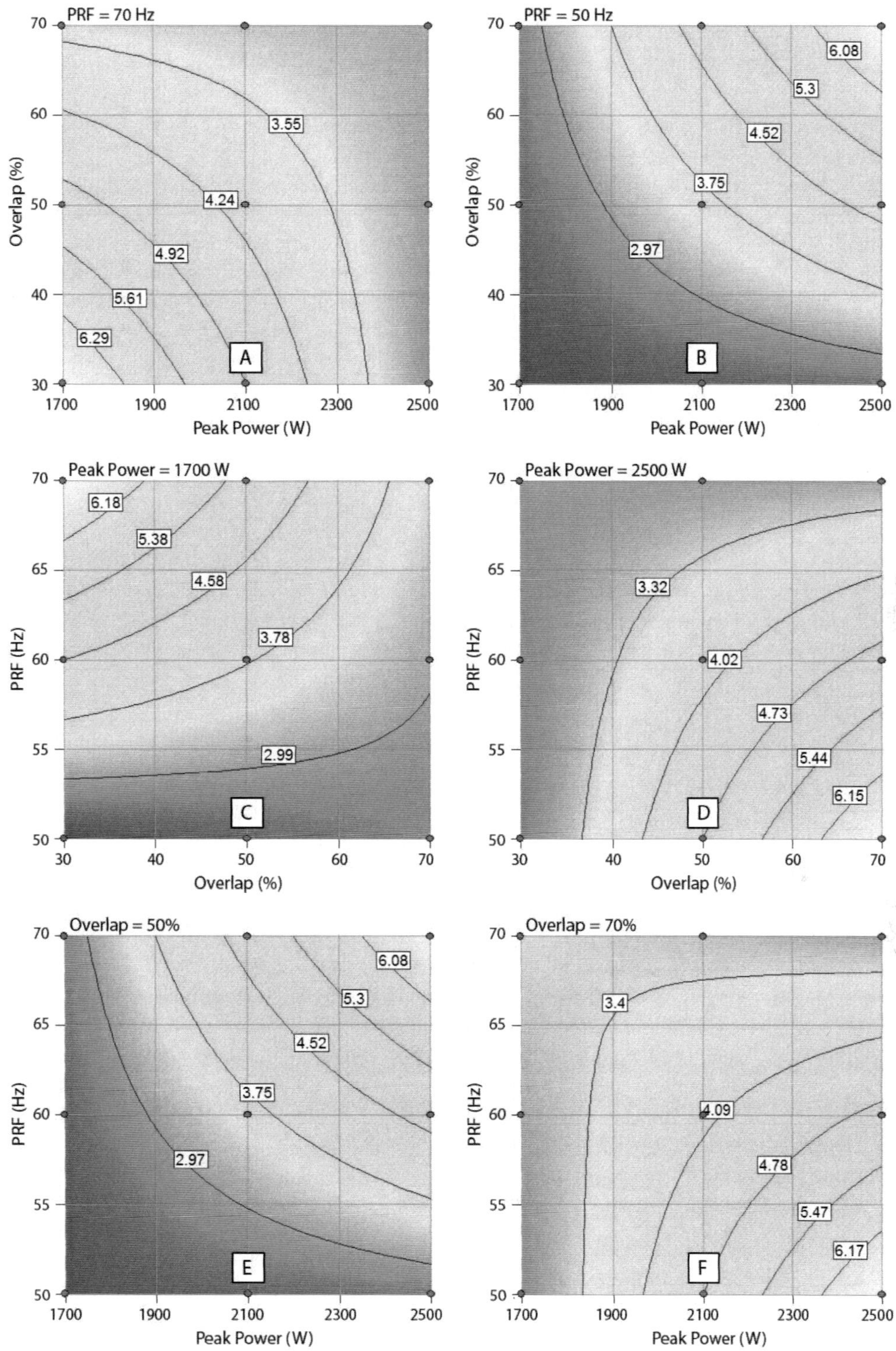

Figure 2. The effects of laser parameters on surface roughness (Ra) after thermal fatigue loading.

Final model equation produced by statistical analysis is shown in equation 1.

$$Ra_{tl} = -36.86 + 0.01P_{peak} + 0.87f_{PRF} + 0.07\eta - 2.83P_{peak}f_{PRF} + 1.17P_{peak}\eta - 5.13f_{PRF}\eta \quad (1)$$

Where, Ra_{tl} is the surface roughness after thermal cyclic loading, P_{peak} is the Peak Power (W), f_{PRF} is the Pulse Repetition Frequency, PRF(Hz), and η is Overlap (%)

The average surface roughness of test specimen increase after thermal fatigue test which is from 3.238 μm increase to 3.922 μm. This increasing of surface roughness occurs on laser modified surface due to entrap of oxides and carbides which possibly initiates the thermal fatigue failure [5].

The surface roughness of laser modified AISI H13 tool steel depends on the overlaps efficiency [6]. The overlap rate is the most important factor to affect surface roughness before thermal loading process. However, the surface roughness after thermal loading has different relationship with the laser parameter due to produced thermal wear resistance properties. Sliding action between molten aluminium and laser modified surface at high temperature induced the changes of surface roughness. The wear mechanism involved during the thermal loading experiments is a friction wear at high temperature. Oxide accumulated due to oxidation process on the surface of test specimen led to increase of surface roughness.

4. Conclusion

Surface roughness of laser modified AISI H13 tool steel surface was significantly influenced by laser parameters. Relationship between the any two parameters and surface roughness was found to be inverse when the third parameters at the lower and the higher value. Finding of this study leads to further investigation on the wear mechanism of laser modified sample in thermal cyclic environment.

Acknowledgments

The authors are grateful to Universiti Malaysia Pahang (UMP) for financial support given under RDU160318.

References

[1] Abdulhadi H A, Aqida S N, Ishak M and Mohammed G R 2016 Thermal Fatigue of Die-Casting Dies: An Overview *MATEC Web Conf.* 74 00032

[2] Klobčar D, Tušek J and Taljat B 2008 Thermal fatigue of materials for die-casting tooling *Materials Science and Engineering: A* 472 198-207

[3] Norhafzan B, Aqida S N, Chikarakara E and Brabazon D 2016 Surface modification of AISI H13 tool steel by laser cladding with NiTi powder *Applied Physics A* 122 384

[4] Abdulhadi H, Ahmad S, Ismail I, Ishak M and Mohammed G 2017 Experimental Investigation of Thermal Fatigue Die Casting Dies by Using Response Surface Modelling *Metals* 7 191

[5] Qin Q C, G.X. 2014 Effects of Parameters on Surface Roughness of Metal Parts by Selective Laser Melting *Advanced Materials Research* 834-836 872-5

[6] Fauzun F, Aqida S N, Naher S, Brabazon D, Calosso F and Rosso M 2014 Effects of Thermal Fatigue on Laser Modified H13 Die Steel *Journal of Mechanical Engineering and Sciences* 6 975-80

IOP Conf. Series: Materials Science and Engineering **319** (2018) 012003 doi:10.1088/1757-899X/319/1/012003

Game-Theoretic Models for Usage-based Maintenance Contract

H Husniah[1], R Wangsaputra[2], A Cakravastia[2] and B P Iskandar[2]

[1] Industrial Engineering, Langlangbuana University
Karapitan 116, Bandung 40261,Indonesia
[2] Department of Industrial Engineering, Bandung Institute of Technology
Ganesha 10, Bandung, 40132, Indonesia

Email: hennie.husniah@gmail.com, bermawi@mail.ti.itb.ac.id

Abstract. A usage-based maintenance contracts with coordination and non coordination between two parties is studied in this paper. The contract is applied to a dump truck operated in a mining industry. The situation under study is that an agent offers service contract to the owner of the truck after warranty ends. This contract has only a time limit but no usage limit. If the total usage per period exceeds the maximum usage allowed in the contract, then the owner will be charged an additional cost. In general, the agent (Original Equipment Manufacturer/OEM) provides a full coverage of maintenance, which includes PM and CM under the lease contract. The decision problem for the owner is to select the best option offered that fits to its requirement, and the decision problem for the agent is to find the optimal maintenance efforts for a given price of the service option offered. We first find the optimal decisions using coordination scheme and then with non coordination scheme for both parties.

1. Introduction

In an open pit mining, heavy equipments are used for loading and hauling mining material and these equipments play an important role to support business processes in a mining company. As the equipments deteriorate with age and usage, an effective maintenance is needed to keep the equipments in a good condition which in turn gives a high availability of the equipments. In general, maintenance is not considered as the core competencies of a mining company, and hence it is outsourced to an external agent. Furthermore, the technology of such equipments tends to be complex and having maintenance facilities and maintenance specialists for conducting maintenance activities of such complex equipments in house is far from economical. In many mining companies, maintenance activities (including preventive maintenance (PM) and corrective maintenance (CM)) are outsourced to an OEM (original equipment manufacturer) or an external agent. OEM or an agent can do a partial or full coverage of the maintenance actions (Preventive Maintenance (PM) or/and Corrective maintenance (CM)). Preventive maintenance (PM) actions are done to prevent the excessive degradation (whether it is an age based or condition based maintenance), while corrective maintenance (CM) is performed to restore the failed equipment to the operational state.

From the mining company (as the owner of the equipment) perspective, the decision problem facing is to choose the best option that results in high availability of the equipments. As a result, it needs to control closely the equipment availability as it influences very much the production and revenue of the company. On the other hand, the OEM's (agent's) decision problem is to determine the price of each option offered that maximises its profit.

In this paper, we propose a new maintenance service contract (MSC) characterized by L representing a time period of contract and U_{max}, representing the maximum usage allowed (See Fig. 1). Here, the contract will not terminate when the usage at time t (<L) is greater than U_{max}, but the owner will be charged some additional cost. The contract will terminate only due to the time limit. This contract can be viewed as the extension of the two dimensional service contract by [1] and [2] where the usage limit is no longer acting as the limit of contract but it is only a maximum usage allowed.

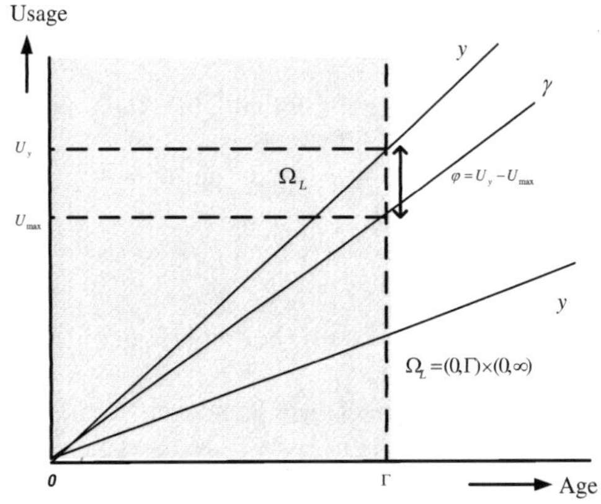

Figure 1. Key elements of maintenance service contract considered

In addition, we study two MSC models based on cooperative and non-cooperative games, where the cooperative case is similar to that studied by [3] but we consider a different maintenance policy (i.e. imperfect PM policy) which fits with the case of dump trucks used in a mining site.

For the cooperative game, it requires two major criteria. First, cooperation should lead to a win-win situation for the owner and OEM, such that the profits of both players become higher compared to independent individual profit [3]. Second, the players should have no incentive to deviate from the non-cooperative solution, i.e., they should modify their profits such that the total maximum solution becomes identical to a Nash equilibrium (for the simultaneous move case) [4].We have allocated the maximum total profit between the players to satisfy the first criterion. In this game, decision makers negotiate over different contract terms .

Next, for the non-cooperative game, we considered: non-cooperative simultaneous move game where the owner and OEM choose their strategies simultaneously. We derive the Nash equilibrium. It is of interest to compare the performance of the non-cooperative contracts and the cooperative alternative. One measure of performance is the difference between the total profit of a non-cooperative contract and that of a cooperative one (which has the maximum total profit).

The remainder of this paper is organised as follows. In section 2 we give model formulation for the service contract studied. The cooperative and the non-cooperative methodologies are described in Section 3. Section 4 analyzes the game solutions, the Nash equilibrium is determined and compares it results with the cooperative solution. Section 5 presents numerical examples to illustrate the MSC models and to investigate their performance. Finally, Section 6 concludes the paper.

2. Model Formulation

The following notations will be used in model formulation.

W, U	: Warranty time and usage limits
X_i	: Downtime caused by the i- th failure and waiting time
ζ	: Total repair time allowed

Υ, L	: Revenue , maintenance contract length
Y	: Usage rate
C_r	: Repair cost done by OEM
C_o	: Preventive maintenance cost per PM
C_v	: Additional cost for level PM created per PM
$\phi_y(O.)$	: Owner profit
$\pi_y(O.)$	: OEM profit
C_b	: The product cost over the contract period
$F(t,\alpha_y)$	: Conditional failure distribution for a given usage rate y
$r_y(t)$, $R_y(t)$	: Hazard and Cumulative hazard functions associated with $r(t)$ and $F(t,\alpha_y)$

2.1. Warranty Policy

In this paper we assume that each dump truck purchased is covered by a two-dimensional warranty. We also assume that to provide more protection to the buyer, the warranty covers PM. Further, every CM is covered by the warranty and rectified at no cost to the buyer. This means, after the warranty ends, the responsibility of the CM and PM actions shifts to the buyer (the owner). To express the warranty coverage we use a rectangle region $\Omega_W = [0,W)\times[0,U)$ where W and U are the time, and the usage limits.For a given usage rate (y) of a dump truck, the warranty ceases at $W_y = W$ for $y \le U/W$, or $W_y = U/y$, for $y > U/W$ (See Fig.2).

2.2. Maintenance Service Contract (MSC)

The OEM proposes a two dimensional MSC for a period of L (e.g. L =2 year) with a fixed price P_G,. Here, the MSC has no usage limit or in other words, the contract coverage forms a region $\Omega_S = \left[W_y, L\right)\times\left[yW_y, \infty\right)$ (See Fig. 2). The contract starts at the end of warranty, W_y. However, since the performance of the equipment degrades with age and usage and often the usage contributes more impact to the deterioration of the equipment, then the owner will be charged an additional cost when the usage rate goes beyond the nominal value. This is considered as a compensation to the OEM due to the increase in repair cost.

In general, the OEM provides a full coverage of maintenance, which includes PM and CM, for each MSC dump truck. In other words, all failures under MSC contract are fixed at no cost to the owner and all PM actions are performed without any charge to the owner. It is also stated that the OEM as a service contract provider assures a minimum down time (repair time and waiting time) for each failure and penalty cost incurred when the down time exceeds the predetermined target.

<u>The MSC is defined as follows.</u>

For a price of P_G , the OEM proposes a two dimensional MSC for a period of L. When the total usage (U_y) at L is greater than U_{max}, an additional cost is charged to the owner. The amount of the additional cost is proportional to $\Delta = U_y - U_{\max}$ given by $C_{na}(U_y - U_{\max})$. The additional cost is viewed as a compensation for the OEM as the total usage > U_{max}. And a penalty cost is born to the OEM when the down timeexcess the down time target.Under this option, the OEM agrees to carry out PM and CM in $(W_y, W_y + L]$ or the maintenance service is full coverage.

We study the MSC proposed from the view points of the OEM and the owner. The objective of the OEM is to minimize the expected maintenance cost according to various usage pattern and the mining operational condition whilst the objective of owner is to maximize the expected profit.

2.2.1. Equipment Failures and Repairs. Let Y be the constant usage rate for a given truck and varies across the trucks. For Y = y, the conditional hazard function for the time to first failure is given by $r_y(t)$ which is a non-decreasing function of t (the age of the truck) and *y*. Here we consider that the degradation of the truck is strongly affected by usage rate of the truck and a land contour of a mining area where the truck is operated. An AFT (accelerated failure time) model is proposed to model the effect of usage rate and the operating condition on degradation of the truck.

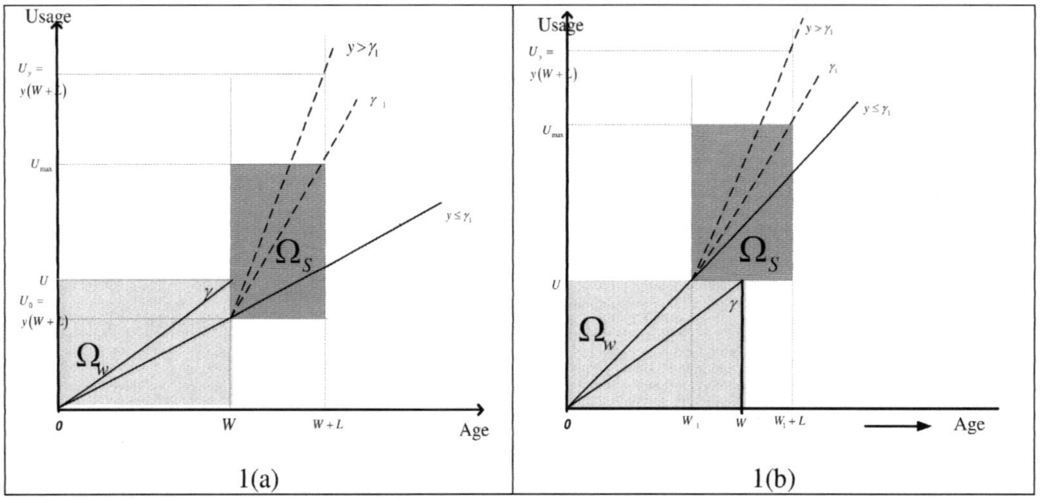

Figure 2. Warranty region Ω_W and *service* contract region Ω_S for (a) $y \le \gamma$ and (b) $y > \gamma$

In AFT model the distribution function for T_y is given by $F(t, \alpha_y)$, with a scale parameter given by $\alpha_y = (y_0/y)^{\rho}\alpha_0$ with $\rho \ge 1$ where ρ is a parameter for the operating condition of a truck. The hazard and the cumulative hazard functions associated with $F(t, \alpha_y)$ are given by $r_y(t) = f(t,\alpha_y)/(1-F(t,\alpha_y))$ and $R_y(t) = \int_0^t r_y(x)dx$ respectively where $f(t,\alpha_y)$ is the associated density function.

2.2.2. Preventive Maintenance Policy. The PM policy is defined as follows. For a truck with usage rate y, the PM policy is characterised by single parameter $\tau_y[\nu_y]$ during Ω_W [Ω_S]. Conditional on $Y = y$, the equipment is periodically maintained at $k\tau_y$, $k = 1,2,\ldots$. This involves k disjoint intervals -$[0,\tau_y)$, …, $[l\nu_y, \Gamma_0)$ in which all failures within PM period are minimally repaired, where k is an integer value. Note $(k+1)\tau_y = W$ $[(\ell+1)\nu_y = W]$ where $k[\ell]$ is an integer value. The effect of PM actions are modelled through the reduction in the intensity function after PM at t_j, $j \ge 1$ is δ. Since any failure occurring between PM is minimally repaired and $\delta_j^y = \delta$, then the expected total number of minimal repairs in $([t_{j-1}, t_j), 1 \le j \le k+1)$ is given by $N = \sum_{j=1}^{k+1} \int_{t_{j-1}}^{t_j} r_{j-1}(t')dt'$.

2.2.3. Modelling Cost. The OEM's expected total cost consists of PM cost, repair cost, and penalty cost (incurred when the down time exceeds the predetermined target). If $J^1(k_y,\delta_y)$, $J^2(k_y,\delta_y)$ and $J^3(k_y,\delta_y)$ are the expected total PM cost, the expected total repair costs and the expected penalty cost over the MSC period $(W_y, W_y + L]$ for a given usage rate y, respectively, then the expected total cost for the OEM, $\Pi^n[k_y,\delta_y]$ is given by

$$\Pi^n[k,\delta_y] = J^1(k,\delta_y) + J^2(k,\delta_y) + J^3(k,\delta_y) \quad (1)$$

Let $C_{pm}(\delta_y)$ and C_r be the cost of the j-*th* PM and the cost of each minimal repair. If $C_{pm}(\delta_y) = C_0 + C_v\delta_y$ as in [5] and [6] then the expected total PM cost over the MSC period $(W_y, W_y + L]$ is given by

$$J^1(k_y,\delta_y)=\sum_{j=1}^{k}C_{pm}(\delta_y)=kC_0+C_v\sum_{j=1}^{k}\delta_y \tag{2}$$

And the expected total minimal repair cost is given by

$$J^2(k,\tau_y,\delta_y)=C_r\left(R_y(L)-\sum_{j=1}^{k}\left(L-j\tau_y\right)\delta_y\right) \tag{3}$$

Let D and $\tilde{D}$ be the down time (consisting repair time and waiting time) for each failure and the down time allowed, respectively.The OEM incurs a penalty cost when $D>\tilde{D}$, and the penalty cost is assumed to be proportional to the excess of down time, $(D-\tilde{D})$. Then, the expected penalty cost is given by $J^3(k,\tau_y,\delta_y)=C_p\bar{G}(S)\left(R_y(L)-\sum_{j=1}^{k}\left(L-j\tau_y\right)\delta_y\right)$ where C_p is the penalty cost per unit time, $\bar{G}(\tilde{D})=\left\{\int_{\tilde{D}}^{\infty}\left(z-\tilde{D}\right)dF(z)\right\}$ is the expected value of penalty, and Z_i (downtime caused by the i-*th* failure) are assumed *i.i.d* with distribution function $F(z)$. After simplification we have the expected total cost of the OEM given by

$$\Pi^n\left[k,\tau_y,\delta_y\right]=\tilde{C}R_y(L)-\left\{\tilde{C}\sum_{j=1}^{k}\left[\left(L-C_v/\tilde{C}\right)-j\tau_y\right]\delta_y-kC_0\right\} \tag{4}$$

where $\tilde{C}=\left(C_r+C_p\bar{G}(\tilde{D})\right)$.

2.2.4. Modelling Profit. We assume that OEM and the owner have the same attitudes to risk, with the utility function , where ϕ is the owner's profit function in order to make the solution reach equilibrium.

Owner Expected Profit
The revenues for the owner consist of the revenue generated from the operation of the equipment plus the penalty cost paid by the OEM. Hence, the expected profit is given by

$$E\left[\phi_y\right]=K\left\{L-E[D_y^n(L)]\right\}+J^3(k_y,\delta_y)-P_G-C_b^n \tag{5}$$

where $E[D_y^n(L)]=\mu N_y$, μ (expected downtime), N_y (expected number of failures), K is the revenue (\$/hour) received by the owner as a result of transporting mining materials from a mining area to a processing unit, and P_G is the MSC price.

OEM Expected Profit
The revenues for the OEM consist of the revenue received as a payment of the MSC contract plus the additional charges paid by the owner. The costs consist of PM cost, repair cost, penalty cost. Hence, the expected profit is given by

$$E\left[\pi_y\right]=P_G+\Phi(y)-\Pi^n\left[k_y,\delta_y\right] \tag{6}$$

The optimal PM interval for the OEM is obtained bymaximizing $E[\pi_y]$ with the respect to τ_y.

3. Model Analysis

3.1 Cooperative Solution [case 1]
We first consider that the owner and OEM follow in a cooperative fashion. It is clear that the owner's objective is to maximize the process' expected uptime, but the OEM's objective is to minimize the total expected maintenance cost. Consequently, an agreement may not be achievable between the two

parties. Here, owner and the OEM want to get a win-win solution in order to have PM interval that maximizes the expected profit both OEM and the owner.

As a result, the terms and condition of the service contract are negotiated and decided jointly (or coordinated) between the two parties involved. In coordination scheme, the strategy set of the OEM and the owner is $Q_y = \{(\delta_y, \tau_y, k_y) | 0 \le \delta_y \le 1, \tau_y \ge 0, k_y \ge 0 \text{int}\}$. Both players choose the set of strategies $q_y^* \in Q_y$ that solves $\max_q E[\Pi_y] = \max_q \{E[\phi_y] + E[\pi_y]\}$ where $E[\phi_y]$ and $E[\pi_y]$ are the total expected profit for the owner and OEM. The coordination solution is obtained by solving the following problem:

$$\max_{\delta_y, \tau_y, k_y} \quad E[\Pi_y] = KL - C_b^n - (\mu K + C_r) R_y(L) - \left[kC_0 - (\mu K + C_r - C_v) \sum_{j=1}^{k} (L - j\tau_y)\delta_y\right] \tag{7}$$
$$s.t. \quad \delta_y > 0, \tau_y > 0, k_y > 0 \text{ and integer}$$

3.2. Non-Cooperative [case 2]

Next, we consider that the owner and OEM follow in a non-cooperative fashion, and the governing optimization equilibrium concept underlying non-cooperative behaviour is that of Nash. Based on Nash Bargaining solution, to find Nash equilibrium, the owner and OEM negotiate over the share and moreover, they equally split the joint profit gain, $E[\Pi_y]$.In this scheme, the strategy set of the OEM and the owner is $Q_y^{nash} = \{(\delta_y, \tau_y, k_y) | 0 \le \delta_y \le 1, \tau_y \ge 0, k_y \ge 0 \text{int}\}$. Both players choose the set of strategies $q_y^* \in Q_y^{nash}$ that solves $\max_q E[\phi_y] = \max_q E[\pi_y]$where $E[\phi_y]$ and $E[\pi_y]$are the total expected profit for the owner and OEM. Next, we determine proper values of the set of strategies $q_y^* \in Q_y^{nash}$ for the owner and OEM by solving the following problem:

$$\max_{\delta_y, \tau_y, k_y} \quad E[\Pi_y^{nash}] = \frac{1}{2}\left\{KL - C_b^n - (\mu K + C_r) R_y(L) - \left[kC_0 - (\mu K + C_r - C_v) \sum_{j=1}^{k} (L - j\tau_y)\delta_y\right]\right\} \tag{8}$$
$$s.t. \quad \delta_y > 0, \tau_y > 0, k_y > 0 \text{ and integer}$$

$q_y^* \in Q_y$ can be obtained by simultaneously solving the two first order conditions for the owner and OEM.

4. Numerical Example

Let us recall that for a given usage rate y, $F(t; y)$ is the time to the first failure which follows the Weibull distribution with $F_y(t; \alpha_y) = 1 - \exp(-t/\alpha_y)^\beta$. Its failure rate function is $r_y(t) = \beta\left(t^{\beta-1}/(\alpha_y)^\beta\right)$ where α_y as in (1). The other parameter values are:$\alpha_0 = 0.4$, $\beta = 2.5$, $W = 12$ (months), $L = 12$ (months), U=24 ($1\text{x}10^4$Km), K= 24 ($1\text{x}10^4$Km) ($\gamma = U/W = 1$), $y_0 = 1$, $\rho = 1.5$ and $C_v = 0.5C_m$, ζ =80 (hours) or 4 (days) .

Table 1 shows optimal solutions for cases 1 and 2. In Case 1, after warranty ends the optimization is carried out jointly for both players. As the usage rate increases, profit obtained for the agent (OEM) increases whilst the profit for the owner decreases.

Table 1. Results for Case 1 and Case 2

Warranty Region (W=12 months)		MSC Region (L=12 months)						
		Coordination				Non-Coordination		
$\bar{y}$	k_y^*; τ_y^*; $E[\text{cost}]$	$\bar{y}$	k_y^*; τ_y^*; δ_y^*	$E[\pi_y]$; $E[\phi_y]$ $(\times 10^5)$	$E[\Pi_y]$ $(\times 10^6)$	k_y^*; τ_y^*; δ_y^*	$E[\pi_y]=E[\phi_y]$ $(\times 10^5)$	P_G $(\times 10^5)$
1.2	6; 1.36; 866.07	1.2	5; 2.19; 1.09	2.04; 9.21	1.13	5; 2.18; 1.09	5.91	6.07
		2.0		2.20; 9.05			5.85	5.96
		3.0		2.30; 8.96			5.81	5.87
		4.0		2.38; 8.88			5.77	5.78

This is as expected since the agent performs a more effective PM (with $\boldsymbol{\delta}^*$ (= 1.09)) and the PM in turn will decrease the number of failures and hence the total maintenance cost and downtime. Compare with the profit resulting from the Nash bargaining solution (Case 2), the profit for the agent (OEM) and for the owner always equal. This is due to the bargaining strategy that maximize both profits. Since the number of failures increase as the usage increases then it in turn decreases the profit for both players. The similar result exists for the MSC prices, it decreases as the usage rate increases.

5. Conclusion

In this paper we study a usage based MSC for dump trucks after the expired of a two-dimensional warranty, where the MSC is characterized by two parameters – age and usage limits which form a region. An imperfect PM (which reduces the age of the equipment) is performed during the MSC and the optimal imperfect PM is obtained by maximizing the expected total profit for the both players. The paper models the contract using the cooperative and non-cooperative game approach with one dimensional approach. One can models using a bivariate approach with multi players, and considers other shapes of a contract region. This is one topic for future research.

Acknowledgements

This work is funded by the Ministry of Research, Technology, and Higher Education of the Republic of Indonesia through the scheme of "PUPT 2017" with contract number SP DIPA-042.06.1.401516/ 2017.

References

[1] Husniah H, Pasaribu U S and Iskandar B P 2014 *Proc. of ICMIT* Performance-based maintenance contract for equipment used in mining industry Singapore.

[2] Husniah H, Pasaribu U S and Iskandar B P 2016 *Proc. of IEEM* Two dimensional maintenance contract with coordination between owner and agent Bali.

[3] Taracki H, Tang K, Moskowitch H and Plante R N 2006 *IIE Transactions* Incentive Maintenance Outsourcing Contracts for Channel Coordination dan improvement **366** 71–684.

[4] Iskandar B P, Husniah H and Pasaribu U S 2014 *Quality Technology and Quantitative Management* Maintenance service contract for equipment sold with two dimensional warranties **11** 3.

[5] Sheu S H, Chang C C, Chen Y L and Zhang Z G 2015 *Rel. Eng.&Sys.Safety* Optimal preventive maintenance and repair policies for multi-state systems **140** 78-87.

[6] Wang Y, Liu Z L and Liu Y 2015 *Rel. Eng.&Sys* Safety Optimal preventive maintenance strategy for repairable items under two-dimensional warranty **142** 326–333.

Assembly Line Efficiency Improvement by Using WITNESS Simulation Software

A S H M Yasir and N M Z N Mohamed

Faculty of Mechanical Engineering, Universiti Malaysia Pahang, 26600 Pekan, Pahang, Malaysia

ir.ahmadshah@gmail.com, nikzuki@ump.edu.my

Abstract. In the nowadays-competitive world, efficiencies and the productivity of the assembly line are essential in manufacturing company. This paper demonstrates the study of the existing production line performance. The actual cycle time observed and recorded during the working process. The current layout was designed and analysed using Witness simulation software. The productivity and effectiveness for every single operator are measured to determine the operator idle time and busy time. Two new alternatives layout were proposed and analysed by using Witness simulation software to improve the performance of production activities. This research provided valuable and better understanding of production effectiveness by adjusting the line balancing. After analysing the data, simulation result from the current layout and the proposed plan later been tabulated to compare the improved efficiency and productivity. The proposed design plan has shown an increase in yield and productivity compared to the current arrangement. This research has been carried out in company XYZ, which is one of the automotive premises in Pahang, Malaysia.
Keywords; Production, Witness software, Line balancing

1. Introduction

In brief, it is vital to have an efficient layout arrangement and material flow path design because material handling needs a massive percentage of the product cost [2]. The problem with assembly system gives increase to the assembly line balancing problem (ALBP), which involves of allocating assembly tasks to some workstations in imperative to enhance a given purpose [3].

In some occasion, other methods in assembly sequence can use and another case; there is no adjustability in the assembly sequence. In an assembly line, each station organised in the sequential system along the assembly line. Each workstation performs given the subset of tasks on each product unit arriving at that location at fixed time intervals. This range is known as cycle time. The formula shown in (Eq. 1) is for the cycle time [6].

$$\text{Cycle time} = \frac{\text{available time}}{\text{desired output}} \qquad (1)$$

Idle time is also one of the essential components that considered in solving assembly line problem. Idle rate measured as the difference between the cycle time and the workload of each station where the station workload described as the total of operation times of jobs that it performs [7]. Assembly system regarded as a single model line. As for the level of automation, it divides two lines, which are manual lines and automated lines. In assembly line, manual material handling implemented, lot-sizing choices can have a significant influence on workload and human performance, which can have a substantial impact on ergonomic parameters and thus worker [9].

Simulation is one of the methods used to figure out assembly line balancing problem. Simulation is a scientific method to not separate study a system without actually disconcerting it, but also to assess ideas that have not been used in the real world [10]. Simulation is used to predict assembly line performance among other machine layouts and scheduling rules, thus find the best performing layout. Suggest that simulation modelling is mostly used to support decision-making in various commercial tasks [11]. The model simulations offer data about the process results, performance and behaviour with diverse process structures [11]. Mostly, a manufacturing approach is a set of manufacturing policies planned to maximise performance among trade-offs among profit standards to meet the manufacturing job determined by a corporate strategy [12].

2. Research background

The XYZ Company was chose after several selection processes has done. Selection of the company is based on permission criteria, which required by the premise before any further research about the company problem addressed. This study was conducted at S-class car assembly line in XYZ Company premise. The existing assembly line is having line-balancing problem, which affected productivity of its output. Total of 10 workstation were involved on S-class car assembly line. This study will collect data of the current design layout, the number of workstation, number of operators, time study and processing time.

3. Methodology

The methodology is created to ensure that this research is well organised to get the precise result and data. Figure 1, show the flow chart of this research methodology.

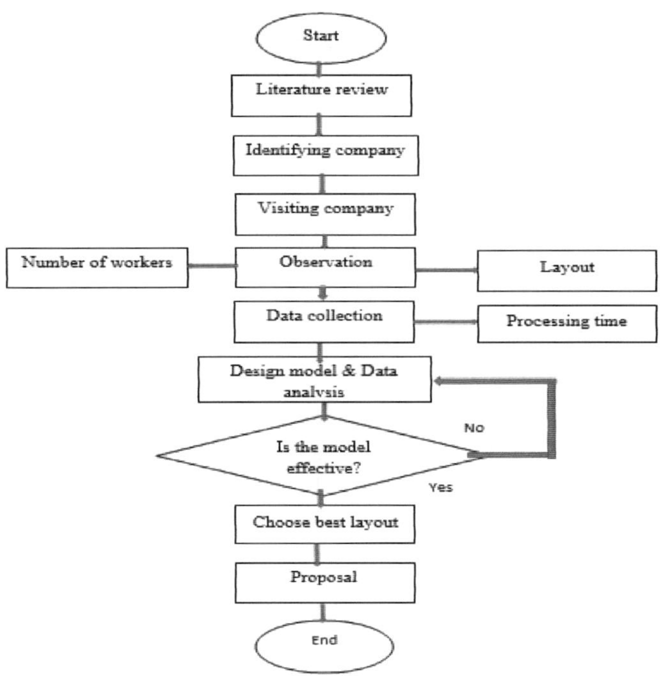

Figure 1. Research methodology flow chart.

Firstly, the research is developing based on literature review method to collect data from the previous study. According to the data obtained, there are many studies about assembly line balancing. Thus, many data can be concluded to develop this research. The suitable company need to be identified to do the research. Next, the researcher visiting the premise is to ensure that the company's condition is known. The observation focus the assembled product's processing time and the current layout design. The data collected has then been analysed and developed onto new layout design that will increase the efficiency of the assembly line. The received data were analysed to get the calculation for standard time, current layout, workstation, number of operators and used material handling system. The collected data will be evaluated using Witness simulation software. The software analyses the processes and makes the optimal choice in a short time. Then, selection of the best arrangement selected from the new layouts, will proposed to the company.

4. Result and Discussion

4.1. Current Layout

Firstly, before the new layout constructed, the current arrangement was analysed using Witness software as in Figure 2. The existing plan is build up using the multi-cycle process on each workstation. All the processing time for each workstation is being filled carefully to avoid any mistakes during analysis.

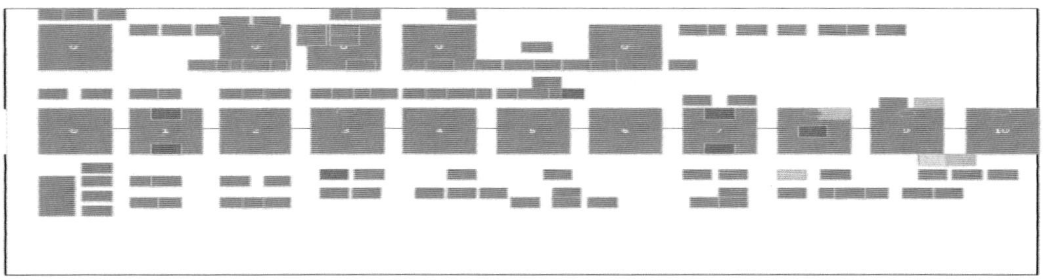

Figure 2. The illustration of the current layout of S-class car.

The present layout of S-class car is in a straight line. The green box in the straight line indicates the workstation while the green box at the side of the workstation is the sub-assembly. The blue square indicates the small parts that will be assembled into the car meanwhile the red square show the lifting equipment to lift the part into the car. In addition, the orange square indicates the machines used such as fuel pump while the light blue square shows the setup machines such as brake bleeding and coolant. After the car reaches workstation number 10, the car then transfers to the testing and quality inspection. The layout begins with workstation preparation and finished in workstation 10. There is five sub-assembly in the current production of S-class car. They are front module, centre console, headlining, cockpit, and rear wall assy. The precedence diagrams show how the work for each workstation progress while the processing time indicates the time used to complete the work. Table 1 shows the overview of all workstation in S-class car assembly line. Figure 3 shows the precedence diagram for all workstation for S-class car. By using all inputs required, Witness Simulation shows, the time required for one car to be finished assemble is 8165 minutes. The time was measured from workstation preparation to workstation 10, including with all the sub-assembly. In a month, the labour-hour is 13200 minutes as the workers only work for 22 days per month and 10 hours per day. After the simulation run for six months, the total number of the car produce is 51. The inputs that were used to measure the productivity is work hours. For six months, the overall work hours is 79200 minutes. The productivity for the current layout for six months is calculated to become the benchmarks for the new plans. Eq. (2) shows the equation for productivity. Table 1 shows the summary of all cycle time for each workstation for the current layout.

$$\text{Productivity} = \frac{\text{units produced}}{\text{inputs used}} \quad (2)$$

$$= \frac{51}{79200}$$

$$= 0.0006439 \text{ units per labour hour}$$

IOP Conf. Series: Materials Science and Engineering **319** (2018) 012004 doi:10.1088/1757-899X/319/1/012004

The productivity for the current layout is 0.0006439 units per labour-hour. Besides productivity, the aspects that also measure are the usefulness of the labours or the effectiveness of the employees. The value of the workers determined by observing the busy and idle time for each worker.

Table 1. The summary of all cycle time for each workstation for current layout

Workstation	Number of workers	Time (min)
Preparation	2	63.13
1	3	62.07
2	3	57.41
3	3	103.15
4	3	64.24
5	3	65.15
6	1	39.54
7	4	58.33
8	3	50.59
9	3	56.29
10	2	41.20

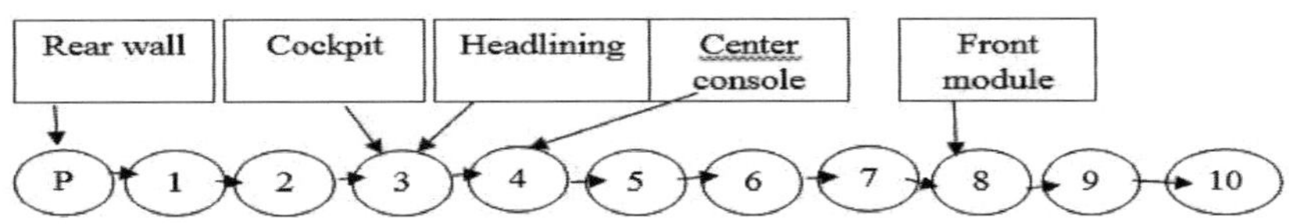

Figure 3. The precedence diagram for all workstation for S-class car

In Appendix A, Figure 7 shows the current layout that was constructed in WITNESS software for workstation preparation until workstation 4 while Figure 8 displays the current design that been built in WITNESS software for workstation 5 until workstation 10. Figure 4, indicates the percentage of the idleness shown is high. To avoid the losses due to labours' inactivity in the current layout, the workers that have too many idle times should be reassigned or fired from the factory.

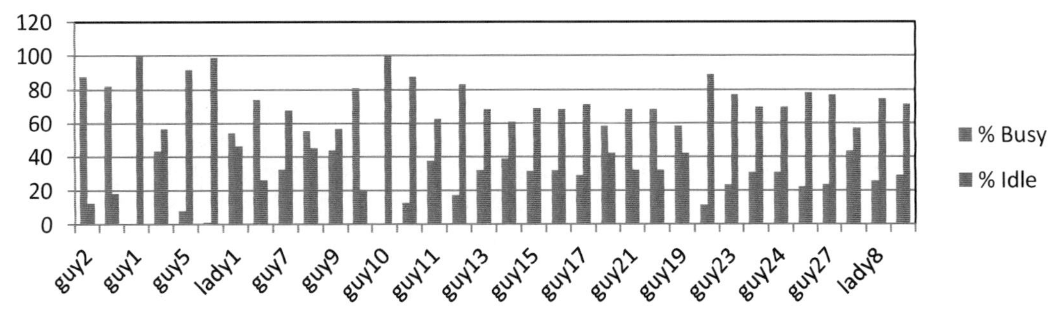

Figure 4. The graph of the effectiveness of each labour for current layout.

4.2. New Layout 1

Simulation using Witness software take place after the New Layout 1 invented as in Appendix A, Figure 9 and Appendix B, Figure 10. The data from Witness simulation show that time demanded one car to assemble is 8119 minutes. Then, the simulation is continued until six months to see the effectiveness of the New Layout 1. The number of products increased by 1, which made it become 52 units in total production. The productivity is measured and compared with the current layout to see the effectiveness of the new design layout. The Productivity is measured using Eq. (2). The Effectiveness of the New Layout 1 was measured using the busy and idle time of each worker as shown detail in Appendix C, Table 3. Figure 5 verifies the Effectiveness of each labour in the New Layout 1. The guy6, guy14, guy20, and lady9 have been reassigned to other department or fired. The total number of workers left is 30 people. The total number of employees reduce because the level of their idleness is high and it does not affect the line.

$$\text{Productivity} = \frac{52}{79200} = 0.0006566 \text{ units per labour hour}$$

The productivity for the New Layout 1 is 0.0006566 units per labour hour. The new layout is compared with the existing layout to measure the effectiveness using Eq. (3).

$$\text{Effectiveness of New Layout 1} = \frac{\text{productivity new layout} - \text{productivity current layout}}{\text{productivity current layout}} \text{ x } 100 \quad (3)$$

$$= \frac{0.0006566 - 0.0006439}{0.0006439} = 1.97\%$$

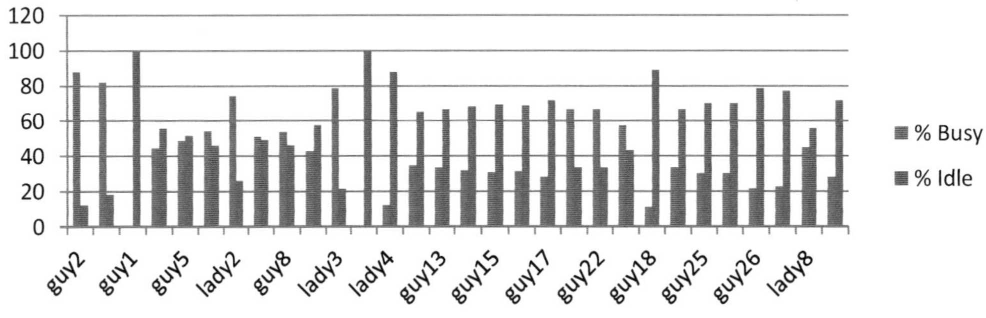

Figure 5. The graph of the effectiveness of each labour in the New Layout 1.

4.3. New Layout 2

The analysis for New Layout 2 is simulated after the design of the layout is completed. The illustration shows in Appendix B, Figure 11 and Figure 12. The time taken for a car to produce is 8077 minutes. After the simulation run for six months, the car units that produced are 53 units. The productivity is calculated to see the effectiveness of the New Layout 2. The productivity is measured using Eq. (2). An alternative way to measure the effectiveness of the new layout is by calculating the efficiency of the labours. Figure 6 demonstrates the graph of the effectiveness of each worker in the New Layout 2, as well as the details as shown by Table 4 in Appendix D. From the data, we can see that lady6 and guy28 are fired or reassigned to another department. The total number of workers in the new layout 2 is 32 people. The other two people that removed in the New Layout 1 reassigned to the new workstation. They have a new workload that can make them efficient instead of having many idle times. The productivity for the New Layout 2 was 0.0006692 units per labour hour, which is higher than the New Layout 1 by 0.0000126 units per labour hour. The effectiveness of the New Layout 2 is measured using Eq. (3).

$$\text{Productivity} = \frac{53}{7920} = 0.0006692 \text{ units per labour hour}$$

$$\text{The effectiveness of the new layout 2} = \frac{0.0006692 - 0.0006439}{0.0006439} \text{ x } 100\% = 3.93\%$$

Table 2 shows the productivity and improved efficiency comparison. Table 2 demonstrated that all new layout shown improvement in efficiency and productivity from the simulation analysis. The New Layout 2 shown the highest productivity of the three layout which is 0.0006692 units/ labour hour. New Layout 2 also improved the efficiency of the current layout by 3.93%.

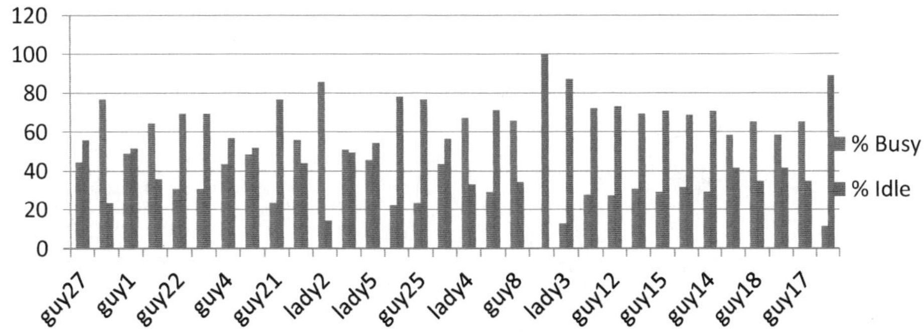

Figure 6. The graph of the effectiveness of each labour in the New Layout 2.

Table 2. Productivity and improved efficiency comparison

Layout	Productivity	Improved efficiency
Current layout	0.0006439 units / labour hour	-
New layout 1	0.0006566 units / labour hour	1.97%
New layout 2	0.0006692 units / labour hour	3.93%

5. Conclusion

The study of assembly lines generates when the contextual analysis arises. The first objective accomplished as the data collected concerning time study, the layout for workstation and number of workers were analysed using Witness software.

The statistic for the current layout examines by identifying the rate of productivity and the effectiveness of each employee. Besides, the second objective also achieved which, the two new layouts are designed to improve the amount of productivity and the efficiency of each worker. After the design of the layouts, the simulation conducted to analyse the effectiveness of the new layout and the productivity of each employee.

The data analysis simulated using Witness software, New Layout 2 is better than the New Layout 1 and the current company layout. The productivity rate and the effectiveness of the New Layout 2 are higher than the New Layout 1. Layout 2 produce efficiency of 3.93% in contrast with New Layout 1 that only provide 1.97% efficiency. New Layout 2 production rate is 0.0006692 units per labour hour also better than current layout which only produce 0.0006439 units per labour hour. By comparison of the improved effectiveness, New Layout 2 is more efficient as it has higher efficiency rate. As a result, it is better to choose the New Layout 2 to increase the production of the product. The result shows that the New Layout 2 is better than the New Layout 1 and the present layout. The simulation was run using Witness software proven to be useful in analysing the assembly lines problems because it helps improve the efficiency of the assembly line and increase the productivity of the operators. After the current layout is analysed, weakness and strength of each workstation is determined. Due to condition in the factory, the workstation cannot be redesigned. That is because the plant does not have big enough spaces mainly due to company produce C-class car and E-class car in the same time. Thus by changing layout for workstation, it will consume more money. This generally because company have to reinstall the machines according to the new layout. It is not recommended to redesign the workstation. As a result, the only option in designing the new layout is by reassigning the labours or terminate the employees.

IOP Conf. Series: Materials Science and Engineering **319** (2018) 012004 doi:10.1088/1757-899X/319/1/012004

Appendix A

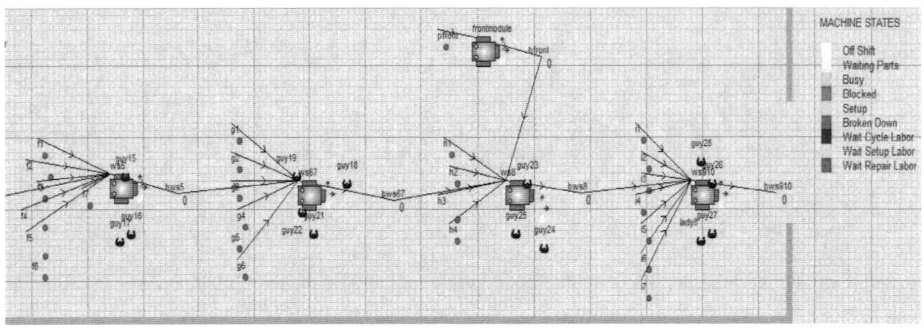

Figure 7. The current layout that been constructed in WITNESS software for workstation preparation until workstation 4

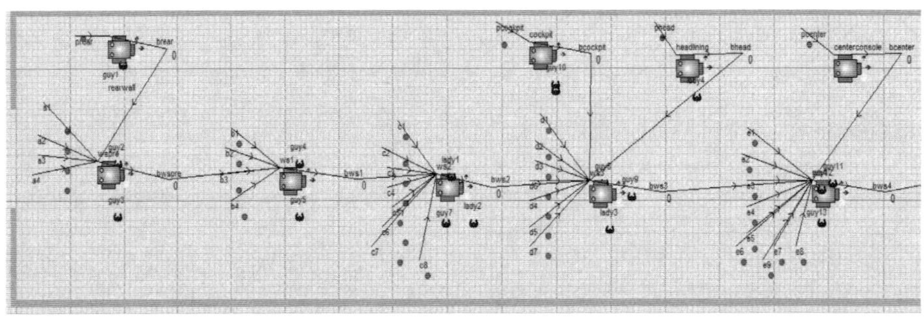

Figure 8. The current layout that built in WITNESS software for workstation 5 until workstation 10

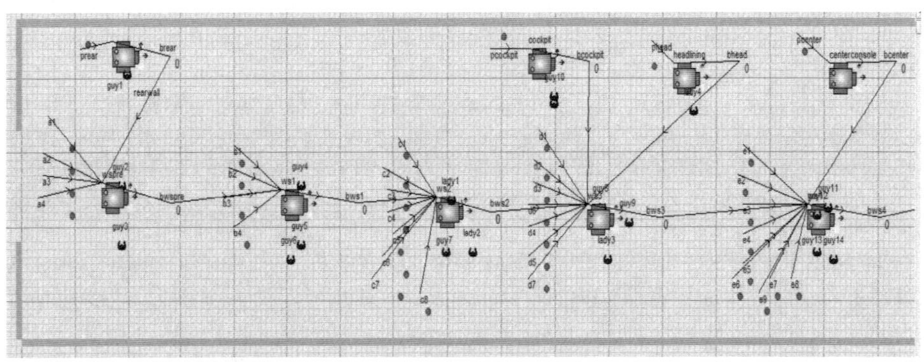

Figure 9. The New Layout 1 of workstation preparation to workstation 4

IOP Conf. Series: Materials Science and Engineering **319** (2018) 012004 doi:10.1088/1757-899X/319/1/012004

Appendix B

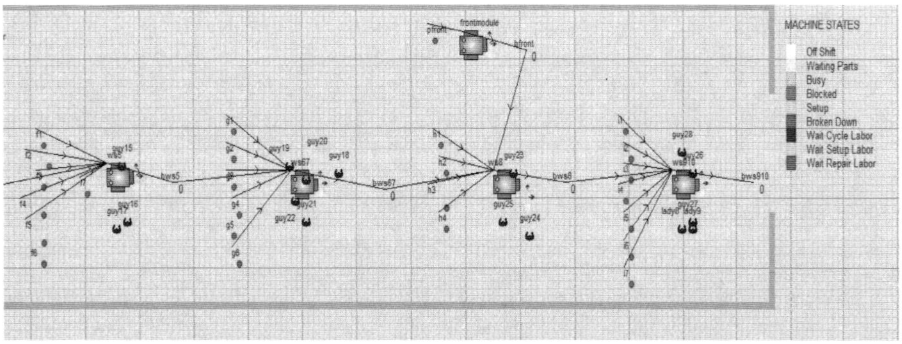

Figure 10. The New Layout 1 of workstation 5 to workstation 10

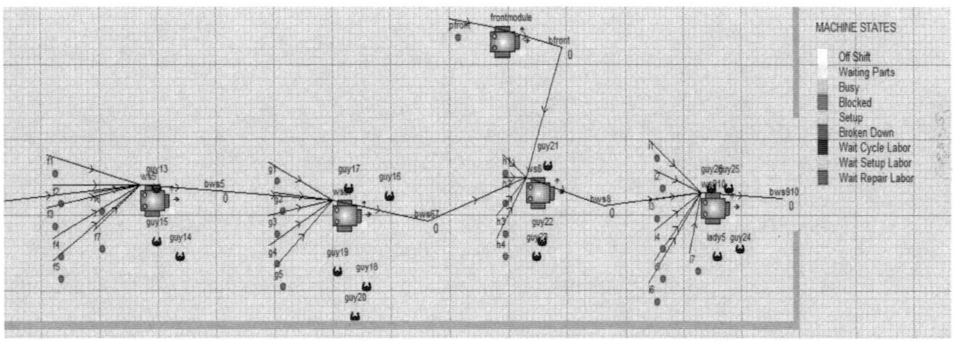

Figure 11. The workstation preparation until workstation 4 for the New Layout 2

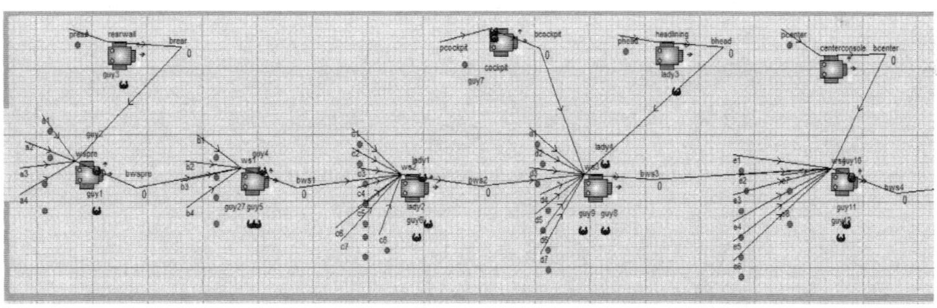

Figure 12. The workstation 5 until workstation 10 for the New Layout 2

IOP Conf. Series: Materials Science and Engineering **319** (2018) 012004 doi:10.1088/1757-899X/319/1/012004

Appendix C

Table 3. The busy and idle time of busy and idle time of each job in the New Layout 1

Name	% busy	% idle
guy2	87.61	12.39
guy3	81.95	18.05
guy1	0	100
guy4	44.34	55.66
guy5	48.41	51.59
lady1	54.03	45.97
lady2	74.03	25.97
guy7	50.92	49.08
guy8	53.91	46.09
guy9	42.6	57.4
lady3	78.63	21.37
guy10	0	100
lady4	12.42	87.58
guy11	34.79	65.21
guy13	33.34	66.66
guy12	31.56	68.44
guy15	30.77	69.23
guy16	31.05	68.95
guy17	28.34	71.66
guy21	66.7	33.3
guy22	66.7	33.3
guy19	57.13	42.87
guy18	11.11	88.89
guy23	33.47	66.53
guy25	29.96	70.04
guy24	29.96	70.04
guy26	21.55	78.45
guy27	22.93	77.07
lady8	44.59	55.41
guy28	28.18	71.82

Appendix D

Table 4. The busy and idle time for each worker in the new layout 2

Name	% busy	% idle
Guy27	44.2	55.8
Guy2	76.73	23.27
Guy1	48.61	51.39
Guy3	64.32	35.68
Guy22	30.72	69.28
Guy23	30.72	69.28
Guy4	43.38	56.62
Guy5	48.31	51.69
Guy21	23.29	76.71
Lady1	55.97	44.03
Lady2	85.66	14.34
Guy6	50.64	49.36
Lady5	45.59	54.41
Guy24	22.08	77.92
Guy25	23.45	76.55
Guy9	43.61	56.39
Lady4	66.98	33.02
Guy26	28.92	71.08
Guy8	65.8	34.2
Guy7	0	100
Lady3	12.7	87.3
Guy11	27.78	72.22
Guy12	26.83	73.17
Guy10	30.73	69.27
Guy15	29.04	70.96
Guy13	31.53	68.47
Guy14	29.3	70.7
Guy19	58.45	41.55
Guy18	65.26	34.74
Guy20	58.45	41.55
Guy17	65.26	34.74
Guy16	11.38	88.62

References

[1] Fleszar, *A new MILP model for the accessibility windows assembly line balancing problem level 2 (AWALBP-L2).* European Journal of Operational Research, 2017. **259**(1): p. 169-174.

[2] Mohamed, N.M.Z.N., et al., *Production Layout Improvement for Steel Fabrication Works.* Journal of Industrial and Intelligent Information, 2014. **3**(2).

[3] Akpinar, S., A. Elmi, and T. Bektaş, *Combinatorial Benders cuts for assembly line balancing problems with setups.* European Journal of Operational Research, 2017. **259**(2): p. 527-537.

[4] Shaaban, S., T. McNamara, and S. Hudson, *Mean time imbalance effects on unreliable unpaced serial flow lines.* Journal of Manufacturing Systems, 2014. **33**(3): p. 357-365.

[5] Riggs, R.J., O. Battaïa, and S.J. Hu, *Disassembly line balancing under high variety of end of life states using a joint precedence graph approach.* Journal of Manufacturing Systems, 2015. **37**: p. 638-648.

[6] Sungur, B. and Y. Yavuz, Assembly line balancing with hierarchical worker assignment. Journal of Manufacturing Systems, 2015. 37: p. 290-298.

[7] Vilà, M. and J. Pereira, A branch-and-bound algorithm for assembly line worker assignment and balancing problems. Computers & Operations Research, 2014. 44: p. 105-114.

[8] Pacauz-Lemoine, M.-P. and P. Millot, *Adaptive Level of Automation fo risk management* 2016: p. 48-53.

[9] Battini, D., et al., *Ergo-lot-sizing: An approach to integrate ergonomic and economic objectives in manual materials handling.* International Journal of Production Economics, 2017. **185**: p. 230-239.

[10] Kavakeb, S., et al., *Green vehicle technology to enhance the performance of a European port: A simulation model with a cost-benefit approach.* Transportation Research Part C: Emerging Technologies, 2015. **60**: p. 169-188.

[11] Orta, E., et al., Decision-making in IT service management: a simulation based approach. Decision Support Systems, 2014. 66: p. 36-51.

[12] Mohamed, N.M.Z.N., et al., Production Layout Improvement for Steel Fabrication Works. Journal of Industrial and Intelligent Information, 2014. 3(2).

IOP Conf. Series: Materials Science and Engineering **319** (2018) 012005 doi:10.1088/1757-899X/319/1/012005

Bridging Technometric Method and Innovation Process: An Initial Study

A A Rumanti[1,2], R Reynaldo[3], T M A A Samadhi[1], I I Wiratmadja[1], A C Dwita[3]

[1]Industrial Engineering Department, Bandung Institute of Technology, Bandung 40132, Indonesia
[2]School of Industrial and System Engineering, Telkom University, Bandung 40257, Indonesia
[3]Industrial Engineering Atmajaya Catholic University of Indonesia, Jakarta, 12930, Indonesia

E-mail: augustinaar@gmail.com, reynaldo.rocky@gmail.com, asamadhi@mail.ti.itb.ac.id, iwan@lspitb.org

Abstract. The process of innovation is one of ways utilized to increase the capability of a technology component that reflects the need of SME. Technometric method can be used to identify to what extent the level of technology advancement in a SME is, and also which technology component that needs to be maximized in order to significantly deliver an innovation. This paper serves as an early study, which lays out a conceptual framework that identifies and elaborates the principles of innovation process from a well-established innovation model by Martin with the technometric method, based on the initial background research conducted at SME Ira Silver in Jogjakarta, Indonesia.

1. Introduction

Human lives will never be apart from technology. Numerous studies have given contribution to the development of technology, and it is justifiable to say that humans and technology have a substantial interdependency, which makes them simply inseparable [1]. Mankind cannot survive without technology and on the other hand, technology will never exist without the hands of human race.

The feeling of dissatisfaction or the notion of "the pursuit of perfection", is in fact a part of human nature, which is naturally invented as a response to their suffering in dealing with survival circumstance. This propensity of improvement becomes an important agent of improvement for technology and science in general [2].

If we analyze further on technology, there is a different perspective that is not as straightforward as we thought. Technology has four essential components i.e. technoware, humanware, infoware, and orgaware [3]. Each component plays a major role in an industry, be it big scale or small and medium enterprises (SME). Nevertheless there is a difference in managing these components that could lead to different level of maximization. Big industry in general has an adequate level of awareness regarding the components in management of technology aside from generous capital and a more structured organization, leading to lesser effort in increasing capabilities. In contrast, SME lacks this kind of wareness due to limited capital and human resources. A study on usage of information technology in Nigeria confirmed that lack of awareness, training, and infrastructure, contributed to underdeveloped fruition [4].

SMEs need to change in order to grow the organization, economy of nearby society, and eventually reduce poverty level. Indonesian SMEs have grown rapidly and been faced with various demand from consumers. In order to serve those diverse market, SMEs need a technology advancement i.e. innovation. The process of innovation is ingrained together with its components and this very process will be examined in this paper by measuring the capability of each component in management of technology, with SME as the research object.

Despite how simultaneous these components are in applying technology, it is very likely to have certain component stands out compared to the other, in terms of having bigger impact to the innovation process. Prior to innovate, SMEs ought to know where they stand with respect to level of sophistication within each component, which can be measured by using technometric method [3]. This method encompasses technology (technoware), organization (orgaware), human resource (humanware), and information related to technology within the organization (infoware), and analyses the dynamic interaction among components to determine the level of sophistication.

1.1. The profile of SME Ira Silver

This reseach focuses on a silversmith in Kotagede, Yogyakarta. Every single touch of silversmith reflects the sheer culture that still exists. The birth of silversmith in Kotagede was in tandem with the genesis of Kotagede as capital city of Matam Islam in the 16th century. There is evidence that shows how silver, gold, and blacksmith have already been known to existence since the 9th century (the era of old Mataram/Hindu) with the finding of insriptions in Central Java. The development of silvermith in Kotagede entered the golden age during 1930 – 1940, with the emergence of new enterprises, quality improvement, and discovery of new patterns.

After experiencing ups and downs, the silversmiths in Kotagede managed to survive and turned themselves into a tourist destination in Yogyakarta, but several problems rose as dollar became unstable causing high price for silver yet the selling price for used-silver remained low. This led to dwindling silver enthusiasts that put pressure on silver industries, including SME Ira Silver. Founded by Mr. Lesmanu in 1990, Ira Silver works through a form of cooperation between a group of silversmiths like cooperation Apikri, that extended their wings to cities even countries.

The workers who were starting to alter their job toward becoming factory labor became a problematic part of silver industries. Moreover, the rising price of dollar affected negatively in Ira Silver's sales. Mr. Lesmanu then started to make some adjustments by producing silver-coated copper and combining silver with stones, all with a hope to reduce the usage of silver. He hoped that with his innovation, the quantity of goods sold would rise, however his limitation on technology advancement diminished the effectiveness ot the innovation itself. The applied changes did not bring substantial impact as he had expected because the change occurred only on product level without paying attention to other components.

The outcome could then be used to increase the capability of promising components to reach their maximum potential i.e. which components contribute significantly in the innovation process by using technometric method. Moreover the research should be extended to generate strategies that would improve the selected component(s), with the ultimate goal of increasing sales.

1.2. Research objective

Based on the circumstance of Ira Silver and theoretical framework, this research aims to lay out a preliminary study with respect to the relation between technometric and innovation process.

2. Theoretical Framework

2.1. Definition of technology

Technology can be defined as knowledge, product, process, equipment/tool, method, and system applied to generate goods and provide service [5]. In short, technology is how one does something to attain her goal. Technology is equally a practical implementation of knowledge to help humans in their work. The common perception of technology is hardware, machines, computer, or any sophisticated devices. In fact, technology is more than just a machine. There are several technology entity beside hardware such as software and human capability.

Hardware is a physical structure and logical arrangement of a device or machine-related things used to complete a task; Software is the knowledge on how to use the hardware; while Brainware is defined as the thinker, who creates and provides reasoning to use technology in a meticulous manner. The fourth aspect is know-how, which translates to knowledge or technical skill required to ensure the task is being done precisely and correctly, incorporates experience, knowledge transfer, and direct training.

Technology and knowledge are two spheres tied together strongly. Any application that entails technology advancement, shall follow the up-to-date knowledge. Knowledge is not information, but the result of thought process behind the information [5]. The vast development of technology information has broadened the spread of information across the world, resulting in what is called as knowledge explosion and astonishing progress in technology.

2.2. *Technometric*

The standard definition according to the Economic and Social Commission for Asia and Pacific (ESCAP) of United Nation classifies technology into four simultaneous components [6]:

- Technoware (T) = object-embodied technology = physical facility = technical device: includes tools, equipment, machines, vehicles, factory, and physical infrastructure, with which humans operate a transformation.
- Humanware (H) = person-embodied technology = human abilities = human resources' capability: includes knowledge, skills. Wisdom, creativity, achievement, and experience of on e or a group of persons in leveraging available natural or technology resources.
- Infoware (I) = document-embodied technology = document fact = information device: includes process, procedure, technique, method, theory, specification, design, observation, manual, and other facts revealed through publication, documentation, and blue print.
- Orgaware (O) = institution-embodied technology = organizational frameworks = institutional device: essential to accommodate physical facility, human skills, facts, and includes management practices, and organizational management to reach a positive result.

Technometric method as an analysis for the level of technology contain, aims to measure aggregate contribution of all four technology components within transformation process from input to output, which then can be described as technology contribution. Technology Contribution Coefficient (TCC) can be formulated in a multiplicative function as follows:

$$TCC = T^{\beta_t} * H^{\beta_h} * I^{\beta_i} * O^{\beta_o} \tag{1}$$

T, H, I, O reflect contribution of Technoware, Humanware, Infoware, and Orgaware; βt, βh, βi, βo reflect the intensity of contribution for each component to TCC. There are five steps that serve as procedure to estimate the value of T, H, I, O, β:

- Estimate the degree of technology sophistication
  - ❖ Data collection: degree of sophistication for technology component
  - ❖ Qualitative observation on T, H, I, O
  - ❖ Evaluation and give score to T, H, I, O
  - ❖ Define top and bottom limit of available technology sophistication
- Conduct assessment for State of The Art (SOTA)
  - ❖ Determine status of technology component against SOTA with the formula (2), (3), (4), and (5):

$$ST_i = \frac{1}{10} x \left[\frac{\sum k_t T_{ik}}{k_t}\right];\ k = 1,2,..,k_t \tag{2}$$

$$SH_j = \frac{1}{10} x \left[\frac{\sum i_h H_{ji}}{i_h}\right];\ i = 1,2,..,i_h \tag{3}$$

$$SI = \frac{1}{10} x \left[\frac{\sum m_f m}{m_f}\right];\ m = 1,2,..,m_f \tag{4}$$

$$SO = \frac{1}{10} x \left[\frac{\sum n_o n}{n_o}\right];\ n = 1,2,..,n_o \tag{5}$$

- ❖ Capability: technical knowledge, in relation with the best condition within transformation facility.
- ❖ Criteria: scoring from 0 (lowest specification) to 10 (highest specification).
- ❖ Between score: interpolation approach.

- Determine component contribution
  - ❖ Define contribution by using the limitation of sophistication degree and SOTA.
  - ❖ T_i , H_j = contribution of each item i technoware and item j humanware. The formulas to determine component contribution are shown below (6), (7), (8), (9):

$$T_i = \frac{1}{9}x\,[LT_i + ST_i(UT_i - LT_i)] \quad (6)$$

$$H_j = \frac{1}{9}x\,\left[LH_j + SH_j\left(UT_j - LT_j\right)\right] \quad (7)$$

$$I = \frac{1}{9}x\,[LI + SI(UI - LI)] \quad (8)$$

$$O = \frac{1}{9}x\,[LO + SO(UO - LO)] \quad (9)$$

- Assessment on the intensity of component contribution β, whose format is a coupled comparison matrix.
- Calculate TCC using formula (1). TCC is the total contribution of operating technology transformation vis-à-vis output made by SME.

As TCC indicates the gap between technology content between input and output, therefore if there are two TCC values but one with higher output, then it can be said that the one with higher output will have the upper hand. The maximum value of TCC is 1, and TCC value represents the level of technology in an organization as shown in table 1 [7].

Table 1. Qualitative assessment of TCC.

TCC Value	Classification
0,1	Very low
0,3	Low
0,5	Normal
0,7	Good
0,9	Very good
1	Highly Sophisticated

After following the aforementioned five steps, one can then position the organization's technology level through a THIO Diagram. TCC value cannot be 0 as it translates to the absence of even the slightest advancement in each component contributing collectively. TCC value will then be compared to table 2 and 3, which are the more elaborated versions of table 1.

IOP Conf. Series: Materials Science and Engineering **319** (2018) 012005 doi:10.1088/1757-899X/319/1/012005

Table 2. Quantitative assessment based on TCC interval

TCC score	Classification
$0 \leq TCC \leq 0{,}1$	Very low
$0{,}1 \leq TCC \leq 0{,}3$	Low
$0{,}3 \leq TCC \leq 0{,}5$	Normal
$0{,}5 \leq TCC \leq 0{,}7$	Good
$0{,}7 \leq TCC \leq 0{,}9$	Very good
$0{,}9 \leq TCC \leq 1$	Highly sophisticated

Table 3. TCC-based technology level.

TCC score	Classification
$0 \leq TCC \leq 0{,}1$	Traditional
$0{,}3 \leq TCC \leq 0{,}7$	Semi Modern
$0{,}7 \leq TCC \leq 1{,}0$	Modern

2.3. Innovation

Innovation is defined as new discovery that differs from previous or existing one [8]. One can be deemed as innovative if she continues to make improvement and provide differentiation, which are paramount virtues to be a successful entrepreneur.

The process of innovation is a complex activity through a transformation from ideas and knowledge to physical form and tangible application. This process of actualization turns knowledge into advantageous products and service leading to a positive value for social and economy aspect of society. Within the process of technology innovation, there are several types of elasticity:

- Technology elasticity: shows the effect of change in quality of technology with respect to demand change over the technology
- Absolute elasticity: shows the responsiveness of total market demand due to dimension improvement in technology quality.
- Relative elasticity: shows the shift in market due to better introduction from competitor.

The advancement of innovation technology has several steps described as the Model of Technology Innovation Process [9].

- Scientific invention: an initial step in the innovation process where theoretical findings or new knowledge can support the success of innovation. Having said that, not all innovation came from this step as it requires huge amount of capital and long period of time.
- Engineering development: this step represents the creation of product, which also oversees the proficiency and anything related to manufacturing process.
- Entrepreneurship: innovation process in this step indicates how innovation has to be able to attract opportunity of new or more exciting subjects, which can also be associated with how to get consumers through marketing or fulfilling their desires. An entrepreneur plays an important role in innovation as she is the driving force for the business that has a vision to a better future state. Equally so, an entrepreneur can create jobs and boost nation's economy. Entrepreneurial activities have a lot to do with innovation process because not only do they involve a set of management skills e.g. problem analysis or decision-making, they also require ideas, thoughts, and creativity in order to have the upper hand over the competitor.
- Management: this step shows the importance of management to achieve the goals through several activities: planning, organizing, actuating, and controlling, in order to have a structured, smooth, and sustainable business.
- Recognized social need: innovation process in this step examines the reason behind one's need for a product and where it belongs within Maslow's hierarchy of needs.
- Supportive environment: this is the last step of the innovation process where environment serves as an important supporting factor.

IOP Conf. Series: Materials Science and Engineering **319** (2018) 012005 doi:10.1088/1757-899X/319/1/012005

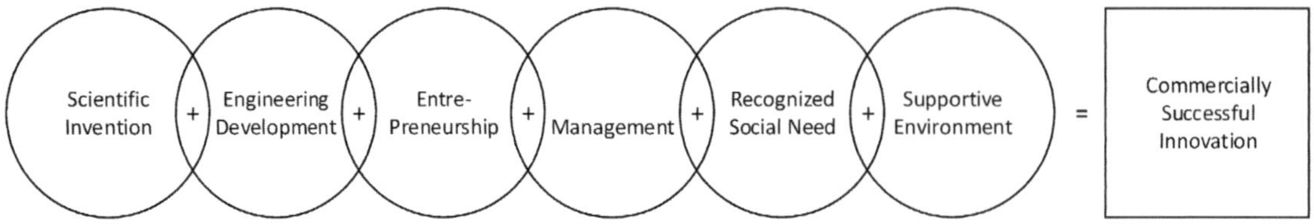

Figure 1. Model of innovation process (Martin, 1994).

3. Research methodology
The flow of thinking in this research is presented through the schemes below:

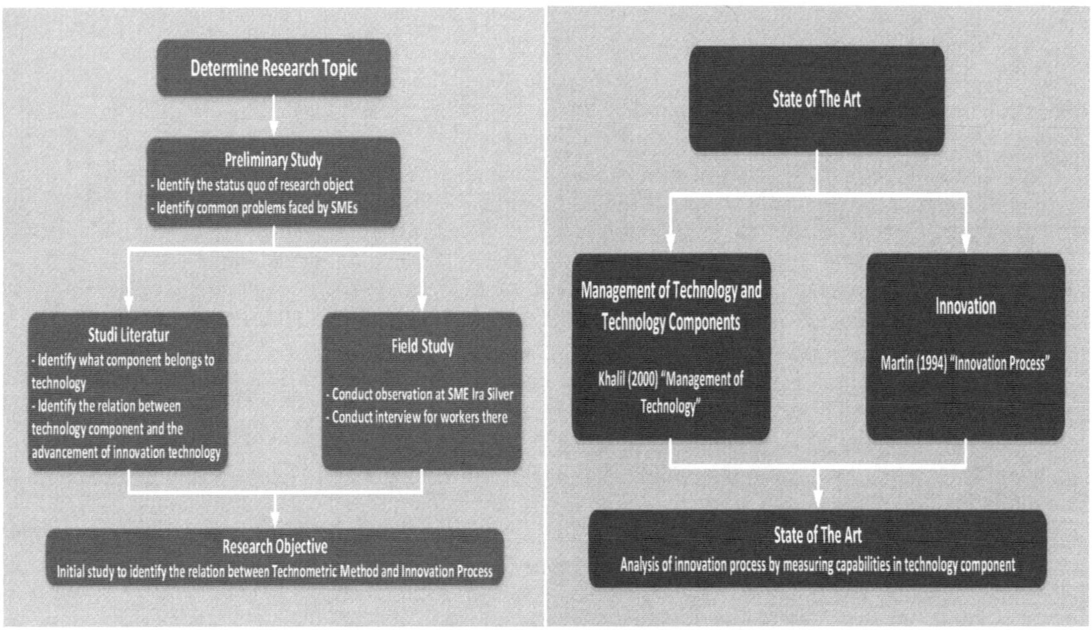

Figure 2. Introduction and state of the art

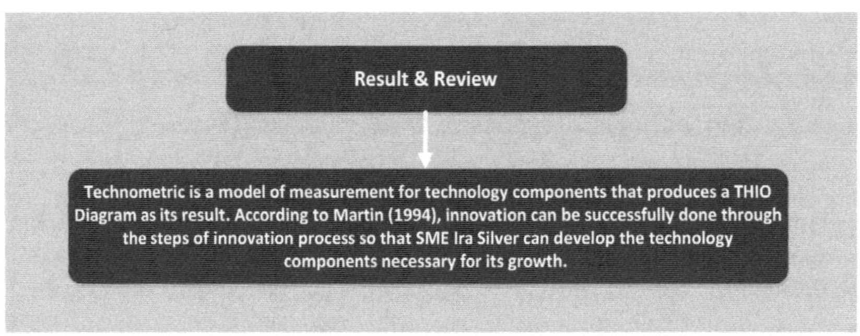

Figure 3. Result and review.

4. Result and review
Technometric model aims for measuring aggregate contribution of four technology components within a transformation process from input to output by firstly analysing the level of technology content in each component i.e. technoware, humanware, infoware, and orgaware. This calculation then will be presented in a form of THIO diagram, reflecting the position of each technology advancement. Visually, this diagram will portend which component that needs to be improved to stimulate innovation process. The aggregate technology score can also be classified into categories to mark the

state of an organization at that time, which can then be re-measured over time in a form of longitudinal study to examine the progress of technology enhancement.

Through his innovation model, Martin indicated the importance of having scientific, engineering, entrepreneurial, and management skills, which resonates with the elements assessed in technometric method i.e. human, technology, organization, and information. This resemblance is perceived as a justification on why the authors proposed to combine Martin's model with this method. Moreover, the additional two spheres i.e. need and socio-politic environment, brings an interesting external aspect of innovation that is not covered in technometric, to the whole conversation with regards to enabling the innovation process itself. A well-rounded innovation road map can then be synthesized by leveraging the technometric assessment on how SME Ira Silver can work on elevating each innovation stage and equally engage with the externalities to realize a fruitful innovation.

5. Conclusion

Technometric method can be used to identify to what extent the level of technology advancement in a SME is, and also which technology component that needs to be maximized in order to significantly deliver an innovation. Furthermore, this study justified the collaboration between technometric and Martin's innovation model through their clear similarity with regards to innovation scope, as well as additional aspects that can be used to extend the discussion to include the external state of the entity per se. As a continuation of this study, a questionnaire can then be utilized to measure the level of sophistication for each component, before extracting them into TCC value. Another important part is to have a better traceability for each innovation stage, which can be mitigated by establishing a strategy roadmap synthesized with the joint-inputs from both authors and SME owner. This approach will increase the feasibility aspect of the whole innovation effort.

References

[1] Skarzauskiene A, Tamosiunaite R and Zaleniene I 2013 Defining Social Technologies: Evaluation of Social Collaboration Tools and Technologies, *Electronic J. Information System Evaluations* **16** 232-241.
[2] Marsen S 2008 Becoming More Than Human: Technology and The Post-Human Condition Introduction, *J. Evolution and Technology* **19** 1-5
[3] Rumanti, A A and Wirawan H 2015 Organization Culture Transformation towards Management of Technology *J. Economics, Business, and Management* **3** 999-1003
[4] Afolayan A, White G, Plant E, Jones P and Davis P B 2015 Information Technology Usage in SMEs in A Developing Economy: *Strategic Change* **24** 483-498
[5] Khalil T 2000 *Management of Technology: The Key to Competitiveness and Wealth Creation* (USA: McGraw-Hill)
[6] Sunarharum W B and Santoso I 2012 Analisis Kontribusi Teknologi pada Industri Susu Pasteurisasi di KUD "Dau" Malang *Jurnal Teknik Pertambangan* **5** 141-150
[7] Nazaruddin 2008 *Manajemen Teknologi* (Indonesia: Graha Ilmu)[
[8] Hamid M A A, Balwi M K and Othman M F 2010 *Reka Cipta dan Inovasi dalam Perspektif Kreativiti* (Malaysia: UTM)
[9] Fayolle A 2011 *Entrepreneurship and New Value Creation* (UK: Cambridge)

IOP Conf. Series: Materials Science and Engineering **319** (2018) 012006 doi:10.1088/1757-899X/319/1/012006

Development of hydropower sustainability assessment method in Malaysia context

Faiz Mohd Turan[1], Kartina Johan[1] and Nur Atiqah Omar[1]

[1]Faculty of Manufacturing Engineering, Universiti Malaysia Pahang, 26600 Pekan, Pahang, Malaysia

E-mail: faizmt@ump.edu.my

Abstract. Nowadays, sustainability is becoming one of the crucial requirement to business success today. This requirement is strongly supported by Bursa Malaysia. In their webpage, they stated that an entire way to business management, incorporating economic, environmental, social and governance considerations alongside financial ones, will serve as a sound business model that supports business continuity and long term value creation for stakeholders and society at large (Bursa Malaysia website, 21th April 2016). This proved that companies need to take sustainability as one of their aspect performance as well as an energy company. Apart from that, energy companies in Malaysia are facing problems as there is still no systematic assessment of sustainability. Before this, Malaysia energy companies assess their large projects based on Environmental Impact Assessments (EIAs) requirement. However, the EIAs mostly covers the environmental issues related to the projects. The EIAs give less attention to the social aspects and economical aspects. In addition, there are still not many companies comply all the three aspects together. So, this study is to help the energy companies to discover the systematic assessment of sustainability. In developing sustainable project, they need to include many criteria that cover the environmental, economic and social aspects at all stages. Thus, the new version of Systematic Sustainability Assessment (SSA) that apply the Hydropower Sustainability Assessment Protocol (HSAP) is used as a guideline to achieve sustainability in Malaysia energy companies. This tool will guide the energy company on how to assess the sustainability in their project and see the performance of the project.

1. Introduction

Sustainability refers to the long-term maintenance of systems according to environmental, economic and social consideration [1-4]. It can be divided into 3Ps which is planet, people and profit [5]. But, recently there is another two P's added which are process sustainability and product sustainability [6]. Both of these two aspects are important for achieving sustainability. Currently, the hydropower sustainability is developed in Malaysia. This is because hydropower is used as power generation. Besides, hydropower is a renewable energy source that has less negative impact to environment as it produces negligible amounts of greenhouse gases.

The development of hydropower will lead to the emergence of environmental issues and social issues [7]. As the hydropower sustainability is getting attention from governments and industry, it is vital for Malaysia to have sustainability assessment method that covers all stages and issues in the hydropower projects. However, the situation in Malaysia currently is there no systematic evaluation of hydropower because the one that they use which is Environmental Impact Assessments (EIAs) only focused on environmental aspects and cover the early stage of the project [8].

In order to help them with these problems, this study introduces new Systematic Sustainability Assessment (SSA) [9] method. This new SSA will provides a guideline to the governments and hydropower industry on how to attain hydropower sustainability.

2. Methodology

The general framework of the approach is as portrayed in Figure 1.

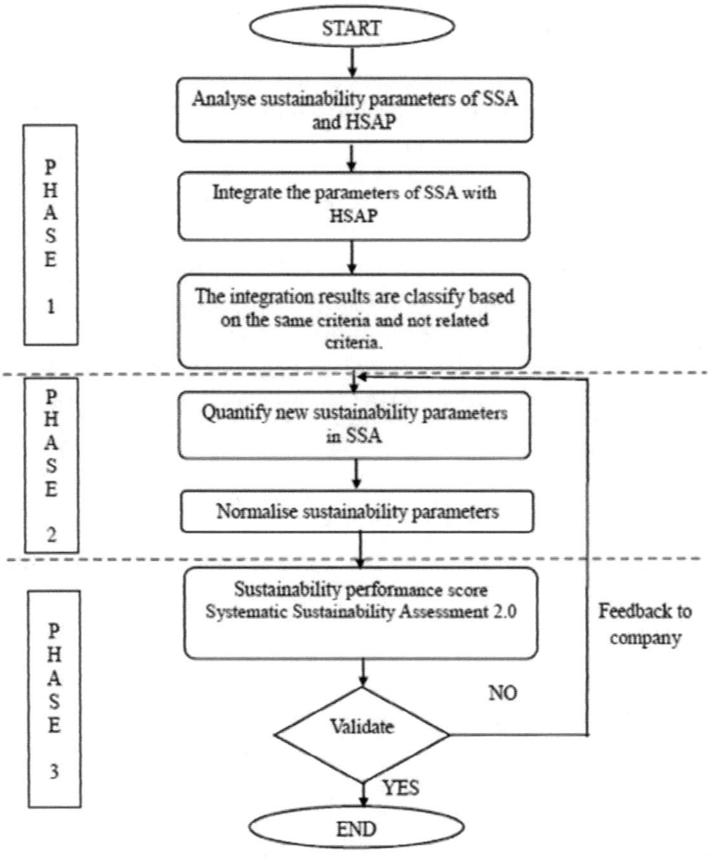

Figure 1. Flow chart of the research

2.1. Phase 1

In this phase, twenty-three criterions of HSAP [7] are classified based on P5 concept integration matrix. The results of this phase are classification of these criterions based on the same criteria and not related criteria between GPM P5 with HSAP. Subsequently, the result of both the same criteria and not related criteria will be used as parameter in SSA version 2.0.

2.2. Phase 2

The scale between 0 to 6 was developed to ease the respondents' group for rating the evaluation criteria. For this phase, an energy company in Pahang is visited to collect some data. There are five departments that are being chosen to do the data collection process. The departments involved are Department of Compliance and Enforcement, Department of Project Construction, Department of Technical, Department of Environmental Quality and Department of Quality. Table 1 describes the scale of "Weighting criteria" in more detail.

Table 1. Scale of "Weighting Criteria" from 0 to 6

Numerical Rating	0	1	2	3	4	5	6
Description	Negative high impact	Negative medium impact	Negative low impact	Neutral	Positive low impact	Positive medium impact	Positive high impact

As the data is gathered from the company, this phase is proceeds with quantifying and normalizing the data gathered according to with "Functional Based" and "Criteria Based". For the "Functional Based", the data is being summarized based on the "Same Criteria" and "Not Related Criteria" that is equated to the Product and Process factors. Meanwhile, the "Criteria Based" is being illustrated according to each of the criterions. For each criterion, the minimum, average and maximum values is computed. The standard deviation of each criterion is determined to identify the sustainability compliance level in the company.

2.3. Phase 3

Result of sustainability compliance ratio of each sustainability parameters are proposed to be ranked as shown in Table 2 below. All the departments of the company sustainability score will be compared to proposed sustainability ranking to recognize they are at which level of sustainability.

Table 2. Proposed ranking of sustainability compliance

Sustainability Score	Sustainability Impact
0-2	Negative impact
3	Neutral
4-6	Positive impact

3. Results

3.1. Criteria integration matrix

There are twelve criterions of GPM P5 and twenty- three criterions of HSAP. These twenty- three criterions of HSAP are integrated with the twelve criterions of GPM P5 according to the P5 concept integration matrix. The results of the integration process are shown in Table 3 and Table 4 below. Based on Table 3, there are ten criterions of HSAP that are related with GPM P5. Meanwhile, the not related criteria of HSAP with GPM P5 consists of thirteen criterions.

Table 3. Same characteristics between GPM P5 with HSAP

GPM P5	HSAP
1. Labour practices & decent work	1. Labour & working conditions (C1)
2. Human rights	2. Affected Communities (C2) 3. Resettlement (C3) 4. Indigenous People (C4)
3. Society & customers	5. Public health (C5)
4. Materials & procurement	6. Procurement (C6)
5. Water	7. Water quality (C7)
6. Return on investment	8. Economic viability (C8)
7. Economic stimulation	9. Project benefits (C9)
8. Business Agility	10. Financial viability (C10)

Table 4. Not related characteristics between GPM P5 with HSAP

GPM P5	HSAP
1. Ethical behaviour 2. Energy 3. Transport 4. Waste	1. Communication & consultation (C11) 2. Governance (C14) 3. Demonstrated need & strategic fit (C12) 4. Siting & design (C13) 5. Environmental & social management (C15) 6. Integrated project management (C16) 7. Hydrological resource (C17) 8. Infrastructure safety (C18) 9. Cultural heritage (C19) 10. Biodiversity & invasive species (C20) 11. Erosion & sedimentation (C21) 12. Reservoir planning (C22) 13. Downstream flow regimes (C23)

3.2. Sustainability rating score for each criteria

In order to achieve a sustainable hydropower project, this assessment need to cover all these twenty-three criterions which including environmental, economic, social and technical elements. The criterion is scored from zero to six by the respondents. The mean score of product with process is determined to evaluate the overall performance of sustainability for each criterion. The result is illustrated in Figure 2 below.

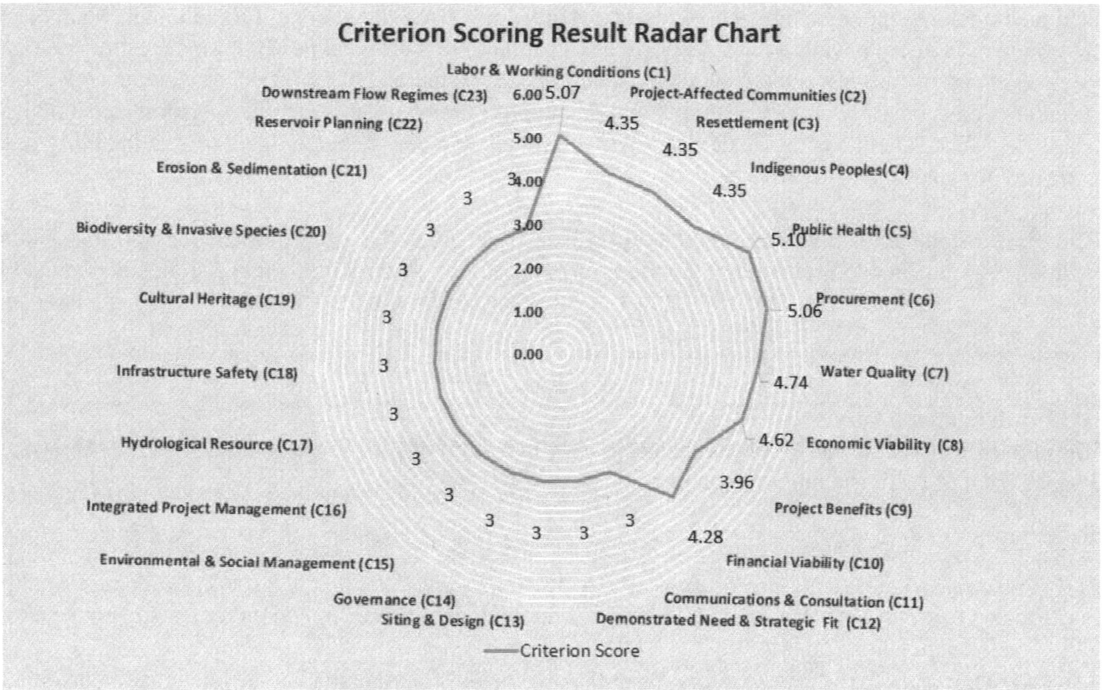

Figure 2. Overall criterion scoring result radar chart

3.3. Sustainability performance score

The result of sustainability performance obtained by using SSA calculator that based on GPM P5 for each department are being summarized as in Table 5 below. Then, this calculated result is compared with the proposed ranking of sustainability compliance.

Table 5. Comparison of sustainability performance with sustainability ranking of each department

Department	Sustainability Performance	Sustainability Ranking
1. Department of Compliance and Enforcement	3.82	Neutral
2. Department of Project Construction	3.71	Neutral
3. Department of Technical	3.75	Neutral
4. Department of Environmental Quality	3.84	Neutral
5. Department of Quality	3.86	Neutral

4. Conclusions

Consequently, this research project shows a correlation between sustainability level of hydropower project in Malaysian context and the improvement of SSA tool at one of the Malaysian energy company. Based on the results and analysis that have been done, Malaysian sustainability level is still low as it not achieved the low positive impact of sustainability yet. But, the results implied that Malaysian hydropower industry having a lot of opportunities to improve their weaknesses in some criterions that being covered in SSA especially the criteria that relates to the technical aspects. Conclusively, this research project not only provide a quality and quantitative report of sustainability performance but also act as Self-Assessment Report (SAR) to provide roadmap to achieve greater level of sustainability in a company for continuous improvement.

There are recommendations that could be useful in order to further improve the method and attain desirable result. Choose a right company to perform the survey. If possible, select a company that listed in the United Nations (UN) Global Compact as most of the UN Global Compact participants have an idea what sustainability is all about.

Acknowledgments

The authors would like to give special thanks to Research & Innovation Department, Universiti Malaysia Pahang, Malaysia for funding this research project (RDU150120).

References

[1] Sabaghi, M., Mascle, C., Baptiste, P., & Rostamzadeh, R. (2016). Sustainability assessment using fuzzy-inference technique (SAFT): A methodology toward green products. *Expert Systems with Applications*, *56*, 69-79.
[2] B. Malaysia, 2015. *Sustainability Reporting Guide*. Available: http://www.bursamalaysia.com/
[3] UN. (2013). Indicators of sustainable development; guidelines and methodologies. United Nations Commission on Sustainable Development.
[4] Global Reporting Initiative. (2015). Linking the SDGs and GRI, 6.
[5] F. M. Turan, K. Johan, " Assessing Sustainability Framework of Automotive-related Industry in the Malaysian Context based on GPM P5 Standard," *ARPN Journal of Engineering and Applied Sciences,* vol. 11, pp. 7606 – 7611, 2016.

[6] The GPM Global Standard P5 of sustainability in Project Management, first edition, *GPM Global*, 2014. Available www.greenprojectmanagement.org
[7] Hydropower Sustainability Assessment Protocol (November 2010), International Hydropower Association.
[8] Ministry of Science, Technology and Environment, 1987. *Environment Quality Order*. Available: http://www.doe.gov.my/eia/wp-content/uploads/2012/03/Environmental-Quality-Prescribed-Activities.pdf
[9] F. M. Turan, K. Johan, " Assessing Sustainability Framework of Automotive-related Industry in the Malaysian Context based on GPM P5 Standard," *ARPN Journal of Engineering and Applied Sciences,* vol. 11, pp. 7606 – 7611, 2016.

A new framework for sustainable hydropower development project

Kartina Johan[1], Faiz Mohd Turan[1] and Nur Syazwani Abdul Gani[1]

[1]Faculty of Manufacturing Engineering, Universiti Malaysia Pahang, 26600 Pekan, Pahang, Malaysia

E-mail: kartina@ump.edu.my

Abstract. This project studies on the establishment of a new framework for sustainable hydropower development. A hydropower development is listed as one of the prescribed activities under the Environmental Quality Order 1987. Thus, Environmental Impact Assessment (EIA) guidelines must be referred to comply with the Department of Environment (DoE) requirements. In order to execute EIA, an assessment tool that will be utilized in the final evaluation phase must be determined. The selected assessment tool that will be used is Systematic Sustainability Assessment(SSA) which is a new integrated tool to evaluate the sustainability performance. A pilot run is conducted in five different departments within the Energy Company to validate the efficiency of the SSA tool. The parameters to be evaluated are constructed aligned with the Sustainable Development Goals (SDG) to maintain the sustainability features. Consequently, the performance level of the sustainability with respect to People, Planet and Profit (3P's) is able to be discovered during evaluation phase in the hydropower development for continuous improvement.

1. Introduction

Hydropower is a globally acknowledged source of clean and green energy which responsible to supply the energy through the world. The increasing demand towards energy and global climate change led many countries include Malaysia to focus on the hydropower development in their energy sectors. However, the rapid development of the hydropower can affect the communities near the project site, the environment engaged with the hydropower development and also economy of the company. Thereon, this research is focusing on maintaining the sustainability performance within the hydropower development with respect to People, Planet and Profit factors in order to maximise the benefits of the new developments and prevented the negative impact towards environmental, social and economic aspects.

Nowadays, every single company that under the auspices of BURSA Malaysia is required to yield the sustainability reporting [1]. Thus, Systematic Sustainability Assessment (SSA) is designed in the advancing of sustainability reporting for promoting sustainability practices [2]. Most of the companies in Malaysia have implemented green practice in their organization management. However, the green practice only emphasizes the environmental aspect, and that causes other important aspects within the company seem to have overlooked [3]. Hence, Green Project Management (GPM P5) standard is introduced as one of the sustainability assessment tool to measure the sustainability practices performance comprehensively by taking into account the aspects of planet, people, profit, process and product [4].

The primary objective in this research is to evaluate the sustainability performance level within the hydropower development project in Malaysia. The specific targets to be accomplished in this research are as follows:

- To investigate the existing environmental assessments for hydropower development project.
- To design a framework for hydropower development project that complies with the environmental and social laws in Malaysia.
- To assist an organization to execute a self-assessment on sustainability in hydropower project during evaluation phase.

This study of SSA is to be explored in the area of the energy sectors to validate the efficiency of the established framework embedded with SSA tool in order to evaluate the sustainability performance of the hydropower development project. This research provides the guidelines to the Energy Company to integrate the sustainability assessment as part of the management and project development within the energy sectors.

2. Literature review

2.1. P5 Integration Matrix

Sustainability assessment is recognized as a powerful and important tool to measure the performance of sustainability in a company or industry [5]. There are various initiatives exists on tools for sustainable development. However, most of the sustainability measurement tools emphasize on environmental, economy and governance aspects. Some of the companies also implement different of sustainability indicators to evaluate the performance of economy, social and environmental separately. Thus, GPM P5 standard is introduced as an advanced alternative to measure sustainability performance thoroughly within an organization. GPM P5 standard is an integrated sustainability tool comprises of people, planet, profit, process and product. GPM P5 standard is designed to assist an organization to produce a comprehensive sustainability report regarding the sustainability performance to improve their strategies and productivity for a long-term success [4].

2.2. EIA

In Malaysia, EIA is necessary for each activity that is listed under the Environmental Quality Order 1987 [6]. Thus, EIA guidelines must be followed to assist an environmental planning of new or existing project. EIA used to minimize the environmental impacts of a proposed project by determine, predict and evaluate the potential impact zones within the project. The main aspects to be considered are the project concept and the site selection. The project proponent shall not oppose any policies or development plans of the Government of Malaysia. The site selection is investigated to ensure any development that is built on the site is not giving huge damage to the communities and the environments. A site visit is conducted to measure the potential impacts on the socioeconomic and environment aspects. Basically, the EIA report is expected at the end of the investigation which comprises of the details on the impact assessment and also the proposed measures that need to be addressed to avoid and minimize the impact on environment.

3. Methodology

The general framework for Sustainable Hydropower Development is as portrayed in Figure 1.

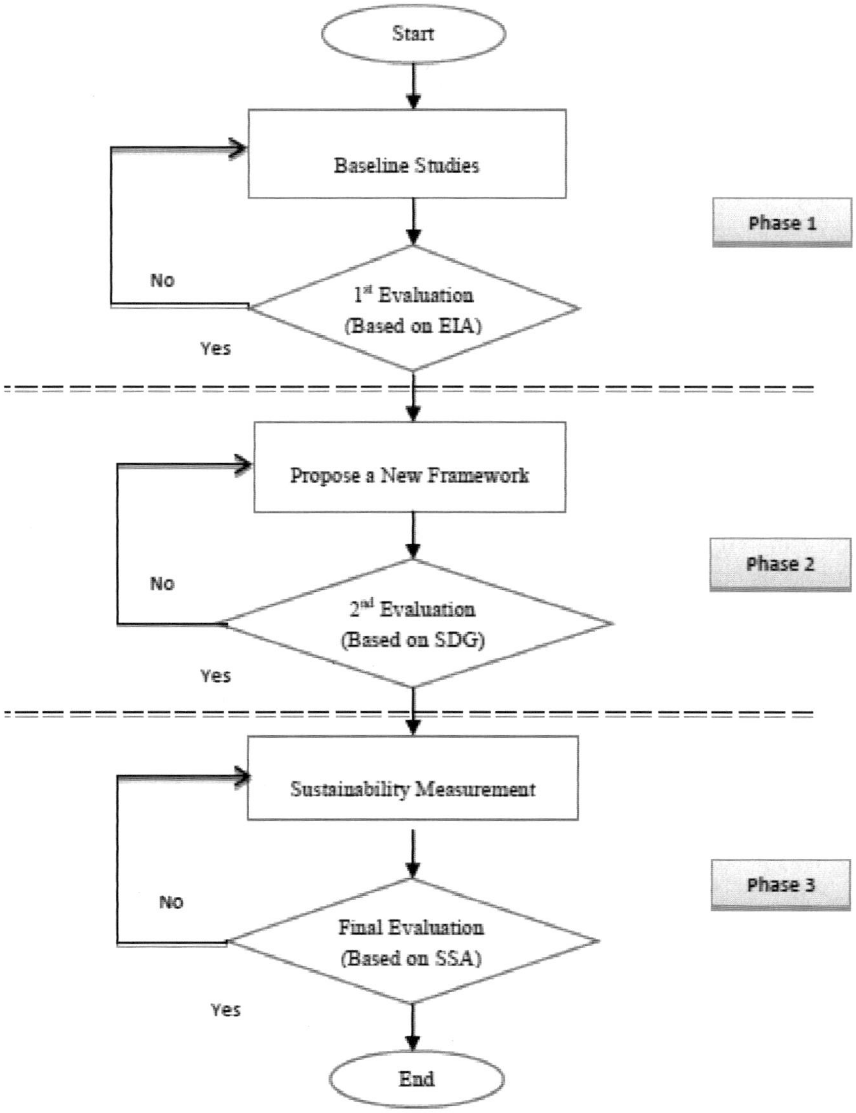

Figure 1. Framework for Sustainable Hydropower Development

The framework for sustainable hydropower development is segregated into three main phases (see Figure 1). Phase 1 consists of the baseline studies, and Phase 2 requires a proposal for a new framework. Meanwhile, the last phase which is Phase 3 relates to the sustainability measurement that will be conducted at the end of the research. Each and every phase has a different evaluation phase with different approaches such as EIA, SDG and SSA. The sustainability parameters are evaluated by scoring method where each parameter is ranked with the most suitable rating value starting from 0 to 6 (see Table 1).

Table 1. Weighting Criteria Scale from 0 to 6

Numerical Rating	Description
6	High Positive Impact
5	Medium Positive Impact
4	Low Positive Impact
3	Neutral
2	Low Negative Impact
1	Medium Negative Impact
0	High Negative Impact

4. Results

4.1. 3P's Distribution for the Attribute
Generally, the 3P's distributions for the attribute within the entire departments are corresponding to each other. The pie chart displays the percentage of the sustainability compliance towards People, Planet, and Profit factors which are 35%, 33% and 32% (see Figure 2). The distinction of the percentage depicted in the pie chart is narrow, thus, it signifies the sustainability approach for product attribute has been complied evenly with respect to 3P's within the Energy Company.

Despite of the evenly distributions of the 3P's for the attribute; it does not imply that the sustainability approach is compiled thoroughly within the Energy Company. Overall, the Energy Company has achieved a total score of 4.46 for the mean score average which is equivalent to 74% for the mean percentage average. The average score indicates that the performance level of sustainability towards People, Planet, and Profit factors for product attribute is ranked in positive sustainability impact or precisely, lower positive sustainability impact (refer Table 1).

Although it is suggested that the sustainability approach towards 3P's has been distributed evenly, however, it does not guarantee the performance level of the sustainability compliance within the Energy Company. Altogether, the total score acquired by the Energy Company is 4.6 for the mean score average which is equivalent to 77% for the mean percentage average. The average score specifies the rank of the performance level of the sustainability towards 3P's factors for process attribute. In this case, the sustainability performance is ranked in positive sustainability impact or specifically in lower positive sustainability impact (refer Table 1).

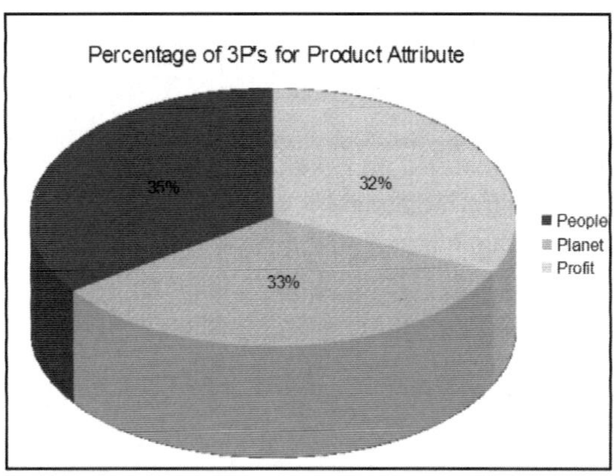

Figure 2. 3P's Distribution for the Attribute

4.2. Control Chart for the Attribute

The range of each criterion is plotted on the R chart in order to measure the consistency of the sustainability variation. The R chart of the attribute is in statistical control (see Figure 3). This signifies that variation of range for each criterion is consistent between each other. On the contrary, if the point on the R chart is out of control, hence, the control limit on the X-bar chart may have tendency to be inaccurate. Although the chart is in statistical control, yet, the result does not imply the best impression as there are several points such as C1, C4, C6, C10, C11 and C12 that have a great difference between maximum and minimum values of the sustainability performance. Thus, the sustainability compliance for the criteria mentioned above must be reviewed for all five departments in order to minimize the range values for continuous improvement.

On the other hand, the average of each criterion is plotted on the X-bar chart aims to evaluate the consistency of the sustainability performance average. The X-bar chart of the Product attribute is not in statistical control as there are several points that are out of control (see Figure 3). This implies that inconsistent variation of the average value does occur between the criteria. In other words, the practice of the sustainability does not evenly distribute among the twelve criteria. The criteria such as C3, C5, C8 and C11 gain high priority to be revised in order to stabilize the sustainability compliance towards 3P's factors within the Energy Company.

Furthermore, the R chart of Process attribute also acts in the same manner as for the attribute which is in statistical control that demonstrates the consistency of range variation between each criterion (see Figure 5). However, the R chart is not in the most desirable state as there are certain points such as C2, C3, C4, C8 and C12 which have a huge gap between maximum and minimum values of the sustainability performance. Consequently, a review is required to examine the sustainability compliance for the recognized criteria within all departments to reduce the range values for a more desirable state.

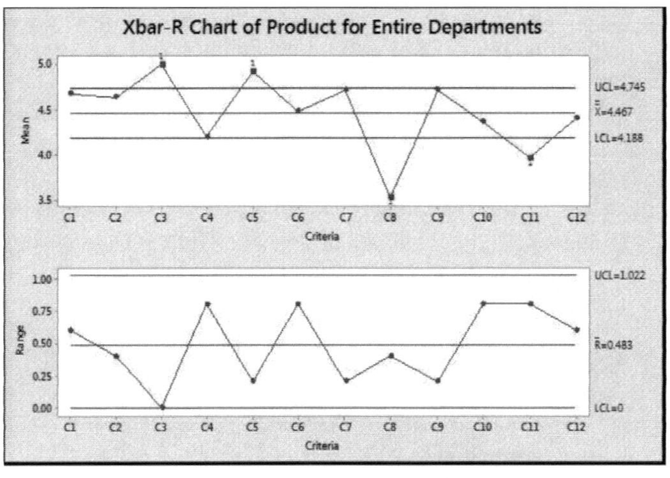

Figure 3. X-bar and R Chart for the Attribute

5. Conclusions

The outcomes from the statistical analysis portrays that the sustainability practice is not implemented thoroughly within the departments of the Energy Company. Although the company declares that all the departments within it comply with the sustainability practice, yet the statistical analysis shows oppositely. Basically, the sustainability performance of the entire departments is ranked at the positive impact of sustainability. However, further analysis shows the sustainability practice have not been adapted appropriately as the sustainability compliance is not distributed evenly among the criteria. In other words, the sustainability assessment is focusing on some criteria only and the other criteria have been neglected.

Acknowledgments
The authors would like to give special thanks to Research & Innovation Department, Universiti Malaysia Pahang, Malaysia for funding this research project (RDU160308).

References

[1] B. Malaysia, 2015. *Sustainability Reporting Guide*. Available: http://www.bursamalaysia.com/
[2] F. M. Turan, K. Johan, W. N. S. W. Lanang, N. H. M. Nur, " Development of Systematic Sustainability Assessment (SSA) for the Malaysian Industry," *IOP Conf. Series: Materials Science and Engineering,* vol. 160, pp. 012047, 2016.
[3] F. M. Turan, K. Johan, " Assessing Sustainability Framework of Automotive-related Industry in the Malaysian Context based on GPM P5 Standard," *ARPN Journal of Engineering and Applied Sciences,* vol. 11, pp. 7606 – 7611, 2016.
[4] The GPM Global Standard P5 of sustainability in Project Management, first edition, *GPM Global*, 2014. Available www.greenprojectmanagement.org
[5] Eastwood, M. D., Haapala, K. R., " A unit process model based methodology to assist product sustainability assessment during design for manufacturing," *Journal of Cleaner Production,* vol. 108, pp. 54 – 64, 2015.
[6] Ministry of Science, Technology and Environment, 1987. *Environment Quality Order*. Available: http://www.doe.gov.my/eia/wp-content/uploads/2012/03/Environmental-Quality-Prescribed-Activities.pdf

IOP Conf. Series: Materials Science and Engineering **319** (2018) 012008 doi:10.1088/1757-899X/319/1/012008

Characteristic and Competency Measurement Instrument Development for Maintenance Staff of Mechanical Expertise with SECI Method: A Case of Manufacturing Company

P A Mahatmavidya[1], R P Soesanto[1], A Kurniawati[1,2], L Andrawina[1]
[1]Industrial Engineering Faculty, Telkom University, Bandung 40257, Indonesia
[2]Industrial Engineering Department, Bandung Institute of Technology, Bandung 40132, Indonesia

E-mail: aryamahatmavidya@gmail.com, raysoesanto@telkomuniversity.ac.id, amelia.kurniawati@gmail.com, luciana@telkomuniversity.ac.id

Abstract. Human resource is an important factor for a company to gain competitiveness, therefore competencies of each individual in a company is a basic characteristic that is taken into account. The increasing employee's competency will affect directly to the company's performance. The purpose of this research is to improve the quality of human resources of maintenance staff in manufacturing company by designing competency measurement instrument that aims to assess the competency of employees. The focus of this research is the mechanical expertise of maintenance staff. SECI method is used in this research for managing knowledge that is held by senior employees regarding employee competence of mechanical expertise. The SECI method converts the knowledge of a person's tacit knowledge into an explicit knowledge so that the knowledge can be used by others. The knowledge that is gathered from SECI method is converted into a list of competence and break down into the detailed competency. Based on the results of this research, it is known that 11 general competencies, 17 distinctive competencies, 20 indicators, and 20 item list for assessing the competencies are developed. From the result of competency breakdown, the five-level instrument of measurement is designed which can assist in assessing employee's competency for mechanical expertise.

1. Introduction

The most important factors in the existence of the company are the competencies within the human resource. The role of human resource in the company is considered as important in order to gain competitive advantage. Thomas [1] stated that with better competencies, an organization can be more effective, efficient, competitive, and easy to achieve high performance. Besides of competencies, knowledge in an organization is considered as important. An organization can obtain great profit if investing in knowledge asset, this can be done because knowledge is one of the un-depreciated assets, on the contrary knowledge, can be increased as long the knowledge is being used by another [2]. According to Nonaka and Takeuchi [3], knowledge can be divided into explicit and tacit knowledge. Explicit knowledge is the knowledge that can be codified and transmitted into the formal and systematic language. Tacit knowledge is the knowledge that consists of a mental model, belief and hard to articulated.in an organization, the tacit knowledge must be converted into explicit knowledge in case the individual in an organization is leaving the organization. Knowledge lies within people, there is a potential problem when the employee in the organization is retired because the form of knowledge in people usually considered as tacit knowledge. Tacit knowledge can't be transferred completely into formal language, so it is hard to save the tacit knowledge electronically [4].

X company is a manufacturing company that focused on manufacturing part of automotive and aircraft. In order to maintain the productivity in optimum level, there is maintenance department which focused on maintain the machine in the company. Figure 1 shows the number of an employee from maintenance department in company X.

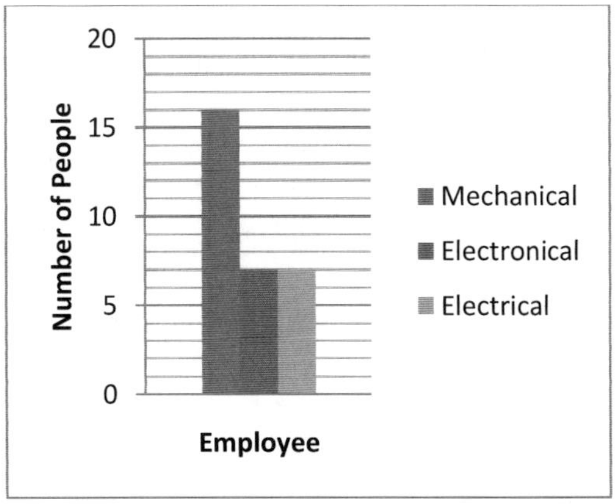

Figure 1. Number of Employee in Maintenance Department Company X,

From Figure 1, it is known that the mechanical expertise employee is the largest in the maintenance department. It is important to have good competencies and skill from the maintenance department employee because it is needed a specific set of skill to maintain a machine. According to the head of the maintenance department, it is known that there is enough employee to do the maintenance job, but there is no standardization about how to assess the employee according to the current skill and competencies. The assessment of maintenance operator usually just intuition from the operator supervisor. The standardized assessment and assignment process is considered important because in 5 years from now a lot of maintenance operator retired.It is known that 37% of maintenance department employee will be retired. The needs of measurement for assessing employee competencies is high. The purpose of this research is to improve the quality of human resources of maintenance staff in manufacturing company by designing competency measurement instrument that aims to assess the competency of employees. SECI method developed by Nonaka and Takeuchi [3] is used in this research for managing knowledge that is held by senior employees regarding employee competence of mechanical expertise. SECI model is widely used for extracting knowledge, the research from Nurunisa, Kurnaiwati, Soesanto, and Hediyanto [5] focused on using SECI model for extracting knowledge for machine maintenance process, the research from Atma, Soesanto, Kurniawati, and Hediyanto [6] focused on utilizing the SECI model for corrective maintenance process. Another research that used SECI model is the research from Rehman [7] which focused on using SECI model for human resources.

2. Theoretical Background

2.1. Knowledge

Knowledge is a main essential for any companies because, in the knowledge-based era, the knowledge itself became the source of competitive advantage [8]. Knowledge is built on individual belief about cause and effect relationship [9]. Knowledge is defined as a fluid mix of experiences, values, information and expert insight that provides some kind of scheme that can evaluate new experience and information [10].

2.2. SECI Model

Nonaka and Takeuchi [3] describe the four phase of converting knowledge which is socialization, externalization, combination, and internalization. Socialization is defined as sharing process and creation of tacit knowledge through interaction and direct observation. Externalization is defined as an articulation of tacit knowledge into explicit knowledge through dialogue and reflexion. The combination is defined as conversion process of explicit knowledge into new explicit knowledge through systemization and information. Internalization is defined as a learning process and knowledge acquisition by organization member towards explicit knowledge that spread into the organization through self-experience so it can be tacit knowledge of organization member. Figure 2 shows the SECI method by Nonaka and Takeuchi [3].

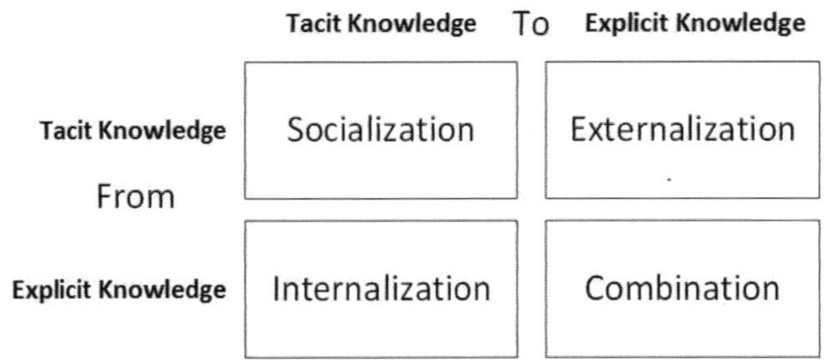

Figure 2. SECI Method

According to Rehman [7], SECI model can be used for Human Resource (HR) Practice. The difference between general SECI model and the HR based SECI model is in the content of each phase. In socialization phase by applying several HR practices, the sharing knowledge process can be extracted. In externalization phase, The codification is in the form of documents, manuals, procedures, methods, etc., but in HR practices the focused is on the work design. During the internalization phase, the main focus is the internalization of the knowledge gain from the HR practices.

3. Research Methodologies

The purpose of this research is to improve the quality of human resources of maintenance staff in manufacturing company by designing competency measurement instrument that aims to assess the competency of employees. The focus of this research is the mechanical expertise of maintenance staff. SECI method is used in this research for managing knowledge that is held by senior employees regarding employee competence of mechanical expertise. The SECI method converts the knowledge of a person's tacit knowledge into an explicit knowledge so that the knowledge can be used by others. The knowledge that is gathered from SECI method is converted into a list of competence and break down into the detailed competency. In socialization phase, tacit knowledge is transformed into another tacit knowledge, in this case, tacit knowledge will be converted into knowledge about the detailed breakdown of general competencies for mechanical expertise. Figure 3 shows the breakdown process for mechanical expertise.

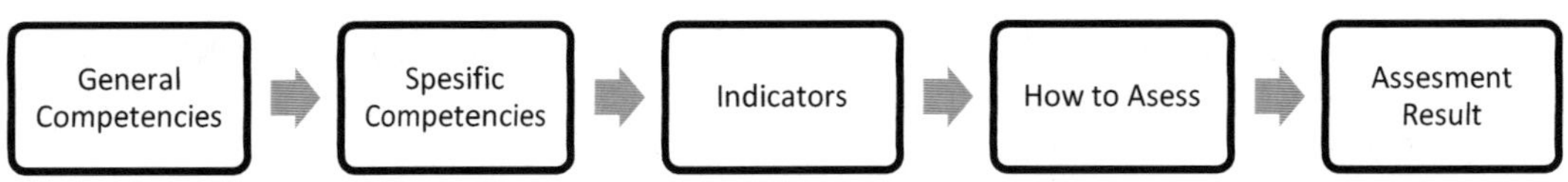

Figure 3. Breakdown Process for Mechanical Expertise Characteristic

From Figure 3, it is known that general competencies are the basic competencies that all operator should have in order to maintain the machine. Specific competencies are the detailed competencies from general competencies. Indicators are the detailed breakdown for specific competencies that are used as the basis for assessment. How to measure explained how to measure each indicator. After the process is break down the next step is to interview the respondent to gain knowledge about the measurement. The data is gathered by interviewing the senior employee. The senior employee will give their tacit knowledge regarding general characteristic of mechanical expertise. During externalization phase, the previous knowledge about the competencies and the interview result is documented. Previous tacit knowledge then converted into explicit knowledge in form of review result. In combination phase, explicit knowledge is converted into new explicit knowledge through a combination of knowledge. From the interview of three respondent, the information will merge into one complete information. All breakdown competencies from the socialization result are merged into one best competencies measurement.

1. Specific Competencies Merge.
   The merge of specific competencies is done based on the knowledge of the respondent. Specific competencies are the detailed breakdown of the general competencies.
2. Indicators Merge.
   The merge of indicators is done based on the knowledge of the respondent. Indicators are used as a basis to measure the competencies.

During internalization phase, the competencies measurement design is shared through entire department. Focus group discussion is conducted to share the new explicit knowledge.

The selected competencies need a scale that can be used to measure the value of each indicator. The measurement scale can be used to assess the employee competencies. The scale that is used in this research is 5 scale where 5 is the highest score.

4. Result and Discussion

4.1. Socialization

The breakdown process of competencies measurement is discussed in the methodology section. After the process is break down the next step is to interview the respondent to gain knowledge about the measurement. 3 respondent selected for the interview process, the respondent was selected from maintenance department that is experienced in mechanical expertise for maintenance process. The tacit knowledge from respondent then will be converted into new tacit knowledge through the interview. The problem from interview process is that the respondent not fully aware of the competencies needed and the answer is biased, but this problem is handled by using three respondent to gain a better perspective about the mechanical expertise in machine maintenance.

Table 1. Interview Result

Respondent	Interview Result
A	• Explanation about specific competencies • Explanation about indicators of each competency • Explanation about how to measure
B	• Explanation about specific competencies • Explanation about indicators of each competency • Explanation about how to measure
C	• Explanation about specific competencies • Explanation about indicators of each competency • Explanation about how to measure

4.2. Externalization

The previous knowledge about the competencies and the interview result is documented. Previous tacit knowledge then converted into explicit knowledge. Table 2 shows the example of externalized knowledge form respondent. From table 2 it is known that the explicit knowledge is gained from the previous phase. The general competencies then derived into specific competencies, the specific competencies then derived into the indicator, lastly, the instrument is complete by adding how to measure of each indicator.

Table 2. Example of Externalization Process

General Competencies	Specific Competencies	Indicators	How to Measure	Measurement Result
Have basic mathematics skill	Know the concept of trigonometry	Able to calculate great angle of triangle	Ask the employee how to measure great angle of triangle	$sin\ A = \frac{opposite}{hypotenuse} = \frac{a}{c}$ $cos\ A = \frac{adjacent}{hypotenuse} = \frac{b}{c}$ $tan\ A = \frac{opposite}{adjacent} = \frac{a}{b}$
	Know the concept of Pythagoras	Able to calculate the angle of triangle	Ask the employee how to measure the angle of triangle	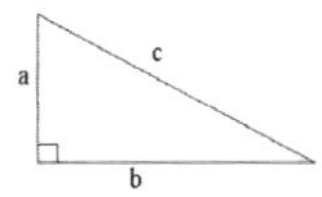 $c^2 = a^2 + b^2$

4.3. Combination

In combination phase, the externalized knowledge from 3 respondent merges into one new externalized knowledge. If 2 out of 3 respondent include the item then it will be used in the new competencies measurement breakdown. The externalized knowledge is used for the design of competencies measurement of mechanical expertise, this will help the supervisor or the stakeholder in assigning, evaluating, and developing competencies of each operator. In designing the competencies measurement, 11 general competencies with 17 specific competencies, 20 indicators, and 20 lists of how to measure each indicator is developed. Table 3 shows the example of combination process. From table 3 it is known that the combined process is done by comparing each answer from the respondent.

Table 3. Example of Combination Process

Competencies	Respondent A	Respondent B	Respondent C	Combined Spesific Competencies	Notes
Able to use the machine's accuracy gauge	Able to use dial indicator	Able to use dial indicator	Able to use dial indicator	Able to use dial indicator	Abilities to use dial indicator and spirit level to measure machine accuracy.
	Able to use spirit level	Able to use spirit level	Able to use spirit level	Able to use spirit level	
English Competencies	Able to understand machine manual book	Able to understand English idiom in machine	Able to understand machine manual book	Able to understand Basic English	This competency related to the ability of the employee to understand the manual book to operate the machine because all machine in the company using English as language.

4.4. Internalization

Focus group discussion (FGD) is conducted in this phase to share the new knowledge with all department employee. FGD is done by discussing the competencies measurement instrument. From FGD result it is known that the explicit knowledge conversion from the previous step is as expected and will be used as basis for evaluating the competencies in the company. Table 4 shows the FGD result. From table 4, it is known that all the instrument that is developed in this research can be used as a basic instrument. The instrument can help the maintenance department to assess current competencies of employee and further can use to improve the skill and competencies of maintenance operators. The knowledge lies within the instrument is rich enough for understanding the basic skill for doing the mechanical task in machine maintenance.

Table 4. FGD Result

No.	Focus Group Discussion Result
1	The proposed competencies measurement of mechanical expertise can be implemented in the company, especially maintenance department
2	The proposed measurement can be a basic instrument to assess employee competencies regarding the mechanical expertise in maintenance department
3	The proposed instrument can help the maintenance department to assess current competencies of employee and can be basic step to improve their skill and competencies
4	The knowledge lies within the instrument is enough to be guidance for next generation to understand the basic work of maintenance department, especially the mechanical expertise.

5. Conclusions

The purpose of this research is to improve the quality of human resources of maintenance staff in manufacturing company by designing competency measurement instrument that aims to assess the competency of employees. SECI method is used in this research for managing knowledge that is held by senior employees regarding employee competence of mechanical expertise. The knowledge about the competencies measurement breakdown is well documented in the form of explicit knowledge. The measurement of employee competencies can be done by using the proposed instrument. The competencies for mechanical or in general in maintenance can be different according to the change in technology or method, another research can be done to break down each characteristic according to the company needs,

References

[1] A. Thomas, Coaching for Staff Development, Wiley-Blackwell, 1995.

[2] A. Kurniawati, T. A. Samadhi, I. I. Wiratmadja and R. P. Soesanto, "The Impact of Source, Recipient, and Tacit Knowledge Characteristics on Tacit Knowledge Transfer Effectiveness," Taipei, 2016.

[3] I. Nonaka and H. Takeuchi, The Knowledge-Creating Company, New York: Oxford University Press, 1995.

[4] R. S.-d. Alwis, E. Hartmann, and H. G. Gemunden, "The role of tacit knowledge in innovation management," Copenhagen, 2004.

[5] S. Nurunisa, A. Kurniawati, R. P. Soesanto and U. Y. S. K. Hediyanto, "e-Learning Application for Machine Maintenance Process using Iterative Method in XYZ Company," *IOP Conference Series: Materials Science and Engineering, Volume 114, conference 1,* 2016.

[6] S. Atma, R. P. Soesanto, A. Kurniawati and U. Y. K. S. Hediyanto, "Best Practice Kegiatan Corrective Maintenance untuk Kerusakan Bearing pada Mesin Millac 5H 6P Berdasarkan Knowledge Conversion," in *Satelit 2017*, Batu, 2017.

[7] W. U. Rehman, "Knowledge creation (seci) through Human resource practices and rapport: a conceptual model from social perspective," *The Business & Management Review,* vol. 4, no. 1, pp. 155-163, 2013.

[8] R. P. Soesanto, L. Andrawina, I. R. Pertiwi and A. Kurniawati, "Relationship of Knowledge Management Process and the Performance from Human, Customer, and Organizational Perspective," in *KMICe*, Chiang Mai, 2016.

[9] G. Probst, S. Raub and K. Romhardt, Managing Knowledge Building Blocks for Success, Chichester: John Wiley & Sons, 2000.

[10] T. Davenport and L. Prusak, Working Knowledge, Cambridge: Harvard Business School Press, 1998.

[11] D. Siagian and Sugiarto, Metode Statistika: Untuk Bisnis dan Ekonomi, Jakarta: Gramedia, 2006.

Success Measures Evaluation for Mobile Commerce Using Text Mining based on Customer Tweets

A A Habib[1], R Govindaraju[1]

[1] Department of Industrial Engineering, Bandung Institute of Technology, Indonesia
Email: habibahmaditb@gmail.com

The customer acceptance of mobile commerce application systems is one of the driving factors of a successful mobile commerce implementation. However, there is an increasing number of companies that fail to implement. This shows the need to measure the success in implementing mobile commerce. One of the approach to determine success measures of mobile commerce can use customer perspective through user feedback regarding the application performance. The development of technology allows customers to share their feedback about mobile commerce services through online media without the limitation of time and place, for example using Twitter. Any tweets that contain comments or questions about a service is called a customer tweet. Customer tweets, processed using text mining, facilitate the company to get information about the success measures of a particular service based on customer point of view. Based on previous literature study, research on success measures of mobile commerce from the customer perspective is still in the early stages. Therefore, there are several reasons why this research is important. First, every mobile service has different characteristics, especially in the measurement of success. This shows that many explorations can be done regarding the success measures of mobile commerce. Second, the availability of customer tweets that contain comments about mobile commerce has allowed the data to be collected and processed into information to analyse the success measures of mobile commerce systems. Third, there is very little or no studies that use text mining method to model the success measures for mobile commerce based on customer tweets. Thus, the study reported in this paper aims at developing the success measures for mobile commerce based on customer tweets using text mining approach. The results from data processing indicate that some constructs that represent the success measures for mobile commerce include System Quality, Information Quality, Process Quality, Service Quality, Use, User Satisfaction, Individual Benefit and IT Infrastructure.

1. Introduction

The globalization era has forced enterprises to face complex competition and rapid changes in order to improve productivity effectively and efficiently. The improvement of efficiency and effectiveness in a company can be achieved by adopting the role of Information Systems (IS) and Information Technology (IT) appropriately and optimally [1]. It will encourage companies to increase their investment in Information Technology (IT), especially the ones using mobile phone technology in helping them to face business competition. Currently, mobile phone is used for various actions that support business activities such as text messaging, taking pictures, web surfing, downloading ring-tones, playing games and payment transactions [2]. The transaction processes, ownership transfer process as well as the right to use goods and services, which are initiated and completed using mobile access to computer networks with the help of electronic devices, are called mobile commerce [3]. Based on a research conducted by Lila Rajabilon in 2015, the number of mobile commerce transactions in the world increase year to year from 498 billion in 2006 to 4.8 billion in 2010 with average transaction increase of $ 7 in 2006 to $ 13 in 2010 [4].

Some companies use mobile commerce as one of media to communicate with customer, such as Uber, Grab, Lyft, etc. Although the number of companies that are using mobile commerce technology is increasing, but there are still many companies that fail in implementing mobile commerce[5]. Many mobile commerce applications failed to get user acceptance from the customer perspective.

The success measures of mobile commerce implementation from customer perspective can be determined using various methods. One of which is based on the final evaluation feedback provided by users about the information systems performance [5]. The development of technology makes it easier for companies to get feedback about certain services, including reviews and online product reviews [6]. The development of technology allows customers to be able to provide online product reviews with no boundary of time and place, one of which is through social media, for instance, Twitter. Twitter is a website service of microblog, a type of blog that limits the size of each post. Twitter facilitates users to write a text for a maximum140 characters. Twitter is one of the most easy-to-use social networks, because it only takes a short time to spread information instantly and widely.

Customer tweets on Twitter can be processed using text mining method. The method can be used to perform the process of classification, clustering, information extraction and information retrieval [7]. Text mining is a mining implemented by a computer to get novel insights as well as to discover implicit information from different text data sources [6]. In this context, text mining can facilitate companies in getting information about the success measures of particular services based on the customer's point of view. In the prior research, customer evaluation was conducted by implementing questionnaire survey in order to obtain a product review. This method requires a lot of time and cost. Moreover, the quality of data depends on the willingness of the respondents to participate and their ability to understand the questions as well as the length and complexity of survey questions given [8].

The research conducted by Chevalir & Mayzlin in 2006 stated that 24% of internet users that previously gave product reviews offline were starting to provide their product reviews online [9]. Through online review, customers are able to share their reviews more freely and voluntary. It makes customers have the highest authority [6] to decide the content of review that they want to share.

Based on several comparative literature studies, the current model to measure the success of mobile commerce using text mining method has not commonly been used. On the other hand, the importance of knowing the success measures of an information system, especially mobile commerce from the perspective of the customers, becomes a challenge in mobile commerce industry. Therefore, this study aims to explain the success measures of mobile commerce from the customers perspective based on customers tweets, using text mining method.

2. Implementation

This section is a preliminary study in order to identify indicator variables that best describe the success measures in mobile commerce. The identification process is performed using text mining technique, aspect extraction and apriori algorithm, with the help of Java, Python and RapidMiner. The steps are presented in Figure 1.

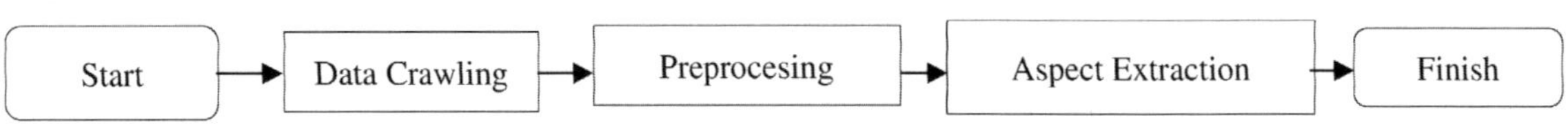

Figure 1. Aspect Identification Process

2.1. Data Crawling

Crawling is a process or automated script with a certain method of scanning or "crawl-ing" numerous pages in the internet to create index of the searched data. In this research, the data crawling is done by using RapidMiner software that connected to Twitter API. Using this software, all tweets that contain keywords such as "Uber", "Grab" and "Lyft" are collected as the dataset.

2.2. Data Preprocessing

Before performing the aspect extraction process, the product review dataset must go through preprocessing stage. The steps of data preprocessing are performed as follows:

IOP Conf. Series: Materials Science and Engineering **319** (2018) 012009 doi:10.1088/1757-899X/319/1/012009

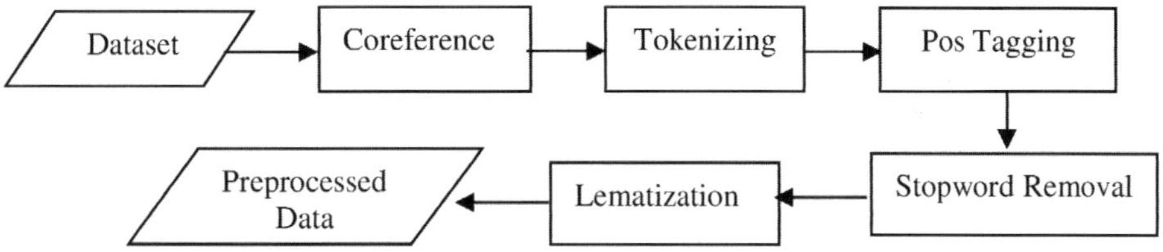

Figure 2. Preprocessing Process

Based on the Figure 2, data preprocessing consists of 5 stages namely coreference, tokenization, PoS tagging, stopwording and lemmatizing. The explanations are as follows:

2.2.1 Coreference Resolution

Coreference Resolution is the process of changing pronouns in English into the real subject. In this process the system will use the type of coreference resolution in the form of anaphora, cataphora, and co-referring noun phrase. An example of conference resolution is presented in Figure 3.

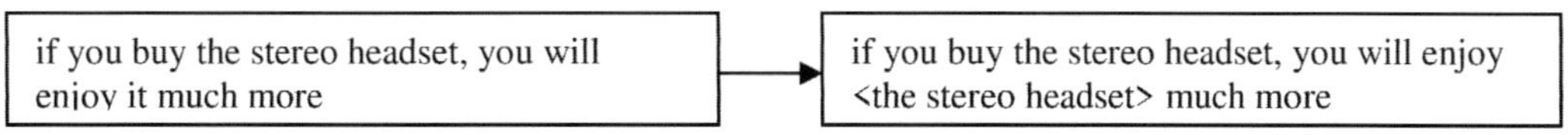

Figure 3. The Example of coreference resolution

2.2.2 Tokenizing

This process is performed by cutting the string based on each words. The result of this process is the snippet of words. An example of tokenizing is presented in Figure 4.

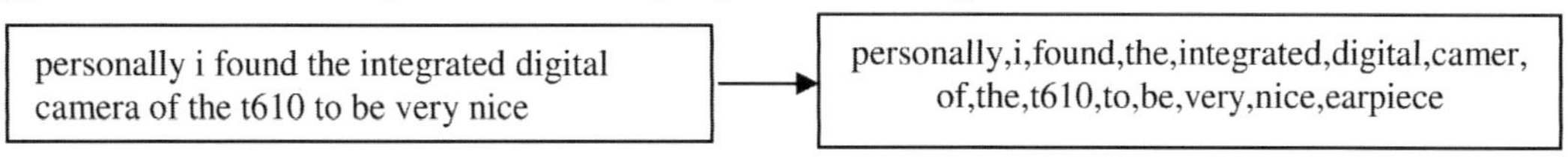

Figure 4. The example of tokenizing

2.2.3 Pos Tagging

PoS Tagging is process which select nouns that are found in the sentence. An example of tagging is presented in Figure 5.

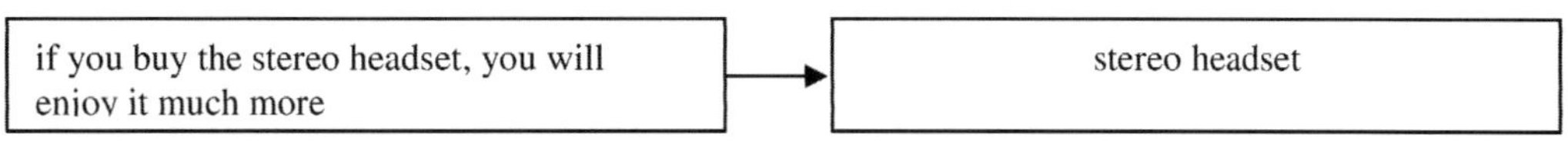

Figure 5. The example of pos tagging

2.2.4 Stopword Removal

The result of the pos tagging process then carried out to the process of removal of words that are unimportant and have no meaning. This filtering process is called stopword. An example of stopword removal is presented in Figure 6.

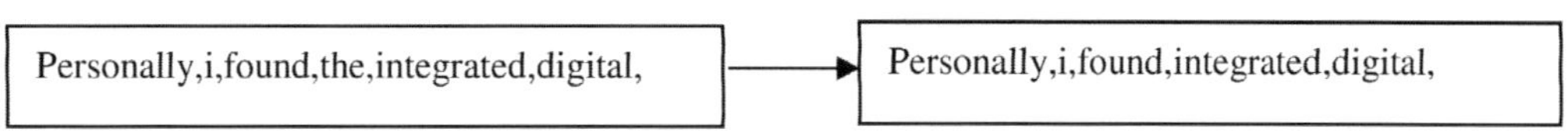

Figure 6. The example of stopword removal

2.2.5 Lemmatization

In this process, each words of the result of stopword removal will be changed into its basic form. An example of lemmatization is presented in Figure 7.

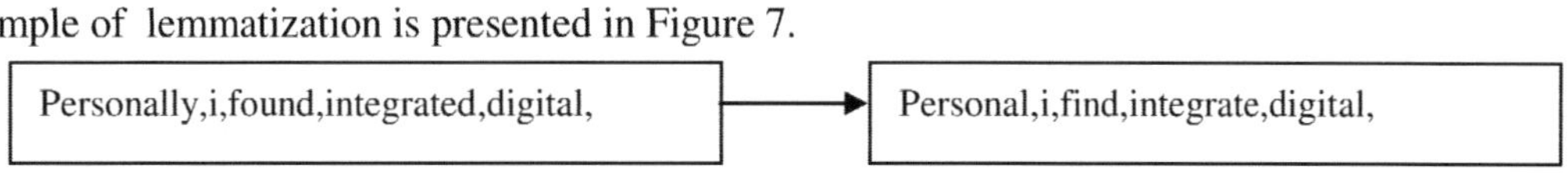

Figure 7. The example of lematization

2.3. Aspect Extraction

After data preprocessing, the next step is the identification of "*aspects*". In this context, "*aspects*" refer to the important words that are identified as indicators for construct of mobile commerce success measures. Nouns or noun phrases that are often mentioned by costumers in product reviews become candidates of potential aspects. Extraction of these aspects can be done using apriori algorithm to find frequent *itemsets* representing the aspects of product. The steps are presented in Figure 8.

Apriori algorithm produces a set of words that often appear alone or simultaneously in a sentence, regardless of the proximity of the words. So it is possible to produce a combination of some words that have no meaning. Elimination of aspects can be performed by calculating the distance between the two words that appear, if the distance is too far and it passes certain threshold then the aspect will be eliminated.

The threshold in this algorithm is called the pure support. Pure support is the number of sentence that only contains aspect which is not a superset of other aspects. If the pure support is less than the minimum pure support and the identified aspect is the subset of the other aspect phrases, then the aspect is eliminated. The apriori process that use Stanford University library package is implemented by choosing all types of nouns, including singular or mass noun, noun plural, proper noun, proper singular, proper plural. Then the coding process is performed based on each words in the dataset. Each words is coded which then will be used as input data for apriori library. The library then displays the output of each result of apriori processing.

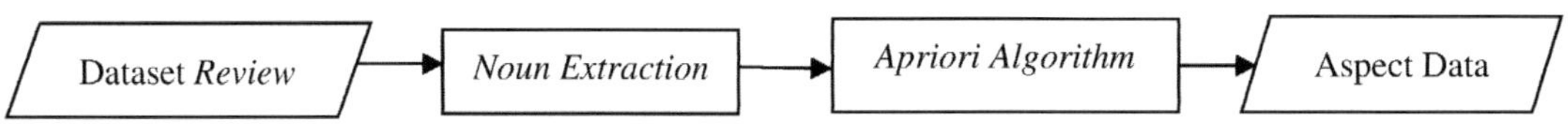

Figure 8. Aspect extraction with Apriori

Here is the example of extraction process performed by the system. Assumed, there is a dataset as with minimum support of 2 as shown in Table 1:

Table 1. The example of Apriori process

Sentence	List of Sentence
Sentence 1	The Application is amazing, the quality is very wonderfull, I like the function too
Sentence 2	Zoom very clear, the quality more than everything, it has good feature
Sentence 3	This Application is bad, I don't like the zoom and the quality, its feature is not good
Sentence 4	Wow I interest with the zoom and feature
Sentence 5	This Application is fantastic. It has good quality and I like the features

From the existing data, it will generate all types of nouns from the PoS tagging process and will continue to the process of numbering each noun data generated. The example is listed in Table 2.

Table 2. The numbering process on Apriori Algorithm

Sentence	List Of Noun	Numbering Apriori
Sentence 1	Application, Quality, Function	1 3 4
Sentence 2	Zoom, Quality, Feature	2 3 5
Sentence 3	Application, Zoom, Quality, Feature	1 2 3 5
Sentence 4	Zoom, Feature	2 5
Sentence 5	Application, Quality, Feature	1 3 5

Then each word is calculated to count the its occurrence as shown in Table 3. *Itemset* {4} does not meet the minimum support of 2 so it will be eliminated and not be processed (Table 4).

Table 3. One itemset frequent

Itemset	Support
{1}	3
{2}	3
{3}	4
{4}	1
{5}	4

Table 4. The data result after selection with Minsup 2 and one itemset

Itemset	Support
{1}	3
{2}	3
{3}	4
{5}	4

The second *itemset* generation process will then be carried out with the combination of 2 aspects that is performed by Brute Force method, as follows (Table 5 and Table 6):

Table 5. Two itemset Frequent

Itemset	Support
{1,2}	1
{1,3}	3
{1,5}	2
{2,3}	2
{2,5}	3
{3,5}	3

Table 6. The data result with minsup 2 and two itemset

Itemset	Support
{1,3}	3
{1,5}	2
{2,3}	2
{2,5}	3
{3,5}	3

Table 7. The result of Apriori 3 Itemset

Itemset	Whether or Not
{1,2,3} {1,2},{1,3},{1,3},{2,3}	No
{1,2,5}{1,2},{1,5},{2,5}	No
{1,3,5} {1,3},{1,5},{3,5}	Yes
{2,3,5}:{2,3},{2,5},{3,5}	Yes

Then it is proceed to the combination of 3 itemsets as shown on Table 7. If the itemset of combination of 3 meets the minimum support threshold, then it is considered as the result of apriori. Based on the above rule, only the itemset with the occurrences of {1,3,5} and {2, 3,5} become the final results of the apriori process. The final results refer to the word “Application”, “Zoom”, “Quality” and“Feature”.

3. Result

RapidMiner is used in this research to crawl customer tweets related to mobile commerce companies which are Uber, Grab and Lyft. From this crawling process, 7000 tweets related to Uber, 1500 related to Grab, 1200 related to Lyft are obtained. Using Apriori Method in factor extraction process, it is found that the best minimum support was 0.02 or the minimum occurrence frequency was at least 2% of the total terms. After the extraction of aspects for the model is obtained, then the clustering process is processed based on expert judgement. The results of the clustered aspects of success measures are shown in Table 8.

Based on Table 8, it was found that the success measures of mobile commerce based on customer tweets consist of several aspects, those are System Quality, Information Quality, Service Quality, Use, User satisfaction and Individual Benefit. This is in accordance with the measurement model of information systems produced by DeLone and McLean in 2003 [10]. In addition, the Proces Quality are also produced in accordance with recent research conducted by Legner et al. in 2016 [5], and IT Infrastructure Services conducted by Younis et al. in 2013 [11]. From Table 8, with the help of apriori algorithm, the more tweets are included in the processed, the more aspects appear. Tweets related to Uber produced 43 aspects, tweets related to Lyft produced 32 aspects, and tweets related to Grab produced only 26 aspects. The amount and quality of data affect the numbers of aspects extracted from the model and quality of data affect the result of the extraction aspect on the model of mobile commerce success measures.

Table 8. The result aspect based on Apriori Algorithm

Uber	Grab	Lyft
System Quality: code promo, email, account, apps, password, email link, refund system, problem application, using credit card, error message, application quality. **Information Quality:** details information, contact, address information. **Process quality:** check order process, request time application, trip process. **Service Quality:** driver, promo, time access, car, credit for new users, application support, awareness, service center, time to reply, customer service, resolve the issue, respons application. **Use:** direct massage, feature, noted feature, payment. **User Satisfaction:** feeling happy, good experience. **Individual Benefit:** helpfull experience **IT Infrastructure:** phone number, connection, phone quality.	**System Quality:** app, application, code. **Information Quality:** news, details information **Process quality:** Join application **Service Quality:** time access, support center, driver, car, promo code, rides, offer, service center. **Use:** payment, feature, **User Satisfaction:** love, enjoy, good, nice. **Individual Benefit:** free apps. **IT Infrastructure:** phone, connection.	**System Quality:** code, enter apps, app quality. **Information Quality:** Credit, information guests, users. **Process quality:** join application, fastest apps. **Service Quality:** driver, ride, promo, coupon, taxi, ride, cab. **Use:** type feature, payment. **User Satisfaction:** Deal, feeling happy. **Individual Benefit:** Free, special offer, lyft discount. **IT Infrastructure:** download, internet, phone.

4. Conclusion

This study found that text mining with Apriori Algorithm using review to mine the the customer tweets data together with the empirical testing with statistics on the results of the text mining method can be used to identify the success measure mobile commerce. Using this method it was found that with the best minimum support of 0.02, factors that measure the success of mobile commerce in the case of Uber, Grab and Lyft are System Quality, Information Quality, Process Quality, Service Quality, Use, User Satisfaction, Individual Benefit and IT Infrastructure. Based on the data processing results it was found that Service Quality has the highest frequency compared to the other measures.

5. References

[1] Park, Suh, H. J., & Yang, H. D. (2007). Perceived absortive capacity of individual users in performance of enterprise resource planning (ERP) usage: the case of korean firms. *Information and Management*, (pp. 300-312).

[2] Smith, E. D. (2005). Marketers Go the Mobile Route," Knight-Ridder Tribune Business News ".

[3] Tiwari, R., Buse, S., & Herstatt, a. (2006). From Electronic To Mobile Commerce Opportunites Through Technology Convergence For Business Services. *Asia Pacific Tech Monitor*, 38-45.

[4] Rajabion, L. (2015). Critical Factors for Adoption of Mobile Commerce Service. *International Journal of Business and Social Science*, 26-35.

[5] Christine, L. D., Nils, U. P., & Christoph, N. D. (2016). Mobile Business Aplication For Service and Maintenance Process: Using Ex Post Evaluation by End-Users as Input for Iterative Design. *Information and Management.*

[6] Zhu, & Zhang. (2010). Impact of online customer review on sales: The moderating role of product and customer characteristics. *Journal of Marketing Research*, 133-148.

[7] Berry, W. M., & Kogan. (2010). *Text Mining Aplication and theory.*

[8] DeeLeew, & DeHeer. (2002). Trends in household survey nonresponse: A longitudinal international perspective. Survery Nonresponse. New York. 41-54.

[9] Chevalir, J., & Mayzlin, D. (2016). The effect of word of mouth on sales: Online Book Review. *Journal of Marketing Research*, 345-354.

[10] DeLone, W., & McLean, E. (1992). Information systems success: The quest for the dependent. *Information Systems Research,* 60–95.

[11] Alsabawy, A. Y., Cater-Steel, A., & Soar, J. (2013). IT infrastructure services as a requirement for e-learning system success. *Computer and Education*, 431-451.

IOP Conf. Series: Materials Science and Engineering **319** (2018) 012010 doi:10.1088/1757-899X/319/1/012010

Investigation of roughing machining simulation by using visual basic programming in NX CAM system

Mohamad Hafiz Mohamad and Muhammed Nafis Osman Zahid

Faculty of Manufacturing Engineering, Universiti Malaysia Pahang
26660, Pekan, Pahang, Malaysia

nafis@ump.edu.my

Abstract. This paper outlines a simulation study to investigate the characteristic of roughing machining simulation in 4th axis milling processes by utilizing visual basic programming in NX CAM systems. The selection and optimization of cutting orientation in rough milling operation is critical in 4th axis machining. The main purpose of roughing operation is to approximately shape the machined parts into finished form by removing the bulk of material from workpieces. In this paper, the simulations are executed by manipulating a set of different cutting orientation to generate estimated volume removed from the machine parts. The cutting orientation with high volume removal is denoted as an optimum value and chosen to execute a roughing operation. In order to run the simulation, customized software is developed to assist the routines. Operations build-up instructions in NX CAM interface are translated into programming codes via advanced tool available in the Visual Basic Studio. The codes is customized and equipped with decision making tools to run and control the simulations. It permits the integration with any independent program files to execute specific operations. This paper aims to discuss about the simulation program and identifies optimum cutting orientations for roughing processes. The output of this study will broaden up the simulation routines performed in NX CAM systems.

1. Introduction

In conventional approach, Computer Numerical Control (CNC) requires a significant human intervention and efforts to execute the cutting processes [1]. Machining operator need to understand the 3D design in Computer Aided Design (CAD) and develop the process planning in terms of tool selection, toolpath planning, machining strategies and fixturing method. After all the desired information has been obtained, the data will be entered into the Computer Aided Manufacturing (CAM) system to produce the CNC machine code. In CNC machining, the machined parts features is important to be defined first, in order to create a set of processes to obtain a geometry according to the required accuracy [2]. In today's competitive industry, automation in process planning becomes crucial with aim to enhance the efficiency reduce lead time and improve productivity and quality. Current methods of machining process planning is highly depend on the machinist experience [3,4]. A demand to increase productivity is the main challenge in global manufacturing and each of the companies will compete to achieve the targets. When dealing with new machining process, the trial and error approach is one of the common methods. Although the trial and error approach seamlessly applicable, but it will cause a major impact on productivity in terms of machining time and cost [5].

There are few ways to improve the current approach with the new and practical one. The development of computer graphics has improved the simulation capabilities. For example, in the generation of NC codes from machining processes. Instead of using the conventional practice, there are few other methods that can be employed by the integration of advanced tool and programming codes. Lack of technical knowledge and integration capability of CAM system is one of the reasons why most manufacturers suffered in their production. Therefore, an independent Computer Aided Process Planning (CAPP) has been introduced as a communication platform between CAD and CAM systems. Computer-Aided Process Planning (CAPP) is the system where computers are used to assist or replace human intervention in order to produce a better process plans in optimum time [6]. The basic fundamental concept of CAPP is to automate the process planning, where the time and efforts of process planning will be reduced and high process consistency achieved.

2. Overview of CNC Machining Simulation

The development of Rapid Prototyping by using CNC Machining (CNC-RP) is one of the effective technique in implementing the machining simulation processes [7]. With this method, process plans for machined component is allowable to be automated. A 3-axis milling with fourth axis indexer devices was employed in CNC-RP and act as a clamping device that permits the parts to be rotated in one rotation axis. The aim of CNC-RP method is to machine all parts surface within specific set of orientation without re-fixturing [8]. **Figure 1** shows the indexer setup for 1 axis rotation. This approach allows a round stock of material to be clamped between 3-jaw chucks and rotary indexer will rotate the material between the operation sequences. This method is able to propose an optimal cutting angle within set of orientation until real part shape is visible.

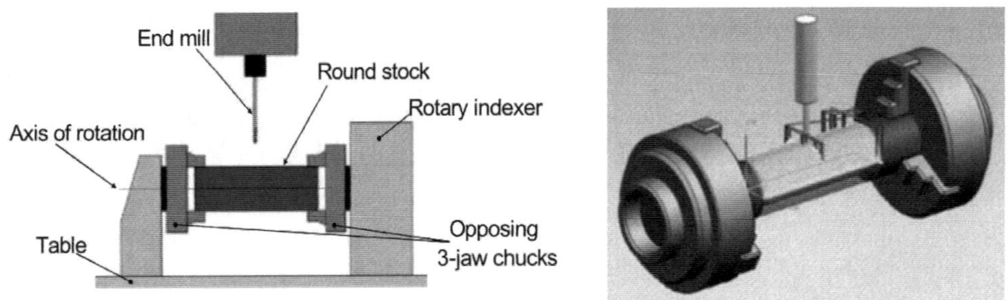

Figure 1. Rotary Indexer Setup

A recent study has proposed several approaches to further improve the roughing machining operation in CNC-RP (Nafis et al. 2014). It was successfully executed by simulating a set of orientation with 3 orientation (0°-120°-240°) (0°-135°-225°) and 4 orientation (0°-90°-190°-270°). Cutting angle with set of 4 orientations has proven to be able to execute roughing operation more efficiently in terms of machining time [9]. This approach improves the previous method by splitting roughing orientation from finishing orientation and incorporating with other orientation. For each orientation, roughing and finishing operation are performed in sequence with different value of cutting angle and parameter.

3. Roughing Simulation Methodology

4. In this study, process planning simulation is executed by customized program using Visual Basic programming in NX CAM system. Machining operation in the form of automation can be developed in NX CAM system where it is equipped with a programming tool that corresponds to the proposed simulation. In NX CAM system, advanced tool called "Journaling" able to record and translate the command or instruction into Visual Basic, JAVA and C++ language [10].

Figure 2 shows a roughing simulation GUI and simulation algorithm. It was developed in Visual Basic Studio using a windows form application template. The GUI has several text boxes and drop-down lists where user can input selected parameter, workpiece diameter, cutting parameter, set of orientation and cutting tool size. The "Activate Simulation Analysis" box needs to be checked in order to enable the simulation mode. Simulation algorithm has been programmed behind the system to execute the simulation based on user input.

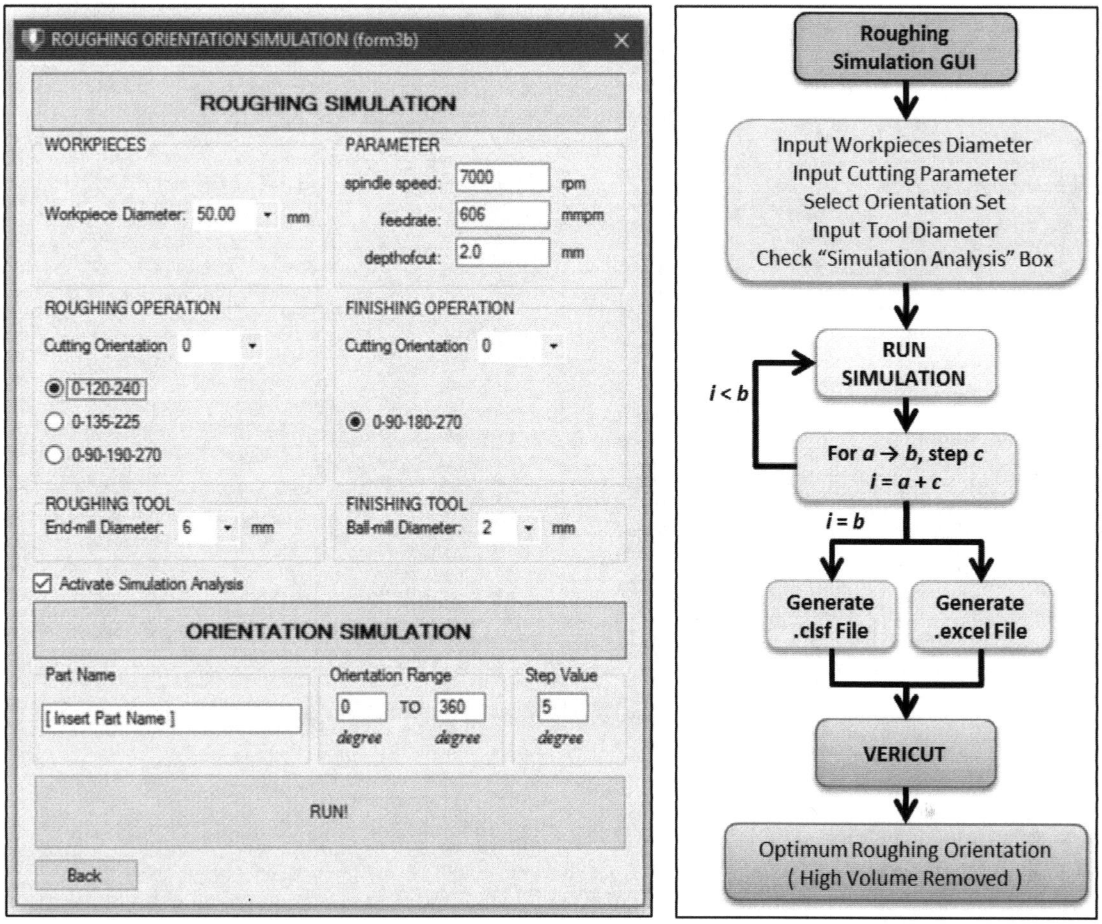

Figure 2. (a) Roughing Simulation Graphical User Interface (GUI) (b) Simulation Algorithm

In this simulation study, the orientation range is set from 0°to 360° with 5° step value. Therefore, for each set of orientations, 72 results will be obtained and analysed by the system which will then be exported to cutter location source file (.clsf) files and excel files, as shown in

Figure 3. CLSF files works as a communication tool between NX and VERICUT software to validate the simulation output [11]. In this study, CGTech VERICUT 8.0 applications have been used to analyse the excess volume left on parts after machining [12].

IOP Conf. Series: Materials Science and Engineering **319** (2018) 012010 doi:10.1088/1757-899X/319/1/012010

Orientation	Cutting Length (mm)	Cutting Time (min)	Volume Removed (mm³)	Orientation	Cutting Length (mm)	Cutting Time (min)	Volume Removed (mm³)
5	44,615.94	88.52	102,671.91	185	44,563.60	88.42	102,653.14
10	44,793.58	88.88	102,686.30	190	44,282.67	87.86	102,665.71
15	44,202.80	87.70	102,813.63	195	44,357.32	88.01	102,650.43
20	44,331.95	87.96	102,625.67	200	44,302.97	87.90	102,615.58
25	44,302.93	87.90	102,657.07	205	44,648.03	88.59	102,650.14
30	44,504.48	88.30	102,661.47	210	44,610.28	88.51	102,639.85
35	44,568.62	88.43	102,692.05	215	44,699.62	88.69	102,704.97
40	44,526.92	88.35	102,727.98	220	44,445.90	88.19	102,717.90
45	44,428.58	88.15	102,756.10	225	44,533.82	88.36	102,708.93
50	44,891.33	89.07	102,710.44	230	44,216.88	87.73	102,684.85
55	44,762.72	88.81	102,689.57	235	44,229.82	87.76	102,651.03
60	44,703.75	88.70	102,677.43	240	45,617.22	90.51	102,633.58
65	44,392.85	88.08	102,639.12	245	44,725.78	88.74	102,613.52
70	44,157.27	87.61	102,663.77	250	44,706.07	88.70	102,646.83
75	44,034.72	87.37	102,663.85	255	44,112.15	87.52	102,645.98
80	44,283.86	87.86	102,674.29	260	44,325.41	87.95	102,658.02
85	44,609.89	88.51	102,693.82	265	44,698.42	88.69	102,680.07
90	44,444.84	88.18	102,707.70	270	45,374.25	90.03	102,641.06
95	44,211.51	87.72	102,703.09	275	44,677.47	88.65	102,685.25
100	44,391.86	88.08	102,633.35	280	44,625.52	88.54	102,657.82
105	44,380.86	88.06	102,766.54	285	44,170.72	87.64	102,660.09
110	44,627.53	88.55	102,654.40	290	44,195.77	87.69	102,668.54
115	44,527.38	88.35	102,669.13	295	45,659.53	90.59	102,169.68
120	44,259.97	87.82	102,667.65	300	44,680.72	88.65	102,682.30
125	44,446.12	88.19	102,691.31	305	44,848.46	88.99	102,690.15
130	44,314.89	87.93	102,691.92	310	45,082.05	89.45	102,755.27
135	44,452.77	88.20	102,718.44	315	44,844.79	88.98	102,779.74
140	47,464.88	94.18	102,674.81	320	45,155.24	89.59	102,724.19
145	44,307.07	87.91	102,662.33	325	44,872.89	89.03	102,712.00
150	44,587.81	88.47	102,651.96	330	44,478.79	88.25	102,663.54
155	44,510.04	88.31	102,615.23	335	44,279.26	87.86	102,636.97
160	44,445.34	88.19	102,646.86	340	44,632.35	88.56	102,717.53
165	44,666.56	88.62	102,616.24	345	44,431.49	88.16	102,645.96
170	44,446.16	88.19	102,658.91	350	44,568.19	88.43	102,675.54
175	44,554.92	88.40	102,677.87	355	44,535.70	88.36	102,683.22
180	44,603.31	88.50	102,640.50	360	44,349.27	87.99	102,716.84

Figure 3. Sample of information extracted from developed program

5. Results and Discussion

Virtual simulation was conducted by using a developed program. Experiments are carried out on four part models with 3 simulations for each model; Baby Toy, Lego, Fan Blade and Gear. Model selection is based on characteristics and geometry, consisting of flat surfaces and non-flat surfaces**. Table 1** summarizes the simulation results for the selected models. For each model, first and second column represents the result of three roughing orientation approach which is (0-120-240) and (0-135-225). The third column represents the result of four roughing orientation approach, (0-90-190-270). Comparisons were made through all the simulation approach that differs in roughing orientation set.

A value of cutting angle proposed by the developed program is dependent by machined parts geometry. Each part has different geometry features which influence the value of cutting angle within orientation set. It is based on the region that can be accessed by cutting tools to complete machine the cylindrical workpieces. The result shown in Table 1 portray that the developed program is capable to adapt with various geometrical parts, where the "stickiness" function inside the program is customized into a free-features function. This customization allows flexibility to the developed program in order to executing the simulation with various parts. At the same time, the developed program manages to analysis and extracts a large number of information from the simulation within less planning processes. In conventional practices, there is no doubt that virtual machining simulation will take a long planning processes as each simulation for each parts needs to be repeated several times manually, in order to obtain a required results. By utilizing this program, the orientation values for each simulation are changing automatically, until the entire desired cutting angles have been analyzed.

Table 1. Simulation Results

Result	BABY TOY			LEGO		
	3 orientation (0-120-240)	3 orientation (0-135-225)	4 orientation (0-90-190-270)	3 orientation (0-120-240)	3 orientation (0-135-225)	4 orientation (0-90-190-270)
Roughing Orientation (degree)	155° 275° 35°	100° 235° 325°	165° 255° 355° 75°	300° 60° 180°	315° 90° 180°	90° 180° 280° 0°
Roughing Cutting Time (min)	81.95	79.82	102.15	71.88	69.12	83.77
Finishing Cutting Time (min)	72.92	75.95	59.15	34.3	36.65	28.8
Machining Cutting Time (min)	154.87	155.77	161.30	106.18	105.77	112.57
Roughing Volume Removed (mm^3)	168,888.45	168,470.44	172,530.24	146,978.76	147,848.89	149,660.64
Total Volume Removed (mm^3)	187,249.22	187,052.31	187,346.71	162,307.53	162,331.04	162,338.57
Percentage of Volume Removed (roughing)	90.2%	90.1%	92.1%	90.6%	91.1%	92.2%

Result	FAN BLADE			GEAR		
	3 orientation (0-120-240)	3 orientation (0-135-225)	4 orientation (0-90-190-270)	3 orientation (0-120-240)	3 orientation (0-135-225)	4 orientation (0-90-190-270)
Roughing Orientation (degree)	250° 10° 130°	25° 160° 250°	225° 315° 55° 135°	155 275° 35°	340° 115° 205°	225° 315° 55° 135°
Roughing Cutting Time (min)	34.17	34.27	47.20	48.88	48.92	58.82
Finishing Cutting Time (min)	22.65	22.93	19.10	96.28	91.03	84.72
Machining Cutting Time (min)	56.82	57.20	66.30	145.16	139.95	143.54
Roughing Volume Removed (mm^3)	67,676.94	67,596.11	68,501.70	83,540.09	84,001.73	87,764.52
Total Volume Removed (mm^3)	73,008.13	72,893.77	73,103.93	120,266.06	120,254.71	120,250.19
Percentage of Volume Removed (roughing)	92.7%	92.7%	93.7%	69.5%	69.9%	73.0%

6. Conclusion

In this research paper, the outcomes of the simulation are generated through the application of Visual Basic Programming, NX CAM and VERICUT tools. A developed roughing simulation program offers another solution for machining simulation routines in CAM environment. Integration of Visual Basic programming language in NX CAM system manages to optimize the process planning and able to execute multiple simulation routines effectively. This approach is considered as an alternative way to determine an optimum cutting orientation, in terms of high volume removed in roughing operation before proceed to finishing processes. As conclusions, the developed program managed to improve the efficiency of simulation routines and shorten the processing time in NX CAM system.

Acknowledgments

We acknowledge with gratitude to Ministry of Higher Education Malaysia for providing a financial support under Research Acculturation Grant Scheme (RDU151406), which realize this research project. We also thankful to Universiti Malaysia Pahang for providing outstanding research facilities and constant encouragement throughout the years of the project execution. Certainly, thanks to all parties involved in this project.

References

[1] Turley S P et al., 2014. Automated Process Planning and CNC-Code Generation. *2014 Industrial and Systems Engineering Research Conference*, 2138(November 2015), pp.2138–2145.

[2] Frank, M.C., 2004. Rapid Planning for CNC Milling A New Approach for Rapid Prototyping. *Journal of Manufacturing Systems*, 23(3), pp.242–255

[3] Xu, X., Wang, L.H. & Newman, S.T., 2011. Computer-aided process planning: a critical review of recent developments and future trends. *International Journal of Computer Integrated Manufacturing*, 24(1), pp.1–31. Available at: http://dx.doi.org/10.1080/0951192x.2010.518632

[4] Relvas, C. & Simoes, J.A., 2004. Optimization of computer numerical control set-up parameters to manufacture rapid prototypes. *Proceedings of the Institution of Mechanical Engineers, Part B: Journal of Engineering Manufacture*, 218(8), pp.867–874.

[5] El-Mounayri, H., Kishawy, H. & Tandon, V., 2002. Optimized CNC end-milling: A practical approach. *International Journal of Computer Integrated Manufacturing*, 15(5), pp.453–470

[6] Anderberg, S., 2009. *A study of process planning for metal cutting*. Chalmers University of Technology

[7] Frank, M., 2007. Implementing Rapid Prototyping Using CNC Machining (CNC-RP) Through a CAD/CAM Interface. *Proceedings of the Solid Freeform Fabrication ...*, pp.112–123.

[8] Boonsuk, W. & Frank, M.C., 2009. Automated fixture design for a rapid machining process. *Rapid Prototyping Journal*, 15(2), pp.111–125

[9] Nafis, M., Case, K. & Watts, D., 2014. Optimization of roughing operations in CNC machining for rapid manufacturing processes. *Production & Manufacturing Research*, 2(1), pp.519–529. Available at: http://dx.doi.org/10.1080/21693277.2014.938277

[10] Moi, M.B., 2013. *Web Based Customized Design*. Norwegian University of Science and Technology

[11] Nafis, M., Case, K. & Watts, D., 2015. End mill tools integration in CNC machining for rapid manufacturing processes: simulation studies. *Production & Manufacturing Research*, 3(1), pp.274–288. Available at: http://www.tandfonline.com/doi/full/10.1080/21693277.2015.1094756

[12] CGTech, 2016. VERICUT 8.0: Release Notes

[13] Nafis, M., Case, K. & Watts, D., 2016. Cutting orientations for non-complex parts in 4axis machining

PVA/Graphene Nanocomposite: Morphology and its Thermal Properties

Abu Hannifa Abdullah[1]*, Zulhelmi Ismail[2], Anis Sakinah Zainal Abidin[1], and Fadwa Sameeha Ismail[1], Kamal Yusoh[1].

[1] Faculty of Chemical & Natural Resources Enginnering, Universiti Malaysia Pahang, Lebuhraya Tun Razak, 26300 Kuantan, Pahang, Malaysia.
[2] Faculty of Manufacturing Engineering, Universiti Malaysia Pahang, 26600 Pekan, Pahang, Malaysia.

kamal@ump.edu.my

Abstract. Graphene is known as a wonder materials that can be used to enhance the properties of nanocomposites. In this work, PVA/Graphene nanocomposite was fabricated using simple solution method. The photograph of the nanocomposite samples shown the transparency of the sample reduced as the graphene content increase. The photograph also shown the PVA and the modified graphene are miscible and compatible. The XRD of the samples proved the exfoliation of graphene in the nanocomposite and the result of the thermal property improvement for the sample is confirmed by the TGA.

1. Introduction

Polymer nanocomposites have found widespread application in automotive, aerospace and even medical field[1]. Besides being the material that is easy to process, it also has high strength to weight ratio[2]. Water soluble polymer such as PVA is always used widely as a host for many types of nanofillers [3, 4]. Recently, the use of graphitic materials as fillers has become a spotlight for many researchers in nanocomposite field[5]. Graphite is a layered structure consists of carbon atoms bonded on two-dimensional honeycomb pattern. The layers are bound by weak van der Waals forces which enable it to be separated into platelets with few layers or even single layer with nanometre scale thickness. That material is called graphene. After the innovative experiment by Sir Andre Geim and Konstantin Novoselov[6], graphene has become a focus in many research area primarily due to its extraordinary electronic and thermal properties. Due to the great properties owned by graphene, many approaches are made to improve other materials such as polymers by producing graphene based materials. Although graphene is known to improve the thermal properties of nanocomposite, agglomerated graphene can cause its properties to worsen. Therefore the morphology of graphene in the nanocomposite is important to ensure the properties improvement of the nanocomposite[7]. Graphene is hydrophobic by nature thus making it difficult to be blended with PVA as they are not miscible[8]. The preparation of PVA/graphene nanocomposite usually involves a reduction process of graphene oxide/PVA solution using hydrazine at 100°C at about 24 hours of stirring[9, 10]. Here, we show the simple method to prepare PVA/graphene nanocomposite by direct solution intercalation method using only water as surfactant. The effects of morphology to thermal properties of the nanocomposite were also studied by varying the weight percentage of the filler to the nanocomposite.

2. Methodology

2.1. Materials
Water dispersible functionalised graphene (1 mg/ml) was obtained from Graphene Production and Nanomaterial Research Group Universiti Malaysia Pahang and used without further purification. PVA powder were bought from Sigma Aldrich (M_w 89,000-98,000, 99+ %). Ultra-pure water was used as the solvent.

2.2. Preparation of PVA/Graphene Nanocomposite
The method to prepare PVA/Graphene nanocomposite films is given as follows. 1 g of PVA powder was slowly added to 100 ml distilled water at 90°C and stirred for 1 hours by using magnetic stirrer at 700 rpm. Desired amount of graphene dispersion was dropped gradually by using glass dropper into the PVA solution and stirred for 1 hour. The mixed solution was then poured into a 10.5 diameter petri dish and dried at 60°C for 24 hours. The dried PVA/graphene nanocomposite was carefully peeled off from the petri dish and stored in a desiccator for further use. The weight content of graphene was varied to 0.01, 0.05 and 0.1 wt % to highlight the effect of the filler content to the properties of the nanocomposite. These formulation is chosen to prevent the possibility of agglomeration as it will reduce the thermal properties of the nanocomposite. High filler content will surely lead to the agglomeration of filler in the nanocomposite[11].

2.3. Characterization
The X-ray diffraction (XRD) test was carried out using Rigaku MiniFlex benchtop x-ray diffractometer with Cu Kα radiation source of (λ= 0.154 nm). The machine is operated at 30kv and 15mA. The Thermogravimetric Analysis was performed by using Mettler Toledo thermogravimetric analyser. The sample was dried in a desiccator for several days before the test was run. The weight loss of the samples was recorded from 35°C to 700°C with the heating rate of 10°C/min in nitrogen atmosphere.

3. Results and Discussion

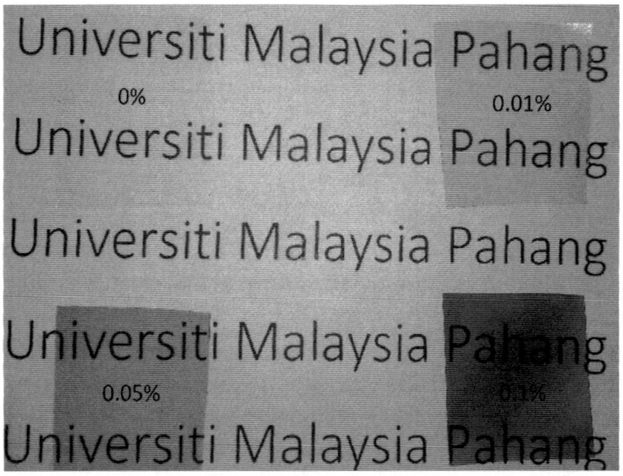

Figure 1. Photograph of PVA/Graphene nanocomposite samples.

Figure 1 shows the photographic image of PVA nanocomposite film samples with different graphene loadings. It shows that the nanocomposites with higher graphene loading tend to decrease the transparency of the PVA which indicates that the functionalised graphene and PVA have good interaction. On the other hand, nanocomposite with higher graphene loading shown ununiformed graphene distribution that may have cause by agglomerated graphene in the nanocomposite.

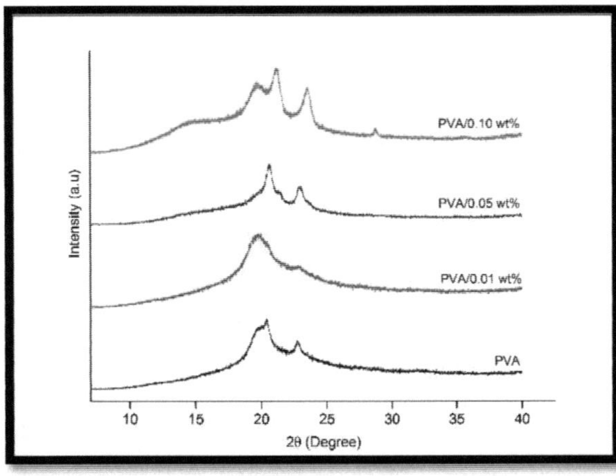

Figure 2. XRD diffractogram of PVA/Graphene nanocomposite with different filler content.

The discussion of the results begins with the XRD diffractogram of nanocomposite with different graphene content. As observed from the Figure 2, the diffractogram of the sample PVA shows the typical polyvinyl alcohol peak at 19.7, and 22.4 which represent the crystalline phase of the PVA. The broadening of peak 19.7 is observed in PVA/0.01 wt%. This finding highlights that the graphene have exfoliated in the polymer matrix at the molecular level [9]. It causes the PVA crystallinity to decrease and thus influencing the properties of the nanocomposites[12, 13]. It has been theorised that the change in crystallinity of the nanocomposite caused by the "molecule restriction effect" to the PVA chains caused by graphene [10]. On the other hand, PVA/0.05 wt% and PVA/0.10 wt% have not shown any peak broadening at 19.7.

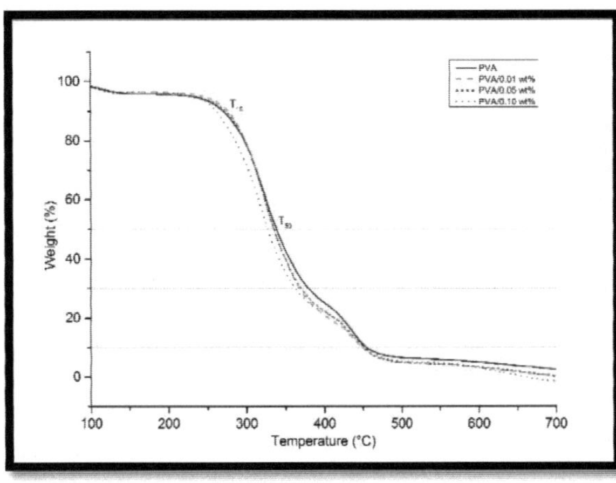

Figure 3. TGA thermogram of nanocomposites with different graphene content.

Table 1. Summary of T10 and T50 of PVA/Graphene Nanocomposites.

Sample	T_{50}	T_{10}
PVA	338	270
PVA/0.01 wt%	335	277
PVA/0.05 wt%	335	274
PVA/0.10 wt%	327	262

The thermal stability of PVA and graphene nanocomposites were measured according to the value of the temperature at 10% weight loss (T_{10}) and 50% weight loss (T_{50}). TGA thermogram Figure 3 shown that the thermal properties of the nanocomposites varied by the addition of graphene. All samples start to lose weight around 120°C due to presence of water in the samples. The T_{10} of PVA/0.01 wt% and PVA/0.05 wt% are higher by 7°C and 4°C respectively compared with PVA. The increase of T_{10} in nanocomposite is due to the dispersion level of graphene in the polymer matrix that form a barrier labyrinth effect. The stability of the graphene structure also acts as a physical barrier to the nanocomposite [14-16]. Graphene are known to have a good thermal conductivity and in this case, that property of graphene itself tends to speed up the heat transfer and eventually causes the degradation temperature of nanocomposite to deteriorate further on as showed in the T_{50}. The weakening of the thermal degradation temperature for PVA/0.1 wt% is also caused by the thermal conductivity of graphene in an agglomerated graphene nanocomposite where it causes heat to focus in that certain area and thus speeding the degradation process.

4. Conclusion

In conclusion, this study set up to figure out the relationship between the morphology of PVA graphene nanocomposite and its thermal degradation properties. PVA/graphene nanocomposites were successfully produced by using solution method. The result found out that the T_{10} value of PVA/0.01 wt% has slightly increased. The XRD diffractogram has confirm a peak broadening at 19.7° for PVA/0.01 wt% which represent the exfoliation of fillers in the nanocomposite. Based on the results, the influence in thermal properties of the PVA/graphene nanocomposite can be attributed to the arrangement and dispersion level of graphene in the PVA polymer matrix.

References

[1] Feldman D 2016 Polymer nanocomposites in medicine *Journal of Macromolecular Science, Part A* **53** 55-62

[2] Atif R, Shyha I and Inam F 2016 Mechanical, thermal, and electrical properties of graphene-epoxy nanocomposites—A review *Polymers* **8** 281

[3] Chang J Y, Godovsky D, Han M, Hassan C, Kim J, Lee B, Lee Y, Peppas N, Quirk R and Yoo T 2000 *Biopolymers· PVA Hydrogels Anionic Polymerisation Nanocomposites* vol 153: Springer Science & Business Media)

[4] Vijaya Kumar R, Elgamiel R, Diamant Y, Gedanken A and Norwig J 2001 Sonochemical preparation and characterization of nanocrystalline copper oxide embedded in poly (vinyl alcohol) and its effect on crystal growth of copper oxide *Langmuir* **17** 1406-10

[5] Geim A K and Novoselov K S 2007 The rise of graphene *Nature materials* **6** 183-91

[6] Novoselov K 2011 Nobel lecture: Graphene: Materials in the flatland *Reviews of Modern Physics* **83** 837

[7] Ismail Z, Abdullah A H, Zainal Abidin A S and Yusoh K 2017 Application of graphene from exfoliation in kitchen mixer allows mechanical reinforcement of PVA/graphene film *Applied Nanoscience* **7** 317-24

[8] Hong G, Han Y, Schutzius T M, Wang Y, Pan Y, Hu M, Jie J, Sharma C S, Müller U and Poulikakos D 2016 On the mechanism of hydrophilicity of graphene *Nano letters* **16** 4447-53
[9] Yang X, Li L, Shang S and Tao X-m 2010 Synthesis and characterization of layer-aligned poly (vinyl alcohol)/graphene nanocomposites *Polymer* **51** 3431-5
[10] Bao C, Guo Y, Song L and Hu Y 2011 Poly (vinyl alcohol) nanocomposites based on graphene and graphite oxide: a comparative investigation of property and mechanism *Journal of Materials Chemistry* **21** 13942-50
[11] Liang J, Huang Y, Zhang L, Wang Y, Ma Y, Guo T and Chen Y 2009 Molecular-level dispersion of graphene into poly (vinyl alcohol) and effective reinforcement of their nanocomposites *Advanced Functional Materials* **19** 2297-302
[12] Zhao X, Zhang Q, Chen D and Lu P 2010 Enhanced mechanical properties of graphene-based poly (vinyl alcohol) composites *Macromolecules* **43** 2357-63
[13] Xu Y, Hong W, Bai H, Li C and Shi G 2009 Strong and ductile poly (vinyl alcohol)/graphene oxide composite films with a layered structure *Carbon* **47** 3538-43
[14] Jiang L, Shen X P, Wu J L and Shen K C 2010 Preparation and characterization of graphene/poly (vinyl alcohol) nanocomposites *Journal of Applied Polymer Science* **118** 275-9
[15] Kashiwagi T, Grulke E, Hilding J, Harris R, Awad W and Douglas J 2002 Thermal degradation and flammability properties of poly (propylene)/carbon nanotube composites *Macromolecular Rapid Communications* **23** 761-5
[16] Patole A S, Patole S P, Jung S-Y, Yoo J-B, An J-H and Kim T-H 2012 Self assembled graphene/carbon nanotube/polystyrene hybrid nanocomposite by in situ microemulsion polymerization *European Polymer Journal* **48** 252-9

IOP Conf. Series: Materials Science and Engineering **319** (2018) 012012 doi:10.1088/1757-899X/319/1/012012

Optimization of PID Parameters Utilizing Variable Weight Grey-Taguchi Method and Particle Swarm Optimization

Nur Iffah Mohamed Azmi[1], Kamal Arifin Mat Piah[1], Wan Azhar Wan Yusoff[1], Fadhlur Rahman Mohd Romlay[2]

[1]Faculty of Manufacturing Engineering, Universiti Malaysia Pahang, 26600 Pekan, Pahang, Malaysia.

[2]Faculty of Mechanical Engineering, Universiti Malaysia Pahang, 26600 Pekan, Pahang, Malaysia

nuriffah87@yahoo.com, kamalarifin87@gmail.com, wawy.engineer@gmail.com, fadhlur@ump.edu.my

Abstract. Controller that uses PID parameters requires a good tuning method in order to improve the control system performance. Tuning PID control method is divided into two namely the classical methods and the methods of artificial intelligence. Particle swarm optimization algorithm (PSO) is one of the artificial intelligence methods. Previously, researchers had integrated PSO algorithms in the PID parameter tuning process. This research aims to improve the PSO-PID tuning algorithms by integrating the tuning process with the Variable Weight Grey-Taguchi Design of Experiment (DOE) method. This is done by conducting the DOE on the two PSO optimizing parameters: the particle velocity limit and the weight distribution factor. Computer simulations and physical experiments were conducted by using the proposed PSO-PID with the Variable Weight Grey-Taguchi DOE and the classical Ziegler-Nichols methods. They are implemented on the hydraulic positioning system. Simulation results show that the proposed PSO-PID with the Variable Weight Grey-Taguchi DOE has reduced the rise time by 48.13% and settling time by 48.57% compared to the Ziegler-Nichols method. Furthermore, the physical experiment results also show that the proposed PSO-PID with the Variable Weight Grey-Taguchi DOE tuning method responds better than Ziegler-Nichols tuning. In conclusion, this research has improved the PSO-PID parameter by applying the PSO-PID algorithm together with the Variable Weight Grey-Taguchi DOE method as a tuning method in the hydraulic positioning system.

1. Introduction

Hydraulic positioning system has been used in modern industrial applications. This is due to the capabilities of the larger driving force, fast response and high power-to-weight ratio. Hydraulic positioning system is generally categorized in the stabilization and tracking control. This system will always be the driving force to reduce energy consumption and to improve the efficiency and accuracy of the motion control. However, the nonlinearity of the system causes the difficulty for accurate control. Recently, researchers have focused on dynamic characteristics of a hydraulic positioning system for the control purpose. Some of them designed and implemented a suitable controller proper operation with dynamic and nonlinear system [2].

Hydraulic positioning system uses proportional-integral-derivative controllers (PID). This controller is the most famous controller among industrial player because of the simplicity of PID structure and easy to implement in various industrial applications. However, PID controller deals have some issues that limit their application. The main issue is the difficulty to tune the PID controller. This is significant when the controlled system is highly dynamic. It will be hard for the system to achieve the desired performance when the system is nonlinear and time-varying.
Recently, particle swarm optimization (PSO) has been implemented in various applications. PSO algorithm is an evolutionary computational technique. This algorithm is designed to solve optimization problems and easy to be implemented [1]. The issue with PSO-PID algorithm is the random assignment of parameter values. For example, there is no clear guideline on the values of the particle velocity limit and the weight distribution factor implemented in the PSO-PID algorithm. Therefore, the goal of this study is to implement PSO-PID tuning algorithm together with the Variable Weight Grey-Taguchi Design of Experiment (DOE) so that an optimal performance of hydraulic positioning system is achieved. The role of the Variable Weight Grey-Taguchi DOE is to determine the proper values for the particle limit velocity and the weight distribution factor.
The paper is organized as follows: Section II describes the system dynamic model. Section III presents an overview of PSO-PID algorithm. Section IV reviews the Variable Weight Grey-Taguchi DOE method. Section V presents the system with PSO-PID tuning algorithm together with the Variable Weight Grey-Taguchi Design of Experiment (DOE) as well as the Ziegler Nichols tuning algorithm. Finally, the results of the PSO-PID tuning algorithm together with the Variable Weight Grey-Taguchi Design of Experiment (DOE) and Ziegler Nichols tuning algorithm are presented in Section V.

2. PSO-PID Algorithm

[5] suggested new performance criterion based on the time domain that is easier and simpler. This performance criterion is used as evaluation function in the PSO algorithm and he has named the algorithm as PSO-PID tuning algorithm. The evaluation function proposed by him is established in the following equation:

$$W(K)=\left[1-e^{(-\beta)}\right](M_p+E_{ss})+\left[e^{(-\beta)}\right](T_s-T_r) \qquad (1)$$

A set of good control parameters K_p, K_i and K_d can produce a good system response by minimizing the evaluation function. The evaluation function includes the settling time (T_s), rise time (T_r), maximum overshoot (M_p) and steady-state error (E_{ss}). [5] applied PSO-PID algorithm to improve the step transient response of an AVR system.PSO algorithm is used to determine the three PID controller parameters: K_p, K_i and K_d. The controller parameters represent an individual K by $K\equiv[K_p, K_i, K_d]$; therefore there are three members in an individual. These members are allocated as real values. In the process of developing the PSO-PID tuning algorithm, [5] replaced few terms in the PSO algorithm developed by [7]. The term "individual" is used to replace the term "particle" and "population" replaces the term "group". If there are n individuals in a population, then the dimension of a population is n x 3. For the population number, [5] and [10] suggested to use 50 population ($n = 50$).

[5] also proposed the Routh-Hurwitz criterion to limit the evaluation value for each individual in the population within a suitable range. According [5], the Routh-Hurwitz test shows the closed-loop control system stability and each individual is suitable to be implemented. It will produce a small value of $W(K)$. The evaluation function f in (2) is the result of the reciprocal of the Eq.(1).

$$f=\frac{1}{W(K)} \qquad (2)$$

This evaluation function will be used in the PSO-PID algorithms as the performance evaluation function. In addition, [5] used inertia weight ω as proposed by [12] to provide a balance between global and local exploration. Inertia weight can reduce the average iterations number and can adequately obtain optimal results. Thus, [5] agreed with [3] and [8] that the most appropriate inertia weight value is in the range of 0.9 to 0.4. [5] introduces Eq.(3) in his research to allow a balance between global and local exploration. The equation involves inertia weight maximum and minimum values $(\omega_{\max}, \omega_{\min})$ the maximum iteration number $(iter_{\max})$ and the current iteration number $(iter)$. The equation is:

$$\omega = \omega_{\max} - \frac{\omega_{\max} - \omega_{\min}}{iter_{\max}} \times iter \quad (3)$$

Next, there are nine steps used by Gaing in the PSO-PID tuning algorithm. In order to find the best value of K_p, K_i and K_d, He used the following nine steps.

3. Variable Weight Grey-Taguchi DOE method

The proposed tuning algorithm includes the Variable Weight Grey-Taguchi method in the process of determining the optimum setting of maximum velocities for each parameter in order to obtain optimum values of K_p, K_i and K_d.

Step 1: Determine the Factors
First step is the definition of the problem. In this step, the control factors and response factors are identified and selected. The control factors are based on parameters that give significant effect to the output. This study focuses to manipulate two PSO parameters: Limit of change in velocity (V_{max}) and weighting factor (β).
V_{max} works as a constraint to control the global exploration ability of a particle swarm. It determines the resolution, or fineness, with which regions between the present position and the target position are explored. If V_{max} is too high, particles might fly past good solutions. However, if V_{max} is too small, on the other hand, particles may not explore sufficiently beyond locally good areas. The weighting factor β on the evaluation function $W(K)$ can be adjusted to meet specific requirements. The settings of β can also be considered as the manipulated variable to determine the optimal parameter values. Therefore, Variable Weight Grey-Taguchi method is applied to determine the best value of V_{max} for each particle and the weighting factor, β. V_{max} and β are set as the control factors in the Variable Weight Grey-Taguchi method.

Step 2: Identify the Level
Second step is designing and identifying the level of Orthogonal Arrays (OA) to determine the suitable number of experiments. By using OA, the number of experiments can be reduced and the calculation is presented in a few steps. The first step is to determine the Degree of Freedom (DOF) using Eq.(4) below.

$$DOF = L_i - 1 \quad (4)$$

whereby L_i is the number of levels. Second step is to design the suitable OA using Eq.(5) below.

$$OA = (DOF \times N) + 1 \quad (5)$$

N is the number of factors. For the same number of factors and levels using OA required at least nine experiments. However, this study needs larger OA because L_9 may not fulfil the requirement for higher resolution. Then, the selection of L_{27} with 27 experiments is appropriate for higher resolution.

Step 3: Conduct the PSO Tuning Algorithm
In this step, the nine steps of PSO algorithm are applied to determine the response factors. The simulation process will be conducted by referring to the rows of selected OA. The control factors combine with levels respectively in each row.

Step 4: Analyse the Data
The next step is to analyse the data based on the Grey-Taguchi algorithm. The normalized raw data and GRC values in each factor are then calculated. This transformation is called Grey Relational Grade (GRG). Using this GRG values the optimal parameter value is determined by the highest values.
The normalized value lies between zero and one whereby better performance is reached when the normalized value is larger. The ideal value is equal to one. The grey relational coefficient (GRC) is used to determine the relationship between ideal and actual normalized response. The average calculation of GRC for each factor called Grey Relational Grade (GRG). Eq.(6) shows the average calculation of GRC.

$$GRG = \frac{GRC}{i} \tag{6}$$

whereby, i is the number of response factors.

Step 5: Select the Optimum Parameters
Then, optimum control factors are obtained from the average GRG at each parameter. The higher values of average GRG yield a better combination of parameters. So, the GRG values which are closer to 1.0, is closer to the ideal response.

Step 6: Determine the System Response
The next step is to verify the results obtained from an optimum set of factors. All the obtained parameters of maximum velocities will be run in PSO-PID tuning algorithm to get the PID parameters and system response values. The following nine steps of PSO algorithm will be repeated in this procedure.

Step 7: Verify the Result
The last step is to validate the obtained PID parameter values by comparing both simulation and physical experiment. Comparison between both results are conducted based on the system responses: transient response and steady-state response.

4. Simulation of the system
The DOE data set is implemented using the Variable Weight Grey-Taguchi in MATLAB program. The calculation of normalized value, GRC and GRG for 20%, 2% and 0.2% of the variable dynamic range on each dimension using Variable Weight Grey-Taguchi algorithm are recorded.
Therefore, the optimum set of PSO parameters is determined by picking the highest values of average GRG at each particle. The levels and values selected as an optimum set of PSO parameters are shown in Table 1.

Table 1. PSO parameters from optimum PSO parameters

PSO Parameters	Level	Value
V_{max}, K_p	3	2.00
V_{max}, K_i	3	0.20
V_{max}, K_d	2	0.02
Weighting factor, β	1	1.00

Then, the optimum sample is investigated using the PSO algorithm to determine the value for the performance characteristics. Table 2 shows the performance characteristic of the optimum PSO parameters.

Table 2. Performance characteristics of the optimum PSO parameters

Performance Characteristics	Weight	Value
Rise time, $T_r\,(s)$	0.01	0.083000
Settling time, $T_s\,(s)$	0.01	0.144000
Maximum overshoot, M_p	0.90	-1.188.E-12
Evaluation value, f	0.80	44.56200

The PID controller parameters are tuned by two methods. The first method is the proposed optimization method. For confirmation, the classical method (Ziegler-Nichols method) is used to compare the performance response for both methods [6]. Table 3 shows the PID parameters values for both methods and performance criteria of both methods.

Table 3. PID parameters values

Methods	Ziegler-Nichols	PSO-PID DOE
K_p	6.34156	8.77385000
K_i	0.20000	0.00861000
K_d	0.05000	0.13136500
$T_r\,(s)$	0.16000	8.3000.E-02
$T_s\,(s)$	0.28000	1.4400.E-01
M_p	0.23720	1.188.E-012

Table 4. Performance response for different set point of step response

Set Point (cm)	Ziegler-Nichols tuning algorithm			The proposed PSO-PID tuning algorithm		
	$T_r\,(s)$	$T_s\,(s)$	M_P	$T_r\,(s)$	$T_s\,(s)$	M_P
10	0.3865	0.5287	1.3514	0.3723	0.5093	0.4886
9	0.3477	0.4652	0.4190	0.3215	0.4587	1.3651
8	0.3100	0.4266	0.4857	0.3128	0.5125	1.4821
7	0.2738	0.3762	0.8082	0.2677	0.4990	1.5306

Table 4 shows performance results at different set points for both methods. The performance of both methods clearly shows that the modified algorithm improves the rise time, settling time and maximum overshoot.

5. Discussion

The purpose of this paper is to model and to improve the system performance of hydraulic positioning system. The paper proposed a system model using a mathematical transfer function to simulate actual system behaviours. This paper also proposed a PSO-PID tuning algorithm in combination with the Variable Weight Grey-Taguchi DOE in order to determine the optimum PID parameters.

6. Conclusion

In conclusion, the PSO-PID tuning algorithm with the Variable Weight Grey-Taguchi DOE successfully obtained the optimum PID parameters. The simulation results verified that the proposed tuning algorithm improved transient response compared to Ziegler-Nichols tuning algorithm. The transient response results from the modified PSO-PID tuning algorithm improved 48.13% of rise time and 48.57% of settling time. The physical experiment results verified that the proposed PSO-PID tuning algorithm improved the system performance.

References

[1] Allaoua, B., Gasbaoui, B. & Mebarki, B. (2009). Setting Up PID DC Motor Speed Control Alteration Parameters Using Particle Swarm Optimization Strategy. *Leonardo Electronic Journal of Practices and Technologies*, 19-32.

[2] Dechrit Maneetham, N. A., Maneetham, D. & Afzulpurkar, N. (2009). Modeling, simulation and control of high speed nonlinear hydraulic servo system. *World Journal of Modelling and Simulation,* 6, 27-39.

[3] Eberhart, R. C. & Shi, Y. (2000). Comparing inertia weights and constriction factors in particle swarm optimization. *In:* Proceedings of the 2000 Congress on Evolutionary Computation. CEC00 (Cat. No.00TH8512), 2000. IEEE, 84-88.

[4] Eberhart, R. C. & Shi, Y. (2001). Particle swarm optimization: developments, applications and resources. *In:* Proceedings of the 2001 Congress on Evolutionary Computation (IEEE Cat. No.01TH8546), 2001. IEEE, 81-86.

[5] Gaing, Z.-L. (2004). A Particle Swarm Optimization Approach for Optimum Design of PID Controller in AVR System. *IEEE Transactions on Energy Conversion,* 19, 384-391.

[6] Haugen, F. (2010). Ziegler-Nichols ' Open-Loop Method. *TechTeach.*

[7] Kennedy, J. & Eberhart, R. (1995). Particle swarm optimization. *Proceedings of ICNN'95 - International Conference on Neural Networks,* 4.

[8] Kennedy, J., Eberhart, R. C. & Shi, Y. (2000). Swarm intelligence. *In:* Proceedings of the 2000 Congress on Evolutionary Computation. CEC00 (Cat. No.00TH8512), 2001 San Francisco. Morgan Kaufmann, 84-88.

[9] Kilian, C. T. (2001). *Modern Control Technology: Components and Systems,* United States, Delmar Thomson Learning.

[10] Mat Yahya, N. (2009). *Peningkatan Prestasi Sistem Pemacu Suap Balik Dengan Mengoptimumkan Parameter Pengawal Pid Menggunakan Algoritma Pengoptimuman Kawanan Partikel.* Master, Universiti Malaysia Pahang.

[11] Rao, R. & Yadava, V. (2009). Multi-objective optimization of Nd:YAG laser cutting of thin superalloy sheet using grey relational analysis with entropy measurement. *Optics & Laser Technology,* 41, 922-930.

[12] Shi, Y. & Eberhart, R. (1998). A modified particle swarm optimizer. *In:* IEEE International Conference on Evolutionary Computation Proceedings. IEEE World Congress on Computational Intelligence (Cat. No.98TH8360), 1998. IEEE, 69-73.

Design for Warehouse with Product Flow Type Allocation using Linear Programming: A Case Study in a Textile Industry

M S A Khannan[1], L Nafisah, D L Palupi

Industrial Engineering Department, Universitas Pembangunan Nasional Veteran Yogyakarta, Indonesia

Email: shodiq@upnyk.ac.id, lailanafisah71@gmail.com,diknalindapalupi@gmail.com

Abstract. Sari Warna Co. Ltd, a company engaged in the textile industry, is experiencing problems in the allocation and placement of goods in the warehouse. During this time the company has not implemented the product flow type allocation and product placement to the respective products resulting in a high total material handling cost. Therefore, this study aimed to determine the allocation and placement of goods in the warehouse corresponding to product flow type with minimal total material handling cost. This research is a quantitative research based on the theory of storage and warehouse that uses a mathematical model of optimization problem solving using mathematical optimization model approach belongs to Heragu (2005), aided by software LINGO 11.0 in the calculation of the optimization model. Results obtained from this study is the proportion of the distribution for each functional area is the area of cross-docking at 0.0734, the reserve area at 0.1894, and the forward area at 0.7372. The allocation of product flow type 1 is 5 products, the product flow type 2 is 9 products, the product flow type 3 is 2 products, and the product flow type 4 is 6 products. The optimal total material handling cost by using this mathematical model equal to Rp43.079.510 while it is equal to Rp 49.869.728 by using the company's existing method. It saves Rp6.790.218 for the total material handling cost. Thus, all of the products can be allocated in accordance with the product flow type with minimal total material handling cost.

1. Introduction

As there is a rapid development of the industry and followed by an increasement of sophisticated technology, it can cause problems to the existance in the manufacturing industry. One of the problems encountered with the utilization of warehouse space for placing the product. A warehouse is a place to store the goods, both raw materials that will be the manufacturing process and finished goods ready for marketing. A good warehousing system is capable of utilizing the warehousing system for the storage space effectively in order to increase the utility of space and minimize material handling cost.

Sari Warna Co. Ltd in Boyolali, Central Java, Indonesia is a company engaged in the field of textile industry. One department in the company is a Weaving Department II which produces semi-finished yarn into fabric (gray). Gray cloth production is made to order that the amount of production tailored to consumer demand. In a gray warehouse at the weaving department there the product is streamed before the product is shipped to the customer. Based on the observations that have been made that the

[1] To whom any correspondence should be addressed.

Department of Weaving II is still experiencing problems in the placement of the product stream for each product. In May 2016 there were placements of stock of products that did not match the products in the warehouse stock plan. Data stock products in May 2016 showed that there is a kind of fabric with conditions that are not in accordance with the stock plan products in the warehouse, i.e. the type of fabric R188, C247, R155 / 63, and R197L. The fabric types do not need to have a stock of fabrics which are stored in the warehouse based on interviews with the head of the warehouse. This leads to reduced utilization of space which will result in the fact that many products can not be accommodated in the shed. In other conditions warehouse clerks also had difficulty to locate and retrieve gray cloth, because the drafting and structuring of gray fabric warehouse that does not comply with the product stream. This leads to inaccuracy in the time of delivery of goods / products to consumers, the high total cost of the movement of products, and the products exceed the storage capacity. So, the current company's warehouse condition is needed to be redesigned on the product flow type allocation policy case.

In the next section, we discuss the theoretical background about warehousing in section 2. The mathematical model that is used in this problem is presented in section 3. Research Methodology in section 4. A numerical example and computational result is presented in section 5. And the paper concludes with section 6.

2. Theoretical Background

Warehouse is a place with the duty to keep the items to be used for the production, until the goods are required under production schedule [1]. Here are some essential warehouse functions [2]:

1. *Provide temporary storage of goods.* To achieve economies of scale in production, transportation, and handling of the product, it is often necessary to store products in warehouses and issued to *customers* when there is a demand.
2. *Put together customer* **orders.** To save the *order* from the *customer* in large quantities.
3. *Serve as a customer service facility.* Because the warehouse is a storage place the product is to be shipped to the *customer,* the warehouse can provide services to the *customer* facility and handle the change as well as damage to the product, conduct market surveys, and provide *aftersales* services.
4. *Protect goods.* Warehouses are usually equipped with a circuitry security system, so that goods manufacturing results can be protected from hazards such as theft, fire, flood and other weather problems.
5. *Segregate hazardous or Contaminated materials.* The security code does not allow hazardous materials near the production floor, so that the warehouse be a good place to separate and put a dangerous and contaminated material.
6. *Perform value-added services.* Some warehouses serve as packing product value addition, preparation of *customer orders* based on *customer* specific requirements, material inspection and testing of products.
7. *Inventory.* Because it is difficult to forecast product demand, in some businesses, it is important to conduct an inventory and *safety stock* to maintain an unexpected request.

To carry out these functions, a *warehouse* area is usually divided into *reserve storagearea,forward area,* and *cross-docking area.Warehouse* configuration including a part of the five functional areas such as the reception area, delivery, *cross-docking,reserve,* and *forward.* Here is a possible flow of material in a *warehouse* [2]:

a. Flow 1: *Receiving* (Reception) - *cross-docking* - *shipping* (delivery)
b. Flow 2: *Receiving* (Reception) - *reserve* - *shipping* (delivery)
c. Flow 3: *Receiving* (Reception) - *reserve* - *forward* - *shipping* (delivery)
d. Flow 4: *Receiving* (Reception) - *forward* - *shipping* (delivery)

3. Mathematical Model

The Model used in this paper is the same as the reference model [3]. This model can be used to find optimal warehouse flow type. The objective function is minimizing the total material handling cost by several flow type allocation.

The following notation used is as follows:

Parameter

i : Number of products; i = 1, 2, 3, ..., n
j : Type of product flow; j = 1, 2, 3, 4
λ_i : Average annual demand of each product
A_i : The cost of production of each product
P_i : The selling price of each product
p_i : The average percentage of time the store in a *reserve* area, if the flow of products in flow type 3
d_i : Comparing the size of the product in the area of *the reserve* and *forward*
q_{ij} : The capacity of the product stack height in each functional area
a B C a = *cross-docking*, b = *reserve*, c = *forward*
r : Cost savings and inventory
H_{ij} : The cost of the movement of products in the product stream type j
C_{ij} : Cost savings per year of products in the product stream type j
S_{i}: :The space required to store each product
TS : Total warehouse capacity available
Q_i : The number of booking products
T_i :average shelf life of each product within one year
$_{CD}LL$ $_{CD}UL$:minimum and maximum capacity for functional areas cross-docking
$LL_{R,}$ UL_R :minimum and maximum capacity for functional area *reserve*
$LL_{M,}$ UL_F: :minimum and maximum capacity for functional areas *forward*

The decision variables

Xij :Proportion of storage space for each functional area
α =*Cross-docking*, β=*reserve*, γ=*forwardCross-docking* area

Objective Function

$$\text{Minimize } Z = 2\sum_{i=1}^{22}\sum_{j=1}^{4} q_{ij} H_{ij} \lambda_i X_{ij} + 0{,}5\sum_{i=1}^{22}\sum_{j=1}^{4} q_{ij} C_{ij} Q_i X_{ij} \quad (1)$$

Constraint

$$\sum_{j=1}^{4} X_{ij} = 1 \quad \forall i \quad (2)$$
$$0{,}5\sum_{i=1}^{n} Q_i S_i X_{i1} \le a\alpha TS \quad \forall \quad i \quad (3)$$
$$0{,}5\sum_{i=1}^{n} Q_i S_i X_{i2} + \sum_{i=1}^{n} p_i Q_i S_i X_{i3} \le b\beta TS \quad (4)$$
$$0{,}5\sum_{i=1}^{n} (1-p_i) Q_i S_i X_{i3} + 0{,}5\sum_{i=1}^{n} Q_i S_i X_{i4} \le c\gamma TS \quad (5)$$
$$\alpha + \beta + \gamma = 1 \quad (6)$$
$$_{CD}LLUL \le a \le TS_{CD} \quad (7)$$
$$LL_R \le b \le UL\,TS_R \quad (8)$$
$$LL_F \le c \le UL\,TS_F \quad (9)$$
$$X_{ij} = 0 \text{ or } 1 \quad \forall \quad i, j \quad (10)$$

Based on the cost of the annual product demand (A_{i}),) the selling price of each product per unit *(Pij)*, and the cost of transporting supplies *(r)*, then to find economic order quantity (EOQ) with the formula:

$$Q_i = \sqrt{\frac{2\,Ai\,\lambda i}{r\,Pi}} \quad (11)$$

IOP Conf. Series: Materials Science and Engineering **319** (2018) 012013 doi:10.1088/1757-899X/319/1/012013

4. Numerical Test

Based on the observations and data gathering, Data related to the cloth type at the Department of Weaving II of Sari Warna Co. Ltd such as current Product flow, The average annual demand of each product, Product flow coefficient, The costs of moving each product in the product flow type *j*, Cost savings of each product in the product flow type *j* can be shown in Table 1.

Table 1. Data related to the cloth types

Cloth types	Product flow	The average annual demand of each product	Product flow coefficient	The costs of moving each product in the product flow type *j*	Cost savings of each product in the product flow type *j*	Optimal product Quantity (Q_i)
S120	1	4832421	1	0.4	0.04	8225
R153	3	4459346	1.0077	0.66	3.29	7901
R152	2	1.24267 million	1	0.53	3.29	4171
R188	1	3573292	1	0.4	0.04	7073
T98	3	3734592	1.0075	0.66	3.29	7231
R155	2	1.07698 million	1	0.53	3.29	3883
T90	3	8.27788 million	1.0077	0.66	3.29	10765
R180	3	633 679	1.0099	0.66	3.29	2979
T88	3	6692585	1.0113	0.66	4.61	9680
S121	3	363 748	1.0077	0.66	3.29	2257
C29	3	302 026	1.0077	0.66	3.29	2056
C247	4	679 216	1	0.32	0.11	3084
T88L	3	1975653	1.0146	0.66	4.61	5259
T140L	3	27 981	1.0146	0.66	4.61	626
R155 / 63	1	178 582	1	0.4	0.11	1581
R180 / 62	3	520 111	1.0097	0.66	3.29	2698
R197L	4	20 574	1	0.32	0.11	537
R169L	3	10 698	1.0097	0.66	3.29	387
T83	3	983 712	1.0209	0.66	7.68	3711
S100	2	2300448	1	0.53	7.68	5675
POT 001	4	151 735	1	0.32	0.11	2783
POT 002	4	192 061	1	0.32	0.11	3113

The data related to flow type as seen in Table 1 is used to be input parameter in the mathematical model and solved by LINGO software. The result of the allocation of each product to the optimal flow type can be seen in Table 2. Cloth type S120 is allocated in to flow type 2, R152 is allocated in to flow type 1, and so on.

IOP Conf. Series: Materials Science and Engineering **319** (2018) 012013 doi:10.1088/1757-899X/319/1/012013

Table 2. Results of the allocation of product

No.	Cloth type	Product allocation			
		flow 1	flow 2	flow 3	flow 4
1	S120	0	1	0	0
2	R153	0	1	0	0
3	R152	1	0	0	0
4	R188	0	1	0	0
5	T98	0	1	0	0
6	R155	1	0	0	0
7	T90	0	1	0	0
8	R180	1	0	0	0
9	T88	0	1	0	0
10	S121	0	0	0	1
11	C29	0	0	0	1
12	C247	1	0	0	0
13	T88L	0	1	0	0
14	T140L	0	0	0	1
15	R155 / 63	0	0	0	1
16	R180 / 62	1	0	0	0
17	R197L	0	0	0	1
18	R169L	0	0	0	1
19	T83	0	1	0	0
20	S100	0	1	0	0
21	POT 001	0	0	1	0
22	POT 002	0	0	1	0

There are four flow types in the current flow type based policy as the same as in the proposed policy. The result of the allocation of product shown in Table 2 above is different with the current allocation of product (shown in Table 1). LINGO *software* output results stating the optimal solution or not and details of the optimization result can be seen on *the Status Solver* shown in Figure 1.

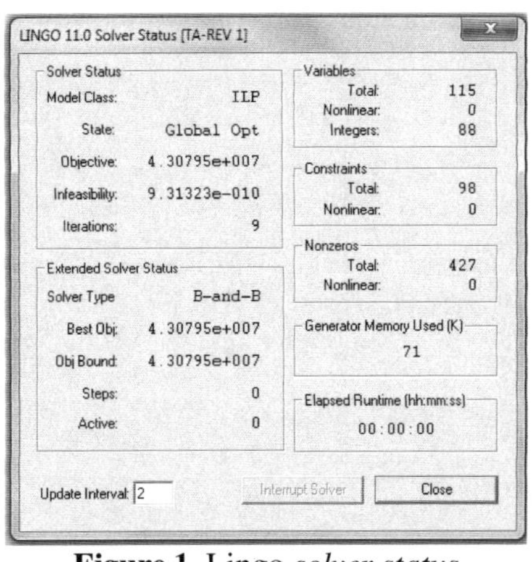

Figure 1. Lingo *solver status*

Figure 1 shows that the optimization model is *the Global Optimum* obtained by *Integer Linear Programming* class model. It indicates that optimization results already show *Global Optimum.*

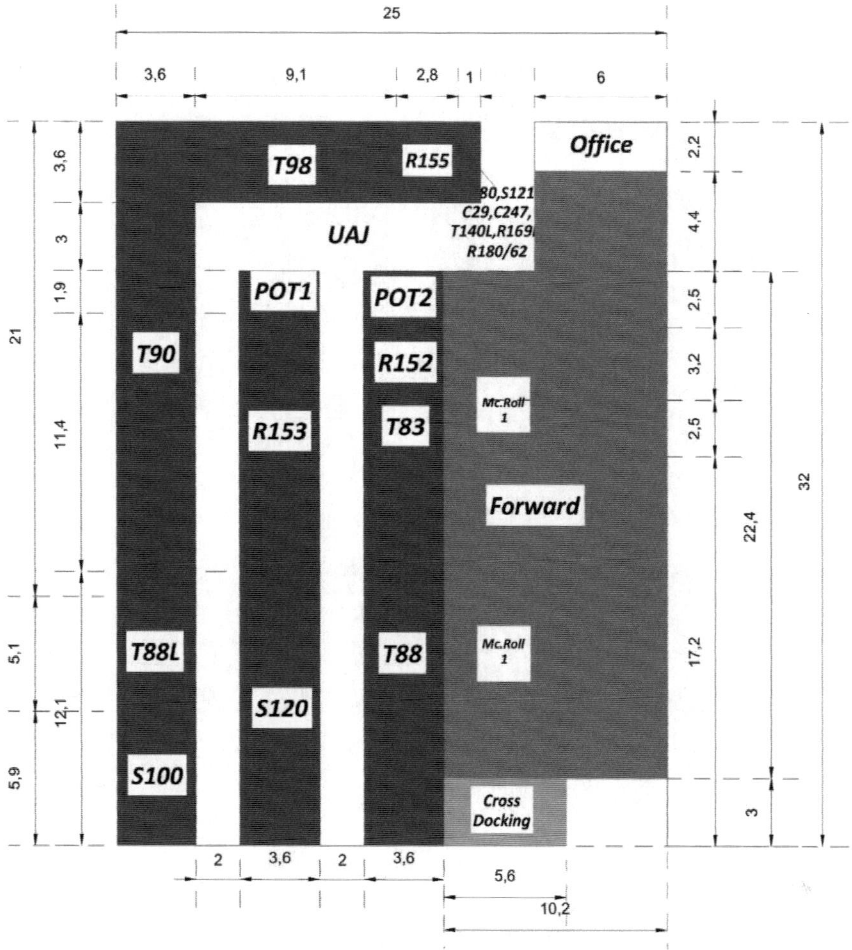

Figure 2. The current layout

From Figure 2 above, we can see that Reserve area is the greatest functional area with 55.9% followed in the second one is Forward area 41.36%, and the smallest one is crossdocking area with 2.72%.

reserve capacity calculation	= Reserve 1 + Reserve 2 + Reserve 3+Reserve 4
	= 279.36 m^3+ 691.2m^3+ 548,64m^3+548,64m^3
	= 1530 m^3
forward area capacity calculation	= *(l x w x h) + (l x w x h)*
	= *(10,2 x 22,4 x 6) + (6 x 4,4 x 6)*
	= *1,530* m^3
cross-docking capacity calculation	= length x width x height
	= 5,6 m x 3 m x 6 m
	= 100.8 m^3

Based on the output generated from the software LINGO it can be determined the proportions of each functional area that is Cross-docking area (α) : 0.0734, Reserve area (β) : 0.1894 , Area forward(γ) : 0.7372. This result can be used to generate the proposed layout as shown in Figure 3.

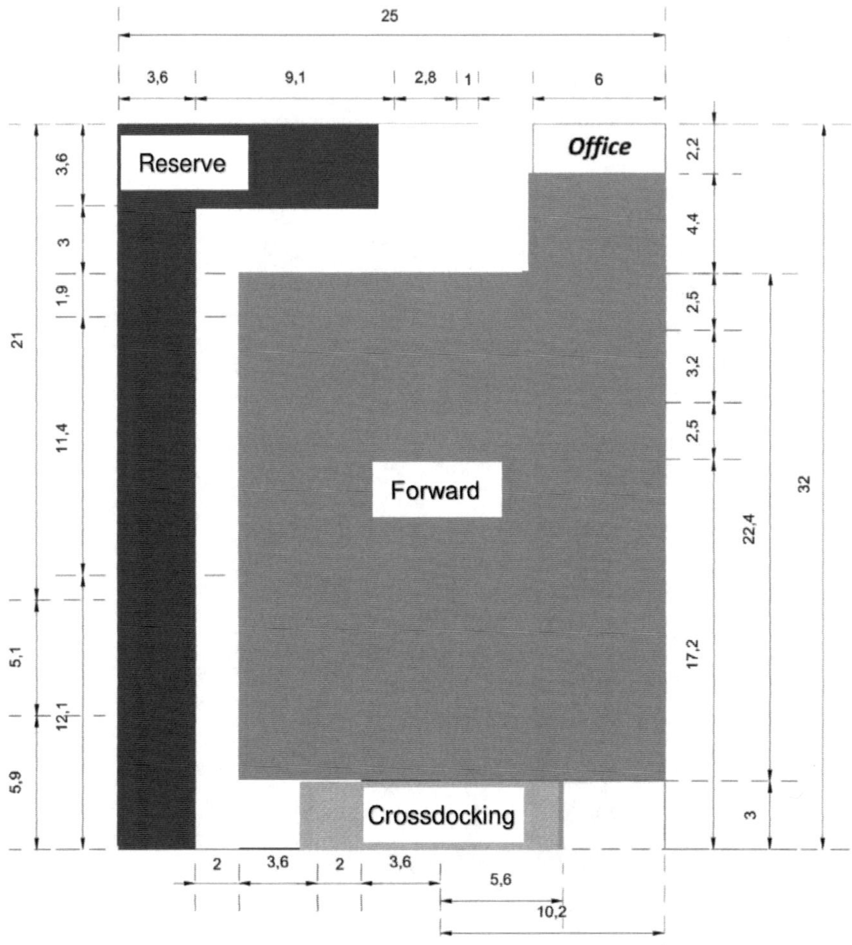

Figure 3. The Proposed Layout

Based on the Figure 3 above, it can be seen that forward area is the greatest functional in the proposed layout, followed by reserve area and crossdocking area is the smallest one. Cross docking is almost double from the current value while reserve area is decrease become the third of the current value while forward area almost double from current value. The full comparison between each fucntional area in the current layout and the proposed lay out can be seen in Table 3.

Table 3. Capacity comparison between each functional area in the current layout and the proposed lay out

Area fungsional	Current Layout	percentage	Proposed Layout	percentage
Cross-docking (α)	100.8 m^3	2.72%	271,48 m^3	7.34%
Reserve (β)	2.067.84 m^3	55.90%	700,52m^3	18.94%
Forward (γ)	1.530 m^3	41.36%	2726,64m^3	73.72%

The total material handling cost before the model applied is Rp 49.722.747. This result found using the mathematical model with current layout parameter. Total material handling cost after the model applied based on the output of LINGO *software* is Rp 43.079.510. It takes 12 times iteration process to obtain optimal results. Based on the calculation above, it is known that the company can reduce the total cost by switching to the proposed model by Rp 6.643.237.

5. Conclusion

In the current work we have applied a mathematical model for design warehouse flow type allocation using linear programming in a textile industry. The objective function of the mathematical model is minimizing the total material handling cost. Based on the output generated from *the software* LINGO it can be determined the proportions of each functional area that is *Cross-docking* area *(α)* : 0.0734, *Reserve* area *(β)* : 0.1894 , Area *forward(γ)* : 0.7372. All kinds of products have been allocated to each type of product flow in the warehouse, with the distribution of products included in the product stream 1 number 5 product, the product stream 2 are 9 product, the product stream 3 number 2 product, and the product stream 4 number 6 products. The total material handling cost before the model applied is Rp49.722.747. Total material handling cost sfter the proposed model applied based on the output of LINGO *software* is Rp.43.079.510. It takes 12 times iteration process to obtain optimal results. Based on the calculation above, the company can reduce the total cost by switching to the proposed model by Rp.6.643.237.

Suggestions needed for further research in order to improve further research are: (1) Subsequent research can be done by adding or considering new constraint function in the mathematical model of optimization that already exist (2) Need to do a review of the production and scheduling of taking goods from the warehouse to the costumer because it is directly related to the capacity and the flow of products in the warehouse.

6. References

[1] Apple J M 1990 *Plant Layout and Material Handling*, 3rd Ed, (Bandung: Institut Teknologi Bandung)
[2] Heragu S S 2008 *Facilies Design,* 3rd ed. (CRC Press)
[3] Heragu S S, Du L, Mantel R J, Schuur P B 2005 Mathematical model for warehouse design *Int. Jrnl. of Production Research*, 43 (2) pp 327-338

IOP Conf. Series: Materials Science and Engineering **319** (2018) 012014 doi:10.1088/1757-899X/319/1/012014

Development of Product Availability Monitoring System In Production Unit In Automotive Component Industry

Rachmad Hartono[1,2], Sri Raharno[1], Yatna Yuwana Martawirya[1], Bagus Arthaya[3]

[1]Department of Mechanical Engineering, Institut Teknologi Bandung, Bandung, 40132, Indonesia
[2]Departement of Mechanical Engineering, Pasundan University, Bandung, 40153, Indonesia
[3]Department of Mechatronics Engineering, Faculty of Industrial Engineering, University of Parahyangan, Bandung, Indonesia

rachmad_hartono@unpas.ac.id

Abstract. This paper described a methodology to monitor the availability of products in a production unit in the automotive component industry. Automotive components made are automotive components made through sheet metal working. Raw material coming into production unit in the form of pieces of plates that have a certain size. Raw materials that come stored in the warehouse. Data of raw each material in the warehouse are recorded and stored in a data base system. The material will then undergo several production processes in the production unit. When the material is taken from the warehouse, material data are also recorded and stored in a data base. The data recorded are the amount of material, material type, and date when the material is out of the warehouse. The material coming out of the warehouse is labeled with information related to the production processes that the material must pass. Material out of the warehouse is a product will be made. The products have been completed, are stored in the warehouse products. When the product is entered into the product warehouse, product data is also recorded by scanning the barcode contained on the label. By recording the condition of the product at each stage of production, we can know the availability of the product in a production unit in the form of a raw material, the product being processed and the finished product.
Key words: record; data base

1. Introduction

Lean Production System (LPS) is a production system that is characterized by a lean and agile condition. The basic principle of LPS is (1) making only products that can be sold in Just In Time, (2) only make the product with good quality, (3) making products in a cheaper method, and (4) building a strong industry quality that can respond to fluctuations in demand and environmental change.

The basic principle of the first LPS is to create only products that can be sold in Just In Time. This means that the product created must be in accordance with the order and completed within the specified time. To realize this need to know the quantity of products at any time either in the form of finished products, WIP, or still in the form of raw materials. In order for the quantity of products can be known at any time easily, need to be made a system that can monitor the quantity of products that are in a production system.

IOP Conf. Series: Materials Science and Engineering **319** (2018) 012014 doi:10.1088/1757-899X/319/1/012014

To know the quantity of products, WIP, and raw materials easily at any time, need to be made a data base that can reflect the quantity of products, WIP, and raw materials. The data base should be updated every time there is a change in the quantity of products, WIP, as well as raw materials. Updates to the quantity of products, WIP, or raw materials shall can be made by each person involved in production activities and such updates can be made easily. With the data base that always follow the changes (up to date), the quantity of products, WIP, and raw materials can be known easily whenever needed. Thus production planning will become easier to do.

2. Problem-solving approach

This research is the first step of LPS design that will be applied to automotive supporting industries. The industry that became the object of research is the industry that produces automotive components with the base plate material. The industry has two production units separated by a considerable distance. The first production unit (location in Cibolerang) is tasked for preparing pieces of plates with a certain width. The second production unit (location in Cimahi) is tasked with processing the pieces of the plate into automotive components through several stages of the manufacturing process. Pieces of the plate from the first production unit are delivered to the second production unit by truck. The material flow in this industry can be seen in Figure 1.

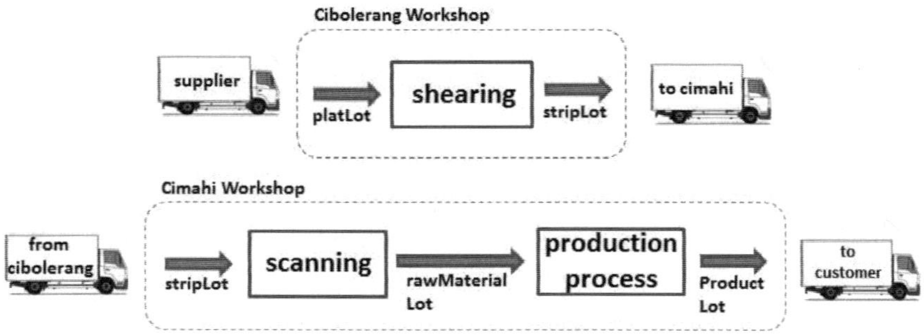

Figure 1. Material flow

The production unit in Cibolerang receives a lot of plate from raw material suppliers. A lot plates from suppliers are then labeled according to the type and size of the plates. When raw materials are needed in the form of pieces of plates of a certain type, thickness and width, plates of a certain type and thickness are cut into sections. These pieces of plates are grouped according to the type, thickness, and width of the pieces and each group is labeled. The pieces of plates that have fulfilled the specified amount are sent to the production unit in Cimahi.

Pieces of plates that have arrived at the Cimahi production unit are stored in raw materials warehouse. Data pieces of the plate should be recorded and entered into the data base. The data recording is done by scanning the barcode contained on the label affixed to each batch of plates. The pieces of the plate are then processed through several stages of the process until it becomes a product.

2.1. Class model to describe material flow

Changes due to material flows should be monitored easily and accurately. Therefore, every component involved in the flow of material needs to be modeled in the virtual world. Since the industry that became the object of research has two production units with different data bases, the model is made up of models that describes the production unit in Cibolerang and the model that describes the production unit in Cimahi.

The classes to describe the condition of production units in Cibolerang consist of suppliers, materialSpec, platLot, stripLot, shearingType and productType. The supplier class represents suppliers who supply raw materials. The materialSpec class represents the material specification of the plate. The platLot class represents a lot of plate material supplied by the supplier. The stripLot class represents pieces of plates which are the base material for automotive components. The shearingType class represents the piece type of plate. The productType class represents the type of automotive

IOP Conf. Series: Materials Science and Engineering **319** (2018) 012014 doi:10.1088/1757-899X/319/1/012014

component. The classes to describe the condition of production units in Cibolerang can be seen in Figure 2.

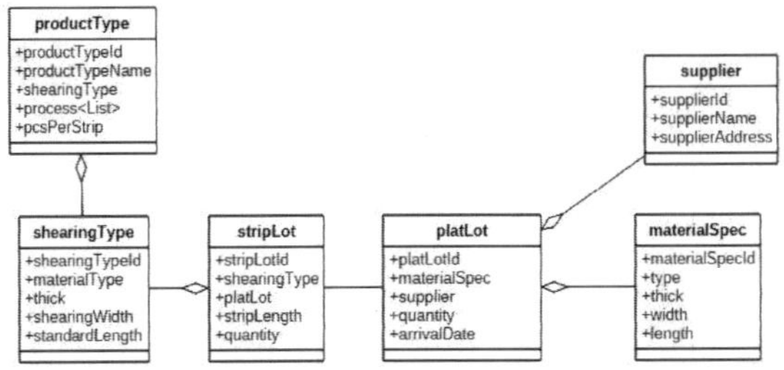

Figure 2. Classes in the Cibolerang production unit

The classes to describe the condition of production units in Cimahi consist of rawMaterialLot, productLot, productType, shearingType, process, customer, order, and orderProduct. The rawMaterialLot class represents the raw material in the form of plate pieces coming from the production unit in Cibolerang. The rawMaterialLot class is a transformation of the lotStrip class. The productLot class represents automotive components that are being made or finished in production units in Cimahi. The customer class represents the customer ordering certain types of products in a certain quantity. Order class represents orders from customers. The order Product class represents the specific product type ordered by the customer. The classes to describe the condition of production units in Cimahi can be seen in Figure 3.

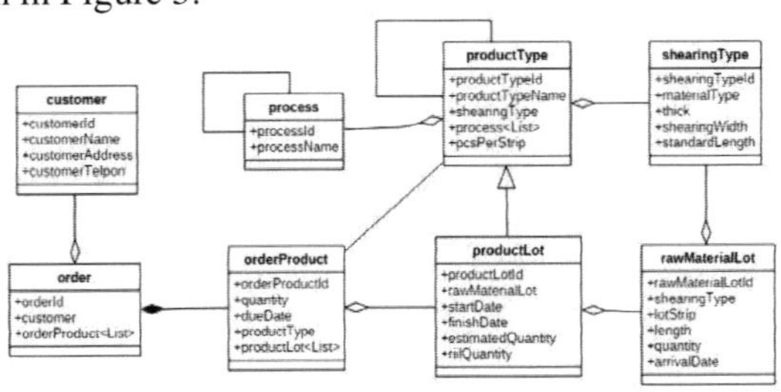

Figure 3. Classes in the Cimahi production unit

2.2. Programs that have been created

The main activity of the material flow from the receiving of plate material from the supplier to the delivery of the product to the customer consists of (1) receiving of plate material (platLot) from suppliers, (2) delivery of strips (stripLot) from production units in Cibolerang to production units in Cimahi, (3) receiving of pieces of plate (stripLot) in Cimahi production unit, (4) retrieval of plate pieces (rawMaterialLot) by the operator from raw material material warehouse, (5) product acceptance (productLot) in the product warehouse, And (6) delivery of products to customers. Activity (1) and (2) are conducted in the Cibolerang production unit, while activities (3) through (6) are conducted in the Cimahi production unit.

The program has not been able to record all data changes that occur on the activity (1) until the activity (6). This is because the information obtained from the industry being studied is still incomplete. The activities successfully recorded by the program to date are activities (2), (3), (4), and (5).

Programs are created using a programming language that supports data base applications and web applications. The data of each object of the class that has been depicted in figure 2 and 3 is stored in a data base. The data stored in the data base is obtained from several different locations. Data obtained in several different locations is sent to the server by utilizing web applications.

Recording of data that occurs on the activity (2) is performed by the administrative staff in charge of making documents delivery of goods. The administrative staff recorded all the pieces of plates (stripLot) to be delivered to the production unit in Cimahi. This record activity generates goods

IOP Conf. Series: Materials Science and Engineering **319** (2018) 012014 doi:10.1088/1757-899X/319/1/012014

delivery documents (SURAT JALAN PENGIRIMAN) and some labels (KARTU INPUT RAW MATERIAL) affixed to each group of plates.

Recording of data occurring on the activity (3) is performed by the operator receiving the lot of pieces of plate (stripLot). Recording is done by scanning the barcode contained on the label affixed in each lot of metal strip. The operator enters the quantity of metallic lot pieces received. This activity produces a labell (KARTU LOT RAW MATERIAL) that will be attached to each appropriate lot raw material.

Recording that occurs on the activity (4) is performs by the operator who takes the lot of raw materials in raw materials warehouse. Operators enter the number of strips taken from the stack of raw materials that match the type of product to be made. This activity produces labels (KARTU LOT PRODUK) that will later be affixed to the stack of strips that will undergo several stages of the manufacturing process.

The recording that occurs in the activity (5) is performed by the operator receiving the completed product in the product warehouse. Recording is done by scanning the barcode contained on the label affixed on the lot of the product concerned. The operator enters the quantity of products received.

3. Testing the program

Programs that have been created need to be tested to determine the ability of the program to record data every time a material flow occurs. Testing the program includes data recording testing on activity (2), data recording test on activity (3), data recording test on activity (4), and data recording test on activity (5).

3.1. Testing of data recording on activity (2)

One of the interface of the program to record the data stripLot can be seen in Figure 4 (left). At the interface, the administrative staff enters the car license, delivery id, product type id, and quantity of stripLot. The data stripLot that has been listed can be seen in the table located at the bottom of the program interface. This program generates goods delivery documents (figure 4, center) and label (KARTU INPUT RAW MATERIAL) (figure 4, right) which amount corresponds to the number of stack of material shipped.

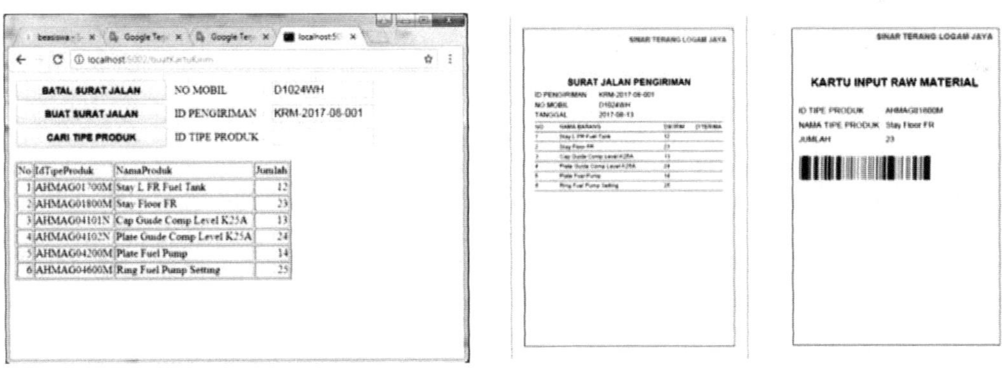

Figure 4. The program interface to record stripLot data, goods delivery document, and KARTU INPUT RAW MATERIAL

3.2. Testing of data recording on activity (3)

One of the interface of the program to record rawMaterialLot data can be seen in figure 5 (left). The interface appears after the operator receiving rawMaterialLot scans the barcode contained on KARTU INPUT RAW MATERIAL. The operator enters the quantity of metallic lot pieces received. This program generates KARTU LOT RAW MATERIAL (figure 5, right).

IOP Conf. Series: Materials Science and Engineering **319** (2018) 012014 doi:10.1088/1757-899X/319/1/012014

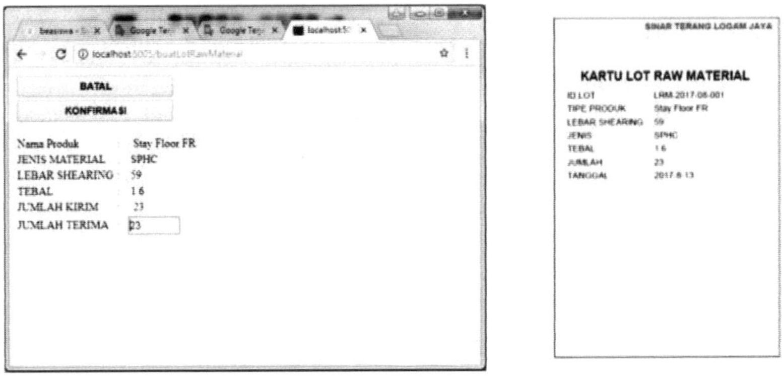

Figure 5. The program interface to record rawMaterialLot data and KARTU INPUT RAW MATERIAL

3.3. Testing of data recording on activity (4)

One of the interface of the program to record productLot data can be seen in figure 6 (left). The interface appears after the operator entered the identity of the product type to be created. In the interface there are some rawMaterialLot that can be taken by the operator to create the desired product that has been sorted according to the date of arrival. The operator selects the desired rawMaterial according to the order of arrival date. This program generates KARTU LOT PRODUK (figure 6, right).

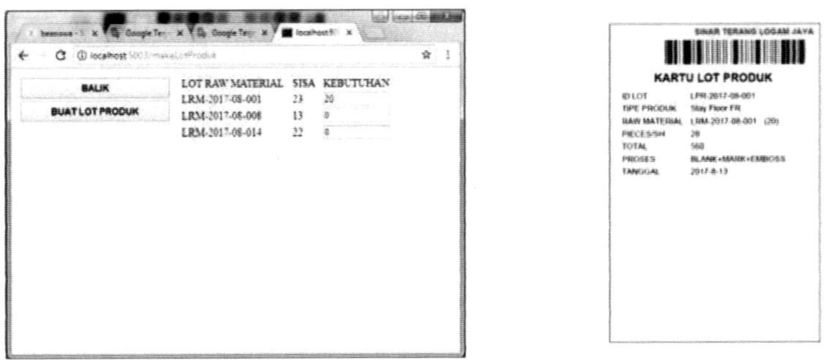

Figure 6. The program interface to record productLot data and KARTU LOT PRODUK

3.4. Testing of data recording on activity (5)

One of the interface of the program to record productLot data can be seen in figure 7. The interface appears after the operator scans the barcode contained on KARTU LOT PRODUK. The operator enters the quantity of product received on the product quantity input box. The quantity of products entered according to the real amount of product received.

Figure 7. The program interface to record productLot data

4. Discussion

stripLot shipped from the Cibolerang production unit to the Cimahi production unit is automatically recorded in the existing data base in the Cibolerang production unit. In addition to data stripLot, delivery time can also be recorded automatically. rawMaterialLot which is a transformation of stripLot can also be recorded automatically on the existing data base system in Cimahi production unit. In addition to rawMaterialLot data, the reception time can also be recorded automatically. By comparing the acceptance time of rawMaterialLot with stripLot delivery time it is possible to estimate the time interval required to deliver the stripLot from the production unit in Cibolerang to the production unit in Cimahi.

The rawMaterialLot data can be compared with the useRawMaterialLot data so that the remaining rawMaterialLot can be obtained. The remainder of each rawMaterialLot can reflect the amount of product reserves of a certain type which is still a basic material.

When data recording on activity (4) is performed, productLot data is automatically recorded in the data base. Data finishDate and realQuantity still empty. productLot whose data finishDate and riilQuantity are still empty describe WIP of a certain type of product.

When data recording on activity (5) is performed, the data finishDate and the realQuantity of the lotProduct will be filled automatically. The productLot for which the finishedDate and realQuantity data are already filled describes the number of completed products of a certain type of product.

5. Conclusions

From the results of tests that have been done, it can be concluded several things:

1. Monitoring system that has been made can record the number of products that have been made, WIP, as well as raw materials to make each automotive component. Thus the number of products, WIP, or raw materials for each type of component can be known easily and accurately.
2. The monitoring system that has been created can also record the time required to make each type of automotive component. By knowing the timing of each type of component, it is possible to estimate the targets that can be obtained at specified intervals.

Monitoring system that has been made still needs to be developed further. The development that will be done is:

1. Making a program that can record the arrival activity of raw materials.
2. Create a program that can record the product delivery activity.
3. Develop an existing monitoring system to monitor machine activity, operators, or use of tools required for production activities.
4. Adding the scheduling function and control function on the system that has been created.
5. Adding a function to overcome the interference on the production system.

References

[1] Jainury, Suhartini Mohd., Ramli, Rizauddin., Rahman, Mohd Nizam Ab., dan Omar, Azhari (2014). Integrated Set Parts Supply system in a mixed-model assembly line, Computers & Industrial Engineering, 75, 266-273.

[2] Jana, Tarun Kanti., Bairagi, Bipradas., Paul, Soumen., Sarkar., dan Saha, Jyotirmoy (2013): Dynamic schedule execution in an agent based holonic manufacturing system, Journal of Manufacturing Systems, 32, 801-816.

[3] Simao, Jean Marcelo., Stadzisz, Cezar Paulo., dan Morel, Gerard (2006). Manufacturing execution systems for customized production, Journal of Materials Processing Technology, 179, 268-275.

[4] Lee, C. H. K., Choy, K. L., Law, K. M. Y., dan Ho, G. T. S. (2014). Application of intelligent data management in resource allocation for effective operation of manufacturing systems, Journal of Manufacturing Systems, 33, 412-422.

[5] Valckenaers, Paul., Brussel, Van Hendrik., Verstraete, Paul., Germain, Bart Saint., dan Hadeli (2007). Schedule execution in autonomic manufacturing execution systems, Journal of Manufacturing Systems, 26, 75-84.

IOP Conf. Series: Materials Science and Engineering **319** (2018) 012014 doi:10.1088/1757-899X/319/1/012014

[6] Zhong, Ray Y., Huang, George Q., Lan, Shulin., Dai, Q. Y., Zhang, T., dan Xu, Chen (2015). A two-level advanced production planning and scheduling model for RFID-enabled ubiquitous manufacturing, Advanced Engineering Informatics, xxx, xxx-xxx.

IOP Conf. Series: Materials Science and Engineering **319** (2018) 012015 doi:10.1088/1757-899X/319/1/012015

Modeling of Geometric Error in Linear Guide Way to Improved the vertical three-axis CNC Milling machine's accuracy

Widiyanti Kwintarini[1,2], Agung Wibowo[1], Bagus M Arthaya[3] and Yatna Yuwana Martawirya[1]

[1] Departement of Mechanical Engineering, Faculty of Mechanical and Aerospace Engineering, Institut Teknologi Bandung, Jl. Ganesha No.10, Bandung 40132.

[2] Departement of Mechanical Engineering, Faculty of Engineering, University of Pasundan, Jl. Setiabudhi No.153, Bandung 40143.

[3] Departement of Mechatronics Engineering, Faculty of Industrial Engineering, University of Parahyangan, Jl. Ciumbuleut No. 94, Bandung 40141.

Email: widyantikwintarini@unpas.ac.id

Abstract. The purpose of this study was to improve the accuracy of three-axis CNC Milling Vertical engines with a general approach by using mathematical modeling methods of machine tool geometric errors. The inaccuracy of CNC machines can be caused by geometric errors that are an important factor during the manufacturing process and during the assembly phase, and are factors for being able to build machines with high-accuracy. To improve the accuracy of the three-axis vertical milling machine, by knowing geometric errors and identifying the error position parameters in the machine tool by arranging the mathematical modeling. The geometric error in the machine tool consists of twenty-one error parameters consisting of nine linear error parameters, nine angle error parameters and three perpendicular error parameters. The mathematical modeling approach of geometric error with the calculated alignment error and angle error in the supporting components of the machine motion is linear guide way and linear motion. The purpose of using this mathematical modeling approach is the identification of geometric errors that can be helpful as reference during the design, assembly and maintenance stages to improve the accuracy of CNC machines. Mathematically modeling geometric errors in CNC machine tools can illustrate the relationship between alignment error, position and angle on a linear guide way of three-axis vertical milling machines.

1. Introduction

With the rapid development of manufacturing industry in the business world, machine tool Computer Numerical Control (CNC) is playing a very important role and widely used in large quantities in the field of modern production such as automotive field to the field of aerospace. Meeting the demands required machine tools with precision and high accuracy and high productivity. One of the key performance criteria for modern production is the ability of CNC machines to produce products with very meticulous dimensions with complex shapes [1]. Similarly, the criteria of accuracy and high accuracy must be owned by the performance of machine tools, especially when the design and assembly of the machine. There are two ways to improve geometric accuracy in machine tools: (1) design and manufacture of precision machines and (2) fault compensation [2].

A decrease in the quality of production results may be due to a decrease in the performance of CNC machine tools caused by various sources of error factors in machine tools. Some types of error sources that contribute to deviations occur in the construction of machine tools, such as geometric errors, thermal errors and servo errors [3]. Among several deviation factors in machine tools, geometric errors have the greatest effect on accuracy and precision in machine tools [1-3].

In the design of CNC machine tools other than construction must be strong, also must be considered in the straightness profile of linear guide way. Linear guide way is one of the most important components in machine tools because it serves as a place where the work table moves along the path that has been determined in accordance with the motion axis of the machine, especially to do the motion straight.

2. Geometric Error Modeling Methodology

The geometric error component consists of position error, alignment error, angle error, as shown in figure 1. Each one axis with translational movement on the machine tool there are six error parameters ie one position error, two parameters error alignment and three angle error is pitch, yaw, roll.

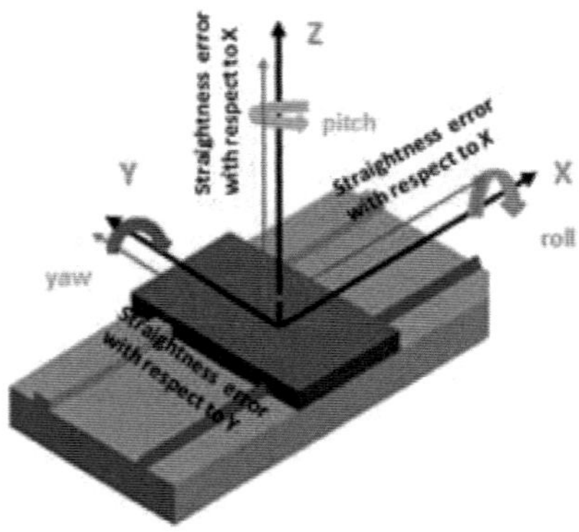

Figure 1. Error parameters of three-axis milling machine.

The source of geometric error will have an effect on the inaccuracy of the parts of the machine so that the position of the cutting tool on the machining process can occur in an undesirable position.

In the three-axis milling machine, there are 21 parameters of geometric errors that come from 3 linear positioning errors, 6 linearity errors, 9 angular errors, and 3 perpendicular errors, as shown in table 1 [4]. Illustration of Figure 1 shows error parameters of three-axis milling machine. Error parameters in figure 1 are elaborated to several geometrical error components for each axis in the three-axis machine. Several geometrical errors are listed in Table 1.

Table 1. Geometrical error components in the three-axis machine

Component errors	Linear position errors	angular errors (roll, pitch, yaw)
axis-X	$\delta_x(X)$, $\delta_y(X)$, $\delta_z(X)$	$\varepsilon_x(X)$, $\varepsilon_y(X)$, $\varepsilon_z(X)$
axis-Y	$\delta_x(Y)$, $\delta_y(Y)$, $\delta_z(Y)$	$\varepsilon_x(Y)$, $\varepsilon_y(Y)$, $\varepsilon_z(Y)$
axis-Z	$\delta_x(Z)$, $\delta_y(Z)$, $\delta_z(Z)$	$\varepsilon_x(Z)$, $\varepsilon_y(Z)$, $\varepsilon_z(Z)$
Squarness error	φ_{xy}, φ_{yz} , φ_{zx}	

3. Linear Guide Ways

Linear guide way is one of the most important components in machine tools and delivers smooth movement on machine tools, so it can produce high accuracy[5]. The main function of the linear guide way is to ensure that the work table or motion element can move along a predetermined path or in accordance with the axis of the machine tool. In every guide way has two Linear Motion (LM).

In a linear guide way there are six parameters of geometric error as shown in the figure 2, and for each linear motion can have six geometic errors, as shown in figure. 3. Geometric error that occurs in one linear guide way will affect the accuracy of machine tools or performance machine tools[6].

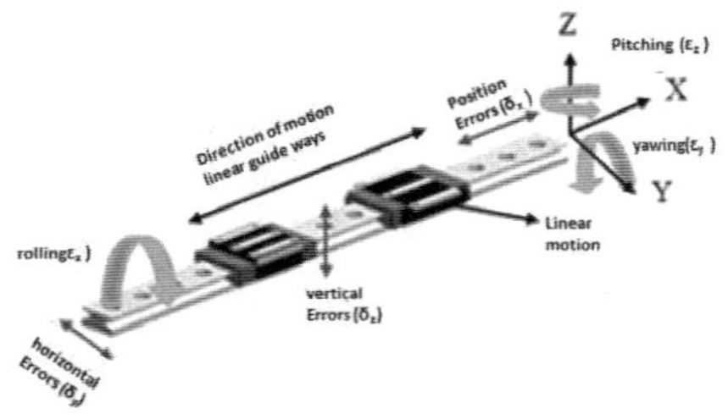

Figure 2. One linear guide way with two linear motion (LM) has six geometric error parameters.

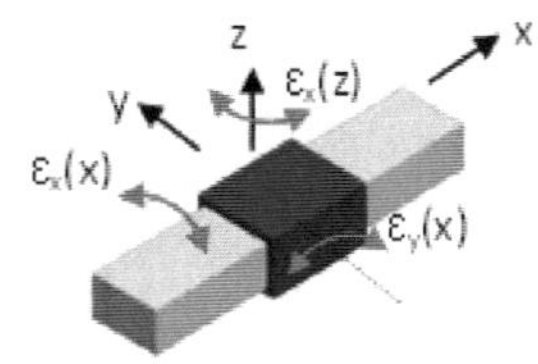

Figure 3. Geometric error in one linear motion at one linear guide way.

3.1. *Angular Error Pitching*

The angular error in linear guide way is pitch error, yaw and roll. The angle error of pitching in one linear guide way can be modeled as shown in figure 4. Point A 'and B', representing each Linear motion that serves as a place to move and support the machine tool workbench, then the equation angle error as follows:

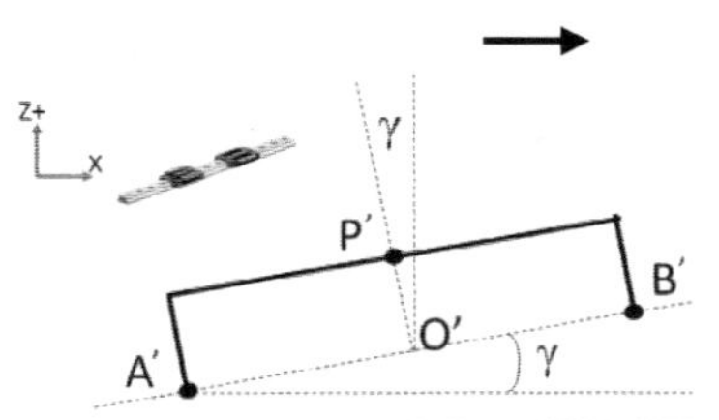

Figure 4. Condition of geometric error occurred in one linear guide way by experiencing angular error (pitching).

$$\gamma = tan^{-1}\left(\frac{(Z_B+\Delta Z_B)-(Z_A+\Delta Z_A)}{(X_B+\Delta X_B)-(X_A+\Delta X_A)}\right) \tag{1}$$

where: z_A , z_B : the linear motion A and B, position of motion A and B moves toward the Z-axis.
x_A, x_B : the linear motion A and B, position of motion A and B moves toward the X-axis.
Δz_A, Δz_B : motion drift position from linear motion A and B toward the Z-axis.
Δx_A , Δx_B : motion drift position from linear motion A and B toward the X-axis.

This one example calculation using mathematical modeling, shown in table 2.

Table 2. Error pitching linear guide way left from Tapping Brother machine.

Z_A(mm)	Z_B(mm)	Y_A(mm)	Y_B(mm)	ΔZ_A(mm)	ΔZ_B(mm)	ΔY_A(mm)	ΔY_B(mm)	pitching(γ^o)
0	0	25	25	-0,0331	-0,0277	0,00	2,60	0,0021

3.2. *Position error and Vertical Error*

Error position in one linear guide way can be modeled through figure 6 with its mathematic modeling written in equation 2, while the mathematical modeling fot vertical alignment error is given in equation 3.

$$X'_P = \frac{(X_B+\Delta X_B)-(X_A+\Delta X_A)}{2} - h\ sin\gamma \tag{2}$$

where: x_A, x_B : the linear motion A and B, position of motion A and B moves toward the X-axis.
Δx_A , Δx_B : motion drift position from linear motion A and B toward the X-axis.
h : high table from the surface of the machine block to the work table surface.
γ : pitching error

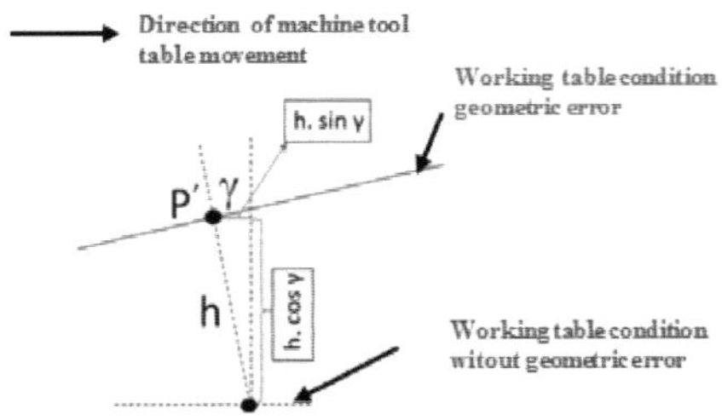

Figure 5. Error position and vertical error in one linear guide way.

Vertical alignment error is a straightness alignment toward the vertical of the linear surface guide ways.

$$Z'_P = \frac{(Z_B+\Delta Z_B+hcos\gamma)-(Z_A+\Delta Z_A+hcos\gamma)}{2} \quad (3)$$

where: z_A , z_B : the linear motion A and B, position of motion A and B moves toward the Z-axis.
Δz_A, Δz_B : motion drift position from linear motion A and B toward the Z-axis.
h : high table from the surface of the machine block to the work table surface.
γ : pitching error

This one example calculation using mathematical modeling, shown in table 3.

Table 3. Error vertical alignment error linear guide way left from Tapping Brother machine.

Za(mm)	Zb(mm)	Ya(mm)	Yb(mm)	ΔZ_A(mm)	ΔZ_B(mm)	h(mm)	γ	cos γ	Z'_P(mm)
0	0	25	25	-0,0331	-0,0277	30	0,0539	0,9985	0,0027

3.3. Angular Yawing Error

The angular error in a linear guide way taking into account the horizontal alignment deviation is called Yawing. Yawing is the rotation that occurs on the vertical axis towards the direction of translation of linear motion guide way. The motion of yawing angle errors can be modeled as shown in fihure 6.

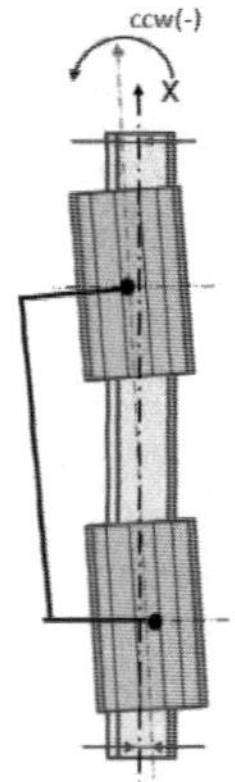

Figure 6. Angular error (yawing) in liner guide way.

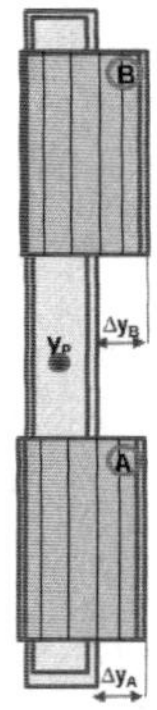

Figure 7. Geometric error linear horizontal error for one linear guide way.

The yawing angular error equation is as follows:

$$\beta_P = tan^{-1}\frac{(y_B+\Delta y_B)-(y_A+\Delta y_A)}{(x_B+\Delta x_B)-(x_A+\Delta x_A)} \quad (4)$$

where : z_A , z_B : the linear motion A and B, position of motion A and B moves toward the Z-axis.
x_A , x_B : the linear motion A and B, position of motion A and B moves toward the X-axis.
Δz_A , Δz_B : motion drift position from linear motion A and B toward the Z-axis.
Δx_A , Δx_B : motion drift position from linear motion A and B toward the X-axis.
This one example calculation using mathematical modeling, shown in table 4.

Table 4. Error yawing error linear guide way left from Tapping Brother machine

X_A(mm)	X_B(mm)	Y_A(mm)	Y_B(mm)	ΔX_A(mm)	ΔX_B(mm)	ΔY_A(mm)	ΔY_B(mm)	yawing(γ^o)
0	0	25	25	-0,2089	-0,5730	0,0039	0	-0,0108

3.4. *Horizontal Error*

The motion of horizontal errors can be modeled as shown in figure 7. The horizontal errors equation is as follows:

$$Y'_P = \frac{(y_B+\Delta y_B)-(y_A+\Delta y_A)}{2} = \frac{1}{2}[(y_B + \Delta y_B) - (y_A + \Delta y_A)] \quad (5)$$

where: y_A , y_B : the linear motion A and B, position of motion A and B moves toward the Y-axis.
Δy_A , Δy_B : motion drift position from linear motion A and B toward the Y-axis.
This one example calculation using mathematical modeling, shown in table 5.

Table 5. Error horizontal alignment error linear guide way left from Tapping Brother machine.

Ya(mm)	Yb(mm)	X_A(mm)	X_B(mm)	ΔX_A(mm)	ΔX_B(mm)	Y'_P(mm)
25	25	0	0	0,2089	0,5730	0,3641

3.5. *Angular Rolling Error*

The motion of horizontal errors can be modeled as shown in figure 8. Shown in figure 9, the direction of rotation of pitch angle error, yaw and roll.

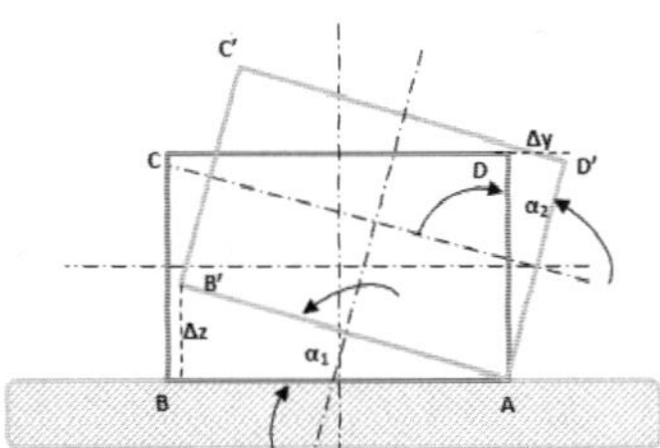

Figure 8. Angular error is a rolling error in linear guide way

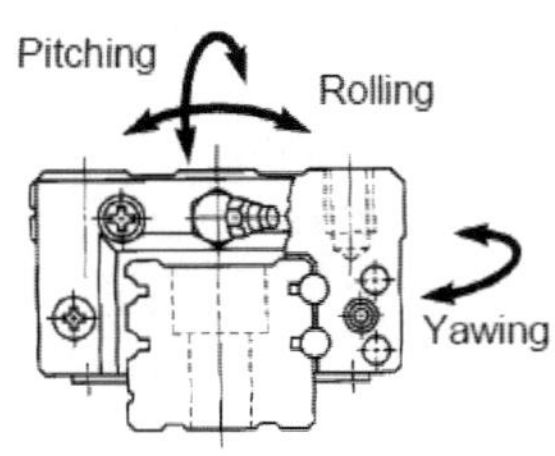

Figure 9. Angular error is a pitching error, yawing error and rolling error in linear guide way.

The rolling errors equation is as follows:

$$\alpha_c = tan^{-1}\frac{1}{2}\frac{\Delta z}{(AB+AD)} \quad (6)$$

where: $\Delta z = \Delta y$: motion drift position from linear motion A and B toward the Z-axi and the Y-axis.
AB and AD: height and width of linear motion.

4. Conclusion

The conclusion of modeling geometric error is mathematically as follows:

1. Pitching error is affected by the magnitude of vertical alignment error toward the direction of translation movement of the work table and position error.
2. The angular yawing error is affected by horizontal alignment error toward the direction of translation movement of the work table and position error.
3. The vertical alignment error is affected by motion deviations in the direction perpendicular to the direction of translation movement of the work table.
4. Rolling angle error is affected by horizontal alignment error and vertical alignment error on the direction of translational movement of the work table.
5. Furthermore, the development of geometric fault modeling will be done by taking into account geometric error to two linear guide way.

References

[1] H Schwenke, W Knapp, H Haitjema, A Weckenmann, R Schmitt and F Delbressine 2008 Geometric error measurement and compensation of machines—An update *CIRP Annals - Manufacturing Technology* vol **57** p 660-675.

[2] WenjieTian, WeiguoGao, DaweiZhang, TianHuang 2014 A general approach for error modeling of machine tools *International Journal of Machine Tools & Manufacture* vol **79** p 17-23.

[3] X B Chen and A Geddam 1997 Accuracy Improvement of Three-Axis CNC Machining Centers by Quasi-Static Error Compensation *Journal of Manufacturing Systems* vol **16** No 5.

[4] Mohsen Soori, Behrooz Arezoo and Mohsen Habibi 2014 Virtual machining considering dimensional, geometrical and tool deflection errors in three-axis CNC milling machines Journal of Manufacturing Systems vol **39** p 1001-1002.

[5] Bhagyesh B Deshmukh 2013 Linear Motion Guideways – A Recent Technology for Higher Accuracy and Precision Motion of Machine Tool *International Journal of Innovations in Engineering and Technology* vol **3**.

[6] T O Ekinci J R R Mayer 2007 Relationships between straightness and angular kinematic errors in machines *International Journal of Machine Tools & Manufacture* vol **47** p 1997-2004.

Preliminary Development of Real Time Usage-Phase Monitoring System for CNC Machine Tools with a Case Study on CNC Machine VMC 250

Herman Budi Harja[1, 2], Tri Prakosa[2], Sri Raharno[2], Yatna Yuwana Martawirya[2], Indra Nurhadi[2], Alamsyah Setyo Nogroho[2]

[1]Department of Manufacturing Engineering, Politeknik Manufaktur Bandung, St. Kanayakan No.21 Dago, Bandung, 40135, Indonesia.

[2]Department of Mechanical Engineering, Institut Teknologi Bandung, St. Ganesha No.10, Bandung, 40132, Indonesia.

Email : herman@polman-bandung.ac.id

Abstract. The production characteristic of job-shop industry at which products have wide variety but small amounts causes every machine tool will be shared to conduct production process with dynamic load. Its dynamic condition operation directly affects machine tools component reliability. Hence, determination of maintenance schedule for every component should be calculated based on actual usage of machine tools component. This paper describes study on development of monitoring system to obtaining information about each CNC machine tool component usage in real time approached by component grouping based on its operation phase. A special device has been developed for monitoring machine tool component usage by utilizing usage phase activity data taken from certain electronics components within CNC machine. The components are adaptor, servo driver and spindle driver, as well as some additional components such as microcontroller and relays. The obtained data are utilized for detecting machine utilization phases such as power on state, machine ready state or spindle running state. Experimental result have shown that the developed CNC machine tool monitoring system is capable of obtaining phase information of machine tool usage as well as its duration and displays the information at the user interface application.

1. Introduction

Flow-shop industry has different production characteristic compared to job-shop industry. The job-shop industry has characteristics of HVLV (High Variety Low Volume), where products have wide varieties but small amounts [1]. Customized product characteristic and limited capacity of workshop equipment induce negative impact on the utilization of production machine, that leads to dissimilar utilization of the machines and frequently result in utilization bottleneck at particular machines although production scheduling has been well planned. A machine with high intensity usage becomes a critical machine candidate which is prone to damage, eventually its quick reliability decreased will affect replacement frequency of machine component. The HVLV characteristics causes any machine tools to be shared and conducting a production process with the dynamic load. Reliability of machine tool will continuously

decrease with time usage but at certain condition it will quickly decrease when the machine tools continuously receives different operating conditions beyond its ideal operating condition [2, 3].
Generally, equipment reliability decrease is due to failure on machine component. It is measured by two factors, (1) life time and usage and (2) life time and number of failure. Maintenance activity for recovering reliability is machine component replacement at certain period, that causes the component is to be ‘as good new’ [4].

This paper studies a monitoring system which utilize CNC component actual usage as inputs for developing maintenance model especially in job-shop industry. This maintenance model applies the concept of Autonomous Distributed Manufacturing System (ADiMS) which emphasizes flexibility and autonomy of each production model element. Hence machine tool level from top until its component level will serve as maintenance object element whose ability and intelligence to define self-maintenance activities. The outcome of maintenance model is in the form of maintenance policy for maximizing production machine availability. One of maintenance policy is maintenance schedule whose activity scope such as cleaning, adjustment and age replacement policy of machine component. Time interval determination in maintenance schedule must consider real condition usage. Therefore the machine tools actual usage data are necessary for accuracy of maintenance model outcomes.

2. Research Problem

Currently, each of the CNC machine controllers such as Fanuc, Siemens, Mori Seiki and other CNC machine controllers have developed a machine monitoring system by their own software with different programming protocols. Therefore the monitoring data of machine utilization cannot be accessed to be taken and entered in maintenance model that is being developed. The examples are software Programming Tool PLC which developed by Siemens controller, and Fanuc controller uses Focas software for monitoring data diagnostic CNC machine. Hence, the focus of the study on this paper is development of a universal CNC machine monitoring system to obtain information about each CNC machine tool components usage in real time approached by component grouping based on its operation phase, and its system can used on various types of CNC machine controllers.

3. Materials and Method

The development study of machine monitoring system begins by studying the access ability factor of machine monitoring data on some CNC machines with different CNC controller. The review results indicate the monitoring data of each CNC machine controller can only be accessed by its own application software and access authority by theirs own maintenance machine maker technician. Hence, maintenance technician of machine tools user cannot access these data.
Generally, the architecture and installation of control and feedback signals on CNC machines are similar. CNC controller is connected to several actuators such as servo motor axis or servo motor spindle through the servo driver interfacing, coolant and lubrication motor pump, lamps, alarm and some other actuators through relay interfacing. The CNC controller is also connected to the feedback sensor devices for accurate of actuator activation. In addition the CNC controller communicates data with computer either through wired or wireless communications [5, 6, 7].

The similarity of control architecture on CNC machine and availability factor of research object, then the experiment stages of study use object research machine is focused on a CNC vertical milling machine VMC 250 with Siemens 802C controller.

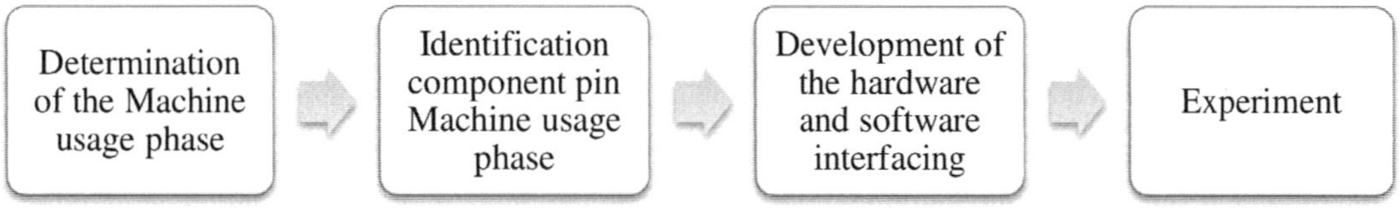

Figure 1. Research stages.

Figure 1 shows the research is consist of four stages of research. The first is defining the machine utilization status, the second is the identification of the component pin which indicate each machine utilization status, the third is development of the hardware and GUI software as an interface and final stage is do the experiment to see the function of monitoring machine device.

3.1. Determination of the machine utilization status

The definition approach of real time or actual usage data of CNC machine component is by component grouping based on its operation phase, which is divided into several operation phase. In Figure 2. The Venn diagram shows three groups machine utilization phase. Power on phase is the status at when the machine's main switch is turned on, Machine ready phase is the status when the machine's actuator system is ready to execute the commands such as setting zero point, machine start-up, or cutting process. And Spindle running phase is the status when the spindle component of CNC machine has been rotating either for start-up or cutting.

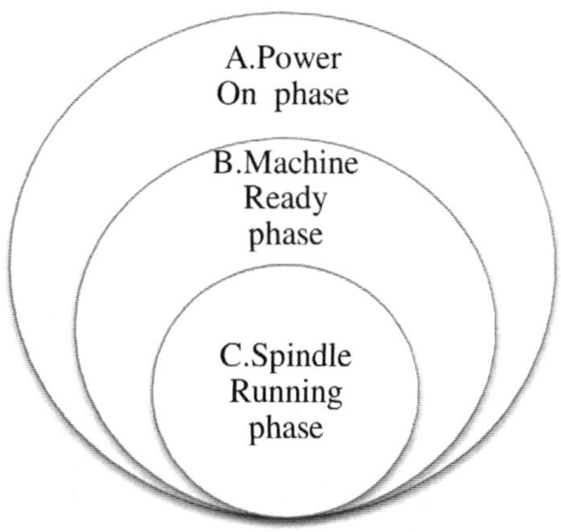

Figure 2. Venn diagram of machine usage status.

In the figure of Venn diagram shows that is:

- $C \subseteq B$: C is subset or set of part B. It is mean that actual usage time for each component in Spindle running group is included as the actual usage time for each component in Machine ready group.
- $B \subseteq A$: B is subset or set of part A. It is mean that actual usage time for each component in Machine ready group is included as the actual usage time for each component in Power on group.
- Because $C \subseteq B$ and $B \subseteq A$ then C is subset or set of part A. It is mean that the actual usage time for each component in Spindle running group is included as the actual usage time for each component in Power on group.

3.2. Identification of the component pin which indicate each machine utilization status

On this stage, Analysis and tracing of control signals on a CNC VMC 250 machine is performed to determine the pins of electronic components of CNC machine that represent the activation of each machine utilization status, such as the CNC controller, adapter and servo driver/ spindle driver. Table 1. Shows the relation between component system and which component pin can be indicator of each machine utilization status.

Table 1. Relation between components system [5, 6, 7].

Machine Usage Phase	CNC Machine Component	Output Name/ Cable	Pin Number	Relay	Pin No. Micro Controller
Power On	Power supply	Volt 24V	P24/ P24	Relay Power	30
Machine Ready	Servo driver Axis-X	Servo Ready	30/ 4116	Relay Axis-x	22
Machine Ready	Servo driver Axis-Y	Servo Ready	30/ 4117	Relay Axis-y	24
Machine Ready	Servo driver Axis-Z	Servo Ready	30/ 4118	Relay Axis-z	26
Machine Ready	Spindle Driver	Spindle Ready	30/ 4108	Relay Spindle Running	28
Spindle Running	Controller 802C	X7 Analog Input	4/ 209 & 37/ 208	-	A0 & GND

3.3. Development of the machine monitoring interface.

Monitoring machine device is consist of two interface that is hardware interfacing and software application as graphical user interface. Hardware monitoring device is build using microcontroller and several relays that schematic diagram of monitoring device and schematic of microcontroller algorithm circuit can be shown in Figure 3. And Figure 4.

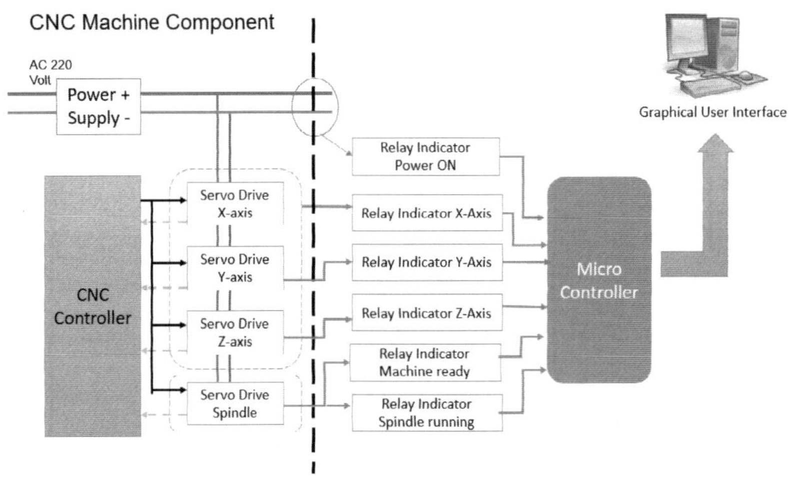

Figure 3. Schematic of monitoring device.

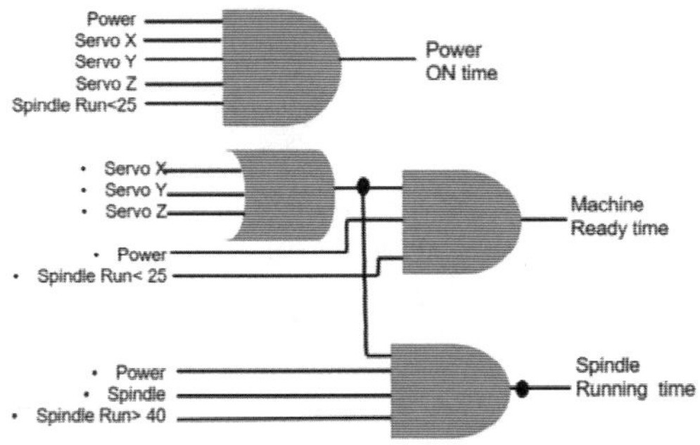

Figure 4. Schematic of microcontroller algorithm circuit.

The monitoring data from hardware device is inputted to user interface (GUI) using the application form as shown in Figure 5. The application form is designed to display the status of machine usage phase and the duration of each phase.

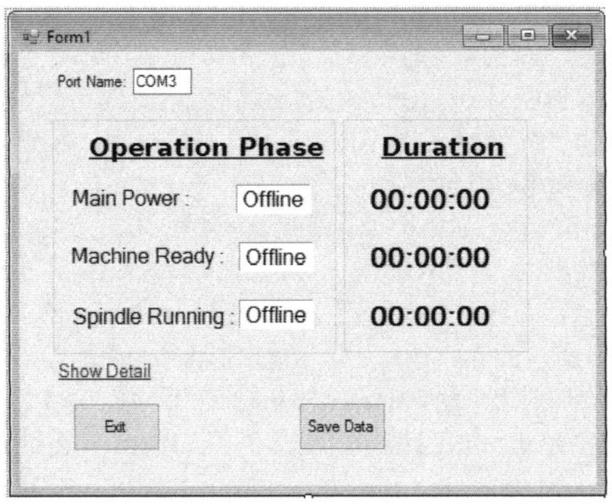

Figure 5. GUI information of operation phase type and its time duration.

4. Result and Discussion

Experimental result have shown that the developed CNC machine tool monitoring system is capable of obtaining phase information of machine tool usage as well as its duration and displays the information at the user interface application. Figure 6 shows display of interface when Power-on status is on.

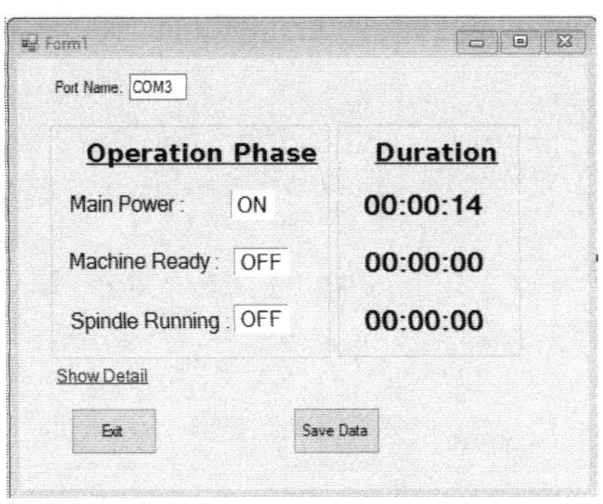

Figure 6. GUI when Power on phase

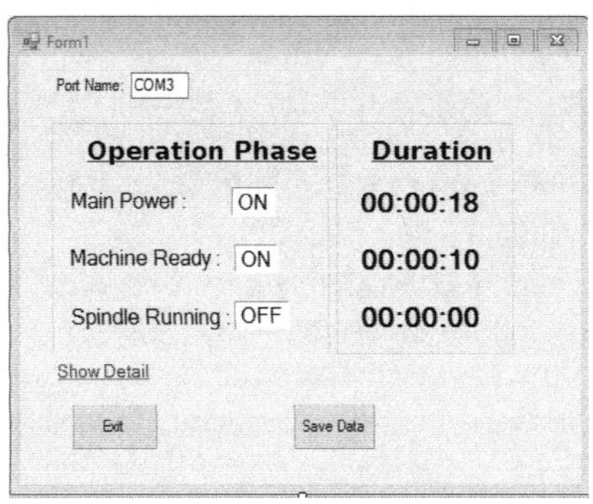

Figure 7. GUI when Machine ready phase

Figure 7 shows GUI when Machine ready phase is on. The figure shows that time duration of Power on is longer than the time duration of Machine ready phase, because Power on phase has previously started since controller booting process.

Figure 8 shows GUI when Spindle running with time duration shorter than other operation phases, because its operation started after Machine ready phase is on.

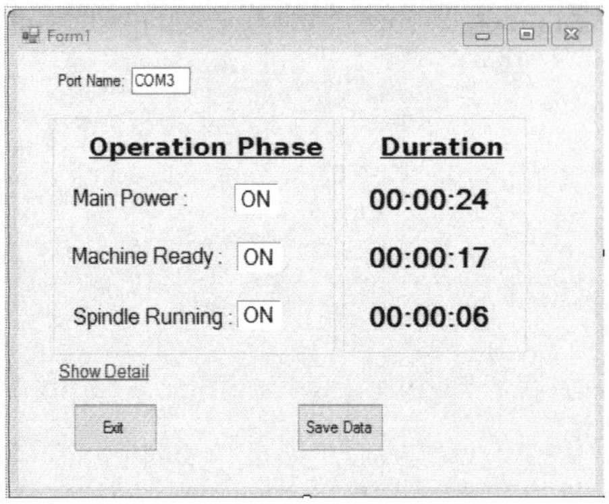

Figure 8. GUI when Spindle Running

5. Conclusion

CNC machine monitoring system based on real time usage-phase has been successfully developed and capable to convey operation phase information status, including its time duration and then displays the information in the user interface application.

Component pins for the machine usage status indicator are determined analyzing and tracing control and feedback signal on CNC machine. It uses signal status from power supply as Power on indicator, signal servo ready status from servo driver as machine ready indicator and spindle ready status from X7 output CNC controller to spindle driver as spindle running indicator.

This research should be continued by involving cutting operation phase as new components group, therefore actual usage for each component is based on real cutting operation phase, not on spindle running phase.

6. Reference

[1] Huang H and Irani S An enhanced systematic layout planning process for high-variety low-volume (HVLV) manufacturing facilities 2003 *Proc. Int. Conf. on Production Research* Virginia Polytechnic Institute and state University

[2] Borris S 2006 *Handbook on the Total Productive Maintenance*, (New York: McGraw Hill)

[3] Budi H, Prakosa T, and Yuwana Y 2015 Review and analysis: Maintenance on job shop industry. *J. Applied Mechanical and Material* ISSN 1662-7482 **842** pp 365-372

[4] Chen M, Mizutani S, and Nakagawa T 2010 Random and age replacement policies *J. International Journal of Reliability, Quality and Safety Engineering* **17**(1) pp 27–39.

[5] Siemens AG 2006 *Handbook on the Sinumerik 802C Baseline Start-Up* (Erlangen: Siemens-Aktiengesellschaft)

[6] Siemens AG 2003 *Handbook on the Simens Micromaster 440 Operating instruction* (Erlangen: Siemens-Aktiengesellschaft)

[7] Yaskawa Electric Corporation 2003 *Handbook on the Yaskawa ∑-II series SGMH/ SGDH user's manual* (Saitama: Yaskawa Electric Coorporation)

IOP Conf. Series: Materials Science and Engineering **319** (2018) 012017 doi:10.1088/1757-899X/319/1/012017

Integrated multi sensors and camera video sequence application for performance monitoring in archery

Zahari Taha[1], Jessnor Arif Mat-Jizat[1*], Muhammad Amirul Abdullah[1], Rabiu Muazu Musa[1,2], Mohamad Razali Abdullah[2], Mohamad Fauzi Ibrahim[3] and Mohd Ali Hanafiah Shaharudin[1]

[1]Innovative Manufacturing, Mechatronics and Sports Lab (iMAMS), Faculty of Manufacturing Engineering, Universiti Malaysia Pahang, 26600 Pekan, Pahang, Malaysia.

[2]Faculty of Applied Social Sciences, Universiti Sultan Zainal Abidin, Gong Badak Campus 21300 Kuala Terengganu, Terengganu, Malaysia

[3]Research and Education Division, National Sports Institute, Kompleks Sukan Negara Bukit Jalil, Peti Surat 7102, 57700 Kuala Lumpur. Malaysia

Email: jessnor@imamslab.com

Abstract. This paper explains the development of a comprehensive archery performance monitoring software which consisted of three camera views and five body sensors. The five body sensors evaluate biomechanical related variables of flexor and extensor muscle activity, heart rate, postural sway and bow movement during archery performance. The three camera views with the five body sensors are integrated into a single computer application which enables the user to view all the data in a single user interface. The five body sensors' data are displayed in a numerical and graphical form in real-time. The information transmitted by the body sensors are computed with an embedded algorithm that automatically transforms the summary of the athlete's biomechanical performance and displays in the application interface. This performance will be later compared to the pre-computed psycho-fitness performance from the prefilled data into the application. All the data; camera views, body sensors; performance-computations; are recorded for further analysis by a sports scientist. Our developed application serves as a powerful tool for assisting the coach and athletes to observe and identify any wrong technique employ during training which gives room for correction and re-evaluation to improve overall performance in the sport of archery.

1. Introduction

Development of innovations in sports sciences and engineering has provided a solution that reinforces the utilisation of inertial measurement units (IMUs) and wearable sensors as a powerful device for the measurement and tracking of athletes in different sports as an alternative to the use of motion analysis laboratory [1]. The IMUs which consist of gyroscopes and accelerometers as well wearables sensors such as Electromyograph (EMG) and Electrocardiogram (ECG) can be used to monitor athletes' performance with respect to speed, position and physiological responses during a sporting performance by integration of the sensor data [2]. The combinations of these sensors have shown to be cost effective and reliable in tracking and assessment of athletes' performance in diverse sporting activities [3,4].

Archery is a static and non-contact sport that requires the archer to be acquainted with the synergies of certain performance variables viz. physiological, mechanical as well psycho-fitness variables which either directly or indirectly influence the outcome of performance in the sport. The previous researchers have described that the capability of the archer to control his or her postural sway, heart rate and activation, as well as relaxation of the relevant muscles, is resulted in shooting consistency and consequently contribute to higher archery scores[5,6]. Nevertheless, for an archer to be acquainted with all the information on the aforementioned performance variables during archery performance, there is a need for an interactive device that can deliver the information in a real time for analysis and evaluation. Moreover, to the best of the authors' knowledge, the development of software and database system for monitoring archery performance proposed in the present investigation is still in infancy. Hence the purpose of the current work is to develop a comprehensive archery performance monitoring system which consisted of three camera views and five body sensors using a single user interface.

2. Materials and Method

The procedures for the development of the present system was undertaken under the following sequential order:

2.1. Sensors development

A total of 2 IMUs and two EMG wearable sensors are embedded into the system to determine the postural balance, movement of the arrow, muscular activations of the muscle flexor digitorum and extensor digitorum and the heart rate of the archer. For the assessment of postural sway, an accelerometer is tied to the pelvic region of the archers which is the centre of gravity that controls the static stability in human. To determine the movement of the bow, a gyro sensor is attached to the hand of the archer gripping the bow in a glove to hold the sensor tightly. Furthermore, bipolar electrodes are connected to the sensor by enabling the heart rate detector to detect the heartbeat of the archers from the wrist. Equally, two sensors are attached to the left muscle extensor digitorum and the right muscle flexor digitorum to obtain Electromyography signals during the performances of the archery related movements. All the data can be streamed in real time at the desired sampling rate and can be transmitted via Bluetooth for further analysis. The sections of all the sensors attachments in the body are shown in figure 1:

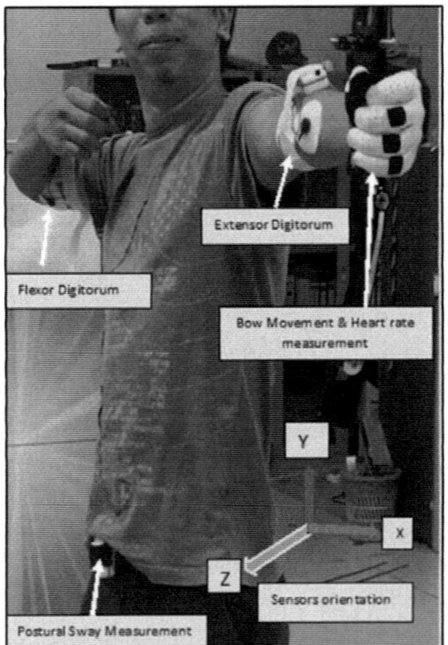

Figure 1: IMUs sensors' location attachments in the archer's body.

2.2. Camera integrations

Cameras has become ubiquitous in monitoring archery training performance. A camera system can help the archer together with their coaches to review their techniques during training. A more advance camera system such as high speed camera system [7] allowed precise technique execution as well as injury prevention such as rotor cuff.

In this paper, a three way camera view is deployed simultaneously to record every archery action during training from three different viewing angle of the archer namely from the front, from the top, as well as from the side. When reviewing the action from this three camera angle, archer and coaches can easily recognize correct form and posture as opposed to the wrong during stance, drawing, aiming and release phase of the arrow. By reviewing the video footage, an improvement can be made for the archer's technique.

The camera system described in this paper also has the capability to capture the archer action in "live" mode. One of the advantages provided by the "live" mode is the ability for the archer to receive feedback immediately from the coaches while standing, drawing, aiming and releasing arrows. This will help the archer to develop muscle memory of proper technique rather than waiting for a special review session after training [8].

2.3. Psycho-fitness assessment

The Athletic Coping Skills Inventory (ACSI), an instrument for assessing athlete's psychological skills, developed by previous researchers was selected [9]. This instrument is considered apt for evaluating the psychological skills of the archers in this study because of its connection to the nature of archery game as an individual sport rather than a team sport. The instrument measured seven basic psychological constructs namely; Coping with Adversity, Coachability, Concentration, Confidence and Achievement Motivation, Goal Setting and Mental Preparation, Peaking Under Pressure, Freedom from Worry.

On the other hand, standard physical fitness evaluation which consists of vertical jump, standing broad jump, stork balance, hand grip, core muscle strength test were implemented. These psycho-fitness assessments are integrated into the system and key in offline before the online measurement of the biomechanical variables.

3. Results and Discussion

The output of this paper is the software development to integrate all the necessary sensors and the camera view into one single performance monitoring interface. The software is developed using Microsoft Visual Studio Community 2017 Version 15.1 in a Windows Forms Apps. The entire programmed is coded using C# language. The software is divided into two main component namely the data entry for psycho-fitness analysis and the camera view with biomechanical analysis

From the instrument described in 2.3, an interface for storing the psycho-fitness analysis for each archer is shown in figure 2. The psycho-fitness variables are assessed offline and key into the system which could predict the level of the psycho-fitness ability of the archer through an integrated equation. The psycho-fitness data give the indication of the performance of the archers given that the psychoanalysis data is gathered consistently once in three months or a week before a tournament. Similarly, the interface is working together with an embedded database features in Microsoft Visual Studio. The manipulation of the data such as insert, update, select, or delete is programmed through SQL Client query inside the database. All the data inserted into the database can be exported into comma-separated values (.csv) form. This .csv form allowed for data sharing between archer and coaches as well as coaches and coaches to track individual archer progress in term of psychological readiness and fitness performance. As reported by Suppiah P K et al. [6], psychological readiness and fitness performance are key components for the archer to prepare for a competition such that in highly rated psychological readiness and highly rated fitness, the archer is expected to perform well in the competition.

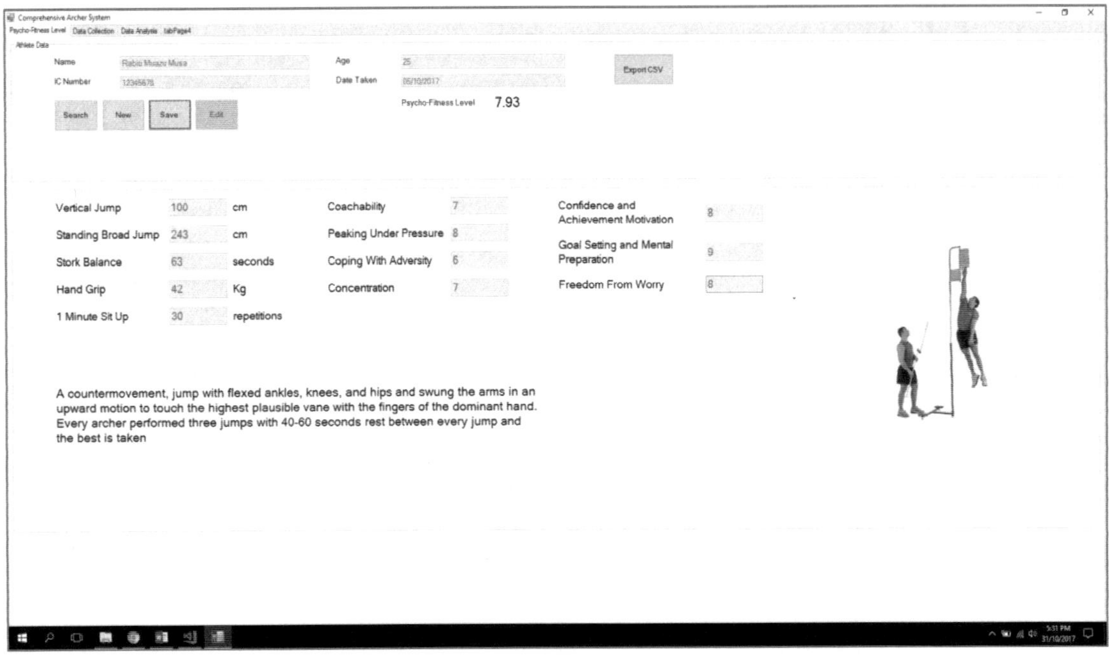

Figure 2: Interface for Psycho-Fitness Analysis Data Entry.

The other main part of the software development is the display of live camera and all measurements from the sensors simultaneously. The live camera displayed is programmed using Emgu.CV library while the sensors are read from Bluetooth. Thus, the software needs a Bluetooth enabled the computer to run. The data streaming from three cameras and five sensors through Bluetooth can put a significant burden on the computer processor to compute in real time. Therefore, the five sensors are programmed in a multithread environment embedded in Microsoft Visual Studio. The term "Threading" in C# programming language refers to a coding section which enables concurrent processing of operation at one time. The data stream from the Bluetooth also being displayed in "live" line chart form in the interface. The interface for the camera view and the chart form is shown in figure 3

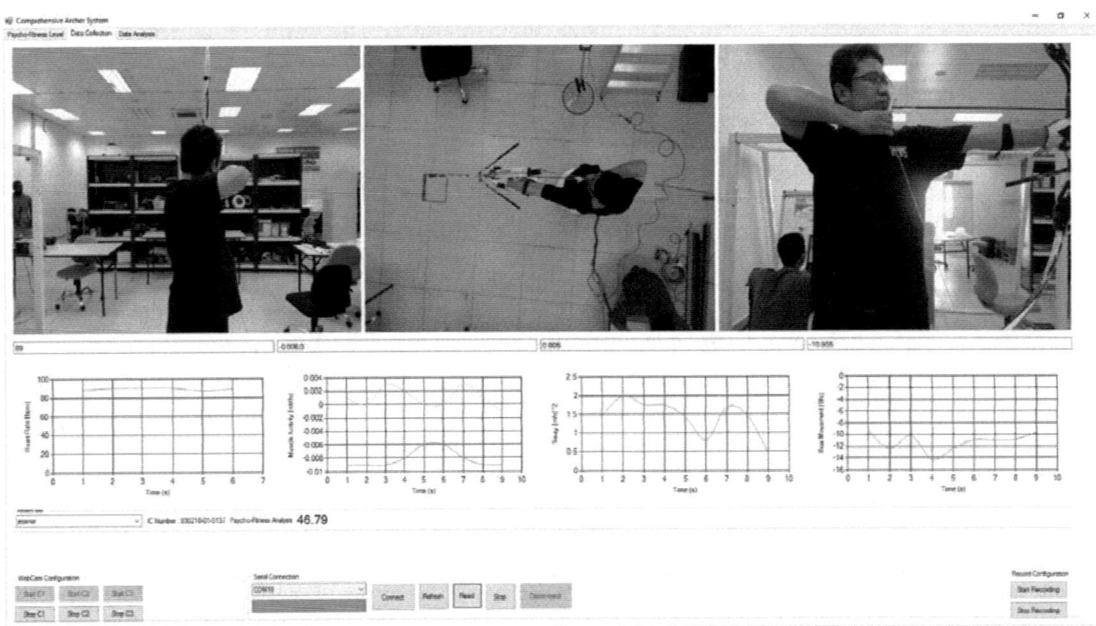

Figure 3: Interface for Camera Live View and Sensors.

The importance of the present developed system is the capability to capture the archer action in real time. This enables the archer to obtain feedback immediately from the coaches during stages of archery performance such as standing, drawing, aiming and releasing arrows. The opportunity for the archers to obtain real time feedback on their performance has been reported by the previous researchers to offer a substantiate benefit in developing muscle memory [8]. It could also serve as an avenue for immediate correction of any detected wrong technique rather than waiting for special review session after training.

4. Conclusion

The comprehensive archery monitoring system developed in this paper has shown ability in assessing postural sway, hand movement, muscular activation, heart rate as well as psycho-fitness variables different connected to the sport of archery as well as archers' technique from three different viewing angle in one user interface. The developed system has shown that a comprehensive sensors integration from physical sensors attached to the archer to three different camera angles able to provide necessary information on every action executed by the archer which can go a long way in facilitating the archers to be conscious of his/her actions and any improper techniques to enhance their own performance. The next part of the system development which is still ongoing is the development of a deployable portable tower in which the cameras can be mounted for recording the archer's performance in real time during training at the archery range.

5. References

[1] Ahmadi A, Mitchell E, Destelle F, Gowing M, O'Connor N E, Richter C and Moran K 2014 Automatic activity classification and movement assessment during a sports training session using wearable inertial sensors *Wearable and Implantable Body Sensor Networks (BSN), 2014 11th International Conference on* (IEEE) pp 98–103

[2] Tan H, Wilson A M and Lowe J 2008 Measurement of stride parameters using a wearable GPS and inertial measurement unit *J. Biomech.* **41** 1398–406

[3] Taha Z, Musa R M, Abdullah M R, Razman M A M, Lee C M, Adnan F A, Abdullah M A and Haque M 2017 The Application of Inertial Measurement Units and Wearable Sensors to Measure Selected Physiological Indicators in Archery *Asian J. Pharm. Res. Heal. Care* **9** 85–92

[4] Pfau T, Witte T H and Wilson A M 2005 A method for deriving displacement data during cyclical movement using an inertial sensor *J. Exp. Biol.* **208** 2503–14

[5] Musa R M, Abdullah M R, Maliki A B H M, Kosni N A and Haque M 2016 The Application of Principal Components Analysis to Recognize Essential Physical Fitness Components among Youth Development Archers of Terengganu, Malaysia *Indian J. Sci. Technol.* **9**

[6] Suppiah P K, Musa R M, Kiet T W K, Abdullah M R, Musawi A B H, Maliki N A K and Haque M 2017 Sensitivity Prediction Analysis of the Contribution of Physical Fitness Variables on Terengganu Malaysian Youth Archers' Shooting Scores *Int. J. Pharm. Sci. Rev. Res.* ***43*** *133-139*

[7] Ahmad, Z., Taha, Z., Hassan, H. A., Hisham, M. A., Johari, N. H., & Kadirgama, K. (2013, July). Biomechanics measurements in archery. In *International Conference on Mechanical Engineering Research (ICMER2013)*(Vol. 1, p. 3).

[8] Bridgman, A. (2013).*Movement Variability in Archery* (Doctoral dissertation, Cardiff Metropolitan University).

[9] Smith R E, Schutz R W, Smoll F L and Ptacek J T 1995 Development and validation of a multidimensional measure of sport-specific psychological skills: The Athletic Coping Skills Inventory-28 *J. Sport Exerc. Psychol.* **17** 379–98

Acknowledgments

The researchers thank the National Sports Institute of Malaysia for providing the research grant for the study (ISNRG: 8/2014-21/2014). The researchers have no any conflict of interest to declare.

IOP Conf. Series: Materials Science and Engineering **319** (2018) 012018 doi:10.1088/1757-899X/319/1/012018

Reverse engineering of wörner type drilling machine structure.

A Wibowo, I Belly, R Ilhamsyah, Indrawanto and Y Yuwana

Faculty of Mechanical and Aerospace Engineering, Institut Teknologi Bandung
Jalan Ganesa 10 Bandung, 40132 Indonesia

Email : a_wibowo_m@yahoo.com

Abstract. A product design needs to be modified based on the conditions of production facilities and existing resource capabilities without reducing the functional aspects of the product itself. This paper describes the reverse engineering process of the main structure of the wörner type drilling machine to obtain a machine structure design that can be made by resources with limited ability by using simple processes. Some structural, functional and the work mechanism analyzes have been performed to understand the function and role of each basic components. The process of dismantling of the drilling machine and measuring each of the basic components was performed to obtain sets of the geometry and size data of each component. The geometric model of each structure components and the machine assembly were built to facilitate the simulation process and machine performance analysis that refers to ISO standard of drilling machine. The tolerance stackup analysis also performed to determine the type and value of geometrical and dimensional tolerances, which could affect the ease of the components to be manufactured and assembled

1. Introduction

Forward engineering is the traditional process of moving from high-level abstractions and logical designs to the physical implementation of a system. In some situations, there may be a physical part/ product without any technical details, such as drawings, bills-of-material, or without engineering data. The process of duplicating an existing part, subassembly, or product, without drawings, documentation, or a computer model is known as reverse engineering [1]. Reverse engineering has several objectives, which one of them is learning to design and make a product with the correct procedure. The reverse engineering process can be performed in seven stages [2]. The seven stages of reverse engineering are the collection of information, the analysis of product functions, dimensional measurement, and modeling, functionality, reliability and manufacturability analysis, product embodiment, testing, evaluation and final improvement. The problem after the completion of the solid CAD model with detailed feature is the accommodation of tolerance and constrains to the dimensional and geometrical values and to the surface features. [3]

As a part of functionality, reliability and manufacturability analysis in reverse engineering, the stack up analysis was perform. The purpose of stack up analysis is to establish the dimensional relationship within a part or assembly. It enables part tolerance to be achieved. One of the most important reason for using stack analysis is that problem can be discovered and solved on paper rather than in the prototype or production, and thus evaluation and modification can be done at early stage of design [4].

The general purpose of this reverse engineering process is to study the reverse engineering by emphasizing how to recognize the precision requirements of machine geometry and the process of establishing tolerances, so the Worner B13 type drilling machine was chosen because it has relatively simple construction and relatively few components. The final objective of this reverse engineering process is to build the drill machine design with the easiness of manufacturing and assembly process, so the machine can be made with simple manufacturing facilities, and limited operator skills. The discussion in this paper will be limited to the process of identifying the requirement for the geometric accuracy of each component of machine structure so that the overall accuracy of the machine can be achieved. The drilling machine structure component consists of worktable, column based, column, and head.

2. Initial information and product function analysis

The Worner B13 type drilling machine can be seen in Figure 1. The initial information of Worner B13 type drilling machine was obtained from the information attached or measured on the machine. The initial information can be seen on the Table 1.

Table 1. Specifications of Worner B13 type drilling machine

Power	0.75 HP
Spindle Speed	560 - 3500 rpm
Spindle stroke	120 mm
Worktable size	330 mm x 430 mm

Figure 1. The Worner B13 type drilling machine

Like other drilling machines, in general this machine is used to make holes. Beside the maximum volume of the workpiece, there are no information is available about the range of cutting operation of this machine. Spindle speed analysis and power analysis are performed to estimate the cutting operation range. Spindle speed analysis is performed for the spindle speed range of 560 rpm. up to 3500 rpm. Based on the table of recommended rotary speed for various tools material [5]. Spindle speed range of 560 rpm. up to 3500 rpm. are suitable for HSS tool with the diameter of 3.175 mm. up to 6.35 mm. Larger tool diameters can be used up to 10.16 mm. in diameter for limited workpiece material. Furthermore, the adequacy of power analysis has been performed to ensure that the power requirements of the cutting process with the variation of tools diameter variations can be supplied by the motor. Based on the power analysis it has been found that with a motor power of 0.75 HP, the range of possible tool diameter variations is only tools of diameter of 3.175 mm. up to a diameter of 6.35 mm.

The main structure of the drilling machine can be seen in Figure 2. Next, the function of each major component of the drilling machine structure will be identified. The drilling machine worktable is a part of the base of the machine. The function of the worktable is to place the workpiece during the cutting process. Another function of machine base is to be the support of the machine column through the column base. The column base connects the column and machine base and supports the column in such a way that the column can move up and down and perpendicular to the worktable surface. Column supports and carries the head up and down. The machine head is the location of the motor, transmission system, and spindle. Spindle system must be able to move up and down (feeding stroke) and the movement is perpendicular to the worktable surface.

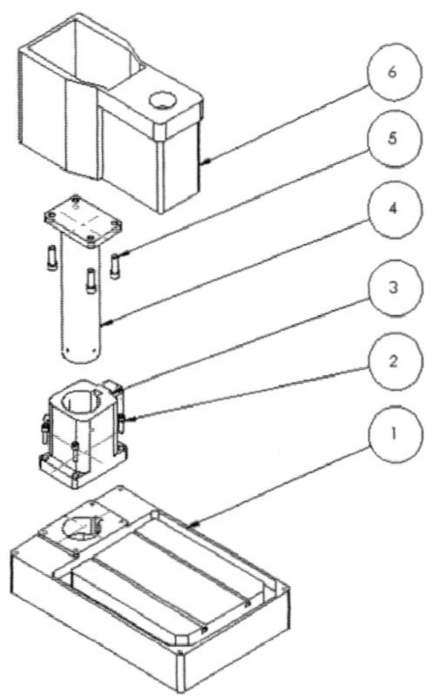

No.	Component name
1.	Machine base with worktable
2.	Column base screw
3.	Column base
4.	Column
5.	Column – Head screw
6.	Head

Figure 2. Exploded drawing of the Drilling Machine Structure

3. Measurement and modeling

The shape of the drilling machine structure component is not too complex, so the geometry and dimensions of the component can be known by measurement using simple dimensional measuring equipment. All information related to the shape and dimensions of each component obtained is used for modeling. The ST42W material is selected for the main components of the drilling machine structure because it has a high machinability. The load simulation due to torque, gravity, and compressive force is carried out on the structure of drilling machine. The simulation results shows that the maximum deformation is about 0.00619 mm as shows in the Figure 3. Figure 4 shows that maximum stress occurs in the fillet area of 1.4 MPa. Furthermore, the simulation results have been used to calculate the angular error of the machine structure in the tolerance analisys.

IOP Conf. Series: Materials Science and Engineering **319** (2018) 012018 doi:10.1088/1757-899X/319/1/012018

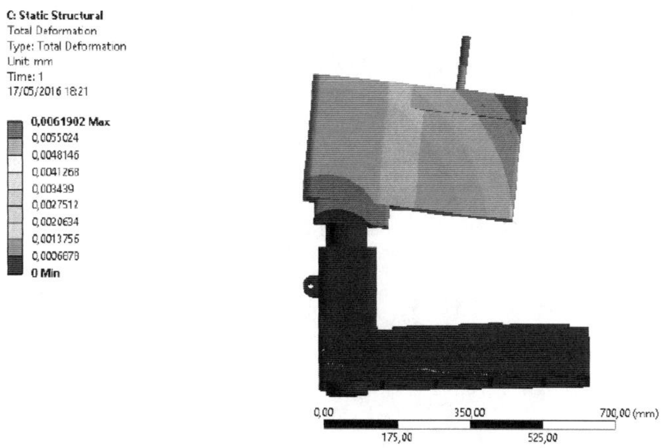

Figure 3. Deformation on drilling machine structure

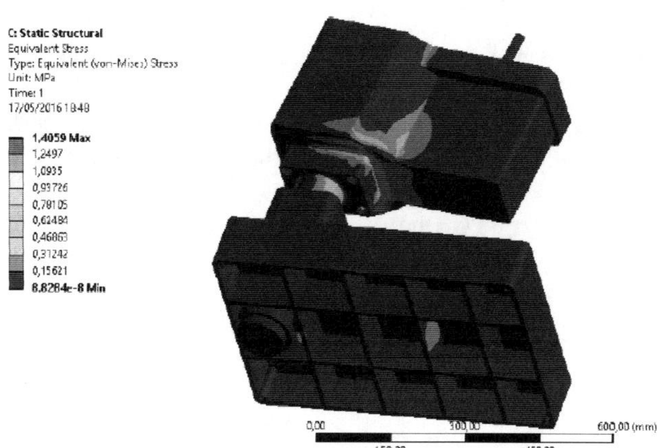

Figure 4. Maximum stress on drilling machine structure

4. Functionality and geometric requirement.

As a machine tool, the accuracy of drilling machines when functioning is very important. In addition to the rigidity of drilling machine structure, geometric accuracy of each component of drilling machine structure play an important role in realizing the overall accuracy of the machine. The geometry accuracy of each component is determined refer to the expected accuracy of the drill machine. The accuracy of the drilling machine cannot be obtained directly from the machine, in this case the geometric accuracy information of the drilling machine is obtained from ISO 2773 [6]. The squareness of spindle axis of drilling machine to the worktable surface is 0.06/300 mm.

The tolerance stackup analysis is performed to determine the accuracy and tolerance of each component so that the final accuracy of the drilling machine can be achieved. The datum reference frame (DRF) is generated for each component based on guides from ASME Y14.5 [7] and becomes a reference when performing a tolerance stackup analysis. At the assembly level, the DRF for the drilling machine is placed at the middle of machine worktable feature as seen in Figure 5.

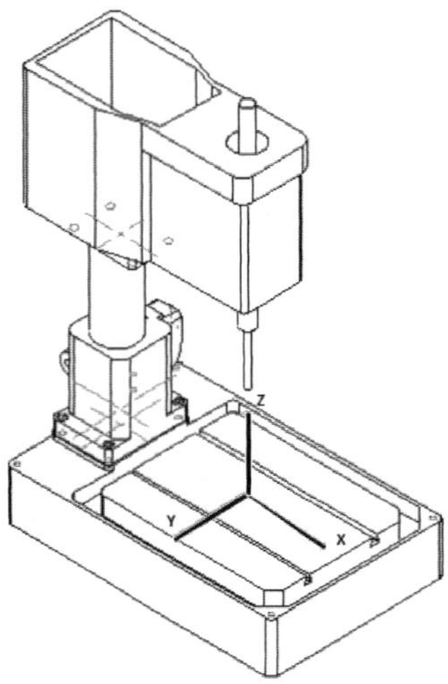

Figure 5. Assembly level datum reference frame

The tolerance stackup analysis has been performed based on two reference planes: X-Z plane and Y-Z plane. The budget tolerance as shown in Table 2, is determined for machine structure, spindle system, and deflection to be used as a tolerance limit on tolerance stackup analysis. The total budget is obtained from the conversion of the squareness tolerance value to the maximum permissible angular error at the location and orientation of the DRF. Only the tolerance stackup analysis for the machine structure to be discussed in this paper. The proportion of the tolerance budget for machine structure subsystem and spindle subsystem was determined based on the number of parts in each subsystems. The number of parts in the machine structure subsystem is 6, whereas in spindle subsystem is 4 so the budget ratio of 6:4.

Table 2. The Budget Tolerance

No.	Tolerance component	Budget for X-Z (E)	Budget for Y-Z (E)
1	Machine Structure	0,0110	0,0070
2	Spindle System	0,0074	0,0041
3	Deflection	0,0006	0
Total Budget		0,019	0,011

The tolerance stackup for machine drilling was perform by using the worst case method. Geometric tolerances as result of the tolerance stackup analysis for drilling machine structure components can be seen in Table 3.

Table 3. Geometric tolerances of structure components

No.	Geometric Feature Correlations	X – Z Plane		Y – Z Plane	
		Sensitivity (E/mm)	Tolerance (mm)	Sensitivity (E/mm)	Tolerance (mm)
1	Parallelism between the column base pad and the worktable	0,41	0,001	0,28	0,002
2	Flatness of each paired surface between column base pad and pad on base	0,41	0,002	0,28	0,002
		0,41	0,002	0,28	0,002
3	Perpendicularity of cylindrical axis and column base pad.	0,29	0,004	0,29	0,002

4	Cylindricity of each paired cylinders on the column base and the column	0,29	0,004	0,29	0,002
		0,29	0,004	0,29	0,002
5	Perpendicularity the head mating surface to the axis of the column cylinder	0,14	0,005	0,14	0,003
6	Flatness of each paired surface between the column head mating surface base	0,32	0,004	0,47	0,002
		0,32	0,004	0,47	0,002
7	Perpendicularity between the main axis of the spindle system and the head mating surface.	0,57	0,003	0,47	0,002

The result of the tolerance stackup analysis becomes a geometric specification on each component of the machine structure, which according to the specification will be determined the production process and the appropriate production equipment. The tolerance value can be compromised by some iteration processes in the tolerance stackup analysis, due to the possibility to looking for the easier manufacturing processes and the simpler production equipments. The result of the tolerance stackup analysis can be used as reference to do some design modification as alternative to meet the strict specifications.

5. Conclusions

The reverse engineering process has been performed on The Worner B13 type drilling machine up to the stage of identifying the requirements for the geometric accuracy of each component of machine structure so that the overall accuracy of the machine can be achieved. The geometric accuracy of machine structure component were obtained by the tolerance stackup analysis with considering the results of the modeling and simulation process. The geometric accuracy for each component can be used as a basis for production planning processes or making some design modifications. By conducting the tolerance stackup analysis, the adjustment of tolerance values is possible according to the existing component relationships, so that the compromises related to the process and equipment to be used can be performed.

References

[1] Raja V, Fernandesh K 2008 *Reverse Engineering* (London: Springer-Verlag).
[2] Rochim T 2012 *Tujuh Tahapan Reverse Engineering* (Indonesia: Teknik Produksi Mesin-ITB)
[3] Jamshidi J, Mileham A R and Owen G W 2006 *Dimensional Tolerance Approximation for Reverse Engineering Applocations* (Internasional Design Conference – Design)
[4] Sahani A K, Jain P K and Sharma S C 2013 *Geometrical Tolerance Stack Up Technologies* (DAAAM Internasional Scientific Book 2013 pp. 875-872 Chapter 52)
[5] Oberg, Erik, Franklin D. Jones, Holbrook L. Horton, Henry H. Ryffel, *29th Edition Machinery's Handbook*, New York. 2012.
[6] ISO 2773/I 1973 *Test Conditions for Pillar Type Vertical Drilling Machines — Testing accuracy — Part I: Geometrical Test* (Switzerland: ISO)
[7] ASME Y14.5 2009 *Dimensioning and Tolerancing* (USA: American Society of Mechanical Engineering)

Effect of Processing Variables on Tensile Modulus and Morphology of Polyethylene/Clay Nanocomposites Prepared in an Internal Mixer

O Ujianto[1], M Jollands[2], N Kao[2]

[1] Center for Polymer Technology, Agency for Assessment and Application of Technology, Indonesia
[2] Chemical Engineering, RMIT University, Melbourne, Australia

Email: onny.ujianto@bppt.go.id

Abstract. A comparative study on effect of internal mixer on high density Polyethylene (HDPE)/clay nanocomposites preparation was done. Effect of temperature, rotor rotation (rpm), and mixing time, as well as rotor type (Roller and Banbury) on mechanical properties and morphology of HDPE/clay nanocomposites were studied using Box-Behnken experimental design. The model was developed according to secant modulus and confirmed to morphology analysis using Transmission Electron Microscopy (TEM). The finding suggests that there is different mechanisms occurred in each rotor to improve the mechanical properties. The mechanism in Roller is medium shear and medium diffusion, while Banbury is high shear and low diffusion. The difference in mechanism to disperse the clay particles attribute to the different optimum processing conditions in each rotor. The settings for roller samples are predicted around mid temperature, mid speed, and mid mixing time. There is no optimum setting for Banbury within the processing boundaries. The best settings for Banbury are at low, high, low settings. The morphology results showed a hybrid composite structure, with some exfoliations and some intercalations. There was a correlation between better mechanical properties and morphology with more exfoliation and thinner intercalated particles.

1. Introduction

Polyethylene (PE) is a commodity polymer that has low tensile modulus compared to most other commodity and engineering plastics. To improve this property, nanoclay is commonly added. However, the preparation of PE nanocomposites is challenging due to polarity difference between polymer and filler. In order to improve the filler-matrix interaction, a compatibilizer can be used. This material enables the polymer to penetrate into the clay interlayer galleries [1, 2] thereby enhancing interfacial adhesion and mechanical properties.

Melt intercalation is one popular method to produce nanocomposites due to wide availability of equipment in industry. A number of studies have shown that the dispersion and delamination degree of nanoparticle produced from this technique was influenced by equipment type [3, 4] and screw design [3, 5-7]. The internal mixer is one of the equipment types commonly used for preparing nanocomposites. It is a mixer which uses various rotor types such as Roller [8] and Banbury [2], with independent setting of temperature, speed and residence time. The Roller rotor is used to process polyolefins, while the Banbury is used to disperse common rubber fillers (such as carbon black, filler with nano scale) in highly viscous rubber blends [9, 10]. However, comparative studies on rotor design are limited. As other studies have shown that polymer nanocomposite properties are influenced by screw design [3, 5, 7], it is of interest to better understand the role of rotor design in polymer nanocomposite preparation with an internal mixer.

Setting conditions are also considered as influencing factors to produce nanocomposites. However, most studies on polyethylene nanocomposites did not explain how processing condition was chosen. A single temperature point was set on mixer at 150 °C [11, 12] and 180 °C [2]. Range of temperature was applied from 130 °C to 190 °C on single screw extruder [13, 14] from 185 °C to 205 °C for twin screw extruder [15-18]. Residence time on extruder was not set because it depends on screw rotation. On the other hand, mixing time in the mixer equipment is independent factor. Lee et al. [2] set screw rotation at 50 rpm for 10 minutes, while another study set 60 rpm for 10 minutes with preheating at 30 rpm for 3 minutes [11, 12]. This fact suggests that the optimum processing condition on polyethylene nanocomposite is still unclear.

This research aims to analyze the effects of processing condition on PE/clay nanocomposites. Various temperatures, rotor rotations, and mixing times were applied in order to observe the effects of mixing conditions and their interactions on tensile modulus. The comparison between Roller and Banbury rotor was done by producing nanocomposites according to Box-Behnken experimental design. Sample morphology were analyzed using transmission electron microscopy (TEM).

2. Materials and Methods

2.1. Materials

All polymer used in this study was a commercially available blow molding grade high density Polyethylene (HDPE) HD5148 donated by QENOS Australia. Filler used in this research was Cloisite 93A (C93A) produced by Southern Clay Products. Maleic anhydride-grafted-polyethylene (HDPE-g-MA, Polybond® 3009) from Chemtura Singapore was used as compatibilizer. SongnoxTM 1010 (SX1010) and SongnoxTM 1680 (SX1680) supplied by SunAce Australia were used as processing stabilizers.

2.2. Nanocomposite preparation

Nanocomposite samples were fabricated according to three factors and three levels of Box-Behnken experimental design (DoE) shown in Table 1. The boundaries were chosen in order to melt the polymer without degrading the materials. Control sample was produced from neat HDPE produced at 180 °C, 30 rpm, and 4 minutes so that the samples have thermal history.

Table 1. Setting conditions applied on sample preparations.

Sample	*temp (°C)*	*speed (rpm)*	*time (min)*	*Sample*	*temp (°C)*	*speed (rpm)*	*time (min)*	*Sample*	*temp (°C)*	*speed (rpm)*	*time (min)*
1	210	80	4	***6***	210	130	10	***11***	180	30	16
2	180	30	4	***7***	180	130	16	***12***	210	30	10
3	180	80	10	***8***	150	30	10	***13***	150	80	4
4	180	80	10	***9***	180	80	10	***14***	180	130	4
5	150	130	10	***10***	150	80	16	***15***	210	80	16

All samples were produced at 2 wt% of clay, 5 wt% of compatibiliser, and 0.05 wt% of SX1010 and SX1680, respectively. The clay was dried in the oven at 80 °C overnight prior to melt compounding. The clay was then hand mixed with HDPE, 5 wt% of HDPE-g-MA, and 0.05 wt% of SX1010 and SX1680, respectively. Haake Rheomix OS R600 equipped with Roller or Banbury rotors was used for mixing.

2.3. Characterizations

Tensile specimens were produced using compression molding at 180 °C according to ASTM D638 type II. Molten materials were pressed into specimen molds at 90 bar pressure, for 2 minutes preheating to obtain desired pressure, and followed by 5 minutes heating and 5 minutes cooling. All specimens were conditioned in a desiccator for 2 days prior to tensile modulus measurement.

The modulus measurements were carried out on minimum four specimens using an Instron 4467 equipped with extensometer. Crosshead rate was set at 5 mm/min for 2% secant modulus. The secant modulus was calculated because the neat HDPE samples did not exhibit any linear region.

Nanocomposite morphology was analyzed using TEM JEOL 1010. The degree of exfoliation was analyzed from 5 different sliced specimens per sample, with thickness ± 50 nm. The specimens were prepared using a Leica microtome with diamond knife at temperatures -110 °C for the specimen and -150 °C for the knife to produce very smooth surface and avoid specimen rolling and wrinkling. The pictures were captured using a Gatan Orius digital camera at 100 keV accelerating voltage.

3. Results and Discussions

3.1. Modelling on Tensile Modulus

Tensile modulus for all nanocomposites and control sample are presented in Table 2. All nanocomposite samples increased with addition of 2% clay in comparison with HDPE, as expected. The increase varied from 2 to 25%. The standard deviation for all nanocomposite samples is 6%. The modulus improvement in this study is similar to other studies in PE nanocomposites [14, 19, 20] with 2 wt% filler loading, but lower than another study which is attributed by different filler modification [21]. The results are also similar to improvements reported in other studies of PE nanocomposites with higher filler loadings, which range from 15% to 26% [1, 2, 22, 23]. These suggest that the mixing undertaken in this study was done effectively.

Table 2. Tensile modulus and strength for samples prepared by Roller or Banbury rotors

Samples	*Filler [wt%]*	*Modulus Roller [MPa]*	*Modulus Banbury [MPa]*
Control	0	745 ± 26	768 ± 22
1	2	897 ± 37 (+20%)	840 ± 23 (+9%)
2	2	761 ± 44 (+2%)	871 ± 9 (+13%)
3	2	917 ± 34 (+23%)	895 ± 21 (+16%)
4	2	913 ± 56 (+23%)	891 ± 44 (+16%)
5	2	777 ± 43 (+4%)	918 ± 27 (+19%
6	2	787 ± 18 (+6%)	831 ± 15 (+8%)
7	2	802 ± 39 (+8%)	815 ± 28 (+6%)
8	2	820 ± 65 (+10%)	874 ± 16 (+14%)
9	2	881 ± 43 (+18%)	891 ± 33 (+16%)
10	2	860 ± 45 (+15%)	951 ± 33 (+24%)
11	2	808 ± 67 (+8%)	938 ± 25 (+22%)
12	2	861 ± 35 (+16%)	918 ± 41 (+19%)
13	2	806 ± 51 (+8%)	960 ± 46 (+25%)
14	2	821 ± 29 (+10%)	898 ± 39 (+17%)
15	2	773 ± 23 (+4%)	875 ± 28 (+14%)
Average 1-15		832 ± 52 (+12%)	891 ± 42 (+16%)
Average at mid-point		904 ± 20 (+21%)	892 ± 2 (+16%)

The complete modulus data were analyzed and interpreted using Minitab 16 software. The data were regressed using a second order polynomial equation as follows:

$$y = \beta_0 + \sum \beta_i x_i + \sum \beta_{ii} x_i^2 + \sum \beta_{ij} x_i x_j \quad \text{.........} (1)$$

where y is the model response, x_i and x_j are independent variables, and $\beta_0\beta_0$, $\beta_i\beta_i$, $\beta_{ii}\beta_{ii}$, $\beta_{ij}\beta_{ij}$ are the coefficients for the constant, linear and quadratic terms, and interactions, respectively. The responses were predicted from generated model using coded unit to (-1, 0, +1) to eliminate effects of measurement units [24]. The equations for each response are shown below:

IOP Conf. Series: Materials Science and Engineering **319** (2018) 012019 doi:10.1088/1757-899X/319/1/012019

$$y_1 = 904 + 6.94\,x_1 - 7.82\,x_2 - 5.32\,x_3 - 28.2\,x_1^2 - 64.4\,x_2^2 - 41.5\,x_3^2 - 7.69\,x_1x_2 - 44.6\,x_1x_3 - 16.5\,x_2x_3 \quad \text{..........(2)}$$

$$y_2 = 892 - 29.9\,x_1 - 17.4\,x_2 + 1.25\,x_3 + 9.46\,x_1^2 - 16.5\,x_2^2 + 4.71\,x_3^2 - 32.8\,x_1x_2 + 11.0\,x_1x_3 - 37.5\,x_2x_3 \quad \text{..........(3)}$$

where y_1 is modulus for Roller and y_2 is modulus for Banbury. The variable x_1 represents temperature, x_2 for rotor rotation, and x_3 for mixing time.

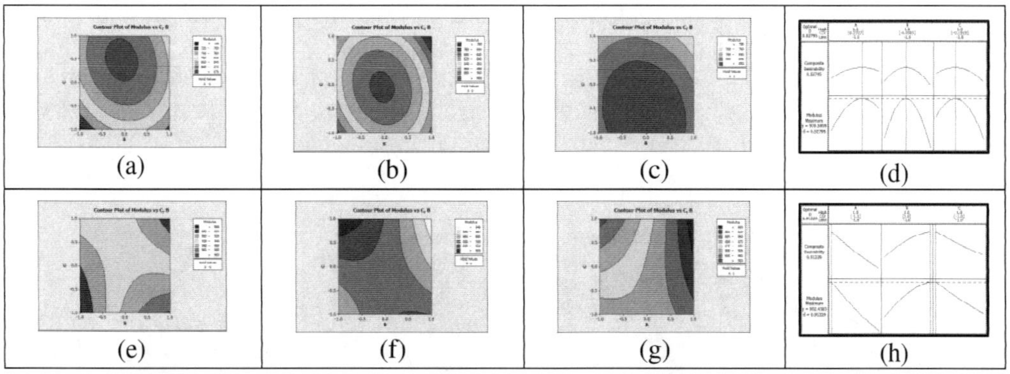

Figure 1. Contour plots for modulus: (a) Roller at low temp, (b) Roller at medium temp, (c) Roller at high temp, (d) Response optimizer Roller, (e) Banbury at low temp, (f) Banbury at medium temp, (g) Banbury at high temp, and (h) Response optimizer Banbury

Figures 1 show interacted factors to influence modulus and response optimizer plots developed form generated model. “A”, “B”, and “C” on the figures represents temperature, rpm, and time, respectively. The different effects of factors and their interactions in the various samples suggest there are several mechanisms influencing platelet dispersion, namely shear, diffusion, and degradation, and these sometimes work together and sometimes work against each other. Generally platelet dispersion is promoted by high shear [8] (low temperature and high speed), high diffusion [3] (long times at medium temperatures), and is reduced by clay surfactant degradation [1, 2, 25] (high temperature or speed for long times).

The contour plots show interactions between factors is complex. The Roller plots (Fig. 1 (a), (b) and (c)) show concentric circles with close contours and a dark green center. This indicates that a maximum modulus value occurs within the DoE boundary. On the other hand, the Banbury samples (Fig. 1 (e), (f), and (g)) show quite different behavior that there is no best modulus within boundaries conditions. That is the high and low modulus positions change with all process conditions.

The results from response optimizer (Fig. 1 (d) and (h)) show that the predicted modulus and corresponding processing conditions depend on rotor design. This shows that optimized processing conditions cannot be transferred from one rotor to another but should be identified for each rotor. The best conditions for Banbury suggested that the shear was a major factor to improve the modulus. This also suggested that Banbury rotor might be beneficial for processing thermal sensitive polymers, but not sensitive to shear. On the other hand, Roller rotor would be useful when it is used for longer mixing time due its consistency to time compare to the other one [26].

Table 3. Comparison of predicted and measured modulus

Rotor	Coded Unit	Actual Settings [°C, rpm, min)	Predicted Modulus [MPa]	Actual Modulus [MPa]	Error [%]
Roller	0.27; -0.05; -0.19	188; 77; 9	906	873 ± 41	-4%
Banbury	-1; 1; -1	150; 130; 4	982	949 ± 26	-4%

Confirmation experiments at best conditions were conducted to test the model accuracy. The results are summarized in Tabel 3. In general, measured modulus was lower than predicted values for both rotors. However, the optimized sample results are higher than the average values modulus by an average of 6% (±1 s.d to 2 s.d). This suggests that optimizing the processing conditions is worthwhile, as it is beneficial to improving modulus.

3.2. Morphology Analysis

TEM analysis was approached from qualitative and quantitative techniques. Quantitative analysis was carried out for 5 images per sample with Gatan Digital Micrograph software. The measurement areas were chosen in region that show no agglomerates (particles over 100 nm) to maximize the number of exfoliated or intercalated structures in the image, thus improving accuracy. The particle thicknesses were divided into 2 (two) groups. All particles with thicknesses less than 10 nm are considered to have an exfoliated structure (that is, a single or double layer), while 10 to 30 nm are considered to be intercalated (that is, several layers). There were very few thicker particles as images without agglomerates were chosen for analysis.

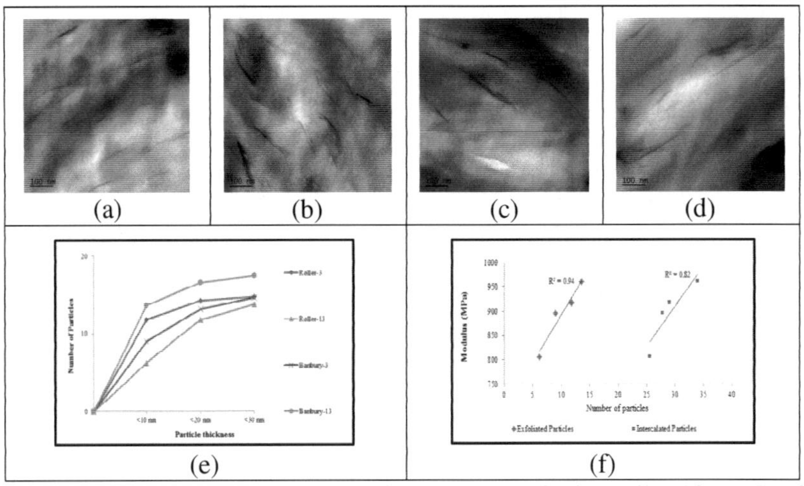

Figure 2. TEM micrographs for selected samples: (a) Roller-3, (b) Roller-13, (c) Banbury-3, (d) Banbury-13, (e) Number of particles thickness on analyzed samples, and (f) Modulus – Number of Particles Correlation

Figure 2 (a – d) presents TEM micrographs for nanocomposite samples Run 3 and 13, representing the highest modulus prepared by each rotor. All TEM micrographs indicate a hybrid structure with some exfoliated and intercalated structures with particle thicknesses ranging between 0 and 30 nm. The Banbury sample run 13 has a higher number of thinner particles than other samples (Fig. 2 (e)). It was produced at high shear and low diffusion settings (150 °C, 80 rpm, and 4 minutes). Interestingly, this setting produced the least number of thinner particles in the Roller sample run 13. For the Roller samples, the highest number of exfoliated particles was produced in run 3 (Roller-3) with medium shear and medium diffusion (180 °C, 80 rpm, and 10 minutes). This further suggests that different dispersion mechanisms occur when different rotors are used.

Further analysis on correlations between number of thin particles and modulus (Fig. 2 (f)) show better linearity for exfoliated particles (<10 nm) than intercalated (10 – 30 nm). This suggests that the modulus improvement is more influenced by exfoliated structures which depend on the rotor type and mechanisms during mixing.

4. Conclusions

Analysis on effect of processing variables on PE/Clay nanocomposites prepared in an internal mixer was done. Analysis on the effects of different rotor design and setting parameters on tensile modulus showed significant improvement on modulus with additional of 2 wt% clay. The results showed that samples produced in an internal mixer using a Banbury rotor had higher improvement compared to a Roller rotor. The modulus improvement in Banbury samples was influenced by high shear (low temperature, high rpm), while in Roller was especially at medium shear conditions (medium temperature, medium rpm). The TEM results showed that the exfoliated and intercalated structures were found in the nanocomposite system produced using either Roller or Banbury rotor at selected conditions. The effect of exfoliated particle was more significant to improve the modulus than the intercalated particles. TEM analysis showed different mechanism needed to produce the exfoliated structures. Medium shear and medium diffusion were the best for Roller rotor, while high shear and low diffusion for Banbury that confirm results from modulus measurements.

5. Recommendation for Future Works

Process variables were focused on this research as formulations have been reported by many studies. However, to minimize the cost and environmental impact, both process and material composition are important. Hence further work on synergy between process and material composition to produce economic material but less environmental effect is recommended.

Acknowledgement

Authors would like to thank to all Rheology and Material Processing Laboratory members, especially Mr. Mike Allan and Dr. Muthu Pannirselvam, as well as all staff in RMIT Microscopy and Microanalysis Facility (RMMF) for their help during this research.

References

1. Dintcheva N T *et al* 2009 *Polym. Degrad. Stabil.* **94** (9) 1571-1588.
2. Lee J H *et al* 2005 *Compos. Sci.Technol.* **65** (13) 1996-2002.
3. Dennis H R *et al* 2001 *Polym. J* **42** (23) 9513-9522.
4. Özen İ *et al* 2012 *Polym.Eng.Sci.* **52** (7) 1537-1547.
5. Jiang G and Huang H X 2008 *J. Mater.Sci.* **43** (15) 5305-5312.
6. Lertwimolnun W and Vergnes B 2007 *Polym.Eng.Sci.* **47** (12) 2100-2109.
7. Villanueva M P *et al* 2009 *Polym. Test.* **28** (3) .
8. Lertwimolnun W and Vergnes B 2005 *Polym. J*, **46** (10) 3462-3471.
9. Chae H D *et al* 2012 *Polym. Compos.* **33** (7) 1141-1153.
10. Charman M *et al* 2011 *J.Polym. Sci. Part B: Polym. Phys.* **49** (22) 1597-1604.
11. Filippi S *et al* 2008 *Eur. Polym. J.* **44** (4) 987-1002.
12. Minkova L and Filippi S 2011 *Polym. Test.* **30** (1) 1-7.
13. Besco S *et al* 2013 *Polym. Eng. Sci.* **53** (4) 689-698.
14. Chu D *et al* 2007 *Polym. Compos.* **28** (4) 499-511.
15. Abu-Zurayk R and Harkin-Jones E 2012 *Polym. Eng. Sci.* **52** (11) 2360-2368.
16. Merinska D *et al* 2011 *J. Thermoplas. Compos. Mater.* **25** (1) 115-131.
17. Spencer M W *et al* 2010 *Polym. J.* **51** (5) 1056-1070.
18. Zhong Y *et al* 2007 *Polym. Eng. Sci.* **47** (7) 1101-1107.
19. Babaei A *et al* 2011 *J. Macromol. Sci. B* **51** (3) 481-495.
20. Sarikhani K *et al* 2012 *J. Appl. Polym. Sci.* **124** (2) 1344-1351.
21. Livi S *et al* 2010 *J. Colloid Interf. Sci.* **349** (1) 424-433.
22. Pöllänen M *et al* 2013 *Polym. Eng. Sci.* **53** (6) 1279-1286.
23. Tanniru M *et al* 2006 *Polym. J.* **47** (6) 2133-2146.
24. Malik Z A *et al 2011 J. Eng.Fiber. Fabr.* **6** (4) 46-53.
25. Cervantes-Uc J M *et al* 2007 *Thermo.Acta* **457** (1–2) 92-102.
26. Salahudeen S A *et al* 2011 *Chem. Eng. Sci.* **66** (12) 2502-2511.

IOP Conf. Series: Materials Science and Engineering **319** (2018) 012020 doi:10.1088/1757-899X/319/1/012020

Reliability Centred Maintenance (RCM) Analysis of Laser Machine in Filling Lithos at PT X

M A E Suryono[1] and C N Rosyidi[2]

Department of Industrial Engineering, Universitas Sebelas Maret, Indonesia

angelein22@student.uns.ac.id[1] and cucuk@uns.ac.id[2]

Abstract. PT. X used automated machines which work for sixteen hours per day. Therefore, the machines should be maintained to keep the availability of the machines. The aim of this research is to determine maintenance tasks according to the cause of component's failure using Reliability Centred Maintenance (RCM) and determine the amount of optimal inspection frequency which must be performed to the machine at filling lithos process. In this research, RCM is used as an analysis tool to determine the critical component and find optimal inspection frequencies to maximize machine's reliability. From the analysis, we found that the critical machine in filling lithos process is laser machine in Line 2. Then we proceed to determine the cause of machine's failure. Lastube component has the highest Risk Priority Number (RPN) among other components such as power supply, lens, chiller, laser siren, encoder, conveyor, and mirror galvo. Most of the components have operational consequences and the others have hidden failure consequences and safety consequences. Time-directed life-renewal task, failure finding task, and servicing task can be used to overcome these consequences. The results of data analysis show that the inspection must be performed once a month for laser machine in the form of preventive maintenance to lowering the downtime.

1. Introduction

PT X is one of Lube Oil Blending Plant (LOBP) under PT X Indonesia. PT X uses automated machines in oil production process from blending until filling. The automated machine is used to keep and improve the product efficiency so the company is able to result competitive products in the market. LOBP works for 16 hours per day and five days per week. Therefore, machines maintenance is important to keep the machines always in good condition to reach the target both in term of quantity and quality.

Maintenance is all actions necessary for retaining an item or equipment in or restoring it to, a specified condition [1] Maintenance clearly impacts on component and system reliability; too little maintenance may result in an excessive number of costly failures and poor system performance, and therefore, reliability decrease; excessive maintenance may improve reliability, but the maintenance cost will be sharply increased [2]. Therefore, a maintenance policy is necessary to a process production. One kind of maintenance policy is Reliability Centred Maintenance (RCM). Dehghanian et al [3] explore a comprehensive RCM framework tailored to power distribution systems. Riyanto et al [4] studied about optimization of maintenance interval with RCM that can provide savings for company. Wilmeth et al applied RCM to improve battery maintenance program at SPSS [5]. Therefore, many companies conducted RCM method.

There are two kinds of maintenance activity, namely corrective maintenance and preventive maintenance. In plant operation, PT. X perform corrective maintenance. Corrective maintenance is a maintenance activity do after machine or production facility were failure or interference so it can't function properly [6]. The drawback of corrective maintenance is this kind of maintenance can bring out many economic impacts like repair or replacement cost without any preparation from the company. Besides that, by corrective maintenance, unpredicted failure on the machine will give impact to the whole production process in the form of stopped the whole production lines. On the other side, preventive maintenance is a maintenance that performed on a schedule basic, generally periodic where a number of activities such as inspection and repair, replacement, cleaning, lubrication, adjustment, and equalization are done [7]. Also, the benefits of preventive maintenance are reducing the probability of equipment breakdowns and extension of equipment life [8].
The aim of this research is to determine maintenance tasks according to the cause of component's failure using RCM and determine the amount of optimal inspection frequency that can be done to the machine at filling lithos process. There are 3 kind of filling process: filling lithos, filling drum, and filling Intermediate Bulk Container (IBC). We choose filling lithos process to be the concern of this research is because filling lithos is the most complicated filling line with 11 machines in the line. There are labelling, filling, capering, volume censor, heater, induction seal, laser, robotic, carton sealer, carton erector, and palletizing.

2. Methodology

For the first step plant observation was done to find the problem that affect the production process. PT. X has a critical problem that affect the production process, that is the line production stop caused of machine downtime. In second step, we perform literature study to find out more about theories related to the issues that will be solved. In this research, RCM method is used to find the maintenance task that suitable to each component to minimize machine downtime. In the third step, data collection and processing were conducted. In this step, downtime data were collected to find critical machine. We collect downtime data of all machine we recapped all of the downtime data so we can find a machine that has the longest downtime and the highest downtime frequency. Afterwards, Fault Tree Analysis (FTA) was used to breakdown the cause of machine downtime. FTA was then followed by creating Failure Mode and Effect Analysis (FMEA) to find the detail potential failure mode, failure effect, and cause of failure. The Risk Priority Number (RPN) of each failure mode can be determined. In the last step, maintenance and the optimal frequency for the whole critical machine can be developed.

2.1. *Collecting and processing data*

The data used in this study is limited only downtime data filling lithos process from August 2015 until July 2016. The assumptions used are every machine work for two shifts starting at 07.00 till 23.00, average inspection time assumed for 2 hours and the average repair time is assumed for 4 hours, and the reliability of the machine after maintenance activities amounted to 85% according to the world standards that have been set.

2.2. *Theoretical probability distribution used in this research*

There are several probability distributions that used in research are Weibull distribution and exponential distribution. The Weibull distribution equation used to calculate MTBF and reliability of the critical machine based on the TBF data. The exponential distribution used to determine the optimal frequency of inspection of critical machine based on the downtime data.

- Weibull Distribution

  The reliability equation is [7]:

$$R(t) = e^{-\left(\frac{t}{\theta}\right)^{\beta}} \quad (1)$$

  With average value as follow [7]:

$$mean = \theta\Gamma\left(1 + \frac{1}{\beta}\right) \quad (2)$$

Where Γ is gamma function [9] with equation:

$$\Gamma(s) = \int_0^{\infty} x^{s-1} e^{-x} dx \tag{3}$$

- Exponential Distribution
  The reliability equation is [7]:

$$R(t) = e^{-\lambda t} \tag{4}$$

With average value as follow [7]:

$$mean = \frac{1}{\lambda} \tag{5}$$

2.3. *Reliability centred maintenance (RCM)*

RCM is a method to determine maintenance task to ensure the reliability of a system. RCM serves to overcome the dominant cause of failure which will bring on maintenance decisions that focus on the prevention of these types of failures that occur often. The application of RCM method can provide some advantages such as the safety and integrity of environment which will be becoming more preferably, increasing operational achievements, higher effectiveness of operating and lowering maintenance cost, increasing the availability and reliability of the equipment, giving longer component life, more comprehensive data base, greater individual motivation, and able to improve cooperation among operators of installation [10]. The steps for implementing RCM are as follow [11]:

- Determine the cause of failure using Fault Tree Analysis (FTA) aimed to obtain the probability of failure and determine the critical components that are prone to failure.
- Develop analytical work with Failure Mode Effect Analysis (FMEA), such as determining the priority of equipment that needs to be maintained.
- Classify the level of consequences of failure.
- Make RCM decisions to classify the levels of maintenance needs.
- Implement maintenance decision based on the RCM.
- Conduct an evaluation based on the results of the proposed maintenance.

RCM classify the failure consequences in four groups:

- Hidden failure consequence is a failure that has many impact on the failure of another component and more serious because of hidden or unknown to the operator.
- Safety consequences is a failure that may causes hazards, injured or even killing someone.
- Operational consequence is a failure that have any impact in operating process such as product, output, operating cost, repair cost and also can cause the stopped of production process.
- Non-operational consequence is a failure that have no impact on safety of production, but can impact directly to the cost with relatively small impact.

Several maintenance tasks are exist in RCM [12]:

- Condition-Directed Task
  This type of maintenance task aims to determine the potential failure which can be prevented before the actual failure occurs and detect the potential failure by checking the component. This maintenance task leads to a regular basis diagnostic test or conduct inspections which can compare the condition of pre-maintain components to their standard.
- Time-Directed Life-Renewal Task
  Time-directed life-renewal task is an action that aims to prevent component failure directly to the source which is based on the life time of the component. Time-directed life-renewal task duty is to replace or repair component before it reaches a time in which the probability of failure becomes greater. In this maintenance task, there are two kinds of task. Replacement which an item must be replaced with new item when it has reached the level of wear-out, while restoration which an item can still be repaired so that it can be used again.
- Failure Finding Task
  Failure finding task aims to find hidden failure to the component by implementing the periodic inspection and evaluate the state of equipment or components.

- Run to Failure
  Run to failure is a task in which the component is used until it is broken, because no economical action can be conducted to prevent the failure.
- Servicing Task
  Servicing task is a task of maintenance for additional products or material before it is used during the normal operation,
- Lubrication Task
  Lubricating is a type of maintenance task in terms of doing lubrication on a regular basis.

2.4. The optimal frequency of inspections

The time to failure of the machine can be determine using the downtime data of critical machine. The total downtime per unit time can be constructed as a function of inspection frequency with the construction of the model as follows [13]:

- Equipment failures occur according to the exponential distribution with mean time to failure (MTTF) = $1/\lambda$, where λ is the mean arrival rate of failures. For example, if the MTTF = 0,5 years, then the mean number of failures per year = 2, i.e. $\lambda = 2$.
- Repair times are exponentially distributed with mean time $1/\mu$.
- The inspection policy is to perform n times per unit time. Inspection times are exponentially distributed with mean time $1/i$.
- The objective is to choose n to minimize total downtime per unit time. The function can be written as follows:

$$D(n) = \frac{\lambda(n)}{\mu} + \frac{n}{i} \tag{6}$$

By changing the value of $\lambda(n) = k/n$ and deriving equation (6) to determine the optimal number n, the equation become as follows:

$$D'(n) = -\frac{k}{n^2\mu} + \frac{1}{i} = 0 \tag{7}$$

Therefore,

$$n = \left(\frac{ki}{\mu}\right)^{1/2} \tag{8}$$

3. Result and Discussion

Based on data shown in Table 1 the filling lithos line 2 has the longest downtime. From Table 2 the laser machine on line 2 is the critical machine in filling lithos process at PT. X. After determining the probability distribution function of laser machine's Mean Time Between Failure (MTBF) data, it was found that the data followed Weibull distribution. Based on the MTBF of that probability distribution, then the machine reliability is calculated and found to be 24.54%, and it can be concluded that the machine is under condition of low reliability so it needs to be analyzed further.

Table 1. Downtime machine in filling lithos

Line	1	2	3	4
Downtime (minutes)	16325	24630	18535	17465

Table 2. Downtime machine in filling lithos line 2

Machine	Labelling	Filling	Capering	Sensor Volume	Heater	Induction Seal	Laser	Robotic	Carton Sealer	Carton Erector	Palletizing
Downtime (minutes)	3730	2325	3165	240	0	30	6115	5765	1020	665	1650

The RCM is started by constructing the FTA (Fig.1), the FMEA (Table 3), and then determining the maintenance task (Table 4). The optimal inspection frequency for laser machine is then can be calculated using equation (8). From the FTA (Fig.1), it can be seen that the failures which might be occurred can be classified into two categories based on the effect of the failure: the effect of machine stops from operating and error marking. The effect of machine stops from operating was caused by lens error. The lens error occurred because of the lens is broken or cracked, and the broken lens was caused by human error. From FMEA in Table 3, marking trouble were caused by inappropriate marking result and machine does not perform marking. FMEA for lastube, the potential failure mode is CO gas cylinder's error because of CO gas is discharge. The error of lastube can produce an error in marking of the product. From Table 4 the cause of the failure of laser machine includes the failure of power supply component, lens, chiller, laser siren, lastube, encoder, conveyor, and mirror galvo. The consequences of such failures are hidden failure consequences, operational consequences, and safety consequences. Maintenance task that are suitable for component laser machine is time-directed life-renewal task, while the other components such as power supply, encoder, and laser siren must use failure finding task and lastube and chiller must use servicing task.

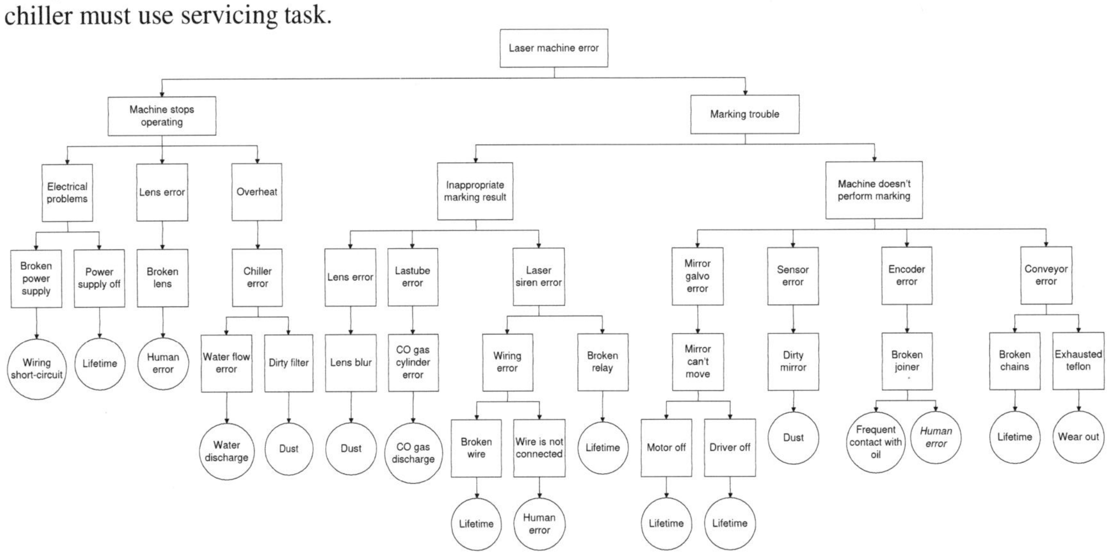

Figure 1. Fault Tree Analysis (FTA)

Table 3. Failure Mode and Effect Analysis (FMEA)

Component	Function	Potential Failure Mode	Failure Effect	Cause of Failure	O	S	D	RPN	RANK
Power Supply	Providing electricity to the power supply or other components in order to work properly	Power supply is broke	Laser machine stops operating	Wiring short-circuit	3	5	8	120	4
		Power supply off		Lifetime	2	5	7	70	11
Lens	As a way to fire a laser beam	Lens blur	Error marking	Dust	8	7	2	112	5
		Lens broke	Laser machine stops operating	Human error	1	5	2	10	15
Chiller	Keeping the condition inside the laser machine remains in proper temperature and not overheat	Error water flow	Laser machine stops operating	Water discharge	7	5	3	105	6
		Dirty filter		Dust	7	5	3	105	6
Laser Siren	Produce a siren sound as a sign that something goes wrong with the machine	Broken wires	Error marking	Lifetime	2	8	3	48	12
		Wire is not connected		Human error	2	8	2	32	14
		Broken relay		Lifetime	2	8	3	48	12
Lastube	A place to store CO which fired to produce writings	Error CO gas cylinder	Error marking	CO gas discharge	7	5	4	140	1
Encoder	Component in charge of determine when sensor fire a laser	Joiner with conveyor broke	Error marking	Frequent contact with oil	3	7	5	105	6
				Human error	3	7	5	105	6
Conveyor	Move the products from one station to another station	Broken chain	Error marking	Lifetime	2	5	1	10	15
		Exhausted Teflon		Wear out	7	7	2	98	10
Mirror Galvo	The components that reflect the laser in order to shoot right at the specified position	Motor is off	Error marking	Lifetime	3	7	6	126	2
		Driver is off		Lifetime	3	7	6	126	2

Table 4. Category consequences of failure and RCM decision of laser machine

Component	Cause of Failure	RPN	Ranking	Category Failure Consequences	Maintenance Task
Lastube	CO gas discharge	140	1	Operational consequences	Time-directed life-renewal task, servicing task
Mirror Galvo	Lifetime of motor	126	2	Operational consequences	Time-directed life-renewal task
Mirror Galvo	Lifetime of driver	126	2	Operational consequences	Time-directed life-renewal task
Power Supply	Wiring short-circuit	120	4	Hidden failure consequences, operational consequences	Time-directed life-renewal task, failure finding task
Lens	Dust	112	5	Operational consequences	Time-directed life-renewal task
Chiller	Water discharge on water flow	105	6	Operational consequences	Time-directed life-renewal task, servicing task
Chiller	Dirty filter	105	6	Operational consequences	Time-directed life-renewal task
Encoder	Frequent contact with oil	105	6	Hidden failure consequences, operational consequences	Time-directed life-renewal task, failure finding task
Encoder	Human error	105	6	Hidden failure consequences, operational consequences	Time-directed life-renewal task
Conveyor	Teflon is wear out	98	10	Operational consequences	Time-directed life-renewal task
Power Supply	Lifetime of power supply	70	11	Hidden failure consequences, operational consequences	Time-directed life-renewal task
Laser Siren	Lifetime of wire	48	12	Hidden failure consequences, safety consequences	Time-directed life-renewal task, failure finding task
Laser Siren	Lifetime of relay	48	12	Hidden failure consequences, safety consequences	Time-directed life-renewal task, failure finding task
Laser Siren	Human error	32	14	Hidden failure consequences, safety consequences	Time-directed life-renewal task, failure finding task
Lens	Human error	10	15	Operational consequences	Time-directed life-renewal task
Conveyor	Lifetime of chains	10	15	Operational consequences	Time-directed life-renewal task

Assuming the firm needs that reliability of the machine be 85% reliable according to world class standard [14], the company must be decreased the downtime from 1,273.76 minutes (before applying the RCM and inspection to the machine) to 17 minutes (after applying the RCM and inspection to the machine).
Based on the downtime data, it can be seen that the average value of failures per month is 1 time and with the assumption of an average inspection time is 2 hours or 0.00625 months and the average number of the repair time is 4 hours or 0.0125 months. Then using equation (8) we can find that the optimal number of inspection is once a month for through inspection on a laser machine and its components.

4. Conclusion
In this research, we calculate total downtime from all machine in line 1-4, then we choose the line 2 with longest downtime and determine that laser machine in line 2 is the critical machine. After that, we applied RCM method to determine the critical component of laser machine and its optimal frequency of inspection for laser machine in line 2 at filling lithos process PT. X. Based on the processing data then it was found that the total downtime of the laser machine is 6,115 minutes with an average downtime 1,273.76 minutes and reliability 24.54%. Then we determine the maintenance task that suitable for each component of laser machine as seen on Table 4. Also, we calculate the optimal frequency of inspection for laser machine is once a month. By assuming that the machine's condition must at 85% after doing maintenance task and inspection, we get the total downtime decreased to 17 minutes.

References
[1] Dhillon B S 2006 *Maintainability, Maintenance, and Reliability for Engineering* (Boca Raton: CRC Press)
[2] Seo J H and Bai D S 2004 An optimal maintenance policy for a system under periodic overhaul *Mathematical and Computer Modelling* **39** 373-380.
[3] Dehghanian P, Firuzabad M F, Aminifar F, Brillinton R 2013 A comprehensive scheme for reliability-centered maintenance in power distribution systems-part II: numerical analysis. *Transaction on Power Delivery* (IEEE) **28** number 2
[4] Riyanto N P, Jauhari W A, Aisyati A, and Rosyidi C N 2014 Developing maintenance policy for solna printing machine with reliability centered maintenance (RCM II) *Proceedings of the International MultiConference of Engineers and Computer Scientists 2014* **2**
[5] Wilmeth R G and Usrey M W 2000 Reliability-centered maintenance: a case study. *Journal of Engineering Management* **12** number 4.
[6] O'Connor and Patrick D T 2001 *Practical Reliability Engineering, Fourth Edition* (England: John Wiley & Sons Ltd)
[7] Ebeling E C 1997 *An Introduction to Reliability and Maintainability Engineering, International Edition* (Singapore: McGraw Hill)
[8] Swason L 2001 Lingking maintenance strategies to performance. *International Journal Production Economics* **70** 237-44
[9] Walpole R. E and Myers R H 1995 *Probability & Statistics for Engineers and Scientists 4th ed* (Bandung.: Bandung Institute of Technology)
[10] Azis M T, Suprawhardana M S, and Purwanto T P 2009 Penerapan metode reliability centered maintenance (RCM) berbasis web pada sistem pendingin primer di reaktor serba guna *Seminar Nasional V SDM Teknologi Nuklir* (Yogyakarta) pp 343-354
[11] Moubray J 1992 *Reliability Centered Maintenance 2nd ed* (New York: Industrial Press Inc)
[12] Sparks R D 2015 *Applying system engineering to improve the main gas turbine exhaust system maintenance strategy for the CG-47 ticonderoga class cruiser* (California: Dudley Knox Library)
[13] Jardine A K S 1973 *Maintenance, Replacement, and Reliability* (England: Pitman Press)
[14] Dal B 2000 Overall equipment effectiveness as a measure of operational improvement *Journal of Operations and Production Management* **20** 1491

IOP Conf. Series: Materials Science and Engineering **319** (2018) 012021 doi:10.1088/1757-899X/319/1/012021

A Goal Programming Optimization Model for The Allocation of Liquid Steel Production

S N Hapsari[1] and C N Rosyidi[2]

Department of Industrial Engineering, Sebelas Maret University, Indonesia

shanellanh@student.uns.ac.id[1] and cucuk@uns.ac.id[2]

Abstract. This research was conducted in one of the largest steel companies in Indonesia which has several production units and produces a wide range of steel products. One of the important products in the company is billet steel. The company has four Electric Arc Furnace (EAF) which produces liquid steel which must be procesed further to be billet steel. The billet steel plant needs to make their production process more efficient to increase the productvity. The management has four goals to be achieved and hence the optimal allocation of the liquid steel production is needed to achieve those goals. In this paper, a goal programming optimization model is developed to determine optimal allocation of liquid steel production in each EAF, to satisfy demand in 3 periods and the company goals, namely maximizing the volume of production, minimizing the cost of raw materials, minimizing maintenance costs, maximizing sales revenues, and maximizing production capacity. From the results of optimization, only maximizing production capacity goal can not achieve the target. However, the model developed in this papare can optimally allocate liquid steel so the allocation of production does not exceed the maximum capacity of the machine work hours and maximum production capacity.

1. Introduction

The steel industry which is often seen as one of the cornerstone for economic growth, has encountered big challenges in recent years [1]. Hence, the steel company company has to maintain and continously improve their service level to satisfy their customers. The company has several factories that produce steel products such as billet steel, slab steel, wire rod, and iron sponge. Along with the increasing competition of national and international steel industry, the company is required to improve its production efficiency so they can improve the productivity and make better profit. Billet steel production process consists of several stages. The first stage of the process is in the form of melting scrap and other raw materials such as iron sponge and limestone in an Electric Arc Furnace (EAF). In the second stage process, the chemical composition of the steel in Ladle Furnace (LF) is determined, and the last stage is the casting process which performed in a Continuous Casting Machine (CCM). This paper concerns with the first stage process. The company faced a problem about how to determine the optimal allocation of liquid steel production in each of four EAF so that the company can satisfy the demand efficiently. The company set several objectives namely maximize production volume, minimize the raw material cost, minimize the maintenance costs, maximize sales revenues, and maximize production capacity. Based on those objectives, the purpose of this paper is to develop an optimization model using goal programming approach to determine the amount of Ton Liquid Steel (TLS) to be produced by four EAF owned by the company. Goal programming is a special type of linear programming which is capable of handling decision situations involving single or multiple goals [2]. The basic approach of goal programming is to

establish a specific numeric goal for each of the objectives, formulate an objective function, and then find a solution that minimize the sum of deviations of the objective functions from their respective goals [3].
Recently, many researchers and practitioners have studied iron and steel production management issues. Solving production operational management problems in Steelmaking-continuous casting (SCC) production through mathematical models and computers is an important research topic and has recently been widely explored [4]. For example, a mathematical programming model has been developed by [5] for a steel production scheduling problem to optimally schedule the slab production through the reheating furnace and the rolling mill. In another study by [6], an optimization model based on EAF production was developed to determine the optimal allocation of semi-finished products to satisfy the final product demand. In research [1], multiple objectives and uncertainties are considered as a part of Integrated Programming (IP) problem. A multi-objective optimization model was developed in the paper with interval-valued objective functions to optimize the throughput of each process. Research [7] used an interactive evolutionary algorithm with decision-maker's preferences to solve the interval multi-objective optimization problems.

2. Methodology

In the first stage of this study several data was collected including liquid steel demand, raw material cost per heat, the sales revenue per TLS, maintenance cost, and production capacity for each EAF. In the second stage. A goal programming model is developed and solved using LINGO 15.0 software.

2.1 Data Collection

Solving the optimization model requires data input which are used as parameters of the model. Table 1-4 shows the data needed to solve the model. Table 1 shows the actual liquid steel demand in 2014. Raw material prices and their corresponding standard usage and cost per heat are listed in Table 2. Table 3 shows standard maintenance cost for each EAF

Table 1. Actual Liquid Steel Demand in 2014

Period	Demand (TLS)	Period	Demand (TLS)
January	16220.7	July	15092.887
February	15793.6	August	14479.642
March	15377.7	September	7958.37
April	10780.151	October	16757.425
May	16454.189	November	15741.045
June	16900.171	December	9829.759

Table 2. Raw Material Costs for Liquid Steel Production

Raw Materials	Price per Ton	Standard Usage of Raw Material per Heat (%)	Material Cost per Heat
Scrap	IDR 2,300,313.00	83	IDR 133,648,194.00
MJIS Sponge	IDR 4,024,364.00	13	IDR 36,621,712,00
DRI Sponge	IDR 4,164,000,00	4	IDR 11,659,200.00
	Total material cost/heat		IDR 181,929,106.00
	Total material cost/TLS		IDR 3,032,152.00

Table 3. Maintenance Costs

Electric Arc Furnace (EAF)	Maintenance costs (in thousand of IDR)
EAF 1	1210
EAF 2	2300
EAF 3	3090
EAF 4	4309

2.2 The Development of Goal Programming Model

In goal programming model, it is necessary to consider the structural or technological constraints and objectives. For each constraint, possible deviations are stated, and for each objective, a target level is set [8]. In this paper, we use a goal programming without priority weight. The notations of index and variables used in this paper are as follows:

X_{ik} : the volume of liquid steel generated in furnace *i* at period *k*
i : the type of liquid steel produced in the furnace, *i* = 1,2,3,4
k : period of production, *k* = 1,2,3
D_k : the volume of demand for liquid steel in period *k*
D_i : the volume of demand for liquid steel in the furnace *i*
B : raw material cost per TLS
X_{ik} : the volume of liquid steel generated in furnace *i* at period *k*
TB_k : material cost target for period *k*
P : the price of liquid steel per ton
TP_k : total revenue target for the period *k*
W_i : average maintenance costs per TLS in furnace *i*
TO_k : total maintenance cost in period *k*
KP_{ik}: total maximum capacity of liquid steel production for furnace *i* in period *k*

2.2.1 The Objectives Function

There are five goals considered in this paper:

1. Maximize the volume of production. To achieve this goal, we have to minimize the positive and negative deviations as shown in Equation (1).

$$Min\ Z_1 = \sum_{n=1}^{7} d_n^- + d_n^+ \quad (1)$$

2. Minimize the cost of raw materials at each period. To achieve this goal, we have to minimize the positive deviations as shown in Equation (2).

$$Min\ Z_2 = \sum_{q=8}^{10} d_q^+ \quad (2)$$

3. Maximize the revenue each period. To achieve this goal, the negative deviation must be minimized as in Equation (3).

$$Min\ Z_3 = \sum_{p=11}^{13} d_p^- \quad (3)$$

4. Minimizes the maintenance cost. To achieve this goal, we have to minimize the positive deviations as expressed in Equation (4)

$$Min\ Z_4 = \sum_{r=14}^{16} d_r^+ \quad (4)$$

Maximize the use of production capacity per furnace. To achieve this goal, we have to minimize the positive and negative deviations.

$$Min\ Z_5 = \sum_{r=17}^{28} d_s^- + d_s^+ \qquad (5)$$

2.2.2 The Model Constraints

1. Constraints of product demand. The product demand constraint for each period can be expressed as in equation (6), while the product demand for each furnace can be expressed as in equation (7).
$$\sum_{i=1}^{4} X_{ik} = D_k \qquad (6)$$
$$\sum_{k=1}^{3} X_{ik} = D_i \qquad (7)$$

2. Constraint of raw material cost. Equation (8) states the constraint for each period.
$$\sum_{i=1}^{4} BX_{ik} \leq TB_k \qquad (8)$$

3. The revenue constraint from sales for each period can be expressed in equation (9).
$$\sum_{i=1}^{4} X_{ik}P \geq TP_k \qquad (9)$$

4. The maintenance cost constraint for each furnace can be expressed in Equation (10).
$$\sum_{i=1}^{4} W_i X_{ik} \leq TO_k \qquad (10)$$

5. The maximum capacity of liquid steel production for each furnace can be expressed in Equation (11).
$$\sum_{i=1}^{4} X_{ik} = KP_{ik} \qquad (11)$$

3. Results and Discussions

3.1 The Optimization Results

The results of the optimization can be seen in Table 4. The model was solved using LINGO 15.0 software.

Table 4. Summary of Optimal Results

Period	Type of Furnace				Total Production (TLS)
	EAF 1	EAF 2	EAF 3	EAF 4	
1	328	11520	4372,7	0	16220.7
2	0	328	7475.3	7990.3	15793.6
3	11520	0	0	3857.7	15377.7

From table 4, we can see that the total of liquid steel production based on the optimal solution in each period is 16220.7 TLS, 15793.6 TLS, and 15377.7 TLS. The allocation of liquid steel production in EAF 1 for period 1 is 328 TLS, EAF 2 for the period 1 is 11520 TLS, EAF 3 for the period 1 is 4372.7 TLS, EAF 4 did not use to produce the demand in period 1. The allocation of liquid steel production in EAF 1 did not use to produce the demand in period 2, EAF 2 for period 2 is 328 TLS, EAF 3 for period 2 is 7475.3 TLS, EAF 4 for a period 2 is 7990.3 TLS. The allocation of liquid steel production in EAF 1 for period 3 is 11520 TLS, EAF 2 did not use to produce the demand in period 3, EAF 3 did not use to produce the demand in period 3, EAF 4 for period 3 is 3857.7 TLS.

The target achievement is shown in Table 5. From the table we can see whether the optimal outcome of the model satisfies the goals of the company or not. The summary of the multi goals achievement is shown in Table 5.

Table 5. Summary of Multi Goal Achievement

Constraint	Goal	Optimal Solutions	Information
Maximize the volume of production	P1: Min $\sum d_n^- + d_n^+$	0.00	Achieved
Minimize the cost of raw materials each period	P2: Min $\sum d_q^+$	0.00	Achieved, with TB 1 = IDR 49,183,600,000.00 TB 2 = IDR 47,888,560,000.00 TB 3 = IDR 46,627,490,000.00
Maximize revenue every period	P3: Min $\sum d_p^-$	0.00	Achieved, with TP 1 = IDR 111,217,200,000.00 TP 2 = IDR 108,288,800,000.00 TP 3 = IDR 105,437,200,000.00
Minimizes maintenance Costs	P4: Min $\sum d_r^- + d_r^+$	0.00	Achieved, with TO 1 = IDR 40,404,520.00 TO 2 = IDR 58,283,280.00 TO 3 = IDR 30,562,030.00
Maximize the use of production capacity per furnace	P5: Min $\sum d_s^- + d_s^+$	> 0,00	Not achieved

From the table 5, we can see that the first goal of the model can be achieved. The factory is capable of producing the optimal solution which is indicated by the zero value. It means there is no negative deviation or achievement is less than the production target of liquid steel in each period. Since the results of the model suggested to produce 16220.7 TLS in period 1, 15793.6 TLS in period 2, and 15377.7 TLS in period 2.

The second goal of the model is to minimize the positive deviation. The factory is capable of producing the optimal solution which is indicated by the zero value. In other words, the second goal can be achieved. It shows that the solution does not have deviations which can increase the cost of raw materials in each period. The results of raw material cost from the optimal solution is IDR 49,183,600,000.00 in period 1, IDR 47,888,560,000.00 in period 2, and IDR 46,627,490,000.00 in period 3.

The third goal of the model is to minimize the negative deviation. The factory is capable of producing the optimal solution which is indicated by the zero value, this indicates that the solution does not have any deviations that can reduce the revenue in each period. The sales revenue from the optimal solution amounted to IDR 111,217,200,000.00 in period 1, IDR 108,288,800,000.00 in period 2, and IDR 105,437,200,000.00 in period 3.

The fourth goal of the model is to minimize the positive and negative deviation. The factory is capable of producing the optimal solution which is indicated by the zero value. Based on the optimal results from Lingo's solution report, the value of a positive deviation on this goal is 0, which indicates that the results of the model already optimal yield maintenance cost is IDR 40,404,520.00 in period 1, IDR 58,283,280.00 in period 2 and IDR 30,562,030.00 in period 3.

The fifth goal of the model is to minimize the positive and negative deviation. The optimal result shows that the optimal solution is more than the zero value. It indicates that optimal allocation of liquid steel production exceeds the maximum capacity of liquid steel production in each EAF. The optimal results show some conditions like EAF 1 in period 1 has remaining production capacity 11,192 TLS, EAF 1 in period 2 has remaining production capacity 11,520 TLS, EAF 2 in period 2 has remaining production capacity 11,192 TLS, EAF 2 in period 3 has remaining production capacity 11,520 TLS, EAF 3 in period 1 has production capacity remaining 71,47.3 TLS, EAF 3 in period 2 has remaining production capacity 4,044.7 TLS, EAF 3 in period 3 has remaining production capacity 11,520 TLS, EAF 4 in period 1 has remaining production capacity 11,520 TLS, EAF 4 in period 2 has remaining production capacity still left 3,529.7 TLS, EAF 4 in period 3 has remaining production capacity 7,662.3 TLS. According to the results of optimal production planning above we can see that there are some electric arc furnace with remaining production capacity. It indicates that the electric arc furnace can still be used to produce liquid steel exceeded the demand in this period. The remaining production capacity that is not 0 can occur due to product demand that are less than the production capacity of each electric arc furnace. Not achieving the 5th goal is influenced by some things such as utility engine that is not maximized, the demand is slightly in the period 1 - 3, so it is not proportional to the number of machine capacity. However, based on the achievement of goal to 5, it turns out the production in this plant does not exceed the capacity of the machine. Can be interpreted that the company was able to meet the demand for billet steel.

4. Conclusion

In this paper, a goal programming model was developed to determine the optimal allocation of steel production in EAF of a large steel company in Indonesia. The model was capable to find the optimal solution, so the company can allocate steel production optimally to fulfill the demand. According to the optimal solution, each EAF can achieve four goals for each period except the fifth goal which shows that the electric arc furnace can still be used to produce liquid steel exceeded the demand in this period. The rest of the production capacity with value other than zero can occur due to product demand that are less than the maximum production capacity of each electric arc furnace. This research can be further developed by considering not only the first stage but also second and third stages as an integrated model to give better solutions.

References

[1] Lin, J., Liu, M., Hao, J., and Jiang, S. (2016). A Multi-objective Optimization Approach for Integrated Production Planning under Interval Uncertainties in the Steel Industry. *Computers and Operation Research.*

[2] Rosyidi, C.N., Perdana, A.Y., Priyandari, Y. (2014). An Optimization Model for Production Allocation in A Large Steel Manufacturing Company. *Proceedings of the 2014 International Conference on Industrial Engineering and Operations Management.* p. 2102 – 2107.

[3] Hillier, F. and Lieberman, G. (1994). Introduction to Operation Research. McGraw-Hill.

[4] Tang, L., Liu, J., Rong, A., and Yang, Z. (2000). A Mathematical Programming Model for Scheduling Steelmaking-Continuous Casting Production, *European Journal of Operational Research* ,120 pp. 423-435.

[5] C.M. Petersen, K.L. Sorensen, and R.V.V. Vidal. (1992), Inter-process synchronization in steel production, *International Journal of Production Research* 30 (6) 1415-1425.

[6] Rosyidi, C.N., Perdana, A.Y., Priyandari, Y. (2014). An Optimization Model for Production Allocation in A Large Steel Manufacturing Company. *Proceedings of the 2014 International Conference on Industrial Engineering and Operations Management.* p. 2102 – 2107.

[7] Gong, D.W., Ji X.F., Sun J, Sun X.Y. (2014), Interactive evolutionary algorithms with decision-maker's preferences for solving interval multi-objective optimization problems, *Neurocomputing* 137:241-51.

[8] Adeyeye, A.D. and Owaba, C.O.E. (2008). Goal Programming Model for Production Planning in a Toothpaste Factory. *South African Journal of Industrial Engineering.* Vol 19(2): 197-209.

IOP Conf. Series: Materials Science and Engineering **319** (2018) 012022 doi:10.1088/1757-899X/319/1/012022

Preparation of tin oxide (SnO_2) thin films using thermal oxidation

N Abdullah[1], N M Ismail[2] and D M Nuruzzaman[2]

[1,2]Faculty of Manufacturing Engineering, University Malaysia Pahang 26600 Pekan Pahang Darul Makmur, Malaysia.

[1]Email: nuraini.abdullah89@yahoo.com

Abstract. The present research study deals with the preparation of tin oxide (SnO_2) nanostructures using thermal oxidation method. At first, Sn thin film was deposited on silicon (Si) substrate by thermal evaporation and then, thermal oxidation of the deposited Sn thin film was carried out at the growth temperature of 100°C with growth time of 1 hour in tube furnace. The structural property of SnO_2 nanostructures was investigated by using FESEM and EDX. The FESEM results showed that Sn was successfully grown on Si substrate and the SnO_2 nanoparticles with diameters of 97.5nm to 142nm were recorded. It was observed that the particles were agglomerated to form the SnO_2 particles. The radiation of sunlight illumination was conducted for four consecutive sunny days and the results showed that the highest reading 189.9 W/m^2 was recorded at day two for the daytime temperature 38°C. It was also noticed that the highest solar radiation percentage at day two was measured 18.9%.

1. Introduction

Tin Oxide (SnO_2) thin films are one of the nanostructures semiconducting oxide materials that are being received broad attention due to unique electrical, physical, chemical, structural and optical properties. It is well known that n-type semiconductors with a wide band gap of 3.6-3.8 eV can be used in various fields, such as a window layer in solar cells, heat reflectors in solar cells, various gas sensors, transparent conducting electrodes, photochemical and photoconductive devices in liquid crystal display, gas discharge display, lithium-ion batteries etc. [1-6]. The properties of tin oxide for the performance of the sodium-ion batteries were investigated [7]. Tin oxide thin films are the most attractive films that are used in automotive sector as an anticorrosive surface treatment of a carbonaceous bipolar plate in proton exchange membrane fuel cells [8]. Using a sol-gel process, nanostructured tin oxide powders were prepared successfully [9]. The obtained results showed that crystallization of tin oxide started at about 250°C and the growth continued at higher temperatures as the organics were removed. On the other hand, it was found that nanocrystalline tin oxide powders were formed only after the hydroxyl groups were completely removed at nearly 600°C.

Mixed semiconductor oxide nanostructures of tin dioxide/tin monoxide (SnO_2/SnO) have been produced using hydrothermal method [10]. The morphological, structural and photoluminescence properties of the fabricated nanostructures were investigated. For the growth of tin oxide nanostructure, silicon wafer was used as a substrate and using thermal evaporation method, tin oxide nanowires have been synthesized for gas sensing application [11]. The synthesized tin oxide nanostructures were characterized using XRD, EDX, SEM and TEM. In this investigation, vapor-liquid-solid (VLS) growth of the tin oxide nanowires was confirmed.

Tin oxide thin films have been synthesized using spray pyrolysis and electron beam evaporation methods [12]. The structural, electrical and optical properties were investigated under different operating conditions of solution flow rate, substrate temperature and deposition rate. The obtained results revealed that conductivity of the deposited films was dependent on structure of films and substrate temperature. The doping effect of neodymium (Nd) on the characteristics properties of tin oxide thin films was investigated [13]. The optical, electrical, crystallographic and morphological characterizations revealed that neodymium doping significantly influenced the properties of tin oxide thin films. Due to the high transparency for visible light and good electrical conductivity, tin oxide thin films are used for numerous applications, such as low-emissivity coatings for glass windows and solar cells [14]. Each application needs different properties of the tin oxide thin films. These properties can be varied by adjusting different parameters involved in the chemical vapour deposition (CVD) process by which tin oxide layer can be deposited on the substrate. The growth of tin oxide nanomaterial was carried out and the optimum growth parameters by varying the growth temperature and growth time were investigated [15]. The obtained results showed that samples were grown on silicon substrate and exhibited like nanobelts and nanorods. Moreover, tin oxide nanomaterial displayed fluorescence and photoluminescence signals.

In this research study, SnO_2 nanostructures were prepared on Si substrate following a route of thermal evaporation and then, thermal oxidation. The properties of these SnO_2 nanostructures have been studied using Field Emission Scanning Electron Microscope (FESEM) with Energy Dispersive X-Ray (EDX) analysis. The efficiency or performance of these fabricated SnO_2 samples was also measured in this study.

2. Experimental

Silicon (Si) wafer p-type (100) with dimension 1.5cm x 2.5cm and thickness of 500±50 μm was used as a substrate for the growth of SnO_2 nanostructures. At first, the Si substrate was cleaned using freshly prepared Piranha solution to remove dust and organic mixture or contaminants. A mixture solution of sulphuric acid (H_2SO_4) and hydrogen peroxide (H_2O_2) (3 : 1) was prepared and the cleaning process was conducted using ultrasonic vibrations for 30 minutes. After that, the Si substrate was rinsed with distilled water and dried with air gun. The tin powder (Sn) was weighed at 0.05g which was then deposited on the Si substrates in thermal evaporator. The base pressure in thermal evaporator was set at 1.60×10^{-4} Pa with target-substrate distance of 10.5mm at room temperature, voltage 0.9V and current 20-25A. The deposition time was constant for 30 minutes for each sample. The obtained Sn thin film sample was characterized by using FESEM and EDX analysis. Next, the thermal oxidation was carried out by using tube furnace with the oxidation temperature at 100°C. The heating rate used was 5°C/min and the oxidation time is 1h. Finally, the substrate was cool down to the temperature and then unloaded from the furnace. The morphological studies were carried out by the FESEM JEOL, JSM-7800F with EDX analysis. The obtained SnO_2 sample was also tested for radiation of sunlight illumination by solar power meter.

3. Results and Discussion

3.1. FESEM and EDX

Figures 1(a)-(d) show the characterization results using FESEM and EDX. Figure 1(a) shows the average particle size of the Sn thin film using 5000x magnification. The average particle size varies from 5.88μm to 7.43μm with some of the particles were almost round shape, and other particles were irregular shapes. The obtained results also showed that the sizes of the particles are not uniform which means that the evaporation process was not uniformly distributed over the surface. It suggests that tin was not uniformly evaporated may be because of the Si substrate had a rough surface during the cleaning process. There might be a possibility that the evaporated material attacked the substrate mostly from a single direction. Figure 1(b) shows the cross section of the Sn films. The obtained results showed that thickness of the deposited Sn film varies from around 0.26 μm to 0.36 μm.

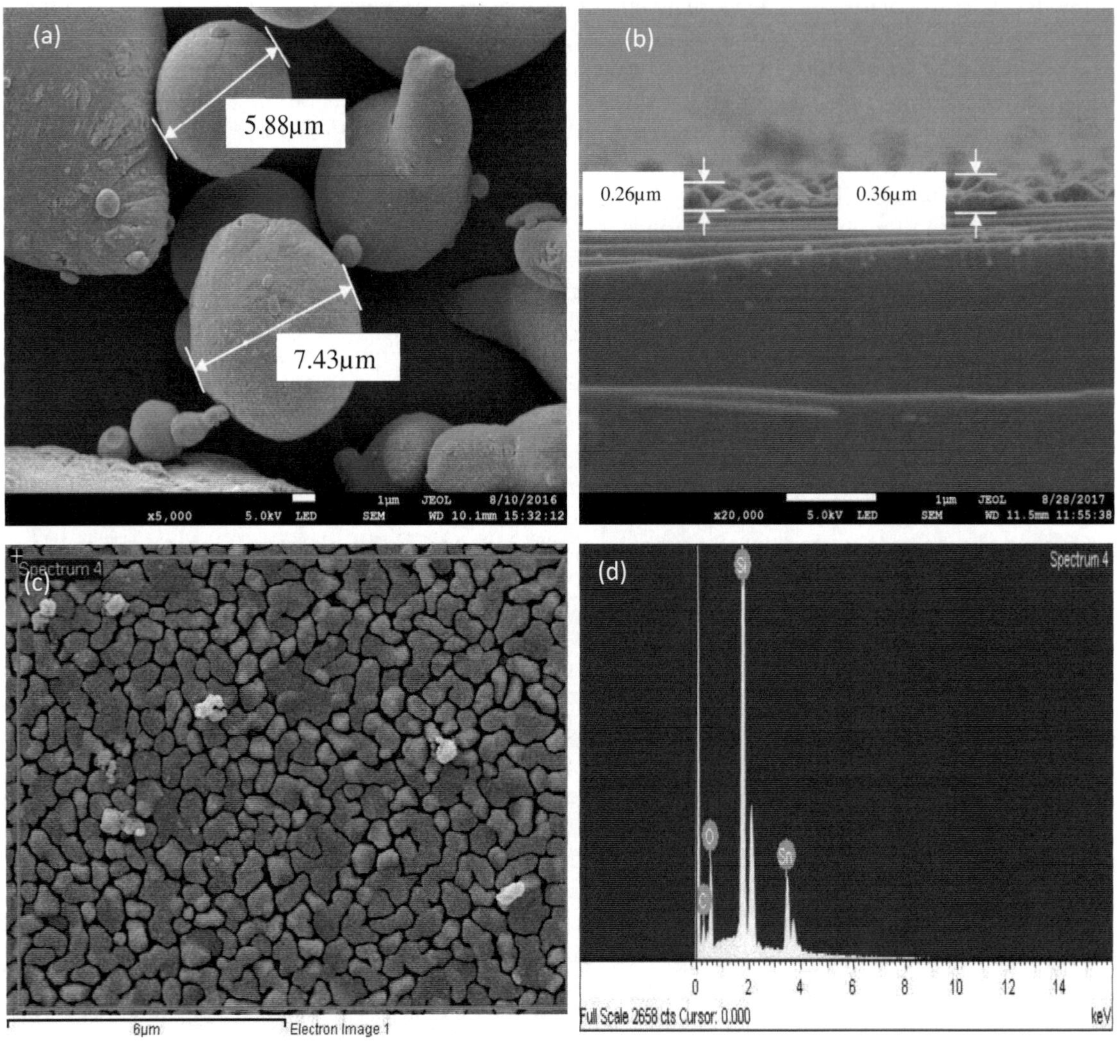

Figure 1. FESEM image of Sn thin film after thermal evaporation (a) the average particle size of Sn and (b) cross section of Sn layer on the Si substrate (c) EDX point surface of Sn layer on Si and (d) EDX spectrum of Sn thin film.

Figures 1(c)-(d) show the EDX results of the Sn film and the elements consist of Si, C, O and Sn meaning that Sn was successfully grown on the Si substrate. Figures 2(a)-(d) show the FESEM and EDX image of SnO_2 at growth temperature 100°C. Figure 2(a) shows the FESEM image of the surface of SnO_2 thin film using x10000 magnification. Figure 2(b) clearly shows that agglomeration of the particles occurred with irregular shapes of tin oxide and the average diameter varies from 97.5nm to 142nm. The tin film was not fully oxidized at this temperature due to the melting point of tin is 232°C. It is believed that grain rotation induced grain coalescence mechanism appears in this growth of the SnO_2 crystals due to the appearance of spherically elongated crystals [16]. Figures 2(c)-(d) show the EDX result of SnO_2 at growth temperature 100°C that confirms the elements consist of Si, Sn, C, and O. The particles sizes were reduced to about 80% after thermal oxidation of Sn thin film.

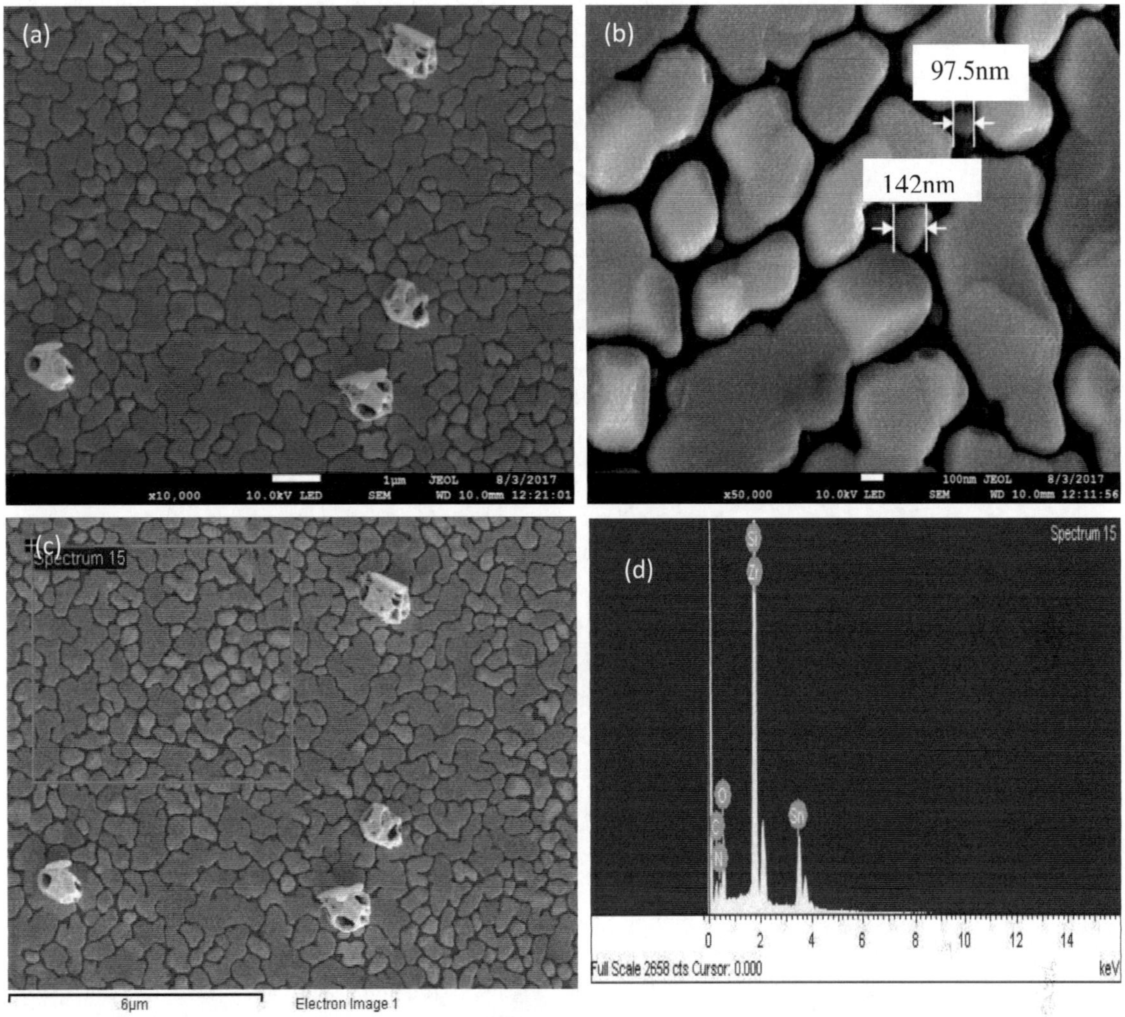

Figure 2. FESEM image of SnO_2 at 100°C after thermal oxidation (a) surface of SnO_2, (b) grains size of SnO_2 particles,(c) EDX point surface of SnO_2 (d) EDX of SnO_2.

3.2 Solar Power Meter (Sunlight Radiation)

Figure 3 shows the variation of sunlight radiation on SnO_2 sample (growth temperature 100°C) for four consecutive sunny days referred to as "peak sun hours" which is solar insolation at a particular location during the sun shining at its maximum level for a certain time. The distance from the surface of the sample parallel to the solar power meter was 65cm. The duration of sunlight radiation measurement for each sunny day was 15 minutes. Since the peak solar radiation is 1 kW/m^2, the percentage is the received solar radiation by the tin oxide sample over the peak solar radiation. The intensity of sunlight radiation at day one was 125.0 W/m^2 with temperature 36°C which results in 12.5% of solar radiation. The intensity of sunlight radiation at day two was recorded 189.9 W/m^2 which resulted 18.9% of solar radiation. The sunlight radiation was also affected by the weather of the day. Temperature at day two was recorded as highest which is 38°C. At day three, the temperature was recorded 34°C and the intensity of sunlight radiation reading showed lower value 69.2 W/m^2 which results in 6.9% of solar radiation. The lowest intensity reading was recorded at day four with radiation

55.5 W/m^2 and the solar radiation percentage of 5.5%. This is due to the fact that at day four, the lowest temperature was recorded 32°C.

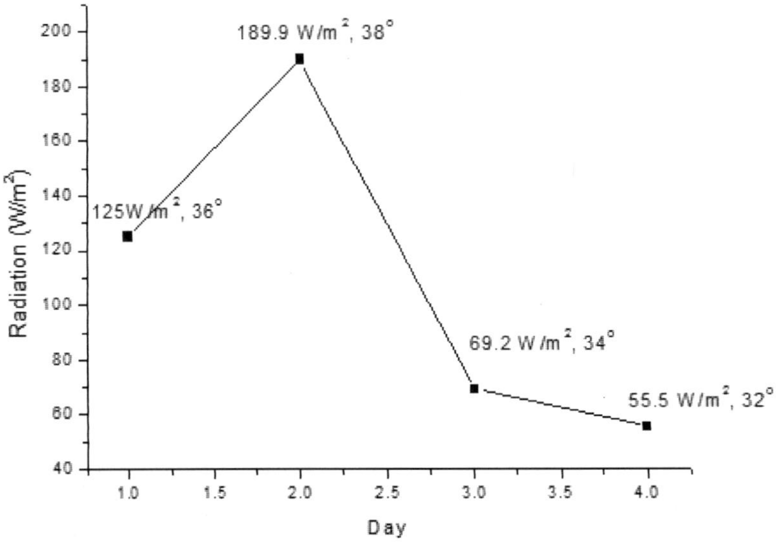

Figure 3. Variation of sunlight radiation on SnO_2 for four sunny days (growth temperature 100°C).

4. Conclusions

SnO_2 nanostructures were successfully prepared using thermal oxidation method at growth temperature 100°C and growth time of 1 hour in tube furnace. The FESEM analyses showed that the average particle size diameter of SnO_2 varies from 97.5nm to 142nm compared to Sn particle size which varies in the range of 5.88μm to 7.43μm. The agglomeration was occurred during the thermal oxidation of the SnO_2 at 100°C oxidation temperature. The sunlight radiation readings on SnO_2 sample (growth temperature 100°C) were measured for four consecutive sunny days and the radiation 189.9 W/m^2 was recorded at day two for the daytime temperature 38°C. At day two, the highest solar radiation percentage was measured 18.9%.

Acknowledgement

The authors are grateful to the RAGS grant #RDU 151410 for financial support to carry out this work. The authors also wish to extend their sincere thanks to all the technical staffs of Faculty of Manufacturing Engineering, University Malaysia Pahang involved related to this research work.

References

[1] Asama N N, Azhar S N and Abdulla M S 2013 Preparation and characterization of SnO_2 nanoparticles, *Int. J. Inno. Res. Sci. Eng. Technol.* **2** 7068-72

[2] Tripti S 2012 Synthesis of tin oxide thick film and its investigation as a LPG sensor at room temperature, *J. Sens. Technol.* **2** 102-8

[3] Arijit C, Divya H, K.Sreenivas and Vinay Gupta 2009 Mechanism of trace level H_2S gas sensing using rf sputtered SnO_2 thin films with CuO catalytic overlayer, *Int. J. Smart Sens. Intell. Syst.* **2** 540-48

[4] Kay A and Gratzel M 2002 Dye-sensitized core–shell nanocrystals: improved efficiency of mesoporous tin oxide electrodes coated with a thin layer of an insulating oxide, *Chem. Mater.* **14** 2930-53

[5] Choudhury S, Betty C A, Girija K G and Kulshreshtha S K 2006 Room temperature gas sensitivity of ultrathin SnO_2 films prepared from Langmuir-Blodgett film precursors, *Appl. Phys. Lett.* **89** 071914

[6] Kim C, Noh M, Choi M, Cho J and Park B 2005 Critical size of a nano SnO_2 electrode for Li-secondary battery, *Chem. Mater.* **17** 3297-3301

[7] Lu Y C, Ma C, Alvarado J, Kidera T, Dimov N, Meng Y S and Okada S 2015 Electrochemical properties of tin oxide anodes for sodium-ion batteries, *J. Power Sources* **284** 287-95

[8] Kinumoto T, Nagano K, Yamamoto Y, Tsumura T and Toyoda M 2014 Anticorrosion properties of tin oxide coatings for carbonaceous bipolar plates of proton exchange membrane fuel cells, *J. power Sources* **249** 503-8

[9] Zhang G and Liu M 1999 Preparation of nanostructured tin oxide using a sol-gel process based on tin tetrachloride and ethylene glycol, *J. Mater. Sci.* **34** 3213-9

[10] Gajendiran J and Rajendran V 2015 A study of the nano-structured aggregated tin oxides (SnO_2/SnO) and their structural and photoluminescence properties by a hydrothermal method, *Mater. Lett.* **139** 116-18

[11] Johari A, Bhatnagar M C and Rana V 2012 Growth, characterization and I-V characteristics of tin oxide (SnO_2) nanowires, *Adv. Mater. Lett.* **3** 515-8

[12] Shamala K S, Murthy L C S and Rao K N 2004 Studies on tin oxide films prepared by electron beam evaporation and spray pyrolysis methods, *Bull. Mater. Sci.* **27** 295-301

[13] Turgut G, Sonmez E and Duman S 2015 Evaluation of an Nd doping Effect on characteristic properties of tin oxide, *Mater. Sci. Semicond. Process.* **30** 233-41

[14] Van Mol A M B, Chae Y, McDaniel A H and Allendorf M D 2006 Chemical vapor deposition of tin oxide: fundamentals and applications, *Thin Solid Films* **502** 72-8

[15] Santos G N C, Salvador A A and Quiroga R V 2011 Temperature and deposition time dependence of the geometrical properties of tin oxide nanostructures, *Int. J. Sci. Eng. Res.* **2** 1-6

[16] Ghanam S and Rajendran V 2010 Luminescence properties of EG-assisted SnO_2 nanoparticles by sol-gel process, *Dig. J. Nanomater. Biostruct.* **5** 699-704

Influence of groove size and reinforcements addition on mechanical properties and microstructure of friction stir welded joints

Ravinder Reddy Baridula [1*], Abdullah Bin Ibrahim[1], Che Ku Mohammad Faizal Bin Che Ku Yahya [1], Ratnakar Kulkarni[1] and Ramgopal Varma Ramaraju[2]

[1]Faculty of Engineering Technology, Department of Manufacturing Engineering Technology, University Malaysia Pahang, Gambang, Malaysia
[2]Faculty of Technology, Department of Mechanical Engineering, Stanford International College of Business and Technology, Toronto, Canada

baridula@gmail.com

Abstract. The butt joints fabricated by friction stir welding were found to have more strength than the joints obtained by conventional joining process. The important outcome of this process is the successful fabrication of surface composites with improved properties. Thus in order to further enhance the strength of the dissimilar alloy joints the reinforcements can be deposited in to the aluminium matrix during the process of friction stir welding. In the present study the multi-walled carbon nanotubes were embedded in to the groove by varying the width during joining of dissimilar alloys AA2024 and AA7075. Four widths were selected with constant depth and optimum process parameters were selected to fabricate the sound welded joints. The results show that the mechanical properties of the fabricated butt joints were influenced by the size of the groove, due to variation in the deposition of reinforcement in the stir zone. The microstructural study and identification of the elements of the welded joints show that the reinforcements deposition is influenced by the size of the groove. It has also been observed that the groove with minimum width is more effective than higher width. The mechanical properties are found to be improved due to the pinning of grain boundaries.

1. Introduction

Welding is a significant process of manufacturing to produce the parts of complex configuration which are difficult to be obtained. It is considered as a secondary process which has to be followed by a primary process and cannot be substituted by any manufacturing process. The aluminium alloys were effectively joined by fusion welding process. However, fusion welding process results in reduced mechanical properties in the weld area due to the formation of intermetallic compounds and coarse grains. Also, the excellence of the weld joint deteriorates due to distortion, hot cracking, porosity and formation of oxide layer [1, 2]. Friction stir welding is an alternative method of joining process which avoids many metallurgical reactions [3]. This solid state fusion process enhances the mechanical properties of the weld joints due to dynamic recrystallization (DRX) in the stir zone and the temperatures yielded are less than the fusion welding process [4-6].In this process the welded joint was obtained by stirring a rotating tool with pin over the surface of the joint which generates the sufficient heat to plasticize the material [7]. This concept was effectively utilized to fabricate the surface composites by friction stir processing to improve the mechanical properties by the addition of reinforcements [8, 9].

Morisada et al. [10] researched the manufacture of surface composites of AZ31 alloy by depositing multi-walled carbon nanotubes of diameter 20-50nm in to a groove size of width 1 mm and depth 2 mm during friction stir processing. The grain refinement was promoted by the nanoparticles which enhanced the microhardness of surface composite to 78 Hv as compared the base metal. Lim et al. [11] investigated the fabrication of aluminium alloy surface composite by depositing the multi-walled carbon nanotubes during friction stir processing. The reinforcements were incorporated in to a groove of size 0.3 mm depth and 2.3 mm width, which enhanced the microhardness of the surface composite. The carbon nanotubes stability and distribution in aluminium surface composites during multi-pass friction stir processing was reported by Izadi and Gerlich [12]. The carbon nanotubes of diameter 30-50nm were deposited in to a groove of 1.8 mm depth and 2.5 mm width. It was found that after three pass the distribution of carbon nanotubes are uniform which restricted the size of the grains and enhanced the microhardness. Liu et al. [13] studied the fabrication of aluminium matrix composites by varying the amount of multi-walled carbon nanotubes during multi-pass friction stir processing. The reinforcements were inserted in to drilled holes of diameter 2,4,6,8 and 10mm and depth of 3.5mm. The results show that the nanoparticles were distributed uniformly which enhanced the microstructure and mechanical properties with the increase in the amount of reinforcements. The maximum tensile strength of 190.2 MPa was obtained at 6vol% of multi-walled carbon nanotubes. Thangarasu et al.[14] investigated the aluminium surface composite by incorporating TiC particles of 2μm in to groove of depth 5mm and varying width of 0.4,0.8,1.2 and 1.6mm during friction stir processing. The reinforcing particles reduced the grain size which enhanced the wear resistance of the composite. Sathiskumar et al. [15] produced a surface composite by depositing B_4C particles of average size 4μm in to the groove of size 0.7mm width and 2.5mm depth. The reinforcing particles were distributed homogeneously at higher rotational while the distribution was poor at low rotational speed, and the composite microhardness found to be decreased with increase in tool rotational speed.

Karthikeyan et al. [16] studied the friction stir welding of Al6351 alloy, by incorporating the SiC particles in to the stir zone. The mechanical properties of the welded joints were found to be improved by 33%, than the joints without reinforcements. This was due to pinning effect of the SiC particles which restricted the growth of the grain boundaries. Abnar et al. [17] contemplated the joining of AA3003-H18 aluminium alloy by inserting the reinforcements in the weld zone. He detailed that Cu and premixed Al-Cu powder were embedded between the two metals without making groove. The mechanical properties were observed to be enhanced more for premixed Al-Cu powder, when contrasted to Cu powder. Kumar et al. [18] investigated the influence of post weld warm treatment and B_4C nano powder addition on hardness, pitting resistance and microstructure of the stir zone during friction stir welding of AA7075 alloy. Abbasi et al. [19] studied the effect of SiC particles on the stir zone during friction stir welding of AZ31 magnesium alloy. The addition of SiC particles influenced the microstructure of the stir zone, which enhanced the corrosion resistance and mechanical properties as compared to the welded joint without reinforcements. Ravinder et al. [20] examined the addition of copper nanoparticles during joining of dissimilar aluminium alloys and found that the mechanical properties of the welded joint enhanced in all the samples.

It is clear from the above literature survey that the researchers have investigated diverse groove sizes for embedding the strengthening particles to fabricate the composites by friction stir processing. Consequently, the present research work concentrates on the influence of groove size and reinforcements addition on mechanical properties and microstructure of dissimilar friction stir welded joints.

2. Experimental Details

The base metals utilized as a part of the present examination are AA2024-T3 and AA7075-T6 having measurements of 100 mm × 50 mm × 4.85 mm. The chemical composition and mechanical properties of the aluminium alloys were shown in Table1 and Table2. The reinforcements used in the present study are multi-walled carbon nanotubes having nominal diameter of 10-20nm and purity more than 90%, which is shown in the Fig. 1. The welded joints of dissimilar alloys with reinforcements were fabricated by using 5T NC FSW machine, and the specimens were fixed in the fixture to avoid their movement. The reinforcements were inserted in to the grooves which are precisely cut in the base metals by a wire cut machine as per the table 3. In order to avoid the sputtering of nanoparticles during the process of joining, a welding tool without pin is moved over the surface of the joint to compact them in to groove. Then the required welded joint with reinforcements was obtained by moving the welding tool with pin over the surface of the metals to be joined. The process parameters selected in the current study are 1400 rpm, 80 mm/min and 1kN. In the present study 4 welded joints with reinforcements were fabricated to understand the mechanical behaviour of the joints. The samples for the tensile test were acquired according to the ASTM-E8 standard by utilizing EDM wire cut machine and the tensile test was done on 12 specimens by INSTRON universal testing machine of 30 kN. The microhardness of the specimens was done by Wilson Vickers hardness tester. The specimens were set up for microstructure examination and etched with Keller's reagent. The microstructures of the stir zone were acquired by optical microscope and identification of elements in weld zone was obtained by FESEM.

Table 1. The chemical composition of the alloys in this work (%wt)

Material	Cu	Mg	Mn	Zn	Si	Fe	Cr	Al
AA2024-T3	4.46	1.42	0.63	0.05	0.04	0.11	0.001	Balance
AA7075-T6	1.7	2.43	0.08	5.93	0.05	0.12	0.001	Balance

Table 2. Mechanical Properties

Material	Yield Strength	Tensile Strength	Elongation	Hardness
AA2024-T3	376Mpa	446Mpa	17%	144
AA7075-T6	489Mpa	573Mpa	13%	172

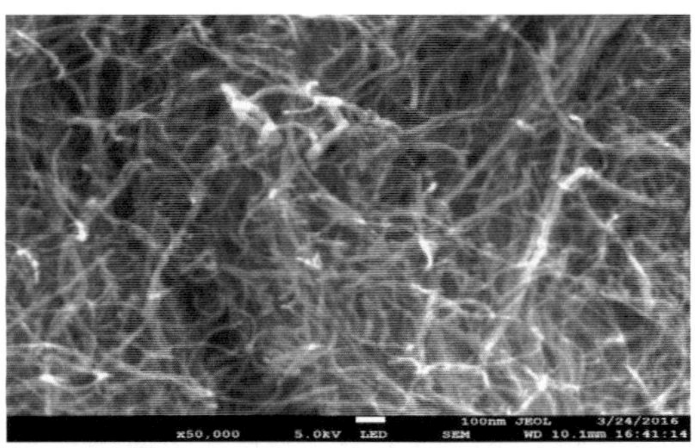

Figure 1. FESEM image of Carbon nanotubes

3. Results & Discussion

The geometry of weld pool and the quality of weld are observed to be influenced by the size of the groove. The imperfections in the welded joints are formed due to difference of velocity field around the welding tool, and also due to the change in the flow of material at low or high tool rotational speeds. It has been observed that the strength of the friction stir welded joints with nanoparticles is influenced by the variation in groove size. This was because of variation in the quantity of reinforcement availability in the microstructure of the stir zone.

3.1 Microstructure

The severe plastic deformation during friction stir welding process at high temperatures results fine grains in the stir zone [10]. The quantity of heat generation during the joining process determines the size of the grains, but the circumstances change as the reinforcements are deposited in the metal matrix [21]. The grain boundaries migration usually occurs at elevated temperatures [22]. Thus the elevated temperatures can be developed by decreasing traverse speed or enhancing the tool rotational speed which yield coarse grains in the weld zone [23]. The addition of reinforcements controls the grain boundary migration by pinning, which results fine grain structure. The microstructures of friction stir welded joints were shown in Fig. 2 and Fig.3. Fig. 2(a) shows the microstructure of the stir zone at groove of width 1 mm. Here the reinforcements are distributed evenly in the metal matrix and the size of the grains was restricted due to pinning effect. Fig. 2(b) shows the microstructure of stir zone for a groove of width 1.6 mm, which shows the combination of fine and coarse grains along with reinforcements. As the groove width was increased it has been observed more amount of reinforcements moved out of groove during the movement of the tool and some accumulation of reinforcements were seen in the microstructure with fine and coarse grains as shown in the Fig. 3(a). Similarly when the width of the groove was enlarged to 3 mm more amount of carbon nanotubes can be deposited, but during the process of welding most of the reinforcements came out of the groove and some accumulation of particles are seen in the microstructure as shown in Fig. 3(b). Thus the compression of nanoparticles by the welding tool without pin was efficient when the groove width was reduced, and enhanced the amount of reinforcements in the weld zone which hindered the dislocations movement and yielded fine grains in the stir zone. This phenomenon depends on volume fraction and size of the particles according to Orowan-Ashby equation [24].

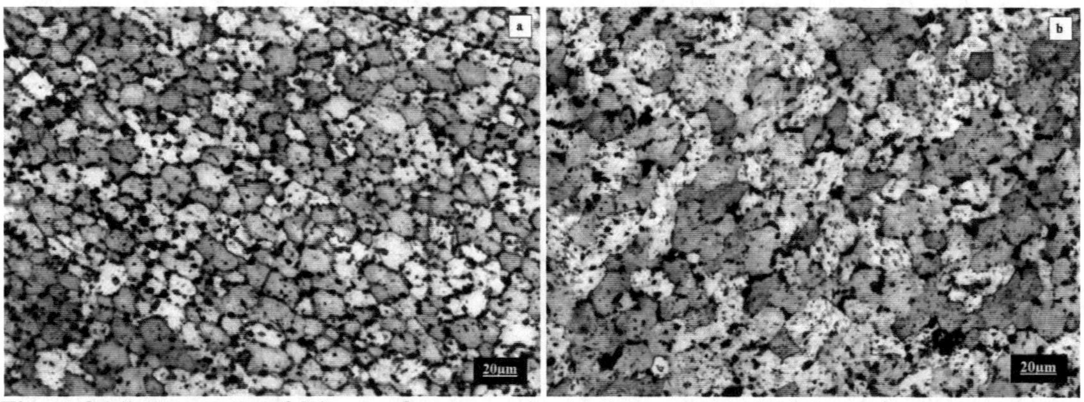

Figure 2. Microstructure of Stir zone for groove size (a) 1 mm×2 mm (b) 1.6 mm×2 mm

Figure 3. Microstructure of Stir zone for groove size (a) 2 mm×2 mm (b) 3 mm×2 mm

3.2 EDX Results

Elemental recognition was acquired by Electron dispersive X-ray analysis (EDX) as the amount of elements distribution in the stir zone was not clear from the study of microstructure. Fig. 3 and Fig. 4 shows the stir zone EDX patterns for the groove of varying width. EDX patterns show the deposition of carbon nanotubes in all the specimens, but there is a deviation in the amount of deposition of carbon nanotubes in the stir zone. This was due to variation in groove size as more amounts of reinforcement can be inserted as the width is increased. Fig. 4(a) shows the EDX pattern for groove of 1 mm width and the amount of carbon nanotubes deposited in the stir zone was 16.49 wt%, and 14.12 wt% for 1.6mm groove width as shown in Fig. 4(b). Thus as the groove width was increased some amount of nanoparticles moved out off the groove during the welding process. The maximum amount of carbon nanotubes of 17.44 wt% were deposited for groove of 2 mm width and 16.38 wt% for the groove of 3 mm width. Hence the groove with minimum width is preferable to deposit the nanoparticles uniformly in the stir zone and if more amount of reinforcements are available the accumulation of particles leads to decrease in the properties of the welded joint.

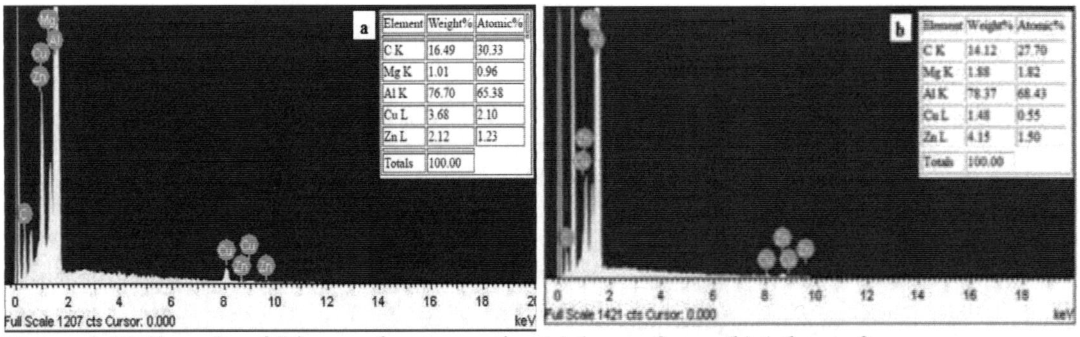

Figure 4. EDX results of Stir zone for groove size (a) 1 mm×2 mm (b) 1.6 mm×2 mm

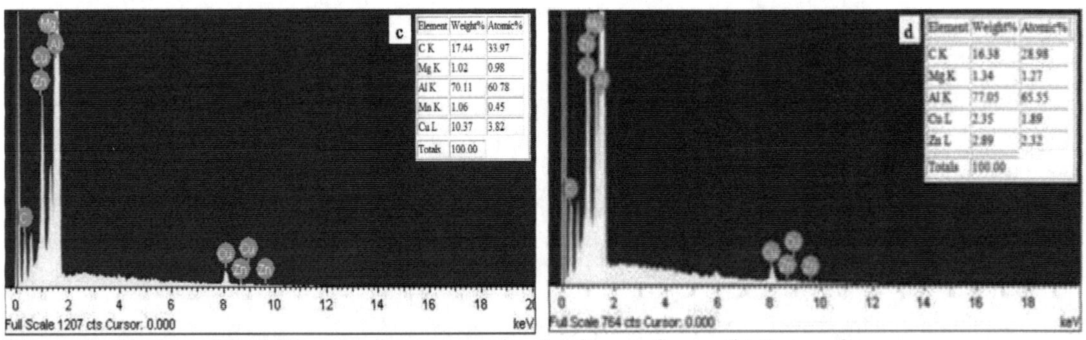

Figure 5. EDX results of Stir zone for groove size (a) 2 mm×2 mm (b) 3 mm×2 mm

3.3 Tensile Strength

Tensile tests were carried out to determine the strength of welded joints with nanoparticles. The mechanical properties are influenced by the factors like dislocation density, the interaction among base metals and reinforcements and grain size [8, 9]. The thermal mismatch was large between the aluminium matrix and carbon nanotubes which generates higher dislocation density at the interface and improves the strength of the matrix by work hardening [25]. Thus the carbon nanotubes in the metal matrix hinders the movement of grain boundaries and dislocations, reduces the grain size and Orowan strengthening mechanisms are the factors for enhancing the strength of the friction stir welded joints[17,19,20]. The mechanical properties of the fabricated dissimilar welded joints were shown in the Table 3. The results show that the highest tensile strength of 449.71 MPa was acquired for the joint with groove dimension of 1 mm width and 2 mm depth. The minimum value of 425.52 MPa was obtained for the joint with groove dimension of 2 mm width and 2 mm depth. This was due to more amount of reinforcement's deposition in the stir zone as compared to minimum groove width. And the accumulation of nanoparticles in the stir zone yielded lower tensile strength. Hence the nanoparticles inserted in to the groove of minimum width enhanced the strength of the welded joint by the homogeneous distribution of the reinforcements in the stir zone.

Table 3. Properties of welded joint

Groove size (mm)	Tensile Strength(MPa)	Microhardness (Hv)
3×2	429.34	186.4
2×2	425.52	192.3
1.6×2	433.34	181.5
1×2	449.71	189.3

3.4 Microhardness

The significant factors which control the metal matrix composite microhardness are heat input, dislocation density and amount of reinforcement distribution and grain size. According to Hall-Petch equation the fine grain size enhances the microhardness [26]. Since the co-efficient of thermal expansion is different for metal matrix and reinforcing phase could generate the dislocations and enhances the microhardness of the specimens with reinforcements [8, 9]. The average microhardness of the samples in the stir zone is shown in Table 3. It is observed that all welded joints exhibited the enhancement of microhardness, due to deposition of nanoparticles in the stir zone. The microhardness was observed to be altering due to change in the quantity of reinforcing particles in the microstructure of stir zone. The maximum microhardness of 192.3 Hv was obtained with a groove of 2 mm width and 2 mm depth, as more amount of reinforcements were present in the stir zone. And the minimum hardness was observed for the joint fabricated with groove of size 1.6 mm width and 2 mm depth. Thus the welded joint fabricated with groove of 1 mm width and 2 mm depth found to be best choice as the nanoparticles were deposited uniformly in the stir zone, which locks the grain boundaries and promotes the fine grains [13, 15].

4. Conclusions

Friction stir welding was effectively utilized to fabricate the dissimilar alloy joints by incorporating the multi-walled carbon nanotubes in to groove of varying width. The nanoparticles were deposited in all the samples irrespective of the groove dimensions. The results show that the groove dimensions play an important role to deposit and distribute the nanoparticles uniformly in the stir zone, which increases the mechanical properties of the welded joint. Tensile strength was observed to be highest for the joint with groove of width 1 mm, as the nanoparticles were distributed uniformly in the stir zone which hindered the progress of grain boundaries and dislocations. The hardness was highest for the joint with groove of 2 mm width, but the strength decreased due to accumulation of nanoparticles in the stir zone.

Acknowledgements

The authors are grateful to Universiti Malaysia Pahang for its steady support during the course of this work. The authors are also gratified to Ministry of Higher Education, Malaysia for conceding FRGS under RDU130103. The first author wants to express gratitude toward University Malaysia Pahang, for granting Doctoral Scholar Ship Scheme.

Reference

[1] Mishra R S and Ma Z Y 2005 Friction stir welding and processing *Material science engineering* **50** 1-78.

[2] Ishak M, Noordin N F M, Razali A S K and Shah L H A 2015 The effect of filler ER4043 and ER5356 on the weld metal structure of 6061 aluminium alloy by Metal Inert Gas (MIG) *International Journal of Engineering Technology and Sciences* **3**

[3] Thomas W M, Johnson K I, Wiesner C S 2003 Friction stir welding recent developments in tool and process technologies *AdvEng Mater.* **5** 485-90.

[4] Li Y, Murr L E, McClure J C 1999 Flow visualization residual microstructures associated with the friction-stir welding of 2024 aluminum to 6061 aluminum *Mater Sci Eng A.* **271** 213-23.

[5] Rhodes C G, Mahoney M W, Bingel W H, Spurling R A and Bampton C C 1997. Effects of Friction Stir Welding on Microstructure of 7075 *Aluminum, Scripta Mater.* **36** 69–73.

[6] Nandan R, Debroy T and Bhadeshia H K D H 2008 Recent advances in friction stir welding – process, weldment structure & properties *progress in material science* **53** 893.

[7] Kurtyka P, Sulima I, Wojcicka W, Rylko N and Pietras A 2012 The Influence of Friction Stir Welding Process on Structure and Mechanical Properties of the AlSiCu/SiC Composites *J. Achiev. Mater. Manuf. Eng.* **55** 339–344.

[8] Alidokht S A, Abdollah-zadeh A, Soleymani S and Assadi H 2011 Microstructure and tribological performance of an aluminium alloy based hybrid composite produced by friction stir processing *Materials and Design* **32** 2727-2733.

[9] Shafiei-Zarghani A, Kashani-Bozorg S F and Zarei-Hanzaki A 2009 Microstructures and mechanical properties of Al/Al_2O_3 surface nano-composite layer produced by friction stir processing. *Mater Sci Eng. A.* **500** 84–91.

[10] Morisada Y, Fujii H , Nagaoka T, Fukusumi M 2006 MWCNTs/AZ31 surface composites fabricated by friction stir processing *Materials Science and Engineering A* **419** 344–348.

[11] Lim D K, Shibayanagi T, Gerlich A P 2009 Synthesis of multi-walled CNT reinforced aluminium alloy composite via friction stir processing *Materials Science and Engineering A* **507(1-2)** 194-199.

[12] Izadi H, Gerlich A P 2012 Distribution and stability of carbon nanotubes during multi-pass friction stir processing of carbon nanotube/aluminum composites. *Carbon* **50** 4744-4749.

[13] Liu Q, Ke L, Liu F,Huang C, Xing L 2013 Microstructure and mechanical property of multi-walled carbon nanotubes reinforced aluminum matrix composites fabricated by friction stir processing *Materials and Design* **45** 343–348.

[14] Thangarasu A , Murugan N, Dinaharan I 2014 Production and wear characterization of AA6082 -TiC surface composites by friction stir processing *Procedia Engineering* **97** 590 – 597.

[15] Sathiskumar R, Dinaharan I, Murugan N, Vijay S J 2014 Influence of tool rotational speed on microstructure and sliding wear behavior of Cu/B4C surface composite synthesized by friction stir processing *Trans. Nonferrous Met. Soc. China* **24** 95-102.

[16] Karthikeyan P and Mahadevan K 2015 Investigating on the effects of SiC particles addition in the weld zone during friction stir welding of Al 6351 alloy *Int J Adv Manuf Technol.* **80** 1919–1926.

[17] Abnar B, Kazeminezhad M and Kokabi A H 2015 The Effect of Premixed Al-Cu Powder on the Stir Zone in Friction Stir Welding of AA3003-H18 *Journal of Materials Engineering and Performance* **24(2)** 1086-1093.

[18] Kumar P V, Reddy G M and Rao K S 2015 Microstructure and pitting corrosion of armor grade AA7075 aluminum alloy friction stir weld nugget zone – effect of post weld heat treatment and addition of boron carbide *Defense Technology* **11** 166–173.

[19] Abbasi M, Abdollahzadeh A, Omidvar H 2016 Incorporation of SiC particles in FS welded zone of AZ31 Mg alloy to improve the mechanical properties and corrosion resistance *Int J Mater Res* **107** 566–572.

[20] Ravinder R B, Ramaraju R V, Ibrahim A B, CheKu M F B 2017 Effect of Nano Particle Deposition on Mechanical Properties of Friction Stir Welded Dissimilar Aluminium Alloys by Taguchi Technique *Trans Indian Inst Met.* **70(4)** 1005-1017.

[21] Faraji G and Asadi P 2011 Characterization of AZ91/alumina nanocomposite produced by FSP *Materials Science and Engineering A* **527** 2320–30.

[22] Humphreys F J and Hatherly M 2004 UK, Elsevier Ltd., University of Manchester, Institute of Science and Technology.

[23] El-Rayes Magdy M and El-Danaf Ehab A 2012 The influence of multi-pass friction stir processing on the microstructural and mechanical properties of Aluminum Alloy *6082 J Mater Process Technol.* **212** 1157–68.

[24] Dieter G E and Bacon D. 1988 *Mechanical metallurgy*. London McGraw-Hill.

[25] George R, Kashyap K T, Rahul R, Yamdagni S 2005 Strengthening in carbon nanotube/aluminum (CNT/Al) composites *Scripta Mater* **53** 1159–63.

[26] Sharifitabar M, Sarani A, Khorshahian S, ShafieeAfarani M 2011 Fabrication of 5052Al/Al_2O_3 nanoceramic particle reinforced composite via friction stir processing route *Materials and Design***32** 4164-4172.

IOP Conf. Series: Materials Science and Engineering **319** (2018) 012024 doi:10.1088/1757-899X/319/1/012024

Design logistics performance measurement model of automotive component industry for srengthening competitiveness of dealing AEC 2015

T G Amran and Mindy Janitra Yose

Magister Industrial Engineering Program
Faculty of Industrial Technology, Trisakti University
Kiyai Tapa no 1. Jakarta 11440. Indonesia

tiena@trisakti.ac.id; amran.tiena@gmail.com

Abstract. As the free trade Asean Economic Community (AEC) causes the tougher competition, it is important that Indonesia's automotive industry have high competitiveness as well. A model of logistics performance measurement was designed as an evaluation tool for automotive component companies to improve their logistics performance in order to compete in AEC. The design of logistics performance measurement model was based on the Logistics Scorecard perspectives, divided into two stages: identifying the logistics business strategy to get the KPI and arranging the model. 23 KPI was obtained. The measurement result can be taken into consideration of determining policies to improve the performance logistics competitiveness.

1. Introduction

As the free trade ASEAN Economic Community (AEC) causes the tougher competition, it is important that Indonesia's automotive industry have high competitiveness as well. A model of logistics performance measurement was designed as an evaluation tool for automotive component companies to improve their logistics performance in order to compete in AEC. The design of logistics performance measurement model was based on the Logistics Scorecard perspectives, divided into two stages: identifying the logistics business strategy to get the KPI and arranging the model. 23 KPI was obtained. The measurement result can be taken into consideration of determining policies to improve the performance logistics competitiveness.

Indonesia supply base is still weak compared to other ASEAN countries, namely Thailand, the number of parts and materials suppliers in Indonesia and Thailand were 709 in 1965 [6][12]. Automotive components industry also still has limited capacity to meet the demand, such as the requirements of Quality, Cost and Delivery (QCD), and the ability of human resources and technology. The unsatisfactory national logistics performance was caused by the high logistics costs and length of delivery time, also the supply chain management and logistics [15]. An effective and efficient logistics system is necessary for competition in the AEC [11]. This research was focused on designing a model of logistics performance measurement for the automotive component industry, which can be a tool of evaluation and guidelines for the improvement of its logistics activities in order to compete in the era of AEC.

The previous researches [5] and [23] developed a method of performance measurement in the application of the supplier relationship management using benchmarking framework. This study was strengthened by [14] who examined the practice of supply chain management and its impact on the performance of a manufacturing company in Ghana. Balanced Scorecard approach [4] and [20] used the in measuring the supply chain performance. Economic Value Added and ABC classification [17] used for evaluating supply chain performance. Others had applied the Supply Chain Operations Reference (SCOR) model [3], [16] and[10] . Anatan conducted based on [13] a research of the effect of implementation of supply chain management practices to the achievement of competitive advantage and performance of the supply chain.

The measurement the competitiveness of some Thailand industries using *Logistics Scorecard* model [18] and [19], which can provide information of how industries should improve its performance to improve its business matters. This resulted a competitiveness index based on KPIs which were determined and compared between industry with other industries, providing guidance for the proposed increase in the performance ratio of logistics industry competitiveness.

This study was aimed to design a logistics performance measurement model for the Indonesian automotive component industry in order to increase competitiveness in dealing with AEC integration. *Logistics Scorecard*'s perspective of [18] and [19] which consists of five perspectives: business strategy, capacity and planning work, the efficiency and productivity of logistics, information technology and supply chain collaboration.

Logistics mission is to meet the needs of the goods corresponding to the right place, at the right time and in the desired conditions, thereby providing benefits for the company [8] . It takes a long time for companies to realize the importance of logistics in order to develop a competitive advantage over competitors [20] and [21]. Research by [9] resulted that the logistics service performance takes effect on customer satisfaction, which is depends on the quality of the management of the flow of goods and services in order to face the free market integration AEC [9].

SCM-Logistics Scorecard (LSC) has been developed since 2001 by Tokyo Institute of Technology with Japan Institute of Logistics System. LSC has been an efficient tool which analyzed relationship between company's supply chain performance and managerial performance [7], investigated the correlation between company's environment and supply chain execution [2], identified influencing factors that determine the performance of the implementation of supply chain management and dan its effect to *bottom-line* financial index [13], evaluation tool for supply chain operational performance to know potential factors which could improve the efficiency of supply chain operational performance [8].

The key index of competitiveness used [19] classified with the perspectives which are: business strategy orientation, capacity planning and execution, logistics efficiency and productivity, information technology implementation, and supply chain collaboration. The five perspectives measurement with 23 KPIs are designed about supply chain practices within the organization by using the rating scale or 1-5 level (Level 1= very bad business processes on the performance, level 5= indicates that the businesses do the best logistics activities). The strategic planning is carried out starting from identification of organizational vision and mission statement, internal environment analysis, analysis of internal and external environment of the organization, determine the opportunities and threats / challenges [2], [10].

2. Research Methodology

The design of logistics performance measurement model is Logistics Scorecard by using the determined indicators business strategy and analysis system. The collected data were including conditions and situations that automotive component industries in the AEC are facing, the measurement model designed, both small-medium scale industries with local ownership and large scale industries with foreign ownership, and also the related agency or institution (PIKKO). The logistics performance measurement using the designed model, for qualitative assessment and quantified methods, and weights for the KPI are determined in order to obtain the output score of overall performance to know how the position of the company's competitiveness.

Designing The Model. The design of the logistics performance measurement model is divided into two stages: (i) identify the logistics business strategy of automotive component company to obtain indicators of logistics performance suitable with the logistics strategy, (ii) create a logistics performance measurement model (Figure 1.Logistics Scorecard) using indicators that have been determined. The performance indicators are identified based on the strategy that had been developed. It is determining the performance indicators along with the targets for each of the strategies that have been grouped in the perspective of *Logistics Scorecard* (Table 1). Those performance indicators validated by using Delphi technique by interviewing five experts consist of the company Director, Head of the Warehouse & Logistics Department, the academics expert in the field of supply-chain logistics, the practitioners of other automotive components, and also the automotive practitioners from distributor. The result summary of the Delphi technique obtained was the five respondents were giving the similar assessment for 'all indicators are required' more than 60% proportional answers [7].

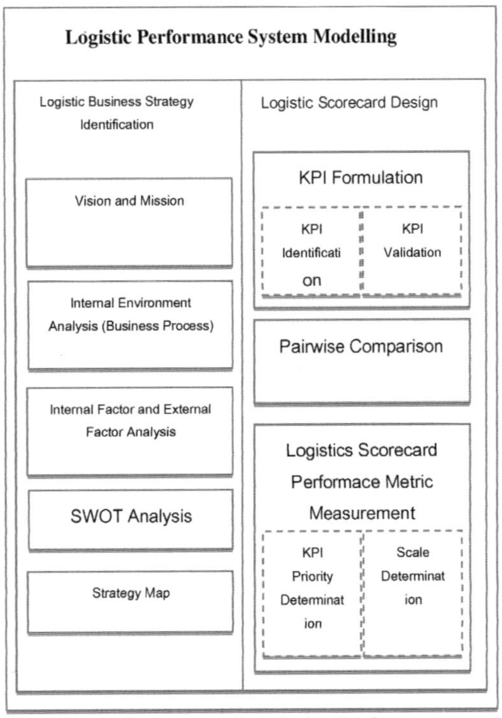

Figure 1. Design of Logistics Performance Measurement Model

Table 1. Strategy, Performance Indicator, Target

Strategy	Performance Indicator	Target
Business-Oriented Strategy Prespective		
Create on point logistic business strategy	Logistis business strategy	Correct and clear business strategy that supports SCM-logistics activity
Contract determination and clear information exchange with supplier	Cotract agreement and information exchange with supplier	A formal written agreement with suppliers, sharing activity and stable supplier planning develonment
Determine contract agreement with a clear information exchange with customer	Contrcat agreement and information exchange with customer	A formal written agreement with customers, sharing activity and stable customer satisfaction planning development
Customer satisfaction system improvement	Customer satisfactory measurement and improvement system	Measuring system enhancement and customer satisfaction improvement, the result will be used as product development and service improvement information
Employee training and evaluation system enhancement	Workers training and evaluation system	Training system and worker evaluation enhancement, knowledge management for knowledge sharing

3. Results and Discussion

Strategic planning begins with identifying vision and mission statement, analyzing the internal and external environment. Then, by using SWOT, the organizational strengths, weaknesses, opportunities and threats are evaluated. The next step is the strategy formulation into five Logistics Scorecard perspectives and arrangement of Logistics Scorecard Strategy Map (case study SME PIKKO, PT. XYZ).

3.1. Designing The Logistics Scorecard

The logistics performances measurement metrics is arranged including the blank column provided to be filled in. KPI priority determination is carried out by sorting the final weight of each KPI from the most high.

Table 2. The Scale of Performance Measurement

Logistic Business Strategy						
Have not yet formulized SCM-logistic strategy, logistic is not considered as important by company	1	2	3	4	5	Clear and precise business strategy, supporting system for SCM-logistic activity is available

Final weights obtained by multiplying the weight of each KPI with the weight of perspective in which KPIs are grouped. Figure 3. shows the KPI priority based on its weight. The measurement scale used is the *descriptive graphic rating scale*, with 1 to 5 range of scale (Table 2).

Below are grouping of the performance score categories based on scale rating [19] :
Level 1 : a very poor logistics performance, in which the company was not involved in logistics activity
Level 2 : a poor logistics performance
Level 3 : a good enough logistics performance, but overall still need much improvement
Level 4 : a good logistics performance
Level 5 : an excellent logistics performance, in which companies do their best in logistics activities
Then the results can be made a diagram to map their performance. This will provide some inputs to the evaluation of the company in determining the logistics performance improvement initiatives [5] (Gong et.al., 2011).

The Logistics KPI Determination based on *Logistics Scorecard* Perspectives. All indicators are really needed in logistics performance measurement of automotive components company since those indicators are able to show in detail the supply-chain logistics business process of the company. The next step is formulate the KPI logistics in the form of a hierarchy to make it easier to see the picture of the logistics performance measurement model based on Logistics Scorecard perspectives. Then the performance measurement results of some companies can be made a diagram to map their performance. This will provide some inputs to the evaluation of the company in determining the logistics performance improvement initiatives. Figure 2 shows the flow of logistics performance measurement system from input to the output. The metric of logistics performance measurement is on the Figure 3.

KPI Final Score

Perspective	Perspective score	Code	Criteria	Criteria score	Final score
Business Strategy Orientation	0.43	BSO1	Logistic business strategy	0.46	0.1984
		BSO2	Contract agreement and information exchange with supplier	0.07	0.0281
		BSO3	Contract agreement and information exchange with customer	0.08	0.0352
		BSO4	Measuring system and customer satisfaction improvement	0.24	0.1042
		BSO5	Worker training and evaluation system	0.15	0.0644
Capacity Planning and Execution	0.27	CPW1	Strategies to optimize logistic resources	0.33	0.0867
		CPW2	Market trend and demand foecast comprehension	0.18	0.0483
		CPW3	Ability to adjust planning according to SCM-logistic	0.17	0.0460

KPI Priority

Code	Criteria	Score
BSO1	Logistic business strategy	0.1984
SCC1	Collaboration on logistic development with similar business partner	0.1072
BSO4	Measurement system and customer satisfaction improvement	0.1042
CPW1	Strategy to optimize logistic system resources	0.0867
BSO5	Employee training system and its evaluation	0.0644
CPW5	Standarization in all business process	0.0541
SCC2	Collaboration on logistic improvement with research and development institution, universities, etc.	0.0529
CPW2	Market trend comprehension and demand forecast	0.0483

							exactness	
		CPW4	Monitoring system ad stock tracking	0.06	0.0168	CPW3	Ability to adjust planning according to SCM-logistic	0.0460
		CPW5	Standarization in all business process	0.20	0.0541	BSO3	Contract agreement and information exchange with customer	0.0352
		CPW6	Logistic department development	0.05	0.0133	LEP1	Logitic activity improvement	0.0331
Efficiency and Logistic Productivity	0.09	LEP1	Logistic activity improvement	0.36	0.0331	ITM3	IT presonnel development related with SCM-logistic	0.0312
		LEP2	Supply cycle and cash-to-cash cycle time	0.09	0.0083	BSO2	Contract agreement and supplier information exchange	0.0281
		LEP3	Customer lead time and load efficiency	0.16	0.0148	CPW4	Supply monitoring and tracking system	0.0168
		LEP4	Product shipment quality and performance	0.15	0.0138	ITM1	Identification standard (code) to for product or process identification	0.0148
		LEP5	Stock management system	0.11	0.0098	LEP3	Customer lead time and load efficiency	0.0148
		LEP6	Working Environment Organization	0.04	0.0138	LEP4	Product shipment's performance and quality	0.0138
		LEP7	Logistic total cost	0.08	0.0077	CPW6	Logistic department development	0.0133
Information Technology Implementation	0.05	ITM1	Identification standard (code) to for product or process identification	0.28	0.0148	LEP5	Supply sistem management	0.0098
		ITM2	Effective computer operation and decising making between company and business partner	0.13	0.0071	LEP2	Stock cycle and cash-to-cash cycle time	0.0083
		ITM3	IT personal development in relation with SCM-logistic	0.59	0.0312	LEP7	Total logistic cost	0.0077
Supply Chain Collaboration	0.16	SCC1	Collaboration on logistic development with same business partner	0.67	0.1072	ITM2	Effective computer usage in operation and decision making between company and business partner	0.0071
		SCC2	Collaboration on logistic development with research institutions, universities, etc.	0.33	0.0529	LEP6	Organization Working Environment	0.0038

Figure 2. KPI Priority based on Final Weight

LOGISTIC PERFORMANCE MEASUREMENT FORM

Company name **Day/Date :**
Person in Charge (Name/Department)

Please circle number 1, 2, 3, 4, or 5 according to the present company's condition

Code	Logistic KPI							Score	Score	Score
	Logistic Business Strategy									
BSO 1	Have not yet formulized SCM-logistic strategy, logistic is not considered as important by company	1	2	3	4	5	Clear and precise business strategy, supporting system for SCM-logistic activity is available		0.1984	
	Logistic development collaboration with similar business partner									
SCC 1	Unaware of the importance of collaboration with business partner	1	2	3	4	5	Formal agreement and clear procedure in the collaboration between business partner for company's business development		0.1072	

IOP Conf. Series: Materials Science and Engineering **319** (2018) 012024 doi:10.1088/1757-899X/319/1/012024

BSO 4	**Measurement system and customer satisfaction improvement**								0.1042	
	No clear definition about customer, complaints were handed temporarily and wisely	1	2	3	4	5	Increase in customer satisfaction measuring and improvement system, in order to develop products and services			
CPW 1	**Logistic system optimizing stategy**								0.0867	
	No improvement strategy, no solution for inefficiency and facility from logistic resources	1	2	3	4	5	Clear strategy for logistic resource optimalization and collaboration system			
BSO 5	**Employee training system and its evaluation**								0.0644	
	Special training program for employee is not available	1	2	3	4	5	Employee's training and evaluation system improvement, knowledge management sustem for knowledge sharing is available			
CPW 5	**Business process standardization**								0.0541	
	Standarization of working methods for several process is not available	1	2	3	4	5	Working method, interface activity, and cooperation in the entire supply chain standardization			
SCC 2	**Logistic development collaboration with research and development institution, universities, etc.**								0.0529	
	Unaware of the collaboration importance in logistic development and research& development institution	1	2	3	4	5	Formal agreement and clear procedure regarding collaboration with research institution for logistic development			
CPW 2	**Market trend comprehension and demand forecast precision**								0.0483	
	Forecast activity in the supply chain process is unavailable	1	2	3	4	5	Demand forecast based on market trend, done together with supply chain partners, and flexible			
CPW 3	**Adjusting abilitu with SCM-logistic planning**								0.0460	
	Plans for sales, addition, and expedition are unavailable	1	2	3	4	5	Improvements in planning ability, sales data availability, and stock and expedition management			
BSO 3	**Contract agreement and information exchange with customer**								0.0352	
	Formal written agreement regarding sharing activity with customer is non-existent. Customer is the only party involved in decision making	1	2	3	4	5	Formal written agreement, established sharing activity and customer development plan			
LEP1	**Logistic activity improvement**								0.0331	
	Logistic activity improvement is unavailable, just-in-time philosophy is not implemented by company	1	2	3	4	5	Sustainable improvement in logistic activities involving customer and supplier			
ITM3	**IT personal development related to SCM-Logistic**								0.0312	
	Company has no concern on IT personnel development regarding SCM-logistic activity	1	2	3	4	5	IT personnel training in SCM-logistic activities			
BSO 2	**Contract agreement and information exchange with supplier**								0.0281	
	Formal written agreement and sharing activity with supplier are unavailable. Decision-making was done independently	1	2	3	4	5	Formal written agreement with all supplier, sharing activity and established customer development planning			
CPW 4	**Monitoring system and supply tracking**								0.0168	
	Supply monitoring and tracking are unavailable	1	2	3	4	5	Transparent supply tracking and monitoring, information sharing in all supply chain			
ITM1	**Identification standard (code) for process or product**								0.0148	
	Standard identification system (code) for product and process is not available anywhere in the company	1	2	3	4	5	Identification standard (code) for all product and process and identifiable by suppliers and customer			
LEP3	**Customer lead time and load efficiency**								0.0148	
	Lead time for order placement is unavailable for ny product	1	2	3	4	5	Continuous effort to lower lead time and increase load efficiency			
LEP4	**Product shipping performance and quality**								0.0138	

	Fixed shipping time and accuracy is unknown, hence company is facing a lot of customer complaints	1	2	3	4	5	On time shipping and order fulfillment accuracy have been accomplished, along with improvements attempts			
CPW 6	**Logistic department development**								0.0133	
	No one is responsible for company's logistic activity	1	2	3	4	5	Personal development for employees (training, recruitment) in logistic department for sustainable improvement			
LEP5	**Stocking management system**								0.0098	
	Good stocking management is unavailable, stock level target is undetermined	1	2	3	4	5	Good stock management system and fulfilled the target lowering stock level deviation			
LEP2	**Stock cycle and cash-to-cash cycle time**								0.0083	
	Stock cycle or cash-to-cash cycle is not measured	1	2	3	4	5	Supply cycle measurement and all products cash-to-cash cycle time is available, sustainable improvement planning			
LEP7	**Total Logistic Cost**								0.0077	
	Logistic cost is unknown for each process	1	2	3	4	5	Logistic cost reduction targeted in each process using predetermined strategy was accomplished			
ITM2	**Computer effective usage in operation and decision making between company and business partner**								0.0071	
	IT devices (e.g. computer) is not implemented anywhere in the business	1	2	3	4	5	TI devices were implemented effectively during operation and decision making process			
LEP6	**Organizational Working Environment**								0.0038	
	Working environment awareness and staff's welfare concern are low	1	2	3	4	5	Iniciative to create condusive, comfortable, and safe working environment. Certified (ISO 14000)			

Figure 3. Logistics Performance Measurement Metric (*Logistics Scorecard*)

3.2. Model Verification and Validation

Measurements were done with the interview to ask how the achievement of the company's logistics business process activities which are specified in each logistics' KPI in the model. The logistics performance score was 3.05 out of 5 scale range, which means the logistics performance is quite good, but still needs a lot of improvement overall. The verification results show that the designed performance measurement model has been verified to be used to measure the companies' logistics performance in accordance with the conditions of business processes-supply chain logistics activities of the company by generating appropriate output. The validation was done using the *face validity* [21] are as follows: conformity of steps during determining KPI, which the KPI were obtained from strategy derived from vision-mission and analysis of internal and external factors. KPI is aligned with the demand to increase the logistics performance competitiveness and become a critical indicator for the improvement of logistics performance automotive components company. Logistics performance measurement model designed can be implemented in automotive components companies in Indonesia.

3.3. Model Implementation

The measurement result is shown on a bar chart in Figure 4. and a radar chart in Figure 5. to show overall logistics performance of each KPI.

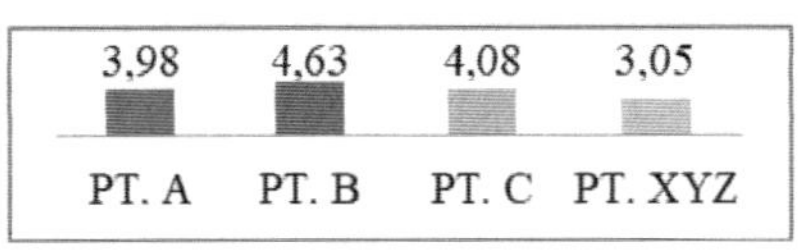

Figure 4. Logistics Performance Score Comparison

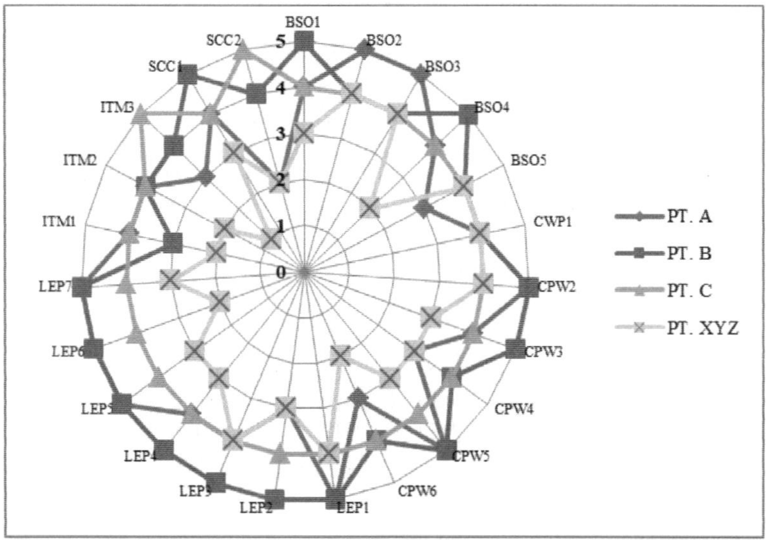

Figure 5. Radar Chart of Logistics Performance Score of The Measured Companies

The small-medium scale automotive components company (PIKKO) was still at a lower level, the average yield of logistics performance appraisal automotive component industry is 3.9, and categorized as a good enough logistics performance. In general, companies have a weakness in the implementation of information technology such as the use of an integrated computer, the lack of IT staffs which related to logistics activities, and also a coding standard products and processes. The weak information technology can have an impact on other KPI, that the information technology has become an important factor which supports the business processes of logistics in both internal and external to the enterprise or business partners. National automotive components industry is also still weak in standardization of the entire business processes and logistics departments which still need to be developed.

4. Conclusions

The logistics performance measurement system is based on the logistics business activities strategy of the automotive component industry in facing the AEC by five Logistics Scorecard perspectives. There are 23 logistics KPI which are 5 for business strategy orientation perspective, 6 for capacity planning and execution perspective, 7 for logistics efficiency & productivity, 3 for information technology implementation perspective, and 2 for supply chain collaboration. KPI are used as a competitive logistics performance benchmarks. The priority order of five perspectives based on the highest weight is the business strategy orientation, capacity planning and execution, supply chain collaboration, logistics efficiency and productivity, and the last is information technology implementation. Priority KPI that has the highest final weight is the logistics business strategies from the perspective of business strategy orientation, before running the logistics activities. Designed model of performance measurement can be used to measure the logistics performance of automotive components company in Indonesia and also for other industries in which KPIs are flexible to be developed. In general, company has a low score on the KPI implementation of IT and business strategy, so that it can be taken into consideration for the improvement in technology, especially information technology and creating a better logistics business strategy. Suggestions related to the results of the study includes the model should be used by the related parties for the evaluation of the automotive component industry in order to improve the competitiveness of logistics performance (PIKKO, GIAMM etc).

Acknowledgments

The author thanks to Universitas Trisakti, the institution which sponsored and supported to attend the seminar and to PIKKO for supporting the author and students to do this research.

References

[1] Anatan, Lina. 2010. Pengaruh Implementasi Praktik-Praktik Manajemen Rantai Pasokan terhadap Kinerja Rantai Pasok dan Keunggulan Kompetitif. *Karisma*, Vol.4(2), pp.106-117.

[2] Armitage, H.M. dan Cameron Scholey. 2006. *Using Strategy Maps to Drive Performance*. Canada: The Society of Management Accountants of Canada, The American Institute of Certified Public Accountants and The Chartered Institute of Management Accountants.

[3] Barnard, James. 2006. A Multi Framework for Defining The Services Supply Chain Using Object Oriented Methodology. Florida: Univ. of Florida.

[4] Brewer, P. dan Thomas Speh. 2000. Using The Balanced Scorecard to Measure Supply Chain Performance. *Journal of Business Logistics*, Vol.21, No.1, pp. 75-93.

[5] Choy, K., Chow, H., Lee, W. and Chan, F. 2007. Development of Performance Measurement System in Managing Supplier Relationship for Maintenance Logistics Providers. *Benchmarking: An International Journal*, Vol. 14 No. 3, pp. 352-68.

[6] Departemen Perdagangan Republik Indonesia. 2009. Buku Menuju Asean Economy Community 2015. http://ditjenkpi.kemendag.go.id/website_kpi/Umum/Setditjen/Buku%20Menuju%20ASEAN%20ECONOMIC%20COMMUNITY%202015.pdf. (Diakses 3 November 2014, pk.20.24)

[7] Eadie, R. et.al. 2010. Identification of E-Procurement Drivers and Barriers for UK Construction Organisations and Ranking of these from the Perspective of Quantity Surveyors. *Journal of Information Technology in Construction*, Vol.15, pp. 23-43.

[8] Gong, J., Ogasawara, T., dan Suzuki, S. 2011. Supply Chain Operational Performance and Its Influential Factors: Cross National Comparison between Japan and China. *Brazilian Journal of Operations & Production Management, 8*, 2nd ser., pp.67-87.

[9] Haryotejo, Bimo. 2015. *Analisis Pengaruh Kinerja Logistik Pemasok terhadap Kinerja Bisnis (Studi pada Bengkel AHASS di Kota Semarang)*. Semarang: Fakultas Ekonomika dan Bisnis Universitas Diponegoro.

[10] Hashemi, Nima F., et.al. 2012. Formulating and Choosing Strategies using SWOT Analysis and QSPM Matrix: A Case Study of Hamadan Glass Company. *Proceedings of The 41st International Conference on Computers & Industrial Engineering*.

[11] Heizer, J dan Barry Render. 2014. Operations Management: Sustainability and Supply Chain Management. Pearson.

[12] Kementrian Luar Negeri Republik Indonesia. 2014. Kerjasama Ekonomi ASEAN. www.kemlu.go.id/Documents/Kerjasama%20Ekonomi%20ASEAN.doc (Diakses 27 Jan 2015, pk.23.10)

[13] Li, S., Nathan, B. R., Nathan, T. S., & Rao, S. S. 2006. The Impact of Supplychain Management Practices on Competitive Advantage and Organizational Performance. *Omega*. 34: 107-124.

[14] Mensah, C., D.Diyuoh, D.Oppong. 2014. Assessment of Supply Chain Management Practices and It Effects on The Performance of Kasaprekko Company Limited in Ghana. *European Journal of Logistics Purchasing and Supply Chain Management*, Vol. 2 No. 1, pp. 1-16.

[15] Mulyadi, Dedi. 2011. Pengembangan Sistem Logistik yang Efisien dan Efektif dengan Pendekatan *Supply Chain Management*. *Jurnal Riset Industri*, Vol. V, No.3, pp.275-282.

[16] Mutakin, A. dan M. Hubeis. 2011. Pengukuran Kinerja Manajemen Rantai Pasokan dengan SCOR Model 9.0. *Jurnal Manajemen dan Organisasi*, Vol.II, No.3, pp.89-103.

[17] Pohlen, T. dan Coleman, B. 2005. Evaluating Internal Operations and Supply Chain Performance Using EVA and ABC. *SAM Advanced Management Journal*, Vol. 70 No. 2, pp. 45-58.

[18] Primiana, Ina. 2012. Logistik dan Daya Saing. Jakarta : LP3E Kadin Indonesia.

[19] Puangchampee, B. dan M.Baramichai. 2010. Thailand Industrial Competitiveness; Enhancing The Logistics and Supply Chain Management Scheme for Thai's Manufacturing. 2010 *International Conference on Management Science and Information Engineering (ICMSIE 2010), UTCC Engineering Research Papers 2010*, pp.203-206.

[20] Saboia, E., L.C.Duclos, C.O.Quandt, A.Souza. 2006. Strategic Management Indicators for Internal Logistics: A Proposal Based on The Balanced Scorecard for An Automotive Sector Company. *XII ICIEOM-Fortaleza, CE, Brasil.*

[21] Sargent, R.G. 2013. Verification and Validation of Simulation Models. *Journal of Simulation*, Vol.7, pp. 12-24.

[22] Sorooshian, S. dan Yin, D.T. 2013. Logistics Evaluation: a Case Study. *International Journal of Engineering & Technology Sciences*, Vol.1, No.4, pp.192-199.

[23] Tracey, M., Lim, J. and Vondrembse, M. 2005. The Impact of Supply Chain Management Capabilities on Business Performance. *Supply Chain Management: An International Journal,* Vol. 10 No. 3, pp. 179-91.

[24] http://www.kemenperin.go.id/artikel/5148/Industri-Komponen-Otomotif-RI-Tertinggal-Jauh-Dari-Thailand (Industri Komponen Otomotif RI Tertinggal Jauh dari Thailand) Diakses 11 Okt 2015, pk.10.55

IOP Conf. Series: Materials Science and Engineering **319** (2018) 012025 doi:10.1088/1757-899X/319/1/012025

PSO-based PID Speed Control of Traveling Wave Ultrasonic Motor under Temperature Disturbance

Kamal Arifin Mat Piah, Wan Azhar Wan Yusoff, Nur Iffah Mohamed Azmi, Fadhlur Rahman Mohd Romlay

Faculty Manufacturing Engineering, University Malaysia Pahang

aviakamal@gmail.com

Abstract. Traveling wave ultrasonic motors (TWUSMs) have a time varying dynamics characteristics. Temperature rise in TWUSMs remains a problem particularly in sustaining optimum speed performance. In this study, a PID controller is used to control the speed of TWUSM under temperature disturbance. Prior to developing the controller, a linear approximation model which relates the speed to the temperature is developed based on the experimental data. Two tuning methods are used to determine PID parameters: conventional Ziegler-Nichols(ZN) and particle swarm optimization (PSO). The comparison of speed control performance between PSO-PID and ZN-PID is presented. Modelling, simulation and experimental work is carried out utilizing Fukoku-Shinsei USR60 as the chosen TWUSM. The results of the analyses and experimental work reveal that PID tuning using PSO-based optimization has the advantage over the conventional Ziegler-Nichols method.

1. Introduction

Proportional Integral Derivative (PID) controller is still widely used in industrial applications even though many new advanced control algorithms have been proposed[1-2]. PID control has a simple structure and easy to be understood by engineers. In practical conditions, PID controller is more reliable compares to new advanced and complex control algorithms. Generally, PID controller consists of three gains: proportional, integral and derivative. The value of each gain needs to be tuned appropriately to get a good or optimal control of the system.

Ziegler-Nichols, Cohen-Coons, Astrom and Haggland tuning methods are some of the early developed tuning methods for PID control. However, tuning results of these methods is insufficient due to problems such as phase lag and high overshoot percentage. Particle Swarm Optimization (PSO) is an evolutionary computation technique approach introduced by Kennedy and Eberhart in 1995[3]. The PSO development is based on observations of animal social behaviour such as bird flocking and fish schooling. In PSO, each particle in swarm interacts in constructive cooperation and share information between them. By using the PSO method, high quality solution can be generated in a short calculation time. The detail of PSO technique is presented later in this paper.

In this study, the PSO tuning of PID controller is implemented to speed system of the ultrasonic motor (USM). An ultrasonic motor is an electric motor driven by piezoelectric vibrations[4,5] Ultrasonic motors have been extensively studied in recent decades due to their key features such as high torque compare to the volume ratio, no demand on gear, fast dynamics, small electromagnetic interference and high stability [4]. However, they are drawbacks in ultrasonic motor such as short life time, low efficiency and temperature-dependence performance [6]. The temperature rise causes shift in resonant frequency and subsequently results in decreasing the vibration amplitude of the stator. The low vibration amplitude degrades the performance of TWUSM, specifically the speed [7]. This is clearly shows that the temperature variations significantly affect the dynamic characteristics of the motor.

The following sections are organized as follows: the model identification of speed system is described, followed by a brief discussion about performance estimation of PID controller. Next, the PSO method and its implementation into the PSO-PID controller are viewed in detail. Further, the simulation and physical experimental methods are explained. Finally, the discussion of the results followed by conclusion of the research is provided.

2. Model Identification of Speed System of TWUSM

The modelling approach in this study is based on the experimental results. The relation between speed and temperature is observed and studied. The mathematical model is developed by assuming the motor temperature as the only dependable variable to the speed performance[6–8]. Before the system model can be identified, experiments were performed to measure the system outputs. The system has one input which is the command voltage to the driver. The outputs of the system are shaft speed and temperature of TWUSM. Commercial ultrasonic motor Fukoku-Shinsei USR60-S2 with its driver circuit D2060 was used in our experiments. The command voltage to the driver is the only user control input. The speed of TWUSM is proportional to the command voltage value. The driver through its voltage-controlled oscillator in the circuitry adjusts the frequency of the applied sinusoidal drive voltage to the TWUSM. The frequency of the applied voltage determines the speed of TWUSM.

A temperature sensor LM35 was attached to the body of the motor, while an incremental rotary encoder (*HENGSTLER Incremental Shaft Encoder*) was used to measure the shaft rotation speed. The ambient temperature was set to be approximately at 25°C when the experiments were performed. Elvis II together with LabVIEW was used to transmit the speed and temperature readings to a personal computer. For each experiment, a constant command voltage was applied to the driver. The speed and temperature were simultaneously logged. The experiments were done for 900 seconds, which is sufficient for observing the temperature related performance of TWUSM. The experiments were repeated for several command voltage settings ranging from 0.4V to 1.4V.

Curve fitting toolbox in MATLAB software was used to get the model equation by fitting the experiments data. The following speed equation was fit to the obtained data

$$s = (a_0 u - a_0 u e^{a_1 t}) + \left(B(T) - B(T)e^{b_0 t}\right) \tag{1}$$

where s is the motor speed, t is the time, T is the temperature, u is the command voltage and $B(T)$ is a function of temperature which can be described as the following equation

$$B(T) = b_1(T - T_0) - b_2 \tag{2}$$

The temperature T of the motor is described by the following equation

$$T = T_0 + [(c_0 u + c_1) - (c_0 u + c_1)e^{c_2 t}] \tag{3}$$

where T_0 is the initial temperature of the TWUSM. In equations above a_{0-2}, b_{0-2} and c_{0-2} are fitting parameters. The equations were chosen based on minimum root mean square error (RMSE) when compared with experimental data. The value of all fitting parameters is shown in Table 1.

Table 1. Value of all fitting parameters

Fitting Parameter	Value	Fitting Parameter	Value
a_0	10.5	b_2	0.6
a_1	-1000	c_0	4
b_0	-0.04	c_1	5
b_1	-0.1545	c_2	-2.096e-3

The model simulation was performed using the equations above. The simulation results were compared with experimental results. The comparison was done to validate the identified model. Figure 1 shows the comparison of simulated and measured temperature with several different command input voltages. The results from experiments show that the temperature of TWUSM rises with time. The rate of temperature in TWUSM depends on the command voltage applied to the motor's driver. The bigger the command voltage, the higher the temperature rise rate. Figure 2 shows the results for TWUSM speed with different command input voltages. The results clearly show that the speed decreases with time with temperature rise. The results also indicate that an increase in command voltage input results in an increase of the motor's speed. The simulation results in both Figure 1 and Figure 2 clearly shows that the model is comparable with the actual experimental results.

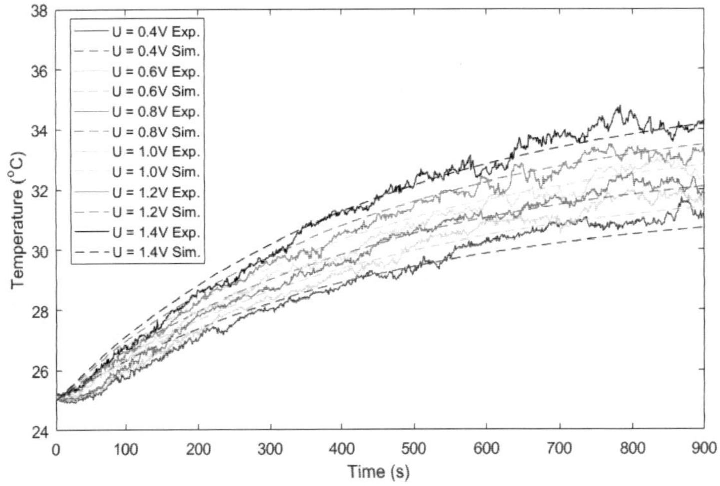

Figure 1. Experimental and simulation results of temperature rise for different command input voltages input.

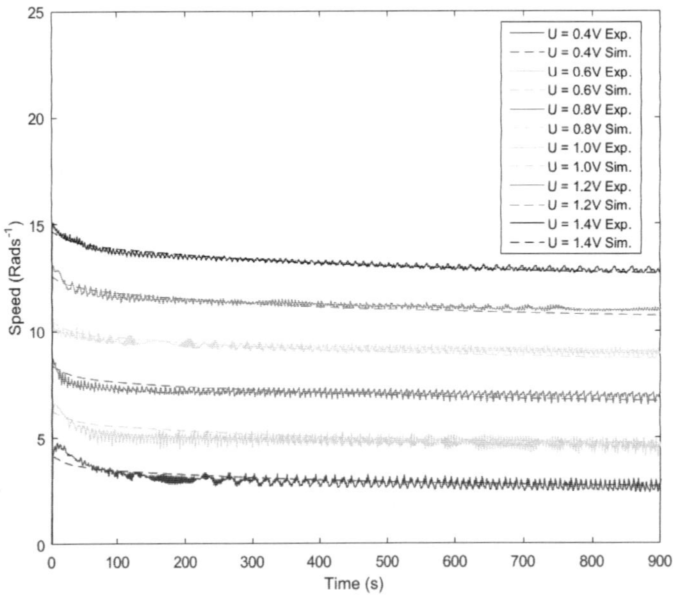

Figure 2. Experimental and simulation results of TWUSM speed for different command input voltages.

3. PSO-PID Algorithm

Gaing [10] proposed the new performance criterion for evaluation PID controller which is denoted as $W(K)$. The criterion $W(K)$ is the combination of several performance criteria. These performance criteria are maximum overshoot M_P, rise time t_r, settling time t_s and steady state error E_{ss}. Therefore, $W(K)$ is defined as

$$W(K) = \left(1 - e^{(-\beta)}\right) \cdot \left(M_p + E_{ss}\right) + e^{(-\beta)} \cdot (t_s - t_r) \tag{4}$$

where K is $\left[k_p, k_i, k_d\right]$ and β is the weighting factor.

The evaluation function f is defined as the reciprocal of $W(K)$. It implies that the smaller the value of $W(K)$ the higher its evaluation value

$$f = \frac{1}{W(K)} \tag{5}$$

By using suitable weighting factor β value in performance criterion W(K) the control performance requirement can be satisfied. Weighting factor β can be set to be larger than 0.7 to reduce overshoot and steady state error. On the other hand, the rise time t_r and settling time t_s can be reduce by setting the value of weighting factor β smaller than 0.7. In this research, the value of β is set in the range of 0.8 to 1.5

In PSO, individuals called as particles represent a potential solution of the problem. Each particle keeps track of its positions as it is treated as flying in g-dimensional search space. The positions are associated with the best solution (evaluating value) it has achieved so far. This value is called *pbest.* The global version of particle swarm optimizer which tracks the overall best evaluating value, and its position obtained by any particle in the group is called *gbest.* The particle size of 50 is used in this research. There are 3 members in each particle to represent three PID parameters.

In PSO concept, each particle changes its velocity toward its *pbest* and *gbest* positions at each time step. As example, the jth particle is represented as $x_j = (x_{j,1}, x_{j,2}, \dots, x_{j,g})$ in g-dimensional space. The best previous position of the jth particle is recorded and represented as $pbest_j = (pbest_{j,1}, pbest_{j,2}, \dots, pbest_{j,g})$. The index of best particle among all particles in the group is represented as $gbest_g$. The rate of the position change (velocity) for j^{th} particle is represented as $v_j = (v_{j,1}, v_{j,2}, \dots, v_{j,g})$. The modified velocity and position of each particle can be calculated using the current velocity and distance from $pbest_{j,g}$ to $gbest_g$ as shown in the following formulas:

$$\begin{aligned} v_{j,g}^{(t+1)} &= w \cdot v_{j,g}^{(t)} + c_1 * rand() * \left(pbest_{j,g} - x_{j,g}^{(t)}\right) \\ &\quad + c_2 * rand() * \left(gbest_g - x_{j,g}^{(t)}\right) \end{aligned} \tag{6}$$

$$x_{j,g}^{(t+1)} = x_{j,g}^{(t)} + v_{j,g}^{(t+1)} \tag{7}$$

$$j = 1,2, \dots, n$$
$$g = 1,2, \dots, m$$

where

n	number of particles in a group;
m	number of members in a particle;
t	pointer of iterations(generations);

$v_{j,g}^{(t)}$	velocity of particle j at iteration t, $v_g^{min} \leq v_{j,g}^{(t)} \leq v_g^{max}$;
w	inertia weight factor;
c_1, c_2	acceleration constant;
$rand()$	random number between 0 and 1;
$x_{j,g}^{(t)}$	current position of particle j at iteration t;
$pbest_{j,g}$	$pbest$ of particle j;
$gbest_g$	$gbest$ of the group

The parameter v_g^{max} determines the resolution, or fitness, with which regions are searched between the present position and target position. If v_g^{max} is too high, particles might fly past good solutions but if v_g^{max} is too low, particles may not explore sufficiently beyond local solutions.

The constant c_1 and c_2 represent the weighting of the stochastic acceleration terms that pull the particles towards $pbest$ and $gbest$. The value of c_1 and c_2 are often set as 2.0 according to experience. The low values of c_1 and c_2 allow particle to fly far from target region before being tugged back while high values result in abrupt movement towards or past target regions [10].

Suitable selection for inertia weight w provides a balance in local and global explorations, thus requiring less iterations on average to find a sufficiently optimal solution. Formula for determining the values of inertia weight w is shown in Equation (8)

$$w = \frac{w_{max} - w_{min}}{iter_{max}} \times iter \tag{8}$$

where $iter_{max}$ is the maximum number of iterations or generations, $iter$ is the current iterations, $w_{max} = 0.9$ *and* $w_{min} = 0.4$

4. Simulation and Experimental Results

The PSO searching algorithms was done in MATLAB and the results for the three PID parameters was applied to PID controller in TWUSM speed system for simulation. The simulation was done using Simulink package in MATLAB. The simulation results with weighting factor $\beta = 1$ and is shown in Table 2.

Table 2. Best solution of PSO algorithms with weighting factor $\beta = 1$

β	1.0
Number of Iteration	50
k_p	0.0519
k_i	0.3678
k_d	0.0181
t_r	0.7792
t_s	1.1618
M_p	0.0646
E_{ss}	0.0053
f	5.4088

Physical experiments were conducted to validate the simulation results. Additionally, to emphasize the advantage of PSO-PID controller, the Ziegler-Nichols-PID control was also implemented is TWUSM speed system. Closed-loop Ziegler-Nichols tuning method was used in determining the parameters for PID control. The experiment results for ZN-PID and PSO-PID are shown in Table 3. The results show that the PSO-PID is more superior method than ZN-PID. The comparison from rise time show that the PSO-PID achieves less time that is 0.5724 seconds compared to ZN-PID that need 0.8991 seconds. Furthermore, the settling time was quicker by using PSO-PID compared to ZN-PID by more than 30%. The results also show that PSO-PID tuning performance is superior than ZN-PID in maximum overshoot and overall evaluation function.

Table 3. Performance comparison between PSO-PID and ZN-PID

Tuning Method	PSO-PID	ZN-PID
k_p	0.0519	0.0480
k_i	0.3678	0.2400
k_d	0.0181	0.0024
t_r	0.5724	0.8991
t_s	0.6671	0.9837
M_p	1.5360	2.5420
E_{ss}	0.1536	0.2542
f	0.9067	0.5660

5. Conclusion

We analytically and experimentally compared the performance of speed control of TWUSM using ZN-PID and PSO-PID tuning methods. The modelling of TWUSM speed system was based on the temperature of the motor. The results of the analyses and experimental work reveal that PID tuning using PSO-based optimization has the advantage over the conventional Ziegler-Nichols method.

Acknowledgement
This work was supported by UMP-RDU grant (RDU095004).

References

[1] Kaya I, Tan N and Atherton D P 2006 A refinement procedure for PID controllers *Electr. Eng.* **88** 215–21
[2] Chang W Der, Hwang R C and Hsieh J G 2003 A multivariable on-line adaptive PID controller using auto-tuning neurons *Eng. Appl. Artif. Intell.* **16** 57–63
[3] Kennedy J and Eberhart R 1995 Particle swarm optimization *Neural Networks, 1995. Proceedings., IEEE Int. Conf.* **4**1942–8 vol.4
[4] Sashida T and Kenjo T 1993 *Introduction to ultrasonic motors* (New York, NY (United States); Oxford Univ. Press)
[5] Mohd Romlay F R, Wan Yusoff W A and Mat Piah K A 2016 Increasing the efficiency of traveling wave ultrasonic motor by modifying the stator geometry *Ultrasonics* **64** 177–85
[6] Zhao C 2010 *Ultrasonic Motor: Application and Theory*
[7] Li S, Ou W, Yang M, Guo C, Lu C and Hu J 2015 Temperature evaluation of traveling-wave ultrasonic motor considering interaction between temperature rise and motor parameters *Ultrasonics* **57** 159–66
[8] Arturo I and Marquez R 2016 *Modeling of Piezoelectric Travelling Wave Rotary Ultrasonic Motors with the Finite Volume Method*
[9] Lu X, Hu J and Zhao C 2011 Analyses of the temperature field of traveling-wave rotary ultrasonic motors. *IEEE Trans. Ultrason. Ferroelectr. Freq. Control* **58** 2708–19
[10] Gaing Z L 2004 A Particle Swarm Optimization Approach for Optimum Design of PID Controller in AVR System *IEEE Trans. Energy Convers.* **19** 384–91

IOP Conf. Series: Materials Science and Engineering **319** (2018) 012026 doi:10.1088/1757-899X/319/1/012026

Six Sigma Approach to Improve Stripping Quality of Automotive Electronics Component – a case study

Noraini Mohd Razali, Siti Murni Mohamad Kadri and Toh Con Ee

Faculty of Manufacturing Engineering, Universiti Malaysia Pahang, Malaysia

norainimr@gmail.com, Sitimurnikadri0@gmail.com, coneetoh@gmail.com

Abstract. Lacking of problem solving skill techniques and cooperation between support groups are the two obstacles that always been faced in actual production line. Inadequate detail analysis and inappropriate technique in solving the problem may cause the repeating issues which may give impact to the organization performance. This study utilizes a well-structured six sigma DMAIC with combination of other problem solving tools to solve product quality problem in manufacturing of automotive electronics component. The study is concentrated at the stripping process, a critical process steps with highest rejection rate that contribute to the scrap and rework performance. The detail analysis is conducted in the analysis phase to identify the actual root cause of the problem. Then several improvement activities are implemented and the results show that the rejection rate due to stripping defect decrease tremendously and the process capability index improved from 0.75 to 1.67. This results prove that the six sigma approach used to tackle the quality problem is substantially effective.

1. Introduction

Six Sigma is an organized methodology that concentrate continuous improvement to minimize the variation and defects of the product. Six Sigma was first developed by Motorola in 1987 and later effectively tested by GE Corp. in 1990. It targeted an invasive goal of 3.4 parts per million (ppm) defects, means that the process is near to "zero defects" or 99.9997% perfect [1]. In other words Six Sigma is a data driven, disciplined approach and methodology. It is based on eliminating the defects in any process. Lo et.al [2] define Six Sigma approach as a method that follow the theories and steps to determine the shifting of standard deviation in between the mean and specification limit and then reduce the defects quantity for a specific process. It is an effective method to determine which part of the process is the most important and determine the weakness of the process [3]. The method offer computable indicators and sufficient data for analysis and became the most outstanding management method for quality improvement [2].

Several studies have been done by the previous researchers on Six Sigma approach applied to manufacturing industries. Valles et al. [4] for instance conducted six sigma project at a semiconductor company, which eliminate 1.88% of nonconforming units. They claimed that the organization management was very supportive and encouraging with the project team and conclude that Six Sigma implementation can be helpful in reducing the nonconforming units or improving the organization quality and personal development. The improvement was a reduction in the electrical failures of around 50%. Gijoa et al. [5] explored how a manufacturing process can use a systematic methodology to move towards world-class quality level. The application of the Six Sigma methodology resulted in reduction of defects in the fine grinding process from 16.6 to 1.19%. The DMAIC methodology had a significant financial impact on the profitability of the company in terms of reduction in scrap cost, man-hour saving on rework and increased output. A saving of approximately US$2.4 million per annum was reported from their project. Jirasukprasert et al. [6] demonstrated the empirical application of Six Sigma and DMAIC to reduce product defects within a rubber gloves manufacturing organization. The study followed the DMAIC methodology to investigate defects, root causes and provide a solution to reduce these defects. A reduction of about 50% in the "leaking" gloves defect was achieved, which helped the organization studied to reduce its defects per million opportunities (DPMO) from 195,095 to 83,750 and thus improve its Sigma level from 2.4 to 2.9. On the other hand Rohit Chandel and Santosh Kumar [7] focused on analysing and elimination of in-process variations causing rejection and rework at different stages of production by using DMAIC approach. By implementing DMAIC technique the rejection rate and rework rate was reduced from 8.79 % to 5.30 % and 12.8 % to 8.2 %, respectively. Also, a significant enhancement in sigma level was obtained from 2.85 to 3.13.

2. Background of the study

This paper presents a case study conducted at one of electronics manufacturing company located in the east coast of Malaysia. The company produce various electronics component for automotive application such as inductor, common mode chock, transformer and others. The manufacturing of components consist of many process steps. The main focus of this study is to produce a good quality at manual mechanical stripping process. The purpose of this stripping process is to remove insulation around wire. Insulation must be removed properly and must not damage the wire to ensure the better connection of wire with terminal. The stripping of the insulation from copper wire is similar to a milling process, but the stripping of insulation is done by rotational tools. The materials needed to undergo stripping process are stripping machine, rotating tool with appropriate wire size and copper wire. The manual stripping process requires experience operators to proceed the operation. All the operators have to be trained before enter the operation in production. The rotating bullet is initially covers with the wires until the whole length of wire is stripped inside the rotating bullet. The operator have to make sure that the wire is always straight when feeding into the bullet, and hitting the centre to feed wire tip to the bullet. At this time, the stripping process created a significant force and the wire is pushed into the bullet. The rest of the wire is usually stay at the end of the stripping part. For the thinner or longer wire, it is required to maintain at the middle of the stripping wire to prevent the deformation of the inserting force.

Based from the process performance history, defects in stripping process accounted for 58.2% of the whole manufacturing process. The defects include bent wire, twist wire, line mark on wire, broken wire, reduced diameter on wire, improper strip and corkscrew on wire. Because of manual operation of stripping process, the defects occur might be due to skills of the operators, tools problem or improper cleaning method at the stripping machine. Figure 1 shows the acceptable wire where it should be as straight as possible and does not peel off, while Figure 2 shows examples of rejected wire in which the wire is ripped off or turned off.

Figure 1. Acceptable wire

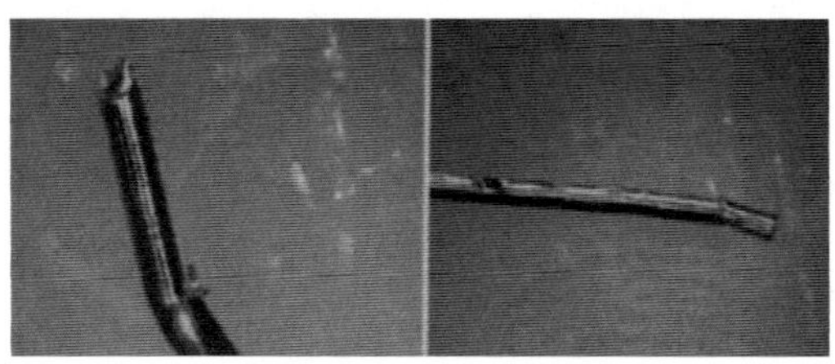

Figure 2. Rejected wire

3. Methodology

The improvement of the stripping process is conducted according to Six Sigma DMAIC steps. DMAIC provides a schedule for project members to identify the problems, solve the problems and make improvement to the process. The objective of the DMAIC method is to ensure the project team follow the steps given to prevent the project straight to the conclusions without searching for more adequate alternative solutions to the related problem. In every phase of the method, several quality tools are used to obtain particular result for each phase [8]. This section discussed in details about the method used to reduce the defects in stripping process of electronic components. It include five steps which is Define, Measure, Analyse, Improve and Control.

3.1. Define Phase

In define phase, the project team start with problem identification. Understanding on the present situation is the important factor for setting the appropriate improvement target. In addition, boundary of the project and team members' responsibility need to be defined clearly. Any related information or data collection such as customer complaint, machine downtime, scrap and rework or process inefficiency can be used to identify the problem statement. There are several common quality tools can be used in this Define Phase such as:

- SIPOC
- Pareto chart
- Process mapping
- CTQ characteristics
- 5W and 1H

In this study, Pareto chart as depicted in Figure 3 was used to prioritize and to identify the major contribution to the problem. By applying Pareto principle also known as the 80/20 rule, it will help to show the most important problem to be solved.

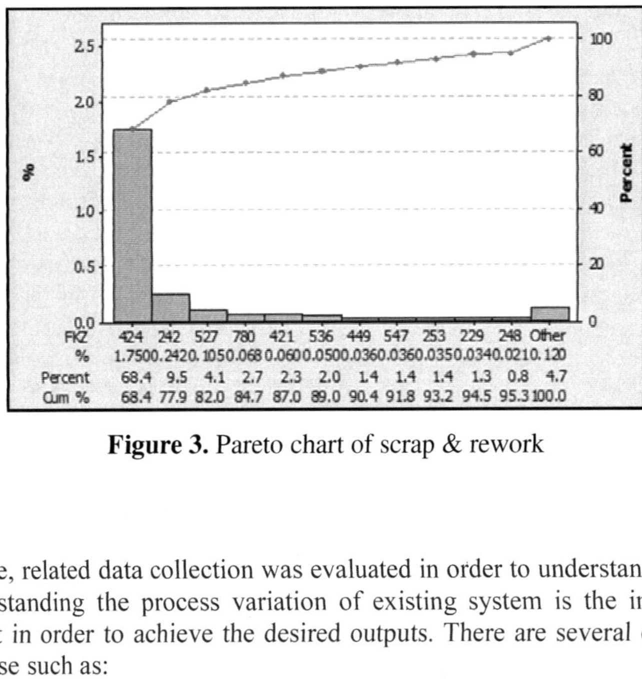

Figure 3. Pareto chart of scrap & rework

3.2. Measure Phase

In the Measure Phase, related data collection was evaluated in order to understand the existing process performance. Understanding the process variation of existing system is the important step for the improvement project in order to achieve the desired outputs. There are several common tools can be used in Measure phase such as:

- Failure analysis
- Process capability study
- Measurement system analysis
- Control chart
- Process analysis

This step will be done by first mapping the overall process and all related parameters. In order to evaluate the present process capability a set of data collection must be done accordingly. The performance of measurement system analysis need to be evaluated to ensure the data is valid and accurate. The important step for improvement project of stripping process is evaluation on present situation. Result from this

IOP Conf. Series: Materials Science and Engineering **319** (2018) 012026 doi:10.1088/1757-899X/319/1/012026

step will be a guideline to identify process variation and to support the understanding of existing system. This will be an important input to the project team to setup a goal of improvement to achieve the desired outputs. Figure 4a shows measurement system analysis while Figure 4b shows a cause and effect diagram for mechanical defect.

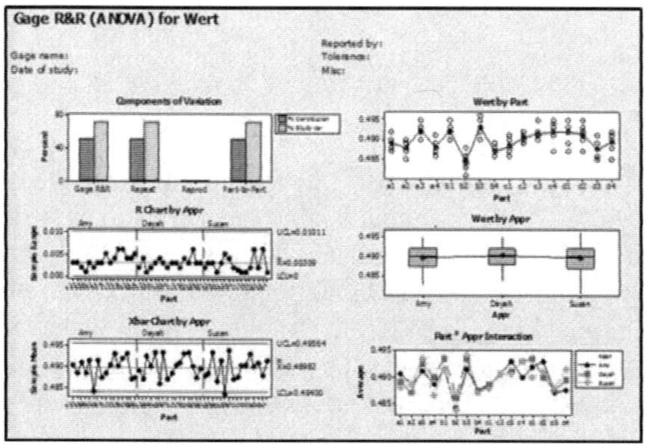

Figure 4a. Measurement system analysis study

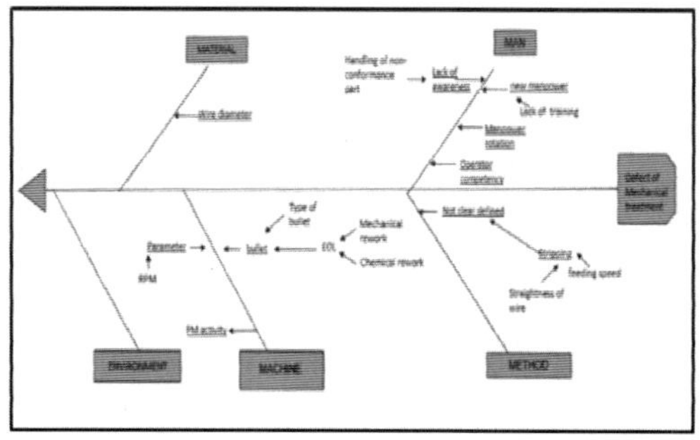

Figure 4b. Cause & effect diagram

Based from the measurement system analysis study it can be conclude that the existing measurement system for checking wire diameter after stripping process is valid and accurate, in which the variability from repeatability and reproducibility is very small. Brainstorming with involvement of multidisciplinary unit has been done to collect all potential root causes. The cause and effect diagram was used in order to identify the real root causes of the defect by mechanical treatment during stripping process. All four elements which is man, machine, material and method were evaluated comprehensively by the team.

3.3. Analysis Phase

In this phase all related analysis or experiment is conducted in order to validate the potential root causes. Result of the analysis will be used to identify the actual root cause of the problem statement. Besides that, this phase can also be used to determine the best setting parameters on the required improvement plan where it is able to determine which inputs parameters are the most important factor for the overall performance. There are several common tools can be used in the Analysis phase:

- Hypothesis Testing.
- Multi-Variance Analysis.
- Cycle-Time Analysis.
- Regression Analysis
- Brainstorming
- Analysis of variance (ANOVA)

In Analysis phase, the current existing system has been analysed to understand the process performance. Data for wire diameter measurement after stripping process were collected and analysed using Minitab Software. The focus in this phase is on the root cause that can be controlled with a reasonable effort. The test results show that a very short insertion speed increases the risk of poor stripping results. It can be conclude that setting the insertion speed slightly slower than the maximum possible speed will give a balance production quality. As a conclusion, bent wire with a diameter greater than 4 mm cannot be removed by manual straightening in a way that they will not influence the stripping process. The straightening of bent wire with diameter greater than 4 mm has a positive effect in stripping process. Therefore, the target is to keep the wires as straight as possible during production process.

IOP Conf. Series: Materials Science and Engineering **319** (2018) 012026 doi:10.1088/1757-899X/319/1/012026

3.4. Improve Phase

Implementation of corrective action is done by selecting the actual root cause of the problem. The validation is done through design of experiment (DOE). After the solution has been validated, the critical factor will be controlled in a way to ensure the effectiveness of corrective action. There are several common tools can be used in this Improve phase:

- Taguchi Method.
- Simulation Method.
- Design of Experiments (DOE.

As an improvement strategy, the team established a prototype for a new stripping machine. A prototype of new stripping machine has been designed with function that is able to strip the bent wire. The machine also has a platform to support the handling of unit during feeding the wire lead-out into bullet stripping. Factors such as insertion angle and feeding speed can be eliminated by this element. Stripping bullet is the main element on this stripping machine. In order, to produce a high quality of mechanical treatment by stripping process a special stripping bullet with rotary spindle blade to grip and remove wire insulation is implemented. The current machine requires the operator to put the jigs on the stripping platform and insert the wire to the stripping point where as for a new design machine, the operator does not need to push the wire to the bullet during wire insertion, the rotation bullet will gently pull the wire inside. This can avoid the wire being twisted and stuck in the bullet of the machine. The advantage of the new machine is that no force needed to push the wire into the stripping blade, thus eliminate 'reduced diameter' and 'improper strip' defects that caused by excess materials and wear blades. Figure 5 shows the design of a new and old stripping machine.

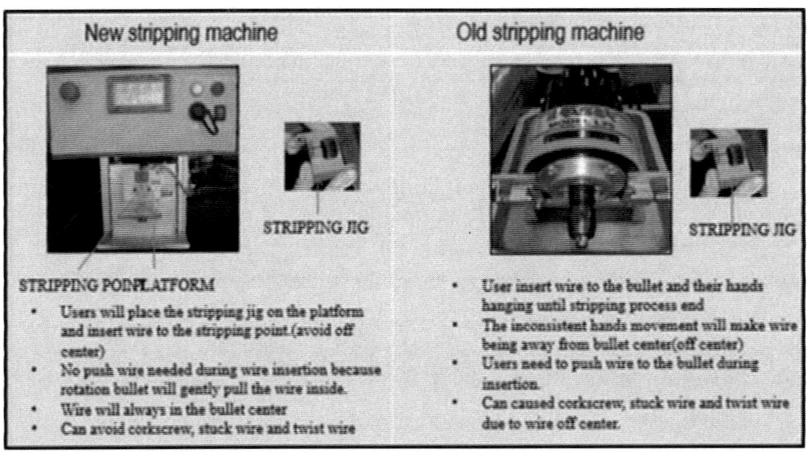

Figure 5. Comparison between new stripping machine and old stripping machine

Validation process is conducted to ensure the best parameter setting for a new machine. In this validation process the 300 pieces sample was run at the new machine using 500 rpm and 3.0 feeding speed. Figure 6a shows the process capability study of wire diameter using current machine, which is before improvement made while Figure 6b shows the process capability study of wire diameter using a new machine. It can be seen that the variability of the process has reduced and the Cpk value has increased from 0.75 to 1.67. Thus, the process with new machine is more capable than the old machine.

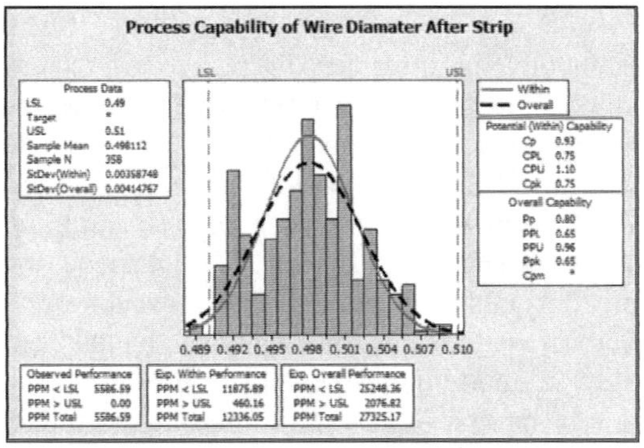

Figure 6a. Process capability study of wire diameter before improvement

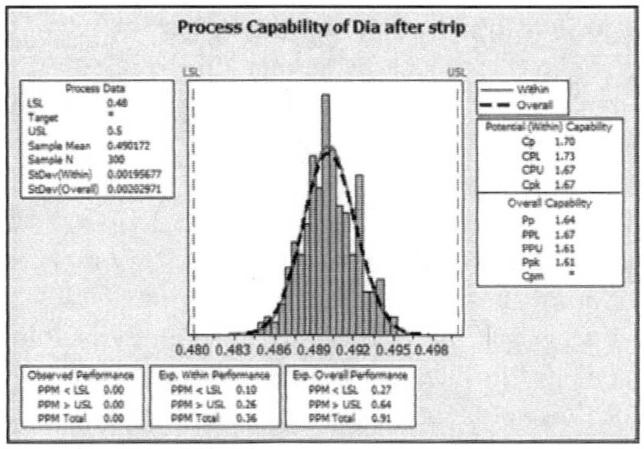

Figure 6b. Process capability study of wire diameter after improvement

3.5. Control Phase

The main objective of this Control phase is to ensure the important aspect is practiced in place. All the improvement plans with introduction of the new setup for man, machine, method and material become part of the normal task and must be maintained all the time. There are several common tools that can be used in Control phase such as:

- Mistake proofing.
- Lean manufacturing.
- Work standardization.
- Preventive maintenance.
- Statistical Process Control (SPC)

In this phase performance of new stripping machine need to be reviewed and monitored. The purpose of monitoring is to understand the positive and negative effect of the improvement project to the manufacturing system. Observation of process performance after implementation of improvement is under process owner responsibility. The data collection was plotted into control chart. In order to maintain the performance of the process, the process should be supervised by control charts over time so that the wire diameter variation can be controlled and meets customer requirements. Figure 7 shows the Xbar and R chart of wire diameter after improvement. Based on both charts, it can be seen that the process are falling within the specification limits. It can be conclude that the process is stable and in control. All the points are within the control limit, and the variation has been reduced. Finally, this project is documented, continuously update and kept into the company directory.

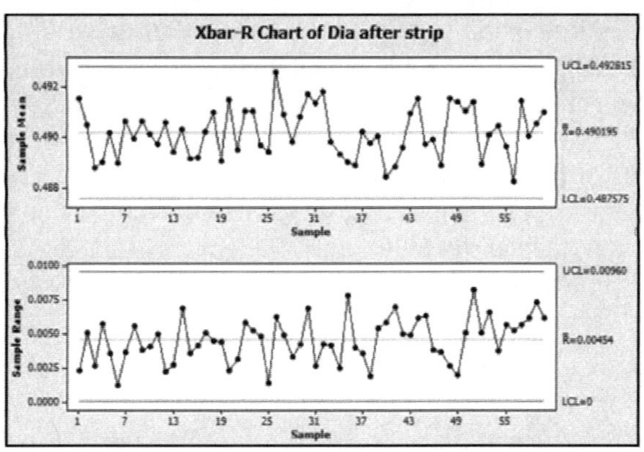

Figure 7. Control chart for wire diameter after stripping process

4. Conclusion and Recommendation

Six sigma approach is a method which aim for continuous improvement to reduce defects in manufacturing process. In this study, implementation of Six Sigma approach with DMAIC steps is carried out to improve the quality of the product and the performance of stripping process. The critical process parameter for this stripping process has been determined. The team agreed to select the wire diameter as a critical control parameters. After implementing the new stripping machine to the stripping process, the process capability index for wire diameter is improved from 0.95 to 1.67. Implementation of control chart to illustrate the sampling result from in-process quality control gives good result for real time feedback. The project involved multidisciplinary approach and the ideas are generated by the team members from various section. There are several improvement opportunities were identified. However selection on the best idea was based on the time limitation of these project. At the end of this project, the team had gain knowledge and better understanding on problem solving technique. As a moving forward, all project team members will be a team leader to establish and setup their own project team to solve another quality problem in future. After implementation of this project, other improvement opportunities can be considered for better performance. This include optimization on winding until cutting process, and optimization of the new stripping machine with fixture to automatically measure the wire diameter after stripping process. All this potential ideas could contribute to better process capability performance.

Acknowledgments
This research is fully funded by the Faculty of Manufacturing Engineering, Universiti Malaysia Pahang under the Grant number RDU170390.

References

[1] Schroeder, R. G., Linderman, K., Liedtke, C., & Choo, A. S. Six Sigma: Definition and underlying theory. Journal of Operations Management, 26(4), 536–554, 2008
[2] Lo, W. C., Tsai, K. M., & Hsieh, C. Y., Six Sigma approach to improve surface precision of optical lenses in the injection-molding process. International Journal of Advanced Manufacturing Technology, 41(9–10), 885–896, 2009
[3] Sokovic, M., Six Sigma process improvements in automotive parts production. Journal of Acheivements in Materials and Manufacturing Engineering, 19(1), 96–102, 2006
[4] Adan Valles, Jaime Sanchez, Salvador Noriega, and Berenice Gómez Nuñez, Implementation of Six Sigma in a Manufacturing Process: A Case Study, International Journal of Industrial Engineering, 16(3), 171-181, 2009
[5] E. V.Gijoa, Johny Scaria and Jiju Antony, Application of Six Sigma Methodology to Reduce Defects of a Grinding Process, Quality and Reliability Engineering International, John Wiley & Sons, Ltd., 2011
[6] Ploytip Jirasukprasert, Jose Arturo Garza-Reyes, Horacio Soriano-Meier, Luis Rocha-Lona, A Case Study of Defects Reduction in a Rubber Gloves Manufacturing Process by Applying Six Sigma Principles and DMAIC Problem Solving Methodology, Proceedings of the 2012 International Conference on Industrial Engineering and Operations Management Istanbul, Turkey, July 3 – 6, 2012
[7] Rohit Chandel, Santosh Kumar, Productivity Enhancement Using DMAIC Approach: A Case Study, International Journal of Enhanced Research in Science, Technology & Engineering ISSN: 2319-7463, Vol. 5 Issue 1, January-2016
[8] Johannsen, F., Leist, S., Johannsen, F., & Leist, S., A Six Sigma approach for integrated solutions, 2009

IOP Conf. Series: Materials Science and Engineering **319** (2018) 012027 doi:10.1088/1757-899X/319/1/012027

The identification of high potential archers based on relative psychological coping skills variables: A Support Vector Machine approach

Zahari Taha[1], Rabiu Muazu Musa[1,2], A P P Abdul Majeed[1], Mohamad Razali Abdullah[2], Muhammad Aizzat Zakaria[1], Muhammad Muaz Alim[1], Jessnor Arif Mat Jizat[1] and Mohamad Fauzi Ibrahim[3]

[1] Innovative Manufacturing, Mechatronics and Sports Laboratory, Faculty of Manufacturing Engineering, Universiti Malaysia Pahang, 26600 Pekan, Pahang, Malaysia
[2] Faculty of Applied Social Sciences, Universiti Sultan Zainal Abidin, 21300 Kuala Terengganu, Terengganu, Malaysia
[3] Research and Education Division, National Sports Institute, Kompleks Sukan Negara Bukit Jalil, 57700 Kuala Lumpur, Malaysia

anwarmajeed1983@gmail.com

Abstract. Support Vector Machine (SVM) has been revealed to be a powerful learning algorithm for classification and prediction. However, the use of SVM for prediction and classification in sport is at its inception. The present study classified and predicted high and low potential archers from a collection of psychological coping skills variables trained on different SVMs. 50 youth archers with the average age and standard deviation of (17.0 ±.056) gathered from various archery programmes completed a one end shooting score test. Psychological coping skills inventory which evaluates the archers level of related coping skills were filled out by the archers prior to their shooting tests. k-means cluster analysis was applied to cluster the archers based on their scores on variables assessed. SVM models, i.e. linear and fine radial basis function (RBF) kernel functions, were trained on the psychological variables. The k-means clustered the archers into high psychologically prepared archers (HPPA) and low psychologically prepared archers (LPPA), respectively. It was demonstrated that the linear SVM exhibited good accuracy and precision throughout the exercise with an accuracy of 92% and considerably fewer error rate for the prediction of the HPPA and the LPPA as compared to the fine RBF SVM. The findings of this investigation can be valuable to coaches and sports managers to recognise high potential athletes from the selected psychological coping skills variables examined which would consequently save time and energy during talent identification and development programme.

1. Introduction

Various researchers have endeavoured to provide insights on the impacts of psychological skills to the advancement of athlete's performance in different sports. Abdullah et al. studied the roles of psychological factors on the performance of elite soccer players and established that although, psychological factors alone could not forecast the performance of elite soccer players, but play an essential part in improving their performance [1]. Likewise, Lazarus [2] and Salim et al. [3] inferred that psychological elements such as stress and worry could be a significant constituent in hindering athletes performance. They further emphasised that athletes who are unable to cope with stress and adversity effectively might likely suffer from performance declining and overall psychological well-being issues. Furthermore, Pasaei et al. [4] stated that in the contemporary archery competitions there are presents of various noises, tensions, and anxiety, and as such selective attention through the application of the psychological coping skills during the competition become paramount for better performance. This implies that the athletes with higher psychological coping skills are likely to outperform their opponents with lesser coping skills.

The employment of machine learning or artificial intelligence has gained popularity in prediction and classification in sports and exercise due to its superiority over conventional means. Bayesian network (BN) analysis was utilised to investigate a semi-professional team's performance and collective efficacy based on a number of psychological features [5]. In a different study, Fuster-Parra et al. [6] also employed BN in correlating negative psychological features with sportive behaviours of young team players in competitive sports.

It is evident from the literature that the employment of machine learning could be a useful tool for prediction as well as classification. Support vector machine (SVM) is a rather popular supervised learning algorithm amongst machine learning methods that have been employed by researchers in a variety of fields for both classification and regression analysis [7–10]. The underlying principle behind SVM is inherently statistical learning theory that was initially proposed by Cortes and Vapnik [11]. The present study aims at correlating the selected psychological coping skill strategies viz. coping with adversity, coachability, peaking under pressure, concentration, freedom from worry, confidence and achievement motivation, goal setting and mental preparation

2. Material and methods

2.1. Participants

A complete number of 50 archers were selected to take part in in this study. The participants consisted of 37 male and 13 female youth archers between the age's range of 13-20 with a mean and standard deviation of (17.0 ± 0.56) drawn from different archery programmes in Malaysia. The archers were under a development program for preparing both at university and the state level and consequently targeted to be preferred to state and national archers respectively. Written consent was obtained, and all the archers signed consent forms. All the procedures, protocol, and equipment for this study were authorised by the Research Ethics Board of the Terengganu Sports Institute (ISNT) with a memo number 04-04/T-01/Jid 2.

2.2. Psychological coping skills inventory assessment

The Athletic Coping Skills Inventory (ACSI), an instrument for assessing athlete's psychological skills, produced by Smith et al. [12] was selected. This instrument is considered apt for evaluating the psychological skills of the archers in this study because of its connection to the nature of archery game as an individual sport rather than a team sport. The instrument was distributed to the archers prior to the shooting test, and their responses were compiled and analysed. A simulated shooting competition area was set up, and all the archers' shoot six arrows (one end) over a distance of 50 meters. All the archers were given trials of four arrows shot before recording the final six arrows scores.

2.3. Data analysis

2.3.1. Clustering: k-means cluster analysis

In the current study, the *k*-means clustering algorithm, which is essentially a type of unsupervised learning, is used to separate the classes of the related performance variables assessed via Orange 2.7. It has been reported in the literature that the k-means clustering method is more reliable than the hierarchical agglomerative clustering in catering large datasets [13]. This is primarily due to the fact that it operates on actual observations rather than the dissimilarity measures employed in hierarchical clustering. The number of clusters, *k* is selected to be two as the data is consists of HPPA and LPPA.

IOP Conf. Series: Materials Science and Engineering **319** (2018) 012027 doi:10.1088/1757-899X/319/1/012027

2.3.2. Classification: Support vector machines (SVMs)
As for classification problems, the SVM is utilised to acquire the optimal hyperplane that correctly classifies or divides the data into two distinct class which in this study, these categories are represented by HPPA and LPPA. Nonetheless, it is worth to note that SVM may also be extended to cater multiclass problems [14]. The acquisition of the aforementioned optimal hyperplane is through the identification of the maximal distant from the classes that in turn minimises the risk of misclassification of both the training and validation data set. For a more in-depth mathematical treatment of the subject matter, the readers are referred to [11]. In this study, two different kernel functions are investigated namely, linear and fine RBF. The scale of the fine RBF is defined by $\frac{\sqrt{P}}{4}$, where P is the number of predictors, i.e. seven. The solver used for the training is the sequential minimal optimisation algorithm.

2.3.3. Model training and testing
A fivefold cross-validation method was utilised in this study. This form of validation technique is desirable as it mitigates the notion of overfitting through partitioning the dataset into a number of folds and estimating the accuracy of each fold. The data (50 observations) is randomly split into five subsets, and for each iteration, one of the five subsets are used as the testing data, whilst the remaining four will be used as the training data. Then, the average performance over all the folds is then computed. The SVM analysis and assessment were conducted by means of MATLAB 2016a (Mathworks Inc., Natick, USA).

2.3.4. Model evaluation
The variations of the SVM employed in this study are evaluated by means of classification accuracy (ACC), sensitivity (SENS), specificity (SPEC), precision (PREC), error rate (ERR) as well as Matthew's correlation coefficient (MCC). The confusion matrix allows the observation of correctly classified and misclassified observations that transpires between the defined classes.

3. Results

3.1. Clustering

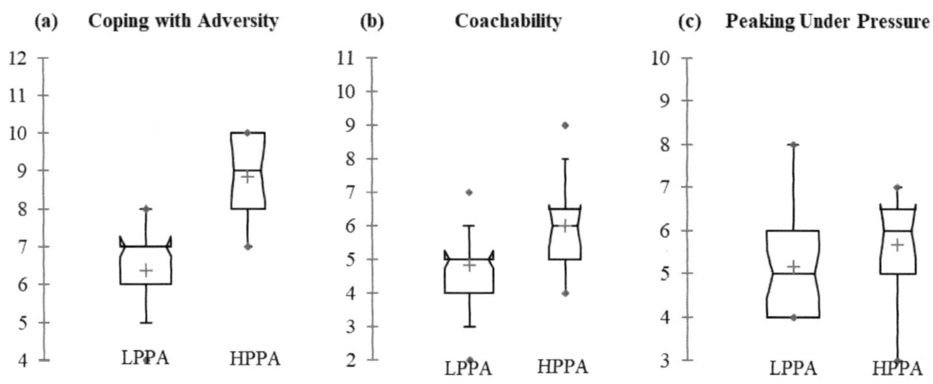

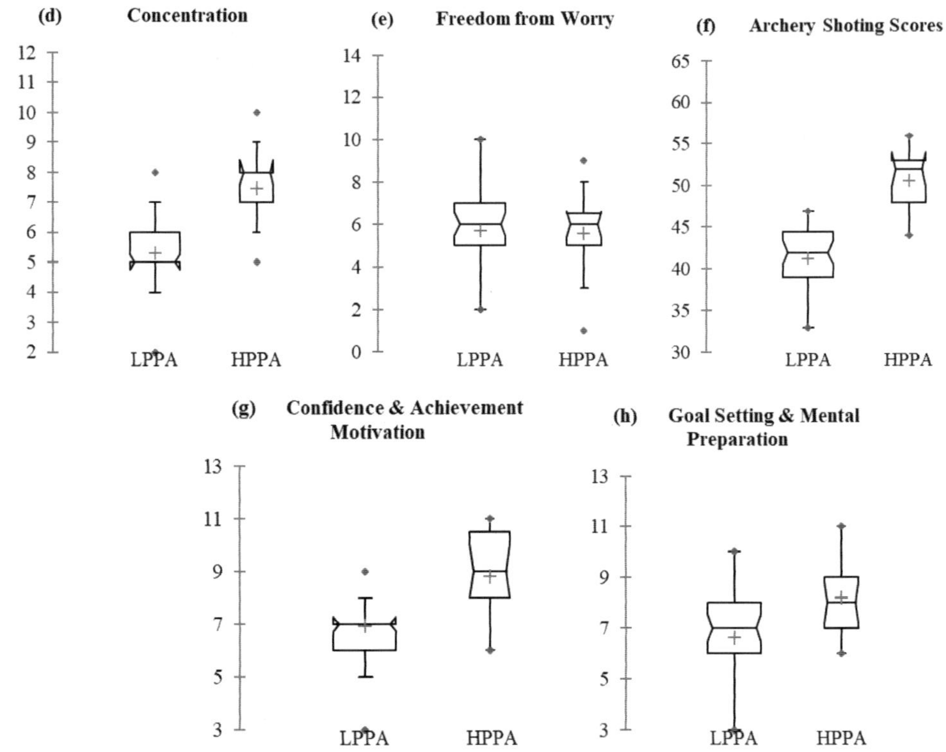

Figure 1. Comparisons of performance differences of the archers based on seven variables evaluated based on ACSI (a) Coping with Adversity; (b) Coachability; (c) Peaking Under Pressure; (d) Concentration; (e) Freedom from Worry; (f) Archery shooting score; (g) Confidence and Achievement Motivation; (h) Goal Setting and Mental Preparation

Figure 1 displays the performance differences of the archers based on the eight performance variables assessed that were clustered via *k*-means. It can be observed from the box plots that the mean performances of HPPA are greater than LPPA across all the seven psychological coping skills measured in the study with the exception of 'freedom from worry' (Figure 1(e)). This observation is rather natural as those who are more worried tend to have less performance. These variables are, therefore, considered as essential attributes that distinguish HPPA from LPPA.

3.2. Classification

Table 1. Model evaluation

Kernel Functions	ACC (%)	SENS (%)	SPEC (%)	PREC (%)	ERR (%)	MCC
Linear	92	100	73.33	89.74	8	0.8112
Fine RBF	70	100	0	70	30	-

It could be observed from the tabulated results that the linear kernel function based SVM model is able to produce exceptional classification through the evaluation of all assessment parameters as compared to the fine RBF. It is apparent that the linear-based model could provide a reasonably accurate classification with an error rate of 8 %. Conversely, the fine RBF-SVM model is rather unsuitable for predicting the correct classification of HPPA and LPPA, as it yields the highest error rate at 30% on top of a poor negative class accuracy demonstrated via the SPEC evaluation that in turn produces an uncorrelated model. Through the present study, it is evident that the linear-based kernel functions

SVM model may be used for the purpose of talent identification for archers based on the predefined psychological coping skills. The confusion matrix s of the evaluated SVMs are depicted in Figure 2.

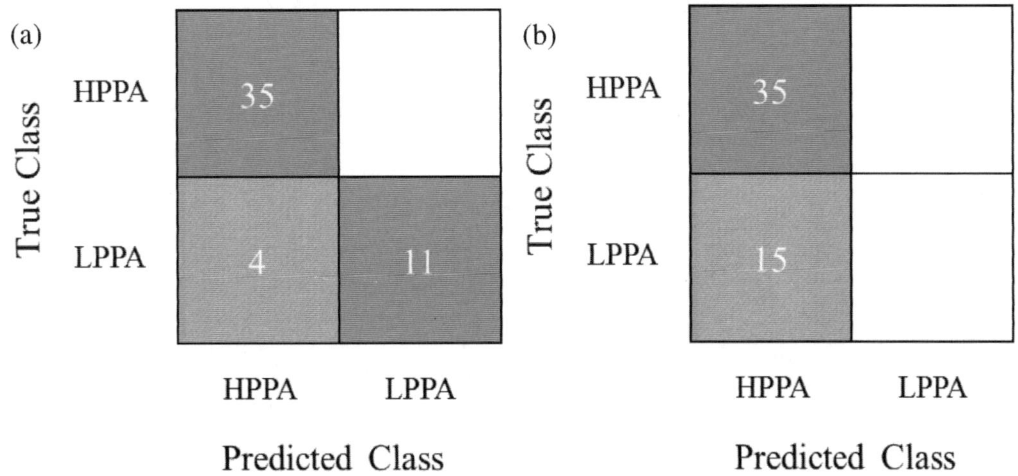

Figure 2. Confusion matrix. (a) Linear SVM; (b) Fine RBF SVM;

4. Discussion

The finding of the present study has indicated that the selected psychological coping skills variables established are able to predict or classify well the performance of the archers, i.e. HPPA and LPPA through the shooting score of the respective archers. The shooting scores enable us to cluster the athletes with respect to measured psychological coping skills via k -means as illustrated in the box plot shown in Figure 1. It was demonstrated that the linear-based kernel function SVM is able to classify the LPPA and HPPA reasonably well.

The finding underscored the necessity for psychological skills such as reliance and self-motivational strategies in the sport. Archery is closed skill as well as an individual competitive sport that calls for the archers to compete against one another. For an archer to obtain a higher score, he or she must be capable of maintaining confidence and be self-motivated to compete. An evolving body of evidence for sports indicates that one's awareness of competence or self-confidence is the crucial factor leading to the success of any chosen sport [15]. Moreover, it has been described that confidence and self-motivation are the symbols of a winner. They are the secret ingredients that all outstanding athletes seem to acquire, irrespective of what level they contest since constructive mental attitude keeps an athlete working hard regardless of how often he/she may fail or how many obstacles get tossed in his/her direction [16]. Confidence and self-motivation can provide a normal athlete or team the determination and focus on defeating a stronger opponent. These psychological skills can inspire an athlete to undertake and achieve the unthinkable. Similarly, lacking confidence and self-motivation could make athletes continually perform way below their potential. Low confidence and self-motivation can wipe out an athlete's indulgence of the sport and turn him/her into a dropout and a consistent loser [1]. This study is somewhat in its preliminary nature, and owing to its limited sample size, the results could not be generalised to other level of archery participation as well as other demography.

5. Conclusion

The current study has successfully demonstrated that some psychological coping skills variables, i.e. coping with adversity, coachability, peaking under pressure, concentration, freedom from worry, confidence and achievement motivation, goal setting and mental preparation does influence the determination the performance class of the archers. The study also has revealed the use of such machine learning methods is nontrivial as it allows coaches to correctly identify high potential athletes in the sport of archery by considering certain psychological coping skills elements. The current findings could be useful to coaches and sports managers in discovering talents which would eventually save cost and energy during talent identification and development programmes. Future study should consider other related performance variables associated with the sport as well as other non-conventional classification methods.

Acknowledgement
This work is funded by the National Sports Institute of Malaysia (ISNRG: 8/2014-12/2014).

References
[1] Abdullah M R, Musa R M, Maliki A B H M B, Kosni N A and Suppiah P K 2016 Role of psychological factors on the performance of elite soccer players *J. Phys. Educ. Sport* **16** 170
[2] Lazarus R S 2000 How emotions influence performance in competitive sports *Sport Psychol.* **14** 229–52
[3] Smith R E and Christensen D S 1995 Psychological skills as predictors of performance and survival in professional baseball *J. Sport Exerc. Psychol.* **17** 399–415
[4] Pasaei N, Abdoli B, Kazem S, Mousavi V and Aslankhani M 2013 Does Predict Personality Introversion-Extraversion Dimension the Archer's Performance? *Middle-East J. Sci. Res.* **13** 760–3
[5] Fuster-Parra P, García-Mas A, Ponseti F J and Leo F M 2015 Team performance and collective efficacy in the dynamic psychology of competitive team: A Bayesian network analysis *Hum. Mov. Sci.* **40** 98–118
[6] Fuster-Parra P, García-Mas A, Ponseti F J, Palou P and Cruz J 2014 A Bayesian network to discover relationships between negative features in sport: a case study of teen players *Qual. Quant.* **48** 1473–91
[7] Azamathulla H M and Wu F-C 2011 Support vector machine approach for longitudinal dispersion coefficients in natural streams *Appl. Soft Comput. J.* **11**
[8] De Yong D, Bhowmik S and Magnago F 2015 An effective Power Quality classifier using Wavelet Transform and Support Vector Machines *Expert Syst. Appl.* **42** 6075–81
[9] Guyon I, Weston J, Barnhill S and Vapnik V 2002 Gene Selection for Cancer Classification using Support Vector Machines *Mach. Learn.* **46** 389–422
[10] Khedher L, Ramírez J, Górriz J M, Brahim A and Segovia F 2015 Early diagnosis of Alzheimer's disease based on partial least squares, principal component analysis and support vector machine using segmented MRI images *Neurocomputing* **151** 139–50
[11] Cortes C and Vapnik V 1995 Support-vector networks *Mach. Learn.* **20** 273–97
[12] Smith R E, Schutz R W, Smoll F L and Ptacek J T 1995 Development and validation of a multidimensional measure of sport-specific psychological skills: The Athletic Coping Skills Inventory-28 *J. Sport Exerc. Psychol.* **17** 379–98
[13] Collins B M, Lydersen J M, Fry D L, Wilkin K, Moody T and Stephens S L 2016 Variability in vegetation and surface fuels across mixed-conifer-dominated landscapes with over 40 years of natural fire *For. Ecol. Manage.* **381** 74–83
[14] Brereton R G and Lloyd G R 2010 Support Vector Machines for classification and regression *Analyst* **135**
[15] Bird A M and Brame J M 1978 Self versus team attributions: A test of the "I'm OK, but the team's so-so" phenomenon *Res. Quarterly. Am. Alliance Heal. Phys. Educ. Recreat.* **49** 260–8
[16] Hardy L, Jones J G and Gould D 1996 *Understanding psychological preparation for sport: Theory and practice of elite performers.* (John Wiley & Sons Inc)

The Identification of Hunger Behaviour of *Lates Calcarifer* through the Integration of Image Processing Technique and Support Vector Machine

Z Taha[1], M A M Razman[1], F A Adnan[1], A S Abdul Ghani[1], A P P Abdul Majeed[1], R M Musa[1], M F Sallehudin[2], Y. Mukai[2]

[1]Innovative Manufacturing, Mechatronics and Sports Lab (iMAMS), Faculty of Manufacturing Engineering, Universiti Malaysia Pahang, Pekan Campus, 26600 Pekan, Pahang, Malaysia.
[2]Kulliyyah of Science, International Islamic University Malaysia, Jalan Sultan Ahmad Shah, Bandar Indera Mahkota, 25200, Kuantan, Pahang, Malaysia.
E-mail: azraai@ump.edu.my

Abstract. Fish Hunger behaviour is one of the important element in determining the fish feeding routine, especially for farmed fishes. Inaccurate feeding routines (under-feeding or over-feeding) lead the fishes to die and thus, reduces the total production of fishes. The excessive food which is not eaten by fish will be dissolved in the water and thus, reduce the water quality (oxygen quantity in the water will be reduced). The reduction of oxygen (water quality) leads the fish to die and in some cases, may lead to fish diseases. This study correlates Barramundi fish-school behaviour with hunger condition through the hybrid data integration of image processing technique. The behaviour is clustered with respect to the position of the centre of gravity of the school of fish prior feeding, during feeding and after feeding. The clustered fish behaviour is then classified by means of a machine learning technique namely Support vector machine (SVM). It has been shown from the study that the Fine Gaussian variation of SVM is able to provide a reasonably accurate classification of fish feeding behaviour with a classification accuracy of 79.7%. The proposed integration technique may increase the usefulness of the captured data and thus better differentiates the various behaviour of farmed fishes.

Keywords: Fish Feeding Behaviour; Support Vector Machine; *Lates Clacarifer*

1. Introduction

There are several factors that one must consider when studying characteristics of a fish, for instance, the behaviour, locomotor activity and sensory cues. Feeding anticipation has been known to increase the locomotor activity either for individual or in a group of fish [1]. The movement and marginalise between the fishes and their ecology will affect the hunger demand as this issues still in an exploring phase amongst the researchers. A study was done to investigate the behaviour between less fed and satiated individuals to measure the cover area scavenging actions within a confined space and demonstrated that individuals in a hungry state display larger coverage area but secluded apart from the confinement walls [2]. Circadian rhythm which is naturally imparted biologically is generally categorised under endogenous rhythms in fishes. It portrays a similar pattern with other mammals sleeping anatomy which may relate to feeding rhythms [3]. The rhythms routinely persist around 24 hours period as some researchers called the free running. With a mannerly scheduled feeding time, hunger behaviour can be modelled accordingly based on the measurements that are required for example the amount needed, capacity that it can withstand and intervals between each feeding time. In this case, to monitor a fish hunger rate, these parameters can be adjusted into manipulating the fish appetite either in famish or satiated state. In most cases, the searching endeavour reduces once the fish is satiated, in turn suggest the lower movement of the fish [4]. In addition, the circadian cycle may fluctuate given that the fish swim as a whole. Another study has implied that the fishes tend to be more adventurous and lurk the area more when they are hungry and consequently alter the fish movement either as individuals or group [5]. A different study has demonstrated that hunger could be influenced through exposing the fish under ultradian rhythm of light and dark environment that replicates day and night moment pulses [6]. They used the self-demand feeder for the fishes to get their food whenever they require it. The present study applies the similar concept that will be further elaborated in the subsequent section.

Self-demand feeder mainly used to ensure the feeding does not exceed the required amount and the time taken for each meal can be measured accurately during the experiment. This echoes the findings of an earlier study where the frequency of growth can be evaluated by an optimal value of feeding for Asian sea bass or *Lates Calcarifer* in net cages [7]. There has been a debate on how the circadian rhythms results were obtained and the reliability of analysis that was acquired since the need of standard of experiment setup is necessary. Vast numbers of variables must be considered such as environmental, social interactions, phenotypic, species amongst others. To date, different type of fish species has been used for such experimentation. Barramundi which is often referred to as the Asian sea bass has shown as a practical specimen for collecting data, where a study conducted has evaluated the growth rate of the fish where the day and night moment are manipulated [8]. Other common species that have been examined are the zebrafish or *Danio rerio* and razor fish or *Xyrithchys novacula* in which the boldness between genders during hunger and circadian rhythms for resting time and active state, respectively are examined [9-10]. The latter study also demonstrates the implementation of using acoustic tracker to evaluate the motion of the fishes by means of hidden Markov models. The concept of presenting valid hunger behaviour representation is adopted into this current study by using image processing for tracking the behaviour and subsequently applying a machine learning technique namely Support Vector Machine (SVM) that will be discussed in the ensuing sections.

Various studies have applied SVM or other classification methods as an analytical tool to support their hypothesis, and it is deemed to be reliable [11-12]. A classification of fish hunger behaviour was done for rainbow trout or *Oncorhynchus mykiss* by evaluating the muscle activity to determine whether the fish are hungry or fed [13]. The present study will evaluate the efficacy of a variation of SVM to classify the state of hunger of *Lates Clacarifer* based on relevant identified paramaters namely center of gravity as well as school density. The structure of the paper is as follows: Section 2 describes the experimental setup employed for the present study as well as the classification method used, the results obtained from the study is presented and discussed in Section 3 and lastly Section 4 draws the conclusion of the study as well as provide some recommendation for future research.

2. Materials and Methods

To incriminate an appropriate behaviour of fishes in group, laboratory conditions must meet several requirements to ensure the results acquired is sufficient and reliable. This section will elaborate on the specimen used, self-demand feeder functions, image recording process and classification method proposed for the study.

2.1 Fish

The fish that was selected for this experiment is the Asia sea bass, as it demands food in a full day cycle regardless either in light or dark environment. The fish are taken from a Fisheries Research Institute, Gelang Patah Johor, Malaysia, and was transported to a laboratory at the International Islamic University Malaysia for safe keeping and experimental setup. A total number of 20 juveniles Asian seabass were taken to form a sizeable school for the purpose of capturing image and to analyse the group size motion.

IOP Conf. Series: Materials Science and Engineering **319** (2018) 012028 doi:10.1088/1757-899X/319/1/012028

2.2 Experiment Setup

The experiment tank area is (92cm x 46cm x 46cm) and filled with 130L of water in which the fishes will be lace in the confined tank. The self-demand feeder were installed above the tank with the infrared sensor suspended from the feeder, and the sensor slightly submerged below the water surface to allow the fish to trigger the sensor. For each time the sensor was triggered, roughly 0.5g of pellets in voume with the diameter of 3mm for each pellet will be dropped. The cycle continues until none of the fish triggers the sensor again. The time taken for the feeder to drop the pellet will be recorded from a microcontroller and sent to the computer. This will determine the time taken for the parameters used for the classification method i.e. Before Feeding (BF), During Feeding (DF) and After feeding (AF). A webcam modeled Logitech, C270h was installed from the front view to record the fish behaviour for a certain period of time. The video was converted into .avi file and was transferred to Roborealm software for evaluation. This program tracks the Centre of Gravity (CoG) of the school fish and the box size. Therefore each parameter had (x, y) coordinates (in pixels). Hence, the position of the school size as well as the school density is determined as the indicator for the fish feeding behaviour. The data obtained that has been clustered with respect to BF, DF and AF is further evaluated for the classification efficacy of SVM.

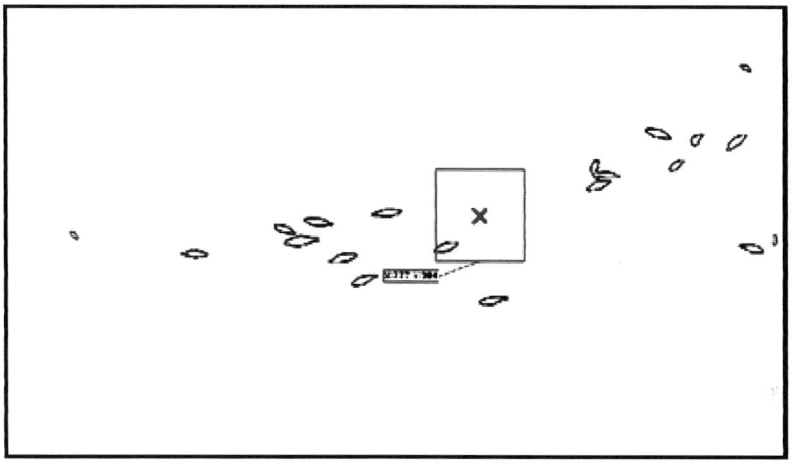

Figure 1. Screenshot from Roborealm for CoG of the box which gives *x* and *y* coordinates for the school.

2.3 Classification method

In this study, six different kernel functions are investigated linear, quadratic, cubic, fine RBF, medium RBF, and Coarse RBF. fine, medium and coarse RBF is defined by $\frac{\sqrt{P}}{4}$, $\sqrt{P}$, and $4\sqrt{P}$, respectively where P is the number of predictors i.e. two. The solver used for the training is the sequential minimal optimisation algorithm. A fivefold cross-validation method was utilised in this study owing to its desirable traits of mitigating the issue of overfitting. The performance of the SVM models were assessed and evaluated via MATLAB 2016a (Mathworks Inc., Natick, USA). The variations of the SVM employed in this study are evaluated by means of classification accuracy (ACC), sensitivity or recall (REC), and precision (PREC).

3. Results and Discussion

It is apparent from the SVM classification results tabulated in Table 1 that the Fine Gaussian kernel was able to provide reasonably accurate classification with a classification accuracy of 79.7 %. Figure 2 depicts the confusion matrix obtained for the Fine Gaussian SVM. Furthermore, it could also be seen that the cubic variation performs the worst amongst all SVM methods. However, it is worth to note that

IOP Conf. Series: Materials Science and Engineering **319** (2018) 012028 doi:10.1088/1757-899X/319/1/012028

the ability of the Fine Gaussian SVM is dependent of the parameters selected namely the centre of gravity as well as the school density, suggesting that the selected parameters are nontrivial or vital in describing the feeding behaviour of *Lates Calcarifer*. The importance of the aforementioned parameters i.e. COG and box size density has been reported in a previous study for understanding fish feeding behaviour [10].

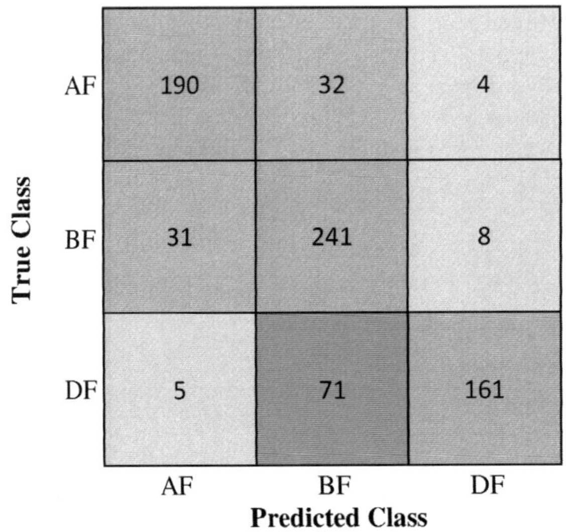

Figure 2. Confusion matrix for Fine Gaussian SVM

Table 1. SVM classification results

SVM Variation	ACC (%)	RECALL (%)			PRECISION (%)		
		BF	DF	AF	BF	DF	AF
Linear	67.0	56.4	77.8	74.1	67.9	63.7	69.5
Quadratic	69.2	57.0	89.3	77.2	80.4	63.3	61.5
Cubic	56.4	49.5	64.1	54.6	38.6	69.2	65.0
Fine Gaussian	79.7	70.1	93.1	84.1	86.1	67.9	88.0
Medium Gaussian	71.9	58.7	90.2	88.0	88.2	62.4	61.5
Coarse Gaussian	69.0	56.6	89.2	81.7	85.4	59.1	59.3

4. Conclusion

It could be concluded from the present investigation that the Fine Gaussian variation of SVM is able to provide a reasonably accurate classification of fish feeding behaviour with regards to the abovementioned parameters examined viz. centre of gravity and school fish density. Further study could be carried out by considering other relevant parameters that may explain the hunger behaviour of *Lates Calcarifer* as well as the effectiveness of other forms of classification techniques.

5. References

[1] Cavallari N, Frigato E, Vallone D, Fröhlich N, Lopez-Olmeda J F, Foà A, Berti R, Sánchez-Vázquez F J, Bertolucci C and Foulkes N S 2011 A Blind Circadian Clock in Cavefish Reveals that Opsins Mediate Peripheral Clock Photoreception *PLOS Biology* **9** e1001142

[2] Hansen M J, Schaerf T M and Ward A J W 2015 The effect of hunger on the exploratory behaviour of shoals of mosquito fish *Behaviour* **152** 1659-77
[3] Beale A, Guibal C, Tamai T K, Klotz L, Cowen S, Peyric E, Reynoso V H, Yamamoto Y and Whitmore D 2013 Circadian rhythms in Mexican blind cavefish Astyanax mexicanus in the lab and in the field **4** 2769
[4] Priyadarshana T, Asaeda T and Manatunge J 2006 Hunger-induced foraging behavior of two cyprinid fish: Pseudorasbora parva and Rasbora daniconius *Hydrobiologia* **568** 341-52
[5] Chapman B B, Morrell L J and Krause J 2010 Unpredictability in food supply during early life influences boldness in fish *Behavioral Ecology* **21** 501-6
[6] F. Javier Sanchez-Vázquez J A M a S Z 1995 ircadian Rhythms of Feeding Activity in Sea Bass, Dicentrarchus labrax L.: Dual Phasing Capacity of Diel Demand-Feeding Pattern *Journal of Biological Rhythms* **10** 11
[7] Biswas G, Thirunavukkarasu A R, Sundaray J K and Kailasam M 2010 Optimization of feeding frequency of Asian seabass (Lates calcarifer) fry reared in net cages under brackishwater environment *Aquaculture* **305** 26-31
[8] Harpaz S, Hakim Y, Barki A, Karplus I, Slosman T and Tufan Eroldogan O 2005 Effects of different feeding levels during day and/or night on growth and brush-border enzyme activity in juvenile Lates calcarifer reared in freshwater re-circulating tanks *Aquaculture* **248** 325-35
[9] Ariyomo T O and Watt P J 2015 Effect of hunger level and time of day on boldness and aggression in the zebrafish Danio rerio *Journal of Fish Biology* **86** 1852-9
[10] Josep Alós M M-B a A C-C 2017 Repeatability of circadian behavioural variation revealed in free-ranging marine fish *Royal Society Publishing* **4** 11
[11] Cortes C and Vapnik V 1995 Support-vector networks *Machine Learning* **20** 273-97
[12] Dutta M K, Sengar N, Kamble N, Banerjee K, Minhas N and Sarkar B 2016 Image processing based technique for classification of fish quality after cypermethrine exposure *LWT - Food Science and Technology* **68** 408-17
[13] Cubitt K F, Williams H T, Rowsell D, McFarlane W J, Gosine R G, Butterworth K G and McKinley R S 2008 Development of an intelligent reasoning system to distinguish hunger states in Rainbow trout (Oncorhynchus mykiss) *Computers and Electronics in Agriculture* **62** 29-34

IOP Conf. Series: Materials Science and Engineering **319** (2018) 012029 doi:10.1088/1757-899X/319/1/012029

Failure Behaviour of Aluminium/CFRP Laminates with Varying Fibre Orientation in Quasi-static Indentation Test

N K Romli[1], M R M Rejab[1], D Bachtiar[1], J Siregar[1], M F Rani[2], Salwani Mohd Salleh[1], M N M Merzuki[1]

[1]Structural Materials & Degradation Focus Group, Faculty of Mechanical Engineering, Universiti Malaysia Pahang, 26600 Pekan, Pahang, Malaysia
[2]Faculty of Engineering, DRB-Hicom University of Automotive Malaysia, 26607 Pekan, Pahang, Malaysia
Phone: +6094246255; Fax: +6094246222

*E-mail: khaleedaromli@gmail.com

Abstract. The response of the aluminium/carbon laminate was examined by an experimental work. The investigation on fibre metal laminate behaviour was done through an indentation test in a quasi-static loading. The hybrid laminate was fabricated by a compression moulding technique and used two types of carbon fibre orientations; plain weave and unidirectional. The plain weave orientation is dry fibre, and unidirectional orientation is prepreg type fibre. The plain weave carbon fibre and aluminium alloy 2024-0 was laminated by using thermoset epoxy while the unidirectional carbon fibre was pressed by using a hot press machine and cured under a specific temperature and pressure. A compression moulding technique was used for the FML fabrication. The aluminium sheet metal has been roughening by a metal sanding method which to improve the bonding between the fibre and metal layer. The main objective of this paper is to determine the failure response of the laminate under five variation of the crosshead speeds in the quasi-static loading. Based on the experimental data of the test, the result of 1 mm/min in the plain weave CFRP has lower loading than unidirectional fibre which the value of both was 4.11 kN and 4.69 kN, respectively.

1. Introduction

A fibre metal laminate (FML) or also known as a hybrid laminate contains metal layers and composite materials [1]. The fibre metal laminate can be applied to many applications because of the fibre metal laminate can withstand the impacts, loadings and harsh environments for a long period. The combination of the metal layers and composite fibres produce superior characteristics such as fatigue and fracture. [2]. The fibre metal laminate started and applied in military application after the Second World War [3]. From the positive feedbacks of the application, the fibre metal laminate began the application widely in aircraft structures such as fuselage and wings [4]. The example of the application is in the Airbus A380 due to the improvement in fatigue, resistance to corrosion and impact and weight reduced nearly 794 kg [5, 6]. The other organisations such as NASA, Bombardier and EMBRAER interested in the substitution of the aluminium in a fibre metal laminate. The high strength and stiffness are a good combination and resulted in a great performance in space applications [7].

The debonding process between a sheet metal layer and composite fibre layer, delamination of the composite layers, fracture due to a plasticity behaviour of a metal commonly happen to a fibre metal laminate structure which because of the impact of a load and depended on the layup configurations [8, 9]. The process of curing by using an autoclave will take long hours of processing and required intensive of labour cost for the whole production [2].

The purpose of this paper is to compare the strength of two types of carbon fibre orientations under quasi-static indentation test.

2. Experimental work

The behaviour of a fibre metal laminate was studied based on the aluminium alloy 2024-0 sheets and unidirectional carbon fibre prepreg. The prepreg fibre was bonded to the two sheet metal layers. The surface of the aluminium alloy was treated with a metal sanding. The grit size of sandpaper is 80. The purpose of the metal sanding is for surface roughness. The surface roughness will increase the bonding strength of the sheet metal and glass fibre. The layup configuration of the FML is 2/1. The size and thickness of the FML are 150 mm x 150 mm and 2.00 mm. The FML was fabricated by a compression moulding technique. The glass fibre and aluminium alloy sheet metal alloy laminated by using the thermoset epoxy. A pressure was applied by placing a moderated load on it and cured in at room temperature for 24 hours. The experiment was setup as in Figure 1. The shape and size of the indenter are hemispheres and 12.7 mm as in Figure 2.

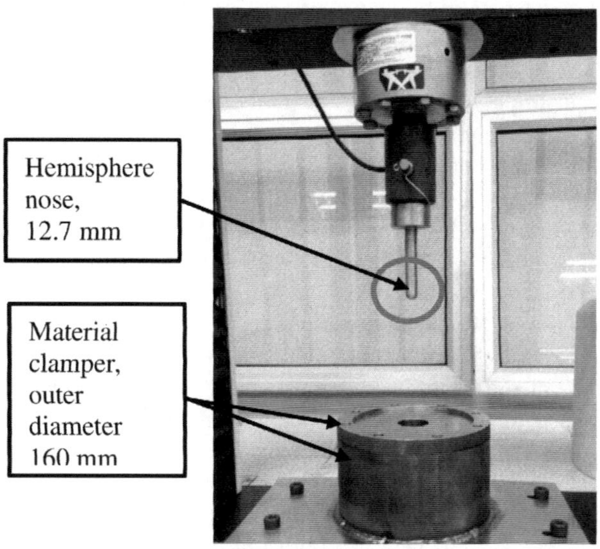

Figure 1. The setup for quasi-static test.

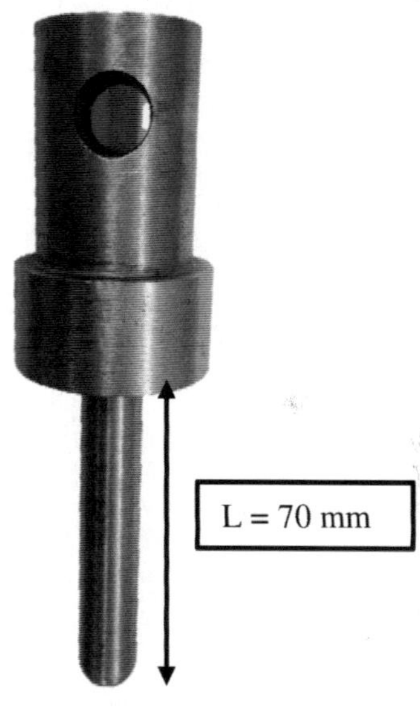

Figure 2. Hemispherical indenter.

3. Results and discussion

The damage area of the laminate depended on a cross speed of the quasi-static loading. The crosshead speed and the constituent materials, the 2024-0 aluminium alloy and carbon fibre reinforced polymer was in the present case. From the graph in Figure 3, it showed that the trending of failure on the fibre metal laminate became decreased when the cross speed was increased from 1 mm/min to 100 mm/min.

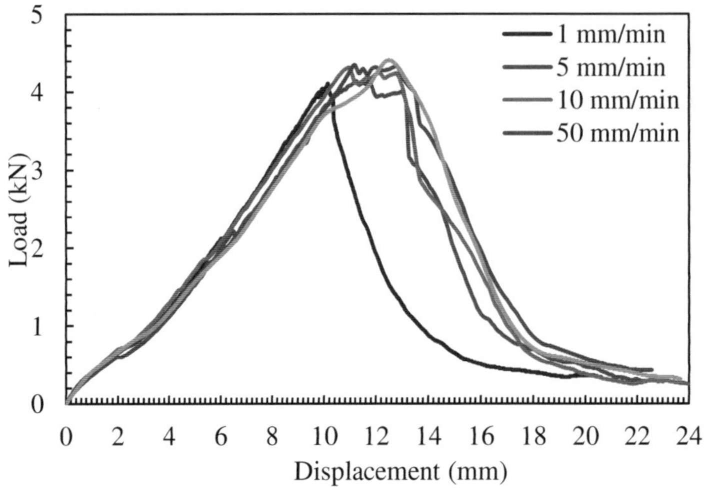

Figure 3. Graph of load against displacement of Aluminium/CFRP FML.

The failure process of the quasi-static indentation on the FML was investigated by examining the front surface and rear-surface of the impact damaged samples. The initial fracture onto the 2/1 FML by 1 mm/min cross speed. The experiment was handle by using Instron 3369 Universal Testing Machine. The 1 mm/min impact shown in a Figure 4a. The damaged area has been taken through an indentation stage by the indenter at a top surface of the aluminium alloy. At the second stage, there was a localize crack on the matrix. The failure of the matrix because of the preparation of layup process. The laminate start to bend because of the maximum load that acted on it and the laminate has a maximum bending. After that, there was a fibre breakage because of the maximum bending of the loading and the stiffness of the fibre. Then, there was an indentation on the bottom ply, aluminium alloy. The perforation of the laminate happened at the third stage of failure. The passage of the indenter through the target and produce a clean hole with a diameter that similar to the indenter. From the graph, the perforation involved a local ductility.

The higher cross speed resulted in a decrease of composite structure degradation. Besides that, there was a number of transverse cracks which increases the delamination of the laminate especially at the middle area of the specimen. The matrix crack also caused delamination of the laminate. When the testing was carried out with the highest cross speed of quasi-static loading it resulted in an extensive delamination at the middle part and bottom ply. When the cross speed increased, the deformation of the fibre metal laminate happened at the same time. The matrix of the laminate damaged caused by impact and the connection between matrix-fibre became degradation. The epoxy resin is brittle and has low resistance to the crack propagation. The crack happened at a high transverse shearing stress which connected with the contact force. The bending crack occurred easily if the volume of the matrix is higher than the volume of the fibre.

(a) (b) (c)

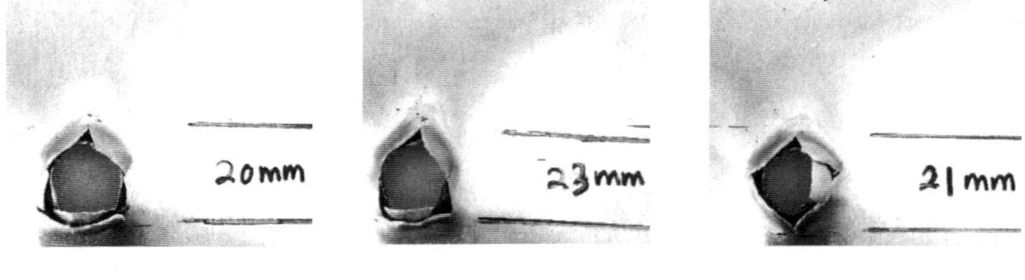

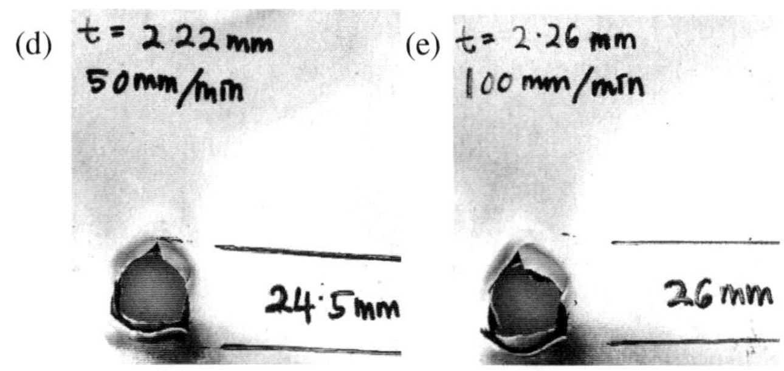

Figure 4. The visual result FML penetration based on different cross speeds.

In the Figure 4 showed a surface area of damages in the aluminium/carbon fibre laminate which based on different crosshead speeds of the indentation. The evaluation of a composite structure based on the damaged area. A delamination between a sheet metal and fibre layer normally found after cracking of the matrix and fibres. The damaged area on the FMLs was analysed and followed by increasing the crosshead speed of the indentation. The values of the damaged area were determined through an experimental. The values were compared based on the crosshead speed and a thickness of the laminate. The shape of damage for all specimens is same which called as petal pattern.

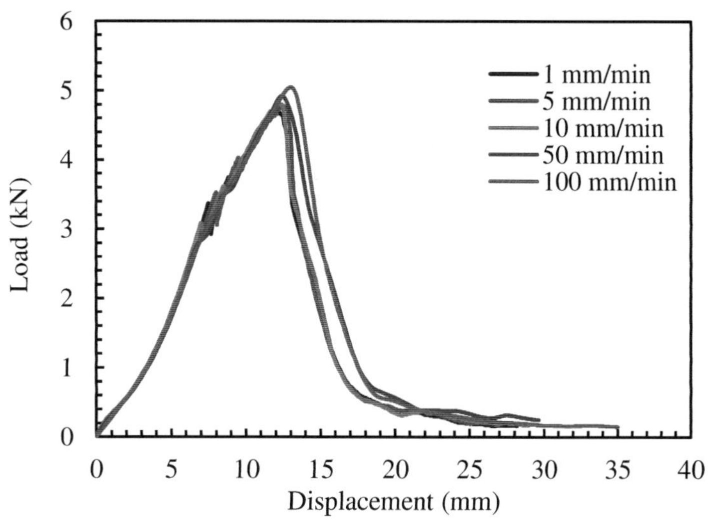

Figure 5. Graph of load against displacement in Aluminium/UD CFRP.

The failure process during the quasi-static indentation was recorded. The damaged samples were investigated by measuring the length of deflection and examined through a load-displacement curve. There was an initial fracture of the 2/1 FML by 1 mm/min crosshead speed. The testing was done by using a compression machine which the compression plate was replaced with an indenter. The damaged area has been taken through an indentation stage by the impactor at a first ply of the aluminium alloy. Based on the graph above, specimen 1 and 5 mm/min almost have same graph pattern. At the first stage, a load was acted on the FML until reached the maximum load. The load was caused the matrix started to appear and the sound of cracking could listen. When the continued load reached the second stage, the fibre started to delaminate. From the continued load, it was caused a sheet metal of first play started to crack. The indenter started to perforate the metal alloy at first ply and the maximum bending happened. The maximum bending caused the fibre tend to break. This happened at the third stage. After the load reached to ultimate value, the indenter started to penetrate the FML and happened at the fourth stage. At this stage, the penetration of a sheet metal also depended on its ductility.

According to Figure 5, the higher crosshead speed resulted in an increase of composite structure degradation. Besides that, there was a number of transverse cracks which increases the delamination of the laminate especially at the middle area of the specimen. The matrix crack also caused delamination of the laminate. When the testing was carried out with the highest crosshead speed of quasi-static loading it resulted in an extensive delamination at the middle part and bottom ply. When the crosshead speed increased, the deformation of the fibre metal laminate happened at the same time. The matrix of the laminate damaged caused by impact and the connection between matrix-fibre became degradation. The epoxy resin is brittle and has low resistance to the crack propagation. The crack happened at a high transverse shearing stress which connected with the contact force. The bending crack occurred easily if the volume of the matrix is higher than the volume of the fibre.

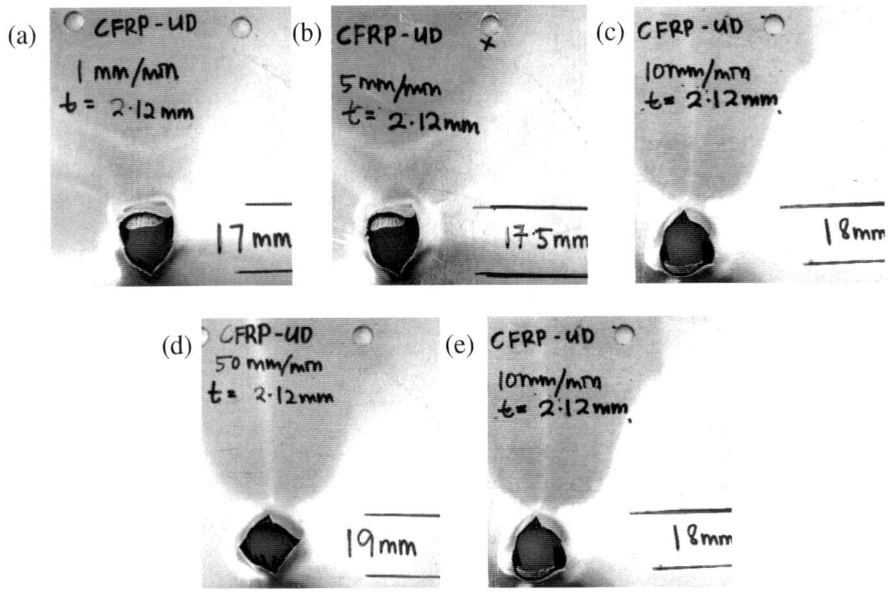

Figure 6. The visual results of FML penetration based on the different crosshead speeds

Based on the Figure 6 showed the damaged surface area of aluminium/unidirectional carbon fibre laminate by various crosshead speed of the indentation loading. The evaluation of the damage FML depended on the surface area. The delamination between a sheet metal layer and a fibre layer was found after a cracking process in the matrix. These FMLs were tested by increasing the crosshead speeds. The measurement of the damaged area has been taken and recorded. The results were compared and analysed the reasons for failure. The pattern of failure looked like a petal. The maximum force values were evaluated by the experimental work under the quasi-static load. Through experimental records that have been made, the records were accompanied by the increasing of quasi-static crosshead speed and laminate thickness. The specimens of aluminium/unidirectional carbon fibre have the same thickness. The internal damage of the FML has been studied and concluded by this author, [10] the penetration and perforation energy was depending on the laminate thickness and cross-section size. Area of damage is a necessity to evaluate the composite laminate. The cracks formation and fibre breakage were recognised as a form of failure to the FML [11].

The cross-section of the laminates has been compared between of them. Based on the observation that has been made, damage contour in the sheet metal and composite fibre layers were mainly followed by a lower layer in a particularised composite interface. The diverse of fibre orientation could cause bending stiffness happened in different phases. The differences in bending stiffness referred to the anisotropy of mechanical properties in composite laminates. These have been related to Young's modulus in the longitudinal and perpendicular direction of the fibre arrangement.

Greenhalgh has made a conclusion that the delamination would happen in the area with higher bending stiffness [12].

4. Conclusion

The current study describes the comparison of a 2/1 fibre metal laminate between plain weave CFRP and unidirectional carbon fibre. Both laminates have been fabricated in a different way. The aluminium/CFRP laminate was manufactured by using epoxy and hardener. Then it was cured at a room temperature for 24 hours. While the aluminium/unidirectional laminate was fabricated by using a hot press machine and it was cured at the specific condition. The technique of manufacturing for both laminates was same, compression moulding. These laminates have been tested under a quasi-static loading with difference crosshead speeds.

On the basis results that have been obtained through the experimental work, the behaviours of the laminates were investigated through a load-displacement curve. The high value of maximum load stipulated to the high resistance of the fibre metal laminates to the quasi-static indentation. The relationship between indentation crosshead speeds and the thickness of the laminate have been confirmed by the experimental research.

The evaluation of the damaged fibre metal laminate it is possible to compare the damaged surface areas by an experimental research. It has been found that increasing of crosshead speed would increase the damaged surface area. A detail observation on the cross-section of the fibre metal laminates was revealed by the presence of matrix cracking and delamination between composite layers as well as delamination at the metal-composite interface.

The laminate thickness was lead to degradation degree. The lower thickness of laminates would have greater deformation and lower metal layer cracking. A laminate increased with thickness was accompanied by decreasing the degree of laminate deformation and occurring delamination between the individual composite layers and delamination on the metal-composite interface.

Acknowledgement

The authors are grateful to the Ministry of Higher Education and Universiti Malaysia Pahang (www.ump.edu.my) for funding this research under RDU 1603108.

References

[1] Vlot, A. and M. Krull, *Impact damage resistance of various fibre metal laminates.* Le Journal de Physique IV, 1997. 7(C3): p. C3-1045-C3-1050.
[2] Hu, Y., et al., *Application of laser peen forming to bend fibre metal laminates by high dynamic loading.* Journal of Materials Processing Technology, 2015. 226: p. 32-39.
[3] Hillermeier, R. and J. Seferis, *Interlayer toughening of resin transfer molding composites.* Composites Part A: Applied Science and Manufacturing, 2001. 32(5): p. 721-729.
[4] Zenkert, D., *An introduction to sandwich structures.* Student edition, 1997.
[5] Zhang, Z., et al., *An experimental investigation into pin loading effects on fatigue crack growth in Fibre Metal Laminates.* Procedia Structural Integrity, 2016. 2: p. 3361-3368.
[6] Vlot, A. and J.W. Gunnink, *Fibre metal laminates: an introduction.* 2011: Springer Science & Business Media.
[7] Edson Cocchieri Botelho, R.A.S., Luiz Cláudio Pardini, Mirabel Cerqueira Rezende, *A Review on the Development and Properties of Continuous Fiber/epoxy/aluminum Hybrid Composites for Aircraft Structures. Materials Research*, 2006. 9(3): p. 247-256.
[8] Abrate, S., *Impact on composite structures.* 2005: Cambridge university press.
[9] Su, Z. and L. Ye, *Identification of damage using Lamb waves: from fundamentals to applications.* Vol. 48. 2009: Springer Science & Business Media.
[10] Evci, C., *Thickness-dependent energy dissipation characteristics of laminated composites subjected to low velocity impact.* Composite Structures, 2015. 133: p. 508-521.

IOP Conf. Series: Materials Science and Engineering **319** (2018) 012029 doi:10.1088/1757-899X/319/1/012029

[11] Romli, N.K., et al., *The behavior of Aluminium Carbon/Epoxy fibre metal laminate under quasi-static loading*. IOP Conference Series: Materials Science and Engineering, 2017. 257(1): p. 012046.

[12] Greenhalgh, E., et al., *Characterisation of impact damage in skin-stringer composite structures.* Composite Structures, 1996. 36(3): p. 187-207.

IOP Conf. Series: Materials Science and Engineering **319** (2018) 012030 doi:10.1088/1757-899X/319/1/012030

Fleet Sizing of Automated Material Handling Using Simulation Approach

Radinal Wibisono, The Jin Ai*, Deny Ratna Yuniartha

[1]Department of Industrial Engineering, Universitas Atma Jaya Yogyakarta, Indonesia

*jinai@staff.uajy.ac.id

Abstract. Automated material handling tends to be chosen rather than using human power in material handling activity for production floor in manufacturing company. One critical issue in implementing automated material handling is designing phase to ensure that material handling activity more efficient in term of cost spending. Fleet sizing become one of the topic in designing phase. In this research, simulation approach is being used to solve fleet sizing problem in flow shop production to ensure optimum situation. Optimum situation in this research means minimum flow time and maximum capacity in production floor. Simulation approach is being used because flow shop can be modelled into queuing network and inter-arrival time is not following exponential distribution. Therefore, contribution of this research is solving fleet sizing problem with multi objectives in flow shop production using simulation approach with ARENA Software
Keywords: Automated material handling, Fleet sizing, ARENA Software, Queuing theory, Flow shop, Simulation

1. Introduction

Currently, manufacturing company has to face tight competition with many competitors. Company try to have lower price for their products compared to another competitor but still maintain or even higher profit level. The ultimate way to reach the objective is lower the cost to perform an enterprise in form of manufacturing and administration activity. It is well known that lean principle can be applied to response this issue, i.e. eliminating seven waste in the company [1], including transportation waste.

A case study is presented here, in which a manufacturing company tried to minimize its material handling cost by automating material handling in production floor. Automated Guided Vehicle (AGV) is the most popular solution in automated material handling in manufacturing due to many advantages over human power in material handling activity [2-4]. However, this company decided to implement a simple type AGV, which is called AGV towing vehicles. The principle is actually the same as AGV in term of driverless material handling [5] and consists of many parts, such as: DC Motors, sensors to ensure the safety, wireless communication, and energy sources in form of rechargeable battery [6]. The difference between them, AGV towing vehicle is simplification of common AGV which are not fully automated and not flexible as AGV system. It has only two stop points, one is considered as start point and another is considered as end point. Term usage of start and end point is based on material flow in production. Start point is stop point to load material that ready to be processed to the next station and end point is stop point to unload material in the arrival station.

AGV towing vehicle still needs operator to load and unload container manually. The purpose of AGV towing vehicle is to reduce headcount and prevent human operator fatigue due to far distances and repetitive transportation activity. This objective is driven by the increasing labor cost, so that

IOP Conf. Series: Materials Science and Engineering **319** (2018) 012030 doi:10.1088/1757-899X/319/1/012030

manufacturing company is forced to reduce workers in production floor and tend to have automated or semi-automated system in the production activity. Another consideration of using AGV towing vehicle rather than common AGV is because cheaper investment. Cheaper investment is caused by not all features in AGV used in AGV towing vehicle system.

In this research, real case study used in manufacturing company which AGV towing vehicle will be implemented and already in prototype phase. AGV towing vehicle will be implemented to transport material consecutively from supermarket to X station and from X station to Y station. The layout map for case study in AGV towing vehicle systems implementation can be seen in Figure 1.

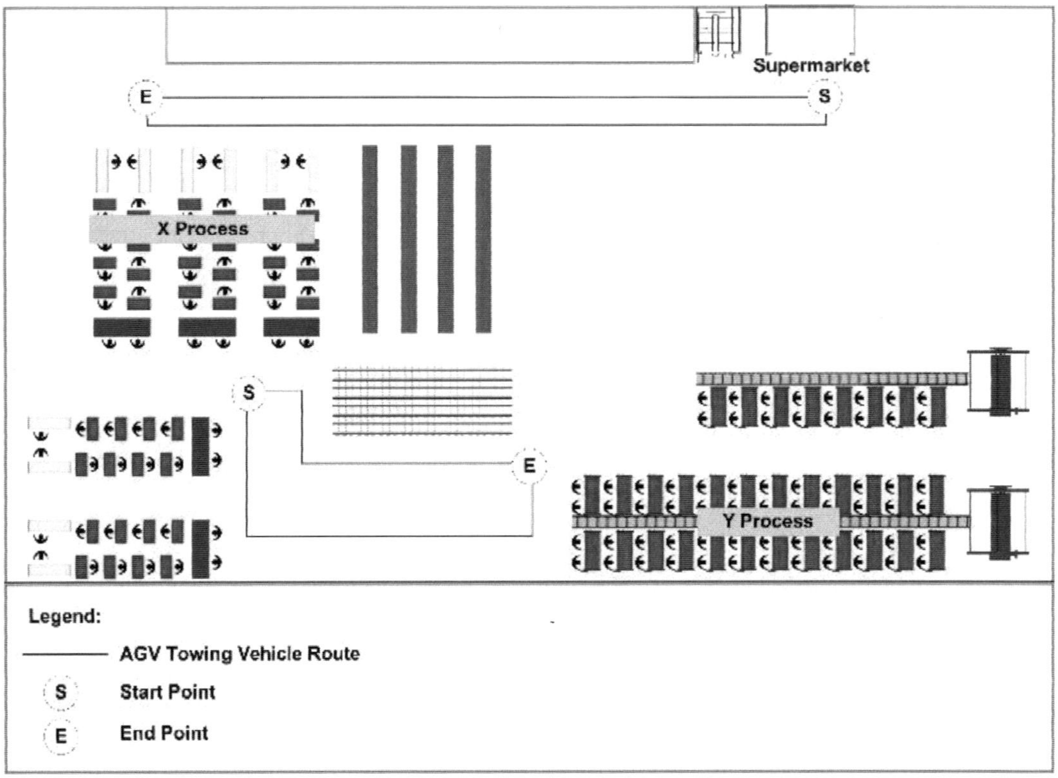

Figure 1. Layout Map

AGV towing vehicle can be said as automation in manufacturing process in term of material handling. As automation, designing the system is a critical phase to ensure that production system is optimized. It is critical because main reason of automation is actually to improve previous process in term of cost spending and ensure that AGV towing vehicle usage is to eliminate and not adding any waste in previous process. AGV towing vehicle also should not become a bottleneck in production or make production capacity reduced. This research is important because appropriate number of AGV towing vehicle can have minimum flow time of material.

2. Simulation Model

In order to imitate how the system works, a simulation model is developed using ARENA software. The description of the model is presented below, while the whole model is available in Figure 2. It is noted that the key decision variables to be taken in this situation are number of AGV towing vehicle to be placed in phase one (between supermarket and X station) and number of AGV towing vehicle to be placed in phase two (between X station and Y station).

IOP Conf. Series: Materials Science and Engineering **319** (2018) 012030 doi:10.1088/1757-899X/319/1/012030

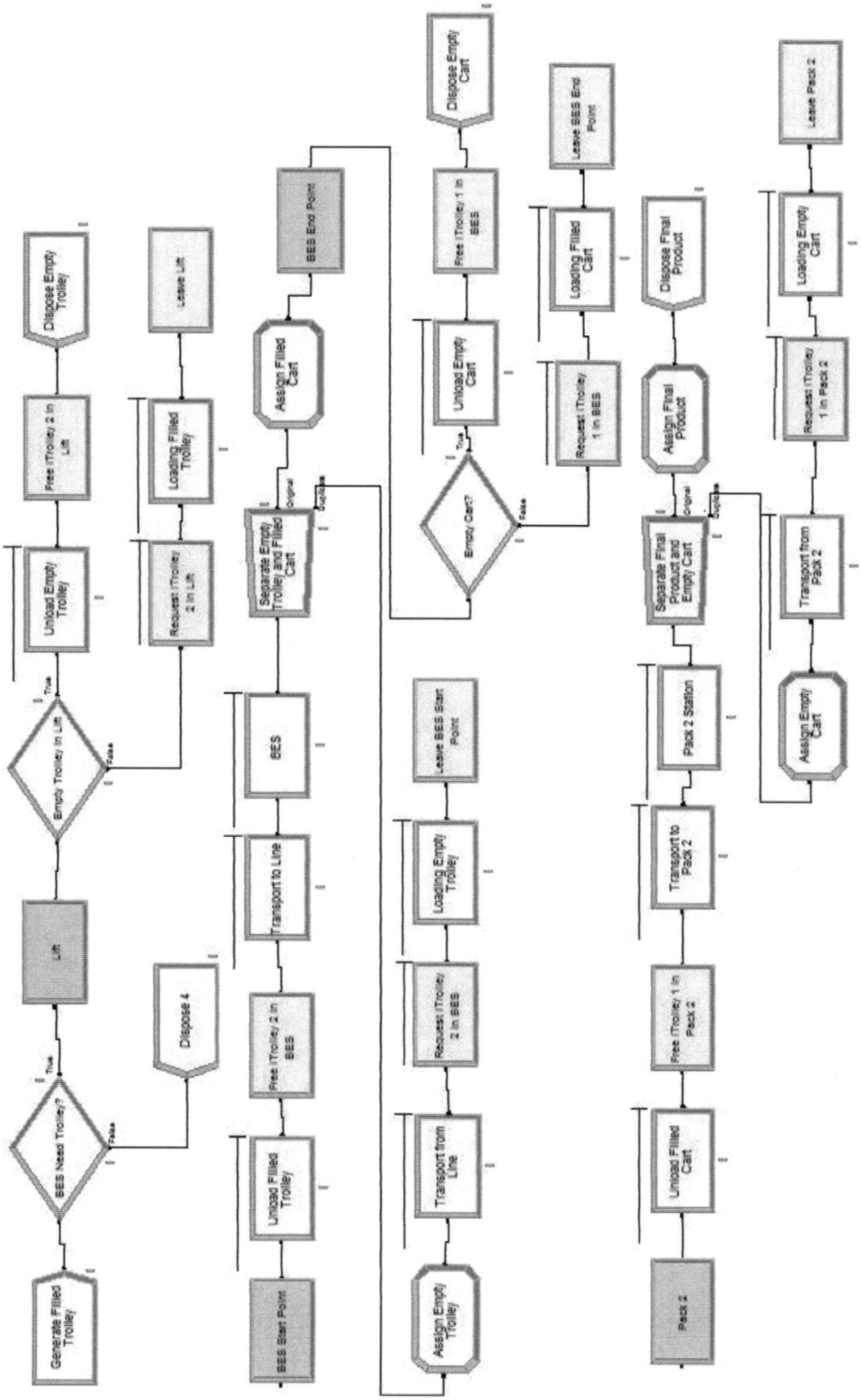

Figure 2. Complete Model in ARENA Software

IOP Conf. Series: Materials Science and Engineering **319** (2018) 012030 doi:10.1088/1757-899X/319/1/012030

In the phase one, start point is supermarket and end point is X station. There will be two entities involved in start point, which are empty entities and filled entities. If entity that arrived is empty, operator should unload it first from AGV towing vehicle before it ready to transport another entity. After that empty entities will disposed or exit the system. If it is filled entities arrived, AGV towing vehicle will be requested first before operator can load entities. Filled entities will be departing to X station. End point in phase one is X station. Filled entities will come first from supermarket to X start station and operator need to unload filled entities before AGV towing vehicle ready to use again. After filled entities being unloaded, it will be transported to X station and processed in the station. If filled entities already processed, two entities will be occurred which are empty entities to reuse in previous process and filled entities to be transported to Y station. Empty entities will be transported to X station start point. After that, AGV towing vehicle is requested and operator load it to AGV towing vehicle after it is available. In the end, empty entities leave X station start point and go to supermarket. Another model made in phase one is to generate filled entities to the system. Assumption of supermarket concept in lift should be considered because systems only generate new filled entities when X station needs it. Unfortunately, create module in ARENA Software cannot generate new entities based on conditions. It is only based on schedule or expression of time distribution. Therefore, decide module is placed after create module to create appropriate model to represent pull system between supermarket to X station.

In phase two, start point is X station and end point is Y station. There will be two entities in X station which are filled entities that ready to be transported to Y station and empty entities that will be reuse in X station. If empty entities arrived to X station, operator will unload the empty entities so AGV towing vehicle will be ready to transport again. Empty entities will be disposed so X station can reuse it again. If it is filled entities arrived, AGV towing vehicle will be requested. After AGV towing vehicle requested, operator load filled entities and AGV towing vehicle will leave to Y station. End point in phase two is Y station. There are three entities involved in end point, which are empty entities, filled entities, and final product. Filled entity is arrived in Y station and operator should unload from AGV towing vehicle first before it is ready to transport another entity. After filled entities being unloaded, operator transports it to Y station line and being processed in Y station. The outputs are final product and empty entities. Final product will exit the system or disposed and empty entities will be transported first to end point of AGV towing vehicle system. Empty entities will be transported to X station again for reuse. Therefore, AGV towing vehicle will be requested first and operator load empty entities to AGV towing vehicle before it leave Y station to X station.

Time data are collected from the system in order to qualitatively model the system, which are the operator time to: (1) load AGV towing vehicle, (2) unload AGV towing vehicle, (3) transport to Y station, (4) transport from Y station, (5) transport to X station, (6) transport from X station. The mean of these times are (1) 6.60 seconds, (2) 5.56 seconds, (3) 20.94 seconds, (4) 20.15 seconds, (5) 32.46 seconds, and (6) 30.00 seconds, respectively. Moreover, the processing cycle time at X and Y stations are also being observed. It is statistically proven that the distribution of processing cycle time at X station is triangular distribution with minimum value is 5 seconds, mode is 5.12 seconds, and maximum value is 136 seconds. While, the distribution of processing cycle time at Y station is beta distribution with parameter $\alpha = 0.655$ and $\beta = 1.48$. Experiment from the AGV towing vehicle prototype is showed that the cycle time for AGV towing vehicle for phase one and two are 315 seconds and 220 seconds, respectively. For the AGV towing vehicle, cycle time means time needed for the vehicle departs from start point and arrives again in start point.

3. Result and Discussion

Preliminary simulation runs indicate that 80 replications are required for fulfilling desired precision. The alternative of number of AGV towing vehicle to be placed in phase one is 6 – 9 vehicles and number of AGV towing vehicle to be placed in phase two is 8 – 9 vehicles. Therefore, 8 combination of scenario are tested. Flow time and daily production volume are recorded as the system performances. Average performances of each scenario from all replications are presented in Table 1.

IOP Conf. Series: Materials Science and Engineering **319** (2018) 012030 doi:10.1088/1757-899X/319/1/012030

Table 1. Simulation Results

Scenario	Phase One Vehicles	Phase Two Vehicles	Flow Time (minutes)	Production Volumes (units)
1	6	8	18.8	1676
2	6	9	17.9	1671
3	7	8	18.8	1674
4	7	9	18.5	1677
5	8	8	18.7	1678
6	8	9	18.0	1676
7	9	8	18.1	1680
8	9	9	17.9	1676

It is shown in Table 1 that the results are quite similar among scenarios, i.e. increasing number of AGV towing vehicles is not having any significance difference in flow time and production volume. Therefore, statistical analysis is required in order to ensure that the performance among scenarios can be considered as statistically same. One-way ANOVA is applied here, with the null hypothesis: all the mean flow time (or the mean production volume) are similar, versus the alternate hypothesis: at least one mean is different. The results of the ANOVA test for flow time and production volume in MINITAB software are presented in Figure 3 and 4, respectively. Since the p-value is greater than α (i.e. α=0.05), therefore, in both hypothesis testing we failed to reject null hypothesis. In other words, there are no statistical evidence that the flow time and production volume are different among scenarios.

Based on these performance analysis, it is recommended that the company use only 6 vehicles in the phase one and 8 vehicles in the phase two. By using this recommendation, the company can save the investment for acquiring the vehicles, while the performance of the system is not significantly different with the system at higher number of vehicles.

Results for: Flow Time.MTW

One-way ANOVA: Scenario 1, Scenario 2, Scena

```
Source   DF       SS     MS     F      P
Factor    7    85.00  12.14  1.85  0.075
Error   632  4138.50   6.55
Total   639  4223.50

S = 2.559   R-Sq = 2.01%   R-Sq(adj) = 0.93%
```

Figure 3. ANOVA Results on Flow Time

Results for: Production Volume.MTW

One-way ANOVA: Scenario 1, Scenario 2, Scenario 3

```
Source   DF      SS   MS     F      P
Factor    7    4573  653  1.64  0.122
Error   632  252081  399
Total   639  256654

S = 19.97   R-Sq = 1.78%   R-Sq(adj) = 0.69%
```

Figure 4. ANOVA Results on Production Volume

4. Conclusion

In order to solve the optimization problem for determining optimum number of AGV towing vehicle used in production floor, a simulation model is developed using ARENA software. Using standard simulation methodology, i.e. input analysis, experiment, and output analysis, the situation of smaller

IOP Conf. Series: Materials Science and Engineering **319** (2018) 012030 doi:10.1088/1757-899X/319/1/012030

material flow time and higher production volume is reached with six AGV towing vehicles in the phase one and eight AGV towing vehicles in the phase two.

References

[1] Marchwinski C and Shook J 2003 *Lean Lexicon: a Graphical Glossary for Lean Thinkers*
[2] Lin L, Shinn SW, Gen M and Hwang H 2006 *Journal of Intelligent Manufacturing* **17** 465–77
[3] Shneor R, Berman S and Edan Y 2006 *Applied Soft Computing Technologies: The Challenge of Complexity* 749–60.
[4] Leite LFV, Esposito RMA, Vieira AP and Lima F 2015 *Independent Journal of Management & Production* **6** 269–85
[5] Fazlollahtabar H and Saidi-Mehrabad M 2015 *Journal of Intelligent & Robotic Systems* **77** 525–45
[6] Zajac J, Slota A, Krupa K, Wiek T, Chwajol G and Małopolski W 2013 *Applied Mechanics and Materials* **282** 59–65

IOP Conf. Series: Materials Science and Engineering **319** (2018) 012031 doi:10.1088/1757-899X/319/1/012031

Analysis of Container Yard Capacity In North TPK Using ARIMA Method

Sirajuddin, Cut Gebrina Hisbach M, Ratna Ekawati, Ade Irman SM

Industrial Engineering, University of Sultan Ageng Tirtayasa, Indonesia

sirajd_udin@yahoo.com, gebrinahisbachms@gmail.com, ratna.ti@untirta.ac.id, irman@untirta.ac.id

Abstract. North container terminal known as North TPK is container terminal located in Indonesia Port Corporation area serving domestic container loading and unloading. It has 1006 ground slots with a total capacity of 5,544 TEUs and the maximum throughput of containers is 539,616 TEUs / year. Container throughput in the North TPK is increasing year by year. In 2011-2012, the North TPK container throughput is 165,080 TEUs / year and in 2015-2016 has reached 213,147 TEUs / year. To avoid congestion, and prevent possible losses in the future, this paper will analyze the flow of containers and the level of Yard Occupation Ratio in the North TPK at Tanjung Priok Port. The method used is the Autoregressive Integrated Moving Average (ARIMA) Model. ARIMA is a model that completely ignores independent variables in making forecasting. ARIMA results show that in 2016-2017 the total throughput of containers reached 234,006 TEUs / year with field effectiveness of 43.4% and in 2017-2018 the total throughput of containers reached 249,417 TEUs / year with field effectiveness 46.2%.

1. Introduction

As a maritime country, Indonesia uses sea transportation as one of the most important modes of transportation. Sea transportation will be smooth as it is supported by good port conditions. Port serves key role in trading between countries, industries, and becomes locomotives of economic growth. One port operator that operates and dominates port business in Indonesia is Indonesia Port Corporation (IPC). IPC was previously known as PT. Pelindo II has developed into a more efficient port service company with modern equipment in every aspect of its operations [1] [2] [7].

IPC has contribution to providing port services to support the creation of cheaper, efficient and safe logistics costs. North TPK is a container terminal located at Port One IPC and has a responsibility to load and unload the domestic container [3]. Port performance depends on the activities in the yard and the availability of the facility in the wharf. If the activities on the yard goes smoothly, the performance of the port will be raise. Because of the activities on the yard has a high impact on the performance of the port, so the activities become the critical point to get higher port performance. The activities on the yard will be go smoothly when the demand of the port in still under the capacity of the yard. This research is expected to be an object to forecast the demand of Indonesian Port, Tanjung Priok, on North TPK yard by using Box-Jenkins ARIMA method [4] [5] [6].

The problems discussed in this research is the flow of the Ships and Containers in Terminal Operation II PT. Tanjung Priok Port, and the level of Yard Occupation Ratio in North TPK at PT Pelabuhan Tanjung Priok. This research is assumed the YOR that applied for the analysis is 80%. The Dwelling Time at container yard North TPK is three days. The maximum Tier on North TPK is 5.5 [6] [8] [9].

IOP Conf. Series: Materials Science and Engineering **319** (2018) 012031 doi:10.1088/1757-899X/319/1/012031

2. Methodology

2.1. Arima method

Autoregressive Integrated Moving Average (ARIMA) first time developed by George Box and Gwilym Jenkins in 1976, therefore ARIMA also called as Box-Jenkins models. ARIMA Model is also used for time series analysis and data forecasting. ARIMA is highly accurate for short term forecasting, but not so good for long term forecasting. Usually it will show flat or constant data forecasting for long period. ARIMA is a combination from Autoregressive (AR) and Moving Average (MA). ARIMA usually noted in the form of p, d, and q. Form p as degree of process ordo, d as differential ordo, and q as degree of MA process [10] [16][19].

ARIMA Model is a model that fully ignores the independent variable to make a prediction. ARIMA use the past and present dependent value to predict the value in the future accurately. ARIMA can be applied if time series observation statistically connected to each other (dependent). There are four steps in ARIMA Box-Jenkins Method i.e. identification, estimation, diagnostic check and forecasting [10] [11] [12] [13] [14].

2.2. Accuracy and forecasting error

Error calculation can be used to compare between a forecasting method to the other methods. And, it is also used to measure its confidence level. This calculation can be used to make sure the forecasting is going well (Heizer and Render 2009:177) in Savira M, Moeliono Nadya NK, (2014) [11]. The error test in this research using Mean Absolute Percentage Error (MAPE) as the indicator. MAPE is an indicator to show the performance or accuracy of the forecasting. MAPE counted as the mean of absolute differentiation between the prediction and actual value, the result is served as actual value percentage [15] [16]. The formula is shown below:

$$\text{MAPE} = \frac{\Sigma\, Absolute\ percent\ error}{n} \quad (1)$$

2.3. Formatting author affiliations

This is the flow chart of our research methodology (figure 1), and the flow chart of ARIMA Box-Jenkins method (figure 2): showing a few steps has to be done.

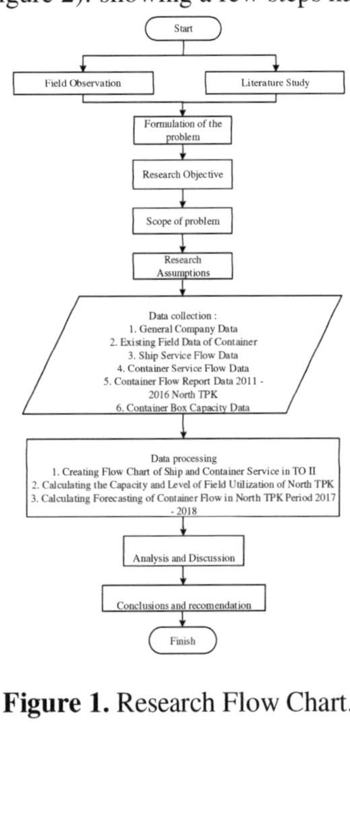

Figure 1. Research Flow Chart.

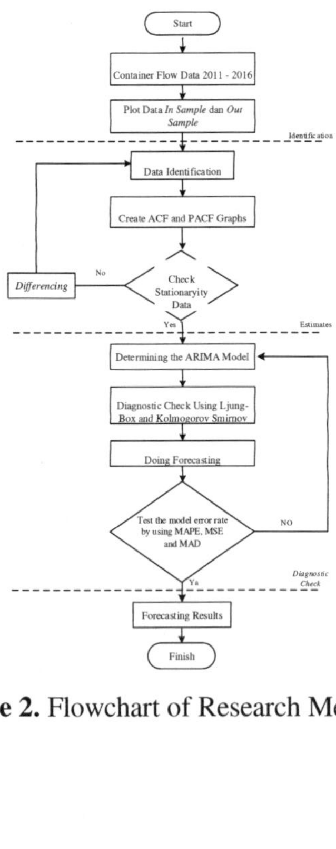

Figure 2. Flowchart of Research Method.

3. Result & discussion

This forecasting uses 60 data of throughput on North TPK since March 2011 until April 2016 shown in table 1.

Table 1. Data Throughput of North TPK

Month	Period				
	2011-2012	**2012-2013**	**2013-2014**	**2014-2015**	**2015-2016**
April	13,252	12,023	13,204	14,855	14,402
Mei	13,481	12,892	14,841	15,533	17,191
June	12,945	11,384	15,913	18,299	15,183
July	12,115	13,589	13,394	17,169	9,041
Augustus	14,782	12,723	14,636	9,655	18,138
September	15,529	14,323	7,622	14,751	19,297
October	12,719	9,759	17,524	14,557	20,508
November	9,492	16,875	17,429	14,218	19,753
December	16,487	17,269	16,389	18,578	18,864
January	12,718	15,091	15,859	18,448	20,870
February	15,373	16,219	16,778	19,528	19,050
March	16,187	17,689	15,670	19,945	20,850
TOTAL	**165,080**	**169,836**	**179,259**	**195,536**	**213,147**

The first step is data identification. The 5 data was deleted because throughput gets down so fast makes the data cannot use. The data is separated in two, 33 data in sample and 22 data out sample. Figure 3 shown data plot of throughput on TPK Utara. Data plot showing un-stationer data, it needs data differencing, shown in figure 4.

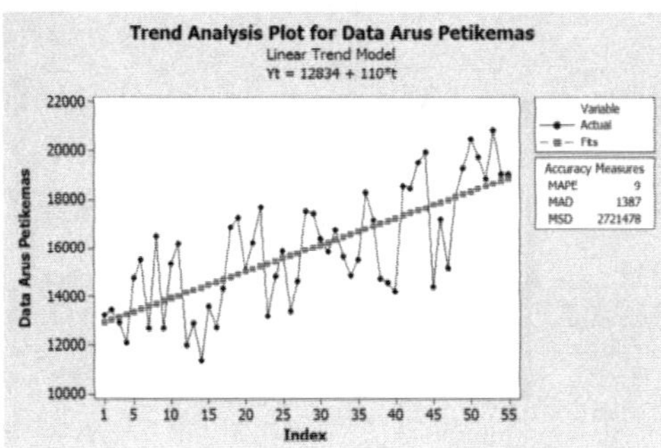

Figure 3. Graphic Trend Analysis Data Throughput.

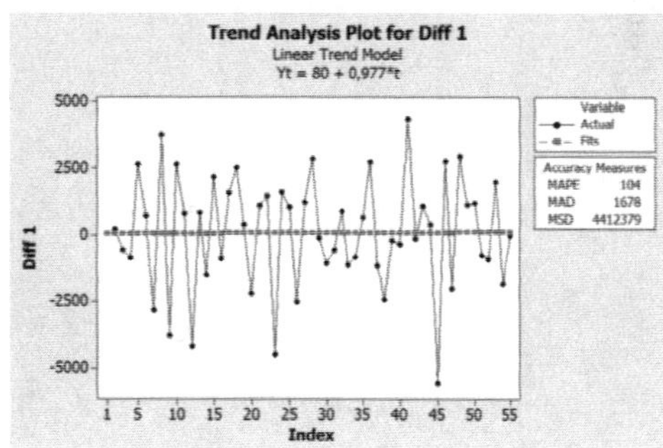

Figure 4. Graphic Trend Analysis Data Throughput Differencing.

After data differencing the data plot has been stationer, and then we make autocorrelation graphic, FAK, in figure 5 and partial autocorrelation graphic, FAKP, in figure 6. From FAKP graphic differencing 1, we can see lag 1, lag 2, lag 10 and lag 14 is cut through white noise line, but the graphic doesn't goes down but following Cygnus graphic.

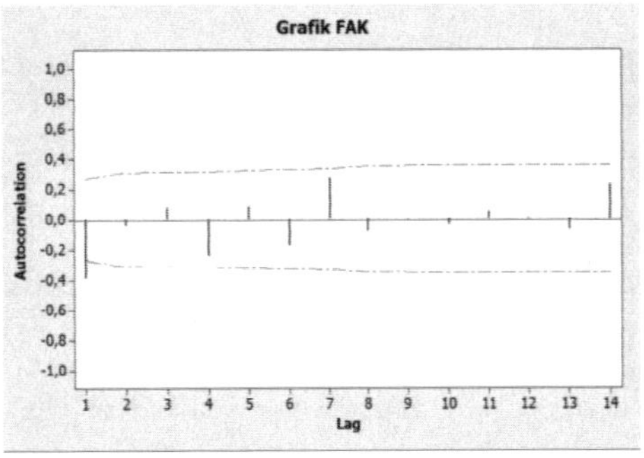

Figure 5. Graphic Autocorrelation (FAK).

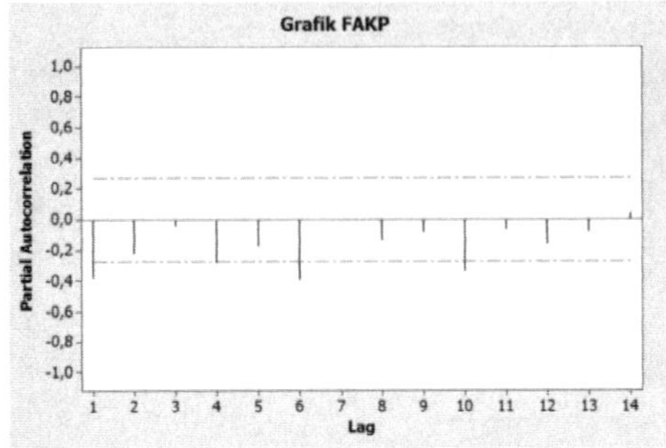

Figure 6. Graphic Partial Autocorrelation (FAKP)

The model that predicted by FAK and FAKP graphic is ARIMA (1, 1, 1). AR (1) because FAK is disconnected on the first lag, integrated (1) because the data get stationer after 1 time differencing and MA (1) because FAKP is disconnected on the first lag. On parameter estimation steps there are three model prediction of ARIMA i.e. (0,1,1), (1,1,0) and (1,1,1). We took the model with the smallest Mean Square, because the smaller the mean square the more accurate the prediction. The smallest MS is model (1,1,1) with 3,366,598. This is the output of Minitab model (1,1,1):

ARIMA Model: Container Throughput

```
Estimates at each iteration
Iteration        SSE          Parameters
        0  678501129   0.100  0.100  115.992
        1  544473884  -0.050  0.250  126.375
        2  522551991   0.045  0.400  114.821
        3  496792953   0.124  0.550  104.998
        4  466664540   0.189  0.700   95.678
        5  432124621   0.247  0.850   82.912
        6  396756049   0.292  1.000   62.311
        7  381891489   0.142  0.995   93.893

Unable to reduce sum of squares any further
Final Estimates of Parameters

Type       Coef  SE Coef     T      P
AR   1  0,2366   0,1920  1,23  0,223
MA   1  0,8227   0,1160  7,09  0,000

Differencing: 1 regular difference
Number of observations:  Original series 55,
after differencing 54
Residuals:    SS =  175063102 (backforecasts
excluded)
              MS =  3366598  DF = 52

Modified Box-Pierce (Ljung-Box) Chi-Square
statistic
Lag              12      24      36      48
Chi-Square     11,6    21,5    37,5    52,2
DF               10      22      34      46
P-Value       0,312   0,492   0,313   0,246
```

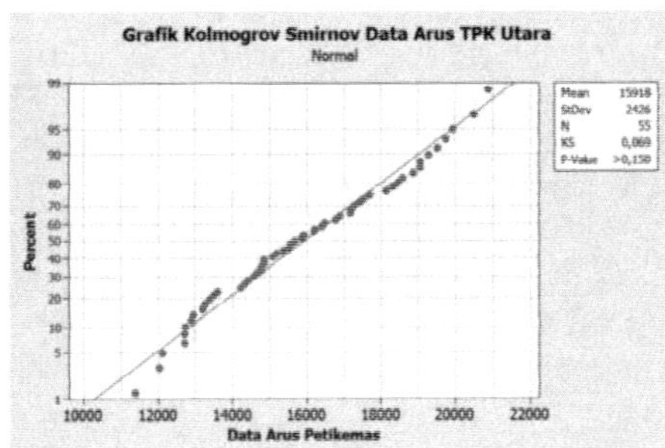

Figure 7. Graphic Kolmogorov Smirnov

Because all of the parameter on significant and residual model is meet the terms of white noise and distributed normally, ARIMA model (1,1,1) is the right model to the forecasting. Forecasting result showing total throughput in 2016-2017 reach 214,911 TEUs/Year and in 2017-2018 reach 229,125 TEUs/Year. The Table above showing the Yard Occupancy Ratio on North TPK in 2016-2017 reach 39,8% and in 2017-2018 reach 42,5%.

4. Conclusion

Container yard on North TPK is able to accommodate 5,544 TEUs container and able to cater the throughput up to 539,616 TEUs/Year. Forecasting Model that we used to predict the throughput in 2016-2017 and 2017-2018 is ARIMA (1,1,1) with MS value 3,366,598 MAPE value 3.36972 with total throughput forecasted 214,911 TEUs in 2016-2017 with YOR 39,8% and 229,125 TEUs in 2017-2018 with YOR 42,5% these forecasting result in North TPK is showing that North TPK is still able to cater the throughput.

References

[1] Amiron HSB, Shahdan. 2009. *Analysis of Feasibility of Size of Docks Length, Warehouse of loading and unloading of cargo case of The Dock of Ujung Baru Port of Belawan Harbor*. (Essay). Medan: Faculty of Civil Engineering University of North Sumatra.

[2] Anita Firlani. 2016. *Due Certain Goods Delivery Time (DT) in the Lifting of Goods by Sea Based on Sales Contract* (Sales Contract). (Essay). Bandar Lampung: Faculty of Law University of Lampung

[3] Arlius, Farendy. 2014. *Forecasting Rate of Usage of Container Crane (CC) and Utility Container Crane (CC) and Rubber Tyred Gantry (RTG) at Banjarmasin Container Terminal.* (Thesis). Surabaya: Faculty of Marine Technology Sepuluh November Institute of Technology.

[4] Aurelia. 2015. *Dynamic System Simulation to Minimize Waiting Time of Ship Using Powersim In Banten CDG Port.* (Essay). Cilegon: Industrial Engineering University of Sultan Ageng Tirtayasa

[5] Hadijah. 2013. *Forecasting Operational Reservation with Minitab Program Using ARIMA Approach PT Surindo Andalan*. Journal The Winners. Vol **14**, No: 1 p. 13-19

[6] Indayani, Eka Febri. 2009. *Forecasting Number of Train Passengers by Using Box-Jenkins Method (Case Study at PT Kereta Api (Persero) DAOP IV Yogyakarta*). (Essay). Yogyakarta: FMIPA State Islamic University Sunan Kalijaga

[7] IPC. 2015. *Annual Report IPC Revisi 24 Juli 2015. Annual Report*. Jakarta Utara : Indonesia Port Corporation

[8] Iriani. 2010. *Measurement of Work Efficiency with Data Envelopment Analysis Approach*. (Essay). East Java: Industrial Engineering UPN "Veteran"

[9] Narindra RAA, Al Musadieq M, Supriono. 2016. *Analysis of the Effect of Dwelling Time on Revenue (Study on Semarang Container Terminal 2011 - 2015).* Journal of Business Administration (JAB) Vol. **41**, No: December 1, 2016.

[10] Panjaitan L, Tarigan G, Bagun P. 2013. *Forecasting of Aluminum Bars Production at PT Inalum by ARIMA Method.* Saitia Mathematic. Vol. **1**, No: 1 pp 1-10. North Sumatra: FMIPA Universitas Sumatera Utara

[11] Savira M, Moeliono Nadya NK. 2014. *Analysis of Generic Drug Sales Forecasting Prologue (OGB) at PT Farma Indonesia*. (Essay). Bandung: Faculty of Economics and Business Telkom University

[12] Shintya, Wike, Mas'ud Effendi. 2013. *Analysis of Product Distribution Efficiency by Data Envelopment Analysis Method (DEA) Case Study at UD Sabar Jaya Malang*. (Essay). Malang: University of Brawijaya

[13] Siswadi. 2005. *Assessment of Container Loading Equipment Loading at Semarang Container Terminal (TPKS).* (Thesis). Semarang: Civil Engineering Diponegoro University

[14] Soeharto. 2003. *Review of Equipment Facilities of Loading and Unloading of Goods at Container Terminal at Tanjung Emas Port.* (Thesis). Semarang: Civil Engineering Diponegoro University

[15] Soludale, Anastasia M.N. 2013. *Assessment of Load Loading Container Performance Facility at Port L. Say Maumene.* (Thesis). Yogyakarta: Civil Engineering Atmajaya University

[16] Sukma Arini. 2011, *World Statistics - Method Kolmogorov Smirnov for Normality Test*. April 22, 2011. [Cited: January 18, 2017]. Available from URL:http://arini2992.blogspot.co.id/2011/04/metode-kolmogorov-smirnov-untuk-uji.html

[17] Sutarti, 2009. *Use of Time-Based Analysis Method with Minitab Relief 11 for Window for Forecasting Textil Production at PT Primatexco Indonesia Batang Regency*. (Thesis). Semarang: FMIPA Semarang State University

[18] Vacca, I., Salani, M., Bierlaire, M. 2010. *Optimization of Operations in Container Terminals.: Hierarchical vs Integrated Approaches.* Lausanne : Ecole Polytechnique Federal de Lausanne

[19] Viara Novita Rizkiyah, 2015. *Normality Test Using Chi Square Test, Kolmogorov Smirnov Test (KS) and Liliefors Test*. May 7, 2015. [Cited: January 18, 2017]. Available from URL: http://viararizkiyah.blogspot.co.i2015/05/ujinormalitas-menggunakan-uji.html

[20] Wibowo, Harmaini. 2010. *Analysis of Factors Affecting Waiting Time for Ships in Port Tanjung Emas Semarang.* (Thesis). Semarang: Civil Engineering Diponegoro University

IOP Conf. Series: Materials Science and Engineering **319** (2018) 012032 doi:10.1088/1757-899X/319/1/012032

Investigation on the effect of Friction Stir Processing Parameters on Micro-structure and Micro-hardness of Rice Husk Ash reinforced Al6061 Metal Matrix Composites

N Fatchurrohman[1], N Farhana[1] and C D Marini[1]

[1]Faculty of Manufacturing Engineering, Universiti Malaysia Pahang, Kuala Pahang, 26600 Pekan, Pahang, Malaysia

Corresponding email: fatchurrohman@ump.edu.my

Abstract. Friction stir processing (FSP) is an alternative way to produce the surface composites of aluminium alloy in order to modify the microstructure and improve the mechanical properties. In this experiment, Al6061 aluminium alloy has been chosen to be used as the matrix base plate for the FSP. Al606 has potential for the use in advanced application but it has low wear resistance. While, the reinforced used was rice husk ash (RHA) in order to produce surface composites which increased the micro hardness of the plate composites. The Al6061 was stirred individually and with 5 weight % of RHA at three different tool rotational speeds of 800 rpm, 1000 rpm and 1200 rpm. After running the FSP, the result in the distribution of particles and the micro hardness of the specimens were identified. The result showed that Al6061 plate with the existing 5 weight % of RHA reinforced at the highest of tool rotational speeds of 1200rpm has the best distribution of particles and the highest result in average of micro hardness with 80Hv.

1. Introduction

In the field of surface modification, friction stir processing (FSP) is used to improved surface properties. Basically, FSP helped so much in improvising the surface properties of metal alloys such as aluminium alloys, to improve its ductility, strength, increase the hardness and formability and fatigue life without changing the bulk metal properties [1]. FSP use a non-consumable tool which plunge into the work piece and moved in direction of interest [2]. The friction between tool and work piece generates the heat. The material become soft and flows around the work piece when it is being heated [3].

FSP also is a technique for creating a composite surface and this method does not deal with melting. This new method which is being derived from the FSW is a novel solid state processing technique for microstructural modification. By referring to the material condition, the dispersion of reinforcement particles on metal surface and the control of its dispersal are more difficult to attain by conventional surface modification technique [4-5].

Methodically, this project works involves the process of the intense of plastic deformation and the mixing by the pressure heat. In addition, it also can improve the surface properties such as abrasion resistance, hardness, toughness, corrosion resistance. With reference to this method, this project is a way to combine between Al6061 aluminium alloy with rice husk ash (RHA). It is intended to produce a composite surface to identify the level of particle distribution, analyse the surface structure and hardness property. It will also adopt a number of parameters such as rotation speed and transverse speed of tool movement to achieve the objectives of the project study.

2. Methodology

Friction stir processing (FSP) is an alternative used to improve the mechanical properties of material especially for alloy. This section provides the procedure in conducting the experiment systematically as shown in figure 1.

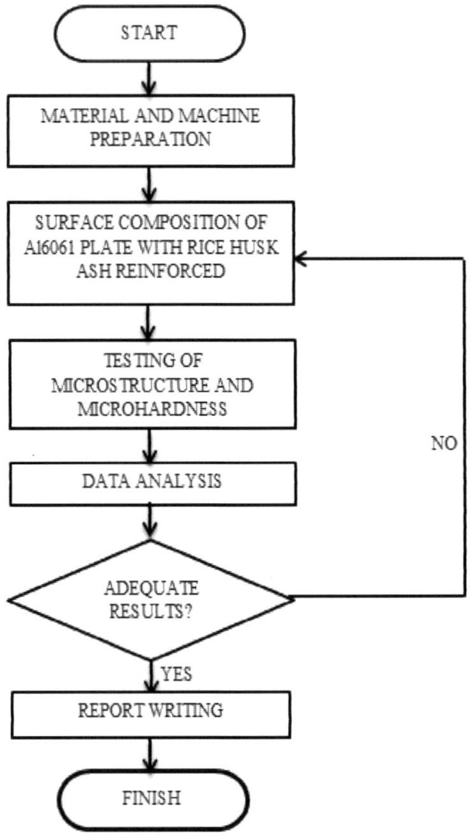

Figure 1. Flow chart of the research.

The experiment was designed for preliminary studies on the effect of RHA addition to Al6061 alloy, additionally on the effect of three different tool rotation speeds. The tool material used in this experiment of FSP is made from steel H13. This H13 tool steel is a versatile chromium-molybdenum hot work steel that is widely used in hot work and cold work tooling applications. The hot hardness (hot strength) of H13 steel resists thermal fatigue cracking which occurs as a result of cyclic heating and cooling cycles in hot work tooling applications. The variable parameter used in this research was tool rotational speed and the constant parameter is transverse speed. Tool rotational and transverse speeds will specify the amount of heat generated in the work piece. Interaction of rotating tool with work-piece generates temperature rise due to friction and plastic deformation. For this investigation, the matrix base material Al6061 alloy and the chemical composition is presented in table 1. Rice husk as (RHA) was utilized as the reinforcement, its composition is clarified in table 2. RHA has high silica contents (87–97 weight % SiO2), high porosity, lightweight and very high external surface area which is a valuable material for industrial applications. RHA contain high silicon content as potential reinforcement for MMC and CMCs [6]. The RHA was heat-treated in order to reduce carbonaceous and volatile substance. In this experiment the amount of tool rotational speed was varied to 800rpm, 1000rpm and 1200rpm while, the traverse speed was constant at 100mm/min.

IOP Conf. Series: Materials Science and Engineering **319** (2018) 012032 doi:10.1088/1757-899X/319/1/012032

Table 1. The chemical composition of Al6061 [7].

Element	Mg	Fe	Si	Cu	Mn	V	Ti	Al
Percentage	1.08	0.17	0.63	0.32	0.52	0.01	0.02	Balance

Table 2. The oxide composition of rice husk ash (RHA).

Oxide composition	**Percentage**
SiO_2	88.32
Al_2O_3	0.46
Fe_2O_3	0.67
CaO	0.67
MgO	0.44
Na_2O_3	0.12
K_2O	2.91
LOI	5.81
Others	2.11

3. Results and discussions

The results obtained from the experiments of friction stir processing for Al6061 without reinforced and with reinforced at different rotation speed of 800rpm, 1000rpm and 1200rpm with constant tool travel speed at 100mm/min. The grain distribution of these specimens and the micro-hardness will be discussed on this section. The heat generates between the plates Al6061 with the tool rotational speed of FSP would causes the plate to obtain its melting point and mixture the reinforced with the plate. This lead to the distribution of the reinforced within the NZ area and would increase the grain growth. Figure 2 show the optical micrograph of Al6061 with RHA reinforced and Al6061 without RHA reinforced at 800 rpm. Figure 3 show the optical micrograph of Al6061 with RHA reinforced and Al6061 without RHA reinforced at 1000 rpm. While, figure 4 show the optical micrograph of Al6061 with RHA reinforced and Al6061 without RHA reinforced at 1200 rpm.

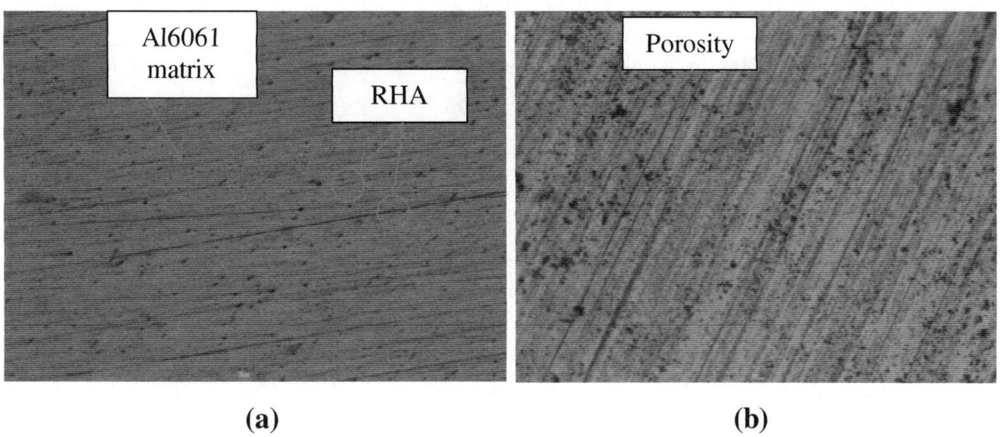

Figure 2. Optical microstructures (10x) of the surface a) Al6061 with RHA b) Al6061 without RHA at 800 rpm

(a) **(b)**

Figure 3. Optical microstructure (10x) of the surface a) Al6061 with RHA b) Al6061 without RHA at 1000 rpm.

(a) **(b)**

Figure 4. Optical microstructure (10x) of the surface a) Al6061 with RHA b) Al6061 without RHA at 1200 rpm

Grain size is an important material characteristic. The material becomes stronger when the average size decreases and opposite effect on strength occurs when the grain size increases. In general, ductility increases with grain size but strength decreases it is happening because the smaller the grains, the shorter the distance for the dislocations to be move. Therefore the smallest the grain size is better for the properties of the material. The plate with RHA reinforced were observed, it shows that the particles dispersed uniformly in NZ area because it was affected by rotating the tool that gives sufficient heat generation and force the particles to distribute the reinforcement particles in wider area. FSP plate without RHA reinforced in surface shows that the porosity occupied the area. After, RHA reinforced in been stir in Al6061 plate, it shows that the porosity has been fully occupied with the reinforcement.

The average micro hardness of Al6061 at three different stage of tool rotational speed with constant traverse speed was observed. The value of tool rotational speed is 800 rpm, 1000 rpm and 1200 rpm while the value of traverse speed is 100 mm/min. The value of micro hardness has been taking at three different areas which are at the surface of FSP, at depth of 2.5mm and at depth of 5mm. Figure 5 show the chart of average micro hardness value when there is existing 5 weight % of RHA reinforced and without RHA reinforced at the surface.

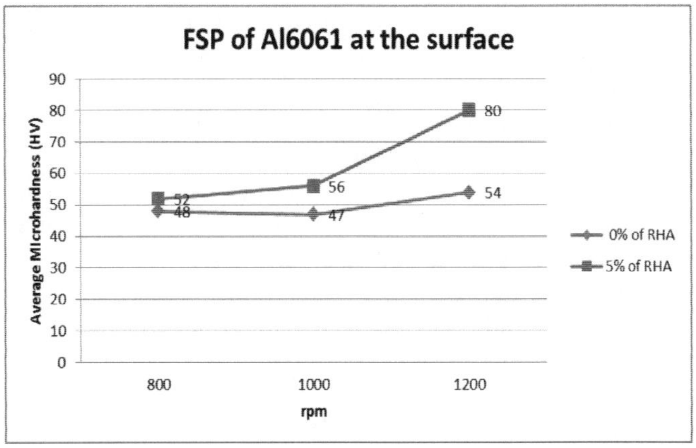

Figure 5. The chart of average micro hardness of specimens at the surface for three different tool rotational speeds.

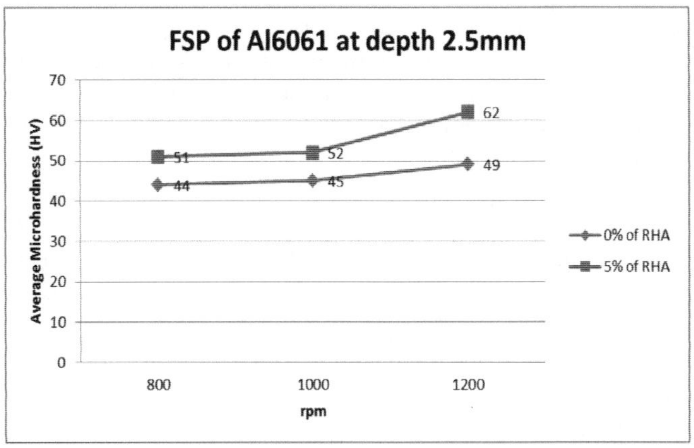

Figure 6. The chart of average micro hardness test of specimens at the depth of 2.5mm for three different tool rotational speeds.

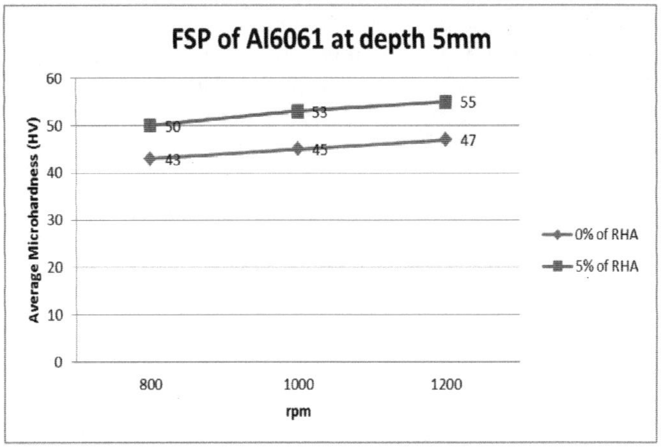

Figure 7. The chart of average micro hardness test of specimens at the depth of 5mm for three different tool rotational speeds.

In all specimens of FSP with RHA, it shown that the micro hardness is significantly increased with reduction of grain size and presence of RHA particles in Al6061 matrix. The overall result shows that the micro hardness value for the Al6061 at the surface and at the depth of 2.5 mm until 5 mm with existing of 5 weight % RHA were higher compared to the value of specimens without RHA reinforcement. In addition, the tool rotational speed also affecting the micro hardness value of the Al6061. By increasing the tool rotational speed, the micro hardness value also increased with or without the existing of RHA. Here the parameter plays an important role. With the increase of the micro hardness value, the size of grain decrease. This is because, when the tool stirs the surface of Al6061 it causes the Al6061 melt and deformation occurred. The distribution of RHA reinforced is uniform and the presence of RHA particles in FSP effected layer compared to the other region.

4. Conclusion

The objectives were successfully achieved which was to fabricate the Al6061 composites reinforced RHA. This was performed using friction stir processing (FSP) with three different tool rotational speed at 800 rpm, 1000 rpm and 1200 rpm with constant feed rate of 100mm/min. Plain specimen undergo FSP also been produced with the same tool rotational speed of 800 rpm, 1000 rpm and 1200 rpm with constant feed rate of 100mm/min. From the experiment, the microstructure of Al6061 with 5 weight % RHA reinforced show the uniform distribution at 1200 rpm compared to other speeds. In addition, the micro hardness of Al6061 with 5 weight % RHA reinforced was higher compared to the micro hardness of Al6061 no RHA reinforcement. The result from the micro hardness test at three different regions of surface, depth of 2.5mm and depth of 5mm using the micro hardness test were analysed. At the surface region showed the higher reading of micro hardness for the rotation speed 1200 rpm with 80Hv in average.

Acknowledgement

The authors gratefully appreciate Universiti Malaysia Pahang for the financial support through *Tabung Fakulti Kejuruteraan Pembuatan* Fund and Internal Research Grant RDU150371.

References

[1] Sharma V, Prakash U and Kumar B V M 2015. A review. *J. Mat. Pro. Tech.* **224** 117–134
[2] Devaraju, A, Kumar A, Kumaraswamy and Kotiveerachari B 2013. *Mater. Des.* **51** 331–341
[3] Palanivel R, Mathews P K, Murugan N and Dinaharan I 2012 *Mater. Des.* **40** 7-16
[4] Salih O S, Ou H, Sun W and McCartney D G 2015 *Mater. Des.* **86** 61-71
[5] Sharifitabar M, Sarani A, Khorshahian S and Afarani M S 2011 *Mater. Des.* **32** 4164-417
[6] Soltani N, Bahrami A, Pech-Canul M I and González L A 2015 *Chem. Eng. J.* **264** 899–935
[7] Vacaero.com/information-resources/metallography georgevander-voort/1217-metallography-and-microstructure-of-aluminum-and-alloys.html. Access date: 22/04/2016

IOP Conf. Series: Materials Science and Engineering **319** (2018) 012033 doi:10.1088/1757-899X/319/1/012033

A feasibility study in adapting Shamos Bickel and Hodges Lehman estimator into T-Method for normalization

N Harudin[1,2], K R Jamaludin[1], M Nabil Muhtazaruddin[1], F Ramlie[1], Wan Zuki Azman Wan Muhamad[1]

[1] Razak School in Engineering & Advanced Technology Universiti Teknologi Malaysia, 54100 Kuala Lumpur, Malaysia
[2] Department of Mechanical Engineering, Universiti Tenaga Nasional, 43000 Kajang Selangor, Malaysia

nolia.harudin@gmail.com, bkhairur.kl@utm.my

Abstract. T-Method is one of the techniques governed under Mahalanobis Taguchi System that developed specifically for multivariate data predictions. Prediction using T-Method is always possible even with very limited sample size. The user of T-Method required to clearly understanding the population data trend since this method is not considering the effect of outliers within it. Outliers may cause apparent non-normality and the entire classical methods breakdown. There exist robust parameter estimate that provide satisfactory results when the data contain outliers, as well as when the data are free of them. The robust parameter estimates of location and scale measure called Shamos Bickel (SB) and Hodges Lehman (HL) which are used as a comparable method to calculate the mean and standard deviation of classical statistic is part of it. Embedding these into T-Method normalize stage feasibly help in enhancing the accuracy of the T-Method as well as analysing the robustness of T-method itself. However, the result of higher sample size case study shows that T-method is having lowest average error percentages (3.09%) on data with extreme outliers. HL and SB is having lowest error percentages (4.67%) for data without extreme outliers with minimum error differences compared to T-Method. The error percentages prediction trend is vice versa for lower sample size case study. The result shows that with minimum sample size, which outliers always be at low risk, T-Method is much better on that, while higher sample size with extreme outliers, T-Method as well show better prediction compared to others. For the case studies conducted in this research, it shows that normalization of T-Method is showing satisfactory results and it is not feasible to adapt HL and SB or normal mean and standard deviation into it since it's only provide minimum effect of percentages errors. Normalization using T-method is still considered having lower risk towards outlier's effect.

1. Introduction

The theory of Mahalanobis Taguchi System (MTS) was inspired from a statistical tool called Mahalanobis Distance (MD) which introduced by a famous Indian statistician named Dr. Prasanta Chandra Mahalanobis in 1936. T-Method is one of various tool govern under MTS theory which specifically developed to calculate an overall prediction based on the signal to noise ratio to predict the future value based on old and current data together. Unique characteristic of T-method is the application of signal to noise ratio as measurement for the function robustness of target system. MTS including T-Method considered to be an ad hoc in the sense that they are not developed using any underlying statistical and probability theory as well as not been compared to any other multivariate statistical method. MTS is not based on probability and distribution theory since the theory developed only measure the descriptive statistic as agreed by [1]–[3]. The normalization stage in T-method for instance, is relying on the differences between unit space and signal data only without considering any statistical theory even the mean and standard deviation. The advantages of T-Method are its simplicity to be understood as well as its ability to do a prediction with a very limited sample size. It is very important for the user of T-Method to clearly understand the data since the method is not considering the effect of outliers within the population data.

In most cases, assumption that receives much attention from most of the statistician is that the regression analysis must be free from the effect of outliers. The errors and distribution was assumed to be normally distributed, observations are random, independent and identically distributed and equally reliable with no outlier in the data. Hodges-Lehmann estimator (HL) is defined as the median of the pairwise Walsh averages and been introduced since 1963 in various application such [4]–[11]. The main advantage of the HL estimator is that it is robust against outliers in a sample. It has a breakdown point of 0.29 (29%) is the least portion of data contamination needed to derive the estimate beyond all bounds). If the underlying distribution for the data is normal, then the asymptotic relative efficiency (ARE) of the HL estimator relative to the sample mean is 0.955 otherwise, it is often greater than unity[12]. It is well known that the classical standard deviation is not robust to outliers, where even a single outlier can have a severe effect[13]. The Shamos Bickel (SB) is an analogous scale estimator to the location estimator of Hodges-Lehmann. The square of the SB estimator can be used as an estimate for variances. Of course, SB is not unbiased or median unbiased for standard deviation, SB^2 is not unbiased or median unbiased for $\boldsymbol{\sigma}^2$. But they give good approximations to what they are estimating [14]. HL and SB are comparable method to normal mean and standard deviation which been applied in various area with intention of considering the outliers effect. The application of HL and SB is very well understood in control chart improvement as discussed by many researchers [12], [14]–[16].

This study was conducted to see the feasibility of embedding robust statistical estimators into T-Method for enhancing the accuracy of normalization steps especially towards outliers which currently T-Method is not relying on any statistical inference analysis. This study is limited to 2 different case studies with different sample size and number of unknown data for prediction. The result is compared based on the prediction of unknown data towards the actual value. The range and mean of the error percentages will be the comparison parameters as well as the R-Squared value. The conclusion driven from this study is mainly applied for these 2 case studies only since more examples are needed for firm fact in determining the robustness of T-Method.

2. Methodology

The T-Method concept is briefly explained in this chapter as well as the Hodges-Lehmann estimator and Shamos Bickel estimator. Most of the literatures are normally pronounced both estimator as Hodges–Lehmann location estimator and Shamos Bickel scale estimator which make them comparable to the theory of mean and standard deviation. The study had been conducted and analysed using two different case studies on prediction output.

2.1. T-Method

T-Method calculation procedure is a combination of Mahalanobis Distance concept with S/N Ratio (SNR). The theory behind T-Method is still focusing on three important elements of SNR which are sensitivity, linearity and variability. Considering on the dynamic environment adding some value added towards the theory developed. T-Method 1 as well applying the concept of reference-point proportional equation which created a linear regression that passes through origin of the graph generated. Unit group and signal group are used to establish and validate forecasting mode. Average value of unit group are subtract from each member of signal group for the normalization.

To find out which of the items (parameter) will be useful for prediction and estimation, an item by item computation of the proportional coefficient (β) and SN ratio (η) is performed. In T-Method, proportional coefficient (β) and SN ratio (η) are calculated between the output value (M_i) and the item value. The computation of the proportional coefficient (β) and SN ratio (η) is performed item by item with the use of normalized data (Xij) and normalized output value (M_i). The SN ratio η and proportional coefficient β are found for each item respectively. If the SN ratio η turns out to be negative, the value will be treated as zero. The integrated estimate output value for Signal Data is found using the proportional coefficient β and SN ratio η, item by item. The higher the SN ratio of an item, the greater the degree of its contribution to general output estimation. Integrated estimate output value $\hat{M}$ can be found using the following formula:

$$\hat{M}_i = \frac{\eta_1 \times \left(\frac{X_{i1}}{\beta_1}\right) + \eta_2 \times \left(\frac{X_{i2}}{\beta_2}\right) + \cdots . + \eta_j \times \left(\frac{X_{ij}}{\beta_j}\right)}{\eta_1 + \eta_2 + \cdots + \eta_j} \quad (2.1)$$

Formulas to calculate the SN ratio η (db) for integrated estimate values are shown below which is used to calculate the degree of contribution:

$$\text{Linear equation, L} = M_1\hat{M}_1 + M_2\hat{M}_2 + \cdots + M_i\hat{M}_i \quad (2.2)$$

$$\text{Effective Divider, r} = M_1{}^2 + M_2{}^2 + \cdots + M_i{}^2 \quad (2.3)$$

$$\text{Total Variation}, S_T = \hat{M}_1{}^2 + \hat{M}_2{}^2 + \cdots + \hat{M}_i{}^2 \quad (2.4)$$

$$\text{Variation of proportional term}, S_\beta = \frac{L^2}{r} \quad (2.5)$$

$$\text{Error variation}, S_e = S_T - S_\beta \quad (2.6)$$

$$\text{Error Variance}, V_e = \frac{S_e}{n-1} \quad (2.7)$$

$$\text{SN ratio}, \eta = 10 \log\left(\frac{(S_\beta - V_e)}{r\, V_e}\right) \quad (2.8)$$

Through the computations performed so far, output values have been normalized in terms of the average Unit Space value. Therefore, in order to find the integrated estimate value ($\hat{Y}$) of the actual output, the average value (M_0) of the output of the Unit Space is added to the normalized value ($\hat{M}$). For instance, the integrated estimate value ($\hat{Y}_1$) for the output of Signal Data No.1 is found as follow (similarly for unknown data):

$$\hat{Y}_1 = \hat{M}_1 + M_0 \quad (2.9)$$

2.2. Hodges Lehman estimator and Shamos Bickel estimator

The Hodges-Lehmann (HL) estimator was proposed by Hodges and Lehmann in 1967 as an estimator for the point of symmetry θ of a continuous and symmetric distribution. It is used for the estimation of the location parameter in one-sample and two-sample models[15]. In this study, only one-sample model is used since the intention is to compare HL to the mean of the distribution data.

For that case, let us assume a random variables *of* X_j $(1 \le j \le m)$ and Hodges-Lehmann estimator (HL) of the expected value $E(X) = E(X_i)$ *(the one-sample problem)* is follow:-

$$E^{HL}(X) = median\left(\frac{x_j + x_k}{2}\right) \quad (2.10)$$

where $1 \le k \le m$ [4]

To make the standard deviation (SD) comparable to Shamos Bickel scale estimator (SB), we consider the normalized SBSE as

$$SB(x) = Median\left(\frac{|x_j - x_k|}{0.9538726}\right) \quad (2.11)$$

Both the HL and SB will be used as a replacement of unit space and signal data deduction steps in calculating the normalization. The normal mean and standard deviation also been applied as part of comparison purposes.

2.3. Data Collection, equipment and tool setup

Data on power consumption prediction is taken from a real case study conducted at nuclear agency Malaysia which involved 15 variables and 14 sample size. The second data is taken from the [17] reference case study on yield% prediction which involving very minimum sample data (6 variables and 7 sample size). In order to test the robustness of each method, extreme outliers' data are purposely added to the population data to see the impact of each method compared. The prediction of unknown data for power consumption involved 13 sample data while yield% prediction is only for 1 data.

As for a comparable study, the formulation of T-Method including the effect of HL, SB, normal mean and standard deviation had been constructed using Matlab R2013A application software. HL and SB is calculated using R programming to get the output prior adapting the result into Matlab.

3. Result and Analysis

In this feasibility study, the main idea is to see how robust T-Method is when it comes to an existence of outliers in the data population. It is believe that by knowing the data trend very well, the effect of outliers might not exist at all since a crucial decision might done earlier. However a deep understanding on the data series is required for that purpose which might be a risk for certain people and can be considered not to be robust enough. As stated earlier, the analysis is involving 2 difference set of case study. Each of the case studies is then differentiate into 2 difference case of error% analysis which is with and without extreme outliers' consideration. The actual data of the unknown trend is taken for a comparison purposes.

Table 1 shows that without adding the extreme outliers into the data, the accuracy shown by HL and SB is much better compared to T-Method , Normal mean and standard deviation which on average error percentages of 4.67%. However, when extreme outliers were added, T-Method is showing higher accuracy compared to others with average error percentages of 3.09%. The R squared value for all method is not vary much and still showing high correlation between predicted and actual data. If compare the range of the error percentages for each method in table 1, T-method is showing minimum range compared to others. This highlighted that the prediction accuracy is having lower gap among samples.

The second case study on yield percentages prediction which summarize in table 2, are showing that HL and SB is having minimum error with extreme outliers case while T-Method is the minimum for case without extreme outliers. This is completely different compared to prediction of power consumption summarize in table 1. There is no range percentages analysis on this since the prediction is involving only 1 unknown data. A few things need to be pondered and question further is the total number of sample size considered in both cases. T method claim to be a good prediction tool for a minimum sample size data which expect no extreme outliers in it and it proved from the yield% prediction case. Minimum sample size can easily trace the dispersion of the data, so normality of it is easily controlled as well as the outliers. By adding extreme outliers on it purposely might be irrelevant to normal practise, however if its involved huge sample size, outliers might be not easily been controlled and normality of the data as well is hardly defined unless the data is really normal or else adjustment and assumption on normality is the usual practice. As for this study, it shows that T-Method is more accurate in doing prediction in case of minimum sample size as well as moderate sample size compared to Hodges Lehman and Shamos Bickel. Since the result is relying mainly on this study, future analysis on data with higher sample size is needed for better comparison.

Table 1. Prediction Vs. Actual data of power consumption

Normalization Method	Error% range for predicted and actual data (without extreme outliers)	R-Squared predicted Vs actual (without extreme outliers)	Error% range for predicted and actual data (with extreme outliers)	R-Squared predicted Vs actual (with extreme outliers)
T-Method (unit space)	Range:0.98%~8.85% Average: 6.50%	75.65%	Range:0.67%~7.29% Average: 3.09%	71.61%
HL and SB estimator	Range: 0.35% ~ 9.75% Average: 4.67%	77.67%	Range:0.13%~9.67% Average: 4.48%	69.85%
Normal mean & SD	Range:0.52% ~ 10.83% Average: 5.42%	77.27%	Range: 0.63%~11.19% Average: 5.48%	72.17%

Table 2. Prediction Vs. Actual data of yield % prediction

	Normalization Method	Estimated	Actual	Error%
Yield% prediction (Without extreme outliers)	T-Method	75.13	73.3	2.50%
	normal mean & SD	82.4725		12.51%
	HL & SB	82.3043		12.28%
Yield% prediction (With extreme outliers)	T-Method	83.7827	73.3	14.30%
	normal mean & SD	80.4		9.69%
	HL & SB	80.0028		9.14%

4. Conclusion

Comparing results in both Table 1 and Table 2 and by considering the irrelevant of adding the extreme outliers on the minimum sample size in yield percentage prediction case, it shows that T-Method is perform better than other 2 methods. However, relying on these 2 case studies might not be sufficient enough to summarize the overall effectiveness of robust normalization in T-method. This result is proved to be effective mainly on these 2 case studies only. Further analysis should be done to several other prediction cases so that the robustness of normalization in T-Method can be summarized in more accurate manner. As for this study, T-method was summarized to be more accurate in doing prediction compared to HL &SB and Normal mean & SD. Therefore, it is not feasible to adapt the Shamos Bickel and Hodges Lehman into T-method for the normalization analysis within this study. However it is very useful to have HL and SB as a comparable method dealing with normalization process. Normalization using T-method is still considered having lower risk towards outlier's effect.

Acknowledgments

Authors wish to acknowledge Mr. Yusnan from nuclear agency Malaysia for his willingness in supporting this research by providing guidance and permission in data collection stage. Special thanks to ministry of higher education (MoHE, Malaysia) and Universiti Tenaga Nasional for their financial support throughout the author research duration period. This research paper publication is mainly supported by Tier 1 Research University Grant No: 15H43 (Universiti Teknologi Malaysia).

References

[1] K. Tsui, T. Sukchotrat, and V. C. P. Chen, "A comparison study and discussion of the Mahalanobis-Taguchi System Seoung Bum Kim," *Int. J. Ind. Syst. Eng.*, vol. 4, no. 6, pp. 631–644, 2009.

[2] W. H. Woodall, R. Koudelik, K. L. Tsui, S. B. Kim, Z. G. Stoumbos, and C. R. Carvounis, "A review and analysis of the Mahalanobis-Taguchi system," *Technometrics*, vol. 45, no. 1, pp. 1–15, 2003.

[3] S. Taguchi, R. Jugulum, G. Taguchi, and J. O. Wilkins, "Discussion," *Technometrics*, vol. 45, no. 1, pp. 16–21, 2003.
[4] R. Duchnowski and Z. Wiśniewski, "Accuracy of the Hodges–Lehmann estimates computed by applying Monte Carlo simulations," *Acta Geod. Geophys.*, 2016.
[5] C. H. Park, S. Lee, and J. H. Chang, "Robust closed-form time-of-arrival source localization based on α-trimmed mean and Hodges-Lehmann estimator under NLOS environments," *Signal Processing*, vol. 111, pp. 113–123, 2015.
[6] I. A. Canay and T. Otsu, "Hodges-Lehmann optimality for testing moment conditions," *J. Econom.*, vol. 171, no. 1, pp. 45–53, 2012.
[7] E. Inference and D. Adminl, "Inaccuracy rates and Hodges-Lehmann large deviation rates for parametric inferences with nuisance parameters," vol. 8, 1995.
[8] Y. Y. Nikitin, "Hodges-Lehmann and Chernoff efficiencies of linear rank statistics," *J. Stat. Plan. Inference*, vol. 29, no. 3, pp. 309–323, 1991.
[9] J.N.Adichie, "Estimates of Regression Parameters Based on Rank Test," *Ann. Math. Stat.*, vol. 38, no. 3, pp. 894–904, 1967.
[10] Hodges Jr, J.L. and Lehmann, E.L., "Estimates of location based on rank tests," *Ann. Math. Stat.*, p. pp.598-611, 1963.
[11] P. j. B. E.L.Lehmann, "Descriptive statistics for nonparametric models. III. Dispersion," *Ann. Stat. ® www.jstor.org*, vol. 4, no. 6, pp. 1139–1158, 1976.
[12] H. Z. Nazir, M. Riaz, R. J. M. M. Does, and N. Abbas, "Robust CUSUM control charting," *Qual. Eng.*, vol. 25, no. 3, pp. 37–41, 2013.
[13] G. Tarr, N. C. Weber, and S. Müller, "The difference of symmetric quantiles under long range dependence," *Stat. Probab. Lett.*, vol. 98, pp. 144–150, 2015.
[14] M. O. Abu-shawiesh, M. B. Abdullah, Q. Methods, and S. Arabia, "A New Robust Bivariate Control Chart for Location," *Commun. Stat. Comput.*, no. March 2012, pp. 37–41, 2007.
[15] M. O. Abu-Shawiesh and M. B. Abdullah, "New Robust Statistical Process Control Chart for Location," *Qual. Eng.*, vol. 12, no. 2, pp. 149–159, 1999.
[16] A. A. Aly and Aydin ozturk, "Hodges-Lehmann quantile-quantile plots," *Comput. Stat. Data Anal.*, vol. 6, pp. 99–108, 1988.
[17] S. Teshima, Y. Hasegawa, and K. Tatebayashi, *Quality Recognition and Prediction: Smarter Pattern Technology with the Mahalanobis-Taguchi System*. 2012.

Occupational Noise Reduction In CNC Striping Process

Kamarulzaman Mahmad Khairai[1,2], Nurul Shamime Salleh[2] and Ahmad Razlan Yusoff[2,3]

[1]TT Technologies Corporation Sdn Bhd, Jalan Tanjung Api,26050 Kuantan, Pahang
[2]Faculty of Manufacturing Engineering, Universiti Malaysia Pahang, 26600 Pekan, Pahang, Malaysia

E-mail[3]: razlan@ump.edu.my

Abstract. Occupational noise hearing loss with high level exposure is common occupational hazards. In CNC striping process, employee that exposed to high noise level for a long time as 8-hour contributes to hearing loss, create physical and psychological stress that reduce productivity. In this paper, CNC stripping process with high level noises are measured and reduced to the permissible noise exposure. First condition is all machines shutting down and second condition when all CNC machine under operations. For both conditions, noise exposures were measured to evaluate the noise problems and sources. After improvement made, the noise exposures were measured to evaluate the effectiveness of reduction. The initial average noise level at the first condition is 95.797 dB (A). After the pneumatic system with leakage was solved, the noise reduced to 55.517 dB (A). The average noise level at the second condition is 109.340 dB (A). After six machines were gathered at one area and cover that area with plastic curtain, the noise reduced to 95.209 dB (A). In conclusion, the noise level exposure in CNC striping machine is high and exceed the permissible noise exposure can be reduced to acceptable levels. The reduction of noise level in CNC striping processes enhanced productivity in the industry.

1. Introduction

Occupational noise, as a common occupational hazard, generally refers to noise at work or noise in the workplace. For the record, 16% of disabling hearing loss in adults is attributed to occupational noise worldwide [1]. For a manufacturing plant to evolve, an additional challenge appears such as requirements the Occupational Health and Safety problems (OHS) to limit the noise impact of manufacturing plants on operators [2] Among the aggression undergone by operators in a machining workshop, noise is a critical phenomenon since it affects them daily without any obvious short-term impact on hearing [3].

In most industrialized nation nowadays, noise becomes one of the major sources of environmental pollution [4]. Noise at CNC area come from several source such as machine itself or any additional element that add into the process. In industry it is very important to control the level of noise so that it will not harm workers. Extreme noise may harm workers when they are exposing to if for a long time as their working condition is not properly monitor by management. It is management duty to provide safe and comfortable working areas to the workers. That is also enforced by Department of Safety and Health (DOSH) about regulation regarding safety condition to the workers, any company that refuse to follow the regulation will be fined or the worst can be force to stop their operation. Increasing numbers of modern people are suffering hearing loss, and undue exposure to noise is one of the important reasons [5].

In most CNC area, workers become uncomfortable due to noise that produces from the machine. Some place chooses to use personnel protective equipment (PPE) to protect their workers as an option. The company must offer individual protection to the operators and to test their hearing capacity annually if the noise reading between 80 and 85 dBA. In the manufacturing industry the critical daily noise exposure is often exceeded 85 dBA level [6]. Exceed noise exposure to workers will harm them in many ways especially their hearing ability. Several workers undergoing specific tasks in this XYZ company was observed to be working in high noise zones. They may be exposed to noise levels exceeding equivalent continuous sound level of 85dB (A) which is set as the action level in Malaysia. Some of the high noise area are identified were from the air compressor and CNC Machine area [7]. Within the range that the noise are produce in CNC Machine, the workers are need to wear a proper Personal Protective Equipment (PPE) especially earmuffs to cover their hearing [8,9]. In additional, the high noise also causes the behavioral effect and physiological effect for the workers [9-11].

In this paper, it is initially needed to identify source of noise at CNC area by measuring using sound level meter. To analyse a safe and comfortable working condition to the workers in CNC area after implement noise reduction method, noise level need to be measured the effectiveness of current methods.

2. Experimental details

Figure 1 shows the noise monitoring plant layout for data collection. Data are collected during working hours from CNC striping process area. By using CNC machine area plant layout, noise monitoring method was developed in detail on where the 11 points to be placed and repeated the instrument sound level meter. The instrument used to measure the noise level in this experiment is CEN0053 Heavy Duty Data Logger Sound Level Meter. It can measure minimum 30dB and maximum 130dB. There are several data were measured in this study during idle time, without operating conditions and during operating conditions. After that, current CNC machines were improved their noise level by several methods. All of this method is used to meet DOSH requirement on safety working condition on noise exposure to the workers. Most company will prefer low cost method and in the same time pass requirement set by DOSH. Finally, the similar data collection of noise was conducted to measure the effectiveness of current method to reduce noise level at CNC striping process.

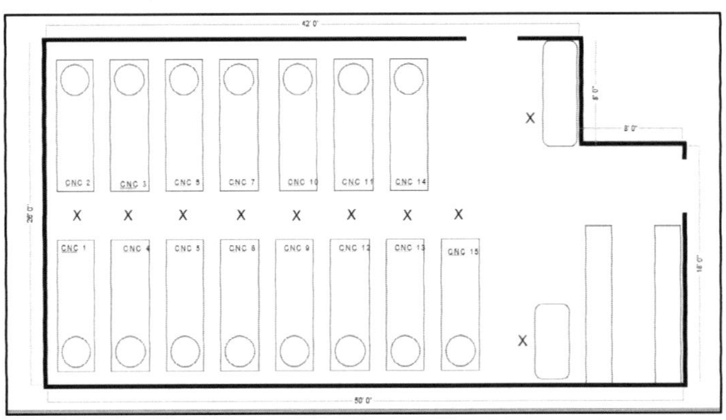

Figure 1. Noise monitoring plant layout

3. Result and Discussion

Figure 2a shows the noise monitoring at the idle time. The maximum noise level are in point 6 with 69.3 dB (A) and 49.7 dB (A) for the minimum noise level. The average between the maximum and minimum are 58.8 dB (A). Figure 2b shows the noise mapping for the noise monitoring at the idle time. Noise level average when the entire machines at CNC Machine area switch off is 53.214 dB (A). Source of noise is important to determine because the noise level produce affecting the whole area. There is a few spot that are potential to affect the noise level after all the machine without operations. There are four machine involve in this experiment which is CNC3, CNC6, CNC9 and CNC14. There are four spot that produce noise exactly after the machine are off.

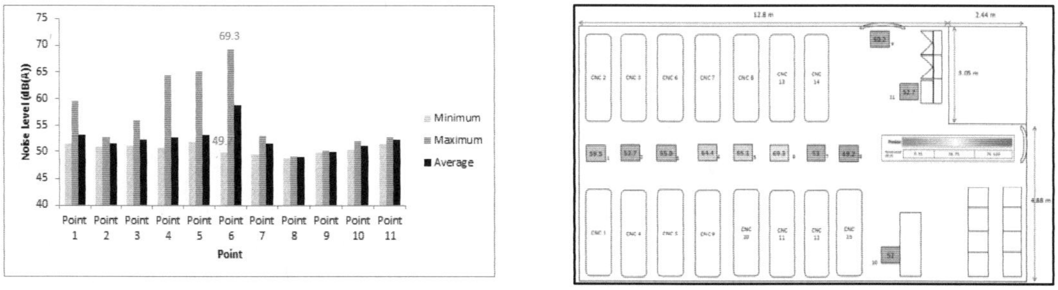

a) Noise monitoring b) Noise level mapping

Figure 2. Noise monitoring at idle time

Figure 3a shows point 1 at machine CNC3. This is because of the airs that blow out from spot at red circle. Even the machine are running the process, the air will blow to push the air coil product. Therefore, this point affects the high noise level monitoring with 98.2 dB (A). Figure 3b also shows the point 2 and point 3 cause the noise sound at machine CNC6 and CNC9 respectively. These points are same as previous point caused the noise from pneumatic. The points are where the stripping process taking place. Consequently, the air pump blow the dust from the coil and cause a noise level at 93.5 dB (A) and 91.2 dB (A), respectively. Figure 3c shows the point 4 at machine CNC14. The air pressure blows out from the red circle about two minutes after the machine are off. The air pressure at this spot is 105 dB (A), which is very high level.

a) CNC3 b) CNC6 c) CNC9 d) CNC14

Figure 3. Noise source at four points at CNC

Figure 4a shows the noise monitoring when machine not operate. The maximum noise level are in point 7 with 106.8 dB (A) and 103.2 dB (A) for the minimum noise level. The average between the maximum and minimum are 105.0 dB (A). Figure 4b shows the noise mapping for the noise monitoring without machine operation. Noise level average when the entire machines at CNC machine are switch-off is 95.797 dB (A).

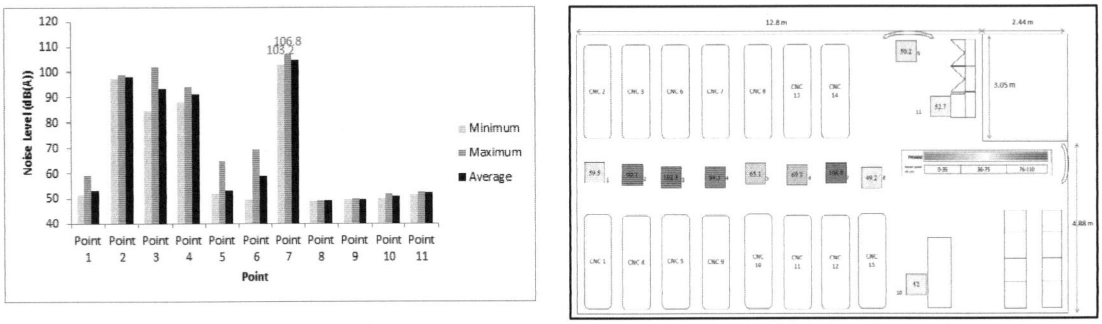

a) Noise monitoring b) Noise level mapping

Figure 4. Noise monitoring when machine without operating condition

Results are now presented for the entire CNC machine during operating conditions. There are 11 points that used to be measured the noise levels. Figure 5a shows the noise level for 11 points in this room during operating conditions. The maximum noise level are in point 7 with 118.2 dB (A) and 112.4 dB (A) for the minimum noise level. The average between the maximum and minimum are 115.3 dB (A). Figure 5a shows the noise mapping for the noise monitoring at operating machine condition. The noise levels average is 95.797 dB(A). The output indicated that there are several points producing very high noise. It is influenced by the machine running the stripping process. Even though the without operating machine, there are still a noise produced for a certain time and this conclude that the noise caused by striping process.

4.

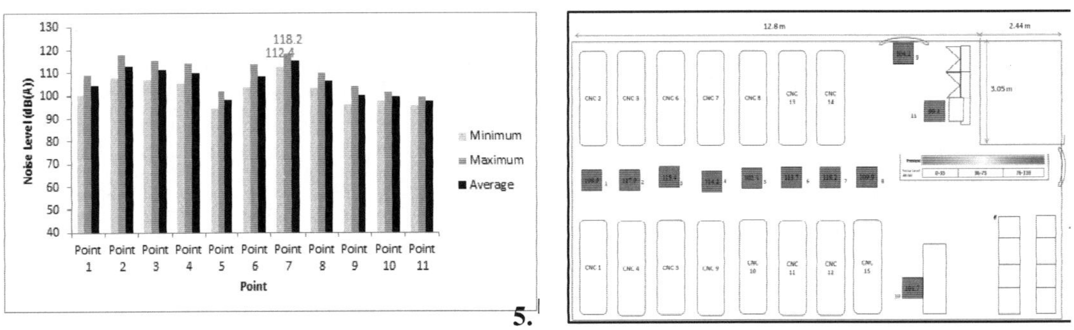

5.

a) Noise monitoring b) Noise level mapping

Figure 5. Noise monitoring when machine in operating condition

First, the high noise at CNC area problem is pneumatic leakage. From the noise monitoring, there are four CNC Machine contribute to pneumatic leakage which are CNC3, CNC6, CNC9 and CNC14. The problem was solved by installation of pneumatic muffler. Figure 6a shows pneumatic muffler used to solve pneumatic leakage. After considering some causes of high noise at CNC area and also how operators and technician doing their job in that area, a suitable method needs to be selected. For this case, plastic curtain was installed to reduce the noise. Initially, all CNC machines with stripping process were moved to one side of CNC area to be be covered by plastic curtain. It is make separation between the high and low noise from the machine. It is also seems to be most suitable for maintaining job. This plastic material can help reduce noise about 10db to 20db when apply at noise exposure area like CNC Area [7]. So for this case it is suitable to be used as it will not disturb routine work and maintenance on that area. Figure 6b and 6c shows condition after plastic cover was installed.

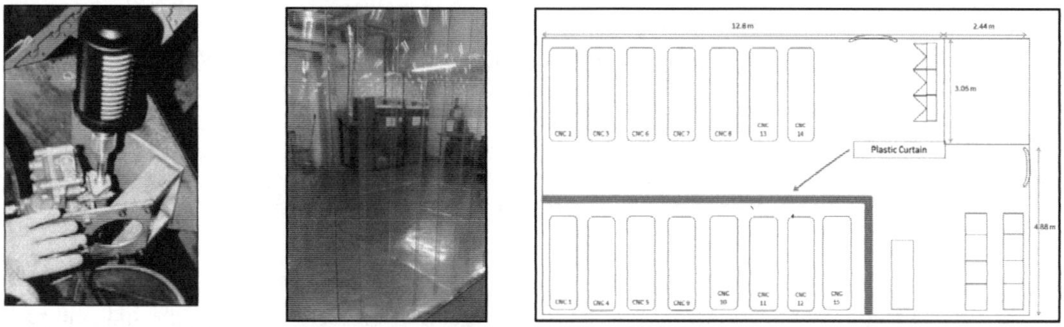

a) Pneumatic muffler b) Plastic Curtain c) CNC area with plastic curtain

Figure 6. Improvement made for current high noise levels

Figure 7a shows comparison of noise level reading without machine working after solving the pneumatic leakage problem at CNC3, CNC6, CNC9 and CNC14. The noise levels decrease to 40.28 dB (A) with 42% improvement on noise reduction. At point 2, 3, 4 and 7 on before improvement, the noise is very high due to pneumatic leakages. After pneumatic leakage was solved by pneumatic muffler installed, the noise was greatly reduced. Figure 7b shows the comparison between noise level reading at machine during operating conditions after installed with plastic curtain at certain area. The noise level are decrease to 14.131 dB (A). It is about 12.9% improvement on noise reduction. Noises at all points are reducing consistently after plastic curtain was installed. After improvement were made, the noise during operating conditions reduce from 109.340 dB (A) to 98.246 dB (A).

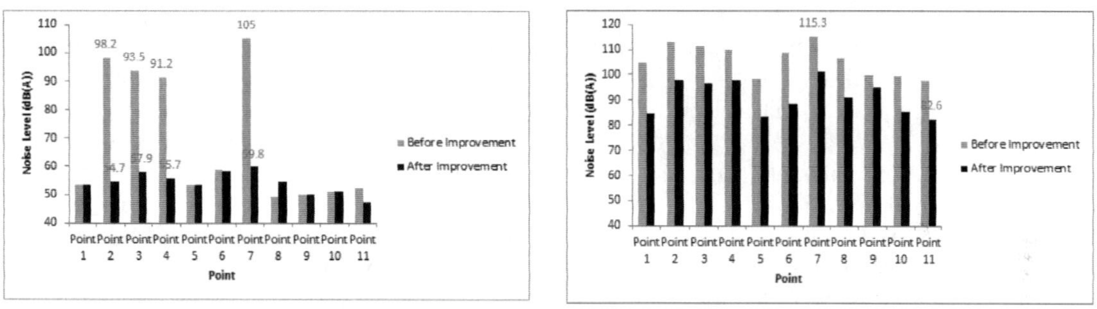

a) Without operation condition b) During operation condition

Figure 7. Noise monitoring reduction

The machine OEE was a data recorded monthly by production department to show on CNC machine performance. It is generally affected by CNC machine downtime and output by the machine. When the downtime was low and output was high the machine OEE increased. In this case, technician plays a role to reduce machine downtime and operators on increasing machine output. Table 1 shows the comparison of machine OEE performance before and after improvement was made. Obviously, output was improved on CNC area. Output, unplanned downtime and actual operating time was improved after improvement, it shows that machine work efficient. This condition happens when workers on that area have better working environment by reducing noise at CNC Area. Over limit noise that expose to workers can reduce their efficiency while working and harm them after a long time. By reducing noise at CNC Area, technician can focused more to do machine maintenance so that machine downtime can be lower and operators can work more efficient to make sure machine producing higher output with machine OEE increase by 12.10%.

Table 1 Comparison of machine OEE performance

Performance	Before Improvement	After Improvement
Output	1,779,000 pcs	2,340,000 pcs
Unplanned Downtime	630 hours	350 hours
Actual Operating Time	6,060 hours	7,210 hours
OEE	54.9%	67.0%

The other recommendation is to provide the employee with training about hearing damage and protection. This is important to the employee that exposed to the noise for a long time as 8-hour shift. In addition, the company should provide the correct hearing protection due to the high noise level present. Even though, the recommendation is for a short term because the noise level may be high in time. Hence, the long term solution should be implementing in that area to reduce more noise from time to time. As for pneumatic leakage problem, technicians have to plays an important role to do proper preventive maintenance and predictive maintenance to the CNC machine. It is to ensure stripping mechanism that using pneumatic run in a good condition and did not contribute much to the high noise at CNC area. A proper setting during installation also needed to avoid any pneumatic leakage after machine running. Permanent action can be done to solve pneumatic problem is to change from pneumatic mechanism to motor mechanism. By using motor, the noise produced lower noise compare to pneumatic. But, it is need to use dust proof motor as the stripping process produce a lot of dust that can affect motor performance.

6. Conclusions

For the first condition, there are four major points that produce the noise. Most of the noise point produce by the pneumatic pressure. The pressures are quite high and affect the noise level. The average noise level at the first condition before improvement was made is 95.797 dB (A) and this cause at the part when the coil have stripping process using pneumatic system with leakage. Then pneumatic leakage problem was solved, after improvement data was taken and the noise was reduced to 55.517 dB (A). For the second condition, there are about six CNC machines that have stripping process with it when running so these six machines contribute most in high noise. The average noise level at the second condition before improvement was made is 109.340 dB (A). By gather all six machine at one area and cover that area with plastic material, noise reading was taken. After improvement data shows the noise reduce to 95.209 dB (A). From noise monitoring conditions, the noise level exposure is the CNC machine room area is high and exceeds the permissible noise level exposure regulation. After some modification and improvement made, it can be reduced maximum 42 percent to acceptable levels. The reduction of noise level in CNC striping processes enhanced productivity in the industry about 12 percent.

Acknowledgements

The authors are grateful to Universiti Malaysia Pahang (UMP) for financial supported by grant RDU 170372.

References

[1] Nelson, D.I., Nelson, R.Y., Concha-Barrientos, M., Fingerhut, M. (2005). The global burden of occupational noise-induced hearing loss. Am. J. Ind. Med. 48 (6), 446-458.

[2] Rech, J., Dumont, F., Le Bot, A., Arrazola, P.J. (2017) Reduction of noise during millinh operations CIRP Journal of Manufacturing Science and Technology, 18, 39-44,

[3] Bollinger, J.G., (1973) Noise – An Industrial Pollutant of International Concern, CIRP Annals, 22(2), 197–202.

[4] Barron, R.F. (2003) Industrial Noise Control and Acoustic. First ed., Marcel Dekkel AG, New York.

[5] Bell, D.H. & Bell, L.H (1994). Industrial Noise Control, Fundamental and Application. Second Edition, Revise and Expand, Marcel Dekkel AG, New York.

[6] Bley, H., Gunther, G., Haeusler, J, Noe, E.L., (1980), Machine Concentration and Noise Annoyance in the Workshop, CIRP Annals-Manufacturing Technology, 29(1), 269–273.
[7] Mendonc, C., Freitas, E., Ferreira, J.P., Raimundo, I.D. and Santos, J.A. (2013), Noise abatement and traffic safety: The trade-off of quitter engines and pavements on vehicle detection Accident Analysis and Prevention 51, 11– 17.
[8] Fetoni AR, De Bartolo P, Eramo SL, Rolesi R, Paciello F and Bergamini C (2013), Noise induced hearing loss (NIHL) as a target of oxidative stress-mediated damage: cochlear and cortical responses after an increase in antioxidant defense. Journal of Neuroscience, 33(9), 4011–4034.
[9] Fernandez, M.D., Quintana, S., Chavarría, N., Ballesteros, J.A., (2009). Noise exposure of workers of the construction sector. Appl. Acoust. 70 (5), 753-760.
[10] Dyn, J. An Analysis of a Pneumatic Servo System and Its Application to a Computer-Controlled Robot, Journal of Dynamics Systems, Measurement and Control, 110(3), 228-235.
[11] WHO (1999), Guidelines for Community Noise, Edited by Birgitta Berglund, World Health Organization, Thomas Lindvall, and Dietrich Schwela, Geneva, April 1999.

Optimization of Surface Roughness and Wall Thickness in Dieless Incremental Forming Of Aluminum Sheet Using Taguchi

Zamzuri Hamedon[1], Shea Cheng Kuang[1], Hasnulhadi Jaafar[2] and Azmir Azhari[1]
[1]Faculty of Manufacturing Engineering, Universiti Malaysia Pahang, Malaysia
[2]School of Manufacturing Engineering, Universiti Malaysia Perlis, Malaysia
E-mail: zamzuri@ump.edu.my

Abstract. Incremental sheet forming is a versatile sheet metal forming process where a sheet metal is formed into its final shape by a series of localized deformation without a specialised die. However, it still has many shortcomings that need to be overcome such as geometric accuracy, surface roughness, formability, forming speed, and so on. This project focus on minimising the surface roughness of aluminium sheet and improving its thickness uniformity in incremental sheet forming via optimisation of wall angle, feed rate, and step size. Besides, the effect of wall angle, feed rate, and step size to the surface roughness and thickness uniformity of aluminium sheet was investigated in this project. From the results, it was observed that surface roughness and thickness uniformity were inversely varied due to the formation of surface waviness. Increase in feed rate and decrease in step size will produce a lower surface roughness, while uniform thickness reduction was obtained by reducing the wall angle and step size. By using Taguchi analysis, the optimum parameters for minimum surface roughness and uniform thickness reduction of aluminium sheet were determined. The finding of this project helps to reduce the time in optimising the surface roughness and thickness uniformity in incremental sheet forming.

1. Introduction

Since the number of vehicles increasing over the last two decades, new model with new design are introduce to market more often. The requirement remain to have a faster to market whilst without neglecting the safety issue became crucial. New materials was introduce either for lighter and yet stronger vehicles or fuel efficient vehicles [1]. Continuous researches are carried out in optimizing the design of vehicle to obtain efficient energy vehicle. The used of aluminum and high strength steel became equal important. Hamedon et al study a new method in joining a high strength steel and reported that the fatigue strength of hemming of high strength is stronger compare to the spot welding [2-3]. Although strength and safety are important, manufacturing lighter vehicles is also become a priority to high end car manufacturer. Since the aluminum is most demanding material for lighter vehicle, the quickest and easiest technique in producing body parts of this low volume vehicles is by using incremental sheet forming (ISF).

Incremental sheet forming (ISF) is a versatile sheet metal forming process where a sheet metal is formed into its final shape by a series of localized deformation. Generally, the process can be carried out on a CNC machine, where the perimeter of the sheet metal is clamped in a special blank holder. While the forming tool is attached to the CNC machine, it is usually round-ended with a diameter of 5 to 20mm, moving along a designed tool path and continuously indent the sheet following the contour until the final part is formed. In a typical ISF process, a general round-ended forming tool is moved along the NC controlled tool path, the tool moves downwards, indents the sheet by a specific depth, causing localized deformation in the sheet, then draws a contour on a horizontal plane, and then makes a step downwards, draws the next contour, makes the next step downwards, and so on [4]

Many issues and breakthrough finding were identified by researchers in the ISF. Tool diameter is one of the significant process parameters in ISF as it will not only affects the formability but also the surface finish of the sheet. Duflou et al showed that increase in tool diameter will increase the required force for forming [5]. Kim and Park found that increasing of tool diameter will increase the forming depth due to increase of contact zone [6]. However Park and Kim's [7] work showed that crack is easily occurred due to biaxial mode of deformation. Oleksik's work revealed that smaller tool diameter along with larger vertical step size will increase the maximum thickness reduction of the sheet metal [8]. Moreover, Malwad and Nandedkar did some experiments to understand to influence of the wall angle and they had observed that larger wall angle will result in a higher thickness reduction [9] . While effect of feed rate was studied by Hamilton and Jeswiet and reported at the feed rate lower than 2540 mm/, characteristic thinning occurred [10]. Beside all above mentioned issued, Echrif and Hrairi suggested a smaller step size will gave surface waviness and very smooth surface [11]. Malwad and Nandedkar's reported that surface roughness increased along with the step size [9].

Although ISF is considered a promising and feasible technology in forming sheet metal products, many researches are still undergoing to improvise the process such as improving the formability, improving the accuracy, eliminating springback, optimizing surface roughness etc. One of the most common research is the optimization of process parameters [12]. The aim of the project is to optimise the wall angle, feed rate, and step size in the ISF process for aluminium sheet to obtain minimum surface roughness and uniform thickness reduction.

2. Experimental procedures

2.1 Process

Firstly, the part to be formed in ISF process is designed in CAD software (CATIA), with the variance of wall angles in which the optimum one will be determined after the experiment. Then, the tool path will be generated on a CAM platform (CATIA), which is fixed throughout the experiment. The generated G-code for the ISF process will then ready to transfer to the 3-axis CNC milling machine.

The forming tool for ISF process is a mild steel rod, was fabricated using a 3-axis CNC turning machine with a ball end and a diameter of 10 mm which is kept constant for all experiments. Figure 1 shows the forming tool of ISF with rough surface to increase the friction between arbour and tool thus preventing slip during the process. Sheet material is aluminum 6061.

Figure 1: Forming tool for ISF

IOP Conf. Series: Materials Science and Engineering **319** (2018) 012035 doi:10.1088/1757-899X/319/1/012035

2.2 Experimental setup

The jigs for the ISF process is fabricated using mild steel rectangular hollow bars by welding each of it into a square shape. The entire ISF experiment was carried out on a 3 axis CNC milling machine. Figure 2 shows the experimental setup with complete structure of the jigs and set up on the CNC machine.

Figure 2: Experiment setup for ISF on CNC milling machine

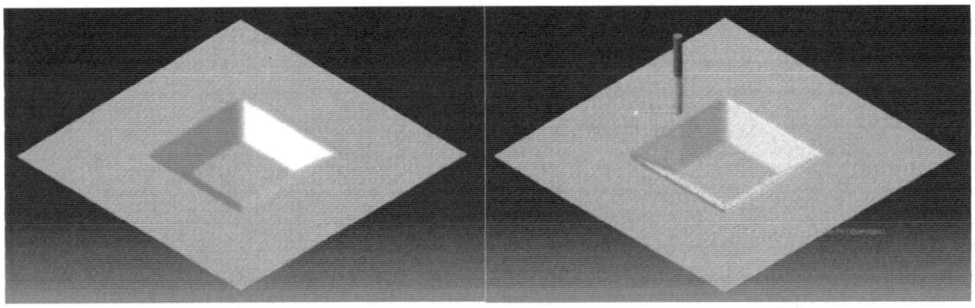

(a) (b)

Figure 3: (a) Pyramid frustum shape, (b) Tool path generation in CATIA

In this experiment, wall angle, feed rate, and step size are going to be investigated with 3 levels each. Table 1 shows the experimental setting for each process parameter, and Table 2 shows the different combinations of every process parameters carry out in this experiment. Figure 3(a) shows the shape of aluminium sheet going to form in ISF, which is a pyramid frustum. The forming depth is fixed at 24 mm throughout the experiment. While Figure 3(b) shows proposed shape and tool path of ISF generated in CATIA which is inward helical along the contour only. Spindle speed was kept constant at 1500 rpm.

Table 1: Process parameters and level descriptions

Parameters	**Level 1**	**Level 2**	**Level 3**
Wall Angle	35°	45°	55°
Feed Rate (mm/min)	700	900	1100
Step Size (mm)	0.25	0.50	1.00

Table 2: Design of experiment plan

Experiment	Wall Angle	Feed Rate (mm/min)	Step Size (mm)
1	35°	700	0.25
2	35°	900	0.50
3	35°	1100	1.00
4	45°	700	0.50
5	45°	900	1.00
6	45°	1100	0.25
7	55°	700	1.00
8	55°	900	0.25
9	55°	1100	0.50

3. Results

The formed aluminium sheets were analysed for average surface roughness on four sides of internal slopes and four different radius size of internal corner. Meanwhile, the average thickness reduction of the sheet was measured and compare with the result obtained from the sine law, the uniformity of thickness reduction was investigated as well. The respective areas to be analysed were labelled as shown in Figure 4.

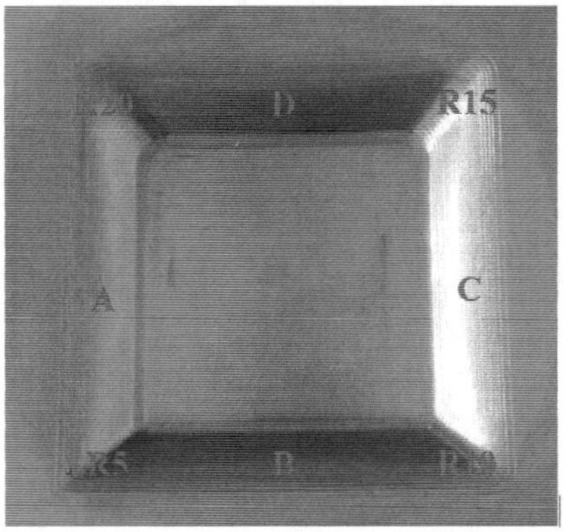

Figure 4: Area for location of measurement taken

Table 3 showed that Experiment 4 has the lowest average surface roughness while Experiment 9 produced the highest average surface roughness. For wall angle of 35°, surface roughness decreased when the feed rate and step size increased. However, the surface roughness increased when the feed rate increased for 45° and 55° of wall angle. Figure 5 shown surface conditions of aluminium sheet for different parameters from experiment 5 and experiment 6.

Table 3: Summary of average surface roughness of aluminium sheet in ISF

Experiment	1	2	3	***4***	5	6	7	8	***9***
Average Surface Roughness (μm)	6.966	6.315	5.126	***2.333***	3.062	6.578	2.772	5.757	***7.178***

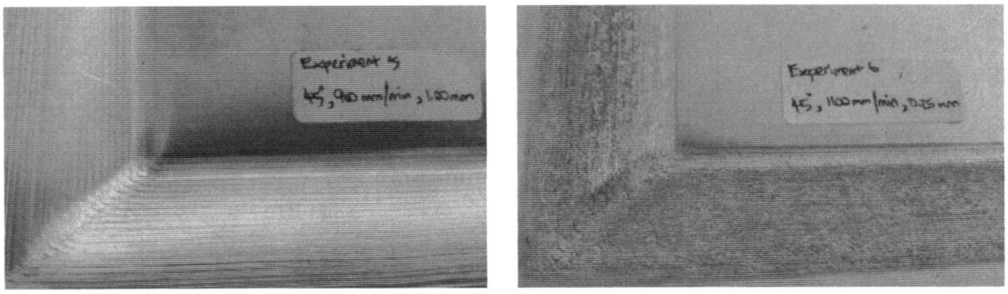

(a) (b)

Figure 5: Surface of aluminium sheet for different parameters (a) experiment 5 and (b) experiment 6

Table 4: Summary of thickness reduction of aluminium sheet in ISF

Experiment	Thickness Reduction (mm)					Standard Deviation
	Area A	Area B	Area C	Area D	Average	
1	0.8322	0.8265	0.8300	0.8254	0.8285	0.0031
2	0.8355	0.8495	0.8616	0.8584	0.8513	0.0117
3	0.8970	0.8691	0.8639	0.8384	0.8671	0.0240
4	0.7337	0.6900	0.7299	0.6965	0.7125	0.0225
5	0.7400	0.6959	0.7440	0.7373	0.7293	0.0224
6	0.7300	0.7379	0.7228	0.7179	0.7272	0.0087
7	0.6084	0.6152	0.5779	0.5762	0.5944	0.0203
8	0.6185	0.5731	0.5648	0.5712	0.5819	0.0247
9	0.5698	0.5644	0.5898	0.6112	0.5838	0.0213

Table 4 shows the standard deviation of average thickness reduction of aluminium sheet for each experiment. A smaller value of standard deviation indicates that the thickness from each area are closer to the mean, which means that the thickness reduction is more uniform.

3.1 Parameters Optimisation for Surface Roughness

In order to minimise the surface roughness of aluminium sheet in ISF, optimisation of parameters was done from the results of the 9 experiments using Minitab software, response graphs for means and S/N ratios were generated. Figure 6 displayed the response graph for S/N ratios and means for surface roughness of aluminium sheet. Generally, a smaller-is-better S/N ratio.

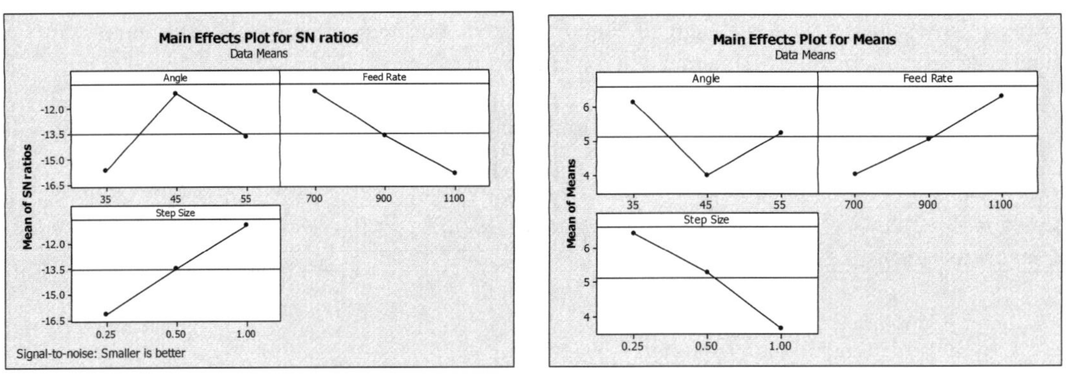

Figure 6: Response graph for S/N ratios and means for surface roughness

3.2 Parameters Optimisation for Uniform Thickness Reduction

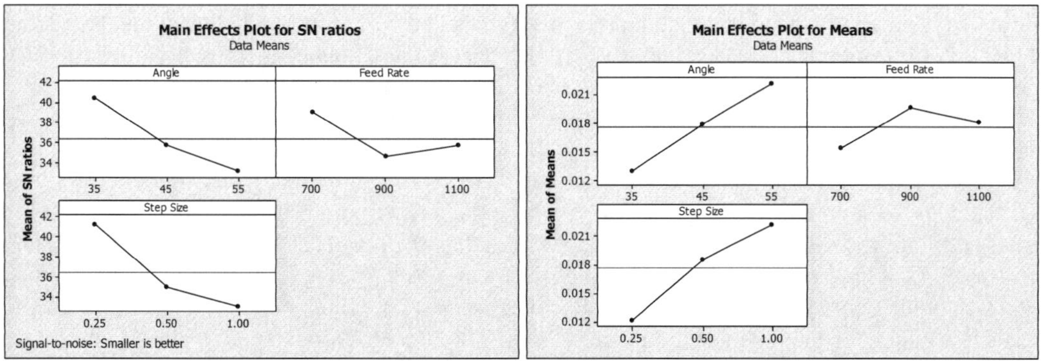

Figure 7: Response graph for S/N ratios and means for thickness uniformity

From the response graph of mean of means (thickness uniformity) as shown in figure 7, increase in wall angle and step size caused the thickness of aluminium sheet less uniform. On the other hand, the thickness uniformity decreased at 900 mm/min and rose at 1100 mm/min.

A confirmation test was carried out to verify the results of parameters optimisation. The optimised parameters for surface roughness and thickness uniformity was summarised in Table 5. However, the optimised parameters for thickness uniformity was exactly same with Experiment 1, therefore a confirmation test was not required. The result of average surface roughness with optimised parameters was revealed in Table 6.

Table 5: Optimised parameters for each outcome

Optimisation	**Angle**	**Feed Rate (mm/min)**	**Step Size (mm)**
Surface Roughness	45°	700	1.00
Thickness Uniformity	35°	700	0.25

Table 6: Comparison of actual surface roughness with the predicted value

Area	**Surface Roughness (μm)**			**Error (%)**
	Prediction	**Actual**	**Difference**	
Overall	2.469	1.705	0.764	30.9437
R5	1.332	0.885	0.447	33.5586
R10	1.525	1.564	0.039	2.5574
R15	1.491	2.290	0.799	53.5882
R20	2.85	2.122	0.728	25.5439

From Table 6, it can be observed that the difference between the predicted and actual surface roughness of aluminium sheet showed a relatively high error percentage except for area R5.

4. Discussions

Durante et al had proposed that the presence or absence of tool rotation will affects the value of surface roughness within the range of 10% [13]. Initially, these experiments was intended to carry out with a spindle speed of 1500 rpm as suggested by Echrif and Hrairi where they produced the best result for surface roughness [11]. Unfortunately, major scratching was found on the surface of the aluminium sheet with the indicated spindle speed in these experiments, causing undesirable surface finish to be produced. Therefore, the surface roughness test was neglected. In order to eliminate or reduce the unwanted scratches on the aluminium sheet surface, tool rotation was disabled for all experiments. In addition, lubricant was applied to the tip of forming tool in the beginning of each experiment. The scratches were greatly reduced and better surface finish was produced. It was believed that tool rotation generated more friction and surface contact between the tool and sheet, moreover no lubrication was involved, and hence it caused rough scratching to occur along the sheet.

The thickness uniformity of aluminium sheet was measured in term of standard deviation of the thickness, where a lower value indicates that the thicknesses are closer to the mean, therefore it means a more uniform thickness. It was observed from the analysis of results that increasing of step size and wall angle will reduce the thickness uniformity, where the former parameter influenced the most to the thickness uniformity followed by the latter one. Besides that, the results clearly revealed that the surface roughness and thickness uniformity were varied inversely. This finding has proved that better surface roughness does not come along with uniform thickness reduction, which can be explained by the effect of surface waviness as proposed by Echrif and Hrairi [11]. In addition, step size played an important role in the formation of surface waviness. The gap of punch between the paths which did not contact the sheet area is not deformed during ISF. A lower step size will reduce the size of the surface waviness and lead to better surface roughness, but the number of non-deformed area will increased, which caused the sheet thickness to be less uniform. On the other hand, forming tool with larger diameter can help to decrease the size of surface waviness.

5. Conclusions

In this project, the optimised parameters for minimum surface roughness and uniform thickness reduction in ISF has been determined via Taguchi analysis. From the design of experiments, the optimum parameters for minimum surface roughness was 45° wall angle, 700 mm/min feed rate, and 1.00 mm step size. On the other hand, uniform thickness of aluminium sheet was optimised at 35° wall angle, 700 mm/min feed rate, and 0.25 mm step size. Besides that, the effect of each parameters to the surface roughness and thickness uniformity has been investigated. Increase in step size and decrease in feed rate will improve the surface roughness of aluminium sheet in ISF. While decrease in wall angle and step size will produce a sheet with better thickness uniformity. Further analysis carried out using ANOVA, (not discussed in this paper) showed that the step size was the most significant parameter to both the surface roughness and thickness uniformity of aluminium sheet.

Acknowledgement

This work was supported by the Universiti Malaysia Pahang Research Grant Scheme RDU1703214. The authors would also like to acknowledge the assistance provided by Faculty of Manufacturing Engineering of UMP.

References

[1] Z Hamedon, K Mori, Y Abe (2014), Journal of Materials Processing Technology, 214, 945-950
[2] Z Hamedon, K Mori, Y Abe (2015). Applied Mechanics and Materials 773, 282-286
[3] Z Hamedon, K Mori, Y Abe, (2014). Procedia Engineering 81, 2074-2079
[4] Pohlak M, Majak J, & Küttner R. (2007). Proceedings of the Estonian Academy of Sciences (Engineering), 13(2), 129–138.
[5] Duflou, J, Tunçkol Y, Szekeres A, & Vanherck P. (2007). Journal of Materials Processing Technology, 189(1-3), 65–72.
[6] Kim Y H, & Park J J. (2002). Journal of Materials Processing Technology, 130-131(3), 42–46.
[7] Park J-J, & Kim Y-H. (2003). Journal of Materials Processing Technology, 140(1-3), 447–453.
[8] Oleksik V. (2014). Procedia Engineering, 81, 2280–2285.
[9] Malwad D S, & Nandedkar V M. (2014). Procedia Materials Science, 6 (Icmpc), 1505–1510.

[10] Hamilton K, & Jeswiet J. (2010). CIRP Annals - Manufacturing Technology, 59(1), 311–314.
[11] Echrif SBM, Hrairi M. (2011). Materials and Manufacturing Processes 26 (11), 1404-1414
[12] S A Bagaber, A R Yusoff. (2017). Journal of Cleaner Production, 157, 30-46
[13] Durante M, Formisano A, Langella A, & Capece Minutolo F M. (2009). Journal of Materials Processing Technology, 209(9), 4621–4626.

Work-related musculoskeletal disorders (WMDs) risk assessment at core assembly production of electronic components manufacturing company

N M Yahya and M N O Zahid

Faculty of Manufacturing Engineering, Universiti Malaysia Pahang, 26600 Pekan, Pahang, Malaysia

nafrizuanmy@ump.edu.my

Abstract. This study conducted to assess the work-related musculoskeletal disorders (WMDs) among the workers at core assembly production in an electronic components manufacturing company located in Pekan, Pahang, Malaysia. The study is to identify the WMDs risk factor and risk level. A set of questionnaires survey based on modified Nordic Musculoskeletal Disorder Questionnaires have been distributed to respective workers to acquire the WMDs risk factor identification. Then, postural analysis was conducted in order to measure the respective WMDs risk level. The analysis were based on two ergonomics assessment tools; Rapid Upper Limb Assessment (RULA) and Rapid Entire Body Assessment (REBA). The study found that 30 respondents out of 36 respondents suffered from WMDs especially at shoulder, wrists and lower back. The WMDs risk have been identified from unloading process, pressing process and winding process. In term of the WMDs risk level, REBA and RULA assessment tools have indicated high risk level to unloading and pressing process. Thus, this study had established the WMDs risk factor and risk level of core assembly production in an electronic components manufacturing company at Malaysia environment.

1. Introduction

The consideration of good workstation design is very essential nowadays especially in industry which require manual assembly by operator [1]. Neglecting ergonomic in workstation will result in work-related musculoskeletal disorders (WMDs) injury to the operator [2]. These disorders have caused a considerable human suffering and also economically very costly, because of reduced working capacity and lessoned production. High incidence rate for WMDs of the upper extremities was reported for workers in office work, manufacturing and agriculture, which include numerous materials-handling tasks in various factories [3]. In the electronic components manufacturing company as the example, there are a lot of processes that require manual handling [4], which can contribute to WMD's symptoms.

Nevertheless, until today, there is no research has been performed specifically to explore about the WMDs in electronic components manufacturing company in Malaysia's environment. Thus, this study conducted to assess the WMDs among the workers at core assembly production in an electronic component manufacturing company located in Pekan, Pahang, Malaysia. This study aims to identify the WMDs risk factor related to working posture and categories the WMDs risk level among the workers.

The company manufactures common mode chokes, which are used widely in automotive industry. Basically, the choke is produced from a combination of magnetic embedded core, which is then encapsulated into a casing. The process of producing that common mode choked usually involve a lot of manual processes. Most of the processes are handled by the operator. The machine is semi-automated, which require at least one person to operate it. From initial observation, the worker can easily experience musculoskeletal disorder due to their job routine. The job routine typically involves repetitive motion, standing or sitting for prolong period of time, excessive force and awkward posture.

2. Work-related Musculoskeletal Disorders (WMDs) risk factor

The term musculoskeletal disorder refers to conditions that involve the nerves, tendons, muscles and supporting structures of the body [5]. It also sometimes called ergonomic injuries and illnesses [6]. These occur when physical and psychosocial demands are too great, resulting in discomfort, pain, or functional impairment. WMDs are defined as a group of painful disorders of muscles, tendons, and nerves. Work activities which are frequent and repetitive, or activities with awkward postures cause these disorders which may be painful during work or at rest [7].

WMDs of the upper extremities are painful and potentially disabling conditions that affect the hands, arms, shoulders, and neck. The Malaysian Social Security Organization (SOCSO) has reported the number of WMDs reported to have been increased up 67% until year 2009 [8]. The location of injury involved head, neck, trunk, upper and lower limb and multiple locations; among of these, upper limb the most affected and resulted higher of WMDs report.

A risk factor is described as an attribute or exposure that increases the probability of disease or disorder [9]. There are various jobs that involve with WMDs hazard, which is contributed in increasing the number of WMD hazard. The primary WMDs hazards are force, awkward postures, and repetitions [10].

Force refers to the amount of effort made by the muscles, and the amount of pressure on body parts as a result of different job demands. According to [11], all work tasks require workers to use their muscles to exert some level of force. However, when a task requires them to exert a level of force that is too high for any particular muscle, it can damage the muscle or the related tendons, joints and other soft tissue. Posture is the position of numerous parts of the body during any activity. For most joints, a good or "neutral" posture means that the joints are being used in the middle of the full range of motion [12]. The further joint moves towards either end of its range of motion, or the farther away from the neutral posture, the more awkward or poor the posture becomes and the more strain are put on the muscles, tendons and ligaments around the joint [13].

The risk of developing an MSD increases when the same parts of the body are used repeatedly, with few breaks or chances for rest [14]. Highly repetitive tasks can lead to fatigue, tissue damage, and eventually pain and discomfort. This can occur even if the level of force is low, and the work postures are not very awkward [15].

Table 1: Four different processes in core assembly production

Process	Description
Winding	Winding process involves unloading coil into the winding machine. Then, the operator need to attach the coil strip to winding pin. After that, they need to adjust the setting at the machine according to the desire specification of core. The machine will then proceed to wind the coil strip.
Unloading	Unloading is a process of transferring winding core into tray. The process of transferring is manually done by operator. The operator need to hold a needle full of core and unload each core in the needle into the tray. This process require long period of standing and also repeated motion of arm and elbow.
Encapsulating	Encapsulation & Pressing process involves inserting core into casing. Then, the operator will insert cover on top of the casing and proceed to do pressing process. The process steps require repetitive motion and also excessive force especially at pressing process
Final Testing	Final test is a process of measuring electrical value of finish unit. It is done manually by operator. Operator need to take each pieces of unit and test it with one by one. This job need high repetition hand motion.

IOP Conf. Series: Materials Science and Engineering **319** (2018) 012036 doi:10.1088/1757-899X/319/1/012036

3. Methodology

A set of questionnaires survey based on modified Nordic Musculoskeletal Disorder Questionnaires, which introduced by [16] have been distributed to respective workers to acquire the WMDs risk factor identification at core assembly production. The questionnaire consists of structured, forced, multiple-choice questions and can be used as a self-administered questionnaire or as an interview. The original version consists of several parts: a general questionnaire, and three specific parts focusing on the lower back, shoulders, and neck [17].

The purpose of the modified Nordic Musculoskeletal Disorder Questionnaire is to collect information on the location of discomfort by reference to particular body regions or by use of partial or whole-body diagrams that designate specific regions to be assessed.

In this study, there is a total of 36 respondents who are chosen to participate in the questionnaire. The chosen workers are selected randomly from four different processes as described in Table 1, which are winding process (10 respondents), unloading process (8 respondents), pressing process (10 respondents) and final testing process (8 respondents).

The questionnaire was divided into two parts, which are one part for physical characteristic questions (age, height, weight, duration of the work) and the other parts consisted of a series question with yes or no response questions. It is involved a detailed question on work-related pain or discomfort in different body parts. Work-related pain or discomfort was reported in 12 months. The participants (sample) were interviewed about any kind of discomfort affecting different body parts during every associated with different processes.

Then, postural analysis was conducted in order to measure the respective WMDs risk level. The analyses were based on two ergonomics assessment tools; Rapid Upper Limb Assessment (RULA) and Rapid Entire Body Assessment (REBA).

RULA was developed by [18]. This method is used to investigate work-related upper extremity disorders and to provide a rapid objective measure of musculoskeletal risk [18]. RULA assesses the posture, force and movement associated with the sedentary task such as a task involves computer, manufacturing or retail tasks. However, the method is more relevant for sedentary jobs. The use of RULA results in a score from one to seven, as shown in Table 2, where higher scores signify greater levels of apparent risk [18].

REBA was developed by [19] to provide a quick and easy observational postural analysis tool for whole-body activities. The design of REBA is very similar to RULA method, and special attention is being made to the external load acting on trunk, neck, and legs and to the worker-load coupling, using the upper limbs [19]. The tools consider critical areas of a job and for each task, it assesses the posture factors by assigning a score to each region. The data obtained through assessment of the task is then entered on a scoring sheet which gives an REBA score for each task as in Table 2 that indicates the level of risk and urgency with which action should be taken. [19].

Table 2: RULA and REBA scoring sheet

RULA score	Risk level	REBA score	Risk level
1-2	Negligible	1	None necessary
3-4	Low	2-3	Maybe necessary
5-6	Medium	4-7	Necessary
7	High	8-10	Necessary soon
-	-	11-15	Necessary now

4. Results and discussion

4.1 Modified Nordic Musculoskeletal Disorders Questionnaires Analysis

From the modified Nordic Musculoskeletal Disorder Questionnaire, the WMDs complaints of the workers can be identified, which are 30 (83.3%) workers out of 36 workers have suffered from WMDs.

Table 3 shown the results of WMDs complaints by process. The results show that the unloading process and pressing process workers reported a higher prevalence of WMDs symptoms compare with winding process and final test process workers.

Table 4 shows the distribution of WMDs complaints according to different process. For winding process, low back (56%) were the area largest prevalence of WMDs symptoms, followed by upper back (33%), and elbow (22%). WMDs complaints for the unloading process are regarded to the largest prevalence of WMDs symptom is shoulder (67%), followed by upper back (50%) and low back (33%). Meanwhile, low back (42%) are the higher WMDs complaint in pressing process followed by shoulder, elbows, upper back and 1 or both hips with 22% for all of them. For final test process, neck has the highest WMDs complaints with a total percentage of 50% followed by shoulder (33%).

Table 3: WMDs complaints by process

Process	WMSDs Complaint (%)	
	Yes	No
Winding (n=12)	66.67	33.33
Unloading (n=6)	83.33	16.67
Pressing (n=12)	77.78	22.22
Final Test (n=6)	33.33	66.67

Table 4: Distribution of WMDs complaints by process

Body parts	Process							
	Winding (n=12)		Unloading (n=6)		Pressing (n=12)		Final Test (n=6)	
	Σ	%	Σ	%	Σ	%	Σ	%
Neck	0	0	1	17	1	11	3	50
Shoulder	1	11	4	67	2	22	2	33
Elbows	2	22	1	17	2	22	1	17
Wrists/hands	1	11	1	17	1	11	1	17
Upper back	3	33	3	50	2	22	1	17
Low back	5	56	2	33	6	67	1	17
1 or both hips	1	11	1	17	2	22	1	17
1 or both knees	0	0	0	0	1	11	0	0
1 or both ankles	1	11	1	17	1	11	0	0
	17.3%		25.9%		22.2%		18.5%	

4.2 Analysis of Relation Between Individual Factors and WMDs Complaints

Table 5 shows the outlines associations between prevalent WMDs symptoms and individual variables. There were significant differences between ages of the workers and years of experience towards the prevalence of WMDs symptoms.

From the Table 5, the correlation between individual factors and WMDs complaints can be identified. From the results of the correlation, it can be concluded that there were significant differences between ages of the workers and years of experience. The experience of the workers is the most significant causes for the WMDs complaints. There were no significant differences between WMDs complaint with Body Mass Index (BMI) and duration of works.

4.3 Postural Analysis

The postural analysis was conducted is to analyse the risk level of the working postures of workers for each selected process and to utilize the results for the prevention and treatment of their WMDs. Then, working postures of workers when dealing the job was analysed. Individual postures were assessed and scored according to the waist, leg, elbow, wrist, wrists twisting, load and muscle use, and the assessment ratings were analysed. The results of the RULA and REBA is presented as in Table 6.

The postural risk analysis of four different processes according to RULA and REBA is shown in Table 7. By evaluate the risk using RULA method, two processes; unloading and pressing give score 7, which is a high-risk level. This indicating a postural change is needed immediately as the working postures are vulnerable to risk of WMDs to the workers. In REBA method, the both processes, unloading and pressing have again given score 8 and 9, which are high-risk level and required intervention soon. This probably makes workers having maximum discomfort of body parts at the neck ad trunk which due to high task repetition and manual involvement during handling the machine. High task repetition, when combine with other risk factors such as force or awkward postures can disclose the workers to WMDs definitely.

Table 5: Correlations of the individual factors influence to WMDs complaint

		Ages of the worker	WMDs Complaints	BMI (Kg/m2)	Duration of work per day (hours)	Years of experience
Ages of the worker	Correlation Coefficient	1	**0.500****	0.217	0.125	**0.556****
	Sig. (2-tailed)	-	0.002	0.204	0.468	0.000
	N	36	36	36	36	36
WMDs Complaints	Correlation Coefficient	**0.500****	1	0.108	0.100	**0.336***
	Sig. (2-tailed)	0.002	-	0.529	0.562	0.045
	N	36	36	36	36	36
BMI (Kg/m2)	Correlation Coefficient	0.217	0.108	1	0.027	0.070
	Sig. (2-tailed)	0.204	0.529	-	0.875	0.684
	N	36	36	36	36	36
Duration of work per day (hours)	Correlation Coefficient	0.125	0.100	0.027	1	0.142
	Sig. (2-tailed)	0.468	0.562	0.875	-	0.408
	N	36	36	36	36	36
Years of experience	Correlation Coefficient	0.556**	0.336*	0.070	0.142	1
	Sig. (2-tailed)	0.000	0.045	0.684	0.408	-
	N	36	36	36	36	36

**Correlation is significant at the 0.01 level (2-tailed);
*Correlation is significant at the 0.05 level (2-tailed)

Table 6: Ergonomic assessment for four different processes

		Winding process		Unloading process		Pressing process		Final testing process	
Working posture	Risk factor	RULA	REBA	RULA	REBA	RULA	REBA	RULA	REBA
Neck, trunk and leg analysis	Neck	1	3	3	3	3	2	3	1
	Trunk	2	3	5	3	5	3	3	2
	Leg	1	1	2	1	2	1	1	1
Arm and wrist analysis	Upper arm	4	2	4	4	2	4	3	4
	Lower arm	1	1	1	1	1	1	2	1
	Wrist	1	1	3	1	1	1	2	1
	Wrist twist	1	-	1	-	1	-	1	-
	Muscle use	1	-	1	-	1	-	1	-
	Force/load	1	1	2	1	0	1	1	1
Assessment Score		**5**	**4**	**7**	**9**	**7**	**8**	**4**	**5**

Table 7: Risk analysis using RULA and REBA

Process	RULA			REBA		
	RULA score	Risk level	Maximum discomfort in body parts	REBA score	Risk level	Maximum discomfort in body parts
Winding	5	Medium	Neck and trunk	4	Necessary	Neck and trunk
Unloading	7	High	Neck, trunk and upper arm	8	Necessary soon	Neck, trunk and upper arm
Pressing	7	High	Trunk and upper arm	9	Necessary soon	Trunk and upper arm
Final testing	4	Low	Neck and trunk	5	Necessary	Neck and trunk

5. Conclusion

Thus, this study established the WMDs risk factor and risk level of core assembly production in an electronic component manufacturing company at Malaysia's environment. The core assembly production involved manual handling task, which contributes to WMDs that are caused by awkward postures, excessive force and repetition because of the limited work area, standing and sitting for prolonged period and deal with heavy equipment.

Using modified Nordic Musculoskeletal Disorders Questionnaires, the study has found that 30 respondents out of 36 respondents have been reported to suffer from WMDs risk. The most affected body parts are shoulder, wrists and lower back. The study also indicates that the ages of the workers and years of workers experience to have become the potential of WMDs risk too. Based on RULA and REBA assessment, the unloading process and pressing process are categorized under high-risk and action to reduce the risk need to be taken urgently. However, another two process; winding and final testing shown a medium and low risk that required no fast intervention.

Acknowledgement

This research is funded by Universiti Malaysia Pahang under research grant RDU170373.

References

[1] Hägg, G. M. (2003). Corporate initiatives in ergonomics—an introduction. Applied Ergonomics 34(1), pp. 3-15.

[2] Choobineh, A., Lahmi, M., Shahnavaz, H., Khani Jazani, R. and Hosseini, M. (2004). Musculoskeletal symptoms as related to ergonomic factors in Iranian hand-woven carpet industry and general guidelines for workstation design. International Journal of Occupational Safety and Ergonomics, 10(2), pp.157-168.

[3] Faucet, J. Garry, M., Nadler, D. and Ettare, D. (2002). A test of two training interventions to prevent work-related musculoskeletal disorders of the upper extremity. Applied Ergonomics, 33(4), pp. 337-347.

[4] Tompkins, J.A. (2003). No Boundaries: Break Through to Supply Chain Excellence. Tompkins Press, Nottingham.

[5] Bernard, B. (1997). Musculoskeletal disorders and workplace factors: A critical review of epidemiologic evidence for work-related disorders of the neck, upper extremities and low back. Report no. (NIOSH) 97-141, Public Health Publication.

[6] Islam, M. (2012). Common work related musculoskeletal disorders among the paediatric physiotherapists at CRP, Doctoral dissertation, Bangladesh Health Professions Institute.

[7] Anap, D.B., Iyer, C. and Rao, K. (2017). Work related musculoskeletal disorders among hospital nurses in rural Maharashtra, India: A multi centre survey. International Journal of Research in Medical Sciences, 1(2), pp.101-107.

[8] PERKESO, Pertubuhan Keselamatan Sosial Malaysia. (2010). Laporan Tahunan 2009 PERKESO. https://www.perkeso.gov.my/images/Laporan_Tahunan_2009_Lengkap.pdf, Retrieved on August, 1, 2017.
[9] Last, J. (1983). Dictionary of epidemiology. Oxford University Press, New York.
[10] Buckle, P.W. and Devereux, J.J. (2002). The nature of work-related neck and upper limb musculoskeletal disorders. Applied Ergonomics, 33(3), pp.207-217.
[11] Sjøgaard, G., Lundberg, U. and Kadefors, R. (2000). The role of muscle activity and mental load in the development of pain and degenerative processes at the muscle cell level during computer work. European Journal of Applied Physiology, 83(2), pp.99-105.
[12] Clarkson, H.M. (2000). Musculoskeletal assessment: joint range of motion and manual muscle strength. Lippincott Williams & Wilkins, New South Wales.
[13] Sharma, A.K., Bansal, P. and Kaur, M. (2016). Ergonomics in dental practice: musculoskeletal disorders, approaches and interventions. Pakistan Oral & Dental Journal, 36(2), pp. 349-355.
[14] Sommerich, C.M., McGlothlin, J.D. and Marras, W.S. (1993). Occupational risk factors associated with soft tissue disorders of the shoulder: A review of recent investigations in the literature. Ergonomics, 36(6), pp.697-717.
[15] Gallagher, S. and Heberger, J.R. (2013). Examining the interaction of force and repetition on musculoskeletal disorder risk: A systematic literature review. Human Factors, 55(1), pp.108-124.
[16] Kuorinka, I., Jonsson, B., Kilbom, A., Vinterberg, H., Biering-Sørensen, F., Andersson, G., Jørgensen, K. (1987). Standardised Nordic questionnaires for the analysis of musculoskeletal symptoms. Applied Ergonomics, 18(3), pp. 233-237.
[17] Roquelaure, Y., Ha, C., Leclerc, A., Touranchet, A., Sauteron, M., Melchior, M., Imbernon, E. and Goldberg, M. (2006). Epidemiologic surveillance of upper-extremity musculoskeletal disorders in the working population. Arthritis Care & Research, 55(5), pp.765-778.
[18] McAtamney, L. and Corlett, E.N. (1993). RULA: a survey method for the investigation of work-related upper limb disorders. Applied Ergonomics, 24(2), pp.91-99.
[19] Hignett, S. and McAtamney, L. (2000). Rapid entire body assessment (REBA). Applied Ergonomics, 31(2), pp.201-205.

A Multiple Items EPQ/EOQ Model for a Vendor and Multiple Buyers System with Considering Continuous and Discrete Demand Simultaneously

Jonrinaldi[2], T Rahman[1], Henmaidi[2], E Wirdianto[2] and D Z Zhang[3]

[1] Master Program of Industrial Engineering, Andalas University, 25163, Indonesia
[2] Department of Industrial Engineering, Andalas University, Padang, 25163, Indonesia
[3] Exeter Manufacturing Enterprise Centre (XMEC), College of Engineering, Mathemetics and Physical Sciences, University of Exeter, Exeter, EX4 4QF, United Kingdom

jonrinaldi@ft.unand.ac.id

Abstract. This paper proposed a mathematical model for multiple items Economic Production and Order Quantity (EPQ/EOQ) with considering continuous and discrete demand simultaneously in a system consisting of a vendor and multiple buyers. This model is used to investigate the optimal production lot size of the vendor and the number of shipments policy of orders to multiple buyers. The model considers the multiple buyers' holding cost as well as transportation cost, which minimize the total production and inventory costs of the system. The continuous demand from any other customers can be fulfilled anytime by the vendor while the discrete demand from multiple buyers can be fulfilled by the vendor using the multiple delivery policy with a number of shipments of items in the production cycle time. A mathematical model is developed to illustrate the system based on EPQ and EOQ model. Solution procedures are proposed to solve the model using a Mixed Integer Non Linear Programming (MINLP) and algorithm methods. Then, the numerical example is provided to illustrate the system and results are discussed.
Keywords: EPQ, EOQ, simultaneously, vendor, multiple buyers

1. Introduction

Inventory management is one of the vital competitive factors for industries [1,2,3]. It used inventory model as decision-making tool to control the inventory [4] and have received a lot attention since the last past century till today [5]. Many of inventory models based on Economic Order Quantity (EOQ) or Economic Production Quantity (EPQ) due to their simplicity and robustness [6, 7, 8]. The EOQ model is used to determine an optimal order size by balancing the setup cost and inventory holding cost [9] and EPQ model is used to determine the optimal replenishment lot size to produce [5]. However, due to their own assumptions and conditions, these models are bound to their applicability in real-world situations [1] so that it is extended by several academicians and researchers [10].

The basic EPQ model was proposed five years later after the basic EOQ model was presented and it used the same assumptions with EOQ model [11]. These assumptions sometimes considered as unrealistic situations, such as the continuous replenishment policy for satisfying the demand and the produces items are in perfect quality [5, 12, 13]. In the real life, the discontinuous or multiple delivery policies are commonly adapted [9, 14] and generation of defective items during production process is inevitable [5, 9, 15, 12, 14]. Therefore, there are many researchers or academicians relaxing the assumptions. Tai [16] extended the EPQ model with considering imperfect/deteriorating items with rework process. Taleizadeh *et al* [12] proposed an EPQ model with backordering, scrap products, rework and interruption in manufacturing process. Sarkar *et al* [3] developed an EPQ model for production with random defective rate, rework process and backorders. Jaber *et al* [17] gives options for EPQ model with imperfect items, whether to buy the new ones or repair the imperfect items, Pasandideh *et al* [10] also considering imperfect production in EPQ model and they extended it with warehouse construction cost. Based on the best knowledge of authors, just few researches consider continuous and discrete demand simultaneously in a system.

In fact, many companies today meet the demand for both continuous and discrete situation. The companies fulfill the demand every time or every day and at the same time they also fulfill the demand every certain period. Therefore, this research proposed a multiple items EPQ/EOQ model with considering continuous and discrete demand simultaneously in a vendor-multiple buyers system. The continuous demand can be fulfilled by vendor anytime, otherwise the discrete demand will be fulfilled by using multiple delivery system [19, 20] and an integrated vendor-buyer inventory model has better performance than the non-integrated inventory model [2]. This research adopt Rahman *et al* [19] inventory model that also considering two-types demands (continuous and discrete demand) simultaneously. The objective of this research is to investigate the optimal production lot size and number of shipments for multiple items or products with considering both continuous and discrete demand simultaneously in the vendor-multiple buyers system. In the following sections, mathematical modeling is given in Section 2. Section 3 is numerical example and discussion. Lastly, the conclusion of this research is presented in section 4.

2. Mathematical modeling

2.1. System characteristics

Figure 1 shows that there are three type of items with two buyers with discrete demand. The vendor will be produce product for buyers (continuous and discrete demands) and the level inventory of products for continuous demand will continuous depleted until the next production cycle time. On the other hand, the discrete demand will be wait for several time until all types of product have finished and the discrete demands ready to deliver to the buyers. In the Figure 1, we can see that the number of shipments for the first buyer is three times deliveries and for the second buyer is two times deliveries in one cycle time.

2.2. Notations and assumptions

Notations that used in this model are list in Appendix A. The mathematical model have some assumptions, there are:

a. Infinite planning horizon.
b. Single facility for all types of products.
c. Shortages are not allowed.
d. There are no imperfect products during production.
e. The total of discrete and continuous demand must least or equal to production rate.

$$\sum_j^k D_{ij} + \sum_r^y C_{ir} \leq P_i \tag{1}$$

f. The total ratio of demand and production rate for all items are maximum 1.

$$\sum_{i=1}^{n} \left(\frac{\sum_j^k D_{ij} + \sum_r^y C_{ir}}{P_i} \right) \leq 1 \tag{2}$$

g. The length of total production time can not exceed the length of cycle time

$$TP \leq T \tag{3}$$

2.3. Model formulation

The objective of this mathematical model are determine the optimal production lot size and number of shipments for each buyers with considering both continuous and discrete demand simultaneously in a system of multiple items and multiple buyers. The mathematical model consist of production cost, setup cost, holding cost, and transportation cost.

2.3.1. Production cost. The production cost is calculated by considering total of discrete demands and continuous demands from buyers with production cost per item. The total production cost (TPC) is given by,

$$TPC = \sum_{i=1}^{n} B_i \left(\sum_{j=1}^{k} D_{ij} + \sum_{r=1}^{y} C_{ir}\right) \quad (4)$$

2.3.2. Setup cost. The total setup cost (TSC) is given by,

$$TSC = \frac{\sum_{i=1}^{n} S_i}{T} \quad (5)$$

2.3.3. Holding costs. The total holding cost (*THC*) are divided to five parts that can be seen in Figure 1. The holding cost that formulated in this model are holding cost during its production process (I), holding cost for discrete demand (II), holding cost for continuous demand (III), holding cost of discrete demand during delivery time (IV) and buyer's holding cost (V).

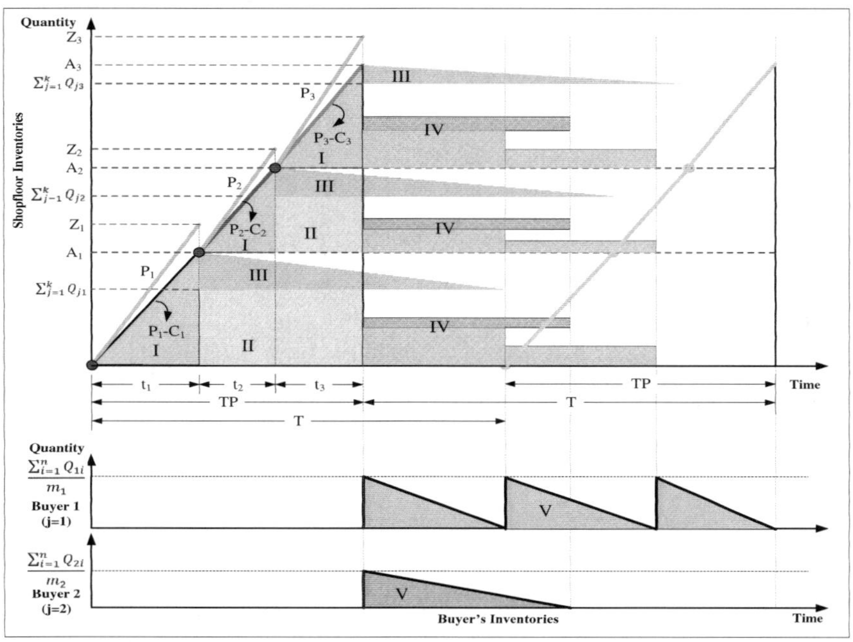

Figure 1. System characteristics with both demand and multiple buyers.

a. The holding cost of item i during its production time.

$$= \frac{1}{2} T \sum_{i=1}^{n} H_i . \left(\sum_{j=1}^{k} D_{ij} + \sum_{r=1}^{y} C_{ir}\right)^2 \frac{(P_i - \sum_{r=1}^{y} C_{ir})}{P_i^{\,2}} \quad (6)$$

b. The holding cost of item i to fulfil its discrete demand during the production time of the other items.

$$= T \sum_{i=1}^{n-1} H_i . \sum_{j=1}^{k} D_{ij} . \sum_{i=i+1}^{n} \frac{\left(\sum_{j=1}^{k} D_{ij} + \sum_{r=1}^{y} C_{ir}\right)}{P_i} \quad (7)$$

c. The holding cost of item *i* to fulfill its continuous demand since the production time of next other items until the next its production time.

$$= \frac{1}{2}\, T \sum_{i=1}^{n} H_i .\ \sum_{r=1}^{y} C_{ir} .\ \left[1 - \frac{\left(\sum_{j=1}^{k} D_{ij} + \sum_{r=1}^{y} C_{ir}\right)}{P_i}\right]^2 \tag{8}$$

d. The holding cost of item *i* to fulfill its discrete demand during delivery periods.

$$= \frac{1}{2} T \sum_{j=1}^{k} \sum_{i=1}^{n} H_i .\ D_{ij} - \frac{1}{2} T\ \sum_{j=1}^{k} \frac{\sum_{i=1}^{n} H_i .\ D_{ij}}{m_j} \tag{9}$$

e. The holding cost of the buyers.

$$= \frac{1}{2} .\ T . \sum_{j=1}^{k} \frac{\sum_{i=1}^{n} D_{ij} .\ L_i}{m_j} \tag{10}$$

2.3.4. Transportation cost. The transportation are consist of fixed transportation cost of each buyers and variable transportation cost. The total transportation cost (TTC) is given by,

$$TTC = \frac{\sum_{j=1}^{k} m_j . F_j}{T} + \sum_{i=1}^{n}\left(\sum_{j=1}^{k} D_{ij} + \sum_{r=1}^{y} C_{ir}\right) . V_i \tag{11}$$

2.3.5. Total inventory-production cost. The total inventory-production cost (TIPC) are given by equation (12) and verified by using dimension test that can be seen in Appendix B. The TIPC is cumulative of TPC, TSC, THC and TTC.

$$\begin{aligned} TIPC = & \sum_{i=1}^{n} (B_i + V_i) \left(\sum_{j=1}^{k} D_{ij} + \sum_{r=1}^{y} C_{ir}\right) + \frac{\sum_{i=1}^{n} S_i}{T} + \frac{1}{2}\ T\ \sum_{i=1}^{n}\ H_i . \Big(\sum_{j=1}^{k} D_{ij} + \\ & \sum_{r=1}^{y} C_{ir}\Big)^2 \frac{\left(P_i - \sum_{r=1}^{y} C_{ir}\right)}{P_i^{\,2}} + \frac{1}{2}\ T \sum_{i=1}^{n} H_i . \sum_{r=1}^{y} C_{ir} . \left[1 - \frac{\left(\sum_{j=1}^{k} D_{ij} + \sum_{r=1}^{y} C_{ir}\right)}{P_i}\right]^2 + \\ & T \sum_{i=1}^{n-1} H_i . \sum_{j=1}^{k} D_{ij} . \sum_{i=i+1}^{n} \frac{\left(\sum_{j=1}^{k} D_{ij} + \sum_{r=1}^{y} C_{ir}\right)}{P_i} + \frac{1}{2} T \sum_{j=1}^{k} \sum_{i=1}^{n} H_i .\ D_{ij} + \\ & \frac{1}{2} T \sum_{j=1}^{k} \frac{\sum_{i=1}^{n} D_{ij}\ (L_i - H_i)}{m_j} + \frac{\sum_{j=1}^{k} m_j . F_j}{T} \end{aligned} \tag{12}$$

The total inventory and production cost will be have optimum solution if Equation $(L_i - H_i > 0)$ and Equation (13) is satisfied. These equations are obtained by using second differential to Equation (12).

$$\left[\left(\frac{2\ \sum_{i=1}^{n} S_i}{T^3} + \frac{2\ \sum_{j=1}^{k} m_j . F_j}{T^3}\right)\left(T \sum_{j=1}^{k} \frac{\sum_{i=1}^{n} D_{ij}\ (L_i - H_i)}{m_j^{\,3}}\right)\right] - \left[\frac{1}{2} \sum_{j=1}^{k} \frac{\sum_{i=1}^{n} D_{ij}\ (L_i - H_i)}{m_j^{\,2}} + \frac{\sum_{j=1}^{k} F_j}{T^2}\right]^2 > 0 \tag{13}$$

2.4. The solution procedures

Algorithm method solves the model step by step. It solves the problem by searching for optimal cycle time first and then determine the optimal number of shipments for each buyers based on the optimal cycle time and total inventory-production cost. There are steps by using algorithm method:

Step 0: Make sure the assumptions of the model are satisfied by doing calculation to equation (1) and equation (2). If the assumptions are satisfied then next to step 1.

Step 1: Determine the optimal cycle time (T^*) by doing calculation with equation (14) and then next to step 2. Equation (14) is given by doing first differential of T to Equation (12). The process to obtain the equation can be seen in Appendix D.

IOP Conf. Series: Materials Science and Engineering **319** (2018) 012037 doi:10.1088/1757-899X/319/1/012037

$$T^* = \left(2\sum_{i=1}^{n} S_i \Big/ \left\{\sum_{i=1}^{n} H_i.\left(\sum_{j=1}^{k} D_{ij} + \sum_{r=1}^{y} C_{ir}\right)^2 \frac{(P_i - \sum_{r=1}^{y} C_{ir})}{P_i^{\,2}} + \sum_{i=1}^{n} H_i.\sum_{r=1}^{y} C_{ir}.\left[1 - \frac{\left(\sum_{j=1}^{k} D_{ij} + \sum_{r=1}^{y} C_{ir}\right)}{P_i}\right]^2 + 2\sum_{i=1}^{n-1} H_i.\sum_{j=1}^{k} D_{ij}.\sum_{i=i+1}^{n} \frac{\left(\sum_{j=1}^{k} D_{ij} + \sum_{r=1}^{y} C_{ir}\right)}{P_i} + \sum_{j=1}^{k}\sum_{i=1}^{n} H_i.\ D_{ij}\right\}\right)^{1/2} \quad (14)$$

Step 2: Determine the optimal numbers of shipment (m_j^*) by doing calculation with equation (15). Equation (15) is given by doing first differential of m_j to Equation (12). If the calculation result is in integer for each buyers (j), then next to Step 3. If not, next to step 4.

$$m_j^* = T\sqrt{\frac{\sum_{i=1}^{n} D_{ij}\ (L_i - H_i)}{2\,F_j}} \quad (15)$$

Step 3: Determine the total inventory-production cost by calculating equation (12).

Step 4: Determine the total inventory-production cost by calculating equation (12) for both roundup and round down of m_j^* from step 2 and for each buyer. Then choose the best combination of the optimal number of shipment for buyers based on minimum total inventory-production cost.

3. Numerical example and discussion

Numerical example is given to illustrate how the proposed mathematical model solve the problem. The data that used in this numerical example consists of six type of items, two buyers with continuous demand and three buyers with discrete demand. The detailed of input data can be seen in Table 1.

Table 1. Input Data

Input		Product (i)						
		i =1	i = 2	i = 3	i = 4	i = 5	i = 6	
D_{i1}		809,500	186,025	1,389,000	432,688	3,218,760	13,868,762	Units / Year
D_{i2}		2,023,750	260,435	1,215,375	605,763	3,755,220	12,607,966	Units / Year
D_{i3}		1,214,250	297,640	868,125	692,300	3,755,220	15,549,824	Units / Year
C_{i1}		0	0	3,069,424	250,000	4,193,700	952,200	Units / Year
C_{i2}		0	0	4,604,136	250,000	2,795,800	1,163,800	Units / Year
P_i		90,720,000	108,864,000	90,720,000	108,864,000	90,720,000	108,864,000	Units / Year
B_i	Rp	3,000	2,300	3,000	2,300	3,000	2,300	/ Unit
S_i	Rp	20,000,000	20,000,000	20,000,000	20,000,000	20,000,000	20,000,000	/ Setup
H_i	Rp	440	440	440	440	440	440	/ Unit.Year
L_i	Rp	880	880	880	880	880	880	/ Unit.Year
V_i	Rp	100	100	100	100	100	100	/ Unit
F_1	Rp			3,0500,000				/ Shipment
F_2	Rp			3,000,000				/ Shipment
F_3	Rp			2,500,000				/ Shipment

Based on the data input, algorithm method gives solution for the problem with length of cycle time is 0.068 year and number of shipments are either 2 or 3 times for both buyer 1 and buyer 2 and either 3 or 4 times for buyer 3. By trying all alternative combinations for the number of shipments, the best solution is 3 times of shipments for buyer 1, buyer 2 and buyer 3 with total inventory-production is Rp 219,341,667,157.62 per year. This result is indifferent with MINLP method, with cycle time is 0.069 year, 3 times of shipments for buyer 1, buyer 2 and buyer 3, and total inventory-production cost is 219,341,200,000.00 per year. Although the differences of the result for both methods is small, the algorithm will take much more time and complexion in calculation, especially when the number of buyer with discrete demand are big, and must calculate the total inventory-production cost with all alternative combination of number of shipments.

4. Conclusions

This research proposes a mathematical model to investigate the optimal production lot size of the vendor and number of shipments policy of orders to multiple buyers with considering continuous and discrete demand simultaneously. The research also proposes two alternative solution model, algorithm method and Mixed Integer Non Linear Programming (MINLP) method. Although both solution model have small differences in problem solving results, the algorithm method will be more complicated to do when the number of buyers are big, since it is used all alternative combination of number of shipments for all buyer and choose the most economic ones. This model still have limitations that can be extended for further research, such as including safety stock, backorder policy or lost sale, considering imperfect product during production process and rework process, and considering delivery capacity for each buyers.

References

[1] Pasandideh S H M and Niaki S T A 2008 A genetic algorithm approach to optimize a multi-products EPQ model with discrete delivery orders and constrained space *Applied Mathematics and Computation* **195** 506-14

[2] Cardenas-Barron L E, Trevino-Garza G, Widyadana A and Wee H-M 2014 A constrained multi-products EPQ inventory model with discrete delivery order and lot size *Applied Mathematics and Computation* **230** 359-70

[3] Sarkar B, Cardenas-Barron L E, Sarkar M and Singgih M L 2014 An economic production quantity model with random defective rate, rework process and backorders for single stage production system *Journal of Manufacturing Systems* **33** 423-435

[4] Mahata G C 2012 An EPQ-based inventory model for exponentially deteriorating items under retailer partial trade credit policy in supply chain *Expert Systems with Applications* **39** 3537-50

[5] Taleizadeh A A, Kalantari S S and Cardenas-Barron L E 2015 Determining optimal price, replenishment lot size and number of shipments for an EPQ model with rework and multiple shipments *Journal of Industrial and Management Optimization* **11** 1059-71

[6] Omar M, Zubir M B and Moin N H 2010 An alternative approach to analyze economic ordering quantity and economic production quantity inventory problems using the completing the square method *Computers and Industrial Engineering* **59** 362-64

[7] Hou K-L 2007 An EPQ model with setup cost and process quality as functions of capital expenditure *Applied Mathematical Modelling* **31** 10-17

[8] Cardenas-Barron L E 2009 Economic production quantity with rework process at a single-stage manufacturing system with planned backorders *Computers and Industrial Engineering* **57** 1105-13

[9] Chiu Y-S P, Chiu S W, Li C-Y and Ting C-K 2009 Incorporating multi-delivery policy and quality assurance into economic production lot size problem *Journal of Scientific and Industrial Research* **68** 505-12

[10] Pasandideh S H R, Niaki S T A, Nobil A H and Cardenas-Barron L E 2015 A multi products single machine economic production quantity model for an imperfect production system under warehouse construction cost International Journal Production Economics **169** 203-14

[11] Garcia-Laguna J, San-Jose L A, Cardenas-Barron L E and Sicilia J 2010 The integrality of the lot size in the basic EOQ and EPQ models: Applications to other production-inventory models Applied Mathematics and Computation **216** 1660-72
[12] Taleizadeh A A, Cardenas-Barron L E and Babak Mohammadi 2014 A deterministic multi product single machine EPQ model with backordering, scraped product, rework and interruption in manufacturing process International Journal Production Economics **150** 9-27
[13] Chiu S W, Lin H -D, Wu M -F and Yang J Ch 2011 Determining replenishment lot size and shipment policy for an extended EPQ model with delivery and quality assurance issues Scientia Iranica E **18** 1537-44
[14] Chiu S W, Chiu Y -S P, and Yang J -C 2012 Combining an alternative multi-delivery policy into economic production lot size problem with partial rework Expert Systems with Applications **39** 2578-83
[15] Chiu S W, Chung C -L, Chen K -K, and Chang H -H 2012 Replenishment lot sizing with an improved issuing policy and imperfect rework derived without derivatives African Journal of Business Management **6** 3817-21
[16] Tai A H 2013 Economic production quantity models for deteriorating/imperfect products and service with rework Computers and Industrial Engineering **66** 879-88
[17] Jaber M Y, Zanoni S and Zavanella L E 2014 Economic order quantity models for imperfect items with buy and repair options **155** 126-31
[18] Al-Salamah M 2016 Economic production quantity in batch manufacturing with imperfect quality, imperfect inspection, and destructive and non-destructive acceptance sampling in a two-tier market Computers and Industrial Engineering **93** 275-85
[19] Rahman T, Jonrinaldi and Henmaidi 2017 A modified EPQ model for multiple items with considering continuous and discrete demand simultaneously Jurnal Optimasi Sistem Industri **16** 1-9 (This reference is in Indonesian)
[20] Oktavia N, Henmaidi and Jonrinaldi 2016 Development of economic production quantity (EPQ) model with synchronizing continuous and discrete demand simultaneously Jurnal Optimasi Sistem Industri **15** 78-86 (This reference is in Indonesian)

IOP Conf. Series: Materials Science and Engineering **319** (2018) 012038 doi:10.1088/1757-899X/319/1/012038

An integer batch scheduling model considering learning, forgetting, and deterioration effects for a single machine to minimize total inventory holding cost

R Yusriski[1], Sukoyo[2], T M A A Samadhi[2], A H Halim[2]
1Department of Industrial Engineering, Universitas Jenderal Achmad Yani (UNJANI), PO. BOX 807 Bandung Indonesia
2 Department of Industrial Engineering and Management, Institut Teknologi Bandung (ITB), Bandung 40132, Indonesia

E-mail: yusarisaki@yahoo.co.id

Abstract. This research deals with a single machine batch scheduling model considering the influenced of learning, forgetting, and machine deterioration effects. The objective of the model is to minimize total inventory holding cost, and the decision variables are the number of batches (N), batch sizes ($Q_{[i]}$, i = 1, 2, .., N) and the sequence of processing the resulting batches. The parts to be processed are received at the right time and the right quantities, and all completed parts must be delivered at a common due date. We propose a heuristic procedure based on the Lagrange method to solve the problem. The effectiveness of the procedure is evaluated by comparing the resulting solution to the optimal solution obtained from the enumeration procedure using the integer composition technique and shows that the average effectiveness is 94%.

1. Introduction

In the manufacturing industry, there are many cases where the processing time of parts could change due to operator learning-forgetting effects, and machine deterioration. The learning effect occurs when an operator work more efficient as the operator has produced identical parts repeatedly (Wright, [1]). On the other hand, if there is an interruption of two repetitive operations, the operator must re-learn the same operation on the next process after an interruption due to forgetting effect (Jaber and Bonney, [2]). Also, it can be observed that the capability of the machine will deteriorate after producing some parts (Kaminskiy and Krivtsov [3]).

Research in Mor and Mosheiov [4], Teyarachakul, et al. [5], Yang [6] shows that on batch scheduling problems the learning, forgetting, and deterioration effects also occur. All the research assume that the processing time of parts in a batch is constant and could change when the parts are scheduled in the next batches. Additionally, the research also assumes that all raw materials have been available since predetermined arrival time (say time zero) and the completed parts will exactly be delivered at the completion time of the batches. Therefore, the planner should consider the learning, forgetting, and deterioration effects when determining the number of batch and batch size to minimize the objective (among other total flow time, total holding cost, and total completion time).

In just-in-time (JIT) production systems, there are the systems (Halim et al. [7]; Halim and Ohta, [8]; Yusriski et al.[9]) where the company is able of managing the parts to be received at the shop at the right time and the right quantities, and the parts must be delivered at a common due date.

For this situation, if the planner does not consider the learning effect, the completion time of the parts could be earlier than the expected time. It leads to a condition where the total inventory holding cost increase as all completed parts must be delivered at a common due date. On the contrary, if the forgetting or deterioration effects are not considered, the completion time of parts will be later than the expected time so that some parts could be finished after the due date, and this cause the company should bear the penalty cost.

2. Problem Formulation

This research discusses an integer batch scheduling model considering learning, forgetting, and deterioration effects in JIT production system to minimize the total inventory holding cost. Suppose that there are n number integer parts of a single item to be processed on a single machine and to be divided into N batches with batch sizes, $Q_{[i]}$ ($i = 1, 2, ..., N$), the completed parts must be delivered at the common due date (d) simultaneously. The batches are scheduled backwardly where the processing time of batch ($T_{[i]}$) is influenced by operator learning-forgetting and machine deterioration effects. A setup time (s) is needed before starting to process a new batch. However, any input material is not needed to accomplish the setup activity so that the parts to be processed can arrive at the shop floor at the same time when the machine is ready to begin processing their parts in a batch ($B_{[i]}$). The objective is to minimize total inventory holding cost (TFC). Two input costs associated with the objective, i.e., the inventory holding cost per unit time for a part in the completed batches (c_1) and in-process batches (c_2). The solve this problem, there are three decisions, i.e., to determine the number of batches, batch sizes, and the sequence of the resulting batches.

3. Inventory Holding Cost

According to Halim and Ohta [8], in batch scheduling problem, the inventory holding cost ($FC_{[i]}$) of the batch ($L_{[i]}$) can be evaluated during the parts in a batch flows in the shop, from its arrival to the common due date. It could be defined as the accumulation of the inventory holding cost for completed parts in batch and the inventory holding cost for in-process parts in batch. This research develops the formula for calculating the total inventory holding cost (TFC) in Halim and Ohta [8] by replacing the static batch processing time, (t) by the dynamic processing time of batch ($T_{[i]}$) which is that processing time influenced by learning, forgetting, and deteriorating. The inventory holding cost function as shown as follows.

$$TFC = \sum_{i=1}^{N}\left(F_{[i]} + I_{[i]}\right) \tag{1}$$

$$\text{where } F_{[i]} = \left\{\sum_{j=1}^{i}\left(s + T_{[j]}Q_{[j]}\right) - s - T_{[i]}Q_{[i]}\right\} c_1 Q_{[i]};$$

$$I_{[i]} = c_2 T_{[i]} Q_{[i]}^2$$

Eq (1) shows that the inventory holding cost in a batch could be calculated by adding the inventory holding cost of completed parts in a batch ($F_{[i]}$) with the inventory holding cost of in-process parts in a batch ($I_{[i]}$). Therefore, the total inventory holding cost (TFC) can be calculated by summing the inventory holding cost in all batches.

4. Learning-Forgetting and Deterioration Functions

There are some functions of learning, forgetting, and deterioration. The reader can find out the functions in (Wright [1]; Jaber and Bonney [2]; Teyarachakul et al. [5]; Janiak et al. [10]; Yusriski et al. [11]; Yusriski et al. [12]). We develop the learning function in Wright [1], the forgetting and the learning-forgetting functions in Jaber and Bonney [2], and the deterioration function by ROCOF function in Kaminskiy & Krivtsov [3] and AFT function in Iskandar and Jack [13]. This research proposes the learning, forgetting, and deterioration functions as shown as follows.

$$T_{[i]} = \max\left\{p\left(E_{[i]}\right)^{-m}, v\right\} \tag{2}$$

where

$$E_{[i]} = 1 + \sum_{i=1}^{N} Y_{[i]}; \qquad Y_{[i]} = \begin{cases} Q_{[i+1]} & , \delta=1 \\ \left[\left(Q_{[i+1]} - Q_{[i+1]}\Lambda(B_{[i]})\right)^{(m+f)/m}\left(X_{[i]}\right)^{-(f)/m}\right] & , 0<\delta<1 \end{cases};$$

$$X_{[i]} = \left[Q_{[i+1]}\left(1-\Lambda(B_{[i]})\right) + \left(s(1-m)/T_{[i+1]}\right)^{1/(1-m)}\right]; \qquad m = -\log\delta/\log 2;$$

$$f_{[i]} = m(1-m)\log\left(Q_{[i+1]}+1\right)\Big/\log\left(1+C_{[i]}\right); \qquad C_{[i]} = t_b / H_{[i]};$$

$$H_{[i]} = \left(T_{[i+1]}/1-m\right)\left(Q_{[i+1]}\right)^{1-m}; \qquad \Lambda(B_{[i]}) = \left(U_{[i]}/\alpha_i\right)^{\beta};$$

$$U_{[i]} = \sum_{i=k}^{N} T_{[i+1]}Q_{[i+1]}; \qquad \alpha_{[i]} = \alpha_0\left[d\Big/\left(d+U_{[i]}\right)\right]^{\gamma};$$

$$\beta > 1,\ \gamma \geq 1; \qquad v = p(q)^{-m}.$$

This research adopts the learn-forget curve model (LFCM) in Jaber dan Boney [2] to develop the learning-forgetting and machine deterioration function. The LFCM function assumes that the processing time of parts influenced by learning and forgetting effect. We assume there is an initial processing time (p) to process parts in batch N-th position (scheduled backwardly). The processing time of parts starting from batch on a position (N-1) to the first position could decreases to be accordance with the experience of operator ($E_{[i]}$) to process the number of parts (as the consequences of the learning effect), and then constant when that achieves a learning threshold (v). The effect of learning could depend on the value of learning rate (δ) where it is between 50% and 100% where δ= 100% implies no learning, and δ= 50% implies maximum learning. It can also be observed that decreasing the value of δ, increases the learning index (m).

The forgetting effect will occur during a setup time between two consecutive batches so that the processing time of parts from batch position (N-1) to the first position could increase in line with the number of batch setup, and that could retard the effect of learning. The effect of forgetting depends on the value of forgetting index ($f_{[i]}$), which associated with the learning index, and the interval time of break for an operator that will cause the operator full forget the operation (t_B). The processing of batch also affected by the machine deterioration ($\Lambda(B_{[i]})$), calculated by ROCOF function. The two parameters, adopted by Weibull distribution, are proposed, that is the scale parameter (α) and the shape parameter (β). The value of β should more than one which to indicate that failure rate increases with time. We assume that the deterioration could occur during an interval [0, α) with a due date (d) is earlier than α. Therefore, the probability of failures during the scheduling period would be less than one ($\Lambda(d) \leq 1$). This indicates that the machine lifetime will end beyond the due date. We also adopt the AFT model (the accelerated failure time) to consider the value of α could decrease as the tool usage.

5. Mathematical Model

This research develops the model in Halim and Ohta [8] by including an additional assumption that the batch processing time is influenced by learning, forgetting, and deterioration effects simultaneously. The function of the batch processing time ($T_{[i]}$) is formulated in Eq. (2). A mathematical model of the problem can be written as follows:

$$\text{Minimize } TFC_{[i]} = \sum_{i=1}^{N}\left(\left\{\sum_{j=1}^{i}\left(s+T_{[j]}Q_{[j]}\right) - s - T_{[i]}Q_{[i]}\right\}c_1Q_{[i]} + c_2T_{[i]}Q_{[i]}^2\right) \tag{3}$$

Subject to constraints

$$\sum_{i=1}^{N} Q_{[i]} = n \tag{4}$$

$$\sum_{i=1}^{N} T_{[i]} Q_{[i]} + (N-1)s \le d \tag{5}$$

$$B_{[1]} + T_{[1]} Q_{[1]} = d \tag{6}$$

$$Q_{[i]} \ge 1 \text{ and integer,} \tag{7}$$

$$1 \le N \le n \text{ and integer,} \tag{8}$$

Eq. (3) expresses the objective of minimizes the total inventory holding cost where the batch processing time is influenced by learning, forgetting, and deterioration effects simultaneously. Constraint (4) is a material balancing that expresses the number of parts in all batches produced equal to demand. Constraint (5) expresses that all batches should be scheduled at the available time, that is, the time interval from the start of scheduling period (time zero) to their common due date. Constraint (6) expresses that the batch scheduled in the first position backwardly should be completed exactly at the common due date. Constraint (7) expresses that batch sizes should be one or more and integer. Constraint (8) expresses that the number of batches should be a positive integer between 1 and the number of demands.

6. Problem Solution

We propose the heuristic solution based on Lagrange method. The step for searching the solution can be found in Halim et al. [7] and yusriski et al. [11]. The formula of the solution can be shown as follows:

$$Q_{[i]} = \max\left\{\left[\left(n - \sum_{k=i+1}^{N} Q_k\right)/i + \{c_1/(2c_2 - c_1)\}\{1/2(i+1)\}(s/T_{[i]}) - \{c_1/(2c_2 - c_1)\}(s/T_{[i]})i\right], 1\right\}, \quad \text{where } i = N,...,1 \tag{9}$$

$$N_{\max} = \min\{1 + \lfloor (d - T_{\min} n)/s \rfloor, n\} \tag{10}$$

Proposition 1. Suppose that there are N batches with batch size, $Q_{[i]}$, (i = 1, 2, ..., N), respectively, and the processing time of a batch is influenced by the simultaneous effects of learning, forgetting, and deterioration. The optimal backward sequence that minimizes the total inventory holding cost is obtained by sequencing the resulting batches in non-increasing order of batch sizes.
$Q_{[1]} \ge Q_{[2]} \ge ... \ge Q_{[N-1]} \ge Q_{[N]}$

Proof. Looking at Eq. (3). The formula can be written as follows:

$$\text{Minimize } TFC = c_1 s \sum_{i=1}^{N} i Q_{[i]} - (1/2) c_1 T \sum_{i=1}^{N} Q_{[i]}^2 + (1/2) c_1 T \left(\sum_{i=1}^{N} Q_{[i]}\right)^2 - s \sum_{i=1}^{N} Q_{[i]} + c_2 T \sum_{i=1}^{N} Q_{[i]}^2$$

The order of batches in a sequence just influenced by the value of the first term on the right-hand side. Relaxing the constant c_1 and s, we can see that since the value of index i (i=1, 2, ..., N) are already in increasing order. Therefore, minimizing the TFC can be obtained by sequencing the batch size, $Q_{[i]}$, in a non-increasing order using the backward scheduling approach ($Q_{[1]} \ge Q_{[2]} \ge ... \ge Q_{[N-1]} \ge Q_{[N]}$). □

7. The Proposed Procedure

Based on Lagrange Relaxation method, we propose the heuristic procedure as shown below:

Step 1: Set input parameters of n, d, p, s, δ, v, w, t_B, μ, α, β, c_1, c_2. Continue to Step 2.

Step 2: Compute N_{max} by Eq. (10). Set j as the index of the number of batches (j=1,2,..., N_{max}). Continue to Step 3.

Step 3: Begin with j= 1. Set $Q_{[1]} = n$. $T_{[1]} = p$. Continue to Step 4.

Step 4: Use Constraint (5) to evaluate the makespan of the job, if it does not exceed the available time, continue to Step 5; otherwise, there is no solution because the demand cannot be scheduled; then STOP.

Step 5: Compute a total inventory holding cost ($TFC_{[1]}$) by Eq. (3) and set the value of $TFC_{[1]}$ as a temporary optimal solution (TFC^*). Continue to Step 6.

Step 6: Set an index of the number of batches $j = j+1$; if $1 < j \leq N_{max}$ then continue to Step 7 otherwise the TFC^* is an optimal solution and the STOP.

Step 7: Set $N = j$. Use Eq. (9) to compute batch size, $Q_{[i]}$, and Eq. (2) to compute batch processing time ($T_{[i]}$) using backward index computation ($i=N,....1$). Continue to Step 8.

Step 8: Use Constraint (4) to evaluate the sum of batch sizes, if that is equal to than the demand, continue to Step 9, otherwise set TFC^* as the optimal solution and then STOP.

Step 9: Use Constraint (5) to evaluate the makespan of the job, if it does not exceed the available time then continue to Step 10; otherwise, the TFC^* is the optimal solution and then STOP.

Step 10: According to Proposition 1, sequence the batches in non-increasing $Q_{[i]}$.The largest $Q_{[i]}$ is closest to the due date. Compute the total inventory holding cost of the batch ($TFC_{[j]}$) by Eq. (3). Continue to Step 11.

Step 11: Compare $TFC_{[j]}$ and TFC^*. If $TFC_{[j]} < TFC^*$ then $TFC_{[j]}$ is seted as a TFC^* otherwise TFC^* is still as a TFC^*. Return to Step 6.

8. Numerical Experience

We demonstrate the ability of the heuristic procedure to solve the following numerical examples. Suppose input parameters as follows: p= 0.5, n= 5, d= 10 , s= 1, v= 0.25, δ= 90%, t_B= 20, w= 1, μ= 1, α= 20, β= 2, c_1= \$3 and c_2 = \$2.5. The result of processing by the heuristic algorithm is presented in Table 1.

Table 1. The heuristic solution of example 1

Nmax	*N*	$Q_{[i]}$, i=1,2,3,5	$T_{[i]}$, i =1,2,3,5	$B_{[i]}$, i =1,2,3,5	*TFC* (\$)
5	1	[5]	[0.5]	[7.5]	31.25
	2	[4, 1]	[0.5, 0.5]	[8, 6.5]	24.25
	3	[3, 1, 1]	[0.42, 0.5, 0.5]	[8.7, 7.2, 5.7]	21.63* optimal
	4	[2, 1, 1, 1]	[0.42, 0.45, 0.5, 0.5]	[9.2, 7.7, 6.2, 4.7]	25.86

Table 1 presents the solution of the procedure. The result of Eq. (10) yields *Nmax* = 5. The search of algorithm solution begins with the number of batches, N, equals 1. The result of *TFC* is 31.25. The search of the solution is continued for N = 2. The result of *TFC* is 24.25. The value of the *TFC* for N= 2 is compared with N= 1, and it is found that the value of *TFC* for N = 2 is cheaper than N = 1. Thus, the value of the total inventory holding cost for N= 2 is a temporarily optimal solution, TFC^*. The search continues for N = 3. It is found that the value of *TFC* is 21.63, and it is found that the value of *TFC* for N = 3 is cheaper than N= 2. Thus, the value of the total inventory holding cost for N =3 is a temporarily optimal solution, TFC^*. The solution for N = 4 found that the value of *TFC* is 25.86, and it is found that the value of *TFC* for N = 3 is cheaper than N = 4. Thus, the value of the total inventory holding cost for N = 3 is an optimal solution, obtained by $Q_{[1]}$= 3, $Q_{[2]}$= 1 and $Q_{[3]}$= 1.

9. The Comparison Test

The comparison test is determined by comparing the solutions of the heuristic procedure to the optimal solution produced by the enumeration procedure using the integer composition technique (see Yusriski et al. [11]). we show the result of the tests for the number of demand between 1 unit and 20 units since the enumeration procedure cannot solve the number of demands more than 22 unit due to out of memory.

Table 2. The effectiveness results

	Proposed Procedure		Enumeration Procedure		Comparison Test Results	
n	*TFC* ($)	Computational Time (second)	*TFC* ($)	Computational Time (second)	Solution Effectiveness (%)	Time-Saving (second)
5	21.63	0.00183	21.63	0.01498	100	0.01315
7	52.58	0.00275	48.40	0.01145	91	0.0087
9	43.82	0.00296	42.31	0.06437	96	0.06141
10	42.12	0.00363	40.43	0.11895	96	0.11532
11	137.62	0.00288	128.60	0.22987	93	0.22699
12	96.90	0.00163	87.95	0.44892	90	0.44729
14	148.28	0.00139	138.58	2.02380	93	202379.9986
16	98.11	0.00302	90.19	10.70594	91	1070593.997
18	113.84	0.00645	107.86	78.24584	94	7824583.994
20	137.85	0.00256	133.49	941.85271	97	94185271
	Mean				94	
	ST Dev				3	

Table 2 shows that although the heuristic procedure does not guarantee an optimal solution. It may produce an outstanding solution with the average effectiveness is 94%. Additionally, the average of time-saving shows that the heuristic procedure is more efficient than the enumeration procedure.

10. Concluding Remarks

The current research deals with single machine batch scheduling problems to minimize inventory holding cost in a just-in-time production system considering that the batch processing time is influenced by learning, forgetting, and deterioration effects simultaneously. We propose a heuristic procedure based on Lagrange relaxation method to solve the problem. The effectiveness of the procedure is determined by comparing the resulting solution with the optimal procedure obtained from the enumeration procedure. The comparison test result shows that the average effectiveness is 94%. The comparison test also shows that the computational time of the heuristic procedure is more efficient than the enumeration procedure. In this paper, we have considered problems with only a single item. Hence, further research may focus on the multi-items problem.

Acknowledgments

The authors wish to thank the editors and the referees for their constructive comments and suggestions. This research was supported by Institute Teknologi Bandung (ITB) and Universitas Jenderal Achmad Yani (UNJANI) Bandung.

References

[1] Wright T P 1936 Factors affecting the cost of airplanes *Journal of Aeronautical Sciences* **3** 122–128.

[2] Jaber M Y and Bonney M 1996 Production break and the learning curve: The forgetting phenomenon *Applied Mathematical Modelling* **20** 162–169.

[3] Kaminskiy M P and Krivtsov V V 2008 An integral measure of aging/rejuvenation for repairable and non-repairable systems *Reliability and Risk Analysis: Theory & Applications* **1** 69–76.

[4] Mor B and Mosheiov G. 2012 Batch scheduling with step-deteriorating processing times to minimize flow time *Naval Research Logistics* **59** 587–600.

[5] Teyarachakul, Chand S and Ward J 2011 Effect of learning and forgetting on batch sizes. *Production and Operations Management* **20** 116–128.

[6] Yang W-H 2009 A batching problem with learning effect considerations *Asia-Pacific Journal Of Operational Research* **26** 307–317.

[7] Halim A H, Miyazaki S and Ohta H 1994 Batch scheduling problem to minimize actual flow times of component through the shop under JIT environment *European Journal of Operational Research* **72** 529–544.

[8] Halim A H and Ohta H 1994 Batch Scheduling Problem to minimize inventory cost in the shop with both receiving Batch Scheduling Problem to minimize inventory cost in the shop with both receiving and delivery Just-In-Time *International Journal of Production Economics* **33** 185–194.

[9] Yusriski R, Astuti B, Sukoyo, Samadhi T M A A and Halim A H 2015 Integer batch scheduling problems for a single-machine to minimize total actual flow time *Procedia Manufacturing* **2** 118–123.

[10] Janiak, Lichtensteind M and Ruson A 2011 Scheduling jobs with a linear model of simultaneous ageing and learning effects *Decision Making in Manufacturing and Services* **5** 37–48.

[11] Yusriski R, Sukoyo S, Samadhi T M A A and Halim A H 2015 Integer batch scheduling problems for a single-machine with simultaneous effect of learning and forgetting to minimize total actual flow time *International Journal of Industrial Engineering Computations* **6** 365–378.

[12] Yusriski R, Sukoyo, Samadhi T M A. A and Halim A H 2016 An Integer Batch Scheduling Model for a Single Machine with Simultaneous Learning and Deterioration Effects to Minimize Total Actual Flow Time *IOP Conference Series: Materials Science and Engineering* **114** 1-10.

[13] Iskandar B P and Jack N 2011 Warranty Servicing with Imperfect Repair for Products Sold with a Two-Dimensional Warranty *Chapter Replacement Models with Minimal Repair, Part of the series Springer Series in Reliability Engineering* 163-175.

IOP Conf. Series: Materials Science and Engineering **319** (2018) 012039 doi:10.1088/1757-899X/319/1/012039

Optimization of Coolant Technique Conditions for Machining A319 Aluminium Alloy Using Response Surface Method (RSM)

S Zainal Ariffin[1*], A Razlan[2], M Mohd Ali[3], A M Efendee[4] and M M Rahman[5]

[1,3,4] Faculty of Manufacturing Engineering Technology, TATI University College, Kemaman, Terengganu, Malaysia
[2,5] Faculty of Manufacturing Engineering, University Malaysia Pahang, Gambang, Kuantan, Pahang, Malaysia;

zainal@tatiuc.edu.my, razlan@ump.edu.my, mohd_ ali@tatiuc.edu.my, efendee@tatiuc.edu.my

Abstract. Background/Objectives: The paper discusses about the optimum cutting parameters with coolant techniques condition (1.0 mm nozzle orifice, wet and dry) to optimize surface roughness, temperature and tool wear in the machining process based on the selected setting parameters. The selected cutting parameters for this study were the cutting speed, feed rate, depth of cut and coolant techniques condition. Methods/Statistical Analysis Experiments were conducted and investigated based on Design of Experiment (DOE) with Response Surface Method. The research of the aggressive machining process on aluminum alloy (A319) for automotive applications is an effort to understand the machining concept, which widely used in a variety of manufacturing industries especially in the automotive industry. Findings: The results show that the dominant failure mode is the surface roughness, temperature and tool wear when using 1.0 mm nozzle orifice, increases during machining and also can be alternative minimize built up edge of the A319. The exploration for surface roughness, productivity and the optimization of cutting speed in the technical and commercial aspects of the manufacturing processes of A319 are discussed in automotive components industries for further work Applications/Improvements: The research result also beneficial in minimizing the costs incurred and improving productivity of manufacturing firms. According to the mathematical model and equations, generated by CCD based RSM, experiments were performed and cutting coolant condition technique using size nozzle can reduces tool wear, surface roughness and temperature was obtained. Results have been analyzed and optimization has been carried out for selecting cutting parameters, shows that the effectiveness and efficiency of the system can be identified and helps to solve potential problems.

Keywords: turning, infrared thermometer, temperature, coated carbide, Response Surface Method.

1. Introduction

An innovative application method of coolant to the cutting point, which is more effective for coolant using 1.0 mm size nozzle orifice cutting application, is increasingly important in high speed and high efficiency machining of difficult to machine materials including not only metal but aluminum alloy 319 as well. In such cases, the experience of the operator plays a major role, but even for a skilled operator it is very difficult to attain the optimum values each time. The objectives of this project are to evaluate the surface roughness, tool wear and temperature of a three machine condition (design 1.0 mm nozzle orifice, dry and wet) of coated cemented carbide tool when machining aluminium alloy 319. This research applies Response Surface Methods (RSM) approach to studying the impact of turning parameters on the tool wear and temperature and also roughness of turned surfaces. RSM analysis method approach is quality control methodology that combines control charts and process control with product and process design to achieve a robust total design. Workpiece material aluminium alloy 319 is very important in choosing tools, machining parameters and cooling strategy. According to [1] [2], is quantified by various parameters, such as, tool life, chip size, achievable surface finish and/or the amount of specific power consumed. It is influenced by a material's alloy chemistry, additives, microstructure, heat treatment, temper and physical and mechanical properties [3]. In addition, according to[4], lead acts as a lubricant, thus reducing the friction coefficient of friction between the chip and the tool. In this study of temperature measurement, the most widely used infrared thermometer laser was selected for the measuring of cutting tool's average temperature during turning operation. The values within certain range of cutting parameters like cutting speed, depth of cut and feed rate were selected and used for building the mathematical model using CCD based Response Surface Method (RSM). The temperature data of cutting tool was obtained by experimentation and the optimization of selected cutting parameters obtained successfully.

2. Experimentation investigations

Experiment were conducted on PUMA 250 type lathe to study the cooling technique condition used in this research that will bring optimum cutting performance like minimum tool wear and reduction in cutting temperature. The photograph of experimental setup is shown in Figure 1. The material used throughout this researched is a cylindrical aluminum alloy 319 rod with an axial length of 360 mm and the Ø 60 mm.

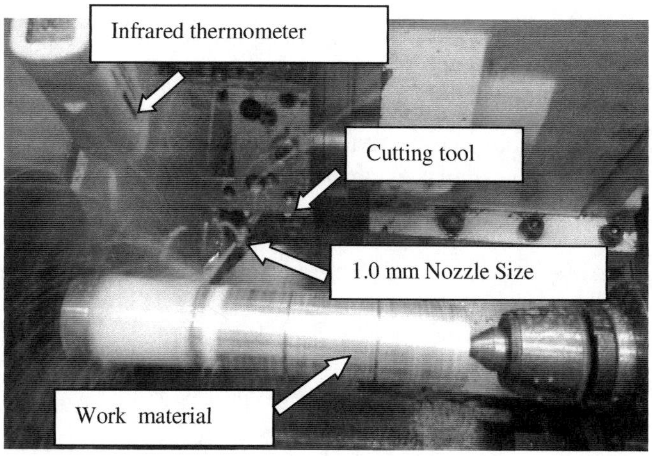

Figure 1. Photographic view of experimental setup.

2.1. Experiment Conditions

The turning process was performed in the following coolant conditions wet cooling, dry cutting and 1.0 mm nozzle orifice. The Central Composite Design (CCD) based RSM was used for the determination of optimum control factors with 3- level factor design with center point six was selected. Values of those factors had been achieved with consideration of the machine tool capacity. There were total 30 combinations of the turning runs were carried out using design expert software to complete experimentation. The experimental conditions with values of level for cutting parameters are shown in table 1 below.

Table 1. Experiment condition with value level for cutting parameter.

	Level 1 (-1)	Level 2 (0)	Level 3 (+1)
Cutting Speed (rpm)	150	210	270
Depth of Cut (mm)	0.08	0.16	0.24
Feed Rate (mm/rev)	0.2	0.06	1.00
Coolant Condition Technique	dry cutting	wet cooling	1.0 mm nozzle orifice

3. Analysis of Results

3.1. Temperature

Central Composite Design (CCD) based temperature and tool wear methodology was selected in this work [5] . RSM require more data compare to the Taguchi method to ensure the optimum condition[6]. RSM relates the independent input variables with output (process response). Design Expert software was used to obtain the set of experimental runs of CCD which would help to investigate the influence of three cutting parameters (cutting speed, depth of cut and feed rate) on (temperature) and surface roughness output.

Again this data had been utilized to analyze through Design-Expert Software. The result of Analysis of Variance (ANOVA) has been carried out and as per Table 2 the value of R-squared (0.7667) and predicted R-squared (0.5489) were obtained. From data temperature analysis, a negative "Pred R-Squared" implies that the overall mean is a better predictor of your response than the current model. "Adeq Precision" measures the signal to noise ratio. A ratio greater than 4 is desirable. The ratio of 8.515 indicates an adequate signal. This model can be used to navigate the design space.

Table 2. Value of R-Squared and Adjusted R-Squared

Std. Dev	4.33	R-Squared	0.7667
Mean	103.43	Adj R-Squared	0.5489
C.V	4.19	Pred R-Squared	-0.1608
PRESS	1399.2	Adeq Precision	8.515

Regression Equation in Un-coded Units:
Temperaturer=+62.94266+0.41185+-196.08684+-4.66598+0.50203-8.56037E004+1002.85416+5.21418+1.68771-0.13021+0.041667+4.16667E-003+11.71875-35.93750-4.29080

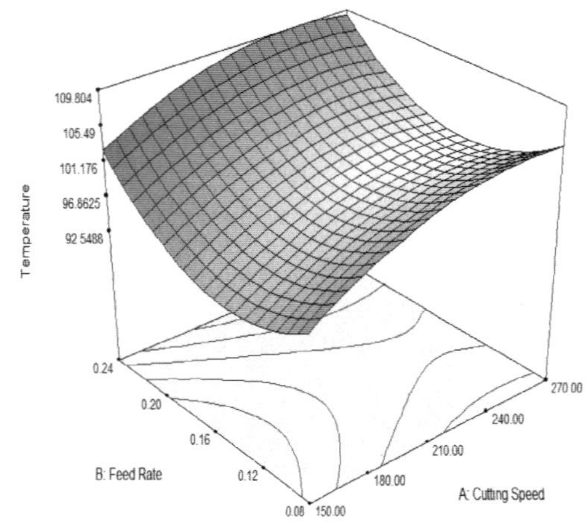

(a) Main and 3D temperature plots of temperture vs Cutting Speed, Feed Rate

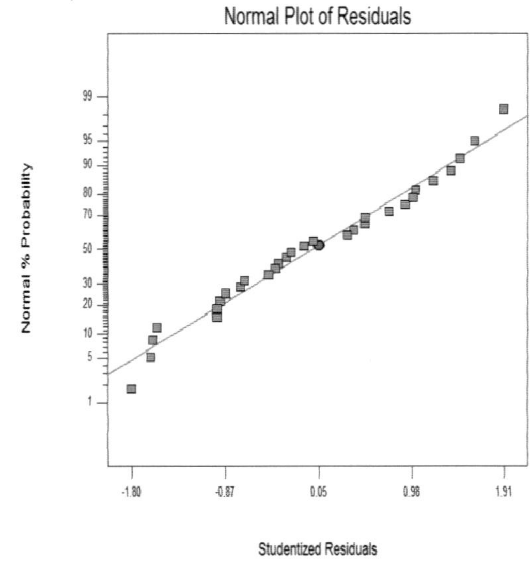

(b) Normal plot of residuals of temperature vs Normal % Probability, Studetized Residuals

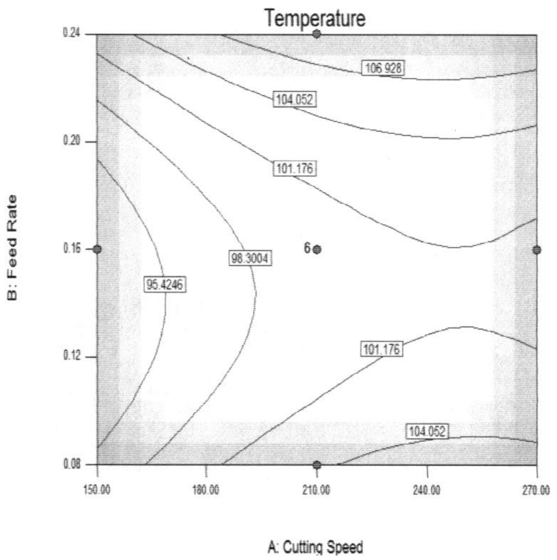

(c) model graphs of temperature Vs Feed rate, Cutting Speed

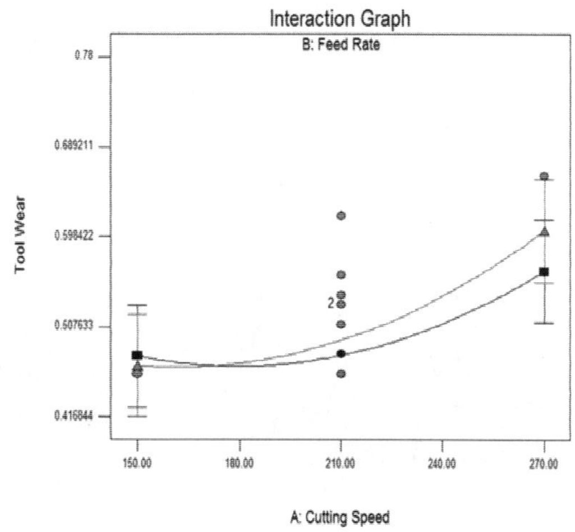

(d) perturbation of temperature vs tool wear, Cutting Speed

Figure 2. (a) Main and 3D temperature plots of temperture (b)Normal plot of residuals of temperature (c) model graphs of temperature (d) perturbation of temperature.

Figure 2a, 2b, 2c and 2d show the 3D surface plot, normal plot of residuals, model graf and perturbation of temperature is presented temperature was minimized when feed rate and depth of cut were at lowest levels. A significant mutual interaction occurred between cutting speed , feed rate, and depth of cut with cutting condition technique.The optimal dearease in the temperature of tool wear was from 0.56 (μm) to 2.45 (μm), where feed rate was within the range of 0.08 - 0.24 mm/rev, depth of cut in the range of 0.8 - 1.0 mm and cutting speed in the range of 150 -270 m/min. These findings were reported by [7] and [3]

3.2. Effect of tool wear

Figure 3a, 3b, and 3c shows the variation in flank wear with machining time under 1.0 mm nozzle orifice, wet coolant and dry machining under coolant technique condition. 1.0 mm nozzle orifice reduces average auxiliary flank wear on auxiliary cutting edge. The temperature also grew very low considerably. It appears on Figure 3 that temperature and flank wear grows quite fast under dry machining due to more intensive temperature and stress at the too-tips. 1.0 mm nozzle orifice technique appeared to be more effective in reducing tool wear and temperature. However, it is evident that coolant technique condition appeared to be effective in reducing tool wear and temperature as well as controlling the deterioration of the auxiliary cutting edge by thermal effect to adhesive wear and built - up edge formation.

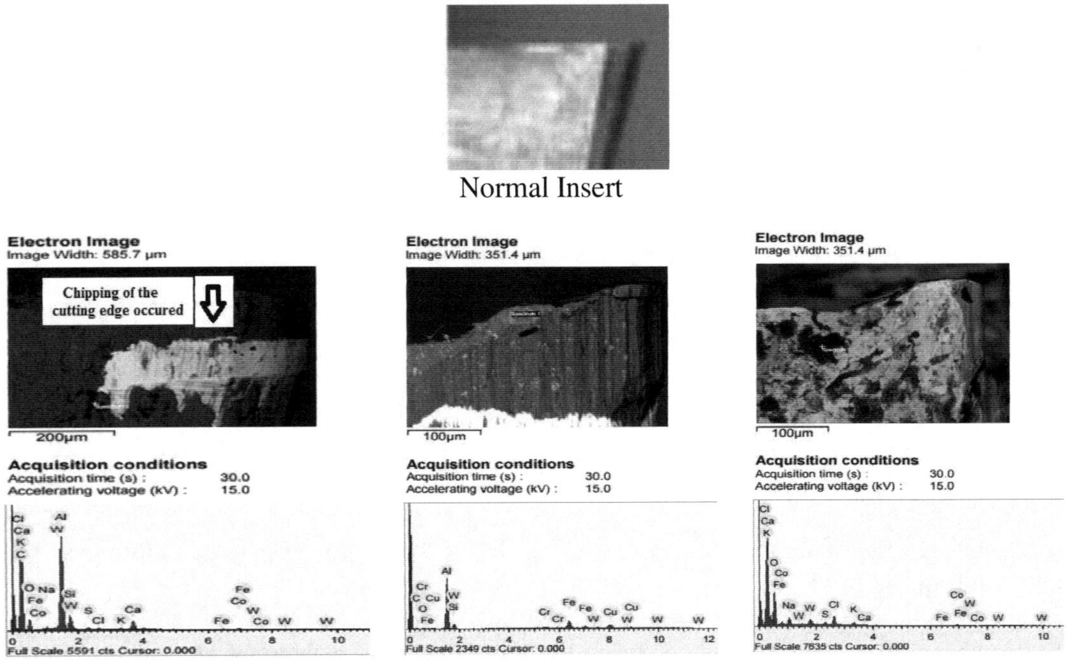

(a) 1.0 mm nozzle orifice (b) wet coolant (c) Dry machining

Figure 3. Schematic illustration between 1.0 mm nozzle orifice ,wet coolant, and dry machining

4. Optimization

In this study, RSM has been utilized for single response optimization. The use of response temperature and tool wear optimization helps to calculate the optimal values of input in order to minimize the temperature and tool wear during the turning process of aluminums alloy 319. The constraints for optimization of cutting parameters are shown in Table 3.

Table 3. Constraints for optimization of parameter

Parameter	Lower limit	Higher limit
Cutting speed (rpm)	150	270
Depth of cut (mm)	0.2	1.0
Feed Rate (mm/rev)	0.08	0.24
Environments Cutting	Dry , Wet and 1.0 mm nozzle orifice	

Table 4 shows the values for the input parameters for minimizing temperature and tool wear. It is clearly seen that obtained optimal value is 94.97 for temperature and 0.439 for tool wear for the respective values of cutting speed, depth of cut and feed rate.

Table 4. Optimization Results

Cutting Speed	Depth of Cut	Feed Rate	Temperature	Tool wear	Desirability
150	0.50	0.05	94.97	0.439	0.488

5. Conclusion

This paper focused on temperature and tool wear tester measurement for measuring the average cutting tool temperature during turning operation on lathe machine with work piece material as round bar aluminum alloy 319 and coated cemented carbide as cutting tool. This paper included the effect of cutting temperature on different parameters such as cutting speed, feed rate and depth of cut. Based on the results following conclusions can be made:

- The cutting performance of 1.0 mm nozzle orifice coolant techniques machining is better than the dry and wet machining with wet cutting fluid supply because this technique is capable of reducing the chip-tool interaction and maintains sharpness of the cutting - edges.
- 1.0 mm nozzle orifice coolant techniques flow provided reduced tool wear, improved tool wear, tool life and better surface finish as compare to dry and wet machining of aluminum alloy 319.
- Surface finish and temperature improved mainly due to reduction of wear and damage at the tool tip by the application of 1.0 mm nozzle orifice coolant techniques.
- Optimized values of cutting parameters have been achieved for minimum temperature with desirability of 0.488%, which is highly acceptable.

Acknowledgement

The author would like to thanks TATI University College (TATIUC) for the full support of TATIUC grant, no. (9001 –1201), University Malaysia Pahang (UMP) for providing the equipments and supervision in making this important research viable and effective.

References

[1] L. Abhang and M. Hameedullah, "Chip-tool interface temperature prediction model for turning process," *International Journal of Engineering Science and Technology,* vol. 2, pp. 382-393, 2010.

[2] B. P. Nedić and M. D. Erić, "Cutting temperature measurement and material machinability," *Thermal Science,* vol. 18, pp. 259-268, 2014.

[3] D. O'sullivan and M. Cotterell, "Temperature measurement in single point turning," *Journal of Materials Processing Technology,* vol. 118, pp. 301-308, 2001.

[4] J. Xie, M. Luo, K. Wu, L. Yang, and D. Li, "Experimental study on cutting temperature and cutting force in dry turning of titanium alloy using a non-coated micro-grooved tool," *International Journal of Machine Tools and Manufacture,* vol. 73, pp. 25-36, 2013.

[5] M. S. Kasim, "Comparison between Taguchi method and response surface methodology (RSM) in optimizing machining condition," in *Proceeding of 1st International Conference on Robust Quality Engineering*, 2013, pp. 60-68.

[6] A. Kus, Y. Isik, M. C. Cakir, S. Coşkun, and K. Özdemir, "Thermocouple and infrared sensor-based measurement of temperature distribution in metal cutting," *Sensors,* vol. 15, pp. 1274-1291, 2015.

[7] Z. A. Selamat, N. B. Baba, M. Alias, A. R. Yusoff, and M. M. Rahman, "Effect of coolant nozzle sizes on turning aluminum alloy AL319," *Indian Journal of Science and Technology,* vol. 9, 2016.

IOP Conf. Series: Materials Science and Engineering **319** (2018) 012040 doi:10.1088/1757-899X/319/1/012040

Development of an Electronic Kit for detecting asthma in Human Respiratory System

Cheow Shek Hong, Ahmad Shahrizan Abdul Ghani, Ismail Mohd Khairuddin

Innovative Manufacturing, Mechatronics, and Sports Lab (IMAMS), Faculty of Manufacturing Engineering, Universiti Malaysia Pahang, 26600 Pekan, Pahang, Malaysia.

cheow93@gmail.com, *shahrizan@ump.edu.my, ismailkhai@ump.edu.my

Abstract. In this paper, a prototype of a carbon dioxide (CO2) measurement device is designed to detect and used to monitor asthma patients. Nowadays, capnogram device is widely used in monitoring asthma and asthma related medical services. However, capnogram is very costly and unaffordable for patient especially those who are in low income household. Thus, the proposed device is cost effective, affordable, and produced to detect and monitor the severity of asthma. Meanwhile, flow meter will cause patient to have chest pain as they needed maximum effort to blow in the device. To overcome these limitations, this prototype electronic kit is easy to use and suitable for all range patients. This prototype electronic kit consists of MH-Z14A carbon dioxide (CO2) sensor to detect the concentration of carbon dioxide from the user exhaled air. Arduino microcontroller is used to process the data while TFT Display shield is applied for data presentation. In addition, HC-06 Bluetooth module is used to communicate with PC for further analysis of the captured graph. This device was tested with 3 asthmatics and 3 normal users. The results showed that asthmatic user has a different graph pattern compared with normal user and these graphs are clearly differentiated on the device TFT screen. Asthmatic user produces "shark fin"-like pattern whereas normal user produces "square wave"-like pattern. This device has successfully produced distinguished-patterns difference between asthmatic and normal user; therefore, it is suitable for asthma monitoring.

1. Introduction

Asthma is a chronic disease involved the airways in the lungs and it causes difficulty in breathing due to respiratory condition marked by the spasms in the bronchi of the lungs. Mc Phee et al. [1] reviewed that asthma exacerbation is a reduction in expiratory flow which is caused by many asthma triggers and the main triggers which will lead to the symptoms of respiratory diseases included viral infection, air pollutions, exercise, cockroach allergen and dust mists.

Furthermore, spirometer and peak flow meter is well known in monitoring asthma. Unfortunately, these two devices have their own limitations in which spirometer has complex procedures as the patient face difficulties on performing multiple criteria when using it and only suitable for age range above 6 years old.

At present, peak flow meter and spirometer are commonly used for monitoring asthma. However, they have their limitations. Based on Zuileka et al. [2], patients whose are tested using peak flow meter will experience chest pain as they need maximum effort to blow into the device. Meanwhile, spirometer is only suitable for patient age 6 and above and not suitable for early-detection of asthma. It also required multiple criteria to fulfill the requirement set by the device. To overcome these problems, capnography is introduced to monitor the severity of asthma without harming patient and suitable for all age patient. Non-invasive continuous analysis of the concentration of carbon dioxide in respiratory cycle is called capnograhy. Infrared technology is used to determine the concentration of carbon dioxide (CO2). Therefore, it can easily use to differentiate normal and asthmatic patient. Figure 1 below shows the difference concentration carbon dioxide between normal and asthmatic patient.

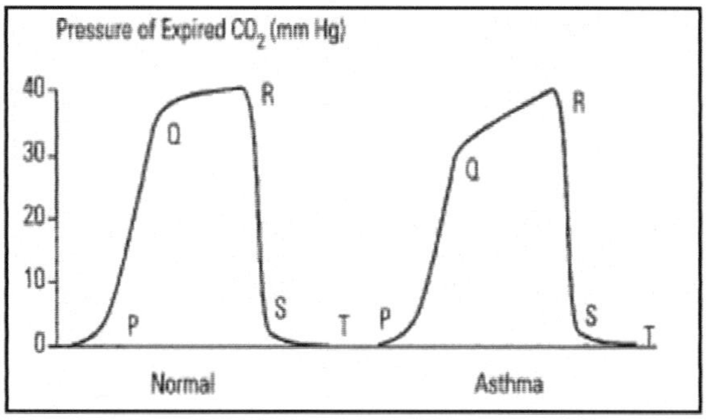

Figure 1. Capnogram between Normal and Asthmatic Patient

There are few sensors considerations in producing this electronic kit. Sensor comparisons were done to select the best sensor. Based on Marani et al. [3], pressure sensor is used for lung sound analysis. It can prevent damages caused by mechanical ventilation which can induce serious damage due to the natural pulmonary elastance. Yu et al. [4] used soft stethoscope for detecting asthma wheeze in young children. Soft stethoscope has been modified and it can be easily used on young children. This sensor can easily use on young children. Normally, soft stethoscope is easily effected by loud sound but this modification of soft stethoscope improves the limitation of this phenomenon whereas high sound amplitude will not affect the soft stethoscope.

On the other hand, Zuleika et al. [2] designed a respiratory carbon dioxide measurement for home monitoring asthma using carbon dioxide (CO2) sensor. This sensor is suitable for all range patient including children below 6 years old. It can also avoid all type of complicated test which leads to the chest pain. However, the reading between normal and asthmatic patient is not much different and hardly differentiated. Kaushal et. al [5] use a pellet sensor based asthma detection system for exhaled breath analysis. This pellet sensor consists of metal oxide semiconductor (MOS) and it is sensitive to exhaled breath. It eliminates the disadvantages of other technique by its low cost and easy operation features. On the other hand, Liu et al. [6] use ultrasonic transducer for asthma pattern identification via continuous diaphragm motion monitoring. It can overcome the limitations of implementing the ultrasound images of diaphragm movement to identify the asthma pattern. In addition, Gatthy et al. [7] used amperometric nitric oxide for asthma detection. It has a detecting limit and sensitivity to changes in relative humidity, response time, flow sensitivity and stability.

Signal transmitter is an electronic device which is capable of generating electromagnetic wave carrying information and signal which lastly received by an electronic receiver. Buechley [8] through multi-purpose user awareness kit for consumer electronic devices used infrared sensor (IR) to receive and process the received signal through electronic devices such as personal computer (PC) and smart phone. Meanwhile, Bjelice et al. [9] used UART communication device in construction of kit for electronic textiles. This kit interface binary intra-processor protocol and data are present by Sky TV application GUI. Next, Ramli et al. [10] developed heartbeat detection kit for biometric authentication system whereas Bluetooth Module is used as signal transmitter. It received and processed by Android platform developed. Captured data will be presented on Android application GUI.

Data processing refers to transformation of collected data into information which is useful in a certain manner for user understanding. There are many types of data processing such as GUI, LCD and oscilloscope. Based on Yu et al. [4], android is used as a platform to collect and present the data. Android operating system is so convenience as this system can be open in any electronic device such as smartphone, PC or even pads everywhere. Rasmli et al. [10] reported that android platform is used as a function of login into the system. Graphical User Interface (GUI) is used to present the result and data to the user [8]. Matlab software is used by Zuileka et al. [2] as a real time data plotter. A real time data logger was developed by sorting the data into the array using the basics of Serial Communication in Matlab by updating the output or data pot according to real-time application.

After comparisons and reviews, carbon dioxide (CO2) sensor has been decided to be used as it can measure the concentration of carbon dioxide (CO2) in a relaxing way and it does not depend on user age unlike some of the sensors used by previous researchers. Arduino microcontroller is used to collect and process the data from the sensor. Meanwhile, Bluetooth Module is used as a signal transmitter to communicate with personal computer. TFT LCD shield is used as a data presentation element. The data will be generated into a graph. Meanwhile Graphical User Interface (GUI) will be build inside the computer for better view of the result. GUI is also used as a platform to analysis the result from the user.

2. Methodology

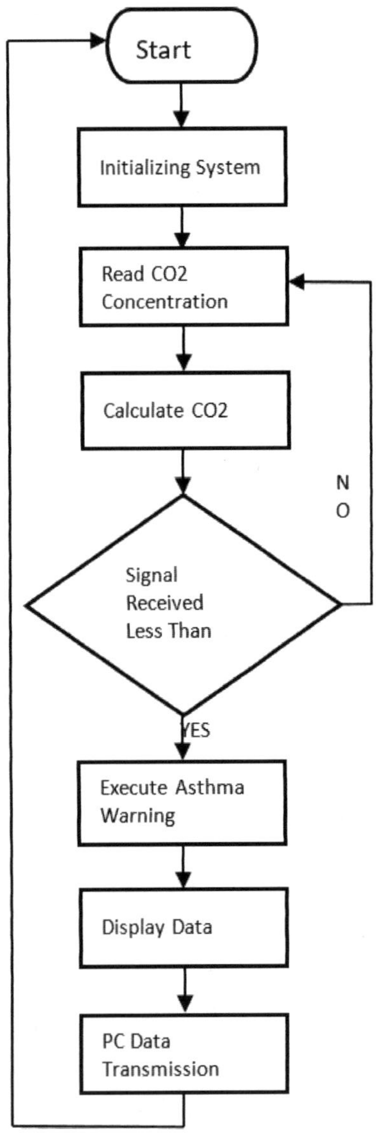

Figure 2 System Flowchart

Figure 2 shows the process flowchart of the electronic kit. Basically, this electronic kit consists of three main parts including electronic, hardware and software parts. The process flowchart will be controlled by the software part. Code is programmed in Arduino microcontroller to process the data collected from the carbon dioxide (CO2) sensor by calculating the CO2 concentration. Next, it will send the data to the TFT LCD shield to display the result in graph form. When a possible asthmatic user is detected, threshold will be activated then warning through LED will be given to the user. Meanwhile, Bluetooth module is used to communicate with personal computer (PC) for a better view of the result.

Figure 3 shows the MH- Z14A carbon dioxide (CO2) sensor which is used in this project. This sensor has three types of output including analogue, PWM and UART communication. In the proposed technique, UART communication is used which consists of host and subsidiary detector. On the other hand, concentration of carbon dioxide (CO2) is calculated based on the Eq. 1 below:

$$Concentration\,of\;CO2 = HC * 255 + LC \qquad (1)$$

where CO2 = Carbon Dioxide; HC = High channel; LC = Low channel

Figure 4 shows the schematic diagram of the electric circuit. This circuit included MH-Z14A carbon dioxide (CO2) sensor as shown in Figure 3, Arduino microcontroller, level logic converter, TFT LCD shield, HC-06 Bluetooth module and LED. The purpose of the level logic converter is to collect the low voltage signal from the sensor and convert to a high voltage signal. TFT LCD shield is used as a tool for data presentation whereas Bluetooth module is used for the communication with PC. In this proposed prototype, LED is used as an indicator to differentiate between asthmatic and normal user. Figure 5 shows the mechanical design for the proposed prototype drawn using CATIA software. In this design, the electronic kit (prototype) included electric junction box, TFT LCD shield, ON/OFF switch, LED, buttons, hose and breather.

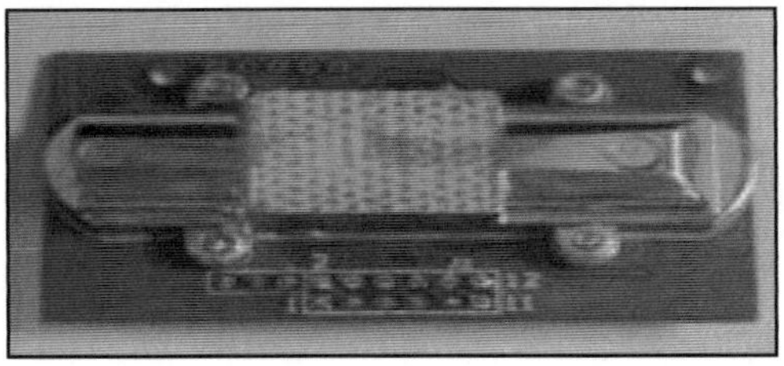

Figure 3. MH-Z14A Carbon Dioxide Sensor

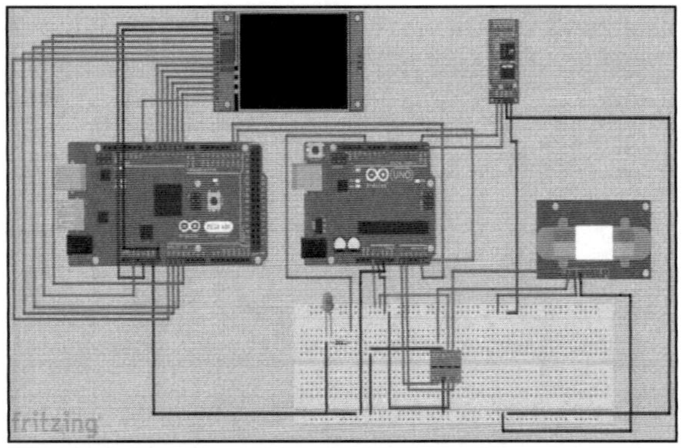

Figure 4. Schematic Diagram for Electric Circuit

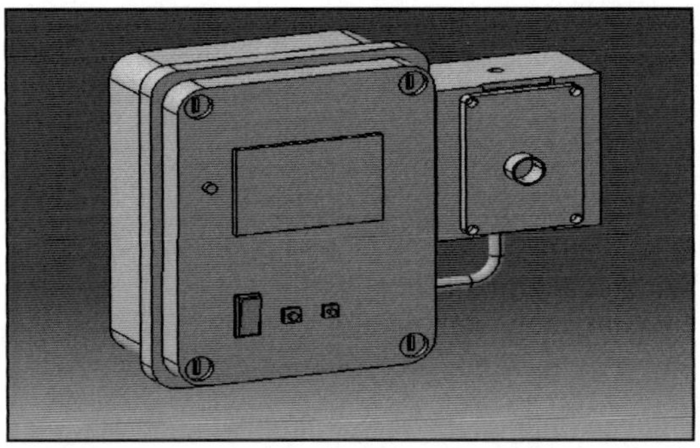

Figure 5. Mechanical Design for Electronic Kit

The sizes of the individual parts are stated in the following table.

Table 1: dimension of the proposed prototype

Part/Model	**Size (mm)**
Cover-front	28.59 x 121 x 161
Cover	3 x 108.5 x 140
Breather	25 x 50 x 76
Breather cover	13 x 50 x 70
Body	50 x 122 x 155

3. Result and Discussion

Figure 6 indicates the sample of output graph of CO2 concentration agains time. As shown in the figure, the concentration graph is divided into 4 phases in which, it is used as main parameter to differentiate between normal and asthmatic patient. Table 2 shows the sample of raw data collected from the normal user. From the table below, there will be at least two same peak value for each test within the range of 40 to 60 seconds. After maintaining the value, CO2 concentration starts to decrease. This shows that, the graph of CO2 concentration for normal user will have flat characteristic which refers to phase 3 in Figure 6. On the other hand, Table 3 shows the CO2 concentration raw data from asthmatic user. From the table, there will be only a peak value before the concentration of carbon dioxide dropped. The peak value happened at 60 seconds then it dropped at least 100ppm. This value could be referred as the peak characteristic of phase 3 in Figure 6. Therefore, from the both table and based on these characteristics, normal and asthmatic user can be differentiated.

Figure 7 refers to the generated graph of carbon dioxide (CO2) concentration obtained from a normal user. From the graph, the concentration of carbon dioxide (CO2) started to rocket up to reach 2880 parts per million (ppm) then it became stable before it decreases to 2264 ppm then back to its original value. Meanwhile, the graph has a "square wave" pattern from time 30s to time 80s. However, the data after that time shows some inconsistent value. This might due to the concentration of carbon dioxide from the surrounding environment.

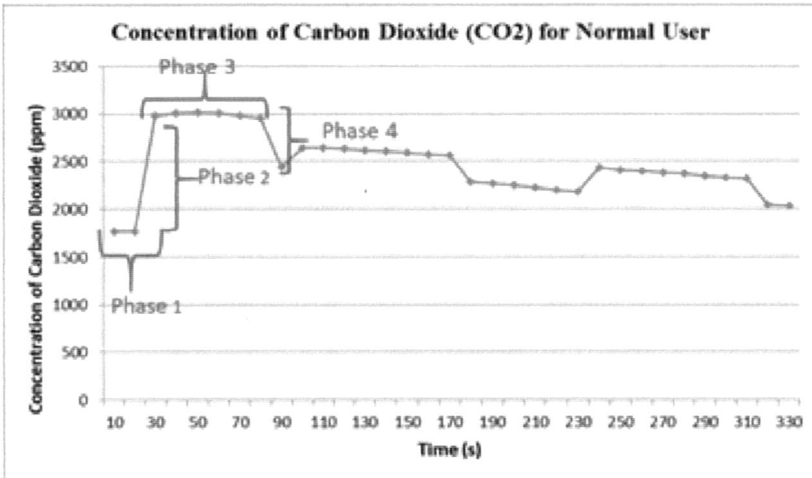

Figure 6. Phase division of CO2 concentration

Based on D'Mello et al. [11], capnogram consists of 4 phase as shown in Figure 6. Phase 1 is also known as Baseline. Baseline is the beginning of exhalation where there is no carbon dioxide presented. As shown in Figure 7, baseline started around 1800ppm due to this sensor measured the concentration of carbon dioxide of the surrounding instead of the human exhalation. When the user exhalation was sensed thus it caused phase 2 to begin. Phase 2 also known as ascending phase, caused by the carbon dioxide from the alveoli begins to reach the upper airway and caused the amount of carbon dioxide to rise rapidly. From Figure 7, the concentration rises from baseline to the peak refers to the phase 2.

Table 2. Raw Data from Normal User

Time (s)	Concentration of CO2 (ppm)		
	Test 1	Test 2	Test 3
10	1764	1724	1754
20	1765	1934	1744
30	2984	2754	3064
40	3004	2774	3400
50	3004	2784	3400
60	3004	2784	3380
70	2984	2774	3150
80	2954	2754	2544
90	2444	2734	2534
100	2640	2714	2524
110	2640	2264	2514

Table 3. Raw Data from Asthmatic User

Time (s)	Concentration of CO2 (ppm)		
	Test 1	Test 2	Test 3
10	1484	1580	1560
20	1484	1580	1570
30	3890	2814	1764
40	4410	4094	4744
50	4400	4360	4744
60	4140	4094	4594
70	3690	3850	4314
80	3670	3620	4284
90	3314	3400	3484
100	2734	2784	3474
110	2724	2714	3710

Next will be phase 3, also known as alveolar plateau. The concentration of carbon dioxide in this phase is stable as shown from the graph above. This is because the carbon dioxide rich alveolar gas constitutes the majority exhaled air. Meanwhile the concentration of carbon dioxide is uniform from alveoli to mouth or nose. Lastly, phase 4 (descending phase) where the carbon dioxide level drops quickly to its lower value. This phase can be seen from the graph above when the concentration starts to drop after the peak level. Figure 8 shows the graph generated from the asthmatic user. The raw data started to rocket up when the sensor senses the concentration of Carbon Dioxide (CO2) from the asthmatic user. Then, the concentration of the Carbon Dioxide (CO2) starts to decrease until it reached the lower (initial) value.

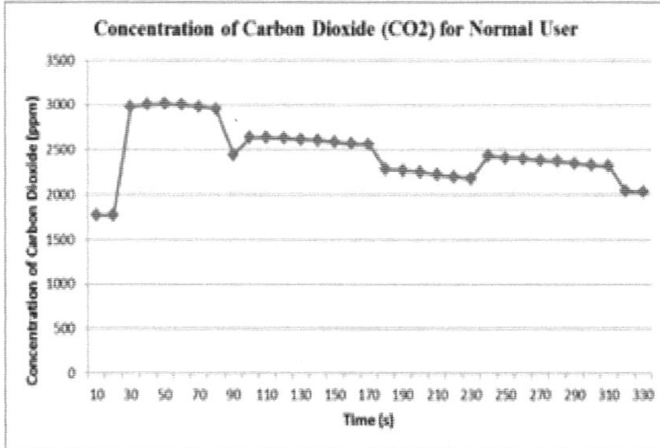

Figure 7. Graph of Concentration Carbon Dioxide for Normal User

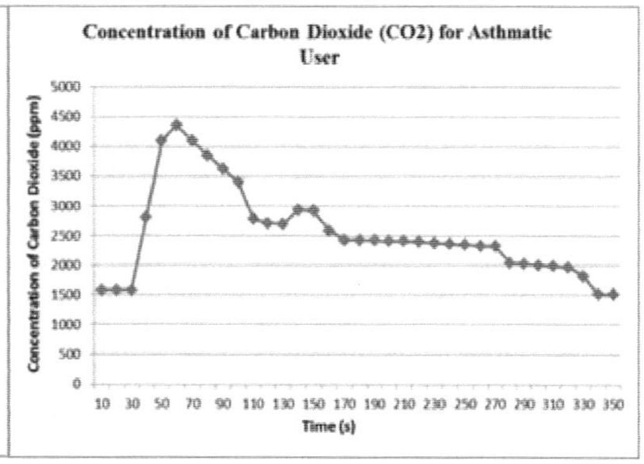

Figure 8. Graph of Concentration Carbon Dioxide for Asthmatic User

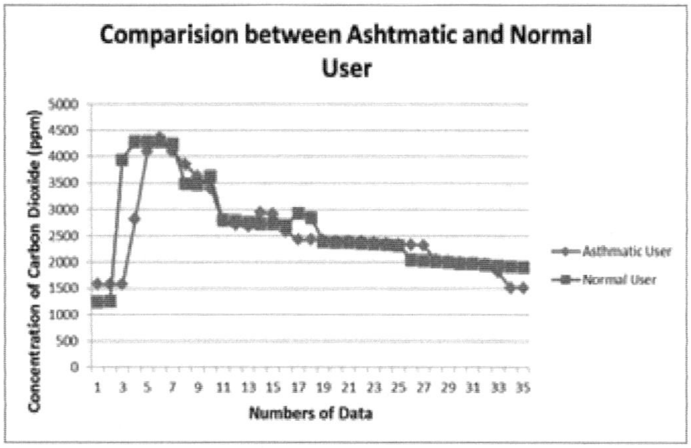

Figure 9. Comparision between Normal and Asthmatic User

For asthmatic user, smaller airway is narrow by the half mark of asthma and caused obstruction to flow within the airways especially during expiration. In asthmatic patient, airway obstruction causes regional decreases in airflow and thus leads to "alveolar ventilation". Therefore, it causes the deformation of the curve as shown in Figure 8. From the graph, it can be observed that the deformation causes the opening angle between the phases 2 and 3.

Figure 9 shows the comparison between the asthmatic and normal user. As shown in the graph, the pattern between the two graphs are slightly difference. At first, the starting point and the peak concentration of carbon dioxide is almost the same. However, when they reach the peak level, the concentration of the normal user stabilizes and maintain for quite some time and formed a "square wave" before it started to decline. Meanwhile, for the concentration of the asthmatic user, it started to decline just after reaching the peak. Then, both of the CO2 concentration stated to fall until they reached a level below 2000ppm.

From Figure 9 above, it is observed that the angle in phases 2 and 3 are slightly different. For asthmatic user, the graph in phases 2 and 3 is wider than the normal user graph due to "alveoli ventilation". Therefore, from the comparison above, the category of the user can be easily categorized either normal or asthmatic.

Figure 10 shows the prototype of electronic kit. The body of the lit was built using electric junction box. Whereas Perspex was used to build the body of the inhaler and the cover was made by 3D printing model. Based on the figure, TFT LCD Shield, ON/OFF button, switches and hose have been included in this prototype.

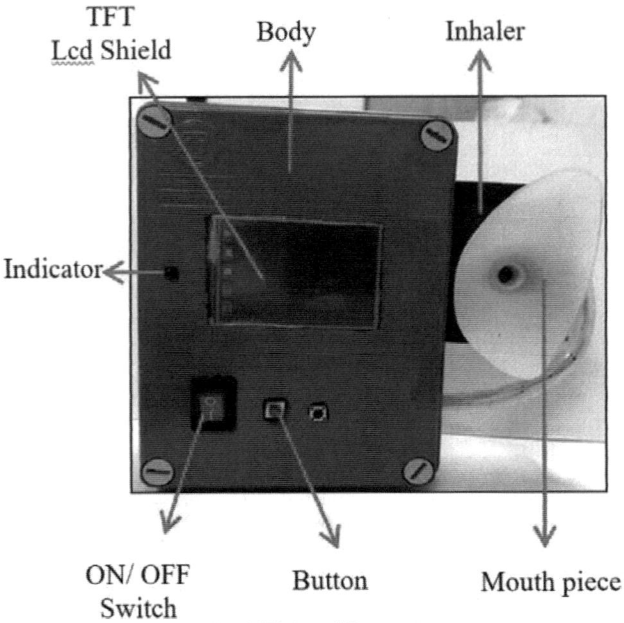

Figure 10 Actual Protoype

4. Conclusion

Use of capnography in monitoring asthma and medical services is increasing; however, capnography is still expensive and unaffordable for patient especially those from low income family for the purpose of asthma self-monitoring application. Therefore, in this research a low cost and portable asthma monitoring devices is proposed. For this device, the main component is carbon dioxide sensor, TFT LCD display and also Bluetooth module. The result from the user can be displayed using TFT LCD display and also PC; meanwhile the result display in a type of graph which can categorize the user either asthmatic or normal. In this research, three asthmatic and normal users are gather to test this device and the result for each of them are gather for analysis. The pattern of the graph for asthmatic user show a significant different compared with normal user. In future, prototype of this device can be tested and validate for the use as a home monitoring device instead of others device such as peak flow meter and spirometer as this prototype device cancelled out the complex procedure for both of the available asthma monitoring devices.

Acknowledgments

The success and final outcome of this project required a lot of guidance and assistance from my project supervisors. In addition, I would like to express my gratitude to En. Suhami Bin Puteh for providing me facilities, laboratory equipment and electric components. This research project is partially supported by Universiti Malaysia Pahang Research Grant RDU170392 entitled "*Dual Image Fusion Technique for Enhancement of Underwater Image Contrast*".

References

[1] McPhee, S. J., & Hammer, G. D. (2010). Pathophysiology of disease: An introduction to clinical medicine (6th ed.). New York, NY: McGraw-Hill.

[2] Zuleika, S., Kazemi, M., & Malarvili, M. B. (2014). Designing a Respiratory CO 2 Measurement Device for Home Monitoring of Asthma Severity, (December), 8–10.

[3] Marani, R. , Perri, A. G. (2013). An Intelligent System for Continuous Blood Pressure Monitoring On Remote Multi-Patients in Real Time. International Journal of Advances in Engineering & Technology, Vol. 5, Issue 2, January 2013

[4] Yu, C., Tsai, T. H., Huang, S. I., & Lin, C. W. (2013). Soft stethoscope for detecting asthma wheeze in young children. Sensors (Basel, Switzerland), 13(6), 7399–7413. https://doi.org/10.3390/s130607399

[5] Kaushal, P. and Mudhalwadker, R.P. 2015. Pellet Sensor Based Asthma Detection System Using Exhaled Breath Analysis. 2015 International Conference on Industrial Instrumentation and Control (ICIC). 28-30 May 2015. Pune, India.

[6] Liu, M., Huang, M.C. 2015. Asthma Pattern Identification via Continuous Diaphragm Motion Monitoring. IEEE Transactions on Multi-Scale Computing Systems. Vol. 1. Issue 2. DOI: 10.1109/TMSCS.2015.2496214

[7] Gatty,H.K., Stemme, G. and Roxhed, N. 2015. A wafer-level liquid cavity integrated amperometric gas sensor with ppb-level nitric oxide gas sensitivity. Journal of Micromechanics and Microengineering. IOP Publishing Ltd. Vol. 25, No. 10.

[8] Buechley, L. (2007). A construction kit for electronic textiles. Proceedings - International Symposium on Wearable Computers, ISWC, 83–90.

[9] Bjelica, M. Z., & Teslic, N. (2010). Multi-purpose user awareness kit for consumer electronic devices. ICCE 2010 - 2010 Digest of Technical Papers International Conference on Consumer Electronics, 239–240.

[10] Ramli, D. A., Hooi, M. Y., & Chee, K. J. (2016). Development of Heartbeat Detection Kit for Biometric Authentication System. Procedia Computer Science.

[11] D'Mello, J., & Butani, M. (2002). Capnography. Indian J. Anaesth, 46(4), 269–278.

IOP Conf. Series: Materials Science and Engineering **319** (2018) 012041 doi:10.1088/1757-899X/319/1/012041

Rigid Polyurethane Nanocomposites Prepared by Direct Incorporation: Effects of Nanoclay, Carbon Nanotubes and Mixing Speed on Physical and Morphological Properties

Benni Ramadhoni, Onny Ujianto and Maxwell Nadapdap

Center for Polymer Technology, Agency for Assessments and Applications of Technology, South Tangerang – Banten, Indonesia 15313

benni.ramadhoni@bppt.go.id

Abstract. Rigid polyurethane (PU) nanocomposites were fabricated via solution mixing of PU, nanoclay and multiwalled carbon nanotubes (MWCNT) according to full factorial DoE. The nanoclay and MWCNT concentration as well as mixing speed were varied. The effects of controlled variables on reduced compressive strength, fire retardancy, hardness and morphological properties were analized. In general, the results showed that incorporation of nanofillers into PU matrix successfully elevated nanocomposites performance. The properties changed from -12% to 45% for reduced compressive strength, 9% to 30% for reduced fire retardancy and -32% to 101% for reduced hardness. The results suggested that the improvements were affected by nanoclay dispersion that acted as nucleating agent which resulted in smaller close cells of PU structures.

1. Introduction

Polyurethane is a polymer commodity that has many applications and products in industry and daily life, including but not limited to flexible and rigid foam, adhesive, coating, thermal insulation and biomedical [1-5]. Among PU products, rigid PU foam contributes to more than 23% of total PU production globally [6]. However, the fact that rigid PU is combustible and has a low compressive strength, inhibits the use thereof for further applications for instance as wall in knock down and lightweight house [7]. Existing flame retardants for PU in the markets generally contain halogenated materials that harmful for health and environment [8]. To address this, a healthy and environmentally benign nanocomposite polymer is designed. Nanomaterials such as CNT and nanoclay are preferable due to their eminence thermal and physical properties at nano size [9]. However, the incorporation of nanomaterials especially CNT into polymer system, encounters agglomeration issue caused by Van Der Waals force [10]. In order to overwhelm the issue, a ternary system is chosen instead of secondary or hybrid polymer/nanomaterial system.

Synergistic dispersion of MWCNT and nanoclay was studied by several groups. Ma Et.al reported that the dispersion of ternary system between ABS/MWCNT/clay was higher as indicated by TEM and XRD result in comparison with hybrid ABS/MWCNT or ABS/clay alone [11]. Peeterbroeck Et.al studied synergical effect of MWCNT and nanoclay on EVA copolymer nanocomposite. Excellence improvement of elongation at break was observed by the incorporation of nanofillers [12]. The effect of simultaneous incorporation of nanoparticles (MWCNT, nanoclay), and nanofibers (polyacrylonitrile, PU) on sound absorption of flexible PU nanocomposite was studied by Bahrambeygi Et.al, indicating better properties due to the concomitant addition of nanoparticles [13].

However, the research about simultaneous incorporation of nanoclay and MWCNT into PU nanocomposite is limited.

The effect of layer by layer assembly of MMT and CNT hybrid flame retardant on the surface of flexible PU was investigated by Pan Et.al. Chitosan wrapped CNT dispersed in acetic acid aqueous solution, MMT suspension and alginate solution were prepared followed by layer by layer deposition of mixture on the surface of flexible PU. 43% reduction on heat release peak was observed [14]. Lee at.al reported that the simultaneous incorporation CNT and MMT in thermoplastics PU via melt mixing slightly increased tensile properties [15]. Madaleno Et.al synthesized PU nanocomposite using hybrid iron modified MMT/MWCNT. Increasing on compressive strength up to 31% compared with neat PU was observed. However, this route required additional step to hybridize MM/MWCNT via CVD technique [16].

This study aims to investigate the effect of direct incorporation of MWCNT and nanoclay as well as mixing speed on compressive strength, fire retardancy, hardness and morphological properties of rigid PU nanocomposite. Instead of modifiying nanoclay and MWCNT, the nanofillers were directly mixed with polyol to produce PU nanocomposite via solution mixing.

2. Experimental

2.1 Materials

Polyether polyol (Bayfit 21 BT 001) and modified diphenylmethane-4,4'-diisocyanate (MDI) (Desmodur 26BM13) from Covestro GmbH were used as polyurethane precursors. Modified montmorillonite nanoclay (Nanomer 1.44PT) from Elementis Specialties Company was chosen as nanofiller due to its eminence thermal stability. MWCNT with diameter 50 nm and length 20 μm were obtained from Cheap Tubes Inc. Polyvinyl butyral (Mowital B60H) used as dispersing agent for MWCNT was supplied by Kuraray GmbH. All materials were used as received.

2.2 Instrumentation

Specimens for testing were cut from cured samples and pre-conditioned at 23°C and 50% for more than 40 hours. The compression strength was measured using Shimadzu AGS 10 kNG Universal Testing Machine, referring to ASTM D1621 with crosshead speed 2.5 mm/min and limited strain to 13%. Horizontal burning rate was determined referring to UL 94. Density was obtained by directly weigthed cubic PU nanocomposite and divided by its volume. Hardness was measured according to ASTM D2240 using shore A durometer. SEM Image was taken using JEOL JSM-6510LA at 10 kV. Reduced compressive strength, burning rate and hardness were calculated by dividing the properties by density. Reduced fire retardancy was calculated by reciprocating the reduced burning rate.

2.3 Procedure

Table 1 lists all variations in PU nanocomposite preparation. Mechanical strirer with speed 1300 rpm and 1900 rpm was used to perform all mixings. Polyether polyol and MDI ratio was fixed at 45:55. Firstly, PU/nanoclay/MWCNT composite was prepared by mixing polyether polyol, MWCNT, PVB and nanoclay at room temperature for 10 minutes, followed by 30 minutes ultrasonication. The amount of nanofillers was varied according to total polyurethane weight, namely 1.5 and 2.5 phr for nanoclay, 0.015 and 0.025 phr for MWCNT. The resulted mixture along with MDI was conditioned at 10°C for 5 minutes prior to further mixing. The chilled materials were mixed for 30 seconds and immediately poured into the mould and left for curing.

3. Results and Discussion

Table 2 shows reduced compressive strength, fire retardancy and hardness for PU nanocomposite samples. The results show that the properties are affected by the incorporation of nanofillers and processing condition. In general, the properties changed from -12% to 45% for reduced compressive strength, 9% to 30% for reduced fire retardancy and -32% to 101% for reduced hardness in comparison with control sample. Our ultimate reduced compressive strength is higher in comparison with another group, suggesting more efficient route to produce PU/Nanoclay/MWCNT nanocomposite [16].

Figure 1 and Figure 2 show the effects of nanoclay and MWCNT concentration as well as mixing speed on reduced compressive strength, fire retardancy and hardness. Figure 1 show that increasing concentration of nanoclay and MWCNT tend to lower the properties. On the other hand, increasing compressive strength and hardness are observed by increasing mixing speed. Figure 3 shows morphological comparison between low and high nanoclay contained PU nanocomposite.

Table 1. Conditions of Nanocomposite Preparations

Runs	Nanoclay (phr)	MWCNT (phr)	Mixing Speed (rpm)
Control	0	0	1900
1	1.5	0.025	1900
2	1.5	0.015	1300
3	1.5	0.025	1300
4	1.5	0.015	1900
5	2.5	0.015	1300
6	2.5	0.015	1900
7	2.5	0.025	1300
8	2.5	0.025	1900

Table 2. Resulted Properties of Nanocomposite Samples

Runs	Density ($g.cm^{-3}$)	Reduced Compressive Strength ($MPa/g.cm^{-3}$)	Reduced Burning Rate ($mm/s.g.cm^{-3}$)	Reduced Fire Retardancy	Reduced Hardness
Control	0.0425	3.4471	93.9675	0.0106	181.2
1	0.0506	4.9743	76.6239	0.0131	191.7
2	0.0511	4.2701	77.2153	0.0130	156.6
3	0.0519	5.0000	71.6525	0.0140	260.1
4	0.0571	4.8284	64.5955	0.0155	364.3
5	0.0545	3.7009	65.7061	0.0152	196.3
6	0.0447	3.9508	85.5787	0.0117	192.4
7	0.0508	3.3720	79.0860	0.0126	196.9
8	0.0561	3.0143	78.0077	0.0128	123.0

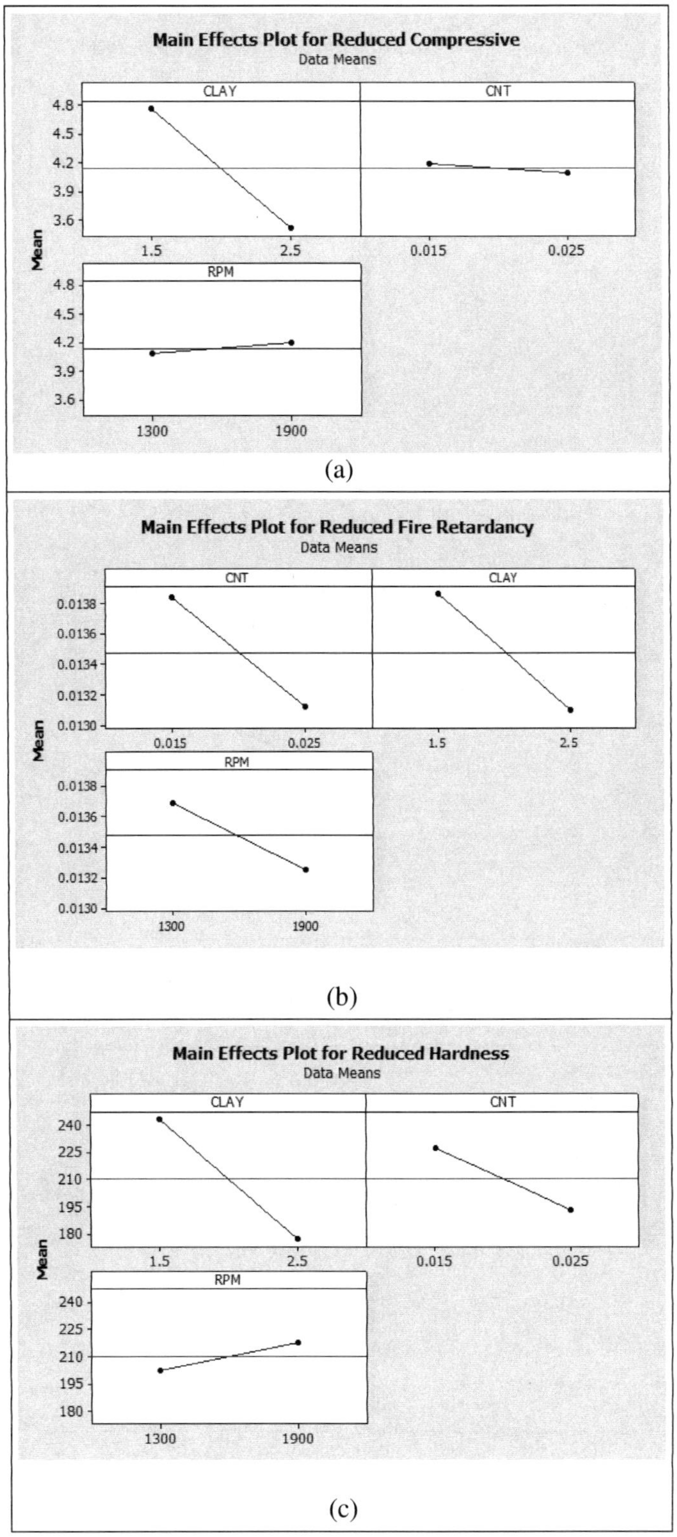

Figure 1. Main Plot of PU/Nanoclay/MWCNT Nanocomposite for Reduced; (a) Compressive Strength ; (b) Fire Retardancy ; (c) Hardness

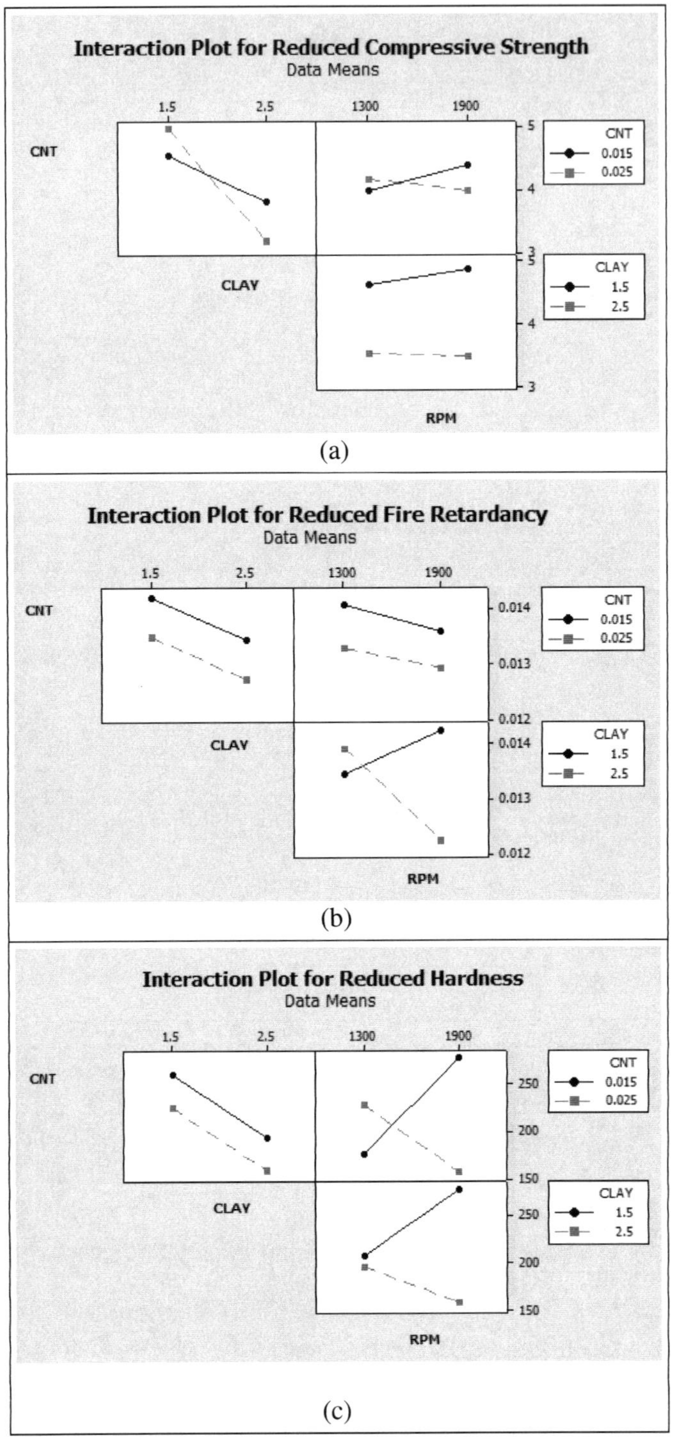

Figure 2. Interaction Plot of PU/Nanoclay/MWCNT Nanocomposite for Reduced; (a) Compressive Strength ; (b) Fire Retardancy ; (c) Hardness

3.1 Compressive Strength and Hardness

Figure 1a shows that reduced compressive strength is less affected by MWCNT and mixing speed probably due to the deficient amount of MWCNT and low mixing speed. On the contrary, nanoclay alter the property. At 1.5 phr nanoclay concentration, the average reduced compressive strength is 4.78 MPa which is 38% higher than the neat PU. This is probably due to the dispersed nanoflakes served as nucleating agent for PU foam that produced small and condense cells [7]. However, the property decreases at higher concentration. At 2.5 phr, only 2% increase is obtained relative to the neat PU. Lower increment at higher concentration is probably due to the agglomeration of nanoflakes that resulted in larger and less populous cells in comparison with the lower one.

Figure 2a also shows the interaction between nanoclay and MWCNT. At low nanoclay concentration, the supplement of MWCNT managed to increase the reduced compressive strength. This is probably due to nanoclay assisted dispersion of MWCNT [17]. However, at higher nanoclay concentration, supplementary of MWCNT further deteriorated the property. Incorporation of agglomerated nanoclay and MWCNT created bigger nanoparticle that served as failure site. Figure 1c shows main effects of nanoclay, MWCNT and mixing speed on hardness. Similar to compressive strength, hardness decreases as nanofillers concentration increases. On the contrary, increasing mixing speed slightly harden PU nanocomposite. Figure 2c shows independency between nanoclay and MWCNT on hardness. However, interaction between nanofillers and mixing speed is observed. At low nanofillers concentration, accelerating mixing speed manage to harden PU nanocomposite. However, at high concentrations, PU nanocomposite is softer in spite of expediting the mixing speed.

3.2 Fire Retardancy

The effects of accretion of nanoclay and MWCNT concentration as well as mixing speed on fire retardancy are shown in Figure 1b and 2b. Figure 1b shows that increasing nanofillers concentrations and mixing speed tended to lower fire retardancy. The average fire retardancy of PU nanocomposite at low nanofillers concentration and mixing speed is 22% higher than the neat PU. While at high states, the average value is decreased 19%. Figure 2b shows less interaction between nanoclay and MWCNT for fire retardancy. However, there is an interaction between nanoclay concentration and mixing speed. At low concentration, rising the mixing speed better dispersed nanoflakes into PU matrix giving a good nucleation on PU foam that resulted in small and packed PU nanocomposite cells [7]. However, at higher concentration and high speed, fire retardancy is decrease due to agglomeration of nanoflakes giving a poor nucleation of PU foam which produced larger and less PU naocomposite cells in comparison with low concentration one yielding shorter and less burning route.

3.3 Morphology

Figure 3 compares the morphology of low and high nanoclay contained PU nanocomposite. It shows that low nanoclay contained PU nanocomposite has smaller, numerous, more condense and structured cells that resulted in better properties in comparison with the high one.

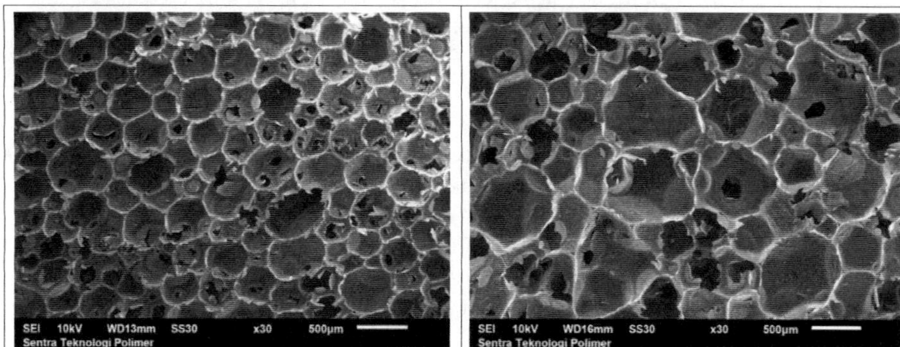

(a) (b)

Figure 3. SEM of PU/Nanoclay/MWCNT Nanocomposite Cell;
(a) Nanoclay 1.5 phr ; (b) Nanoclay 2.5 phr

4. Conclusion

Rigid PU/nanoclay/MWCNT nanocomposite was fabricated successfully via solution mixing of the monomers, nanoclay and MWCNT. The effects of variables on physical and morphological properties were measured. Overall improvements on compressive strength, fire retardancy and hardness compared with neat PU were obtained. The dispersion of nanofillers especially nanoclay that served as nucleating agent for PU matrix yielding small, compact and structured cells was responsible for positive effects on the properties at low nanoclay concentration. However, the agglomeration of nanoclay at high concentration generating larger, less condense and poor structured cells declined the properties compared with the low one.

References

[1] Pillai, P.K.S., Li, S., Bouzidi, L., and Narine, S.S., 2015, Metathesized palm oil polyol for the preparation of improved bio-based rigid and flexible polyurethane foams, *J.Ind. Crops Prod*, 12, 068.

[2] Desai, S.D., Patel, J.V., and Sinha, V.K., 2003, Polyurethane adhesive system from biomaterial-based polyol for bonding wood, *International Journal of Adhesion & Adhesives*, 23, 393-399

[3] Noreen, A., Zia, K.M., Zuber, M., Tabasum, S., and Zahoor, A.S., 2016, Bio-based polyurethane: An efficient and environment friendly coating systems: A review, *Progress in Organic Coatings*, 91, 25-32

[4] Kim, C., and Youn, J.R., 2000, Environmentally friendly processing of polyurethane foam for thermal insulation, *Polymer-Plastics Technology and Engineering*, 39, 1, 163–185

[5] Yang, H., Ye, Q., Zhou, Y., Xiang, Y., Xing, Q., Dong, X., Wang, D., and Xu, W., 2014, Formation, morphology and control of high-performance biomedical polyurethane porous membranes by water micro-droplet inducedp hase inversion, *Polymer*, 55, 5500-5508

[6] Fauzi, N.N.P.N., Majid, R.A., Dzulkifli, M.H., and Yahya, M.Y., 2014, Development of rigid bio-based polyurethane foam reinforced with nanoclay, *Composites: Part B*, 67, 521-526

[7] Cao, X., Lee, L.J., Widya, T., and Macosko, C., 2005, Polyurethane/clay nanocomposites foams: processing, structure and properties, *Polymer*, 46, 775-783

[8] Chen, M.J., Shao, Z.B., Wang, X.L., Chen, L., and Wang, Y.Z., 2012, Halogen-Free Flame-Retardant Flexible Polyurethane Foam with a Novel Nitrogen–Phosphorus Flame Retardant, *Industrial & Engineering Chemistry Research*, 51, 9769-9776

[9] Cha, C., Shin, S.R., Annabi, N., Dokmeci, M.R., and Khademhosseini, A., 2013, Carbon-based nanomaterials: multifunctional materials for biomedical engineering, *ACS Nano*, 7, 4, 2891-2897

[10] Schulz, M.J., Kelkar, A.D., and Sundaresan, M,J., 2006, Nanoengineering of structural, functional and smart materials, CRC Taylor & Francis, London

[11] Ma, H., Tong, L., Xu, Z., and Fang, Z., 2007, Synergistic effect of carbon nanotube and clay for improving the flame retardancy of ABS resin, *Nanotechnology*, 18, 375602-375609

[12] Peeterbroeck, S., Alexandre, M.,Nagy, J.B., Pirlot, C., Fonseca, A., Moreau, N., Philippin, G., Delhalle, J., Mekhalif, Z., Sporken, R., Beyer, G., and Dubois, P., 2004, Polymer-layered silicate–carbon nanotube nanocomposites: unique nanofiller synergistic effect, *Composites Science and Technology*, 64, 2317-2323

[13] Bahrambeygi, H., Sabetzadeh, N., Rabbi, A., Nasouri, K., Shoushtari, A.M., and Babaei, M.R., 2013, Nanofibers (PU and PAN) and nanoparticles (Nanoclay and MWNTs) simultaneous effects on polyurethane foam sound absorption, *J. Polym. Res.*, 20, 72

[14] Pan, H., Pan, Y., Wang, W., Song, L., Hu, Y., and Liew, K.M., 2013, Synergistic Effect of Layer-by-Layer Assembled Thin Films Based on Clay and Carbon Nanotubes To Reduce the Flammability of Flexible Polyurethane Foam, *Industrial & Engineering Chemistry Research*, 53, 14315-14321

[15] Lee, M., Kim, W., Ku, J., Kim, J.D., Min, B.H., and Kim, J.H., 2012, Influence of Nanoclays on Electrical and Morphological Properties of Thermoplastic Polyurethane/Multiwalled Carbon Nanotube/Clay Nanocomposites, *Applied Polymer Science*

[16] Madaleno, L., Pyrz, R., Crosky, A., Jensen, L.R., Rauhe, J.C.M., Dolomanova, V., Timmons, A.M.M.V.D.B., Pinto, J.J.C., and Norman, J., 2013, Processing and characterization of polyurethane nanocomposite foam reinforced with montmorillonite–carbon nanotube hybrids, *Composites: Part A*, 44, 1-7

[17] Etika, K.C., Liu, L., Cox, M.A., and Grunlan, J.C., 2016, Clay-mediated carbon nanotube dispersion in poly(N-Isopropylacrylamide), *Colloids and Surfaces A: Physicochemical and Engineering Aspects*, 489, 19-26

Batch Scheduling for Hybrid Assembly Differentiation Flow Shop to Minimize Total Actual Flow Time

R Maulidya, Suprayogi, R Wangsaputra and A H Halim

Industrial Engineering Faculty, Institut Teknologi Bandung
JL. Ganeca 10 Bandung, 40132, Indonesia

rahmimaulidya@gmail.com, yogi@mail.ti.itb.ac.id,
rachmawati_wangsaputra@yahoo.com, ahakimhalim@mail.ti.itb.ac.id

Abstract. A hybrid assembly differentiation flow shop is a three-stage flow shop consisting of Machining, Assembly and Differentiation Stages and producing different types of products. In the machining stage, parts are processed in batches on different (unrelated) machines. In the assembly stage, each part of the different parts is assembled into an assembly product. Finally, the assembled products will further be processed into different types of final products in the differentiation stage. In this paper, we develop a batch scheduling model for a hybrid assembly differentiation flow shop to minimize the total actual flow time defined as the total times part spent in the shop floor from the arrival times until its due date. We also proposed a heuristic algorithm for solving the problems. The proposed algorithm is tested using a set of hypothetic data. The solution shows that the algorithm can solve the problems effectively.

1. Introduction

In this paper, we study the three-stage production system consists of machining, assembly and differentiation stage called *Hybrid Assembly Differentiation Flow Shop*. The system is producing at least two types of products where every type of product consists of three parts. The first stage is the machining stage and has several parallel unrelated machines to produce each part of product types. The second stage is an assembly operation where all parts of a product is assembled into an assembly part. Each of the assembled parts is delivered to the final process at the differentiation stage. In the differentiation stage, an assembled part is further processed to become a certain type of final product. This system can be found in an electronic industry which has three stages production system and produces several types of products.

Recent research conducted by [1] and [2] has developed the system for job scheduling problem. [1] develop a forward scheduling model with the criteria of total flow time while [2] develop a backward scheduling model with the criteria of total actual flow time. In these research, the setup time is neglected. According to [3], the independent setup is an anticipatory setup for changing tool or cleaning the machine. These setup activity can be shared if jobs have the same characteristic. A reduction in total setup time can be achieved by grouping the jobs into batches [4].

The reason for batching jobs is to gain efficiency with the maximal set of jobs [3]. In job processor, jobs in a batch is processed sequentially so that the processing time of a batch is equal to the sum of the processing time of its jobs [5]. In some literatures, a batch consists of identical processing time of jobs [5-8] but others have develop the non-identical processing time of jobs in a batch [9-10]. The decision of batching and scheduling can be conducted sequentially [5-6],[8],[10] or simultaneously [7].

Some literatures develop the objective of scheduling problem such as minimizations of the makespan [9-10], the maximum lateness [11], the total completion time [13], the total flow time [1] and the multiple objective such as minimization of the makespan and maximum lateness and minimizations of the total flow time and sum of weighted tardiness [12]. These objectives of scheduling problem are completion time based and suitable for forward scheduling. The objective suits to backward scheduling should be due date based. The objective introduced by [5] define the total actual flow time as the interval time of parts in the shop floor started from the arrival time of parts until they finished at their due date. This objectives ensure that the completed product can be delivered to the costumer at their due date and we can control the arrival of parts at the shop at the time when the processes is started. In this paper, we adopt the total actual flow time as the objective of scheduling problem.

This paper deals with a problem of batch scheduling model in a hybrid assembly differentiation flow shop to minimize the total actual flow time. This paper is organized as follows. In Section 2, we formulate our problem into a non-integer linear programming model. In Section 3, we proposed an heuristic algorithm to solve the problem. In order to evaluate the performance of the algorithm, we conduct some computational experiments in Section 4 and we present our conclusions in Section 5.

2. Problem Formulation

There are J jobs ($J=J_1,J_2,\ldots,J_J$) to be processed and are divided into two specific type of jobs JH_h ($JH_1=J_1,J_2,J_3,J_4$ and $JH_2=J_5,J_6,J_7,J_8$). All jobs must be processed in three stages i.e. the machining, assembly and differentiation stage. All jobs in a batch are processed one by one on the machine with its non-identical processing time of job. Each job J has k components (Part#1,…,Part#K) to be processed on k different parallel unrelated machines in the machining stage. When all jobs in a batch are completed, the batch will be transferred to the assembly stage. There is a setup time between the processed of two consecutive batches. In the assembly stage, the parts of a job is assembled. The group of jobs for the three stages production system is in the same order. In the differentiation stage, each of assembled part is finally processed to become a type of finished products. At the end of their common due date, all batches of finished product will be ready to be delivered to customer. Figure 1 shows the system of batch scheduling.

The problem is to determine the number of batches and the batch sizes and to define the production schedule for the three stages production system so that it can minimize the total actual flow time. The following assumptions are considered in the paper.

1. Productions are finished at their common due date.
2. Each batch only consists of a similar type of jobs.
3. The setup operation can be started before the batch arrives on the machine.
4. Transportation times are neglected.
5. No interruption is allowed until all the jobs in a batch are completed.
6. There are unlimited buffers between the stage one and two and the stage two and three

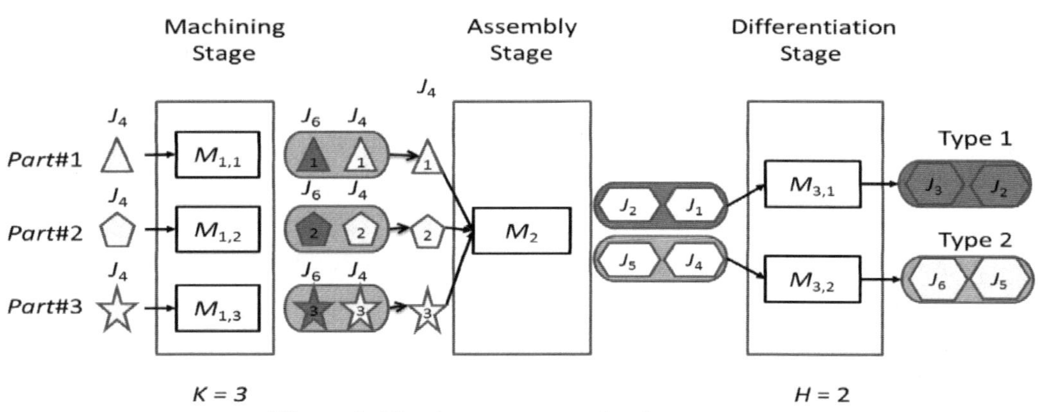

Figure 1. The three stages production system

Halim et.al. [5] define the actual flow time of a job as the time that the job spends in the shop from its starting time for processing until its due date. The actual flow time, $F^a_{[u]}$ can be written as follows.

$$F^a_{[u]} = d - B_{[u]}, \qquad u=1,2,\ldots,U \tag{1}$$

For batch scheduling problems, the actual flow time of a batch is similar to equation (1). The differentiation is that a batch consists of parts that wait in the batch until all parts in the batch are completed. The actual flow time can be written as follows.

$$F^a_{[u]} = \left(d - B_{[u]}\right)Q_{[u]} \tag{2}$$

The machining stage in this paper consists of three parallel machines. So the actual flow time of a batch defines as due date minus the starting time of a batch u in the machining stage ($\mathrm{B}^{(1)}_{[u]}$), times the batch sizes of batch u. The formulation is as follows.

$$F^a_{[u]} = \left(d - \min\left(B^{(1)}_{[u,k]}\right)\right)Q_{[u]} \tag{3}$$

The formulation of the total actual flow time (*TAFT*) is:

$$TAFT = \sum_{\mathrm{u}=1}^{\mathrm{U}}\left(d - \min_{\mathrm{k}\le K}\left(\mathrm{B}^{(1)}_{[\mathrm{u}],\mathrm{k}}\right)\right)Q_{[u]} \tag{4}$$

Notation for this model is introduced below:

Indexes and parameters

j	index of job, j=1, 2,…,J
u_m	index of a batch at stage m, $u_m = 1, \ldots, U_m$, m=1,..,3.
k	index of a machine in the machining stage, k=1, 2,...,K
h	index of a machine in the differentiation stage, and also as the index of product type, h=1, 2,...,H
JH_h	number of processed job at machine h
d	common due date
ϕ	a large number
$M_{1,k}$	machine for processing part k of all the job in the machining stage, k = 1, 2,…,K
M_2	assembly machine in the assembly stage
$M_{3,h}$	dedicated machine for processing product type h in the differentiation stage, h = 1, 2,…,H
$p^{(1)}_{j,k}$	processing time of job J_j on machine $M_{1,k}$ at the machining stage
$p^{(2)}_j$	assembly time of job J_j on assembly machine M_2
$p^{(3)}_{j,h}$	processing time of job J_j on machine $M_{3,h}$ at the differentiation stage, $p^{(3)}_{j,h} > 0$ if $j \in n_h$; 0, otherwise
$A_{j,h}$	binary variable for processing *job j* on machine h, $A_{j,h} = 1$ if $p^{(3)}_{j,h} > 0$; $A_{j,h} = 0$ if $p^{(3)}_{j,h} = 0$.

Decision variables

$W^{(m)}_{j[u_m]}$	binary variable, $W^{(m)}_{j[u_m]} = 1$ if job j is assigned to batch u ; 0, otherwise..
$Q^{(m)}_{[u_m]}$	batch size at stage m
$QQ^{(m)}_{[u_m],h}$	binary variable for processing batch u_m on machine h, $QQ_{[u],h} = 1$ if batch u_m at stage m is assigned to machine h; 0, otherwise.
$\mathrm{B}^{(1)}_{[u_1],\mathrm{k}}$	starting time of processing batch u_1 on machine k in the machining stage (stage 1)
$\mathrm{B}^{(2)}_{[u_2]}$	starting time of assembly operation for batch u_2 in the assembly stage (stage 2)
$\mathrm{B}^{(3)}_{[u_3],\mathrm{h}}$	starting time of processing batch u_3 on machine h in the differentiation stage (stage 3)

The batch scheduling for hybrid assembly differentiation flow shop model can be formulated as follows.

IOP Conf. Series: Materials Science and Engineering **319** (2018) 012042 doi:10.1088/1757-899X/319/1/012042

$$\text{Minimize } TAFT = \sum_{u_1=1}^{U_1} \left(d - \min_{1\le k\le 3}\left(B_{[u_1],k}^{(1)}\right)\right) Q_{[u_1]}^{(1)} \quad (5)$$

Subject to:

$$B_{[u_3],h}^{(3)} = QQ_{[u_3],h}^{(3)} \left(d - \sum_{z_3=1}^{u_3}\left(QQ_{[z_3],h}^{(3)} \sum_{j=1}^{J}\left(W_{j[z_3]}^{(3)} . p_{j,h}^{(3)}\right) + S_h\right) - S_h\right),\ z_3 \le u_3, \forall u_3, \forall h \quad (6)$$

$$B_{[u_2]}^{(2)} + \left(\sum_{j=1}^{J}\left(W_{j[u_2]}^{(2)} . p_j^{(2)}\right)\right) \le \sum_{h=1}^{H} B_{[u_3],h}^{(3)} \frac{X_{[u_2],[u_3]}}{\sum_{j=1}^{J}\left(W_{j[u_2]}^{(2)} . W_{j[u_3]}^{(3)}\right)} + \left(1 - \frac{X_{[u_2],[u_3]}}{\sum_{j=1}^{J}\left(W_{j[u_2]}^{(2)} . W_{j[u_3]}^{(3)}\right)}\right)\phi, \quad \forall u_2, \forall u_3, \forall h \quad (7)$$

$$B_{[u_2+1]}^{(2)} + \left(\sum_{j=1}^{J}\left(W_{j[u_2+1]}^{(2)} . p_j^{(2)}\right)\right) + \sum_{h=1}^{H}\left(QQ_{[u_2],h}^{(2)} . S_h\right) \le B_{[u_2]}^{(2)}, \quad u_2 < U_2, \forall u_2 \quad (8)$$

$$B_{[u_1],k}^{(1)} + \left(\sum_{j=1}^{J}\left(W_{j[u_1]}^{(1)} . p_{j,k}^{(1)}\right)\right) \le B_{[u_2]}^{(2)} \frac{X_{[u_1],[u_2]}}{\sum_{j=1}^{J}\left(W_{j[u_1]}^{(1)} . W_{j[u_2]}^{(2)}\right)} + \left(1 - \frac{X_{[u_1],[u_2]}}{\sum_{j=1}^{J}\left(W_{j[u_1]}^{(1)} . W_{j[u_2]}^{(2)}\right)}\right)\phi, \quad \forall u_1, \forall u_2, \forall k \quad (9)$$

$$B_{[u_1+1]}^{(1)} + \left(\sum_{j=1}^{J}\left(W_{j[u_1+1]}^{(1)} . p_{j,k}^{(1)}\right)\right) + \sum_{h=1}^{H}\left(QQ_{[u_1],h}^{(1)} . S_h\right) \le B_{[u_1],k}^{(1)}, \quad u_1 < U_1, \forall u_1, \forall k \quad (10)$$

$$A_{j,h} = \begin{cases} 1, & \text{jika } p_{j,h}^{(3)} > 0 \\ 0, & \text{jika } p_{j,h}^{(3)} = 0 \end{cases} \quad j = 1,2..J, \forall h \quad (11)$$

$$\sum_{h=1}^{H}\sum_{u_m=1}^{U_m}\left(Q_{[u_m]}^{(m)} . QQ_{[u_m],h}^{(m)}\right) = J, \quad (12)$$

$$\sum_{u_m=1}^{U_m}\left(W_{j[u_m]}^{(m)} * QQ_{[u_m],h}^{(m)}\right) = A_{j,h}, \quad \forall j, \forall h \quad (13)$$

$$\sum_{h=1}^{H} QQ_{[u_m],h}^{(m)} = 1, \quad \forall u_m \quad (14)$$

$$\sum_{u_m=1}^{U_m} W_{j[u_m]}^{(m)} = 1, \quad \forall j \quad (15)$$

$$\sum_{j=1}^{J} W_{j[u_m]}^{(m)} = Q_{[u_m]}^{(1)}, \quad \forall u_m, \forall j \quad (16)$$

$$X_{[u_m],[u_{m+1}]} = \sum_{j=1}^{J}\left(W_{j[u_m]}^{(m)} . W_{j[u_{m+1}]}^{(m+1)}\right), \quad \forall u_m, \forall j \quad (17)$$

$$QQ_{[u_m],h}^{(m)} \in \{0,1\}, \quad \forall u_m, \forall h \quad (18)$$

$$W_{j[u_m]}^{(m)} \in \{0,1\}, \quad \forall u_m, \forall j \quad (19)$$

$$Q_{[u_m]}^{(m)} > 0, \quad \forall u_m \quad (20)$$

$$B_{[u_1],k}^{(1)}, B_{[u_2]}^{(2)}, B_{[u_3],h}^{(3)} \ge 0, \quad \forall u_1, \forall u_2, \forall u_3, \forall h \quad (21)$$

Equation (5) shows the total actual flow time where the actual flow time of each batch is the longest interval between the parts being processed in the machining stage until its common due date. Constraint (6) show that the completion time of every batch processed at the differentiation stage should exactly coincide with the due date. Constraint (7) show that the completion time of batch processed at the assembly stage is less than the starting time of batch processed at the differentiation stage. Constraint (9) show that the completion time of every batch at the machining stage is less than the starting time of the batch at the assembly stage. Constraint (8) and (10) ensures that each batch can only be processed at the same stage after the next position is finished. Constraint (11) define the job that is processed in machine *h* at the differentiation stage has the processing time greater than zero. Constraint (12), (13) and (14) ensure batch u contained only the same type of jobs. Constraint (15) ensure that a job is only assigned to one batch. Constraint (16) accomplishes a material balance in the shop. Constraint (17) defines the batching relation for job *j* that belong to a certain batch. Constraint (18), (19) and (21) define the domain of the decision variables. Constraint (20) ensure that the batch size is greater than zero.

3. Algorithm

The problem is divided into three subproblems, i.e. batching, integrating job and batch, and scheduling. In solving the first problem we define the number of batches and the batch sizes based on [5]. Each stage has a different number of batches and also the batch sizes. So, equation (22) until (25) determine the batching for each stage, except for machining stage there are alternatives to be selected as the batching solution. The second subproblem is solved by sequencing the jobs with SPT based heuristic. The job sequence is grouped into batches and a batch is set only for the same type of product. The job sequence is optimized by variable neighborhood search (Insert move and swap move). The third subproblem is solved by examining the resulting batches to minimize the total actual flow time. The stopping rule is the condition where the iterations give the minimum total actual flow time.

In order to determine the number of batches and the batch sizes, we use equation (22) to (25) based on Halim et.al. (1994) as follows.

$$Q_{[u]} = \frac{J}{N} + \frac{\left(\frac{s}{t}\right).(N+1)}{2} - \left(\frac{s}{t}\right)u \qquad u=1,\dots,N \qquad (22)$$

$$N = \frac{-Z}{s} + \left(\frac{1}{s}\right)\sqrt{Z^2 + 2Jst} \qquad (23)$$

Where:

$$Z = \frac{1}{2}Jt + \frac{Js}{\frac{d}{t}+\frac{s}{t}-J} - \frac{1}{2}d - \frac{1}{2}s \qquad (24)$$

$$N_{max} = \frac{1}{2} + \sqrt{\frac{1}{4} + \frac{2Jt}{s}} \qquad (25)$$

Variable Neighbourhood Search in this paper consists of two neighborhood operators i.e. Insert and Swap move [1]. The Insert move is the process of moving the job in position u to position v, whereas all jobs in position k, with k=u+1,...,v, are shifted one position forward. The swap move is the process of swapping the job in position u and the job in position v. Figure 2 represent the insert and swap move.

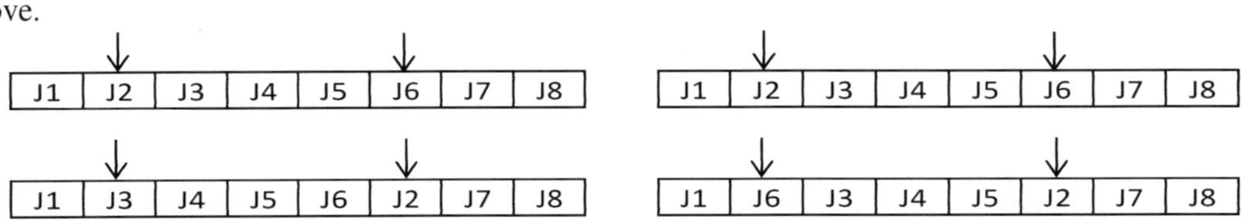

Figure 2. The insert move (left) and swap move (right) for u=2 and v=6

The complete algorithm is as follows.

Step 0: Initialize the problem

Step 1: Define the number of batches and the batch sizes for each machine using equation (22) through (25) following the steps from [5].

Step 2: Define six of job sequences based on SPT based heuristic [1]. Grouped a job sequence into batches of the same type of product and define the total actual flow time. Set the minimum TAFT as the initial solution.

Step 3: Set the maximum iteration. Perform procedure VNS on the initial solution as follows.

- Randomly choose two positions u and v, where $u<v$.
- Move the job in position u to position v, whereas all jobs in position k, with $k = u+1,\dots,v$, are shifted one position forward along solution x. Then a new solution will be x'.
- If TAFT (x') < TAFT (x), then $x \leftarrow x'$. else
- swap the job in position u and the job in position v of solution x. Then a new solution will be x'.
- If TAFT (x') < TAFT (x), then $x \leftarrow x'$. end.
- Repeat this procedure until the maximum iteration.

Step 4: Is the solution better than the best initial solution?
If Yes, output the best solution
If No, go to step 3.

4. Numerical Experience

We use hypothetic data for analyzing the model. The data are: The number of jobs processed (J) is 8 jobs, the common due date (d) is 1000, The setup time for type h (S_h) are 3 and 5, respectively. Table 1 shows the processing time of 2 product types (Type 1 and Type 2). Type 1 consists of jobs J_1, J_2, J_3 and J_4, written as $JH_1=\{J_1,J_2,J_3,J_4\}$ and Type 2 consists of jobs J_5, J_6 ,J_7 and J_8 written as $JH_2=\{J_5,J_6,J_7,J_8\}$. This illustration is for the number of jobs (J)=8, the number of machines in the first stage (K)=3, and the number of machines in the third stage (H)=2. We use processor of Intel Core i7-6500U CPU, 2,50 GHz and 12 GB RAM to run all data. Table 2 represents the resulting number of batches and the batch sizes of the algorithm from step 1.

Table 1. Processing time

Job	Type	$p_{j,1}^{(1)}$	$p_{j,2}^{(1)}$	$p_{j,3}^{(1)}$	$p_j^{(2)}$	$p_{j,1}^{(3)}$	$p_{j,2}^{(3)}$
J_1	1	5	4	3	3	6	0
J_2	1	6	3	4	9	10	0
J_3	1	6	4	10	3	9	0
J_4	1	3	5	5	4	5	0
J_5	2	3	4	5	6	0	8
J_6	2	3	4	6	4	0	7
J_7	2	10	4	6	5	0	6
J_8	2	5	4	3	10	0	4

The number of batches and the batch sizes for every stage can be seen in table 2. $M_{1,1}$, $M_{1,2}$ and $M_{1,3}$ is the unrelated parallel machine and the batching must be selected from N=5 and N=6 so there are no parts left behind in the assembly stage. The resulting batch proceeds to the assembly stage (M_2).

Table 2. The number of batches and the batch sizes

	$M_{1,1}$	$M_{1,2}$	$M_{1,3}$	M_2	$M_{3,1}$	$M_{3,2}$
N_{max}	6	5	6	6	6	4
$Q_{[1]}$	3	4	3	3	1	2
$Q_{[2]}$	2	3	2	2	1	1
$Q_{[3]}$	2	2	2	2	1	1
$Q_{[4]}$	1	1	1	1	1	0
$Q_{[5]}$	1	0	1	1	0	0
$Q_{[6]}$	0	0	0	0	0	0

The initial solution with the minimum $TAFT$=555 is S2 with the resulting batch as follows.

$M_{1,k}$	M_2	$M_{3,1}$	$M_{3,2}$
$Q_{[1]} = J_1, J_3, J_4$	$Q_{[1]} = J_1, J_3, J_4$	$Q_{[1]} = J_1$	$Q_{[1]} = J_6, J_7$
$Q_{[2]} = J_6, J_7$	$Q_{[2]} = J_6, J_7$	$Q_{[2]} = J_3$	$Q_{[2]} = J_5$
$Q_{[3]} = J_5, J_8$	$Q_{[3]} = J_5, J_8$	$Q_{[3]} = J_4$	$Q_{[3]} = J_8$
$Q_{[4]} = J_2$	$Q_{[4]} = J_2$	$Q_{[4]} = J_2$	

Table 3. Variable Neighbourhood Search

u	v	Job sequence	$M_{1,k}$	M_2	$M_{3,1}$	$M_{3,2}$	TAFT
-	-	$J_1, J_3, J_4, J_6, J_7, J_5, J_2, J_8$	$Q_{[1]} = J_1, J_3, J_4$	$Q_{[1]} = J_1, J_3, J_4$	$Q_{[1]} = J_1$	$Q_{[1]} = J_6, J_7$	555
			$Q_{[2]} = J_6, J_7$	$Q_{[2]} = J_6, J_7$	$Q_{[2]} = J_3$	$Q_{[2]} = J_5$	
			$Q_{[3]} = J_5, J_8$	$Q_{[3]} = J_5, J_8$	$Q_{[3]} = J_4$	$Q_{[3]} = J_8$	
			$Q_{[4]} = J_2$	$Q_{[4]} = J_2$	$Q_{[4]} = J_2$		
Insert Move							

2	5	$J_1, J_4, J_6, J_7, J_3, J_5, J_2, J_8$	$Q_{[1]} = J_1, J_4, J_3$	$Q_{[1]} = J_1, J_4, J_3$	$Q_{[1]} = J_1$	$Q_{[1]} = J_6, J_7$	555
			$Q_{[2]} = J_6, J_7$	$Q_{[2]} = J_6, J_7$	$Q_{[2]} = J_3$	$Q_{[2]} = J_5$	
			$Q_{[3]} = J_5, J_8$	$Q_{[3]} = J_5, J_8$	$Q_{[3]} = J_4$	$Q_{[3]} = J_8$	
			$Q_{[4]} = J_2$	$Q_{[4]} = J_2$	$Q_{[4]} = J_2$		
Swap Move							
2	5	$J_1, J_7, J_4, J_6, J_3, J_5, J_2, J_8$	$Q_{[1]} = J_1, J_4, J_3$	$Q_{[1]} = J_1, J_4, J_3$	$Q_{[1]} = J_1$	$Q_{[1]} = J_7, J_6$	555
			$Q_{[2]} = J_7, J_6$	$Q_{[2]} = J_7, J_6$	$Q_{[2]} = J_3$	$Q_{[2]} = J_5$	
			$Q_{[3]} = J_5, J_8$	$Q_{[3]} = J_5, J_8$	$Q_{[3]} = J_4$	$Q_{[3]} = J_8$	
			$Q_{[4]} = J_2$	$Q_{[4]} = J_2$	$Q_{[4]} = J_2$		

The algorithm is set for 50 iterations and there is no new solution. The resulting batch for stage 1 and stage 2 is the same. It means that the part processed in the machining stage will be grouped with the same number of batch and the batch sizes. Figure 3 below shows the Gantt chart for the resulting batch.

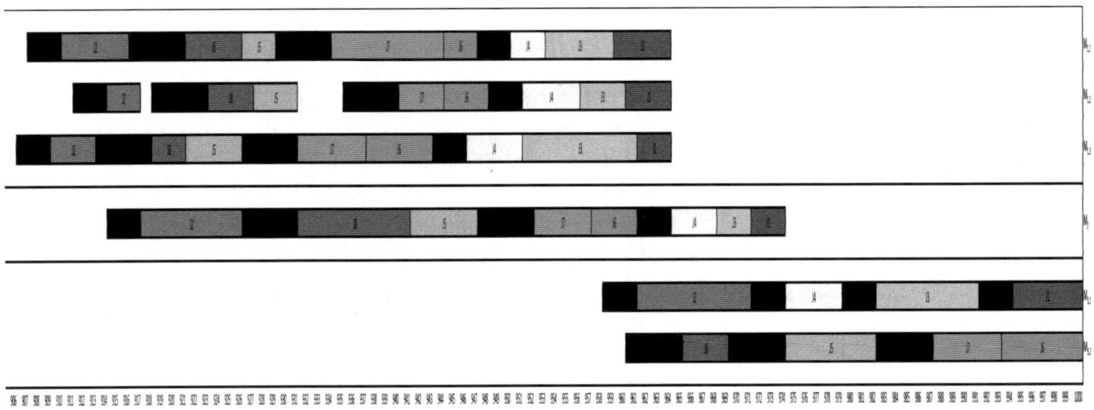

Figure 3. Gantt Chart for $TAFT$=555

5. Conclusions

This paper deals with the batch scheduling problems for a Hybrid Assembly Differentiation Flow Shop, processing H different types of products to minimize the total actual flow time. This research has shown that the non-identical processing time of the same type of jobs can be batched with unnecessarily equal batch sizes. The numerical example of this approach is limited to 8 jobs with the scenario that each stage produces a different number of batches. The proposed algorithm consists of batching, integrating jobs into batches, and scheduling. The final schedule shows that the batches are sequenced from the shortest processing time. But it still needs to be observed how batches are built. The future research is to find the proposed algorithm that can give an optimal solution in short time.

References

[1] Xiong F, Xing K and Wang F 2015 Scheduling a Hybrid Assembly-Differentiation Flow Shop to Minimize Total Flow Time. *European Journal of Operation Research* **240** 338-54

[2] Maulidya R, Suprayogi, Wangsaputra R and Halim A H 2016 Job Scheduling for Hybrid Assembly Differentiation Flowshop to Minimize Total Actual Flow Time. *Proc. of 17th Asia Pacific Industrial Engineering and Management System Conference (APIEMS) Conf.* December 7-10, 2016 (Taipei) 1198-1204

[3] Potts C N and Kovalyov M Y 2000 Scheduling with Batching: A Review. *European Journal of Operational Research* **120** 228–49

[4] Kovalyov M Y, Potts C N and Strusevich V A 2004 Batching decisions for assembly production systems, *European Journal of Operational Research* **157** 620–642
[5] Halim A H, Miyazaki S and Ohta H 1994a Batch-Scheduling Problems to Minimize Actual Flow Times of Parts Through The Shop under JIT Environment. *European Journal of Operation Research* **72** 529-44
[6] Halim A H, Miyazaki S and Ohta H 1994b Lot Scheduling Problems of Multiple Items in the shop with Both Receiving and Delivery Just in Times. *Production Planning and Control* **5(2)** 175-184
[7] Halim A H, Suryadhini P P and Toha I S 2006 Batch Scheduling to Minimize Total Actual Flow Time in a Two-stage Flow Shop with Dedicated Machines in the Second Stage, *Proc. of 7th Asia Pasific Industrial Engineering & Management System (APIEMS) Conf.* December 17-20, 2006 (Bangkok) 1343-49
[8] Halim A H and Yusriski R 2009 Batch Scheduling for the Two-stage Assembly Model to Minimize Total Actual Flow Time. *Proc. of 7th Asia Pasific Industrial Engineering & Management System (APIEMS) Conf. 2009* (Kitakyushu)
[9] Huang T and Lin B M T 2013 Batch Scheduling in Differentiation Flow Shops for Makespan Minimization, *International Journal of Production Research* **51(17)** 5073-82
[10] Shen L and Gupta J N D 2017 Family scheduling with batch availability in flow shops to minimize makespan, *Journal Scheduling*, 1-15 (New York: Springer)
[11] Hariri A M A and Potts C N 1997 Single Scheduling with Batch Set-up Times to Minimize Maximum Lateness, *Annals of Operation Research* **70** 75-92
[12] Nikzad F, Rezaeian J, Mahdavi I and Rastgar I 2015 Scheduling of Multi-component Products in a Two-stage Flexible Flow Shop. *Applied Soft Computing* **32** 132-43
[13] Cetinkaya F C and Kayaligil M S 1992 Unit Sized Transfer Batch Scheduling with Setup Times. *Computers Industrial Engineering* **22(2)** 177-83

Integrating Kano's Model into Quality Function Deployment for Product Design: A Comprehensive Review

Rosnani Ginting, Juliza Hidayati and Ikhsan Siregar

Departement of Industrial Engineering, University of North Sumatra
Jl. Almamater, Kampus USU, Medan, 20155, Indonesia

rosnani_usu@yahoo.co.id, rivaijuliza@yahoo.com, siregar_ix@yahoo.com

Abstract: Many methods and techniques are adopted by some companies to improve the competitiveness through the fulfillment of customer satisfaction by enhancement and improvement the product design quality. Over the past few years, several researcher have studied extensively combining Quality Function Deployment and Kano's model as design techniques by focusing on translating consumer desires into a product design. This paper presents a review and analysis of several literatures that associated to the integration methodology of Kano into the QFD process. Various of international journal articles were selected, collected and analyzed through a number of relevant scientific publications. In-depth analysis was performed, and focused in this paper on the results, advantages and drawbacks of its methodology. In addition, this paper also provides the analysis that acquired in this study related to the development of the methodology. It is hopedd this paper can be a reference for other researchers and manufacturing companies to implement the integration method of QFD-Kano for product design.

Keywords: QFD, Kano, Product Design, Integrated Methodology.

1. Introduction

Product design is very important in the manufacturing sector. Good product design will increase the amount and price, so that it can optimally increase profits. However, product designs fail to produce products that are not sold in the market. It will cause losses not only in design, other areas will also be affected. The design of new or existing products is a major part of all engineering activities. This activity is derived from the perception of human needs, followed by product formation, product design, product development and refinement, and end up with the production and distribution of such products. Product design must meet three important aspects that are often called the triangles of quality product, low cost, and schedule. Next, these aspect are developed to be a requirement in design, the design must be installed, recycled, manufactured, analyzed results, low cost, and exact time. Therefore, in designing a product, it can be done using several product design methods.

Ideally, the company develops products to understand customer needs, so they can develop products to be sold [1]. High quality products is a prerequisite for companies to competitive [2]. In addition, in order to get competitive in the market, it is imperative that products are quickly available in the market and have high reliability at the lowest cost [3]. Therefore, companies need to develop strategic goals, based on achievements [4]. The main factor in product innovation creativity is the development of methodologies, design methods developed for product development and design in various fields [5,6,7,8]. Many academic practitioners and researchers have proposed many design principles and methods to improve the quality of design, and some design methods are implemented as part of the design activities of some manufacturing companies [9]. Quality Function Deployment (QFD) is one of the widely used approaches today. It can drive a product development process from conception into manufacturing. It is a well-structured methodology and technique tool that combines customers' requirements with technical requirements which aids designers and manufacturers to produce better products, enhance their competitiveness in the marketplace, and increase customer satisfaction [10,11,12,13,14].

On the other hand, the various problems faced at several stages of QFD implementation, have been widely reported in the study, in particular the traditional QFD method [15,16]. First, The methodological framework of the conventional QFD method is no longer suitable to meet the design and product development requirements [12,9,17,18]. Second, The QFD matrix is too large and Third, the time required for the matrix sequence deployment is too long and the product time to be marketed is not acceptable [19,20]. Then fourth, QFD is difficult to meet the needs of different customer groups or segments [21]. Fifth, customer voice is still qualitative, cannot be measured and often misleading, it is not systematic and the terms of product function too complex (in this case, engineering process) are not easily determined [22]. Sixth, the customer's requirement translated into engineering terms (technical requirements) obtained from the company is still vague, too subjective, difficult to verify and expressed in linguistic form [23,24,25]. These problems or drawbacks prompted the need for other approaches to be added when applying the QFD method. There are many different methods for generating new ideas and selecting the ideas in order to create a new design or to improve existing ones. Combining QFD with other techniques helps to address these drawbacks and can form the basis of future research. The integrated innovation method, which combines QFD with another technique tool, can precisely solve main contradictory problems in the process from the product demand analysis to the product design, production and application [26].

2. Methodology.

As it could be seen in Fig. 1, shows a flowchart of the methodology, and five majors are taken into account to provide the literature review of this paper. The method and year of publication are the parameters taken into consideration. A screening process to restrict the articles pertaining to combined Kano with QFD. As a result, a total 49 articles were selected from the following nine journals, reviewed and further analysed: Production Research Journal, Institute of Electrical and Electronics Engineers Journal, Journal of Applied Operational Research, Industrial Management Journal, International Journal of Operations and Production Management, Computers and Industrial Engineering Journal, International Journal of Ergonomic, Journal of Engineering Education.

This paper focuses and considers on only journal articles whose goal was to either develop theoretical–concepts work, or reviews of the literature and a case study.

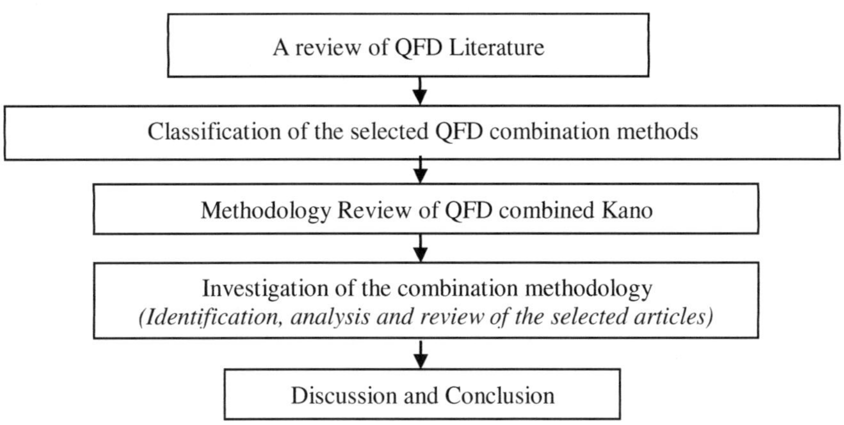

Figure 1. Flowchart of the Methodology

3. Literature Review

3.1. Kano's Model

The Kano method was developed by the professor of Kano in the early 1970s to determine the quality of service based on customer satisfaction. As shown in fig. 2.1, in the Kano method, customer requirements are categorized into three groups, namely the basic element, performance, and quality desired [27]. It based on customer perception and understanding the relationship between product function and customer satisfaction. [28]. [29] said that the Kano's model is generally used to acquire knowledge about the needs of customer[30,31].

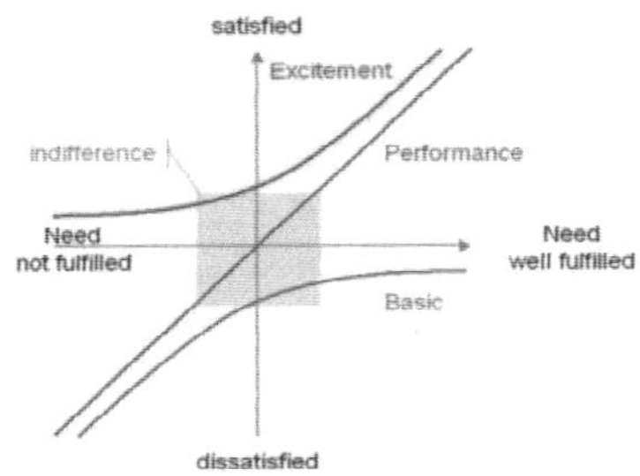

Figure 1. Kano's model of Customer Satisfaction

There are several basic elements of the Kano in product development and customer satisfaction, and the Kano model classifies customer preferences into five categories such as attractive, one-dimensional, must-be, indifferent and reverse. [32]. In the Kano's model, the intent of the quality is attributed to the goal that is actually achieved and these attributes are usually expected. One-dimensional quality features are attributed to quality features that satisfy customers when their expectations are met, and vice versa. Characteristics of quality must be attributed to the attributes that must be met but may result in discomfort if not fulfilled [33].

3.2. Quality Function Deployment (QFD)

QFD was developed in 1996 by Yoji Akao in Japan, and it has been recognized as an effective method for integrated products and process development. QFD is a structured approach to integrate customer voices into design/development products [34]. It is applied in many industries such as automotive, electronics, construction and service sectors. QFD is implemented as a multi-process process, offering the greatest potential to realize significant benefits [35]. The purpose of QFD is to increase customer satisfaction based on customer needs as well as to increase company profits. For the development of new products, QFD is a useful approach to maximize customer satisfaction [36]. In other words, QFD is a tool for changing customer requirements into product design [37]. The common real benefits when the true QFD is used is a 30-50% reduction in engineering changes, a 30-50% shorter design cycle, 20-60% lower start-up costs, and a 20-50% less security claim [38]. Although QFD has been proposed and put in use for several decades, it has been fraught with a number of problems in real application cases. Some of the weaknesses and limitations that have been done by researchers based on the decision of some studies related to the application of the QFD method can be seen in Table 1.

Table 1. Introduction to the Drawbacks of QFD.

References	Drawbacks
[24]	The correlation between technical requirements, the relationship between customer needs and technical requirements is often expressed formally in subjective and vague terms and the Variables are linguistic
[12]	Customer needs are often contrary to technical requirements, conflicts between technical measures, and HoQ are too large and cause excessive confusion.

[25]	Customer voice is dynamic, and inadequate.
[48]	It is not possible to translate individual desires into collective customer demands and the impossibility of translating customer requests into engineering features without the need to violate one or more highly sensitive states
[44]	Determining the exact relationship between customer's desire and technical requirements and amongst technical requirements is done in a subjective. It is because the information, guidelines, and materials that will be used to determine the relationship are often limited and incomplete.
[45]	QFD does not fully guarantee customer satisfaction, as activities such as re-designing a product depend on the designer's interpretation and talent.
[47]	The QFD relies heavily on the results of the study. If the study is designed or done poorly, the QFD implementation will not meet its purpose.

4. Integrated Methodology.

The effort to respond to the needs of customers and to compete competitively, various decision-making methods have been developed [45]. For more accurate understanding and identification of customers' needs, the Kano's model can help QFD in identifying and quantifying customer need. Kano provides an effective approach by classifying product attributes based on customer perceptions and understanding the relationship between product function and satisfaction. [7]. Therefore, in order to better understand and obtain more precise customer requirements, the use of the Kano model has been incorporated into QFD analysis by some researchers. To obtain more accurate, effective and objective customer data or VoC data. [48] suggested that Kano quantitative models be integrated into the QFD method. [6] Argued that an integrated approach involving the Kano and QFD involves a series of activities, from classifying customer traits to prioritizing technical feature priorities. Furthermore [33] said that this combination provide a unique way to differentiate the effects of different customer needs on customer satisfaction at an early stage of product. Table 2 illustrates some relevant literatures in the implementation of the Kano in enhancing QFD capabilities.

Table 2. Selected literature survey on Integrated of QFD – Kano.

References	Methods Applied in	Limitations
[50]	Road mapping technology (TRM) with Flex Cold Start case study	The House of Quality undescribed systematically and integrate Road mapping technologies into product development processes.
[51]	Web site design	This research uncategorized variable that is categorized as must, one-dimensional, simple, and attractive, and do not display conclusions about the responses of schoolchildren to the design system of the workstation and the desired furniture for the design of the repair.
[27]	Notebook Computer Design	The development of a developed integration method has constraints, that is, the method of integration is difficult to apply to a completely new product that has never been introduced in the market. In addition, the product and cost development of each module are not considered in this study, and measures developed by the system are not systematic, structured and gradual, and improvements to SUV car designs through QFD integration with Kano are only performed on phase 1, not forward to the next phase, the product design phase.
[52]	Workstation Design	The combination of both methods requires mathematical

		methods for selecting technical conditions to optimize the product design process
[53]	Design and Development of Pencil products	Do not explained about the validity tests for questionnaire data used and unexplained data sources in determining the technical requirements and conditions of the product.
[54]	Tractor Brake Design	This research does not categorize variables that belong to the category of must, one dimension, simple, and attractive, and do not display conclusions about customer response to the desired brake system for the design of the repair, and only focus on CS correlations and benefits.
[55]	Static Bike Design	Case study with a combination of these Kano-QFD methods does not display quality homes and the implementation of the QFD extension, so that the QFD preparation stage is not detailed in detail.

As seen in Table 2, a number of studies related to the application of Kano's model in the QFD process are widely applied in product design, focusing on customer desires and satisfaction. In this case [50] conducted a combination of QFD method with Kano to choose whether Technology A or Technology B was able to satisfy most of the customer's needs and over achieving today's technological performance. Meanwhile, [27] prioritized the renovation elements implemented in new ergonomic workstation planning successfully identified and given priority based on information from users and technical requirements to develop modified workstations based on ergonomic approaches. Besides, [53] found that Kano divided the needs of customers into QFD into three categories that are coordinated with categories of quality initiative programs such as maintaining, improving and innovating.

A combination of QFD with Kano to provide some obstacles from previous works by using adjusting factors to improve the improvement phase of the QFD Matrix was addressed by [55]. Adaptive factors are based on satisfaction or dissatisfaction. Meanwhile, [54] through his research he argued that combinations of Kano quantitatively with QFD resulted in the optimum product design of notebook computers in maximizing customer satisfaction under constraints of cost and technical constraints. In maximizing customer satisfaction under constraints of cost and technical constraints. Moreover, [52] has successfully developed the innovative design of the new 'static bikes' and 'fitness equipments such as horse riding' have been successfully designed based on QFD combinations with Kano quantitative methods.

Although Kano method is often combined with the QFD method in many studies, it is unclear how Kano should be integrated into the QFD process. Several studies only recommend that the customer's interest is included in the line of matrix of HoQ. Meanwhile, other researchers advised to include three categories of Kano into HoQ, and then make a difference how the functional requirements generated will be disseminated. It seems as if in many years QFD researchers are talking about Kano method, but none of them really explain where Kano categories are included in the QFD process. In addition, to the integration method developed only in phase 1 of QFD (product planning phase), does not extend to phase 2 (product planning). Moreover, the steps of the integration method developed are not stage, unsystematic, unstructured and the analysis is more qualitatively and subjectively.

5. Discussion and Conclusion.

The objective of this paper is hopefully achieved as expected, which is contains the important result, such as success to improve the design of the product case study by applying the selected methods and also develop a product that have maximized value, convenience, suitable and easy to use by the consumer. Customer's need, design requirements, product characteristics, technical operation and producing requirements become the important design procedures. These procedures are organically linked through House of Quality, so they can support the product innovation design processes effectively. The Kano combined QFD model is the most techniques used to explore the customer needs and engineering requirements. From the literature review investigated in this paper, the application system of integrating QFD and Kano's model has some benefits and advantages that are as follows:

- Kano integrated QFD effectively help QFD in the process of the development of product design;
- The combined QFD-Kano is one of the best ways to meet the challenge of satisfying customers' inconstant demands and then to thrive in business.
- The combination of QFD and Kano's model can be a very powerful tool for any manufacturing organization in developing new products or optimizing existing products.
- Manufacturing organizations today need to be able to apply new technology to their products and processes to be successful in the highly competitive global marketplace, and the usage of Kano in combination with QFD can help them meet this objective.
- QFD-Kano combination methodology could provide guidance and relevant information to designers on aspects that should be considered already during the product design and development process.
- The integration of QFD–Kano is one of the most commonly used techniques to deal with incomplete and imprecise information from customer.
- For understanding and more accurate identification of the voice of customer, the use of Kano can help QFD in identifying and determining the required number of customers.

However, shortcomings for this combination methodology should be noticed. First, from point of view this study and investigation of some such case studies, we have found that the respective combination used method in several studies was not tested statiscally. Second, the costs of component product development are not included, and Third, systematic framework in combining methods are not described in detail.

References

[1] Khalid HM, and Helander MG 2004 A Framework For Affective Customer Needs In Product Design *Journal Theoretical Issues in Ergonomics Science. Volume 5, Pages 27-42.*

[2] Augusto, Paulo and Cauchick M 2007 Case Research In Production Engineering: Structure And Recommendations for Its Conduction. *Produção, v. 17, n. 1, p. 216-229, Jan./Abr. 2007.*

[3] Lai X, Tan KC, and Xie M 2007 Optimizing Product Design Using Quantitative Quality Function Deployment: A Case Study *Quality and Reliability Engineering International, 23(1), 45 – 57.*

[4] Felice FD 2015 A Multiple Choice Decision Analysis: An Integrated QFD–AHP Model For The Assessment Of Customer Needs *International Journal of Engineering, Science and Technology ,Volume 2, International Journal of Engineering, Science and Technology, Vol. 2, No. 9, 2010, pp. 25-38.*

[5] Yamashina H, Ito T, Kawada H 2002 Innovative Product Development Process by Integrating QFD with TRIZ. *Journal of the Japan Society for Precision Engineering, 66(11), 1705 – 1710.*

[6] Hsu CH, Chang TM, Wang SY and Lin PY 2007 Integrating Kano's Model into Quality Function Deployment to Facilitate Decision Analysis for Service Quality. *Conference on Mathematics and Computers in Bussiness and Economics, Vancouver, Canada, 19 – 21.*

[7] Yeh CH, Jay, Huang CY and Yu 2011 Integration of Four-Phase QFD and TRIZ in Product R&D: A Notebook Case Study. *Research in Engineering Design, 22(3), 125 – 141.*

[8] Liu HT 2011 Product Design And Selection Using Fuzzy QFD and Fuzzy MCDM Approaches. *Applied Mathematical Modelling. 35 (2011) 482–496.*

[9] Sakao T 2013 A QFD-Centred Design Methodology for Environmentally Conscious Product Design *International Journal of Production Research, 45(18–19), 4143 – 4162.*

[10] Mendoza, Horacio N and Molina A 2003 Case Studies in the Integration of QFD, VE and DFMA during the Product Design Stage *The Proceedings of the 9th International Conference of Concurrent Enterprising, Espoo, Finland, 16-18 June 2003.*
[11] Prasad B 1998 Review of QFD and Related Deployment Techniques *Journal Manufacturing System, 17(3), 221–234.*
[12] Chan LK, and Wu ML 2002 Quality Function Deployment: A Literature Review *European Journal Operation Research, 143, 463–497.*
[13] Lai CJ, Hsu CH, Kuo HM 2012 An Empirical Study of Constructing a Dynamic Mining and Forecasting System for the Life Cycle Assessment- based Green Supply Chain. *Wseas Transactions On Systems, ISSN: 2224-2678.*
[14] Farsi JY and Hakiminezhad N 2012 The integration of QFD Technique, Value Engineering and Design for Manufacture and Assembly (DFMA) during the Product Design Stage *Journal of Advances in Environmental Biology, 6(7), 2096 – 2104.*
[15] Law HW and Hua M 2007 Using Quality Function Deployment in Singulation Process Analysis. *Engineering Letters, 14:1, EL_14_1_6, (Advance online publication: 12 February 2007), pp 1 – 5.*
[16] Mohammadi F, Sadi MK, Nateghi F and Abdullah A 2014 A Hybrid Quality Function Deployment And Cybernetic Analytic Network Process Model For Project Manager Selection. *Journal of Civil Engineering and Management Volume 20, 2014 - Issue 6. pp 795-809.*
[17] Akao Y and Mazur GH 2003 The Leading Edge in QFD: Past, Present And Future *International Journal of Quality & Reliability Management, Vol. 20 Issue: 1, pp.20-35.*
[18] Brad S 2004 Complex System Design Technique (CSDT) *International Journal of Production Research, pp 1 – 28.*
[19] Prasad B 1998 Review of QFD and Related Deployment Techniques *Journal Manufacturing System, 17(3), 221–234.*
[20] Kao, Su E and Wang B 2002 IQFD: A Blackboard-Based Multi-Agent System For Supporting Concurrent Engineering Projects *International Journal of Production Research, 40(5), 1235 – 1262.*
[21] Kim JK, Hun CH, Choi SH, Kim SH 1998 A Knowledge-Based Approach To The Quality Function Deployment *Computers Industrial Engineering Vol. 35, Nos. 1-2,233 – 236*
[22] Law HW and Hua M 2007 Using Quality Function Deployment in Singulation Process Analysis. *Engineering Letters, 14:1, EL_14_1_6, (Advance online publication: 12 February 2007), pp 1 – 5.*
[23] Zhou M 1998 Fuzzy Logic and Optimization Models for Implementing QFD *Computer Industrial Engineering, 35(1–2), 237–240.*
[24] Kim KJ, Moskowitz H, Dhingra A and Evans G 2000 Fuzzy Multi-criteria Models for Quality Function Deployment *European Journal Operation Research, 121(3), 504–518.*
[25] Fung RYK, Chen Y, Chen L and Tang J 2005 A Fuzzy Expected Value-Based Goal Programing Model for Product Planning Using Quality Function Deployment *Engineering Optimization Vol. 37, No. 6,633 – 647*
[26] Ginting R and Amir AY 2016 TRIZ or DFMA Combined With QFD as Product Design Methodology: A Review *Pertanika J. Sci. & Technol. 24 (1): 1 - 25. ISSN: 0128-7680.*
[27] Hashim AMD and Dawal SZM 2012 Kano Model and QFD Integration Approach for Ergonomic Design Improvement *The 2012 Internationl (Summer) Comference on Bussiness Innovation and Technology Management. 57, 22 – 32.*
[28] Hsu CH, Chang TM, Wang SY and Lin PY 2007 Integrating Kano's Model into Quality Function Deployment to Facilitate Decision Analysis for Service Quality *Conference on Mathematics and Computers in Bussiness and Economics, Vancouver, Canada, 19 – 21.*
[29] Seyedi SM, Shirazifar M, Dalvand MR and Zohdi MH 2012 Optimal Examination And Prioritization Of The Factors Affecting Customers' Satisfaction Using Integrated Quality

Function Deployment (QFD) and Kano's model: Case study of Shiraz's Refah bank. *African Journal of Business Management Vol.6 (35), pp. 9762-9772.*

[30] Chen YH and Su CT 2006 A Kano-CKM Model for Customer Knowledge Discovery. *Total Qual Manage 17: 589–608.*

[31] Gerson T 2003 Development of Customer Needs in the QFD Using a Modified Kano Model. *J. Acad. Bus. 27(5):482-500.*

[32] Kano N 2001 Life Cycle And Creation Of Attractive Quality *Paper presented at the 4th International QMOD Conference Quality Management and Organizational Development. Linkopings Universitet, Sweden, pp 1 – 28.*

[33] Raharjo H, Brombacher AV, Goh TN and Bergman B 2009 On integrating Kano's model dynamics into QFD for multiple product design. *Quality and Reliability Engineering International. 351-363.*

[34] Mendoza N, Horacio A and Molina A 2003 Case Studies in the Integration of QFD, VE and DFMA during the Product Design Stage. *The Proceedings of the 9th International Conference of Concurrent Enterprising, Espoo, Finland, 16-18 June 2003.*

[35] Naseri K 2014 Algorithm for Cost-optimized QFD Decision-Making Problem *Evolutionary Intelligence, Volume 9, Issue 1–2, pp 21–36.*

[36] Gupta S and Okudan GE 2012 Computer-Aided Generation of Modularised Conceptual Designs with Assembly And Variety Considerations *Journal of Engineering Design, 19(6), 533 – 551.*

[37] Pourhasomia MH, Khamseh AA and Ghorbanzad Y 2012 A Hybrid of Kano and QFD for ranking customers' preferences: A case study of bank Melli Iran.

[38] Lai CJ, Hsu CHH, Kuo HM 2012 An Empirical Study of Constructing a Dynamic Mining and Forecasting System for the Life Cycle Assessment- based Green Supply Chain *Wseas Transactions On Systems, ISSN: 2224-2678.*

[39] Kao P, Su E and Wang B 2010 A Blackboard-Based Multiagent System for Supporting Concurrent Engineering Projects *International Journal of Production Research (Impact Factor: 1.46), 1235 – 1262.*

[40] Hung HF, Kao HP and Juang YS 2008 An Integrated Information System For Product Design Planning. *Journal Expert Systems with Applications: An International Journal, 35. 338 – 349.*

[41] Lim PC, Nelson KH and Tang 2000 The Development of A Model for Total Quality Healthcar *Managing Service Quality Volume 10 . Number 2, 103 – 111*

[42] Bouchereau V and Rowlands H 2000 Methods and Techniques to Help Quality Function Deployment (QFD) *Benchmarking: An International Journal, Vol. 7 No. 1,8 – 19*

[43] Partovi FY and Corredoira RA 2002 Quality Function Deployment For The Good Of Soccer *European Journal of Operational Research, vol. 137 (3), pp. 642–656.*

[44] Mehrjerdi YZ 2010 Quality Function Deployment and Its Extensions *International Journal of Quality Science merged into International Journal of Quality & Reliability Management. Vol. 27 Issue: 6, pp.616-640.*

[45] Erkarslan, Onder, Yilmaz and Hande 2011 Optimization of The Product Design Through Quality Function Development (QFD) and Analytical Hierarchy Process (AHP): A Case Study in a Ceramic Washbasin *Turkish : Izmir Institute of Technolog*

[46] Jaiswal ES 2012 A Case Study on Quality Function Deployment (QFD) IOSR Journal of Mechanical and Civil Engineering (IOSR-JMCE) 3(6), 27 – 35.

[47] Somadatta B, Karanjekar, Ramesh R, Lakhe and Deshpande VS 2013 Integration of Quality Function Deployment and Value Engineering in Furniture Manufacturing

Industry for Improvement of Computer Work Station *International Journal of Basic and Advanced Research, 124 – 132.*

[48] Matzler K and Hinterhuber HH 1998 How to Make Product Development Projects More Successful by Integrating Kano's Model of Customer Satisfaction Into Quality Function Deployment *Technovation, 18(1), 25 – 38.*

[49] Poel IVD 2007 Methodological Problems in QFD and Directions for Future Development *Research in Engineering Design, 18(1), 21 – 36.*

[50] Paula Jr GA, Trabasso LG, Souza MCF 2009 Combining the Kano model, p-Diagram and Quality Function Deployment (QFD) To Build An Adapted Technology Roadmapping (A-Trm) Process *20th International Congress of Mechanical Engineering Copyright © 2009 by ABCM November 15-20, 2009,pp 1 – 10.*

[51] Chauda A, Jain R, Singh AR 2011 Integration of Kano's Model into Quality Fucntion Deployment *Int. Journal Adv. Manuf. Technol, 53:689 – 698.*

[52] Wu CZ, Wang MT, Liu NT and Pan TS 2015 Developing a Kano-Based Evaluation Model for Innovation Design *Mathematical Problems in Engineering, vol. 2015, Article ID 153694, 8 pages.*

[53] Suef M, Suparno, Singgih ML 2013 Quality Initiatives as QFD-Kano Technical Responses: a Conceptual Model *Proceeding of Industrial Engineering and Service Science. IESS G12-1. 1 – 6.*

[54] Ji P, Jin J, Wang T and Chen Y 2014 Quantification and Integration of Kano's Model Into QFD For Optimising Product Design *International Journal of Production Research Volume 52, 2014 - Issue 21 6335-6348.*

[55] Telang, Samir, Vichoray and Chandan 2004 Development in Agricultural Tractor Brakes through QFD Application-A Conceptual Analysis *IOSR Journal of Mechanical and Civil Engineering (IOSR-JMCE) e-ISSN: 2278-1684, p-ISSN: 2320-334X PP 55-59 International Conference on Advances in Engineering & Technology (ICAET-2014).*

IOP Conf. Series: Materials Science and Engineering **319** (2018) 012044 doi:10.1088/1757-899X/319/1/012044

Solving Assembly Sequence Planning using Angle Modulated Simulated Kalman Filter

Ainizar Mustapa, Zulkifli Md. Yusof, Asrul Adam, Badaruddin Muhammad and Zuwairie Ibrahim

Universiti Malaysia Pahang, 26600 Pekan, Pahang Malaysia

zmdyusof@ump.edu.my

Abstract. This paper presents an implementation of Simulated Kalman Filter (SKF) algorithm for optimizing an Assembly Sequence Planning (ASP) problem. The SKF search strategy contains three simple steps; predict-measure-estimate. The main objective of the ASP is to determine the sequence of component installation to shorten assembly time or save assembly costs. Initially, permutation sequence is generated to represent each agent. Each agent is then subjected to a precedence matrix constraint to produce feasible assembly sequence. Next, the Angle Modulated SKF (AMSKF) is proposed for solving ASP problem. The main idea of the angle modulated approach in solving combinatorial optimization problem is to use a function, $g(x)$, to create a continuous signal. The performance of the proposed AMSKF is compared against previous works in solving ASP by applying BGSA, BPSO, and MSPSO. Using a case study of ASP, the results show that AMSKF outperformed all the algorithms in obtaining the best solution.

1. Introduction

Assembly optimisation in the production planning stage deals with determination of optimum assembly sequence and determination of optimum location of each resource. Solving the Assembly Sequence Planning (ASP) problem is crucial because it will determine many assembly aspects including tool changes, fixture design and assembly freedom. Assembly sequence also influences overall productivity because it determines how fast and accurate the product is assembled.

Assembly sequence planning consists of assembly, operations, existing assembly techniques, and some details of relations between parts. Some researchers have dedicated their work on some important issues related to concurrent engineering analyses on assembly sequence planning. These issues are the representation of a product to be assembled, the generation of assembly sequence plans and the determination of precedence constraints, the representation of resulting assembly sequence plans, and the selection of the optimum assembly sequence planning. In this paper, assumptions for ASP are as follows: (1) setup time and the actual assembly time for each part and component are given, (2) transfer time between workstations is included in set up time, and (3) downtime of machines and workstations is omitted.

The total assembly time is the combination of setup time and actual assembly time. It is assumed that regardless of the assembly sequence, the actual assembly time is constant. A proper tool and setup for each component to be assembled is required. These two items depend on the geometry of the component itself and the components assembled to that point. The setup time for a component can be predicted using the following function in Eq. (1):

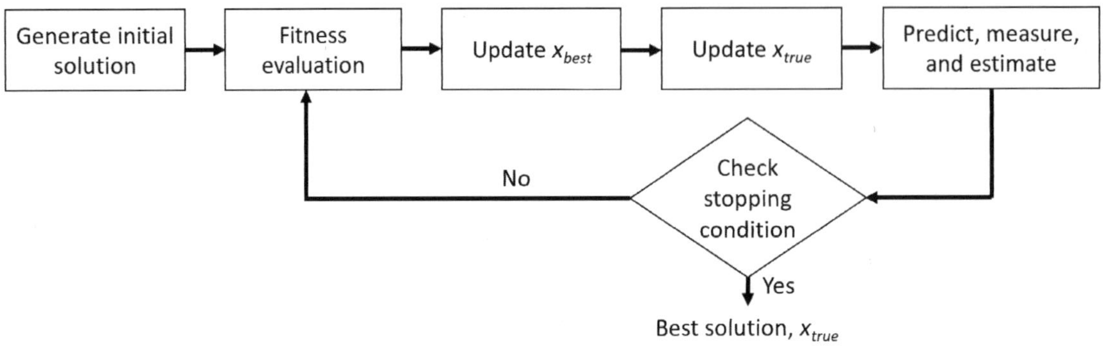

Figure 1. The original simulated Kalman filter (SKF) algorithm [1-2].

$$Time_{\text{Setup}}(a) = p_{a0} + \sum_{b=1}^{c} p_{ab} q_{ab} \tag{1}$$

where a is the number of component to be assembled, p_{a0} is the setup time for product (a) being the first component, p_{ab} is the contribution to the setup time due to the presence of part (b) when entering part a, and $q_{ab} = 1$ if component (b) has already been assembled. Otherwise, $q_{ab} = 0$. The total assembly time is the summation of setup time and actual assembly time. Hence, the objective function for minimizing the assembly time is shown in the Eq. (2).

$$\text{Min } Time_{\text{Assembly}} = \sum_{b=1}^{c} \left(Time_{\text{Setup}}(a) + A_a \right) \tag{2}$$

where A_a is the assembly time for component a.

2. Angle Modulated Simulated Kalman Filter (AMSKF) Algorithm

The simulated Kalman filter (SKF) algorithm [1-2] is illustrated in Figure 1. Consider n number of agents, SKF algorithm begins with initialization of n agents, in which the states of each agent are given randomly. The maximum number of iterations, $t_{\max}$, is defined. The initial value of error covariance estimate, $P(0)$, the process noise value, Q, and the measurement noise value, R, which are required in Kalman filtering, are also defined during initialization stage.

Then, every agent is subjected to fitness evaluation to produce initial solutions $\{\boldsymbol{X}_1(0), \boldsymbol{X}_2(0), \boldsymbol{X}_3(0), \ldots, \boldsymbol{X}_{n-2}(0), \boldsymbol{X}_{n-1}(0), \boldsymbol{X}_n(0)\}$. The fitness values are compared and the agent having the best fitness value at every iteration, t, is registered as $\boldsymbol{X}_{\text{best}}(t)$.

The-best-so-far solution in SKF is named as $\boldsymbol{X}_{\text{true}}$. The $\boldsymbol{X}_{\text{true}}$ is updated only if the $\boldsymbol{X}_{\text{best}}(t)$ is better (($\boldsymbol{X}_{\text{best}}(t) < \boldsymbol{X}_{\text{true}}$ for minimization problem, or $\boldsymbol{X}_{\text{best}}(t) > \boldsymbol{X}_{\text{true}}$ for maximization problem) than the $\boldsymbol{X}_{\text{true}}$. The subsequent calculations are largely similar to the predict-measure-estimate steps in Kalman filter. In the prediction step, the following time-update equations are computed.

$$\boldsymbol{X}_i(t|t) = \boldsymbol{X}_i(t) \tag{3}$$

$$P(t|t) = P(t) + Q \tag{4}$$

where $\boldsymbol{X}_i(t)$ and $\boldsymbol{X}_i(t|t)$ are the current state and current transition/predicted state, respectively, and $P(t)$ and $P(t|t)$ are the current error covariant estimate and current transition error covariant estimate, respectively. Note that the error covariant estimate is influenced by the process noise, Q.

The next step is measurement, which is a feedback to estimation process. Measurement is modelled such that its output may take any value from the predicted state estimate, $\boldsymbol{X}_i(t|t)$, to the true value, $\boldsymbol{X}_{\text{true}}$. Measurement, $\boldsymbol{Z}_i(t)$, of each individual agent is simulated based on the following equation:

$$\boldsymbol{Z}_i(t) = \boldsymbol{X}_i(t|t) + \sin(rand \times 2\pi) \times |\boldsymbol{X}_i(t|t) - \boldsymbol{X}_{true}| \quad (5)$$

The $\sin(rand \times 2\pi)$ term provides the stochastic aspect of SKF algorithm and $rand$ is a uniformly distributed random number in the range of $[0,1]$. The final step is estimate. During this step, Kalman gain, $K(t)$, is computed as follows:

$$K(t) = \frac{P(t|t)}{P(t|t)+R} \quad (6)$$

Then, the estimation of next state, $\boldsymbol{X}_i(t+1)$, and the updated error covariant, $P(t+1)$, are computed based on Eq. (7) and Eq. (8), respectively.

$$\boldsymbol{X}_i(t+1) = \boldsymbol{X}_i(t|t) + \Delta_i \quad (7)$$

$$P(t+1) = \left(1 - K(t)\right) \times P(t|t) \quad (8)$$

where $\Delta_i = K(t) \times (\boldsymbol{Z}_i(t) - \boldsymbol{X}_i(t|t))$. Finally, the next iteration is executed until the maximum number of iterations, $t_{\max}$, is reached.

The angle modulated simulated Kalman filter (AMSKF) algorithm [3] is an extension of simulated Kalman filter (SKF) algorithm. It has been applied as feature selection for EEG signal peak classification [4]. The main idea of the angle modulated approach in solving combinatorial optimization problem is to use a function, $g(x) = \sin(2\pi(x-a)\, x\, b\, \text{x}\, cos(A)) + d$, to create a continuous signal, where A = $2\pi(x-a)$ x c. The region $g(x) > 0$ is called binary 1 region and region $g(x) < 0$ is called binary 0 region. After that sampling based on sampling time, T, is executed to generate a bit string of length n. The required length of the bit string is problem dependent and determined by the size of a combinatorial optimization problem. The main advantage of angle modulated approach is that complex calculation in producing high dimensional bit string can be avoided. The search process in solving a combinatorial optimization problem can be done by tuning the values of a, b, c, and d only. In this work, the tuning is done by the SKF algorithm.

3. Angle Modulated Simulated Kalman Filter (AMSKF) for the ASP

To find an optimal solution, each agent representing feasible assembly sequence must be evaluated to obtain its fitness value. This evaluation of the fitness value is done after initial population is generated and Precedence Matrix (PM), coefficient table and actual assembly time are loaded. These three processes are illustrated along others in the proposed approach based on AMSKF, as depicted in Figure 2. The PM can be referred to Table 1. The coefficient table can be referred in [5] for details. The PM shows relationship betweens 19 components as illustrated in Figure 3. It is worth pointing out that the components of free to be assembled are the components that can be placed regardless of any part of a sequence.

As a result, each agent produces a feasible assembly sequence. The optimum one is then selected from the feasible assembly sequences by evaluating fitness of each agent. The best of the population which is a more optimal until the stopping condition is met. After the stopping condition is met, the performance of the AMSKF can be investigated. To conclude the finding, the performance of AMSKF in solving the ASP problem is compared against some related metaheuristic methods such as Binary Gravitational Search Algorithm (BGSA) [6], Binary Particle Swarm Optimization (BPSO) [7], and Multi-State Particle Swarm Optimization (MSPSO) [8].

IOP Conf. Series: Materials Science and Engineering **319** (2018) 012044 doi:10.1088/1757-899X/319/1/012044

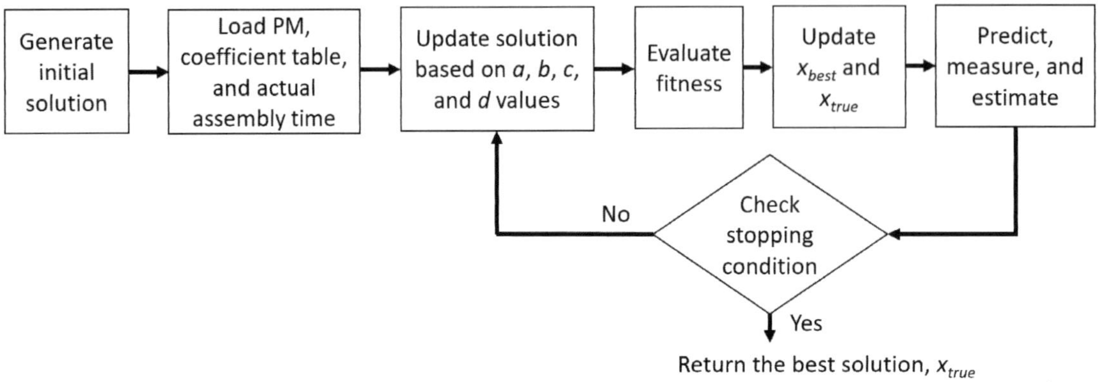

Figure 2. The AMSKF algorithm for the ASP.

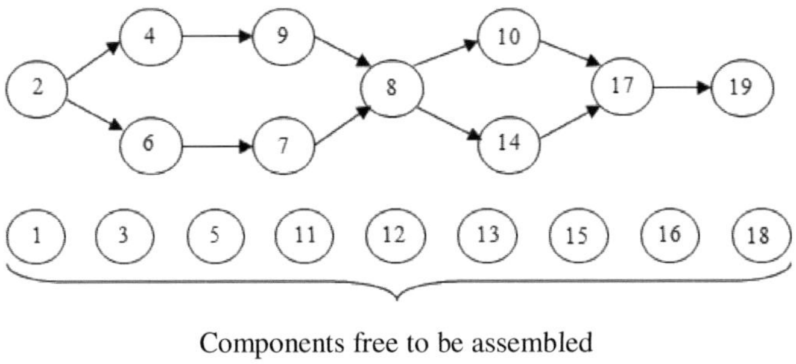

Figure 3. The assembly precedence diagram for the case study.

Table 1. Precedence matrix (PM) for the case study

Comp. (*a*)	Comp. (*b*) 1	2	3	4	5	6	7	8	9	10	11	12	13	14	15	16	17	18	19
1																			
2																			
3																			
4		1																	
5																			
6		1																	
7		1				1													
8		1		1		1	1		1										
9		1		1															
10		1		1		1	1	1	1										
11																			
12																			
13																			
14		1		1		1	1	1	1										
15																			
16																			
17		1		1		1	1	1	1	1				1					
18																			
19		1		1		1	1	1	1	1				1			1		

IOP Conf. Series: Materials Science and Engineering **319** (2018) 012044 doi:10.1088/1757-899X/319/1/012044

Table 2. The parameters and its value used for AMSKF, BGSA, BPSO, and MSPSO experiments

Parameter	AMSKF (proposed work)	BGSA	BPSO	MSPSO
Iteration	1000	500	500	500
Component No.	19	19	19	19
No. of agent	10	50	50	25
Initial error covariance estimate, $P(0)$	100	-	-	-
Process noise, Q	0.5	-	-	-
Measurement noise, R	0.5	-	-	-
Initial temperature	-	100°C	-	-
Cooling rate	-	0.95	-	-
Inertia weight, ω	-	-	0.6	-
Learning factor	-	-	1.42	-
ω Initial	-	-	-	0.9
ω Final	-	-	-	0.4
Coefficient factor, $c1$ and $c2$	-	-	-	2

Table 3. The results of AMSKF against BGSA, BPSO, and MSPSO

Algorithm	Min	Mean	Max	SD	Assembly sequence
AMSKF	503.91	520.19	528.70	4.94	1-2-12-4-3-9-13-15-11-5-16-6-18-7-8-14-10-17-19
BGSA	508.20	523.79	531.60	4.53	2-1-4-9-3-12-5-13-15-18-16-6-11-7-8-10-14-17-19
MSPSO	514.00	530.00	538.50	4.50	2-4-3-1-9-12-5-13-15-18-16-11-6-7-8-10-14-17-19
BPSO	515.80	520.80	523.60	3.10	13-2-3-5-12-15-16-4-1-11-9-18-6-7-8-10-14-17-19

4. Experiments, Results, and Discussion

In this study, the assembly of product consisting 19 components is considered. In experiments, the parameters and its value used for AMSKF, BGSA, BPSO, and MSPSO are presented in Table 2. The quality of results of AMSKF is then measured based on the fitness values of the best solutions in minimizing the total assembly time. For instance, if the number of independent trials on the case study is 50, the quality of results is determined based on the fitness values of 50 solutions.

To simplify the understanding of this work, fitness or objective value and solution is now called total assembly time and feasible assembly sequence, respectively. The average (mean), minimum (min), and maximum (max) of total assembly time of 50 feasible assembly sequences, and the standard deviation (SD) are recorded. Table 3 presents comparison of the result of AMSKF against BGSA, BPSO, and MSPSO. Based on the results given in Table 3, AMSKF outperformed BGSA, BPSO, and MSPSO in minimizing total assembly time and obtaining minimum average time of the ASP problem. The mininum total assembly time obtained by AMSKF is 503.91 unit of time with associated sequence of components suggested by the AMSKF is 1-2-12-4-3-9-13-15-11-5-16-6-18-7-8-14-10-17-19. The average assembly time of AMSKF is 520.19 unit of time and this average value is better than BGSA, BPSO, and MSPSO, which indicate AMSKF's consistency over 50 runs.

It is true that the finding reported in this paper is much dependent on the parameter values used by other algorithms, which are BGSA, MSPSO, and BPSO. However, in this study, there was no attempt to replicate the experiments of BGSA, MSPSO, and BPSO. All the results were taken from the published paper.

5. Conclusion

The ASP is a combinatorial optimization problem with a large-scale candidate solution. In this study, an approach based on a variant of SKF called AMSKF is proposed to solve ASP problem. To evaluate the performance of the proposed approach, a case study of ASP consisting nineteen components is chosen, and the performance of AMSKF is evaluated against three different approaches that uses BGSA, BPSO, and MSPSO as the search engine. Experimental results obtained showed that the proposed AMSKF outperformed the other three approaches.

In future, the AMSKF could be applied to solve ASP problem with different constraints such as assembly stability, machine and workstation assignment, and work load. Perhaps the experiments reported in this paper can be reimplemented with adjusted parameter values to get better result and more convincing comparison.

Acknowledgement

The researchers are indebted to the Ministry of Higher Education of Malaysia for supporting this research by awarding a Fundamental Research Grant Scheme (FRGS) research grant (RDU170106) to Universiti Malaysia Pahang (www.ump.edu.my). This research is also supported using UMP-Research University grant, (RDU1603109).

References

[1] Z Ibrahim, NH Abdul Aziz, NA Ab Aziz, S Razali, MI Shapiai, SW Nawawi, and MS Mohamad 2015 *A Kalman filter approach for solving unimodal optimization problems*, ICIC Express Letters, 9(12), 3415-3422.

[2] Z Ibrahim, NH Abdul Aziz, NA Ab Aziz, S Razali, and MS Mohamad 2016, *Simulated Kalman Filter: a novel estimation-based metaheuristic optimization algorithm*, Advance Science Letters, 22, 2941-2946.

[3] Z Md Yusof, I Ibrahim, Z Ibrahim, KZ Mohd Azmi, NH Abd Aziz, NA Ab Aziz, and MS Mohamad 2016 *Angle modulated simulated Kalman filter algorithm for combinatorial optimization problems*, ARPN Journal of Engineering and Applied Sciences, 11(7), 4854-4859.

[4] A Adam, Z Ibrahim, N Mokhtar, MI Shapiai, M Mubin, and I Saad 2016 *Feature selection using angle modulated simulated Kalman filter for peak classification of EEG signals*, SpringerPlus, 5(1580).

[5] YK Choi, DM Lee YB Cho 2008 *An approach to multi-criteria assembly sequence planning using genetic algorithms*, International Journal of Advanced Manufacturing Technology, 42(1-2), 180-188.

[6] I Ibrahim, Z Ibrahim, H Ahmad, ZM Yusof, MI Shapiai, SW Nawawi, and M Mubin 2014 *An assembly sequence planning approach with binary gravitational search algorithm*. In: Fujita H., Ali M., Selamat A., Sasaki J., Kurematsu M. (eds) New Trends in Software Methodologies, Tools and Techniques, IOS Press, 179-193.

[7] JAA Mukred, Z Ibrahim, I Ibrahim, A Adam, K Wan, ZM Yusof, and N Mokhtar 2012 *A binary particle swarm optimization approach to optimize aseembly sequence planning*, Advance Science Letters, 13, 732-738.

[8] I Ibrahim, Z Ibrahim, H Ahmad, and ZM Yusof 2016 *An assembly sequence planning approach with a multi-state particle swarm optimization*. In: Fujita H., Ali M., Selamat A., Sasaki J., Kurematsu M. (eds) Trends in Applied Knowledge-Based Systems and Data Science. IEA/AIE 2016. Lecture Notes in Computer Science, vol 9799. Springer, Cham.

IOP Conf. Series: Materials Science and Engineering **319** (2018) 012045 doi:10.1088/1757-899X/319/1/012045

Flexure and impact properties of glass fiber reinforced nylon 6-polypropylene composites

N M Kusaseh[1], D M Nuruzzaman[2], N M Ismail[2], Z Hamedon[2], A Azhari[2] and A K M A Iqbal[2]

[1,2]Faculty of Manufacturing Engineering, University Malaysia Pahang, 26600 Pekan, Pahang Darul Makmur, Malaysia.

[1]E-mail: izzathanis92@gmail.com

Abstract. In recent years, polymer composites are rapidly developing and replacing the metals or alloys in numerous engineering applications. These polymer composites are the topic of interests in industrial applications such as automotive and aerospace industries. In the present research study, glass fiber (GF) reinforced nylon 6 (PA6)-polypropylene (PP) composite specimens were prepared successfully using injection molding process. Test specimens of five different compositions such as, 70%PA6+30%PP, 65%PA6+30%PP+5%GF, 60%PA6+30%PP+10%GF, 55%PA6+30%PP+15%GF and 50%PA6+30%PP+20%GF were prepared. In the experiments, flexure and impact tests were carried out. The obtained results revealed that flexure and impact properties of the polymer composites were significantly influenced by the glass fiber content. Results showed that flexural strength is low for pure polymer blend and flexural strength of GF reinforced composite increases gradually with the increase in glass fiber content. Test results also revealed that the impact strength of 70%PA6+30%PP is the highest and 55%PA6+30%PP+15%GF composite shows moderate impact strength. On the other hand, 50%PA6+30%PP+20%GF composite shows low toughness or reduced impact strength.

1. Introduction

Polymers and their composites are gaining popularity in numerous engineering applications such as bearing material, gears, cams, wheels, rollers etc. for improved mechanical properties in relation to industrial requirements. The application of high performance polymer blend composites is a suitable alternative to conventional materials due to their high specific strength and stiffness. Polymer blends of different thermoplastics can be prepared and these polymer blends can be reinforced by different fibrous materials to enhance the mechanical properties. During the past decade, a number of research works were carried out to investigate the mechanical and tribological properties of different types of polymer composites [1-8]. The obtained results showed that the properties were significantly influenced by several operating conditions and the type of reinforced material.

Impact behaviour of E-glass fiber reinforced epoxy and polypropylene matrix composites was investigated and the obtained results show that the type of resin is an important parameter for the impact behaviour of the composites [9]. Flexural strength of polymethyl methacrylate (PMMA) resin reinforced with glass, aramid and nylon fibers were evaluated using 3-point bending test with universal testing machine [10]. The experimental data were compared with the flexural strength of pure PMMA resin and the comparison revealed that reinforced specimens showed better flexural strength than the conventional acrylic resin. Moreover, the specimens reinforced with glass fibers exhibited highest flexural strength, followed by the specimens reinforced with aramid and specimens reinforced with nylon showed the lowest flexural strength. Flexural properties of glass fiber reinforced polyamide 6-6 composites were investigated [11]. The obtained results showed that flexural properties were significantly improved as the fiber content was increased whereas flexural properties were not much improved with the increase in fiber length.

Flexural strengths of epoxy resin based hybrid polymer composites were investigated using 3-point bending test for different thicknesses of the composites [12]. The obtained results revealed that bending strength of the composite was significantly influenced by the thickness of the composite and the grade of the epoxy resin. Mechanical properties of short glass fiber reinforced polyamide 66 and polypropylene (PA66/PP) thermoplastic blend composites were investigated [13]. It was found that flexure and impact strengths of the composites were significantly different as compared to the strengths of the PA66/PP pure polymer blend. Mechanical properties of glass and carbon fibre reinforced polymer composites were investigated at different strain rate conditions [14]. From the experimental data it was concluded that flexural strength of the composite was greatly influenced by the strain rate condition. In addition, experimental data revealed that impact strength of the carbon fiber reinforced polymer composite was much higher than that of glass fiber reinforced polymer composite. For pure nylon polymer, the injection parameters such as pressure, temperature, fill time etc. were theoretically investigated [15]. Mechanical properties of glass and carbon fiber reinforced with polyester resin composites were investigated [16]. The obtained results showed that flexure and impact properties of the composites were significantly influenced depending on the type of composite and percentage reinforcement. Tensile strength and impact strength of glass fiber reinforced nylon were examined and the obtained results revealed that tensile strength and impact strength are greatly influenced by glass fiber content [17].

Within this research, five different types of glass fiber (GF) reinforced nylon 6 (PA6)-polypropylene (PP) composite specimens were prepared successfully using injection molding machine. The effects of glass fiber content on the flexure and impact properties of the composites were investigated.

2.Experimental

Using injection molding process, dog bone shaped specimens of five different compositions such as 70%PA6+30%PP, 65%PA6+30%PP+5%GF, 60%PA6+30%PP+10%GF, 55%PA6+30%PP+15%GF and 50%PA6+30%PP+20%GF were prepared according to ASTM D638 standard. During injection process, suitable process parameters were selected in producing successful specimens. During processing of the molded specimens, the temperature was controlled according to the type of specimen of different compositions in order to fabricate defect free specimens. Flexural tests of glass fiber reinforced nylon 6-polypropylene composites were carried out to investigate the important mechanical properties such as, flexural modulus, flexural yield strength, flexural strength and flexural strain. Three point flexural tests of the composite specimens were carried out using universal test machine Instron (model:3369, maximum load capacity: 50KN) and followed ASTM D790 standard. All the flexure tests were carried out under cross-head speed of 2 mm/min. Impact tests of the composite specimens were carried out to determine the energy absorbed in fracturing the specimens. Impact tests were carried out by using Instron CEAST 9050 Impact Pendulum-Izod Impact (Unnotched) according to ASTM D4812 standard for different types of polymer composites. For impact tests, the hammer energy of 11J was used and 4 N.m torque was applied during the test.

3. Results and Discussion

The flexural modulus of different compositions of glass fiber reinforced PA6-PP composites is shown in figure 1. Flexural modulus is a measure of the stiffness of the glass fiber reinforced PA6-PP composites. The experiments were performed under a cross-head speed of 2mm/min. The data generated by flexure stress-strain curves during the test is shown in Table 1. From figure 1 below, 70%PA6+30%PP neat polymer blend shows a low flexural modulus of 0.75 GPa. The 65%PA6+30%PP+5%GF composite shows an increased modulus of 1.14 GPa which is 52% higher in comparison with neat polymer blend. The 60%PA6+30%PP+10%GF composite reveals an increased modulus of 1.51 GPa which is about 101% higher than that of neat polymer blend. The flexural modulus of 55%PA6+30%PP+15%GF composite was recorded 1.84 GPa which is about 145% higher in comparison with neat PA6-PP blend. Finally, 50%PA6+30%PP+20%GF composite shows a further increased flexure modulus of 2.05 GPa which is nearly 173% higher in comparison with neat PA6-PP.

This proves that the composite shows higher resistance to deformation due to increased glass fiber content, hence in turn increases the stiffness of the PA6-PP-GF composite under flexure loading.

Table 1. Flexure test results of polymer blend composites.

Strain rate: 2 mm/min				
Specimen composition	Flexural modulus [GPa]	Flexural yield strength [MPa]	Flexural strength [MPa]	Flexural strain [%]
70%PA6+30%PP	0.75	25.62	26.07	7.20
65%PA6+30%PP+5%GF	1.14	31.95	32.17	5.81
60%PA6+30%PP+10%GF	1.51	35.58	36.11	4.38
55%PA6+30%PP+15%GF	1.84	36.96	37.24	3.50
50%PA6+30%PP+20%GF	2.05	41.84	42.61	3.53

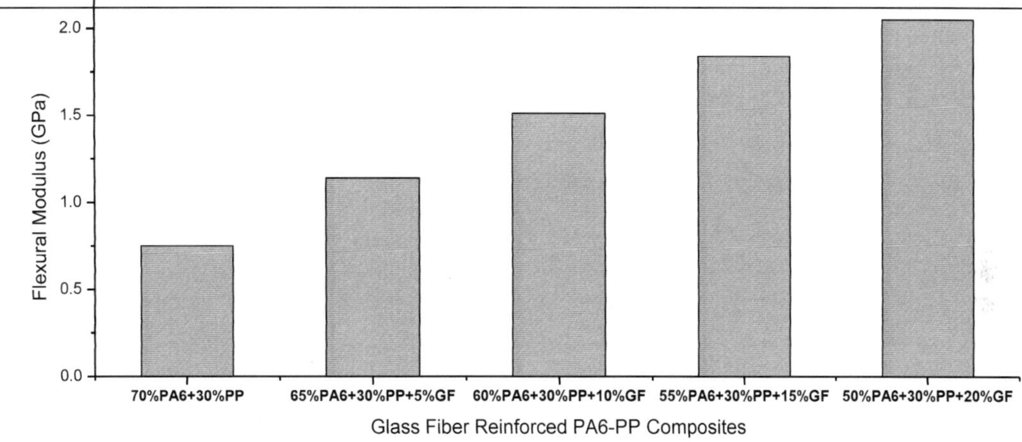

Figure 1. The flexural modulus of different compositions of glass fiber reinforced nylon 6-polypropylene composites.

The flexural yield strength (offset 0.2%) of glass fiber reinforced nylon 6-polypropylene composite is shown in figure 2. The obtained results revealed that the plastic deformation of 70%PA6+30%PP pure polymer blend started at low stress level of 25.62 MPa. Yielding of 65%PA6+30%PP+5%GF composite started at higher stress level of 31.95 MPa which is about 25% higher than PA6-PP blend. The 60%PA6+30%PP+10%GF composite showed increased yield stress level of 35.58 MPa which is almost 39% higher than the neat blend. For further increase of glass fiber

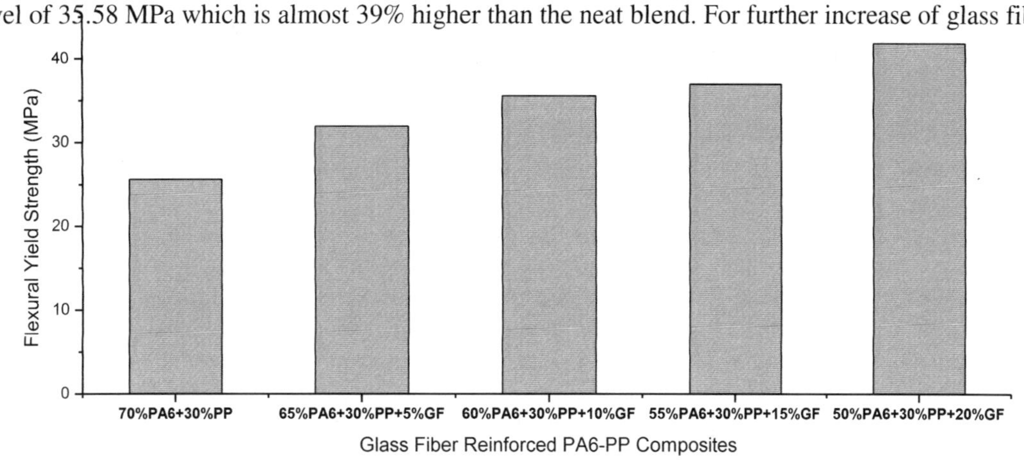

Figure 2. The flexural yield strength (MPa) of different compositions of glass fiber reinforced nylon 6-polypropylene composites.

content to 15%, the flexural yield strength of the composite slightly increases to 36.96 MPa which is 44% higher than that of neat blend. Finally, 50%PA6+30%PP+20%GF composite showed an improved value of flexural yield strength of 41.84 MPa which is nearly 63% higher than the neat blend. It is apparent that the plastic behavior of the blend composite started at higher stress level with the increase in glass fiber content.

The flexural strength of glass fiber reinforced PA6-PP composite is shown in figure 3. Flexural strength is defined as the maximum stress at the outermost fiber on tension side of the specimen under a strain rate of 2 mm/min. The obtained result shows that flexure strength of 70%PA6+30%PP neat polymer blend is 26.07 MPa. The 65%PA6+30%PP+5%GF composite shows the strength of 32.17 MPa which is 23.4% higher than neat polymer blend. Flexure strength of 60%PA6+30%PP+10%GF composite is 36.11 MPa which is nearly 38.5% higher in comparison with neat polymer blend. For 55%PA6+30%PP+15%GF composite, the flexure strength is slightly improved to 37.24 MPa. Lastly, 50%PA6+30%PP+20%GF composite shows further improved strength of 42.61 MPa which is about 63.4% higher in comparison with neat polymer blend. From the obtained results, it is apparent that flexure strength of the composite improves gradually with the increase in glass fiber content.

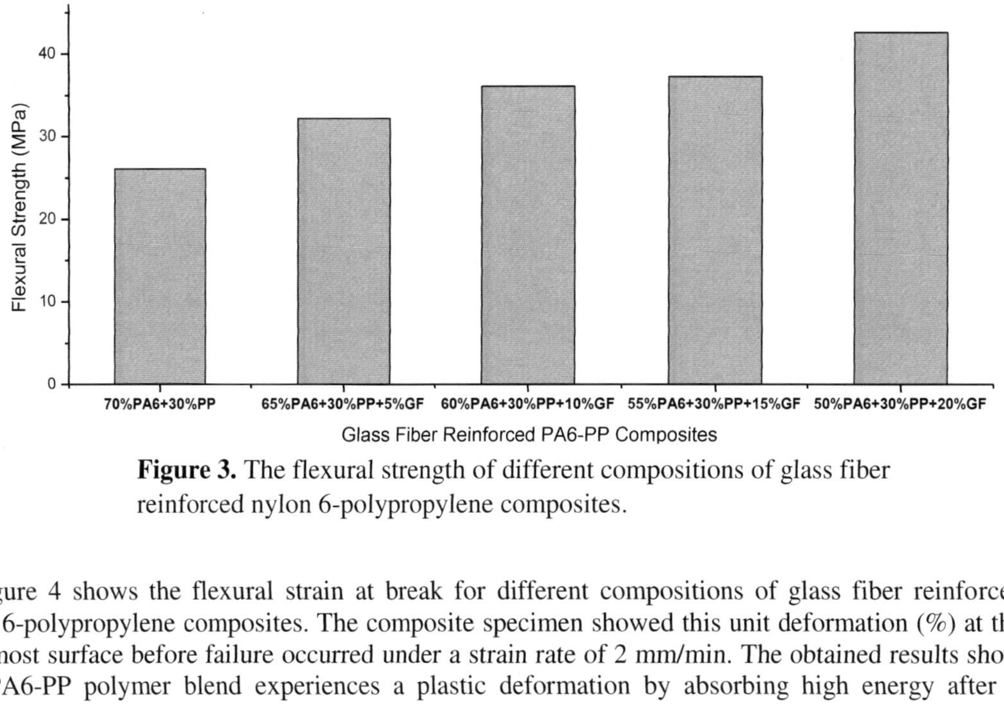

Figure 3. The flexural strength of different compositions of glass fiber reinforced nylon 6-polypropylene composites.

Figure 4 shows the flexural strain at break for different compositions of glass fiber reinforced nylon 6-polypropylene composites. The composite specimen showed this unit deformation (%) at the outermost surface before failure occurred under a strain rate of 2 mm/min. The obtained results show that PA6-PP polymer blend experiences a plastic deformation by absorbing high energy after it reaches to yield point and in this case the flexure strain is 7.20%. The results also show that for the fiber content of 5%, 10%, 15%, and 20%, the composite experiences the strain 5.81%, 4.38%, 3.50% and 3.53% respectively. This shows that in general, the flexure deformation gradually decreases as the glass fiber content increases.

The impact strength of glass fiber reinforced PA6-PP composite is shown in figure 5. All the impact tests were carried out under an applied torque 4 N.m and the hammer energy of 11 J was used. From the obtained results, pure polymer blend 70%PA6+30%PP shows significantly high impact strength of 78.17 kJ/m^2. On the other hand, all the composites show reduced impact strength. The impact strength of 65%PA6+30%PP+5%GF composite is 35.33 kJ/m^2 which is much lower than that of neat polymer blend. The impact strength of the composite is improved to 38.50 kJ/m^2 when the glass fiber content is increased to 10%. For 55%PA6+30%PP+15%GF composite, the impact strength is further improved to 42.73 kJ/m^2. It can be noticed that for the glass fiber content 20%, the impact strength of the composite is decreased to 34.70 kJ/m^2. It is believed that due to the increase in the glass fiber content from 15% to 20%, the polymer blend composite behaves as ductile to brittle transition which results in reducing the impact toughness [17].

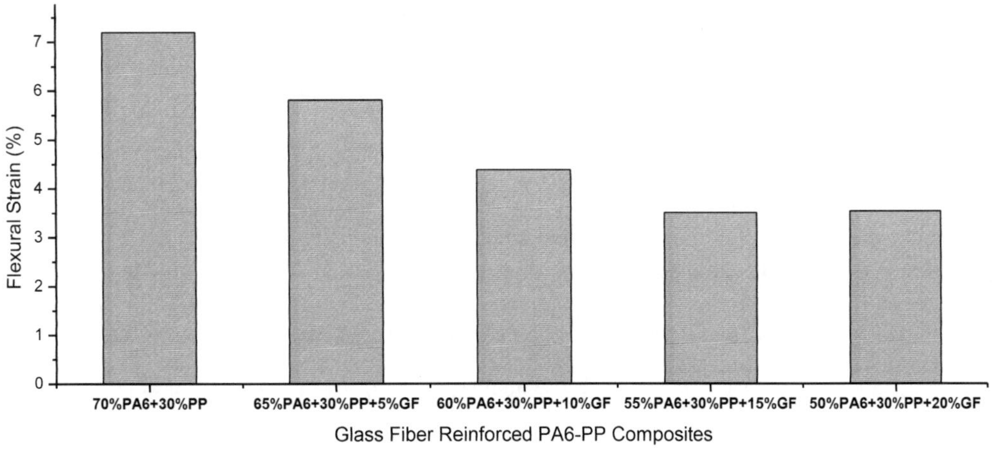

Figure 4. The flexural strain (%) of different compositions of glass fiber reinforced nylon 6-polypropylene composites

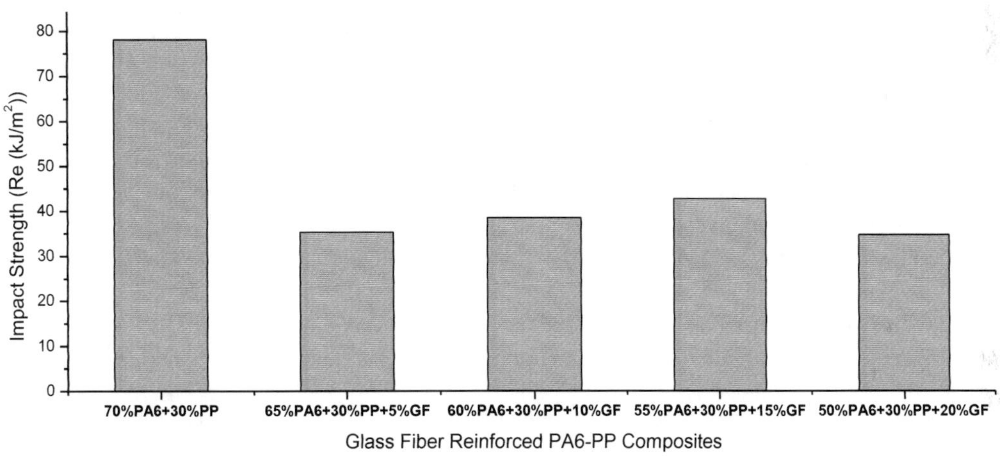

Figure 5. The impact strength, Re (kJ/m^2) of different compositions of glass fiber reinforced nylon 6-polypropylene composites (Torque: 4 N.m)

4. Conclusions

Within this research, five different types of glass fiber reinforced nylon 6-polypropylene thermoplastic blend composites were successfully prepared. The effects of glass fiber content on flexural and impact properties of the composites were investigated. The main conclusions are:

1. The flexural modulus of 70%PA6+30%PP pure polymer blend is the lowest whereas the polymer composite shows increased modulus as the glass fiber content is increased and 50%PA6+30%PP+20%GF composite shows significantly high modulus.
2. Flexural yield strength of pure polymer blend is the lowest and it increases gradually for the polymer composite as the glass fiber content increases. Flexural strength of pure polymer blend shows the lowest value but it improves gradually with the increased glass fiber content.
3. Pure polymer blend shows the highest impact strength and all the composites show reduced impact strength. It is noticed that 50%PA6+30%PP+20%GF composite shows the lowest impact strength.

Acknowledgements
The authors gratefully acknowledge University Malaysia Pahang (UMP) for the financial support through the research grant schemes PGRS160374 and RDU1703131. Sincere thanks to the technical staffs of faculty of manufacturing engineering related to this research work.

References
[1] Mouhmid B, Imad A, Benseddiq N, Benmedakhene and Maazouz A 2006 A study of the mechanical behaviour of a glass fibre reinforced polyamide 6,6: Experimental investigation *Polym. Test.* **25** 544-52
[2] Yuan L, Ma X, Liang G and Yan H 2007 Fibre reinforced organic rectorite/unsaturated polyester composites *Compos. Sci. Technol.* **67** 2311-22
[3] Nuruzzaman D M, Chowdhury M A and Rahaman M L 2011 Effect of duration of rubbing and normal load on friction coefficient for polymer and composite materials *Ind. Lubr. Tribol.* **63** 320-26
[4] Nuruzzaman D M, Rahaman M L and Chowdhury M A 2012 Friction coefficient and wear rate of polymer and composite materials at different sliding speeds *Int. J. Surf. Sci. Eng.* **6** 231-45
[5] Chowdhury M A, Nuruzzaman D M, Roy B K, Samad S, Sarker R and Rezwan A H M 2013 Experimental investigation of friction coefficient and wear rate of composite materials sliding against smooth and rough mild steel counterfaces *Tribol. Ind.* **35** 286-96
[6] Zhou S, Zhang Q, Wu C and Huang J 2013 Effect of carbon fiber reinforcement on the mechanical and tribological properties of polyamide6/polyphenylene sulfide composites *Mater. Des.* **44** 493–9
[7] Braga R A and Magalhaes P A A 2015 Analysis of the mechanical and thermal properties of jute and glass fiber as reinforcement epoxy hybrid composites *Mater. Sci. Eng.,C* **56** 269-73
[8] Unterweger C, Bruggemann O and Furst C 2014 Effects of different fibers on the properties of short-fiber-reinforced polypropylene composites *Compos. Sci. Technol.* **103** 49-55
[9] Arikan V and Sayman O 2015 Comparative study on repeated impact response of E-glass fiber reinforced polypropylene & epoxy matrix composites *Composites Part B* **83** 1-6
[10] John J, Gangadhar S A and Shah I 2001 Flexural strength of heat-polymerized polymethyl methacrylate denture resin reinforced with glass, aramid, or nylon fibers *J. Prosthet. Dent.* **86** 424-7
[11] Lafranche E, Krawczak P, Ciolczyk J P and Maugey J 2007 Injection moulding of long glass fibre reinforced polyamide 6-6: guidelines to improve flexural properties *Express. Polym. Lett.* **1** 456–66
[12] Ahmed M N, Kumar P V, Shivanand H K and Muzammil S B 2013 A study on flexural strength of hybrid polymer composite materials (e glass fibre-carbon fibre-graphite) on different matrix material by varying its thickness *Int. J. Mech. Eng. Technol.* **4** 274-86
[13] Lingesh B V, Rudresh B M and Ravikumar B N 2014 Effect of short glass fibers on mechanical properties of polyamide 66 and polypropylene (PA66/PP) thermoplastic blend composites *Procedia. Mater. Sci.* **5** 1231-40
[14] Elanchezhian C, Ramnath B V and Hemalatha J 2014 Mechanical behaviour of glass and carbon fibre reinforced composites at varying strain rates and temperatures *Procedia. Mater. Sci.* **6** 1405-18
[15] Nuruzzaman D M, Kusaseh N, Basri S, Oumer A N and Hamedon Z 2016 Modeling and flow analysis of pure nylon polymer for injection molding process *IOP Conf. Ser.: Mater. Sci. Eng.* **114** 1-7
[16] Durairaj R B, Mageshwaran G and Sriram V 2016 Investigation on mechanical properties of glass and carbon fiber reinforced with polyester resin composite *Int. J. Chemtech. Res.* **9** 417-23
[17] Nuruzzaman D M, Iqbal A K M A, Oumer A N, Ismail N M and Basri S 2016 Experimental investigation on the mechanical properties of glass fiber reinforced nylon *IOP Conf. Ser.: Mater. Sci. Eng.* **114** 1-7

Microstructure and properties of aluminium-aluminium oxide graded composite materials

F F Kamaruzaman[1], D M Nuruzzaman[2], N M Ismail[2], Z Hamedon[2], A K M A Iqbal[2] and A Azhari[2]

[1,2]Faculty of Manufacturing Engineering, University Malaysia Pahang, 26600 Pekan, Pahang Darul Makmur, Malaysia.

[1]Email: farafazira92@yahoo.com

Abstract. In this research works, four-layered aluminium-aluminium oxide (Al-Al_2O_3) graded composite materials were fabricated using powder metallurgy (PM) method. In processing, metal-ceramic graded composite materials of 0%, 10%, 20% and 30% weight percentage of ceramic concentration were prepared under 30 ton compaction load using a cylindrical die-punch set made of steel. After that, two-step pressureless sintering was carried out at sintering temperature and time 600°C and 3 hours respectively. It was observed that the sintered cylindrical specimens of 30 mm diameter were prepared successfully. The graded composite specimens were analysed and the properties such as density, microstructure and hardness were measured. It was found that after sintering process, the diameter of the graded cylindrical structure was decreased. Using both Archimedes method and rule of mixture (ROM), he density of structure was measured. The obtained results revealed that the microvickers hardness was increased as the ceramic component increases in the graded layer. Moreover, it was observed that the interface of the graded structure is clearly distinguished within the multilayer stack and the ceramic particles are almost uniformly distributed in the Al matrix.

1. Introduction

At present, the demand for multifunctional composite materials is growing increasingly for advanced engineering applications. A new class of graded composite material is suitable for multiple functions and this metal-ceramic graded composite structure can be attained in a single component. Currently, variety of fabrication methods available for processing of these metal-ceramic graded composite materials, namely, centrifugal casting, thermal spraying and powder metallurgy [1-3]. Among these methods, powder metallurgy (PM) is widely used in order to meet the desired requirements and to produce the functional gradients which can also be tailored while providing the best use of each component [4-6]. The most challenging step within powder metallurgy technique is the sintering process. Due to the residual stress caused by the mismatches in thermal expansion and sintering between successive layers, cracks and camber are noticeable during sintering of these multi-layered graded composite materials [7].

In the recent past, several research efforts were carried out to investigate the behaviour of functionally graded materials (FGMs) of different combinations [8-14]. Mechanical responses and the structures of these composites significantly influenced by a number of factors such as powder characteristics, compaction load, metal-ceramic system, sintering temperature, number of layers and the consolidation process. Influence of reinforcements on the mechanical properties of functionally graded aluminium composites was investigated [15]. The obtained results revealed that the outer regions of the composites exhibit higher hardness than the middle and inner regions. The outer region of all composites showed higher tensile strength than the inner one. Very recently, the effects of various atmospheres on the performances of functionally graded hydroxyapatite/titanium (HA/Ti) were investigated [16]. The obtained results revealed that hydroxyapatite and titanium elements reacted differently depending on sintering atmosphere.

The present research work considers the preparation of four-layered aluminium-aluminium oxide ($Al-Al_2O_3$) graded composite specimens using powder metallurgy method. The properties such as densification and shrinkage, microstructure and hardness were analysed.

2. Materials and experimental procedures

In this research, four-layered $Al-Al_2O_3$ graded composite samples were fabricated by powder metallurgy route. The Al and Al_2O_3 powders were mixed with different composition (100:0, 90:10, 80:20, 70:30) according to the design shown in figure 1.

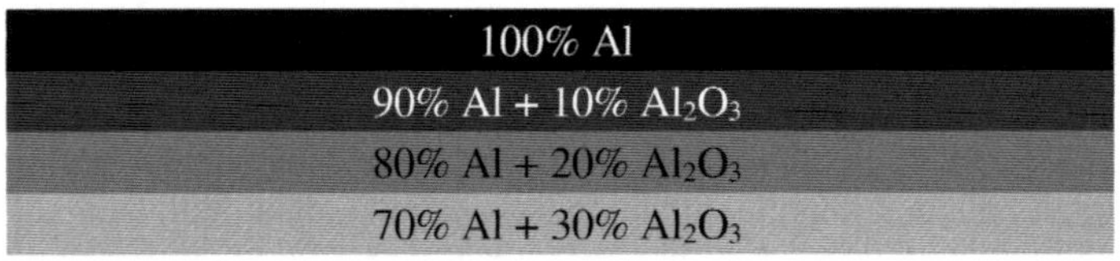

Figure 1. Composition contribution model of $Al-Al_2O_3$ FGM.

Three major steps were involved to process the specimens which are blending, compacting and sintering. At first, the weights of aluminium and aluminium oxide powders were measured based on their molecular weight. After that, these two materials were mixed and blended allowing sufficient time until the good homogenous mixture was achieved. During the stacking process, the powders were filled in order from 70%Al+30%Al_2O_3 and end with 100%Al (figure 1) into a cylindrical steel die with a diameter of 30 mm. The four-layered powder compositions were cold compacted at 30 ton (294.3 kN) using a hydraulic press (TOYO: Model TL30). After the compaction process, the green compacts are fragile and have low cohesive strengths. Finally, all the composite specimens were consolidated using a sintering furnace (Nabertherm: Made in Germany) by pressureless sintering using a two-step heating cycle/map as shown in figure 2.

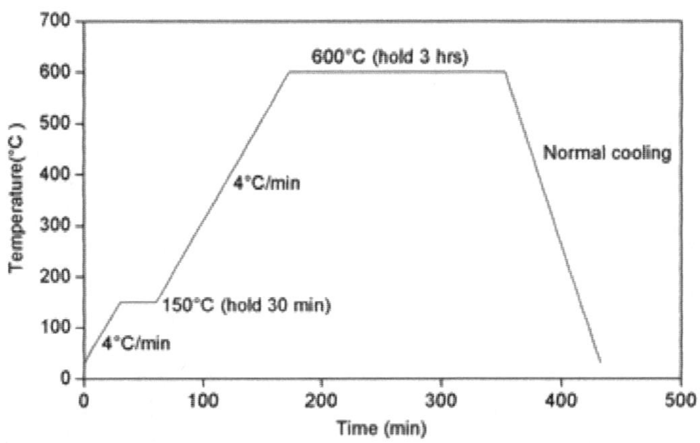

Figure 2. Two-step heating cycle for $Al-Al_2O_3$ FGM synthesis at sintering temperature 600ºC for 3 hours.

After sintering process, the FGM samples were prepared for characterization and hardness testing. Both green and sintered density of FGM samples were measured by using the rule of mixture (ROM) and Archimedes' principle respectively. For microstructural studies, the samples were observed by using metallurgical microscope (OLYMPUS BX51M, Made in Japan). The hardness of the FGM

IOP Conf. Series: Materials Science and Engineering **319** (2018) 012046 doi:10.1088/1757-899X/319/1/012046

samples was measured using micro-vickers hardness tester (Wilson Hardness: Model 402 MVD, Made in USA) under a load of 300 gf (2.94 N) for a dwell time of 15 seconds along a longitudinal axis with average of 10 measurements.

3. Results and discussion

3.1. Densification and shrinkage

Figure 3 shows the fabricated FGM specimen before and after sintering. The linear shrinkage of Al-Al_2O_3 graded composite could be calculated accurately by comparing the specimen diameter before and after sintering. The diameter of FGM sample was decreased from 30 mm to 29 mm due to the sintering shrinkage. Neither any physical damage nor any crack was found on above, below or on the sides of fabricated sample. From the obtained results and according to figure 4, the diameter shrinkage of each layer started from 100%Al to 70%Al+30%Al_2O_3 can be seen decreasing in trend with the increase in Al_2O_3 from the original diameter 30 mm (green compact).

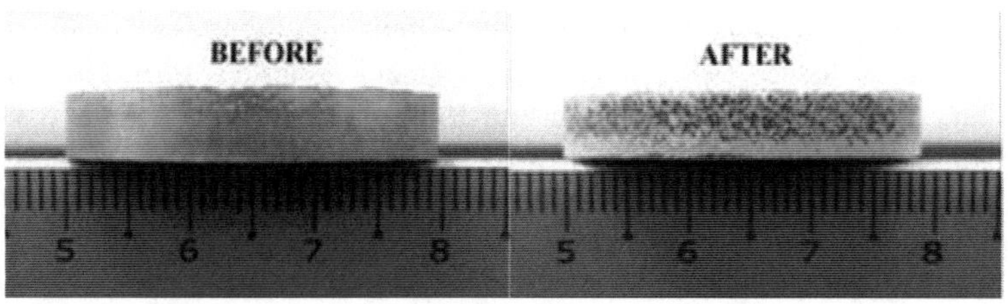

Figure 3. Diameter shrinkage due to pressureless sintering.

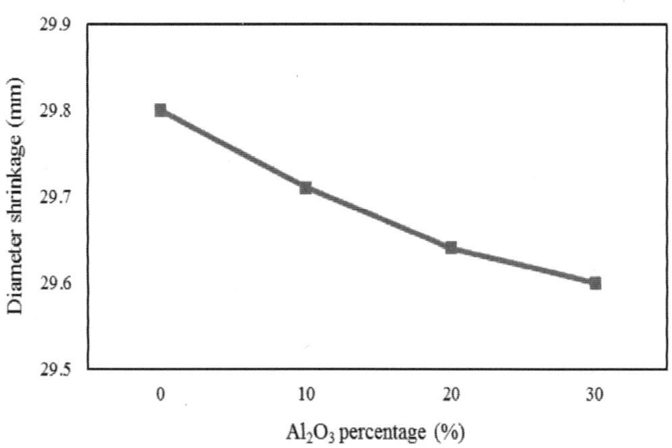

Figure 4. Diameter shrinkage each layer of Al-Al_2O_3.

The densities of the homogeneous layers for different compositions of Al-Al_2O_3 are shown in figure 5. From the obtained results it can be seen that there is a reduction trend of density with increase in Al_2O_3 ceramic percentage. After the sintering process, the maximum density was achieved for pure Al layer. It is clear that the reduction trend of density for each homogeneous layer of Al-Al_2O_3 composite is gradual throughout the system as the ceramic content increases. It is noticed that the sintered density of pure Al layer is slightly higher than the density of green compact. On the other hand, it is apparent that the sintered density of each homogeneous layer of Al-Al_2O_3 is somewhat lower than the density of green compact.

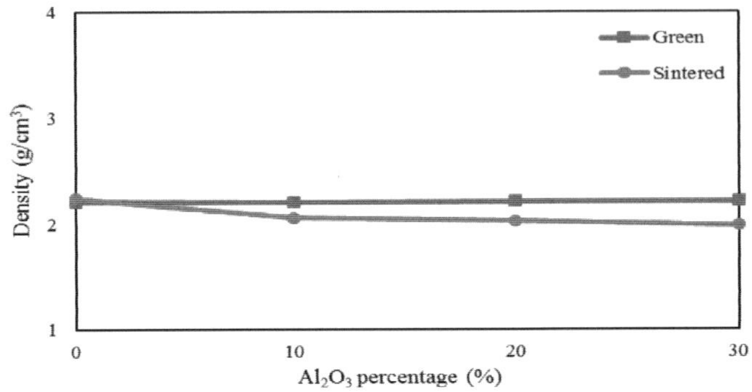

Figure 5. Densities of homogenous layer of Al-Al_2O_3.

3.2. Microstructural analysis

The microstructures of sintered Al-Al_2O_3 graded composites were characterized in order to determine if the particle distribution was consistent and the particle agglomeration occurred. The sintered composites were cut along the cross-sectional direction and mounted using cold mounting process. The mounted specimen was prepared using grinding and polishing. Then the microstructural analysis was carried out using a high resolution optical microscope and the micrographs are shown in figures 6(a)-(d). The microstructure in figure 6 consists of two phases, namely Al and Al_2O_3 are observed in Al-Al_2O_3 graded composite materials. The darker phase is Al_2O_3 and the lighter one is Al matrix. A uniform dispersion of spherical Al_2O_3 phase is clearly evident in the micrographs for the 2nd, 3rd and 4th layers of the graded structure. It can be seen that the weight percentage of ceramic is increased from 10% to until 30%. Figure 7 shows the interfacial region in Al-Al_2O_3 graded structure which represents aa', bb' and cc' as the interface layer. No cracking was observed between the layers which indicates that the FGMs have good strength due to the better metal-ceramic bonding and continuous microstructure at the interfaces of the FGMs, as reported in previous studies [12]. From these observations, it is apparent that the interfaces are almost straight and nearly parallel which confirms that powder stacking processes of the graded composites were well executed.

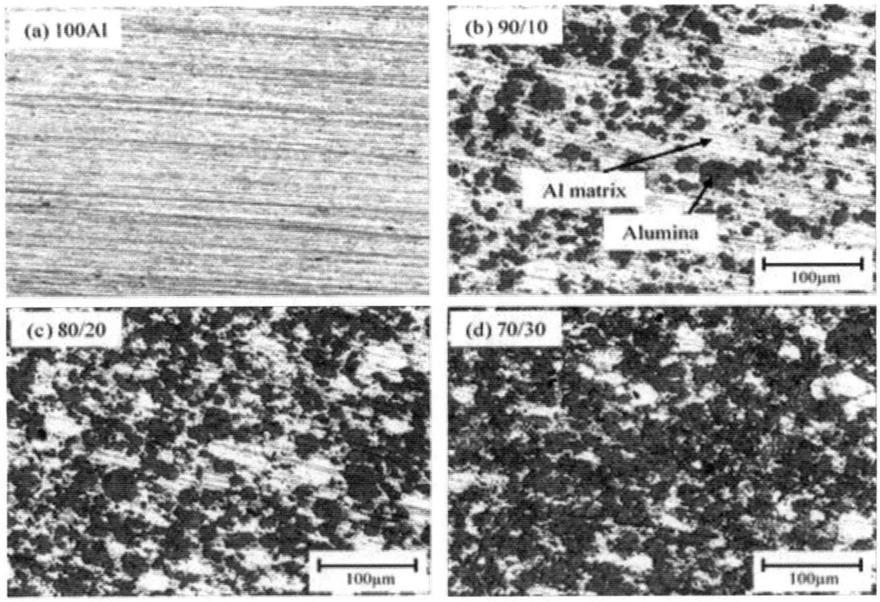

Figure 6. Optical micrographs of Al-Al_2O_3 graded layers (a) 100% Al (1st layer), (b) 90% Al + 10% Al_2O_3 (2nd layer), (c) 80% Al + 20% Al_2O_3 (3rd layer) and (d) 70% Al + 30% Al_2O_3 (4th layer).

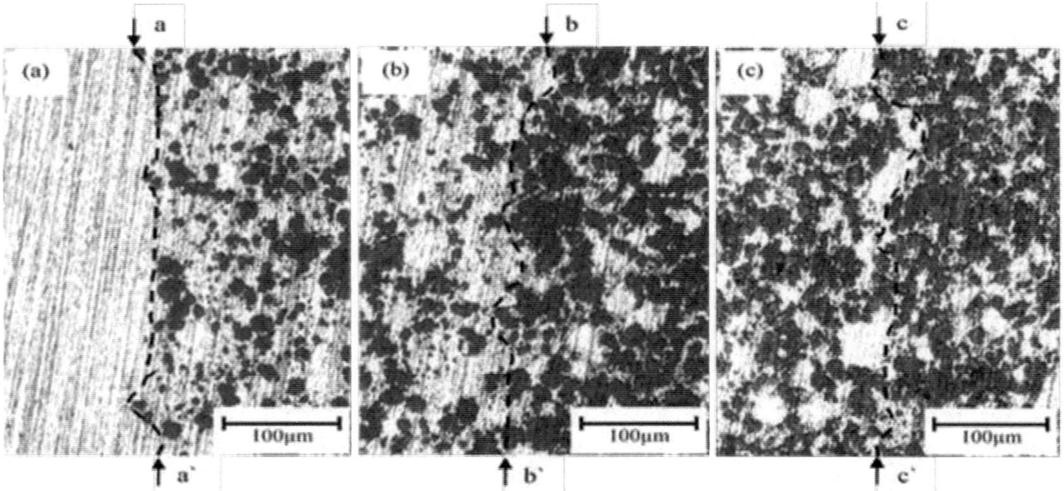

Figure 7. Micrographs of interface between two layers of Al-Al_2O_3 graded composites (a) interface (aa`) between 1st layer and 2nd layer, (b) interface (bb`) between 2nd layer and 3rd layer and (c) interface (cc`) between 3rd layer and 4th layer.

3.3 Microhardness measurement

The variation in hardness across the thickness of graded structure is presented in figure 8. The obtained results revealed that microhardness was increased at each layer with the increase in ceramic content as reported by the researchers in earlier publication [9] and it was noticed that this trend occurred layer by layer almost consistently. From the obtained results, the hardness of layer 1 (100%Al) was measured 29.6 HV. However, at layer 2 (10%Al_2O_3), the hardness was increased to 32.3 HV and further increased to 33.5 HV at layer 3. At layer 4 (30%Al_2O_3), the hardness continued to increase to 35.2 HV as the ceramic content was increased. This increment in hardness was found to be reasonably consistent for dispersion of ceramics in the metal matrix.

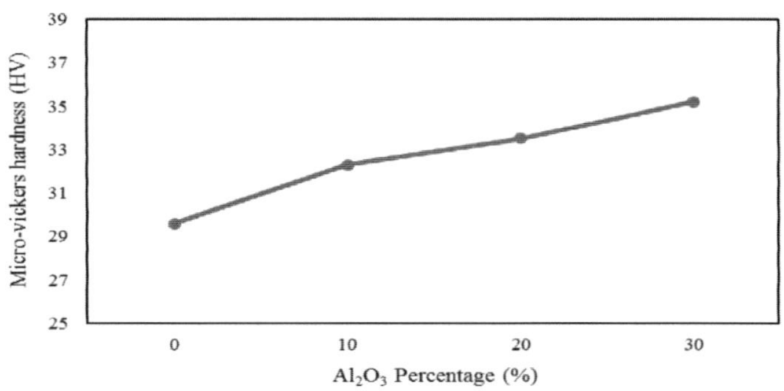

Figure 8. Hardness distribution through the thickness of Al-Al_2O_3 graded composites sample.

4. Conclusions

The reduction trend of density for each homogeneous layer of Al-Al_2O_3 composite is gradual throughout the system as the ceramic content increases. The sintered density of pure Al layer is slightly higher than the density of green compact whereas the sintered density of each homogeneous layer of Al-Al_2O_3 is somewhat lower than the density of green compact. The Al-Al_2O_3 graded composite structure exhibits microstructure with uniform distribution of Al_2O_3 phase in Al matrix. Moreover, the interfacial layers are almost straight and parallel which proves that powder stacking processes and techniques are well executed. The hardness increases with the increase in Al_2O_3 content and this trend occurs layer by layer almost consistently.

Acknowledgement
The authors gratefully acknowledge University Malaysia Pahang (UMP) for the financial support through the research grant PGRS160333. The authors also wish to extend their sincere thanks to all the technical staffs of Faculty of Manufacturing Engineering, University Malaysia Pahang involved related to this research work.

References
[1] Watanabe Y, Hattori Y and Sato H 2015 Distribution of microstructure and cooling rate in Al-Al_2Cu functionally graded materials fabricated by a centrifugal method *J. Mater. Process. Technol.* **221** 197–204
[2] Khoddami A M, Sabour A and Hadavi S M M 2007 Microstructure formation in thermally-sprayed duplex and functionally graded NiCrAlY/Yttria-stabilized zirconia coatings *Surf. Coat. Technol.* **201** 6019–24
[3] EL-Wazery M S, EL-Desouky A R, Hamed O A, Fathy A and Mansour N A 2013 Electrical and mechanical performance of zirconia-nickel functionally graded materials *Int. J. Eng. Trans. A: Basics* **26** 375–82
[4] Shahrjerdi A, Mustapha F, Bayat M, Sapuan S M and Majid D L A 2011 Fabrication of functionally graded hydroxyapatite-titanium by applying optimal sintering procedure and powder metallurgy *Int. J. Phys. Sci.* **6** 2258–67
[5] Wu K, Scheler S, Park H S and Porada M W 2013 Pressureless sintering of ZrO_2-$ZrSiO_4$/NiCr functionally graded materials with a shrinkage matching process *J. Eur. Ceram. Soc.* **33** 1111–21
[6] Rubio W M, Paulino G H and Silva E C N 2012 Analysis, manufacture and characterization of Ni/Cu functionally graded structures *Mater. Des.* **41** 255-65
[7] Sun L, Sneller A and Kwon P 2008 Fabrication of alumina/zirconia functionally graded material: from optimization of processing parameters to phenomenological constitutive model *Mater. Sci. Eng., A* **488** 31–8
[8] Tahvilian L and Fang Z Z 2012 An investigation on thermal residual stresses in a cylindrical functionally graded WC-Co component *Mater. Sci. Eng., A* **557** 106-12
[9] Bhattacharyya M, Kumar A N and Kapuria S 2008 Synthesis and characterization of Al/SiC and Ni/Al_2O_3 functionally graded materials *Mater. Sci. Eng. A* **487** 524-35
[10] Jamaludin S N S, Mustapha F, Nuruzzaman D M and Basri S N 2013 A review on the fabrication techniques of functionally graded ceramic-metallic materials in advanced composites *Sci. Res. Essays.* **8** 828-40
[11] Hosseini S S, Bayesteh H and Mohammadi S 2013 Thermo-mechanical XFEM crack propagation analysis of functionally graded materials, *Mater. Sci. Eng. A* **561** 285-302
[12] Erdemir F, Canakci A and Varol T 2015 Microstructural characterization and mechanical properties of functionally graded Al2024/SiC composites prepared by powder metallurgy techniques *Trans. Nonferrous Met. Soc. China* **25** 3569-77
[13] Latiff M I A, Jamaludin S N S, Basri S, Hussain A, Al-Othmany D S, Mustapha F, Nuruzzaman D M, Ismail N M and Ismail I 2014 Effect of sintering temperature on functionally graded nickel/alumina plate *Appl. Mech. Mater.* **629** 437-43
[14] Jamaludin S N S, Basri S, Hussain A, Al-Othmany D S, Mustapha F and Nuruzzaman D M 2014 Three-dimensional finite element modeling of thermomechanical problems in functionally graded hydroxyapatite/titanium plate *Math. Probl. Eng.* **2014** 1-20
[15] Radhika N and Raghu R 2016 Development of functionally graded aluminium composites using centrifugal casting and influence of reinforcements on mechanical and wear properties *Trans. Nonferr. Met. Soc. China* **26** 905-16
[16] Jamaludin S N S, Basri S, Mustapha F, Nuruzzaman D M, Latiff M I A and Ismail N M 2017 Phase contamination characterization of stepwise-bulit functionally graded hydroxyapatite/titanium (HA/Ti) sintered under various atmospheres *Mater. Sci. Forum.* **889** 90–5

IOP Conf. Series: Materials Science and Engineering **319** (2018) 012047 doi:10.1088/1757-899X/319/1/012047

Design of Work Facilities for Reducing Musculoskeletal Disorders Risk in Paper Pallet Assembly Station

Dian Mardi Safitri, Zahra Arfi Nabila and Nora Azmi

Laboratory of Work System Design and Ergonomics,
Department of Industrial Engineering, Universitas Trisakti
Jakarta, Indonesia

dianm@trisakti.ac.id

Abstract. Musculoskeletal Disorders (MSD) is one of the ergonomic risks due to manual activity, non-neutral posture and repetitive motion. The purpose of this study is to measure risk and implement ergonomic interventions to reduce the risk of MSD on the paper pallet assembly work station. Measurements to work posture are done by Ovako Working Posture Analysis (OWAS) methods and Rapid Entire Body Assessment (REBA) method, while the measurement of work repetitiveness was using Strain Index (SI) method. Assembly processes operators are identified has the highest risk level. OWAS score, Strain Index, and REBA values are 4, 20.25, and 11. Ergonomic improvements are needed to reduce that level of risk. Proposed improvements will be developed using the Quality Function Deployment (QFD) method applied with Axiomatic House of Quality (AHOQ) and Morphological Chart. As the result, risk level based on OWAS score & REBA score turn out from 4 & 11 to be 1 & 2. Biomechanics analysis of the operator also shows the decreasing values for L4-L5 moment, compression, joint shear, and joint moment strength.

1. Introduction

Work is one cause of injury even though it is not the only major factor. Interaction between occupational factors, relationships with co-workers and supervisors, and organizations is a contributing factor to the occurrence of injury [1]. The phenomenon of work-related injury is still happening. This recommends that the risk of work-related injury is still being a significant study. Several studies on the risk of MSD were applied to several sectors such as health [2], banking in correlation with mental workload [3], fishery [4], butchery [5], etc. Study in South Korea is ensued a conclusion that there is a differences risk distribution between men and women in different work sectors [6]. This research focused on paper pallet assembling operator for consumer goods in one of manufacturing industry in Indonesia. In making pallets for consumer goods, the use of stitching machine is not allowed to assist the gluing process. It must be done manually by the operator using glue auxiliaries to avoid damage to the product packaging due to scratches of the adhesive material used and maintain the safekeeping of the product to consumer.

The paper pallet assembling process for consumer goods is divided into six stages of the process i.e. the gluing process for each corrugated paper sheet using a paint roller, multiple corrugated paper gluing process, gluing process on each curve (end of corrugated paper sheet) using carton as sizing tool, the gluing process on each groove using a clamp as a tool to hold the glued portion, the gluing process on each paper pallet element using a paint roller, and the paper pallet element bonding process. Each assembly process involves work movements related with musculoskeletal risk factors such as bending, squatting, repetitive motion, and contact stress. This problem was evidenced by the results of body map surveys on 10 operators in the assembly department which shows the percentage value of 50% upwards are complaints on the location of the upper neck, lower neck, right shoulder, back, upper right arm, waist, left wrist, right wrist, left ankle, and right ankle.

2. Musculoskeletal Disorder Risk Analysis

The purpose of the postural observation method is to identify tasks or parts of tasks that potentially cause physical stress. The two methods available for postural observation are REBA [7], and OWAS [8]. REBA was developed to assess the type of work posture with data collected in terms of posture, strength, movement or action, repetition, and coupling [7]. The final score is used to give an indication of the level of risk and urgency of action to be taken. The measurement results using different methods will result in different levels of risk, in which each method has different measurements and measures of risk levels. OWAS was first developed as an observational technique to evaluate work posture [8], [9]. Measurements use a posture work analysis chart consisting of values for backs, hands, feet, and weights. The Strain Index is a method used to evaluate a job on the operator to see if the operator is exposed to musculoskeletal risk on the distal upper extremity (DUE). DUE itself can be interpreted on the elbows, wrists and hands. Measurements with strain index use predefined criteria rating, and then the final score can be determined by multiplying the six multipliers according to the criteria rating in which each job is analyzed subjectively by the analyst [10].

The OWAS method has limitations in which the OWAS method is intended to analyze the work posture, but the measurement of the upper limb assessment has a very small scoring method of precision. From the results of the whole analysis to the measurement results of the entire assembly process, there is one assembly operator that shows the highest level of risk in measuring OWAS, SI, and REBA. Unlike other assembly operators, the measurement results on the assembly operator one using OWAS, SI, and REBA show that work postures and work looping have a high level of risk. The results of high measurements on the assembly process is caused by inappropriate work posture, there are many changes in posture or movement. In addition, the sheer number of repetitions, fast movements, and the length of time required by the assembly operator to handle the job can also increase the risk level on the work loop problem. These results indicate that the assembly operator one needs a refit immediately. Operator's work posture should be fixed immediately and need another solution to handle the problem on repetitiveness.

3. Ergonomics Improvement for Reducing Musculoskeletal Disorders Risk

Some design improvements on work tool and facility were proposed using the method of Quality Function Deployment (QFD) as the main reference. QFD is used to build a product planning and development framework based on customer needs. QFD consists of a series of "quality tables" that form the design of the voice of customer down to a comprehensive and detail operational level where HOQ is the most important phase of the QFD process [11]. Axiomatic Design is a design method that aims to define the basis of product development by providing a mapping technique between the functions of product requirements and design parameters. In other words, Axiomatic Design can be considered as a more structured, logical, and developed HOQ (House of Quality) version with a focus on design according to product function [12]

Axiomatic House of Quality (AHOQ) can interpret the voice of customers more structured and assist in the development of design with functional requirements, design parameters, and constraints, so decisions can be made with less risk. This method also allows design changes without affecting other design requirements. The Morphological Chart is a table based on functional analysis to support the development of process steps that will allow the morphological chart to be used to generate an integrated conceptual list (interrelated into one) design solution. The purpose of conceptual design is to describe the sketches of the system as a whole, sketches of important sub-systems, and the rough dimensions of the proposed product. CATIA (Computer Aided Three-Dimensional Interactive Application) was developed by French aircraft manufacturer Avions Marcel Dassault which was originally used for home development but has now been used in various industries to develop products [13].

IOP Conf. Series: Materials Science and Engineering **319** (2018) 012047 doi:10.1088/1757-899X/319/1/012047

Proposed improvements for the assembly operator are determined through several stages:

1. *List Customer Attributes* (CAs)
   From the results of measurements, observations, and direct interviews with company owners, production managers, and some operators in the assembly department can be known the current working conditions and needs of the work station. This data is converted into criteria for required and desired products (CAs).
2. *Convert* CAs *into* FRs
   The result of identification to CAs is changed to Functional Requirements (FRs) to create function domain. The output of this stage shows the CA data converted to FRs for the tabletop and paint roller products there is no ambiguous function so decomposition is not necessary to clarify the results to be achieved.
3. *Identifying Constraints*
   Constraints (constraints) are found on table tops and paint products. The size of the work-piece handled has a size large enough, i.e. 1200 mm x 425 mm. The working conditions do not currently have a fixed work place, so this should be considered in product design where these conditions require that the proposed tool be moved easily without disrupting the surrounding activities but can be used according to the conditions and work-piece. For the paint roller, it takes a convenient handle but is able to reach all parts of the work-piece surface. The function constraints for both products look like this:
   1. FR1 = The minimum working ground size is 1200 mm x 425 mm.
   2. FRb = Able to reach all parts of the surface area of 1200 mm x 425 mm.
4. *Formulation of Design Parameters*
   Design Parameters (DPs) is located in the third domain, the physical domain. This domain contains a list of physical elements that aim to meet the needs of each functional. DP1 and DP2 will meet FR1 and FR2, while DPa and DPb will meet FRA and FRb.
5. *Formulating the Design Matrix and Initial Design*
   The relationship between the functional requirements and the design parameters is described using the design matrix. This relationship aims to identify that there is no violation of the function.
6. *Correlation Matrix*
   Correlation in AHOQ (Axiomatic House of Quality) aims to determine the dependencies between design parameters. This correlation is indicated by a sign (+) or (-) on the roof of AHOQ. In both matrices it appears that there is no correlation between design parameters, meaning that one design does not affect changes in other designs.
7. *Listing of Constraints*
   In the identifying constraints stage, it has been found that the function constraint for the table is in the first function while the functional constraint for paint roller is in the second function.
8. *Evaluation of* AHOQ *Model*
   Evaluation of the AHOQ model is used to ensure that the CAs have been met before entering the conceptual design. The AHOQ model for table and paint rollers has something in common. First, the design matrix shows the ideal matrix (uncoupled design), meaning that each function has a relationship with DPs. Second, there is no dependence (positive or negative dependencies) between DPs. Finally, there are constraints to consider in making the best product design.
9. *Development of Conceptual Design*
   At this stage of development, we were using the morphological chart to select some concepts among several alternative concepts that exist. The result of the combination will be used as comparison of competing product. Based on FR and alternative concepts, six alternative combinations of concepts for each product are produced.
10. *Matrix Screening Concept & Matrix Scoring Concept*
    The alternative choice of concept for table and paint roller is done by direct interview to five main responders i.e. company owner, production manager, and three operators in the assembly department who used or have handled the assembly process one. The combined results of these assessments are used for conceptual selection and select some concepts to continue in concept

development. The concept of the selected table to continue is the concept of 1, 2, 3, and 4, while for the paint roller is the concept of C, E, and F. The scoring concept matrix show that the concept of 1 is selected for the table and the F concept is selected for paint roller to be developed.

11. *Concept Testing*
The results of interviews with the respondents were used for the examination of selected concepts. In the AHOQ method it is allowed a change as each DP responds to each FR. Based on the results of concept testing on the table, it is necessary to add plastic material to coat the wood used as the basis of work. This aims to avoid table damage due to glue material that accidentally fell on the table. For the wheel, you need to add the tool as a brake tool so that the table remains stable when used for work. In the concept test for paint roller no change is required, but for handles will take precedence over rubber materials based on recommendations provided by the Canadian Centre for Occupational Health and Safety (2017) on hand tool design. The design of the selected product for the table used as the proposal is the adjustable table as needed and equipped with a medium rubber wheel with a diameter of 75 mm, while the paint roller has a thin nap and handle type foam with three types of grip for close, medium and, far.

Illustrations for working conditions as before and after proposed improvements are represented using CATIA software. From the picture, there is a change of work posture performed by the operator while doing the assembly process. The results of the identification of working conditions after the proposed improvement can be concluded that the proposed improvement provides a decrease in the risk level of musculoskeletal disorders. This is evidenced by the decreasing OWAS and REBA score. The new OWAS and REBA score was become 1 and 2. It could be interpreted as low risk and no action changes on posture are required. Improvements to work posture are also related with a decreased level of risk caused by repetition of work, because one way that can be done to reduce the risk of repetitive motion over a long period of time is to change posture with different movements and allow time for stretching. Biomechanical analyses of operators are also conducted for working conditions before and after repair using CATIA software. This analysis is conducted to determine whether there are changes in L4-L5 Moment, Compression, Joint Shear, and Joint Moment Strength values. The analysis results for L4-L5 Moment and Compression can be seen in Table 1.

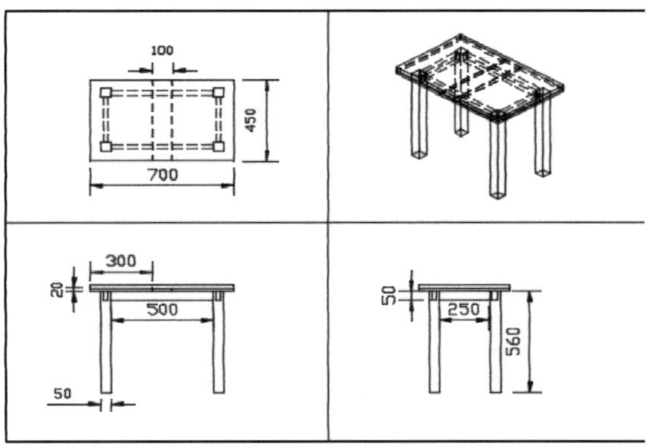

Figure 1. Assembly table

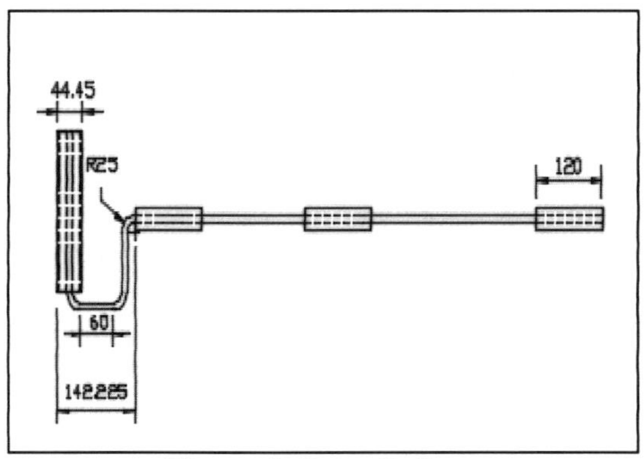

Figure 2. *Paint Roller*

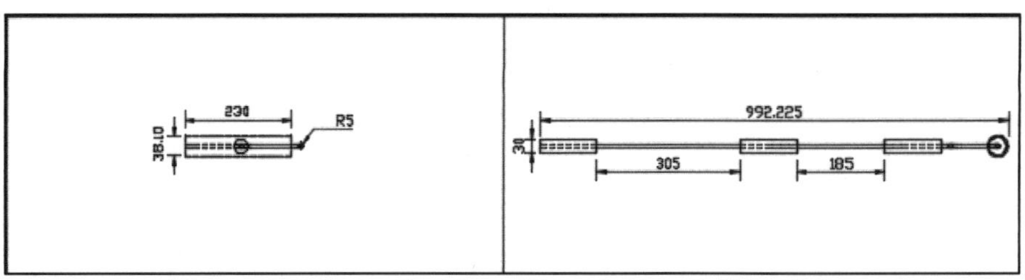

Figure 3. Paint Roller Front View (left) and Side (right)

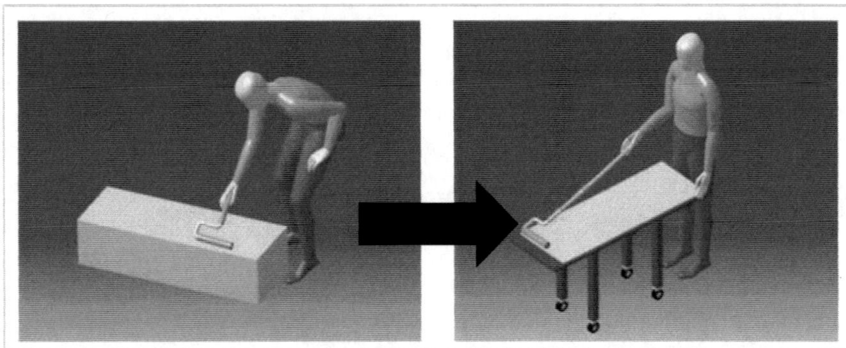

Figure 4. Ilustration of working posture improvement using CATIA

Table 1. L4-L5 Moment, Compression, Joint Shear, and Joint Moment Strength values

Before: Analysis	Value	After: Analysis	Value
L4-L5 Moment [Nxm]	183	L4-L5 Moment [Nxm]	5
L4-L5 Compression [N]	3433	L4-L5 Compression [N]	545
Body Load Compression [N]	222	Body Load Compression [N]	412
Axial Twist Compression [N]	77	Axial Twist Compression [N]	1
Flex/Ext Compression [N]	3055	Flex/Ext Compression [N]	80

From the results of the analysis for moment and compression, it can be seen that there is a decrease in value for both. For compression there is an increase in value for body load compression. This is due to the pressure of pressure between gravity on the body weight and the reaction force of the floor that encourages each other to maintain balance, resulting in a decreased risk for L4-L5.

Figure 5 provides information on whether the values for compression and shear stress are within the limits recommended by NIOSH and the University of Waterloo. There are two defined limits: Action Limit and Maximum Permissible Limit. Based on the graph, it can be seen that there is a decrease in value for compression and shear stress after repair. Both values are within the safe limits, meaning no further action is required for Action Limits and still within the limit of the compressive force set for Maximum Permissible Limit. These data indicate that the after-repair condition has a decrease in population percentage for lumbar strength (L4-L5). This decrease is caused by posture changes, so the work handled does not require lumbar strength (L4-L5). Figure 6 shows the analysis results of joint moment strength for lumbar (L4-L5). It can be compared with population percentage.

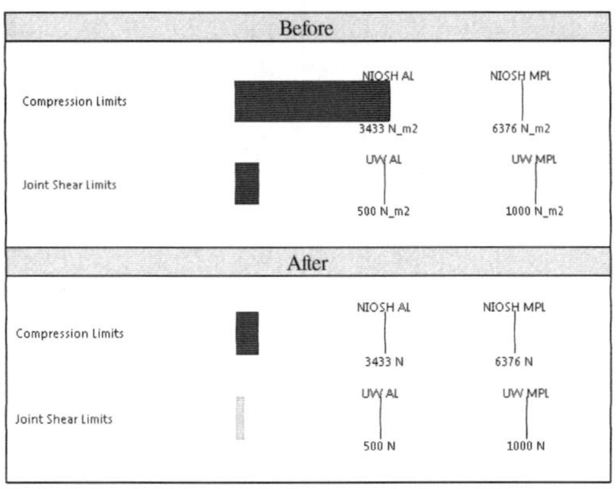

Figure 5. Compression Limits & Joint Shear Limits Chart

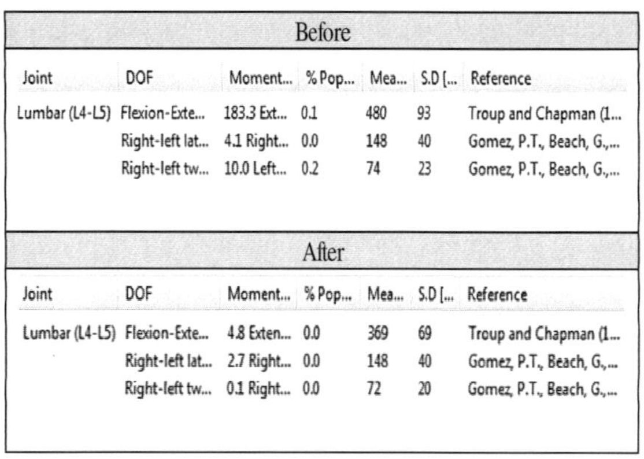

Before

Joint	DOF	Moment...	%Pop...	Mea...	S.D [...	Reference
Lumbar (L4-L5)	Flexion-Exte...	183.3 Ext...	0.1	480	93	Troup and Chapman (1...
	Right-left lat...	4.1 Right...	0.0	148	40	Gomez, P.T., Beach, G.,...
	Right-left tw...	10.0 Left...	0.2	74	23	Gomez, P.T., Beach, G.,...

After

Joint	DOF	Moment...	%Pop...	Mea...	S.D [...	Reference
Lumbar (L4-L5)	Flexion-Exte...	4.8 Exten...	0.0	369	69	Troup and Chapman (1...
	Right-left lat...	2.7 Right...	0.0	148	40	Gomez, P.T., Beach, G.,...
	Right-left tw...	0.1 Right...	0.0	72	20	Gomez, P.T., Beach, G.,...

Figure 6. Joint Moment Strength Data

IOP Conf. Series: Materials Science and Engineering **319** (2018) 012047 doi:10.1088/1757-899X/319/1/012047

4. Conclusion

Posture problems that result in the risk of muscle and skeletal injury can be solved by redesigning work facilities. The proposed of ergonomics improvement for assembling department provide a better condition for workers with a lower risk of musculoskeletal disorder. It can be proven by the decrease of risk level based on OWAS score & REBA score turn out from 4 & 11 to be 1 & 2. Biomechanics analysis of the operator also shows the decreasing values for L4-L5 moment, compression, joint shear, and joint moment strength.

References

[1] I. D. Cameron, "How to manage musculoskeletal conditions: When is 'Rehabilitation' appropriate?," *Best Pract. Res. Clin. Rheumatol.*, vol. 18, no. 4, pp. 573–586, 2004.

[2] H. R. Zaker Jafari and M. H. Yekta-Kooshali, "Work-Related Musculoskeletal Disorders in Iranian Dentists: A Systematic Review and Meta-analysis," *Saf. Health Work*, no. July, pp. 1–9, 2017.

[3] E. Darvishi, A. Maleki, O. Giahi, and A. Akbarzadeh, "Subjective Mental Workload and Its Correlation with Musculoskeletal Disorders in Bank Staff," *J. Manipulative Physiol. Ther.*, vol. 39, no. 6, pp. 420–426, 2014.

[4] Q. Sholihah, A. S. Hanafi, A. A. Bachri, and R. Fauzia, "Ergonomics Awareness as Efforts to Increase Knowledge and Prevention of Musculoskeletal Disorders on Fishermen," *Aquat. Procedia*, vol. 7, pp. 187–194, 2016.

[5] B. Kaka, O. A. Idowu, H. O. Fawole, A. F. Adeniyi, O. O. Ogwumike, and M. T. Toryila, "An Analysis of Work-Related Musculoskeletal Disorders Among Butchers in Kano Metropolis, Nigeria," *Saf. Health Work*, vol. 7, no. 3, pp. 218–224, 2016.

[6] J. Park, Y. Kim, and B. Han, "Work Sectors with High Risk for Work-Related Musculoskeletal Disorders in Korean Men and Women," *Saf. Health Work*, pp. 4–7, 2017.

[7] S. Hignett and L. McAtamney, "Rapid Entire Body Assessment (REBA)," *Appl. Ergon.*, vol. 31, no. 2, pp. 201–205, 2000.

[8] O. Karhu, P. Kansi, and I. Kuorinka, "Correcting working postures in industry: A practical method for analysis," *Appl. Ergon.*, vol. 8, no. 4, pp. 199–201, 1977.

[9] O. Karhu, R. Härkönen, P. Sorvali, and P. Vepsäläinen, "Observing working postures in industry: Examples of OWAS application," *Appl. Ergon.*, vol. 12, no. 1, pp. 13–17, 1981.

[10] J. S. Moore and A. Garg, "The strain index - A proposed method to analyze jobs for risk of digital upper extremity disorders," *AM. IND. HYG Assoc.*, no. 56, pp. 443–458, 1995.

[11] Y. Akao and G. Mazur, "The Leading Edge in QFD: Past, Present and Future," *Int. J. Qual. Reliab. Manag.*, vol. 20, no. 1, pp. 20–35, 2003.

[12] N. Machulenko, "Applying Axiomatic Design Principles to The House of Quality," 2001.

[13] Y. Haik and Shahin. Tamer M., *Engineering Design Process 2nd Ed.* Stanford, USA: CENGAGE Learning, 2011.

Comparison of Two Buyer-Vendor Coordination Models

Ririn Diar Astanti[1*], The Jin Ai[1], Dah-Chuan Gong[2,3] and Hunyh Trung Luong[4]
[1]Department of Industrial Engineering, Universitas Atma Jaya Yogyakarta, Yogyakarta, Indonesia
[2]Department of Industrial and Business Management, Chang Gung University, Guishan Dist., Taoyuan City 33302, Taiwan, ROC
[3]Supply Chain Division, Chang Gung Memorial Hospital, Guishan Dist., Taoyuan City 33302, Taiwan, ROC
[4]Department of Industrial and Manufacturing Engineering, Asian Institute of Technology, Bangkok, Thailand

*ririn@staff.uajy.ac.id

Abstract. This paper develops and compares two mathematical models for describing situation in coordination of buyer and vendor. In this case the vendor which is an Original Equipment Manufacturers (OEMS) of automotive parts, are supplying different type of buyers, i.e. automotive industry, repair shop and automotive dealers. It is well known that automotive industries are operated in Just in Time (JIT) Production Environment, so that the demand behaviour from this buyer has different characteristics than the demand behaviour from other buyers. Two mathematical models are developed in order to depict two different manufacturing strategies as the vendor response dealing with different type of buyers. These strategies are dividing production lot size for each type of buyer and consolidating all buyer's demand in to single production lot size.

1. Introduction and Literature Review

Nowadays, a company needs to perform activities in their business processes efficiently. This is done so that the company can produce products or services with better quality, cheaper price and faster delivery. A company that produce tangible product is called as manufacturing company. Manufacturing companies interacts with external entities such as suppliers. Suppliers have a role in providing raw materials. In addition, manufacturing companies also interact with other external entities i.e. distribution centre. Distribution centre helps company in distributing products to reach their targeted customers. External entities such as suppliers and distributions are usually separate from manufacturing companies. In addition, different parties usually own them. Therefore, each party might have different ways of managing their company.

At the past each company try to focus on how to improve the performance of their internal operations [1]. Because each company is independent, it may occur that the objective of a company is conflicting with the goals of the other company. For example, a manufacturing company has an economical quantity of orders, but this quantity might economical for the supplier side. In the era of Supply Chain Management (SCM), all activities in each supply chain member ranging from suppliers, manufacturers, to distribution centres must be integrated. Therefore, there exists a need to have a coordination mechanism to align objective of each supply chain member to improve the performance of the whole systems [1]. Buyer-vendor coordination is one of activity that can be done in order to improve the performance of the whole system. A buyer can be a manufacturer while the vendor is their suppliers. Or a buyer can be a distribution centre or retail while their vendor is a manufacturer. According to [2] replenishment decision at each member in supply chain can be coordinated in three ways: 1) using quantities discount so that it can attract the buyer to order with the quantity that is profitable from the supplier point of view [3, 4, 5, 6]; 2) synchronizing order from multiple buyers [7]; and 3) information sharing [8, 9]. The first two ways are related to the vendor's perspective coordination model or according to [10] it is called as decentralized model. As opposite of decentralized model is centralized model, where in the centralized model the decision regarding the lot size is determined by both vendor and buyer. However according to [10] this type of coordination model is difficult to apply due to incentive conflict.

In reality there exists a situation where the buyer operates in Just-in-Time (JIT) environment for example in several automotive manufacturers. When the buyer operates in JIT environment usually they prefer that the vendor delivers the item in the smaller quantity and more frequent delivery [11, 12]. To the best of author knowledge, the research on vendor perspective model for decentralized buyer-vendor coordination model where buyer operates in JIT environment was firstly conducted by [5, 6]. The motivation of their research is because when the buyer operates in JIT environment the vendor has to follows buyer's lot size or according to [13] it is said as buyer dominance. In addition, according to [14] sometimes it is un-economical. Therefore, the vendor need to find pricing strategy dealing with this situation to maintain the vendor's target profit. Other form of coordination mechanism in the decentralized model is what it is called as vendor-managed inventory (VMI) [15]. In VMI, vendor takes replenishment decision [15]. This can be happened because the buyer supports the vendor with real time data of inventory [16].

Research on a vendor and multiple buyers coordination models were conducted by several researchers in the past where the buyer can be categorized as homogeneous buyers and heterogeneous buyer. As in reality it is common that each buyer has different characteristics with others such as demand rate, holding cost and order cost, therefore we limit to review the coordination model to a vendor and multiple heterogeneous buyers. Research to determine optimal pricing policies for the case of one vendor multiple heterogeneous buyers was conducted by [17] who discussed about developing discount pricing structure. Other research such as [18] also conducted the research on coordination mechanism for one vendor multiple buyers when a vendor offers a single quantity discount schedule to many buyers.

The research in this paper is motivated by real case study of an automotive part manufacturer that supply a product to:1) automotive manufacturers; 2) repair shop and 3) automotive dealer. Demand rates for automotive manufacturer is different with that of repair shop and automotive dealer. Therefore, according to [1] it is categorized as a manufacturers and multiple buyers coordination models. In addition, the automotive manufacturer operates in JIT environment.

The remainder of the paper is organised as follows: Section 2 presents about Problem Statement and Mathematical Model followed by numerical example in Section 3 to illustrate the applicability of the proposed model. Some concluding remarks is presented in Section 4.

2. Problem and Mathematical Formulation

In the research presented in this paper, a real case study is presented where we consider a manufacturer of spare part who produce the product and sell it to two different types of buyers as it is illustrated in Figure 1. The 1st type of buyer is automotive manufacturer who operates in JIT environment. The 2nd type of buyer is retail which are repair shop and dealer. Recently, the vendor's production lot size is following the production lot size of the automotive manufacturer. This happens because the bargaining power of first type of buyer which is automotive manufacturer is strong. This can happen because the automotive industry can guarantee the continuity of the order to the vendor. For the 2nd type of buyer which is repair shop and dealer, a make-to-order contract is implemented, where in this case there are two alternatives that a vendor can do in order to fulfil the order for the second type of buyer. First, the vendor can do demand consolidation. It means that when the demand from the repair shop or dealer come, the vendor will not process it directly however they will wait for another demand from other retailer or repair shop until it reaches its economic production quantity

(Model 1). Second, vendor will fulfil the order immediately after they receive the order (Model 2). The decision that has to be made by the vendor is that about the selling price that they have to offer for each type of buyer.

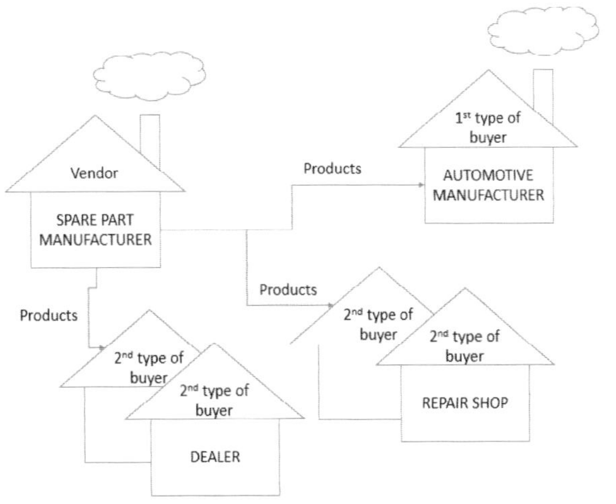

Figure 1. Problem Illustration

Notation:

D_1 : demand parts for the 1st buyer, units/time
D_2 : demand parts for the 2nd type buyers, units/time
Q_1 : production lot size for the 1st buyer, units
Q_2 : production lot size for the 2nd type buyers, units
P : vendor's production rate for fulfilling the demand, units/time
S_1 : 1st buyer's unit setup cost, \$
S_v : vendor's unit setup cost, \$
h_1 : 1st buyer's unit holding cost, \$/unit/time
h_v : vendor's unit holding cost, \$/unit/time
C_1 : vendor's unit selling price for 1st buyer, \$/unit
C_v : vendor's unit production cost, \$
τ_1 : production up time for fulfilling the demand from the 1st buyer
τ_2 : production up time for fulfilling the demand from the 2nd buyer and 3rd buyer

2.1. Model 1

This model presents a situation when a vendor which is an Original Equipment Manufacturer (OEM) supply the product to two type of buyers. They are: 1) automotive manufacture (1st type of buyer); 2) repair shop and dealer (2nd type of buyer). The 1st type of buyer is operated in JIT environment with these characteristics: 1) to meet the demand from the buyer then a vendor produces in a batch of Q_1 at a rate of P unit per time; 2) lot for lot shipment is applied. For 2nd type of buyer (repair shop and dealer), which are not operating in JIT environment, the vendor is free to decide the value of Q_2 (production lot size for the 2nd type of buyer in unit). Vendor produces with 2 separate production lots by assuming that:

$$T_2 = nT_1 \tag{1}$$

subject to:

$$\tau_1 + \tau_2 \le T_1 \tag{2}$$

where:

n : an integer multiplier

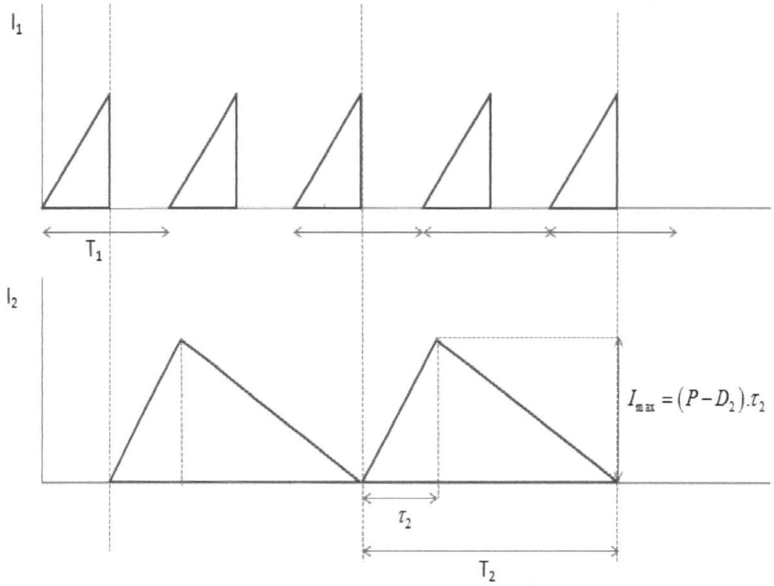

Figure 2. Inventory level of finished goods at vendor (Model 1)

Based on Figure 2 above it can be developed two expression of total cost. They are 1) total production cost for fulfilling demand of the 1st type of buyer (TC_1); 2) total production cost for fulfilling demand of the 2nd type of buyer (TC_2). The expression of TC_1 and TC_2 are derived through this following steps:

1. Finding the lot size in the buyer side for the 1st type of buyer
   To find the lot size in the buyer side, the expression of total cost described by [5] is used:
$$TC_{1B} = D.C_1 + DQ_1^{-1}.S_1 + \frac{1}{2}.Q_1.h_1.C_1 \tag{3}$$
   then the expression of lot size in the buyer side can be derived as follows:
$$Q_1^* = \left(\frac{2.D.S_1}{h_1.C_1}\right)^{1/2} \tag{4}$$
2. Find the expression of T_1 as follows:
$$T_1 = \frac{Q_1^*}{D_1} \tag{5}$$
   Then, subtituting equation (5) to equation (1) the expression of Q_2 can be expressed as:
$$Q_2 = n.\frac{Q_1^*}{D_1}.D_2 \tag{6}$$
3. Find the expression of TC_1
$$TC_1 = \frac{D_1}{Q_1^*}.S_v + \frac{1}{2}.\frac{(P - D_1)}{T_1}\tau_1^2.h_v.C_v \tag{7}$$
$$TC_1 = \frac{D_1}{\left(\frac{2.D.S_1}{h_1.C_1}\right)^{1/2}}.S_v + \frac{1}{2}.\frac{(P - D_1)}{T_1}\tau_1^2.h_v.C_v \tag{8}$$
4. Find TC_2
$$TC_2 = \frac{D_2}{Q_2}.S_v + \frac{1}{2}.I_{max}.h_v.C_v \tag{9}$$

$$TC_2 = \frac{D_1}{n.Q_1^*}.S_v + \frac{1}{2}.(P - D_2)\frac{n.\frac{Q_1^*}{D_1}.D_2}{P}.h_v.C_v \tag{10}$$

2.2. Model 2

In this model, the vendor produces product for the 1st type of buyer and 2nd type of buyer simultaneously. Inventory model of Model 2 is presented in Figure 3. Similar with Model 1, the cycle time T_1 is obtained from the perspective of 1st type of buyer.

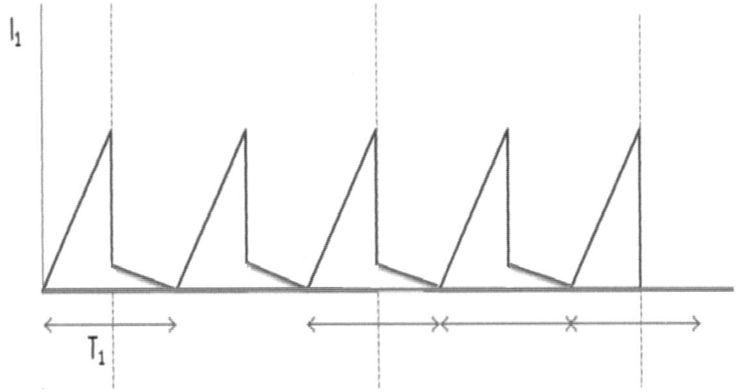

Figure 3. Inventory level of finished goods at vendor (Model 2)

Therefore, the production lot size of this model can be stated as

$$Q = (D_1 + D_2)T_1 \tag{11}$$

and the length of production time in each cycle can be calculated as

$$\tau = \frac{Q}{P} = \frac{D_1 + D_2}{P}T_1 \tag{12}$$

Hence, the total cost of this model can be written as

$$TC = \frac{S_v}{T_1} + \frac{1}{2}\left\{\left(\frac{D_1 + D_2}{P}\right)^2 (P - D_2)T_1 + \left((P - D_2)\frac{D_1 + D_2}{P} - D_1\right)\left(T_1 - \frac{D_1 + D_2}{P}T_1\right)\right\}h_v \tag{13}$$

3. Numerical Example

A numerical example represents the real situation when the vendor which is an OEM that produces the automotive part that has longer end-of-life with following parameters:

D_1 = 10,000 units/year
D_2 = 1,000 units/year
P = 20,000 units/year
S_1 = 50\$
S_v = 500 \$
h_1 = 10%\$/\$ inventory
h_v = 10% \$/\$ inventory
C_1 = 87 \$
C_v = 70 \$

For Model 1, following equation (4) one can obtain Q_1^* = 339.03 units. Therefore, it is obtained T_1^* = 0.033903 year from equation (5) and TC_1 = \$15044.53/year from equation (8). Using iterative procedure, it is obtained that n^* = 10. It is also calculated that TC_2 = \$2602.07/year from equation (10). Therefore, total cost for Model 1 is $TC=TC_1+TC_2$. The value of TC= \$17646.60.

For Model 2, following equation (13) total cost for Model 2 TC = \$15453.91. Therefore, TC Model 2 is less than that of Model 1.

4. Concluding Remarks

The research presented here are mainly about the buyer vendor coordination for an OEM manufacturer that support the component to 2 types of buyers which are: 1) automotive manufacturer and 2) repair shop and dealer. The automotive manufacturers operate in JIT environment. Two model that represents two manufacturing strategies are proposed. They are: 1) dividing production lot size for each type of buyer (Model 1); 2) consolidating all buyer's demand in to single production lot size. Based on the total cost (TC) obtained from those models, the OEM can decide which strategy provides lower operational cost.

Acknowledgment

This work is partially supported by Ministry of Research, Technology, and Higher Education, Republic Indonesia under International Research Collaboration and Scientific Publication Research Grant No. 005/HB-LIT/III/2016 and No. 045/HB-LIT/IV/2017.

References

[1] Li X and Wang Q 2007 *European Journal of Operational Research* **179** 1–16
[2] Zipkin P 2000 *Foundations of Inventory Management.* McGraw-Hill
[3] Chen F, Drezner Z, Ryan JK and Simchi-Levi D 2000 *Management Science* **46** 436–43
[4] Wang Q 2005 *Decision Sciences* **36** 627–46
[5] Banerjee A 1986 *Management Science* **32** 1513–17
[6] Monahan JP 1984 *Management Science* **30** 720–6
[7] Klastorin TD, Moinzadeh K and Son J 2002 *IIE Transactions* **34** 679–89
[8] Lee HL, So KC and Tang CS 2000 *Management Science* **46** 626–43
[9] Moinzadeh K 2002 *Management Science* **48** 414–26
[10] Toptal A and Çetinkaya S 2008 *European Journal of Operational Research* **187** 785–805
[11] Khan LR and Sarker RA 2002 *Computers & Industrial Engineering* **42** 127–36
[12] Wang IC 2010 *Expert System with Applications* **37** 2117–23
[13] Kelle P and Miller PA 1998 *International Journal of Purchasing and Materials Management* **34** 25–30
[14] David I and Eben-Chaime M 2003 *International Journal of Production Economics* **81-82** 361–68
[15] Chakraborty A, Chatterjee AK, and Mateen A 2015 *International Journal of Production Research* **53** 13–24
[16] Nagarajan M and Rajagopalan 2008 *Production and Operations Management* **54** 1453–66
[17] Lal R and Staelin R 1984 *Management Science* **30** 1524–39
[18] Drezner Z and Wesolowsky GO 1989 *European Journal of Operation Research* **40** 38–42

Improving Quality of Shoe Soles Product using Six Sigma

Athalia Jesslyn Wijaya, Wildan Trusaji*, Muhammad Akbar, Anas Ma'ruf and Dradjad Irianto

Faculty of Industrial Technology, Institut Teknologi Bandung, Bandung 40132, Indonesia

E-mail: wildan@mail.ti.itb.ac.id

Abstract. A manufacture in Bandung produce kind of rubber-based product i.e. trim, rice rollers, shoe soles, etc. After penetrating the shoe soles market, the manufacture has met customer with tight quality control. Based on the past data, defect level of this product was 18.08% that caused the manufacture's loss of time and money. Quality improvement effort was done using six sigma method that included phases of define, measure, analyse, improve, and control (DMAIC). In the design phase, the object's problem and definition were defined. Delphi method was also used in this phase to identify critical factors. In the measure phase, the existing process stability and sigma quality level were measured. Fishbone diagram and failure mode and effect analysis (FMEA) were used in the next phase to analyse the root cause and determine the priority issues. Improve phase was done by designing alternative improvement strategy using 5W1H method. Some improvement efforts were identified, i.e. (i) modifying design of the hanging rack, (ii) create pantone colour book and check sheet, (iii) provide pedestrian line at compound department, (iv) buying stop watch, and (v) modifying shoe soles dies. Some control strategies for continuous improvement were proposed such as SOP or reward and punishment system.

Keywords: Six Sigma, DMAIC, Delphi Method, FMEA, 5W1H

1. Introduction

A make-to-order (MTO) manufacture produces some based-rubber products i.e. trim, rice rollers, shoe soles, etc. The manufacture has been rapidly growth since 1995 characterized by an increasing number of production, labours, and product types. The manufacture has a vision to be the number one in Indonesia's market shares. One of its strength is that it has entered the Original Equipment Manufacturing (OEM) market. In 2015, this manufacture created a contract of shoe soles product with a loyal customer who implements tight quality control. Based on secondary data, defect level of this product was 18.08% that needed to be reworked. The rework process caused the manufacture experiencing loss in time of 18.3 days of work and money about 49 million rupiahs.

To reduce the number of rework, the manufacture implement various quality improvement efforts for example recruiting new personnel who expert in quality to perform hardness test, quality analysis, etc. Moreover, this proposed research is also an effort to improve the shoe soles products in the manufacture by implementing six sigma methodology to improve the product quality continuously. This method was chosen because the root cause of higher reject rate was higher variability of product. Therefore, six sigma method which can reduce process variance is suited implementing in this kind of quality problem.

Although six sigma is possible to be implemented in high variance process, but the key factor of this method is the involvement of all stakeholder related to the project [1]. Six sigma methodology implements DMAIC (define, measure, analyse, improve, and control) cycle that is often used for continuous improvement strategy [2].

Six sigma methodology has been successfully implemented in many world class manufacture in the world such as Samsung, Kodak, General Electric, IBM, DEC, Sony, Texas Instrument, Nokia, Philips Electronic, and LG [3]. Six sigma methodology can be also implemented in industry with low level production or service sector. We also found a research which implement sig sigma to improve quality of seal leak test product in a manufacture [5].

2. Methodology

We follow DMAIC cycle to generate quality improvement on shoe soles product. Many tools are possible to be implemented in each phase of six sigma. SIPOC (supplier-input-process-output-customer) diagram, CTQ (critical to quality) matrix, or stakeholder analysis are some tools that can be used in defining of quality project. While statistics quality control such as descriptive or process capability is rapidly used in measure phase of six sigma. There are many usable tools in analyse phase such as cause and effect diagram, FMEA, value stream map, ANOVA, design of experiment, etc. Then, creative technique, 5S, or theory of constraint (TOC) are feasible implemented in improve phase. Finally, in control phase, project documentation or control chart are important tools [4].

In the define phase, our target is to get the quality project aiming to reduce the number of specific reject condition. We interview an expert from the manufacture and get secondary data to identify the product reject history. After discussing, we choose a critical product as an object of this research and characterize its specification e.g. size and colour. SIPOC and flowchart diagrams are used for understanding the production process of the product chosen. In addition, we classify the product reject condition and focus on some critical rejects that contribute to at least 80% of the reject number using Pareto diagram. Finally, we collect critical to quality (CTQ) information by implementing brainstorming and Delphi method to the experts of the manufacture.

Generally, process stability and capability are computed in measure phase of sig sigma. The process stability is mapped using fraction of nonconforming control chart. If the process is not stable, we must find and reduce some assignable causes so the process variance can fulfil the normal assumption. On the other hand, the capability process is used to state the sigma level of the process which can be computed after getting the defect per million opportunities (DPMO) value. If the sigma level is lower than six, some improvements must be applied to the production process.

Finding the correct root cause is the most important thing in the analyse phase. We implement brainstorming method for 4 rounds to identify some critical factors that cause the reject cases. Then, these factors are mapped using fishbone (cause and effect) diagram on several aspects i.e. man, machine, material, method, and environment resulting some root causes. After that, failure mode and effect analysis (FMEA) is applied to rank between the root causes. We choose some root causes to be tackled based on the risk priority number (RPN) that is resulted from FMEA.

To solve the chosen root causes, we design some improvement in the improve phase. Firstly, we generate some solution alternatives considering time and resource constraints. Then, with the experts, we chose the best alternatives as the improvement strategy. The implantation schedule is created in the form of gantt chart. Finally, we implement 5W1H (what, why, who, when, where, how) to make the detail of the solutions.

In control phase, control mechanism is designed to assure that implementation execution is aligned to the implementation plan. We also develop standardization process to make the solution will be implemented continuously. The detail of DMAIC cycle used in this research could be seen in Figure 1.

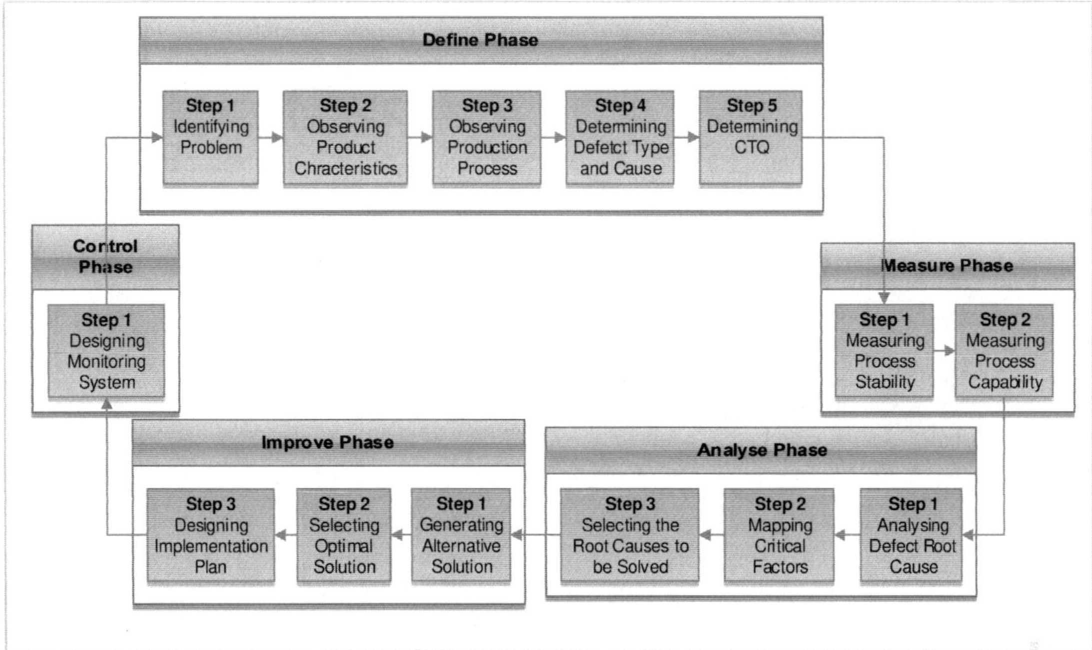

Figure 1. DMAIC Cycle

3. Implementation

This research was conducted at TRC, a manufacture located in Bandung, that produce rubber-based semi-finished and finished products i.e. shoe soles, vulcanization tires, spare part, etc. In the design phase, we chose shoe soles as the research object because the product was requested by the manufacture. It is known that the shoe soles had the second highest of demand with 36.4% of the manufacture's turnover. The customer with highest demand was chosen despite it had the highest number of returned product of 18.08%. The expected loss because of the returned product was about IDR 50 million from December 2015 to Mei 2016. The product characteristics discussed in this study was shoe soles colour (white, gum, and black), size with range of 38 – 44 using European standard, thickness between 0.8 – 1 cm, and hardness between 45 – 50 sore. We also defined supplier, raw materials, and process needed to produce the specific shoe soles ordered by specific customer using SIPOC diagram. We found at least 7 type of defects i.e. (a) contaminated by unwanted colour – 91 cases, (b) hardness is not appropriate – 17 cases, (c) surface eroded – 14 cases, (d) inconsistent colour – 13 cases, (e) contaminated by unwanted objects – 10 cases, (f) deformation – 7 cases, and (g) unsmooth surface – 5 cases. Then, we used Pareto diagram resulting this study only focused on the six first of the defect to tackle 96.82% of all reject cases (Figure 2). From these rejects, we analysed the critical factors by using 4-round Delphi method which delivered 5 CTQ i.e. (a) raw material, (b) operator, (c) press machine, (d) procedure, and (e) work environment.

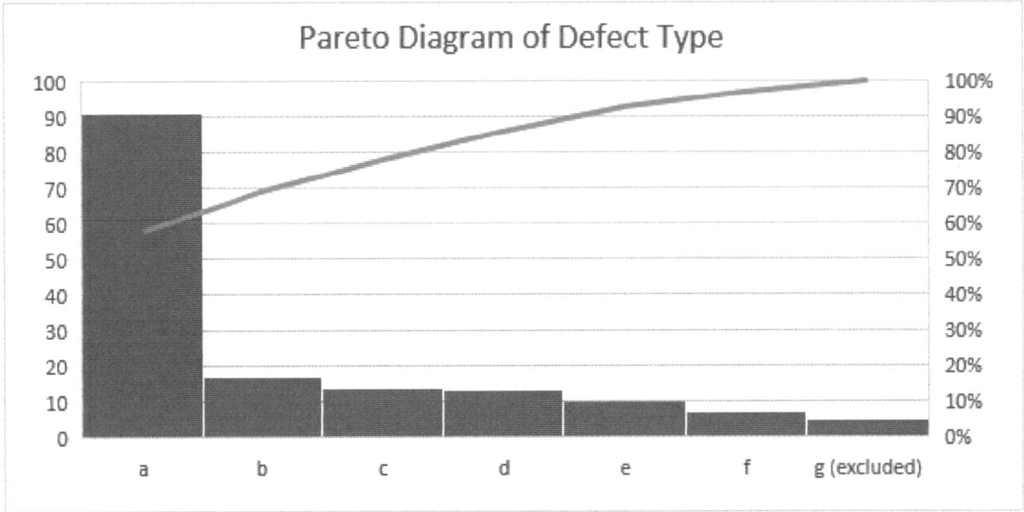

Figure 2. Pareto Diagram of Defect Type

After CTQ had been known, shoe soles production process performance was measured using defect fraction and DPMO to get process stability and process capability respectively. The mapping process was done to two important process linked to all reject types i.e. vulcanization and pressing process. It can be known from historical data that there were some out-of-control events as seen in Figure 3 and the experts identified those as false alarm events resulting the processes can be assumed as normal condition. DPMO of vulcanization and pressing process respectively were 89411.76 and 10359.96 with the sigma score respectively of 2.84σ and 3.81σ. This result shown that some improvement processes were needed to both processes. Moreover, we also computed sig sigma score based on product returns number which was still 3.3σ.

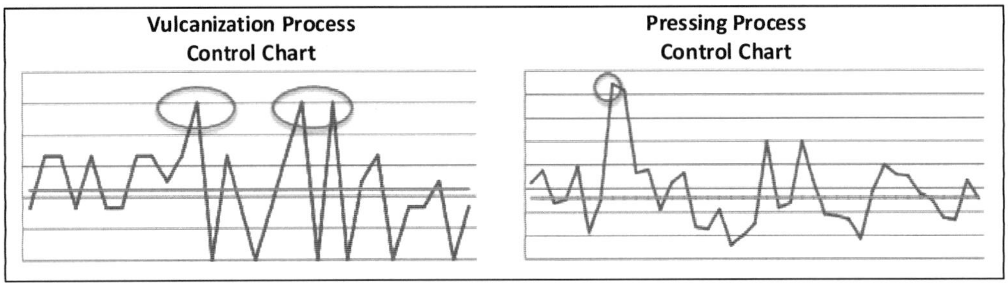

Figure 3. Fraction of Defect Control Chart

In analyse phase, we used brainstorming and Delphi method to find 15 root causes of the six defects. Then, we mapped it to the fishbone diagram for each defect as shown in figure 4. Using FMEA, we computed risk priority number (RPN) considering severity, occurrence, and detectability of each root causes and select the root causes with the RPN score more than 600. So, there were 7 critical root causes to be tackled in this study i.e. (a) hanging shelf near open mill machine, (b) composition of colour pigments, (c) operator on open mill machine, (d) raw material inspection procedure, (e) ripeness inspection process on pressing process, (f) operator on press machine, and (g) shoe soles mould. From these root causes, in improve phase, we generated 6 solutions to be applied at TRC using 5W1H method i.e. (a) hanging shelf modification, (b) pantone colour book, (c) drawing pedestrian line at rubber compound department, (d) make check sheet of pantone colour, (e) buying and using stopwatch, and (f) shoe soles mould modification. Those solutions needed about IDR 15 million to be implemented. The relationship between the root causes and the solutions is shown at Table 1.

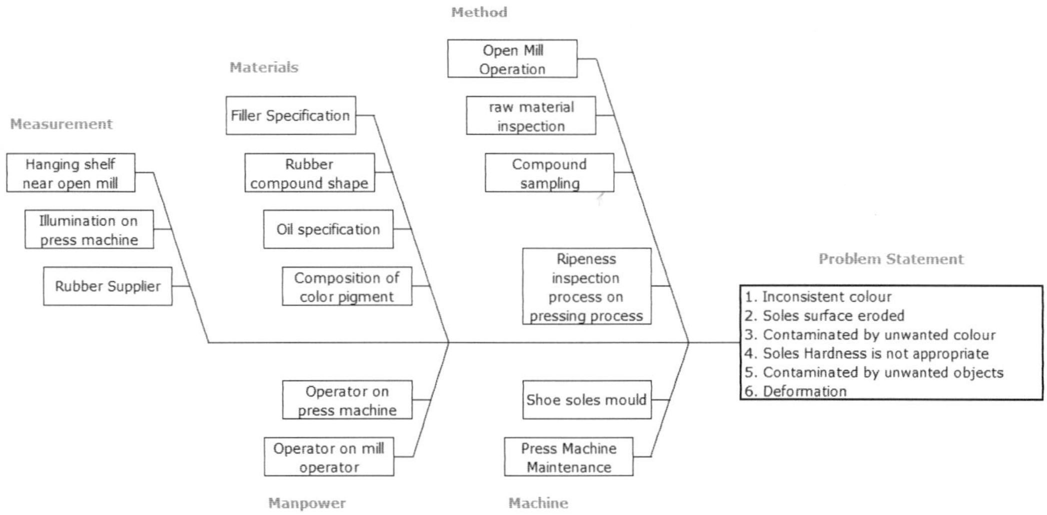

Figure 4. Fishbone Diagram

Table 1. Root Causes and Solution Relationship

Root Causes	Solutions
hanging shelf near open mill machine	hanging shelf modification
composition of colour pigments,	pantone colour book
operator on open mill machine	drawing pedestrian line at rubber compound department
raw material inspection procedure	make check sheet of pantone colour
ripeness inspection process on pressing process	buying and using stopwatch
operator on press machine	shoe soles mould modification
shoe soles mould	

To assure the continuity of the improvement, some monitoring and evaluation strategies in control phase has been proposed which were (a) reward and punishment system for some department heads and open mill machine operator, (b) implementing operator ID card containing pantone colour information, (c) putting warning board near press machine, (d) create SOP of plastic sheet allocation at the bottom of rubber compound, (e) create SOP of raw material inspection and ID card making regulation, (f) create SOP on press machining, and (g) data frequency recording of reject caused by operator on press machine.

4. Conclusion

Shoe soles got the second number as the most reject product at TRC manufacturing. The customer with highest demand of this product with the number of returned product reach 18.08% resulting expected loss about IDR 50 million. To overbear this problem, six sigma methodology consists DMAIC cycle was used. Seven defects were found in the define phase, but we only focus to 6 defects that contributed to 96.82% of cases. There were 5 CTQ that caused the defect i.e. (a) raw material, (b) operator, (c) press machine, (d) procedure, and (e) work environment. In measure phase, DPMO of vulcanization and pressing processes were measured respectively by the value of 89411.76 with the sigma value of 2.84σ and 10359.96 with the sigma value of 3.81σ. After analysis phase, 6 solutions were chosen to tackle 7 critical root causes that had RPN number higher than 600 i.e. (a) hanging shelf modification, (b) pantone colour book, (c) drawing pedestrian line at rubber compound department, (d) make check sheet of pantone colour, (e) buying and using stopwatch, and (f) shoe soles mould modification. Those solutions needed about IDR 15 million to be implemented. To assure the continuity of the improvement, some control strategies were submitted i.e. (a) reward and punishment system for some department heads and open mill machine operator, (b) implementing operator ID card containing pantone colour information, (c) putting warning board near press machine, (d) create SOP

IOP Conf. Series: Materials Science and Engineering **319** (2018) 012049 doi:10.1088/1757-899X/319/1/012049

of plastic sheet allocation at the bottom of rubber compound, (e) create SOP of raw material inspection and ID card making regulation, (f) create SOP on press machining, and (g) data frequency recording of reject caused by operator on press machine.

Acknowledgment
This research was conducted at the Manufacturing System Research Group, funded by P3MI research grant of ITB

References
[1] Kubiak, T., Benbow, D. 2009. *The Certified Six Sigma Black Belt Handbook, 2nd Ed.* Milwaukee: American Society for Quality.
[2] Montgomery, D. C., 2009. *Statistical Quality Control: A Modern Introduction* (7th ed.). Hoboken: John Wiley & Sons.
[3] Park, S. H., 2003. *Six Sigma for Quality and Productivity Promotion*. Tokyo: The Asian Productivity Organization.
[4] Lunau, S., John A, Meran R., Roenpage, O., and Staudter, C. 2006. *Six Sigma+Lean Toolset.* Springer: Berlin
[5] Malik, A.L., Akbar, M., Irianto, D. 2016. IOP Conference Series: Materials Science and Engineering. 114 012054.

IOP Conf. Series: Materials Science and Engineering **319** (2018) 012050 doi:10.1088/1757-899X/319/1/012050

Taguchi experimental design to determine the taste quality characteristic of candied carrot

Y Ekawati and A A Hapsari

Department of Industrial Engineering, Universitas Ma Chung, Malang, Indonesia

yurida.ekawati@machung.ac.id

Abstract. Robust parameter design is used to design product that is robust to noise factors so the product's performance fits the target and delivers a better quality. In the process of designing and developing the innovative product of candied carrot, robust parameter design is carried out using Taguchi Method. The method is used to determine an optimal quality design. The optimal quality design is based on the process and the composition of product ingredients that are in accordance with consumer needs and requirements. According to the identification of consumer needs from the previous research, quality dimensions that need to be assessed are the taste and texture of the product. The quality dimension assessed in this research is limited to the taste dimension. Organoleptic testing is used for this assessment, specifically hedonic testing that makes assessment based on consumer preferences. The data processing uses mean and signal to noise ratio calculation and optimal level setting to determine the optimal process/composition of product ingredients. The optimal value is analyzed using confirmation experiments to prove that proposed product match consumer needs and requirements. The result of this research is identification of factors that affect the product taste and the optimal quality of product according to Taguchi Method.

1. Introduction

Carrot is an agricultural product that needs to be developed to increase its value. The development of carrot products had been studied to find the carrot products which were preferred by consumers [1]. One of the products preferred was candied carrot. The research also resulted in the attributes of consumer needs for the candied carrot. There are nine attributes of consumer needs for candied carrot. The attributes are grouped into two dimensions of quality, taste and texture. The quality dimension assessed in this research is limited to the taste dimension. Five of the nine attributes of consumer needs are the quality of taste, namely no unpleasant taste, no pungent smell, have about the right sweetness level, have about the right sourness level, and no bitter aftertaste. The product design of candied carrot had been developed to find the critical parts in producing candied carrot [2]. Based on the attributes of consumer needs and the product design, then further research was carried out to determine the optimal process and composition of candied carrot. The optimal process and composition need to be found as there is variation in the performance of a product. The variation degrades the product's quality. One of these approaches to determine the optimum is robust design (also known as parameter design or optimization). In the process of designing and developing the innovation product of candied carrot, robust parameter design is done using Taguchi Method. The method focuses in reducing the influence of factors, called noise, that degrade the performance of a product or process [3]. The Taguchi method achieves the robustness to noise through proper settings of certain parameters, called control factors. The improved settings of the parameters are obtained through statistically designed experiments that enable one to study a large number of factors with relatively few experiments.

There are several researches in various food products that had been conducted using Taguchi method [4],[5],[6],[7]. However, most of the researches conducted using Taguchi method for food products used a quality response of products or processes that typically measured quantitatively on interval or ratio scale. The research conducted on candied carrot was based on the attributes of consumer needs which the quality response of the product will be measured qualitatively using organoleptic test. The organoleptic test results data on ordinal scale that needs to be analyzed using an appropriate method to find the robust levels of product and process parameters. The data analysis in this research uses a method proposed by Erdural et al. [8] that offers a simple and effective method for the analysis of categorical response data for robust product or process design.

2. Research Methods

2.1. *Experimental Design Planning*

The planning stage of Taguchi is done by formulating the problem, determining the experimental objectives, choosing quality characteristics, selecting factors that affect the quality characteristics, identifying controlled and uncontrolled factors, determining the number of levels and factor values, and selecting orthogonal arrays to determine the number of experiments to be performed. All these are done by conducting discussions and brainstorming with the help of experts in the field of food technology.

2.2. *Perform Organoleptic Test*

The conducting stage is carried out using organoleptic test. The organoleptic test involves 60 - 90 consumer panelists. The consumer panel is selected based on the commodity's marketing target. The panelist is asked for his/her personal response on the products on a scale of 1 (very unsatisfied) to 5 (very satisfied). The results of these organoleptic tests will then be collected for further processing.

2.3. *Data Processing*

Data processing is done in accordance with the steps of the method proposed by Erdural et al [8]. This research will use data analysis according to the following steps:

1. Generating an appropriate experimental design and collecting data.
2. Fitting an ordinal categorical regression model and calculating event probabilities for each category. Ordinal logistic regression method is used to fit a model that estimates the event probabilities for each category. Ordinal regression model is as follows:

$$Link(P(Y_i \leq j)) = \gamma_j + \beta' X_i \tag{1}$$

$$P(Y \leq j \mid x) = \frac{e^{(\gamma_j + \beta'_1 x_{1i} + \beta'_2 x_{2i} + \ldots + \beta'_k x_{ki})}}{1 + e^{(\gamma_j + \beta'_1 x_{1i} + \beta'_2 x_{2i} + \ldots + \beta'_k x_{ki})}} \tag{2}$$

   where j is the category (j = 0,1,...,J-1), γ_j is the cut point (constant), β is the vector of coefficients and X_i is the vector of the control factors' levels at combination i of the experiment.
3. Estimating expected category for each factor combination. Using factor level combinations in step 1 and estimated event probabilities for each category in step 2, the expected category and the variance for each factor combination i of the experiment are estimated as follows:

$$E(Y_i) = \sum_{j=0}^{j-1} jP(Y_i = j) \tag{3}$$

$$V(Y_i) = E(Y_i^2) - [E(Y_i)]^2 \tag{4}$$

4. Calculating Taguchi's signal-to-noise ratios using $E(Y_i)$ and $V(Y_i)$ calculated in step 3 for each factor combination i. This research uses the S/N Ratio for a larger-the-better type of a response and the formula is:

$$\text{SNR}_i = -10 \times \log\left[\left(\frac{1}{(E[Y_i])^2}\right) \times (1+3) \times \left(\frac{V(Y_i)}{(E[Y_i])^2}\right)\right] \quad (5)$$

5. Determining the optimal factor level using main effect. Finding the optimal factor levels that maximize SNR, hence achieve the minimum variance.

3. Result and Discussion

The attributes of consumer needs for candied carrot used in this research are the attributes to quality of taste, namely 'no unpleasant taste', 'no pungent smell', 'have about the right sweetness level', 'have about the right sourness level', and 'no bitter aftertaste'. These five attributes become the dependent variable in this study. In addition, this research also consider the Indonesian National Standard (SNI) for candied products, namely SNI 01-4443-1998. SNI 01-4443-1998 is an SNI specifically intended for product of candied nutmeg. This standard is used as there is no SNI for candied carrot. There are two important reasons that make the standard composition of the product should be based on SNI. The first reason is the fulfillment of the SNI quality requirement is an obligation to be followed by the products to be marketed in Indonesia. The second reason is that when it comes to food, the quality aspects, the health standard and the safety become the sensitive aspects that consumers pay attention to. Therefore, the fulfillment of food standards can provide a peace of mind to the consumers which results in the increasing trust to the product. So, based on the two reasons, the fulfillment of SNI quality requirements in the process of determining the standard composition of candied carrot is considered necessary. SNI 01-4443-1998 contains references, definitions, quality requirements, sampling methods, testing methods, marking requirements, and packing methods for candied nutmeg. In the quality requirement section, there is a table of quality requirements specifying the standard conditions, moisture content, sugar level, standard of food additives, metal contamination, arsenic content, microbial contents, and mold. According to SNI 01-4443-1998, the minimum sugar level of candied products is 25% with or without food additives permitted in accordance with applicable regulations.

3.1. Experimental Design Planning

Experimental design planning began with determining the production process of candied carrot. The production process used the process resulted from a research of determining the process quality planning for candied carrot [10]. The planning stage continued with discussion and brainstorming sessions with experts. These are the results of the discussion. Factors that affect the taste of candied carrot are materials, processes, machinery and people. The material factors consist of the amount of sugar, the amount of cinnamon, and the amount of citric acid. Process factors consist of the roasting duration and the marinating duration. Engine factor consists of roasting temperature consistency. As for the human factor consists of the expertise of workers. There are three control factors that can affect the taste of candied carrot. These factors are the amount of sugar (A), the marinating duration (B), and the amount of cinnamon (C). While the uncontrolled factor chosen is the age of carrot (D). Factor A and B will each be given 3 levels, while factor C and D will each be given 2 levels. Based on the discussion, the levels for each factor are as follows:

Table 1. Levels for Each Factor

Factors	Levels		
	1	2	3
Amount of sugar (A)	• 15% • 25% • 40% • 50%	• 25% • 40% • 50% • 60%	• 35% • 50% • 60% • 70%
Marinating duration (B)	7 hours	8 hours	9 hours
Amount of cinnamon (C)	10 grams	30 grams	
Age of carrot (D)	3,5 months	5 months	

The attributes of consumer needs included in the experimental design are 'no unpleasant taste' (A1), 'no pungent smell' (A2), 'have about the right sweetness level' (A3), 'have about the right sourness level' (A4), and 'no bitter aftertaste' (A5). The determination of the number of experiments is determined by orthogonal arrays. Based on the orthogonal array it was found that the number of experiments was 18. Based on those combinations, eighteen candied carrot experimental samples are made and tested on 60 consumers.

3.2. Data Processing

The data obtained from the organoleptic test will then be tested using ordinal logistic regression method with the help of Minitab 17. This test is conducted to all five attributes of consumer needs. As an example, the calculation of the attribute A1 (no unpleasant taste) is shown on this paper. The analysis results of the attribute A1 show that factor A and C are statistically significant. The logistic regression equations for the event probabilities of the categories are given as:

$$\text{Logit } P(Y \leq 1 \mid x) = -2{,}9667 - 0{,}09015A + 0{,}129529C$$
$$\text{Logit } P(Y \leq 2 \mid x) = -1{,}08564 - 0{,}09015A + 0{,}129529C$$
$$\text{Logit } P(Y \leq 3 \mid x) = 0{,}396992 - 0{,}09015A + 0{,}129529C$$
$$\text{Logit } P(Y \leq 4 \mid x) = 2{,}47472 - 0{,}09015A + 0{,}129529C$$

For each experiment trial, the event probabilities of each category are calculated using equation 2. Using equations 3 and 4, expected category and variance from this category are obtained. Then, SNR can be calculated by using equation 5. Table 2 shows results for each experiment trial.

Table 2. Estimates for Probabilities of Categories, Expected Category, Category Variance and SNR of the attribute A1 (no unpleasant taste)

Factors			$P(Y_i = j)$					$E(Y_i)$	$V(Y_i)$	SNR
A	B	C	j=1	j=2	j=3	j=4	j=5			
1	1	1	0,05082	0,20912	0,34745	0,31773	0,07488	3,15672	1,00507	8,83666
1	2	1	0,05082	0,20912	0,34745	0,31773	0,07488	3,15672	1,00507	8,83666
2	1	1	0,04664	0,19634	0,34272	0,33293	0,08136	3,20604	0,99885	9,00834
2	2	1	0,04664	0,19634	0,34272	0,33293	0,08136	3,20604	0,99885	9,00834
3	1	1	0,04279	0,18399	0,33689	0,34797	0,08836	3,25511	0,99150	9,17677
3	2	1	0,04279	0,18399	0,33689	0,34797	0,08836	3,25511	0,99150	9,17677
1	3	1	0,05082	0,20912	0,34745	0,31773	0,07488	3,15672	1,00507	8,83666
2	3	1	0,04664	0,19634	0,34272	0,33293	0,08136	3,20604	0,99885	9,00834
3	3	1	0,04279	0,18399	0,33689	0,34797	0,08836	3,25511	0,99150	9,17677
1	1	2	0,05744	0,22818	0,35219	0,29580	0,06638	3,08550	1,01198	8,58442
2	1	2	0,05275	0,21484	0,34915	0,31106	0,07219	3,13511	1,00743	8,76066
3	1	2	0,04843	0,20187	0,34493	0,32631	0,07847	3,18452	1,00171	8,93374
1	2	2	0,05744	0,22818	0,35219	0,29580	0,06638	3,08550	1,01198	8,58442
2	2	2	0,05275	0,21484	0,34915	0,31106	0,07219	3,13511	1,00743	8,76066
3	2	2	0,04843	0,20187	0,34493	0,32631	0,07847	3,18452	1,00171	8,93374
1	3	2	0,05744	0,22818	0,35219	0,29580	0,06638	3,08550	1,01198	8,58442
2	3	2	0,05275	0,21484	0,34915	0,31106	0,07219	3,13511	1,00743	8,76066
3	3	2	0,04843	0,20187	0,34493	0,32631	0,07847	3,18452	1,00171	8,93374

3.3. Determining the Optimal Composition

The optimal level is determined by the highest average of SNR on the overall quality attributes.

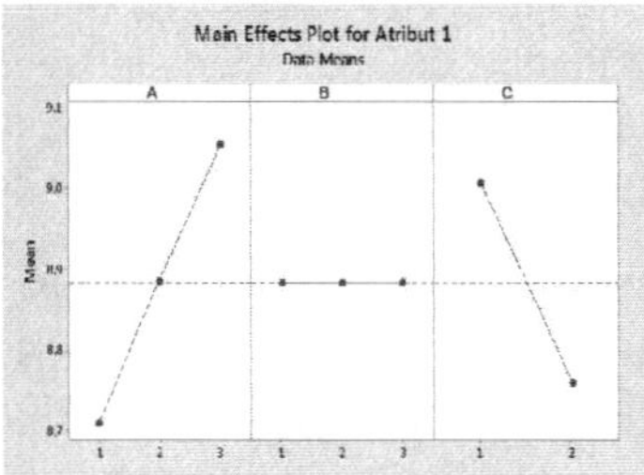

Figure 1. Main effects plot for A1 (no unpleasant taste)

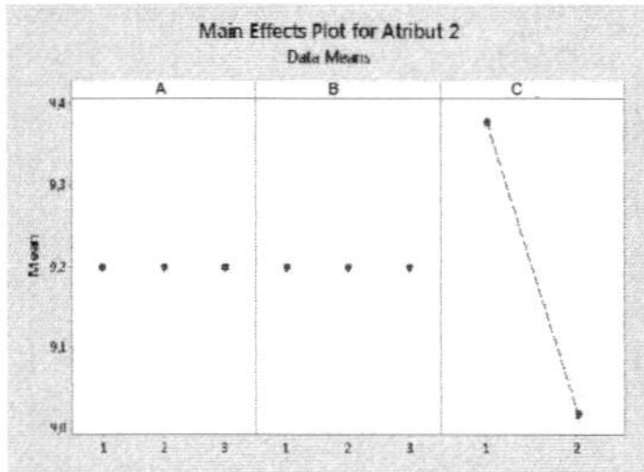

Figure 2. Main effects plot for A2 (no pungent smell)

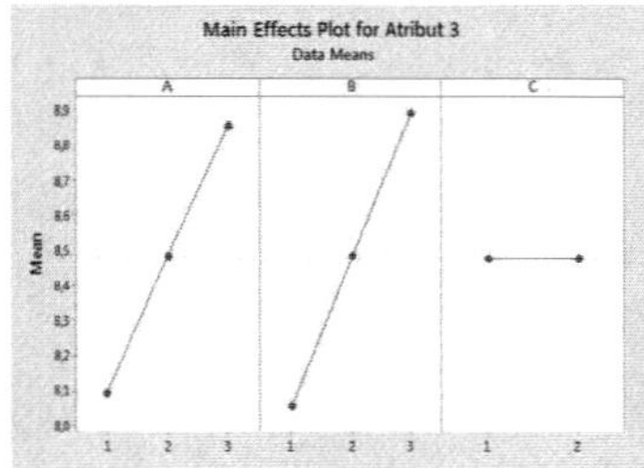

Figure 3. Main effects plot for A3 (have about the right sweetness level)

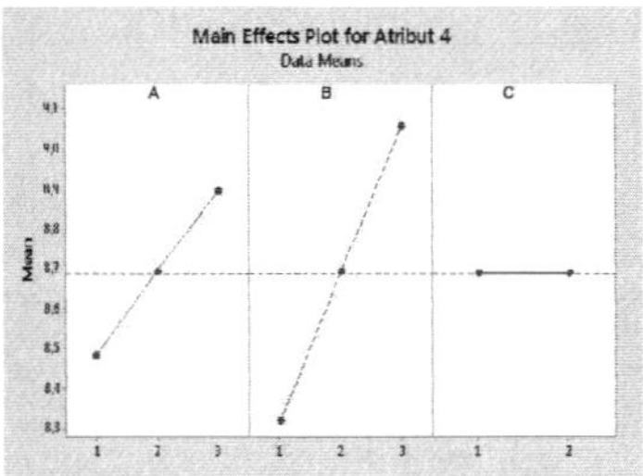

Figure 4. Main effects plot for A4 (have about the right sourness level)

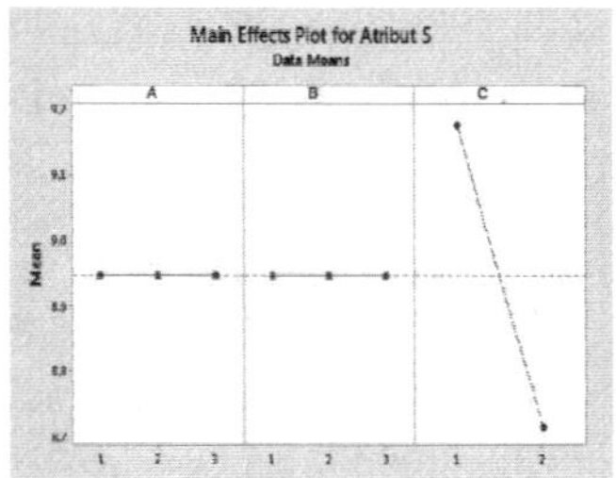

Figure 5. Main effects plot for A5 (no bitter aftertaste)

Based on the optimal level that has been obtained in each attribute, the overall optimal composition then can be determined. The overall optimal composition is determined by the most selected level on each factor. The results are presented in Table 3.

Table 3. The Overall Optimal Composition

Factors	Levels	Composition
Amount of sugar (A)	3	35% (1st marinade) 50% (2nd marinade) 60% (3rd marinade) 70% (4th marinade)
Marinating duration (B)	3	9 hours
Amount of cinnamon (C)	1	10 grams

Table 3 shows that the optimal combination is the ninth combination in the orthogonal array. Amount of sugar influence the level of sweetness, level of sourness, and able to determine whether or not the product has unpleasant taste. Then based on data analysis known that the amount of cinnamon affects the aroma, the aftertaste of the product, and also can determine whether or not the product has unpleasant taste. While the marinating duration can affect the level of sweetness and sourness of the product. This optimal combination is not only preferred by the consumers, but also technically proven to meet the minimum requirement of sugar level regulated by SNI regardless of the age of carrot used in the process. The sugar level of candied carrots aged 3.5 months using this optimal composition is 31.42%. While the sugar level of candied carrots aged 5 months using the same composition is 49.08%. Both are above 25% therefore meet the criteria made by SNI.

4. Conclusion

Based on the experiment, there are three control factors that can affect the taste of candied carrot. These factors are the amount of sugar (A), the marinating duration (B), and the amount of cinnamon (C). While the age of carrot (D) is used as a noise factor that attempted to be controlled in the study. Factor A and B are given 3 levels, while factor C and D are given 2 levels. The determination of the number of experiments is determined by orthogonal arrays. Based on the orthogonal array it was found that the number of experiments was 18. The results of the organoleptic test are analyzed using ordinal logistic regression analysis to determine the control factors that affect each attribute. Overall there are five attributes studied in this research. The optimal combination obtained is technically proven to meet the minimum requirement of sugar level regulated by SNI regardless of the age of carrot used in the process. Hence the optimal combination is robust to noise and therefore able to produce better quality product.

References

[1] Halim V and Ekawati Y 2014 Perencanaan Produk Olahan Wortel Menggunakan Metode Quality Function Deployment. *Symbol,* **1** 57-67

[2] Ekawati Y and Bazarado M 2016 Designing Food Products Based on Carrots Using the Product Design Phase of Quality Function Deployment *ARPN Journal of Engineering and Applied Science,* **11** 3109-16

[3] Dehnad K (ed.) 1989 *Quality Control, Robust Design, and the Taguchi Method* (California: Wadsworth & Brooks/Cole)

[4] Sihombing H, Hafiz M K, Yuhazri M Y and Kannan R 2012 Taguchi's Quality Improvement Analysis of the SME Bread Manufacturing *Global Engineers & Technologists Review* **2** 6-13

[5] Habibi N, Soleimanian-Zad S and Zeinoddin M S 2011 Optimization of Kefir Grains Production by Using Taguchi Technique and Mini-Fermentation *World Applied Sciences Journal* **12** 613-18

[6] Noorwali A 2013 Apply Lean and Taguchi in Different Level of Variability of Food Flow Processing System *Procedia Engineering* **63** 728-34

[7] Naik E K M and Bala E K L 2013 Study on Optimization of Microwave Frying of Potato Slices by Using Taguchi Design *International Journal of Engineering Research & Technology* **2** 141-8

[8] Erdural S, Köksal G and İlk Ö 2006 A Method for Analysis of Categorical Data for Robust Product or Process Design *Proceedings of the 17th Conference of IASC-ERS* 573-80.

[9] Badan Standardisasi Nasional Indonesia. *SNI 01-4443-1998*

[10] Noya S, Ekawati Y and Utami N 2017 Quality Function Deployment for Candied Carrots Process Planning *International Journal of Engineering and Advanced Technology Studies* **5** 1-7

IOP Conf. Series: Materials Science and Engineering **319** (2018) 012051 doi:10.1088/1757-899X/319/1/012051

Performance Analysis of Abrasive Waterjet Machining Process at Low Pressure

M Murugan, MA Gebremariam, Z Hamedon, A Azhari

Faculty of Manufacturing Engineering, Universiti Malaysia Pahang, 26600 Pekan, Pahang, Malaysia

azmir@ump.edu.my, azmir.azhari@gmail.com

Abstract. Normally, a commercial waterjet cutting machine can generate water pressure up to 600 MPa. This range of pressure is used to machine a wide variety of materials. Hence, the price of waterjet cutting machine is expensive. Therefore, there is a need to develop a low cost waterjet machine in order to make the technology more accessible for the masses. Due to its low cost, such machines may only be able to generate water pressure at a much reduced rate. The present study attempts to investigate the performance of abrasive water jet machining process at low cutting pressure using self-developed low cost waterjet machine. It aims to study the feasibility of machining various materials at low pressure which later can aid in further development of an effective low cost water jet machine. A total of three different materials were machined at a low pressure of 34 MPa. The materials are mild steel, aluminium alloy 6061 and plastics Delrin®. Furthermore, a traverse rate was varied between 1 to 3 mm/min. The study on cutting performance at low pressure for different materials was conducted in terms of depth penetration, kerf taper ratio and surface roughness. It was found that all samples were able to be machined at low cutting pressure with varied qualities. Also, the depth of penetration decreases with an increase in the traverse rate. Meanwhile, the surface roughness and kerf taper ratio increase with an increase in the traverse rate. It can be concluded that a low cost waterjet machine with a much reduced rate of water pressure can be successfully used for machining certain materials with acceptable qualities.

1. Introduction

The technology and applications of high pressure waterjet have been studied for many decades (as early as 1960s) [1]. It has been used extensively in various industry-related applications including machining, surface preparation, cleaning, coating removal and surface treatment or waterjet peening [2]. In general, water is compressed to an ultrahigh pressure up to about 700 MPa and discharged from a small orifice typically between 0.2 to 0.4 mm in diameter [3]. The system produces a high velocity water stream up to 900 m/s, with or without abrasive particles causing damage to materials by shearing, cracking, erosion, cavitation, delamination and plastic deformation [3]. With an addition of abrasive particles, the machining capability of the waterjet is significantly improved. A wide range of materials and thicknesses can be cut with good cutting quality and small taper. However, different processing parameters and material properties have to be carefully assessed as to produce the desired cutting qualities [4].

There has been considerable demand for a cost-effective machine tool for various industrial applications. With the advance of research and development in machine tools technology, industrial and consumer products derived from this technology are increasingly becoming commercially available. It is anticipated that the demand for low cost machine tools would increase considerably as their availability would make these products more affordable [5]. These machine tools will be used to produce various components. It is important to develop a low-cost versatile machine tool which is capable of preserving the structural integrity of components. Waterjet machine is one of the tools that have been considered to have potential applications for machining.

The heart of a waterjet machine is the water pump. Due to the low cost waterjet machine, the pump may have a limited capability in generating water pressure typically below 150 MPa. This range of pressure may be suitable for machining thin and soft materials. Summers et al. [6] attempted for drilling and cutting of dolomite rock at a low pressure of 35 MPa. They managed to drill the rock to a depth of 70 cm which proved effective in breaking out blocks in rock. Composite materials were also successfully machined using low-pressure pre-mixed abrasive waterjet machining process [7]. The present study attempts to investigate the performance of abrasive water jet machining process at low cutting pressure using self-developed low cost waterjet machine. It aims to study the feasibility of machining several materials at low pressure which later can aid in further development of an effective low cost water jet machine.

2. Methodology

The commercial waterjet machines available in the market are mainly used for cutting operation since they utilize high pressure mostly between 200 – 600 MPa. The heart of a waterjet machine is the water pump. The currently developed machine has a pump which can generate water pressure of up to about 100 MPa due to its low cost. The type of the pump used is an air driven liquid pump. This type of pump is reliable, compact, robust, and easy to maintain. Most importantly, the overall cost of the pump is way cheaper than the standard electrical or hydraulic type pumps which are used in a commercial waterjet machine. The machine is controlled by a computer numerical control (CNC) system. Table motion is controlled along multiple axes (*X* and *Y*) as well as the nozzle that moves in the *Z*-axis (depth). The cutting head consists of a ruby orifice of 0.15 mm in diameter and a tungsten carbide focusing tube of 0.76 mm and 76.2 mm in diameter and length respectively. The developed CNC waterjet machine is shown in Figure 1.

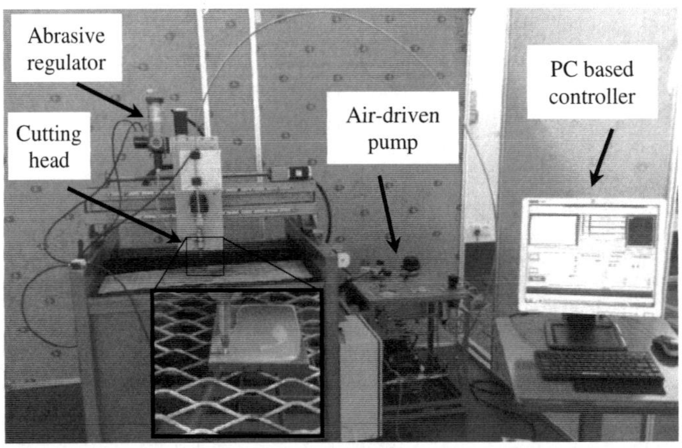

Figure 1. The self-developed CNC waterjet machine (The inset shows the cutting process).

All samples were prepared in wedge shape of 30° for easy measurement of penetration depth. They were cut using the waterjet machine according to the parametric levels as shown in Table 1. All machining procedures were conducted using a single-pass cutting. The machining performance of all cut samples was assessed in terms of depth penetration, kerf taper ratio and surface roughness. Digital Vernier high gauge was used in this experiment to measure the depth of penetration and kerf width. The tapers of the machined surfaces were measured using video measurement system (Econ). Kerf taper ratios were calculated as the ratio of top kerf width to the bottom kerf width. Three different readings for top and bottom kerf widths respectively were taken at each sample and the average readings were calculated as to minimize the error. Finally, a surface roughness measuring device, SURFCOM130A, equipped with a cone-shaped diamond stylus having the diameter of 10 μm and tip angle of 90° was used in this study. The surface roughness was measured over the edge of the cut specimen. It was obtained across the thickness of the test sample surface. All measurements were acquired using 0.8 mm cut-off length. Due to the variability of surface finish data, three different measurements were taken for each surface evaluated so that averages could be calculated.

Table 1. List of parameters and their ranges.

Parameter	Range	Unit
Pressure	34	MPa
Traverse Rate	1 - 3	mm/min
Abrasive	Garnet (80)	Mesh
Abrasive Flowrate	2	g/s
Standoff Distance	5	mm
Workpiece	Mild Steel Aluminium Alloy (6061-T6) Plastic Delrin® (Polyoxymethylene)	

3. Results and discussions

3.1. Effect of low water pressure on machining capability

At such an extremely low pressure (34 MPa), it is expected that the machine may be capable to cut soft materials like plastics. It is known that a higher pressure results in a higher kinetic energy of the waterjet thus giving a greater capability in causing damage for materials [8]. The kinetic energy will be great enough to accelerate the abrasive particles to hit the material and cut to a deeper depth. The depth of materials penetration increases when a higher water pressure is used. At a low pressure, its cutting capability is much reduced. Therefore, it is more suitable for an application other than cutting where a low pressure is needed for example in waterjet peening process (WJP) [2, 9]. WJP is a mechanical surface strengthening process where high-frequent impact of water drops on the surface of metal components, which causes local plastic deformation. The process is not intended to cause full penetration to the surface but rather treating the surface. It is interesting to note that based on preliminary results, using this low pressure proves that it may also cut some hard materials such as metals. However, cutting capability is very much reduced depending on the hardness of the machined materials. The depth of penetration of the materials decreases when the hardness of the material increases. Furthermore, the process has to be properly controlled in order to optimize the cutting capability of the waterjet at a low pressure. The traverse rate has to be as low as possible as to give enough time for the jet to impact the surface thus causing the penetration. Traverse rate is actually the speed of the feed motion in X or Y direction. It plays a bigger role in producing a good cutting quality.

3.2. Effect of traverse rate on penetration depth

The depth of penetration is a measure of the depth that the waterjet machining process could achieve during cutting of a material. Analysing the depth of penetration is very important as it gives an indicator about the machining performance of the process using a particular waterjet machine. The depth of penetration is not an independent issue as it varies according to different parameters. Figure 2 shows the effect of traverse rate on penetration depth for different materials. The depth of penetration increases with a decrease of traverse rate. As the traverse rate increases, the nozzle head moves faster. In this case, the waterjet with the mixture of abrasives does not have enough time to make penetration on the workpiece. The contact time between the waterjet and the material has been reduced [8]. As a result, the waterjet could not fully erode the material while the nozzle moves in a faster rate. The waterjet is only able to cut to a certain depth with a higher traverse rate.

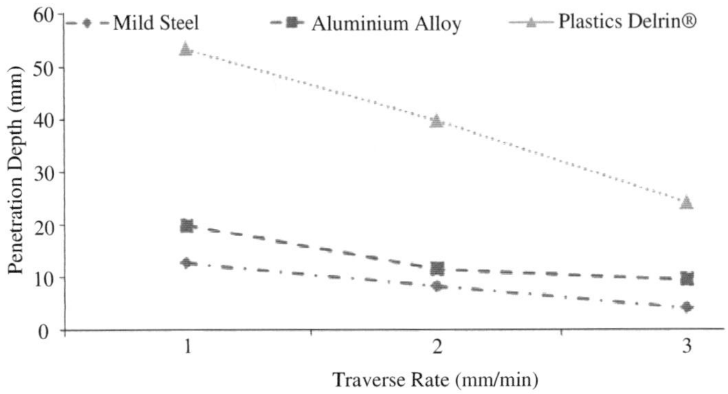

Figure 2. Effect of traverse rate on penetration depth for different materials

3.3. Effect of traverse rate on kerf taper ratio

The effect of traverse rate on the kerf taper for different materials is shown in Figure 3. In general, a higher traverse rate produces a wider kerf taper ratio. The negative effect of traverse rate on the kerf width is due to the fact that a faster passing of abrasive water jet allows fewer particles to strike on the target material and, hence, generates a narrower slot [10]. In other words, the decrease in the exposure time that was caused by increasing the traverse rate resulted to the reduction in both of the kerf top and bottom width. Whereas, the increasing trend of the kerf taper ratio is the result of the more rapidly decreasing kerf width at the bottom than at the top as the traverse rate increases. Hence, the waterjet does not have enough time to penetrate through the material that leads to a wider upper kerf than the bottom kerf. However, at a higher traverse rate, its effect on the kerf taper ratio is not significant as shown in Figure 3. This result is in agreement with Azmir [11] where he found that the traverse rate was not significant in influencing the kerf taper ratio.

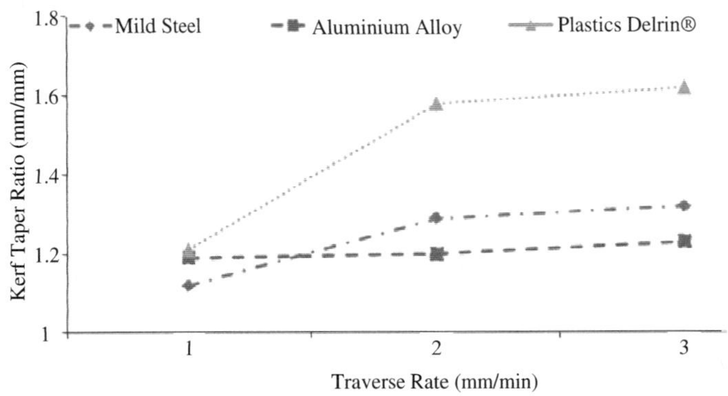

Figure 3. Effect of traverse rate on kerf taper ratio for different materials

3.4. Effect of traverse rate on surface roughness

As shown in Figure 4, it can be seen that increasing the traverse rate produces a higher surface roughness. This is due less overlap machining actions and fewer abrasive particles to impinge the surface at a higher traverses rate thus increasing the roughness of the surface. As a result, the abrasives could not erode the target material properly thus resulting in a rough surface. At a lower traverse rate, the abrasives particles will have enough time to impinge on the target material thus producing a well penetrated cut with a smoother surface. A slower traverse rate also gives a cleaner cut with a smoother surface [1]. The roughness of the cut profiles also changes with traverse rate and it is more obvious at the highest traverse rate [12]. In this case, a lower traverse rate is desirable to produce a better surface finish as shown in Figure 4.

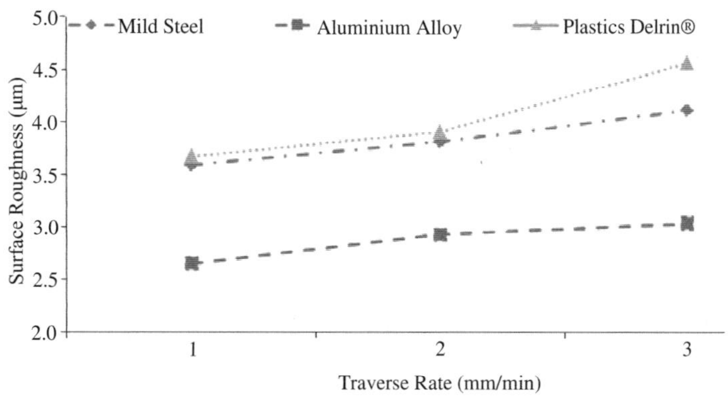

Figure 4. Effect of traverse rate on surface roughness for different materials

3.5. Machining performance for different materials

The machining performance for different materials varies depending on their types. It is known that the material properties (i.e. strength, hardness, ductility, fatigue resistance, etc.) affect their machining performance. In general, mild steel has produced the lowest values of penetration depths and followed by aluminium alloy and plastics as shown in Figure 2. This can be expected since the mild steel has the highest hardness amongst other materials. Therefore, it has a higher resistance to material removal thus producing a lower penetration depth. It is also well understood that plastics has hardness very much lower than metals. As a result, their penetration depths are many times higher than the mild steel and aluminium alloy as illustrated in Figure 2. In case of kerf taper ratio, plastics produce the highest values and followed by mild steel and aluminium alloy as shown in Figure 3. A higher value of kerf taper ratio constitutes to a lower accuracy of geometrical cut since the dimension of top width shows a higher variation with a bottom width. This is possibly due to the nature of plastics in which some fibrous particles were left during machining when the jet made an exit from the samples. Due to this nature, their surfaces are also rougher than other materials as shown in Figure 4. Furthermore, due to low hardness of plastics, its upper kerf width tends to become bigger because the jet impingement can easily erode the material thus making a bigger slot. In case of metals, lower kerf taper ratios were found since they have lower penetration depths with smoother surfaces thus resulting in more accurate cut. Aluminium alloy is softer than mild steel thus producing a higher penetration depth with a smoother surface. Examples of machined samples are shown in Figure 5 which shows their penetration depths and kerf tapers.

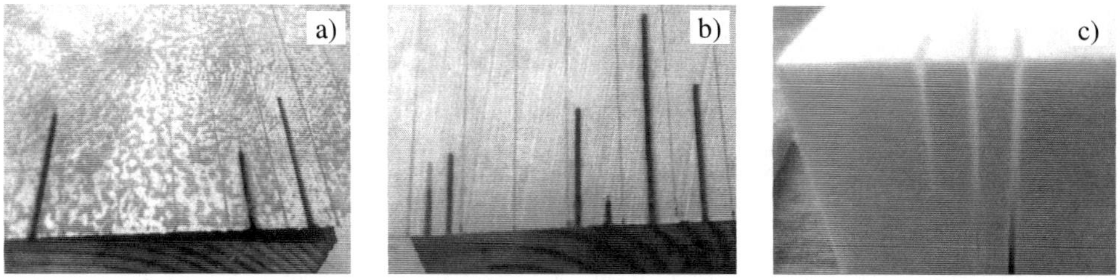

Figure 5. Examples of machined samples, a) mild steel, b) aluminium alloy, and, c) plastics

The performance of the waterjet machine during cutting of mild steel, aluminium alloy and plastics Delrin® at a very low pressure of 34 MPa are satisfactory with acceptable machining qualities. Therefore, it would be interesting to further test its capability by machining other materials. Ceramic tiles and glass were chosen for further experiment. Machining of ceramic tiles at a pressure of 34MPa and traverse speed of 5 mm/min produced a surface roughness of 4.591 μm and penetration depth of 8.7 mm. On the mild steel and ceramic tile surfaces, an obvious wavy pattern or waterjet lagging can be clearly observed as shown in Figure 6. In case of glass machining at low pressure of 34 MPa, the glass surface has experienced slight chipping due to its brittleness. When the pressure had increased to

69 MPa while maintaining the traverse rate, the waterjet was able to cut through the glass without any crack or chipping as shown in Figure 6. This is possibly due to sufficient energy of the jet to cause smooth erosion to the glass.

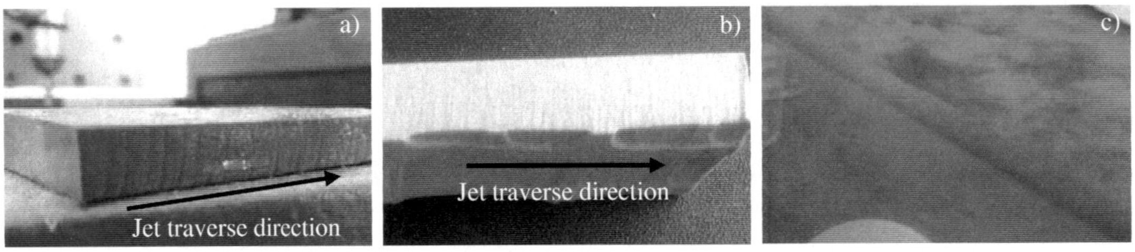

Figure 6. Examples of machined samples, a) mild steel, b) ceramic tile, and, b) glass

4. Conclusions

It was shown that mild steel, aluminium alloy 6061 and plastics Delrin® were successfully machined at low cutting pressure using self-developed low cost waterjet machine. The cutting performance of the waterjet machine is satisfactory with acceptable machining qualities. The depth of penetration decreases with an increase in the traverse rate. Meanwhile, the surface roughness and kerf taper ratio increase with an increase in the traverse rate. It can be concluded that the self-developed waterjet machine with a much reduced rate of water pressure can be successfully used for machining some materials with acceptable qualities.

5. Acknowledgments

Authors would like to gratefully acknowledge the financial support from the Universiti Malaysia Pahang through RDU140128 and RDU150347.

6. References

[1] Folkes J 2009 Waterjet—An innovative tool for manufacturing *Journal of Materials Processing Technology* **209** 6181-9
[2] Azhari A, Schindler C, Kerscher E and Grad P 2012 Improving surface hardness of austenitic stainless steel using waterjet peening process *The International Journal of Advanced Manufacturing Technology* **63** 1035-46
[3] Hashish M, Steele D E and Bothell D H 1997 Machining with super-pressure (690 MPa) waterjets *International Journal of Machine Tools and Manufacture* **37** 465-79
[4] Azmir M and Ahsan A 2008 Investigation on glass/epoxy composite surfaces machined by abrasive water jet machining *Journal of materials processing technology* **198** 122-8
[5] Liu H-T 2010 Waterjet technology for machining fine features pertaining to micromachining *Journal of Manufacturing Processes* **12** 8-18
[6] Summers D, Yao J, Blaine J, Fossey R and Tyler L 1992 Low pressure abrasive waterjet use for precision drilling and cutting of rock *Fluid Mechanics and its Applications* **13** 233-
[7] Zou Z L 2012 Study of Cutting Composite Materials with Low Pressure Abrasive-Water Jet. In: *Applied Mechanics and Materials*: Trans Tech Publ) pp 1480-3
[8] Kovacevic R 1992 Monitoring the depth of abrasive waterjet penetration *International Journal of Machine Tools and Manufacture* **32** 725-36
[9] Azhari A, Schindler C, Godard C, Gibmeier J and Kerscher E 2016 Effect of multiple passes treatment in waterjet peening on fatigue performance *Applied Surface Science* **388** 468-74
[10] Wang J and Wong W 1999 A study of abrasive waterjet cutting of metallic coated sheet steels *International Journal of Machine Tools and Manufacture* **39** 855-70
[11] Azmir M and Ahsan A 2009 A study of abrasive water jet machining process on glass/epoxy composite laminate *Journal of Materials Processing Technology* **209** 6168-73
[12] Ma C and Deam R 2006 A correlation for predicting the kerf profile from abrasive water jet cutting *Experimental Thermal and Fluid Science* **30** 337-43

IOP Conf. Series: Materials Science and Engineering **319** (2018) 012052 doi:10.1088/1757-899X/319/1/012052

Forward and Inverse Predictive Model for the Trajectory Tracking Control of a Lower Limb Exoskeleton for Gait Rehabilitation: Simulation modelling analysis

M A Zakaria, A P P A Majeed, Z Taha, M M Alim, K Baarath

Innovative Manufacturing, Mechatronics and Sport (iMAMS) Laboratory, Faculty of Manufacturing Engineering, Universiti Malaysia Pahang

maizzat@ump.edu.my

Abstract. The movement of a lower limb exoskeleton requires a reasonably accurate control method to allow for an effective gait therapy session to transpire. Trajectory tracking is a non-trivial means of passive rehabilitation technique to correct the motion of the patients' impaired limb. This paper proposes an inverse predictive model that is coupled together with the forward kinematics of the exoskeleton to estimate the behaviour of the system. A conventional PID control system is used to converge the required joint angles based on the desired input from the inverse predictive model. It was demonstrated through the present study, that the inverse predictive model is capable of meeting the trajectory demand with acceptable error tolerance. The findings further suggest the ability of the predictive model of the exoskeleton to predict a correct joint angle command to the system.

1. Introduction

The growing demand for health services such as rehabilitation tends to outstrip the supply in a limited economic setting. Notwithstanding, rehabilitation services are essential to address disabilities that are associated with gait disorder. Gait, is the ability to maintain the walking pattern of a person. Gait abnormalities are the inability of a person to walk that may originate from cerebellar disease, neuromuscular disease, cognitive impairment, stroke, brain or spinal injury [1], [2]. Studies have shown that through the continuous and repetitive motion, the patient's walking pattern may be restored [3]. The assistance from physiotherapists are required to facilitate this form of activity, however, it is worth to note that it is labour and cost intensive. The shortcomings of conventional rehabilitation therapy sessions have led towards the engagement of robotics to mitigate the aforesaid issues.

Various types of methods have been researched previously to address the rehabilitation issues by means of the control system development for rehabilitation device. For instance, a 3-dof device developed by Rahman et. al. uses the integration of Proportional Derivative (PD) control with the integration of neuro fuzzy system [4]. Sliding mode control and nonlinear torque control which is proven to reduce the chattering of the system was employed by the author in [5]. Furthermore, several other methods using AFC-PD were addressed in [6] which provide the adaptility to the patient's movement based on the torque and current feedback to the system.

IOP Conf. Series: Materials Science and Engineering **319** (2018) 012052 doi:10.1088/1757-899X/319/1/012052

This paper presents the mathematical modelling of the lower limb exoskeleton with actuators. The movement of the lower limb exoskeleton requires a reasonably precise positional control to track the prescribed trajectory to rehabilitate the lower limb joints. Contrary to the previous research, the forward and inverse kinematic are developed to predict the correct positional tracking of the exoskeleton. The actuator dynamics is also modelled in order to control the movement of the lower limb exoskeleton. A classical proportional-integral-derivative (PID) control architecture is used as a controller for this robotic system. The advantage of the inverse model control is that the predictive motion that can be estimated by the device without any additional sensors besides the encoders attached to the joint angles. This method is easier to implement and require less burden to the processing unit of the system as the solution of the joint angle is already derived and only require one line of the joint angle command to the system.

2. Methodology

Figure 1 depicts the overall block diagram of the exoskeleton system. The actuator and forward kinematics model are assigned as the plant of the system, whilst the inverse kinematics prediction model provides the solution to the required Cartesian coordinate input to the exoskeleton. The controller block is used to provide the error correction to the evaluated system.

The Denavit-Hartenberg (D-H) algorithm is used to construct the exoskeleton axis and its relevant parameters. Although the D-H method provides a long and complex mathematical solution, nonetheless, it is a systematic method to develop sophisticated robotic modelling. It is worth mentioning that there are other methods that are readily available to develop such model, however, it is beyond the scope of this paper. The readers are encouraged to refer to [7] for further detailed treatment of the other techniques. The methods involved in the present study are as follows:

1. The lower limb exoskeleton is shown in **Figure 2**. From the figure, the D-H parameter may be assigned to the degree of freedom of interest.
2. Then, the D-H parameters are tabulated as in **Table 1** to ascertain its corresponding values.
3. The resultant D-H parameter is then shown in Equation 3.

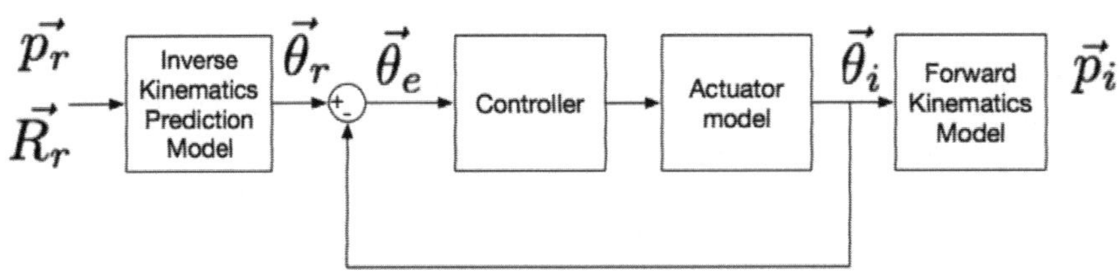

Figure 1 Overall System Block Diagram

IOP Conf. Series: Materials Science and Engineering **319** (2018) 012052 doi:10.1088/1757-899X/319/1/012052

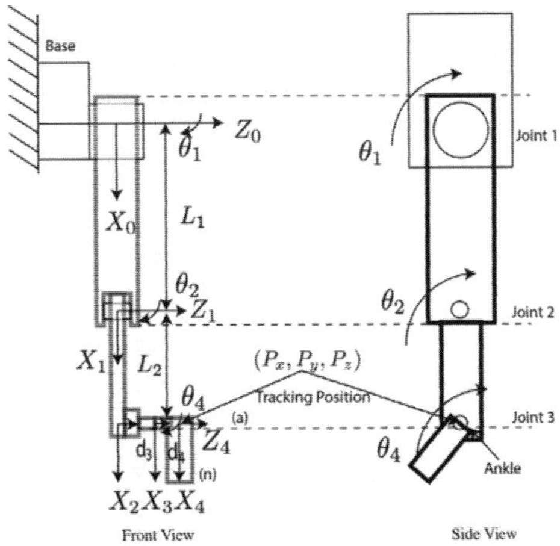

Figure 2 Lower Limb Exoskeleton Coordinate Assignment

Table 1 D-H Exoskeleton Robotic Leg Parameters

Link, i	**Joint angle (degree), θ_i**	**Link offset (meter), d_i**	**Link length (meter), a_i**	**Twist angle (degree), α_i**
1	θ_1	0	1.5	0
2	θ_2	0	0.5	0
3	0	0.2	0	0
4	θ_4	0.2	0	0

2.1 Transformation Matrix

Once the D-H table has been tabulated; the robot transformation matrix can be obtained by substituting the D-H parameter into the Equation 3.

$$A_i^{i-1} = \begin{bmatrix} c\theta_i & -c\alpha_i s\theta_i & s\alpha_i s\theta_i & a_i c\theta_i \\ s\theta_i & c\alpha_i c\theta_i & -s\alpha_i c\theta_i & a_i s\theta_i \\ 0 & s\alpha_i & c\alpha_i & d_i \\ 0 & 0 & 0 & 1 \end{bmatrix} \tag{1}$$

$$T_4^0 = A_1^0 A_2^1 A_3^2 A_4^3 \tag{2}$$

The total transformation matrix is as follows:

$$T_4^0 = \begin{bmatrix} C_{1,2,4} & -S_{1,2,4} & 0 & L_2 C_{1,2} + L_1 C_1 \\ S_{1,2,4} & c\alpha_i c\theta_i & 0 & L_2 S_{1,2} + L_1 S_1 \\ 0 & 0 & 1 & d_3 + d_4 \\ 0 & 0 & 0 & 1 \end{bmatrix} \tag{3}$$

Where $C_{i,j,k} = \cos(\theta_i + \theta_j + \theta_k)$ and $S_{i,j,k} = \sin(\theta_i + \theta_j + \theta_k)$.

2.2 Inverse Kinematics

The inverse must be obtained to derive the relationship between the required position and joint angles relationship. In order to derive this; the forward kinematic equation can be compared with the necessary inputs (position vectors and orientation vectors). The inverse kinematics is essential for the control system development. It allows the user to determine the required joint angles command in order to reach specific position or configuration. The inverse kinematics utilises the ankle required position Px, Py, Pz as inputs to determine the joint angles, θi.

To calculate the desired angle of the exoskeleton, the general orientation, and position of the represented end effector can be defined as below after derivation using trigonometric derivations and rearrange the transformation matrix as in Equation *4*. The further derivation details can be referred to [8]. If the dynamics of the exoskeleton is required, the readers are encouraged to refer to [6].

$$A_3^{2^{-1}} A_2^{1^{-1}} A_1^{0^{-1}} T_4^{0^{-1}} = A_4^3 \tag{4}$$

$$\theta_1 = atan2(-L_2 S_2 P_x + P_y(L_1 + L_2 C_2), P_x(L_1 + L_2 C_2) + P_y L_2 S_2) \tag{5}$$

$$\theta_2 = atan2(n_y, n_x) - \theta_1 - \theta_4 \tag{6}$$

$$\theta_4 = atan2(-n_x S_{1,2} + n_y C_{1,2} n_x C_{1,2}, n_x C_{1,2} + n_y S_{1,2}) \tag{7}$$

2.3 Actuator dynamics

The joints of the exoskeleton are controlled by actuators. Hence, the modelling of the actuators is non-trivial to understand the interaction issues between the control input and the resultant joint angles, namely the transient response of the actuators. More often than not, the torque, current and speed of the driving motor during transient time is studied to observe the performance of the system. The modelling of the actuators can be performed by applying Kirchoff's Voltage Law to the motor model which yields the following equations:

$$\frac{di}{dt} = \frac{V}{L} - \frac{R}{L} i - \frac{K}{L} \omega \tag{8}$$

$$\frac{d\omega}{dt} = \frac{K}{J} i - \frac{b}{J} \omega \tag{9}$$

Where I, V, L, R, K, J, w are the motor current, voltage, inductance, internal resistance, motor constant, load inertia, and motor speed.

PID Control System

In this study, a conventional PID control system is used to move the actuator to the desired angles. The PID control law may be defined as:

$$u(t) = K_p e + K_i \int e\, dt + K_d \frac{de}{dt} \tag{10}$$

Where K_p, K_i, K_d are the PID control gain, *e* is the PID error and $u(t)$ is the control input.

3. Results

To verify the model, the exoskeleton is assumed to be running in all quadrant operations to ensure the correct values are obtained. All quadrant operations are illustrated into the desired input as shown in **Table 2**.

Table 2 Quadrant Working Operations

IOP Conf. Series: Materials Science and Engineering **319** (2018) 012052 doi:10.1088/1757-899X/319/1/012052

Quadrant 3: (negative X, negative Y)		Quadrant 2: (negative X, positive Y)	
$T_d = \begin{bmatrix} 0 & 1 & 0 & -0.5 \\ -1 & 0 & 0 & -0.4 \\ 0 & 0 & 1 & 0.4 \\ 0 & 0 & 0 & 1 \end{bmatrix}$	(11)	$T_d = \begin{bmatrix} 0 & 1 & 0 & -0.3 \\ -1 & 0 & 0 & 0.5 \\ 0 & 0 & 1 & 0.4 \\ 0 & 0 & 0 & 1 \end{bmatrix}$	(12)
Quadrant 4: (positive X, negative Y)		Quadrant 1: (positive X, positive Y)	
$T_d = \begin{bmatrix} 0 & 1 & 0 & 1.0 \\ -1 & 0 & 0 & -0.8 \\ 0 & 0 & 1 & 0.4 \\ 0 & 0 & 0 & 1 \end{bmatrix}$	(13)	$T_d = \begin{bmatrix} 0 & 1 & 0 & 1.5 \\ -1 & 0 & 0 & 0.01 \\ 0 & 0 & 1 & 0.4 \\ 0 & 0 & 0 & 1 \end{bmatrix}$	(14)

The trajectory tracking response is observed in terms of its error and desired angle calculated by the inverse kinematics. In the present study, the Symmetric Mean Absolute Percentage Error (SMAPE) is employed to compute the relative error of the response. The formula is defined in Equation 1.

$$SMAPE(\%) = \frac{100}{n}\sum_{t=1}^{n}\frac{|F_t - A_t|}{\frac{|A_t| + |F_t|}{2}} \tag{15}$$

Where F_t, A_t, n are the desired value, actual value and sample count respectively. The tracking percentage error and performance is shown in Table 3, respectively. From the table, it can be seen that the robotic exoskeleton is able to reach the correct quadrant with less than 5% angular tracking error. The positional tracking error (P_x, P_y, P_z) is acceptable in Quadrant 1, 2, and 3. In the quadrant 4 operation, the exoskeleton contain a larger relative error percentage (18.77%) in P_y position. However, the orientation and direction of the quadrant operation is still in correct position. Furthermore, the angular relative error for the 4th quadrant is still relatively small with no additional correction altough some fine tuning may be required to improved the tracking error percentage. The mean response time of the exoskeleton to reach the steady state is 10.75 seconds which is acceptable.

Figure 3 shows the trajectory tracking error of the exoskeleton control. The error is measured from the distance of the actual exoskeleton position compared with the desired position. From the figure, it can be seen that the error converged to smaller value over time. The largest positional tracking error can be seen in Quadrant for with a steady state error of 0.2 meters offset from the actual position.

Table 3 Trajectory Tracking Performance

Quadrant		P_x	P_y	P_z	θ_1	θ_2	θ_4	Response time, t
1st	Desired	1.5	0.01	0.4	1.604	-3.12	-3.12	
	Actual	1.499	0.0154	0.4	1.676	-3.263	-3.263	7 seconds
	% error	0.07	0.54	0	4.39	4.4	4.4	
2nd	Desired	-0.3	0.5	0.4	147	-146	-146	10 seconds
	Actual	-0.3869	0.471	0.4	-152.5	-151.5	-151.5	
	% error	25.3	5.97	0	3.67	3.69	3.69	
3rd	Desired	-0.5	-0.4	0.4	-151.6	147.5	147.5	14 seconds
	Actual	-0.4237	-0.4242	0.4	-157.3	152.9	152.9	
	% error	16.20	5.87	0	3.69	3.59	3.59	
4th	Desired	1.0	-0.8	0.4	-67.84	65.14	65.14	12 seconds
	Actual	0.8369	-0.9658	0.4	-70.28	67.47	67.47	
	% error	17.75	18.77	0	3.53	3.51	3.51	

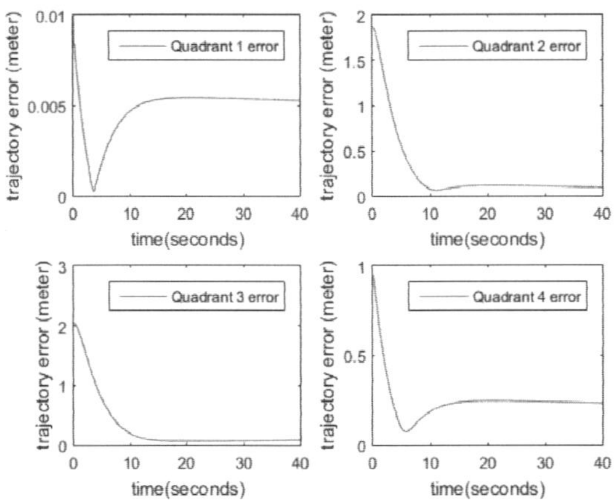

Figure 3 Trajectory Tracking Error for Lower Limb Exoskeleton

4. Conclusion and Future Works

In this paper, the lower limb exoskeleton system model is validated in all quadrants of operations. It has been demonstrated that the PID controller with inverse prediction model is able to track the required position and angular position of the system. However, it is worth to mention that some positioning error does exist, in which improvement in the overall closed loop system is required. This may include the inclusion of an external checking mechanism which periodically monitors whether the minimisation of the positional error has been achieved through an additional forward predictive kinematics in front of the required inputs.

Acknowledgments

The authors would like to thank Universiti Malaysia Pahang for the financial aid under research grant RDU160343 and RDU170731.

References

[1] S. Murray and M. Goldfarb, "Towards the use of a lower limb exoskeleton for locomotion assistance in individuals with neuromuscular locomotor deficits," in *Engineering in Medicine and Biology Society (EMBC), 2012 Annual International Conference of the IEEE*, 2012, pp. 1912–1915.

[2] O. Franch, L. Calandre, J. Álvarez-Linera, E. D. Louis, F. Bermejo-Pareja, and J. Benito-León, "Gait disorders of unknown cause in the elderly: Clinical and MRI findings," *J. Neurol. Sci.*, vol. 280, no. 1, pp. 84–86, 2009.

[3] I. D'iaz, J. J. Gil, and E. Sánchez, "Lower-limb robotic rehabilitation: literature review and challenges," *J. Robot.*, vol. 2011, 2011.

[4] M. H. Rahman, K. Kiguchi, M. M. Rahman, and M. Sasaki, "Robotic exoskeleton for rehabilitation and motion assist," in *Industrial and Information Systems, First International Conference on*, 2006, pp. 241–246.

[5] M. H. Rahman, M. Saad, J. P. Kenné, and P. S. Archambault, "Modeling and control of a 7DOF exoskeleton robot for arm movements," in *Robotics and Biomimetics (ROBIO), 2009 IEEE International Conference on*, 2009, pp. 245–250.

[6] A. P. P. A. Majeed, Z. Taha, A. F. Z. Abidin, M. A. Zakaria, I. M. Khairuddina, M. A. M. Razman, and Z. Mohamed, "The Control of a Lower Limb Exoskeleton for Gait Rehabilitation: A Hybrid Active Force Control Approach," *Procedia Comput. Sci.*, vol. 105, pp. 183–190, 2017.

[7] G. Fu and R. C. Gonzales, "Lee (1987), Robotics: Control, Sensing, Vision and Intelligence." McGraw-Hill International.

[8] M. A. Zakaria, A. P. P. A. Majeed, I. M. Khairuddin, and Z. Taha, "Kinematics Analysis of a 3DoF Lower Limb Exoskeleton for Gait Rehabilitation : A Preliminary Investigation," pp. 4–7.

IOP Conf. Series: Materials Science and Engineering **319** (2018) 012053 doi:10.1088/1757-899X/319/1/012053

Simulation on Effect of Preform Diameter in Injection Stretch Blow Molding

Z Q Tan[1], Nurrina Rosli[1], Muchamad Oktaviandri[1,2]

[1]Faculty of Manufacturing Engineering, Universiti Malaysia Pahang, 26600 Pekan, Pahang, Malaysia

[2]Fakultas Teknologi Industri, Universitas Bung Hatta, Padang, 25143, Indonesia

nurrinarosli@ump.edu.my

Abstract. Polyethylene terephthalate (PET) is the most common material of resin for manufacturing plastic bottle by injection stretch blow molding due to its excellent properties. As various issues of health and environmental hazards due to the PET use have risen, PET bottle manufacture may be improved by minimizing the wall thickness to reduce the PET use. One of the critical qualifications of the manufacturing process which lead to the wall thickness distribution is the initial preform diameter. In this project, we used the ANSYS Polyflow with aim to evaluate the wall thickness distribution of PET bottle for different diameter of initial preform. As a result, only 4 mm preform diameter presented wall thickness below than 1 mm. On the other hand, at least 6 mm preform diameter can permit the wall thickness 1.3 mm i.e. at the shoulder area.

1. Introduction

Polyethylene terephthalate (PET) is among the common thermoplastic polymer resins that widely used in household, medical, food and beverages. The food and beverages industry was the greatest volume consumptions of plastic packaging in statistical report 2014 [1]. The global PET resin demand has been rising continuously over the years and is predicted to grow at a rate of 7.3% annually in 5 years starting from 2014 [2]. The extraordinary mechanical strength and toughness with the ease of processing have prompting the production of PET packaging products. However, as the demand of PET plastic increases significantly among the years, various issues have cropped related to health and environmental hazards [3].

Many approaches have been carried out to minimize those impacts. One of them is to reduce the materials used during the processing stage of PET bottle [4]. By practicing a sustainable design, the bottle manufacture may be improved by minimizing the thickness to reduce the PET use. One of the critical design qualifications of PET bottle which lead to its thickness distribution is the initial preform diameter. The initial preform diameter are extremely important to produce the optimal bottle mass and uniform final wall thickness distribution [5, 6]. Since reports of the wall thickness of PET bottle is still scarce, this project intends to evaluate the wall thickness distribution of PET bottle for different diameter of initial preform, with aim to obtain the potential thinnest wall of the PET bottle.

In order to study the PET bottle manufacturing process, Finite Element Method (FEM) has been widely used to replace the trials through repeating experiments, which can effectively save time and materials consumption. Schmidt et al. [7] through their viscoelastic constitutive equations have successfully estimated the contact kinetic, thickness and stress distribution. Code ABAQUS® has been utilized by Chung [8] to optimize the process of stretch blow-molding for a PET bottle. He found that the sliding at neck area of the bottle made the more material flows towards the side wall and thus causing a thicker side wall by the end. However, the process temperature is assumed to be constant along the process. The ABAQUS model has neglected the effects of elasto-visco-plasticity and thermal behaviours of the PET material. The same approach were also been conducted by Bordival et al. [9] through their works to predict accurately the blow molding kinematics. They have successfully maintained the modelling of fluid-structure interaction presented between the preform and its utilized air flow.

BlowView Software has been proposed by Daver and Demirel [10] in predicting the way of blow molded preform in order to solve the non-isothermal solid mechanics equations of blow molding process. It comprised the thermo-mechanical material models in terms of viscoelasticity and visco-hyper-elasticity for PET and encountered the strain-rate and temperature effects which reverts the real performance of PET. Furthermore, they have used ANSYS Polyflow simulation to address the maximum top load and burst strength that could be affordable by the bottle. Other works that used ANSYS Polyflow were also done by Gupta et al. [11] to solve the problem dealing with the time and stage of PET resin fill-in the cavity of mold. The volumetric mechanical computation allows prediction of wall thickness distribution, force distribution and contact kinetic distribution.

In this project, we used the ANSYS Polyflow due to its great library of finite elements and material models as well as the application of automatic remeshing which is a major contribution to succeed a simulation process.

2. Methodology

In order to analyze the thickness of the bottle formed, Finite Element Analysis (FEA) was performed using ANSYS Polyflow and the data generated was quantified using Simple Digitizer. The method to conduct the ANSYS Polyflow simulation for identifying the thickness, stress and velocities distribution will be explained in detail in the following subsections.

2.1. Simulation procedure

2.1.1. Geometry

The preform and bottle mold were designed as 2D drawing using ANSYS Design Modeler. Because of the complexity of modelling operation in Polyflow, assumptions for procedure were made in general terms. The bottle mold was 350 mm in height (*y*-direction) with a width of 96 mm (*x*-direction).

2.1.2. Meshing

In this project, mapped face shell elements with sizing of 0.0001 m are used. The boundaries were set as shown in figure 1.

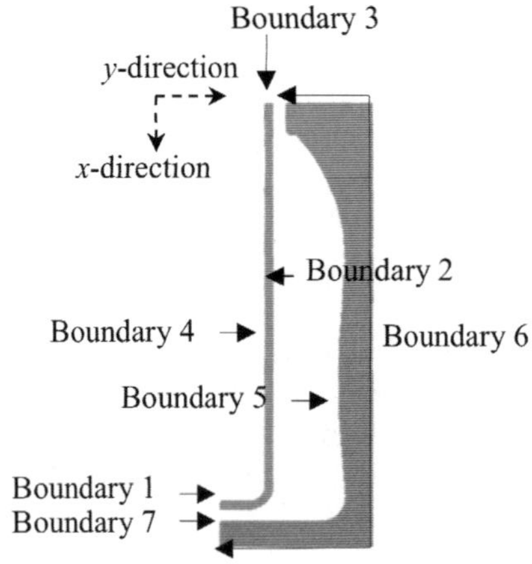

Figure 1. Boundary setup.

2.1.3. Setup
2D axis-symmetry time dependent FEM is defined in Polydata. All the parameters and mathematical expressions were assigned to be used at the setup stage. The sub-sections include models, material data, boundary conditions, remeshing, numerical parameters, outputs and thickness postprocessor.

Table 1. Material properties.

Property	Value
Shear rate dependent	100 000 Pa.s
Density	1 g/cm^3
Gravity	-981 g/cm^3

Table 2. Numerical parameters.

Numerical Parameters	Data
Upper time limit	1.0
Initial value of time-step	1.0e-03
Minimum value of time-step	1.0e-07
Maximum value of time-step	1.0e-03
Tolerance	1.0e-02
Maximum number of successful steps	200
Implicit Euler methods	Enable

Table 1 and table 2 show the material properties and numerical parameters, respectively. Only some material properties were available for the sub-task in Polydata including density, viscosity, inertia term, and gravity. For those irrelevant properties in this sub-task, they will be grayed out straight away when accessing Polydata. A simplified behavior was assumed in this project on account of the accurate characterization for all mechanical behaviors for PET are very complex. Then, boundary conditions were set up by inputting slip coefficient, penalty coefficient as 1.0e+09, normal force as -2e06 (negative indicates the direction of fluid towards mold). Numerical parameters in table 2 was then inputted.

2.1.4. Solution and result
Contours of time steps were then shown graphically to present the relationship between parameters with their dependent variables.

2.2. Data quantification
Simple Digitizer software was used to quantify the data of images collected from the result generated using ANSYS Polyflow. Multiple points were ticked at outer-line and inner-line of bottle in the generated images to produce *x*-coordinates and *y*-coordinates values as the measurement position. The difference between outer-line and inner-line were then calculated to determine the wall thickness.

3. Results and discussion

3.1. Wall thickness distribution
Figure 2 shows the thickness distribution at final time-step. The results of wall thickness fluctuation along the measurement position marked using Simple Digitizer software is shown in figure 3.

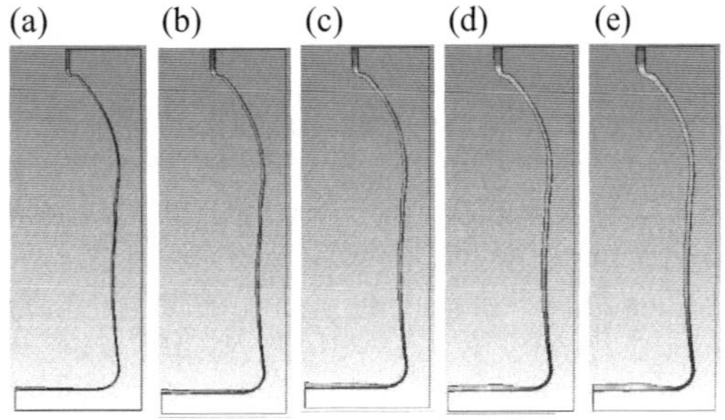

Figure 2. Wall thickness distribution at final-step for (a) 4 mm, (b) 6 mm, (c) 8 mm, (d) 10 mm and (e) 12 mm preform diameter.

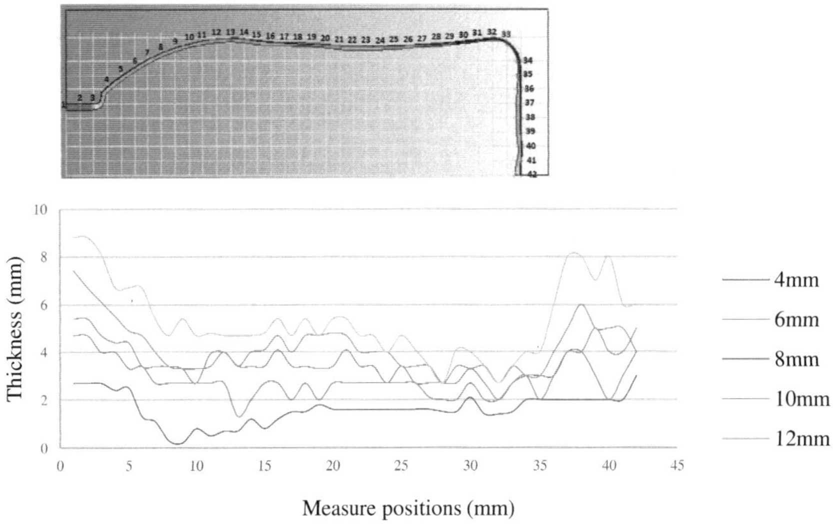

Figure 3. Wall thickness fluctuation at different preform diameter.

As the diameter of the preform increases, the overall thickness for the final bottle formed were found to increase. The wall thickness for all preform diameters were seen the thickest at the finish then followed by the base of bottle. It can be observed clearly that the thickness at the body (middle lower part) is the thinnest. For 4 mm preform diameter, there were some parts formed almost cannot be clearly seen especially at the bottom (corner area). However, for 12 mm preform diameter, the thickness of bottle can be visualized clearly where the thickness was getting thinner from the neck to bottom but growing thicker from bottom to base.

Initially at the finish, the thickness was found to drop rapidly until the shoulder part. It then found to remain constant or drops gradually through the body until bottom and rise significantly throughout the base. The base should be thick satisfactory to eliminate the problem related with drop impact strength. However, if the base is over-thick, it will become fracture and resulting in reduction of drop impact strength. The finish should be thick enough to withstand the pressure come from the blowing nozzle during the blow molding process and the torque used to tighten the bottle cap [12].

Table 3. Summary of wall thickness distribution for different preform diameter.

Preform Diameter (mm)	Percentage of distribution (%)		Minimum wall thickness (mm)
	Over than 1 mm	Less than 1 mm	
4	83.33	16.67	0.2
6	100	0	1.3
8	100	0	2.0
10	100	0	2.7
12	100	0	2.7

Table 3 shows the summary of results collected for each preform diameter. As can be seen, only 4 mm preform diameter presented wall thickness below than 1 mm. Therefore, this value of preform diameter should be to prevent very small thickness of wall in PET bottle. On the other hand, at least 6 mm preform diameter can permit the wall thickness It was found that only 4 mm preform diameter presents wall thickness below than 1 mm. On the other hand, at least 6 mm preform diameter can permit the wall thickness at least 1.3 mm i.e. at the shoulder area.1.3 mm i.e. at the shoulder area. This suggest that preform diameter of 4 to 6 mm can be further investigated to get the smallest 1.0 mm wall thickness.

3.2. Contour of thickness at time

As the fluid preform is moving towards the bottle mold, the thickness is inflated and the color represents the distribution of thickness over the time. The contour of thickness at different time step have displayed clearly the direction of the streching force applied to the preform which included the first step until the final step that form the shape of the mold [11].

As shown in figure 4, it can be observed clearly at the final time step that the thickness was found the smallest at the bottom corner of the bottle where the preform has been the most extended and had a largest plastic stretching in circumferential and longitudinal directions. But the thickness was observed the largest at the top. This can be attributed to the deformation that was less take place in a small diameter.

Furthermore, it can be seen at the final time-step in all figures that the thinnest part was found at the corner, where bottom of the bottles confide fundamentally on the shape and dimensions of the cavity mold. It also lies on shifting the degrees of individual areas where the preform was stretched as well as the contact between preform and the mold at different times. Therefore, the correlation between the final thickness profile of the bottle and preform diameter which acts as the shape of cavity mold is probably fluctuating depends on the desired final shape of bottle.

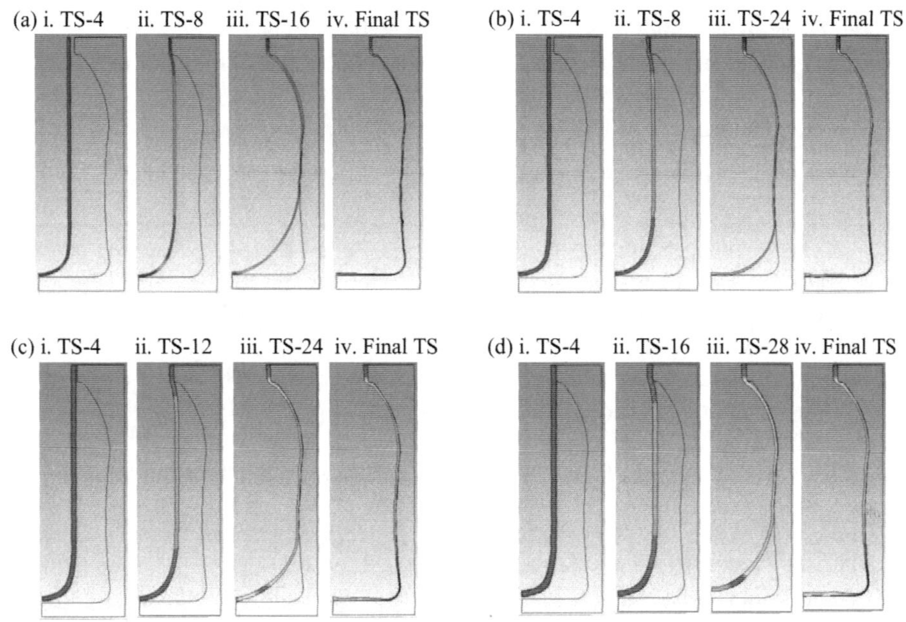

Figure 4. Contour of thickness at different TS (time step) for (a) 4 mm, (b) 6 mm, (c) 8 mm and (d) 10 mm preform diameter.

3.3. Contour of pressure at time-step

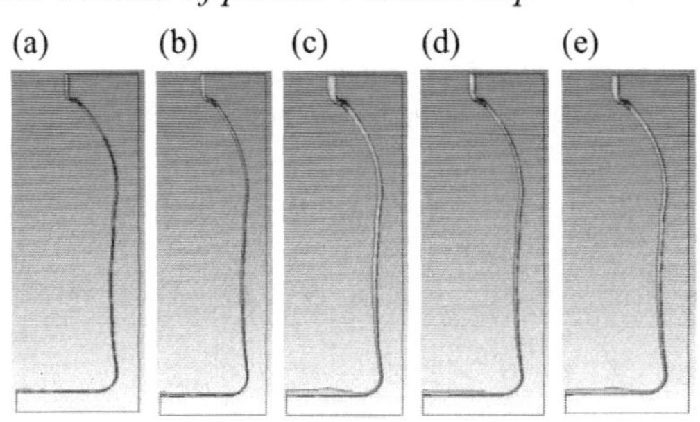

Figure 5. Contour of pressure at final-step for (a) 4 mm, (b) 6 mm, (c) 8 mm, (d) 10 mm and (e) 12 mm preform diameter.

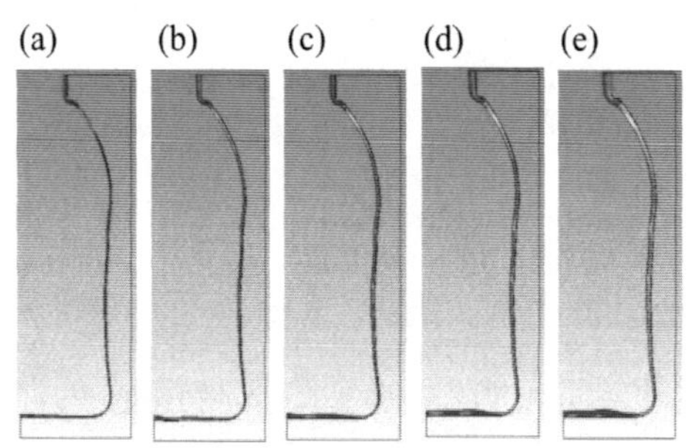

Figure 6. Contour of velocities distribution in *x*-direction at final-step for (a) 4 mm, (b) 6 mm, (c) 8 mm, (d) 10 mm and (e) 12 mm preform diameter.

In order to present the results of polymer pressure field within the mold, the contour plots of pressure at different time steps were extracted. Generally, the magnitude of stress was interrelated to the stretching force at boundary. The stretching force will then affect the processing time directly for forming a complete PET bottle [8].

A general observation to figure 5 suggests that increasing preform diameter increased the pressure at the final step. It was also revealed that the 4 mm diameter preform has the least of stress distribution, where the contour of minimum stress were found. For 12 mm diameter preform, it has the most stress distribution as can be confirmed from the maximum stress indicated in the figure.

4. Conclusion

A simulation study in injection stretch blow molding has been conducted using ANSYS Polyflow. The study was mainly performed to evaluate the wall thickness distribution of PET bottle for different diameter of initial preform. It was found that only 4 mm preform diameter presents wall thickness below than 1 mm. On the other hand, at least 6 mm preform diameter can permit the wall thickness at least 1.3 mm i.e. at the shoulder area. The thickness was found the smallest at the bottom corner of the bottle where the preform has been the most extended and had a largest plastic stretching in circumferential and longitudinal directions. But the thickness was observed the largest at the top.

5. References

[1] Embree K. Global plastics packaging market to hit $375 billion by 2020 | PlasticsTodayhttps://www.plasticstoday.com/packaging/global-plastics-packaging-market-hit-375-billion-2020/23800481624973 (2016, accessed 22 September 2017).

[2] Global PET (Resin) Market Report- 2012 Edition - Research and Markets. *Koncept Analytics* 2013; 53.

[3] Intergovernmental Panel on Climate Change (ed). *Climate Change 2013 - The Physical Science Basis*. Cambridge: Cambridge University Press. Epub ahead of print 2014. DOI: 10.1017/CBO9781107415324.

[4] Hopewell J, Dvorak R, Kosior E. Plastics recycling: challenges and opportunities. *Philos Trans R Soc Lond B Biol Sci* 2009; 364: 2115–26.

[5] Karol Pepliński AM. Ansys-Polyflow Software Use to select the Parison Diameter and its Thickness Distribution on Blowing. *J POLISH CIMAC*; 6.

[6] García-Rejón A, Rapra Technology Limited. *Advances in blow moulding process optimization*. Rapra Technology Ltdhttp://www.chemtec.org/proddetail.php?prod=978-1-85957-050-0 (1995, accessed 12 June 2017).

[7] Schmidt FM, Agassant JF, Bellet M, et al. Viscoelastic simulation of PET stretch/blow molding process. *J Nonnewton Fluid Mech* 1996; 64: 19–42.

[8] Chung K. Finite element simulation of pet stretch/blowmolding process. *J Mater Shap Technol* 1989; 7: 229–239.

[9] M. Bordival, F.M. Schmidt, Y. Le Maoult VV. Optimization of Preform Temperature Distribution for the Stretch-Blow Molding of PET Bottles: Infrared Heating and Blowing Modeling. *Polym Eng Sci* 2009; 783–793.

[10] Daver F, Demirel B. A simulation study of the effect of preform cooling time in injection stretch blow molding. *J Mater Process Technol* 2012; 212: 2400–2405.

[11] Gupta S, Uday V, Raghuwanshi AS, et al. Simulation of Blow Molding Using Ansys Polyflow. *APCBEE Procedia* 2013; 5: 468–473.

[12] NatureWorks. Designing a Bottle for Water Packaging. 2017; 1–3.

Acknowledgments
Authors wish to acknowledge the Research & Innovation Department of University Malaysia Pahang for the financial support given to complete this project (Grant number: RDU1703158).

IOP Conf. Series: Materials Science and Engineering **319** (2018) 012054 doi:10.1088/1757-899X/319/1/012054

Study on the effectiveness of Extreme Cold Mist MQL system on turning process of stainless steel AISI 316

A S Jamaludin[1],*, A Hosokawa[2], T Furumoto[2], T Koyano[2], Y Hashimoto[2]

[1] Faculty of Manufacturing Engineering, Universiti Malaysia Pahang, 26600 Pekan, Pahang, Malaysia
[2] Institute of Science and Technology, Kanazawa University, Kakuma-machi, Kanazawa, 920-1192, Ishikawa, Japan

shahir@ump.edu.my

Abstract. Cutting process of difficult-to-cut material such as stainless steel, generates immensely excessive heat, which is one of the major causes related to shortening tool life and lower quality of surface finish. It is proven that application of cutting fluid during the cutting process of difficult-to-cut material is able to improve the cutting performance, but excessive application of cutting fluid leads to another problem such as increasing processing cost and environmental hazardous pollution of workplace. In the study, Extreme Cold Mist system is designed and tested along with various Minimum Quantity Lubrication (MQL) systems on turning process of stainless steel AISI 316. In the study, it is obtained that, Extreme Cold Mist system is able to reduce cutting force up to 60N and improve the surface roughness of the machined surface significantly.

1. Introduction

Several approaches are proposed to improve the cutting process difficult-to-cut materials, especially in improving the tool life and higher surface finish. It is necessary to understand about the mechanic of heat generation and temperature increment during the machining process. Basically, lubricating process can be considered as successful when the cutting fluid penetrated the sticking-sliding zone of tool-chip contact. It is proven that, cutting process assisted by cutting fluid is able to improve the tribological properties at the interfaces through alteration of normal, shear stresses, and their distribution, and/or by removing the excessive heat, thus improving the machined surface roughness and tool wear etc. [1-8].

Previous studies had shown that dry machining is the optimum solution in most of material removal processes for soft material, but less effective in machining difficult-to-cut material. Thus, a near-dry cutting system is proposed by implementing the Minimum Quantity Lubrication (MQL) method. This MQL system utilizes only a tiny amount of cutting fluid, with flowrate unit of ml/h compared to conventional cutting fluid supplying method. The cutting fluid is turn into microscale droplet and supplied to the cutting [4, 7].

There are various types of cutting fluids available nowadays that can be utilized for MQL system, which are lubrication purpose oil-based cutting fluid, cooling purpose of emulsion-based cutting fluid (made of soluble oil in water and hopefully to be able to reduce oxidation on the surface of material), water or cold air that are mostly for cooling application [4,7]. Various studies had shown that MQL system capability in reducing the frictional contact between tool, chip and workpiece and able to provide efficient heat exchange through conduction [5].

In this study, Extreme Cold Mist system is proposed along with various types of MQL method in turning process of difficult-to-cut materials, stainless steel AISI 316. Stainless steel, AISI 316 is a commonly available stainless steel along with AISI 304, where this stainless steel is austenitic chromium nickel stainless steel, as shown in Table 1 [6]. The molybdenum content of the material elevates the resistance to corrosion and pitting from chloride ion solutions, along with sustaining extreme temperatures. Stainless steel AISI 316 is utilized mostly in industrial sector involving processing chemicals, as well as under-sea (high saline) such as coastal regions and outdoor areas [6]. Thus, it is considered as a valuable effort in improving the knowledge of machining process of AISI 316 effectively.

Table 1. Stainless steel AISI 316 general chemical composition.

Chemical Composition	%
Cr	16-18
Ni	10-14
C	0.08
Mn	2.0
Si	0.75
P	0.045
S	0.03
N	0.10
Mo	2.0-3.0

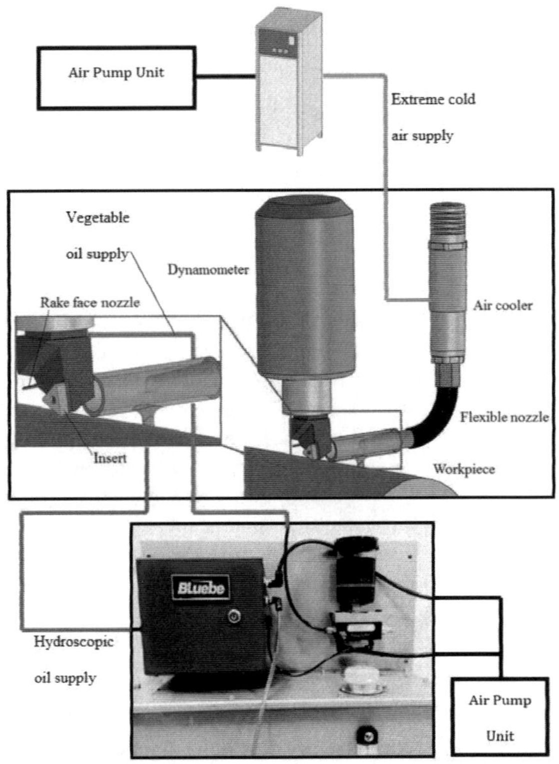

Figure 1. Hybrid lubrication-coolant Minimum Quantity Lubrication (MQL) supplier

2. EXPERIMENTAL PROCEDURES

In the study, a system that capable of supplying heterogeneous Minimum Quantity Lubrication (MQL) is designed, where the MQL mist is supplied to the rake and flank surface of the cutting tool (Figure 1). The specification of MQL systems and cutting fluids applied in the study are shown in Table 2 and Table 3, respectively. In the study, the effectiveness of each MQL system in machining stainless steel AISI 316 is analyzed in terms of cutting force and surface finish.

Table 2. MQL system specification

System Type	MQL type for rake face	Fluid rate (mL/h)	MQL type for flank face	Fluid rate (mL/h)	Total fluid rate (mL/h)
Dry	Dry	-	Dry	-	-
A	Oil A	30	Dry	-	30
B	Oil B	100	Dry	-	100
C	Oil A	30	Cold air	-	30
D	Oil A	30	Oil B	100	130
E	**Oil A**	**30**	**Cold air + Oil B**	**100**	**130**

Table 3. Characteristic of the MQL oil

MQL oil	Oil A (Lubrication)	Oil B (Coolant)
Fluid type	Vegetable oil	Hydroscopic oil
Viscosity	37	
Oil contents (%)	99	30
Oil pressure (MPa)	0.6	
Cold air pressure (MPa)	0.6	
Cold Air temperature (˚C)	-13.6	

In the study, MQL with lubrication characteristic (Oil A) is supplied onto rake face of the tool by rake face nozzle, while hydroscopic mist (Oil B) with coolant characteristic is supplied onto flank face with the application of flexible nozzle (Figure 1) in MQL system Type D. For MQL system Type C and E, the flexible nozzle is connected to an air cooler, which is connected to an air dryer to supply extreme cold air (-13.6°C). MQL system Type E is the proposed MQL system for the study called Extreme Cold Mist system. Material properties and cutting conditions are shown in Tables 4 and 5.

Table 4. Material properties

Materials	TiCN-coated Cermet	AISI 316
Young Modulus, E (GPa)	650	205
Poisson Ratio, ε	0.25	0.30
Thermal Conductivity, k (W/(m·K))	59	17
Density, ρ (kg/m^3)	14900	8070
Volumetric Heat Capacity, c (x 10^6 J/(m^3·K))	15.0	2.78
Hardness Vickers, $HV_{0.3}$ (GPa)	13.7	2.2

IOP Conf. Series: Materials Science and Engineering **319** (2018) 012054 doi:10.1088/1757-899X/319/1/012054

Table 5. Cutting condition

Cutting Speed v_c (m/min)	:	50 ~ 200
Depth-of-cut a (mm)	:	0.5
Feed f (mm/rev)	:	0.3

3. Results and discussions

3.1. Cutting force comparison with the application of various MQL system

During the experimental procedure for turning process of stainless steel (AISI 316) with TiCN-coated cermet cutting tool with the proposed MQL system, principal force F_p, thrust force F_t, and feed force F_f is observed and plots in Figures 2, 3 and 4. Reduction of 50N of the principal force F_p is observed with the application of MQL Type A and E, while reduction of feed force F_f and thrust force F_t up to nearly 30N and 60N, respectively with the application of MQL Type E only.

It is assumed that, the oil mist is able to penetrate the tool-chip and tool-machined surface contact effectively for MQL system Type A and Type E, as observed from the force reduction. However, only MQL system Type E is able to reduce feed force F_f and thrust force F_t with higher effectiveness. It is believed that, reduction of temperature at the tool flank with its extremely low temperature cold mist and higher total fluid rate is occurring effectively during the experiment.

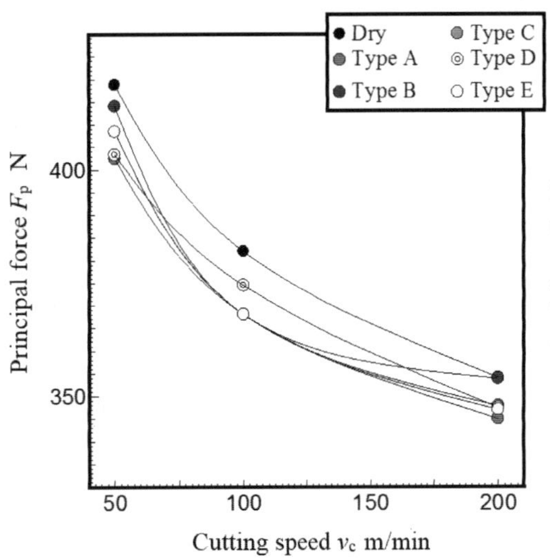

Figure 2. Relationship between cutting speed v_c and principal force F_p for various MQL systems.

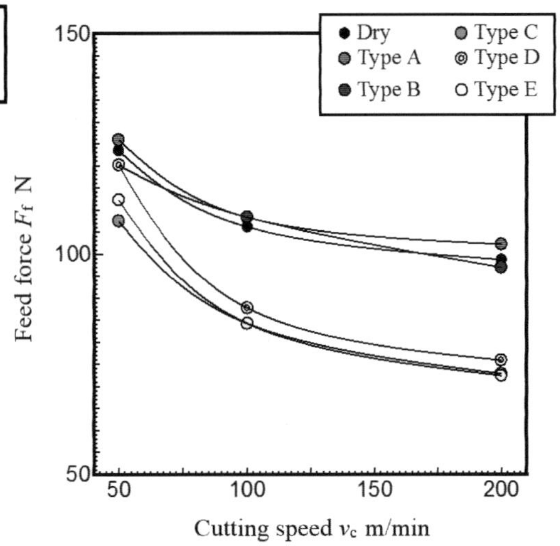

Figure 3. Relationship between cutting speed v_c and feed force F_f for various MQL systems.

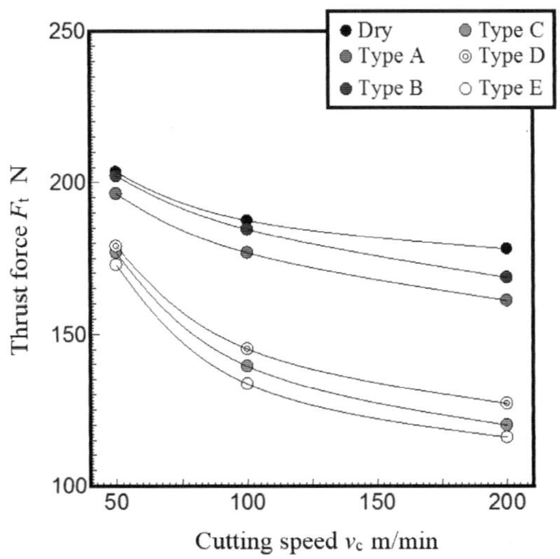

Figure 4. Relationship between cutting speed v_c and thrust force F_t for various MQL systems.

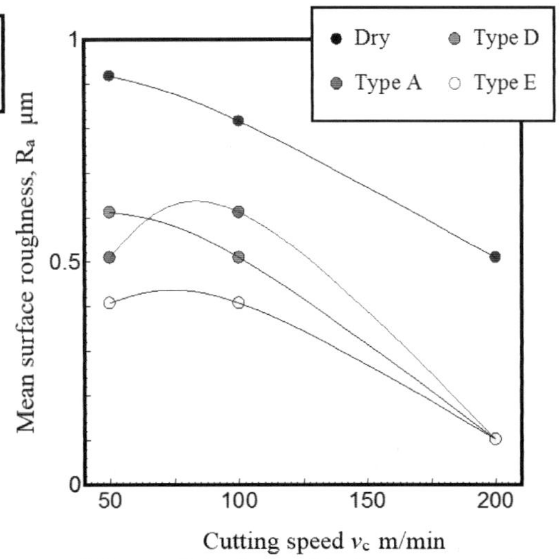

Figure 5. Relationship between cutting speed v_c and mean surface roughness, *R*a

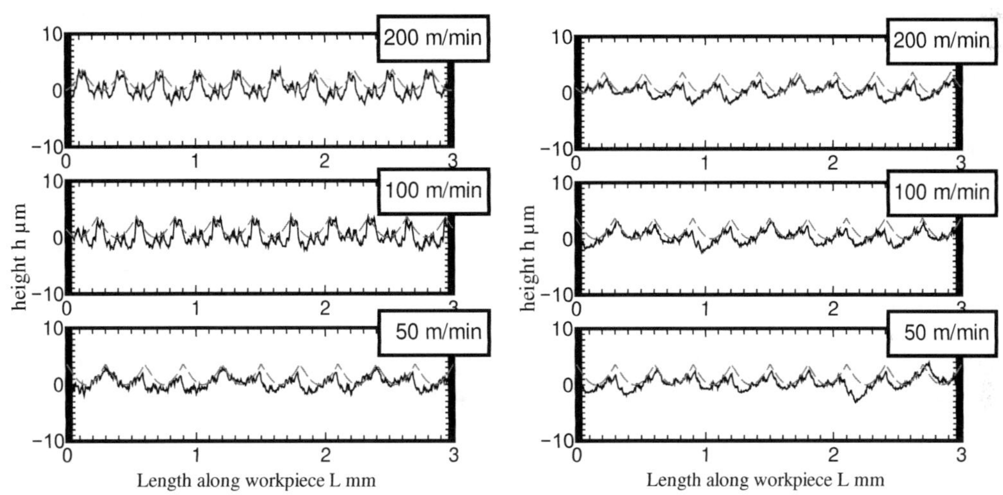

Figure 6. Surface roughness profile comparison between MQL system Type A (left), MQL system Type E (right) and ideal surface roughness profile (red line)

3.2. Surface roughness comparison with the application of various MQL system

During the experiment with the application of MQL system Type A, C and E (depth of cut a=0.5 mm, feed rate f =0.3 mm/rev), it is observed that, the mean surface roughness *R*a decreases as cutting speed increases, but MQL system Type E shows the lowest mean surface roughness *R*a.

It is believed that, at lower temperature, cutting tool integrity is increased, especially with the application of MQL System type E. Plus, the chip is able to flow smoothly by reducing the frictional coefficient through hybrid lubrication-cooling method. It is concluded that Extreme Cold Mist System (MQL system type E) has high effectiveness for cutting process difficult-to-cut material.

4. Conclusions

In this study, turning process experiments were conducted on stainless steel (AISI 316) with the application of various MQL conditions. Cutting force were measured and the effect of cold air was studied. Several conclusions can be made as following.

a. Supplying vegetable oil to the rake face is able to reduce principal cutting force component F_p up to 50N compared to the dry type, whereas the vegetable oil mist able to lubricate the contact surface between tool and chip.
b. By supplying cold air (−13.6°C) + hydroscopic oil (Extreme Cold Mist system) at 100mL/h volume rate, the system is able to decrease feed force F_f and thrust force F_t component up to 30N and 60N compared to dry type, respectively.
c. By supplying cold air (−13.6°C) to the flank face (Extreme Cold Mist system), the finished surface roughness is greatly improved compared with the dry type.
d. It can be concluded at current stage, the Extreme Cold Mist system is the best MQL system in the study. It is needed to study the mechanism of MQL onto cooling process of cutting tool during cutting process in the future.

Acknowledgments
The author would like to express deeply gratitude University Malaysia Pahang for providing the financial support under the internal research grant financial scheme RDU 1703252. The author also wanted to acknowledge and express indebtedness to the NC turning machine provider, OKUMA Co. Ltd.

References

[1] El-Wardany T, Mohammed E, & Elbestawi M, 1996 Cutting temperature of ceramic tools in high speed machining of difficult-to-cut materials. *International Journal of Machine Tools and Manufacture,* **36(5)**, 611
[2] Debnath S, Reddy M M, & Yi Q S, 2014 Environmental friendly cutting fluids and cooling techniques in machining: a review. *Journal of Cleaner Production,* **83**, 33
[3] Shyha I, Gariani S, & Bhatti M, 2015 Investigation of Cutting Tools and Working Conditions Effects when Cutting Ti-6al-4V using Vegetable Oil-Based Cutting Fluids. *Procedia Engineering,* **132**, 577
[4] Priarone P C, Robiglio M, Settineri L, & Tebaldo V, 2015 Effectiveness of Minimizing Cutting Fluid Use when Turning Difficult-to-cut Alloys. *Procedia CIRP,***29**, 341
[5] Agustina B D, Bernal C, Camacho A, & Rubio E, 2013 Experimental Analysis of the Cutting Forces Obtained in Dry Turning Processes of UNS A97075 Aluminium Alloys. *Procedia Engineering,* **63**, 694
[6] Tsisar V, Kondo M, Muroga T, Nagasaka T, & Matushita I, 2012 Morphological and compositional features of corrosion behavior of SUS410 – SUS410, SUS316 – SUS316 and SUS410 – SUS316 TIG welded joints in Li. *Fusion Engineering and Design,* **87(4)**, 363
[7] Ueda T, Hosokawa A, & Yamada K, 2006 Effect of Oil Mist on Tool Temperature in Cutting. *Journal of Manufacturing Science and Engineering,* **128(1),** 130
[8] Rahim E, Ibrahim M, Rahim A, Aziz S, & Mohid Z, 2015 Experimental Investigation of Minimum Quantity Lubrication (MQL) as a sustaitainable Cooling Technique. Procedia CIRP, **26**, 351

IOP Conf. Series: Materials Science and Engineering **319** (2018) 012055 doi:10.1088/1757-899X/319/1/012055

Interpretive Structural Model of Key Performance Indicators for Sustainable Maintenance Evaluatian in Rubber Industry

E Amrina and A Yulianto

Department of Industrial Engineering, Andalas University, Padang, Indonesia

elita@ft.unand.ac.id

Abstract. Sustainable maintenance is a new challenge for manufacturing companies to realize sustainable development. In this paper, an interpretive structural model is developed to evaluate sustainable maintenance in the rubber industry. The initial key performance indicators (KPIs) is identified and derived from literature and then validated by academic and industry experts. As a result, three factors of economic, social, and environmental dividing into a total of thirteen indicators are proposed as the KPIs for sustainable maintenance evaluation in rubber industry. Interpretive structural modeling (ISM) methodology is applied to develop a network structure model of the KPIs consisting of three levels. The results show the economic factor is regarded as the basic factor, the social factor as the intermediate factor, while the environmental factor indicated to be the leading factor. Two indicators of social factor i.e. labor relationship, and training and education have both high driver and dependence power, thus categorized as the unstable indicators which need further attention. All the indicators of environmental factor and one indicator of social factor are indicated as the most influencing indicator. The interpretive structural model hoped can aid the rubber companies in evaluating sustainable maintenance performance.

1. Introduction

Companies today has facing a challenge to implement sustainability in all the business aspects. Sustainability is commonly defined as meeting the needs of the present without compromising the ability of future generations to meet their own needs [1]. Companies that adopt sustainability practices can achieve a better product quality, a higher market share, and an increased profit [2]. Maintenance as one of manufacturing function has become an important part to achieve the sustainable company's status [3]. Besides its benefits in minimizing the operating costs and optimizing the equipment durability, an appropriate maintenance also affects the overall company's performance [4]. Thus, integrating the sustainability into the maintenance activities has become a need.

Sustainable maintenance is a process of continuous development and constant improvement of maintenance processes, increasing efficiency (operational excellence), safety of operations and maintenance of technical objects and installations, and focused on employees [4]. Sustainability has usually evaluated in three aspects known as the triple bottom line of sustainability consist of economic, environmental, and social [5]. Since any change in an aspect will influence the other aspects, these three aspects are interrelated and thus, should be assessed with considering their relationships [6].

This paper develops an interpretive structural model of key performance indicators (KPIs) evaluation for sustainable maintenance evaluation in rubber industry. A literature study is conducted to identify the initial KPIs of sustainable maintenance evaluation in rubber industry. The initial indicators are then validated by academic and industry experts. Finally, a network relationship model is then developed using Interpretive Structural Modelling (ISM) method.

2. Methodology

The methodology has three main stages.

2.1. Identification of KPIs

A literature review is carried out to identify the KPIs of sustainable maintenance evaluation in rubber industry. The initial KPIs are developed using the three factors of sustainability consist of environmental, economic, and social. Furthermore, the three factors are then divided into seventeen indicators as shown in Table 1.

Table 1. The initial KPIs.

Factors	Indicators	References
Environmental	1. Energy consumption	[3], [4], [8], [9], [10]
	2. Water utilization	[3], [10]
	3. Lighting and ventilation	[3], [8], [9], [10]
	4. Emission	[3], [4], [8], [9], [10]
	5. Material consumption	[3], [4], [7], [8], [9], [10]
	6. Working environment	[3], [7], [8], [10]
	7. Land utilization	[4], [9]
Economic	8. Preventive cost	[3], [10]
	9. Failure cost	[3], [10]
	10. Maintenance procedure	[3], [9], [10]
	11. Breakdown rate	[3], [4], [10]
	12. Failure rate	[3], [4], [7], [9], [10]
	13. Product quality	[3], [4], [7], [8], [10]
	14. Availability rate	[3], [7], [8], [10]
Social	15. Occupational health and safety	[3], [4], [7], [8], [9], [10]
	16. Employee involvement	[3], [4], [10]
	17. Training and education	[3], [10]

2.2. Validation of KPIs

The initial KPIs of sustainable maintenance evaluation in the rubber industry are then validated by academic and industry experts. Four academics in maintenance expertise area are consulted to determine the importance of the initial KPIs. Six managers of production and maintenance department from a rubber company located in Padang, Indonesia are also asked to validate the initial KPIs. A five-point Likert scale ranging from 1 (not important at all) to 5 (very important) is used to rate the perspective of academics and managers on the importance level of the initial KPIs. The mean importance values ranged from 2.00 to 4.40 as shown in Table 2.

The results show preventive cost and failure cost are regarded as the most important KPIs. These two costs are the indicators of economic factor which most related to the maintenance cost. It followed by occupational health and safety. On the other hand, four indicators of product quality, material consumption, water utilization, and land utilization are regarded as the least important indicators. Based on the results, the initial KPIs of sustainable maintenance evaluation in rubber industry are then modified. Because of the less importance, four indicators are removed from the initial KPIs. As a result, three factors with a total of thirteen indicators have proposed as the KPIs for sustainable maintenance evaluation in rubber industry as shown in Table 3.

Table 2. The mean importance values of initial KPIs.

Indicators	Mean	Indicators	Mean
Preventive cost	4.40	Lighting and ventilation	3.80
Failure cost	4.40	Working environment	3.80
Occupational health and safety	4.20	Energy consumption	3.60
Maintenance procedure	4.00	Emission	3.60
Breakdown rate	4.00	Product quality	2.80
Failure rate	4.00	Material consumption	2.60
Training and education	4.00	Water utilization	2.20
Availability rate	3.80	Land utilization	2.00
Employee involvement	3.80		

Table 3. The proposed KPIs.

Factors	Environmental	Economic	Social
Indicators	1. Energy consumption	5. Preventive cost	11. Occupational health and safety
	2. Lighting and ventilation	6. Failure cost	12. Employee involvement
	3. Emission	7. Maintenance procedure	13. Training and education
	4. Working environment	8. Breakdown rate	
		9. Failure rate	
		10. Availability rate	

2.3. Development of network structure model

The Interpretive Structural Modeling (ISM) methodology is applied to develop a network structure model of the KPIs for sustainable maintenance evaluation in rubber industry. An ISM questionnaire is then designed to determine the relationships amongst the KPIs. A total of 11 managers and staffs of production and maintenance department from the rubber company located in Padang, Indonesia are consulted. These managers and staffs are carefully selected based on their knowledge and experience in maintenance area. Experts are asked through the questions such as "will indicator *i* affect indicator *j*?" to indicate the direct influence that they believe each indicator on each of the other indicators according to an integer scale ranging from 0 = there is no relationship to 1 = there is a relationship. The details are given in the following section.

3. Development of Interpretive Structural Model

The Interpretive Structural Modeling (ISM) method proposed by Warfield in 1973 is an interactive learning process that enables to develop a map of the complex relationships among many elements involved in a complex problem [11]. ISM aids in developing an interaction map to identify the interrelationships amongst system variables. ISM method provides a better understanding of a system structure and draws up a useful guideline in generating a graphical representation of the structure [12]. It is often used to provide fundamental understanding of complex situations, as well as to find solutions for complex problem. ISM can reduces the complex system interactions into a logically oriented matrix which is useful for the purpose of structure determination. The procedures of ISM methodology consist of five steps as follows:

3.1. Developing structural self interaction matrix (SSIM)

A total of 11 experts are consulted to determine the interrelationships amongst the KPIs of sustainable maintenance evaluation in rubber industry through ISM survey. The answers of each question are averaged. In total, 38 direct relationships are confirmed from the survey. The SSIM of the KPIs is presented in Table 4. Four symbols are used to denote the direction of relationship between the KPIs (*i* and *j*) consist of V for the relation from *i* to *j*, A for the relation from *j* to *i*, X for both directions, relations from *i* to *j* and *j* to *i*, and O if the relation between the indicators does not appear.

3.2. Constructing initial reachability matrix

The SSIM is then transformed into the initial reachability matrix by substituting the symbols of V, A, X, and O into a binary matrix of 1 and 0, where 1 means there is relationship between the KPIs and otherwise, 0 means there is no relationship between the KPIs. The substituting rule is as follows:

1) If (i, j) entry in the SSIM is V, then (i, j) entry in the reachability matrix is 1 and (j, i) entry is 0.
2) If (i, j) entry in the SSIM is A, then (i, j) entry in the reachability matrix is 0 and (j, i) entry is 1.
3) If (i, j) entry in the SSIM is X, then entry for both (i, j) and (j, i) is 1.
4) If (i, j) entry in the SSIM is O, then entry for both (i, j) and (j, i) is 0.

The initial reachability matrix of the KPIs for sustainable maintenance evaluation in rubber industry is presented in Table 5.

Table 4. Structural Self-Interaction Matrix (SSIM) of KPIs.

KPIs	1	2	3	4	5	6	7	8	9	10	11	12	13
1	-	O	V	A	O	O	O	O	O	O	O	O	O
2		-	A	V	O	O	O	O	O	O	A	A	O
3			-	O	O	O	O	O	O	O	A	O	O
4				-	O	O	O	O	O	O	X	O	O
5					-	X	X	A	A	O	O	A	O
6						-	X	A	A	O	O	A	O
7							-	X	X	X	O	A	O
8								-	A	V	O	A	O
9									-	V	O	A	O
10										-	A	A	O
11											-	V	O
12												-	X
13													-

Table 5. Initial Reachability Matrix of KPIs.

KPIs	1	2	3	4	5	6	7	8	9	10	11	12	13
1	1	0	1	0	0	0	0	0	0	0	0	0	0
2	0	1	0	1	0	0	0	0	0	0	0	0	0
3	0	1	1	0	0	0	0	0	0	0	0	0	0
4	1	0	0	1	0	0	0	0	0	0	1	0	0
5	0	0	0	0	1	1	1	0	0	0	0	0	0
6	0	0	0	0	1	1	1	0	0	0	0	0	0
7	0	0	0	0	1	1	1	1	1	1	0	0	0
8	0	0	0	0	1	1	1	1	0	1	0	0	0
9	0	0	0	0	1	1	1	1	1	1	0	0	0
10	0	0	0	0	0	0	1	0	0	1	0	0	0
11	0	1	1	1	0	0	0	0	0	1	1	1	0
12	0	0	0	0	1	1	1	1	1	1	0	1	1
13	0	0	0	0	0	0	0	0	0	0	0	1	1

3.3. Establishing final reachability matrix

The final reachability matrix is developed from the initial reachability matrix by incorporating the transitivities using the following equation:

$$M = M^k = M^{k+1},\ k>1 \quad (1)$$

where k denotes the powers and M is the reachability matrix. Noted that the reachability matrix is under the Boolean operations. The transitivity is a basic assumption of ISM method, which stated that if indicator-A related to indicator -B and indicator -B related to indicator -C, then indicator -A necessarily related to indicator -C [11]. The final reachability matrix of the KPIs for sustainable maintenance evaluation in rubber industry is presented in Table 6.

3.4. Level partitions

From the final reachability matrix, the reachability set and antecedent set [13] for the KPIs can be determined. The reachability set consists of the KPI itself and other KPIs, to which it may relate. The antecedent set consists of the KPI itself and other KPIs, which may relate to it. Then, the intersection of these sets is derived for all KPIs. The KPIs for which the reachability and the intersection sets are the same is placed at the top-level KPIs. After the identification of top-level KPIs, those KPIs discarded from other remaining KPIs. This iteration is continued until the level of all KPIs is obtained. The level partitions are then summarized and shown in Table 7.

Table 6. Final Reachability Matrix of KPIs.

KPIs	1	2	3	4	5	6	7	8	9	10	11	12	13	Driver power
1	1	1	1	1	1	1	1	1	1	1	1	1	1	13
2	1	1	1	1	1	1	1	1	1	1	1	1	1	13
3	1	1	1	1	1	1	1	1	1	1	1	1	1	13
4	1	1	1	1	1	1	1	1	1	1	1	1	1	13
5	0	0	0	0	1	1	1	1	1	1	0	0	0	6
6	0	0	0	0	1	1	1	1	1	1	0	0	0	6
7	0	0	0	0	1	1	1	1	1	1	0	0	0	6
8	0	0	0	0	1	1	1	1	1	1	0	0	0	6
9	0	0	0	0	1	1	1	1	1	1	0	0	0	6
10	0	0	0	0	1	1	1	1	1	1	0	0	0	6
11	1	1	1	1	1	1	1	1	1	1	1	1	1	13
12	0	0	0	0	1	1	1	1	1	1	0	1	1	8
13	0	0	0	0	1	1	1	1	1	1	0	1	1	8
Dependence power	5	5	5	5	13	13	13	13	13	13	5	7	7	

Table 7. Level Partitions of KPIs.

Indicators	Reachability set	Antecendent set	Intersection	Level
1. Energy consumption	All indicators	1,2,3,4,11	1,2,3,4,11	III
2. Lighting and ventilation	All indicators	1,2,3,4,11	1,2,3,4,11	III
3. Emission	All indicators	1,2,3,4,11	1,2,3,4,11	III
4. Working environment	All indicators	1,2,3,4,11	1,2,3,4,11	III
5. Preventive cost	5,6,7,8,9,10	All indicators	5,6,7,8,9,10	I
6. Failure cost	5,6,7,8,9,10	All indicators	5,6,7,8,9,10	I
7. Maintenance procedure	5,6,7,8,9,10	All indicators	5,6,7,8,9,10	I
8. Breakdown rate	5,6,7,8,9,10	All indicators	5,6,7,8,9,10	I
9. Failure rate	5,6,7,8,9,10	All indicators	5,6,7,8,9,10	I
10. Availability rate	5,6,7,8,9,10	All indicators	5,6,7,8,9,10	I
11. Occupational health & safety	All indicators	1,2,3,4,11	1,2,3,4,11	III
12. Employee involvement	5,6,7,8,9,10,12, 13	1,2,3,4,11,12,13	12,13	II
13. Training and education	5,6,7,8,9,10,12, 13	1,2,3,4,11,12,13	12,13	II

The determining process of level partitions for the KPIs consists of three iterations. In the first iteration, six KPIs of economic factor are identified as the KPIs at level I. Then, two KPIs of employee involment and training and education are determined to be placed at level II through the second iteration. Finally, the remaining five indicators are determined into level III. The identified levels of the KPIs can aid in building the digraph and the final model of ISM [11].

3.5. Developing network relationship model

Based on the level partitions of the KPIs, an ISM-based network model is generated . The KPIs are organized in a network structure into three levels as shown in Figure 1. All six indicators of economic factor are regarded as the basic KPIs in evaluating sustainable maintenance in rubber industry. It shows that economic factor still has get more attention from the rubber industry. Level II consists of two indicators of social factor of employee involvement and training and education as intermediate KPIs. The rubber industry must put much efforts to improve these two indicators. One indicator of social factor and all four indicators of environmental factor are indicated to be the leading KPIs in achieving sustainable maintenance in rubber industry. It is not surprisingly since environmental factor is most related to sustainability practices. Of those indicators, employee involvement is regarded as the most influencing KPIs for sustainable maintenance evaluation in the rubber industry.

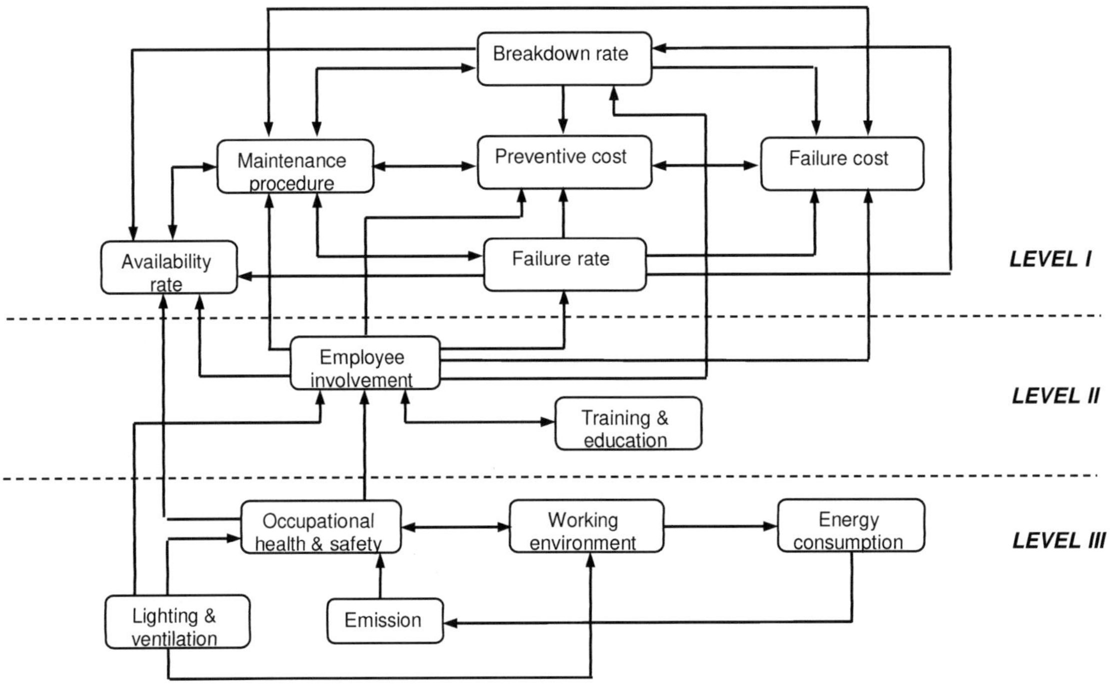

Figure 1. ISM Model

4. Conclusions

This paper has developed an interpretive structural model of KPIs for sustainable maintenance evaluation in rubber industry. The KPIs are structured into three levels. Economic factor are regarded as the basic indicators. On the other hand, environmental factor are indicated as the leading indicators in achieving sustainable maintenance in rubber industry. The rubber industry should put much attention into the indicators of environmental factor. The ISM model provides a better understanding of the relationship amongst KPIs. The model can aid the decision makers with a more realistic representation of relationships amongst KPIs for sustainable maintenance evaluation in rubber industry. Future study will further incorporate the model into Analytical Network Process (ANP) methodology to the development of sustainable maintenance evaluation model in rubber industry.

Acknowledgments

Authors would like to thank to Andalas University and Ministry of Research, Technology, and Higher Education, Indonesia.

References

[1] WCED (World Commission on Environment and Development) 1987 Our common future, (Oxford University Press: Oxford UK)

[2] Nambiar A N 2010 Challenges in sustainable manufacturing *Proc. Int. Conf. on Industrial Engineering and Operations Management* (Dhaka: Bangladesh) p 1

[3] Sari E, Shaharoun A M, Ma'aram A and Yazid A M 2015 Sustainable maintenance performance measures: a pilot survey in Malaysian automotive companies *Procedia CIRP* **26** p 443

[4] Jasiulewicz-Kaczmarek M 2013 Sustainability: orientation in maintenance management - Theoretical Background *EcoProduction and Logistics* ed Golinska P (Springer-Verlag) p 117

[5] Bakshi B R and Fiksel J 2003 The quest for sustainability: challenges for process systems engineering *AIChE Journal* **49** (6) p 1350

[6] Gasparatos A, El-Haram M and Horner M 2008 A critical review of reductionist approaches for assessing the progress towards sustainability *Environmental Impact Assessment Review* **28** p 286

[7] Inwai R C, Diaotrakun R and Thaiupathum T 2013 Key indicators for maintenance performance measurement: the aircraft galley and associated equipment manufacturer case study *Proc IEEE Int. Conf. On Service Systems and Service Management* (Hong Kong: China) p 844

[8] Liyanage J P, Badurdeen F and Ratnayake R M C 2013 Industrial asset maintenance and sustainability performance: economical, environmental, and societal implications *Handbook of Maintenance Management and Engineering* ed Ben-Daya M, Duffua S O, Raouf A, Knezevic J and Ait-Kadi D (London: Springer-Verlag) p 665

[9] Iung B and Levrat E 2014 Advanced maintenance services for promoting sustainability *Procedia CIRP* **22** p 15

[10] Amrina E and Aridharma D 2016 Sustainable maintenance performance evaluation model for cement industry *Proc of IEEE IEEM* p 350

[11] Kannan G, Pokharel S and Kumar P S 2009 A hybrid approach using ISM and fuzzy TOPSIS for the selection of reverse logistics provider *Resource, Conservation & Recycling* **54** p 28

[12] Chen S P and Wu W Y 2010 A systematic procedure to evaluate an automobile manufacturer-distributor partnership *European J. of Operational Research* 205 p 687

[13] Warfield J N 1974 Developing interconnection of complex structural modeling *IEEE Transactions on Systems, Man, and Cybernetics* **4**(1) p 81

IOP Conf. Series: Materials Science and Engineering **319** (2018) 012056 doi:10.1088/1757-899X/319/1/012056

Smart mobile robot system for rubbish collection

Mohammed A H Ali and Tan Sien Siang

Faculty of Manufacturing Engineering, Universiti Malaysia Pahang, 26600 Pekan, Malaysia

E-mail: hashem@ump.edu.my

Abstract. This paper records the research and procedures of developing a smart mobility robot with detection system to collect rubbish. The objective of this paper is to design a mobile robot that can detect and recognize medium-size rubbish such as drinking cans. Besides that, the objective is also to design a mobile robot with the ability to estimate the position of rubbish from the robot. In addition, the mobile robot is also able to approach the rubbish based on position of rubbish. This paper explained about the types of image processing, detection and recognition methods and image filters. This project implements RGB subtraction method as the prior system. Other than that, algorithm for distance measurement based on image plane is implemented in this project. This project is limited to use computer webcam as the sensor. Secondly, the robot is only able to approach the nearest rubbish in the same views of camera vision and any rubbish that contain RGB colour components on its body.

1. Introduction

As the technology keep on growing each and every day, the life of human has become easier and convenient together with the growth of technology. Humans are helped by the technology every day in daily life. Technology today has developed into creating more automatic devices or robots which will help human to save their time to accomplish simple tasks [1]. The growth of technology has been incredibly great that even for simple task like cleaning is included in improvement and development. Cleaning has been a task that requires human energy and time to accomplish. However, people nowadays do not have the interest in cleaning. Hence, the discussion of this paper is about designing a mobile robot with rubbish detection and recognition system and its application [2].

The development of this mobile robot enables human to work at least effort to pick up rubbish. Especially for those people that having problem bending down to the floor to pick up rubbish [1]. There are other types of robotic cleaning machine that have been available in the market, however they did not comes with the application of differentiating the trash into categories. In this project, the smart dustbin is able to differentiate the type of trash it needs to clean by implementing DRI. DRI is the term stands for Detection, Recognition and Identification. Detection is to identify object; recognition is to match the type of object; and identification is specify or categorize the object.

Path planning of robot is one of the most complicated problems that occurs during WMR in outdoor or indoor environments. In path planning approaches, the path trajectory is planned continuously between the start and goal positions with avoiding the obstacles and determine the location of rubbish [3-5]. The controlled of WMR have be done to accomplish WMR rubbish collection task in the right manner [6-8].

2. Overview of the System Architecture

The system design consists of a camera as the sensor, a computer with MATLAB software as the core channel, two MC30 motor driver, an Arduino Uno to control gear motor through motor driver, and 3 wheel mobile robot prototype for the system as shown in Fig. 1, 2.

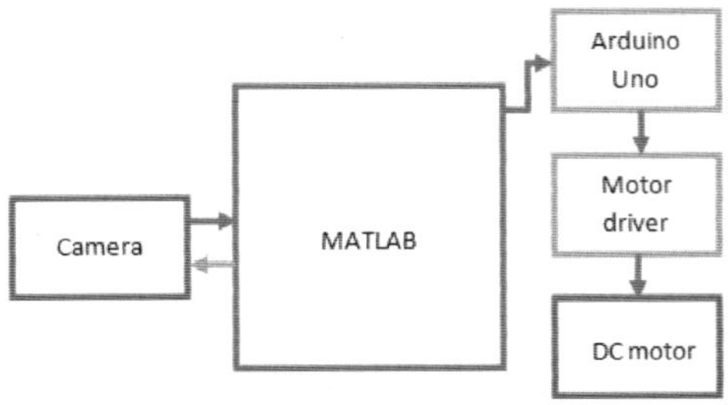

Figure 1. System architecture of Smart Mobile Robot System

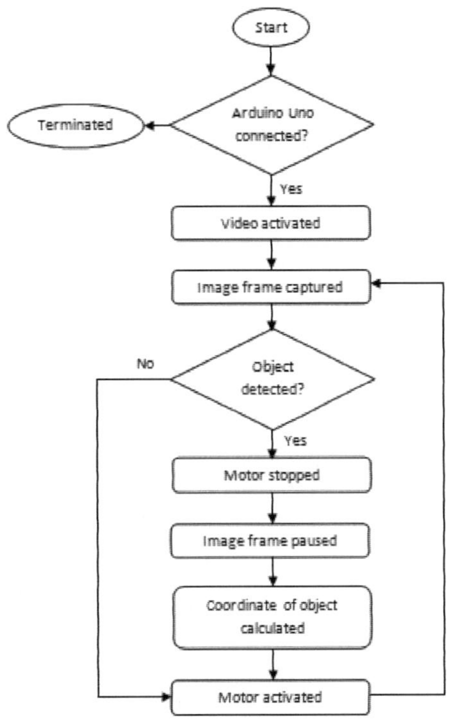

Figure 2. Overall system flow chart

3. Methodology

The methodology activities of this project are as follows:

3.1 Development of Detection and Recognition System

RBG detection is implemented in this project to detect the rubbish around a certain area of view. By retaining the RGB element and adjusting each threshold value, any objects that contain any colour of RGB within the threshold can be recognized and detected.

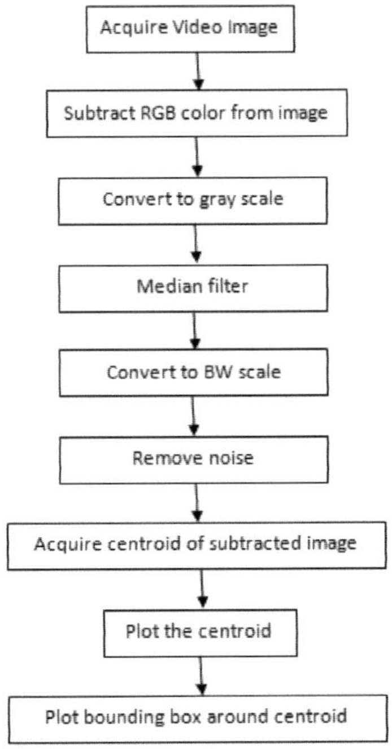

Figure 3. Object detection system process flow

3.2 Image to Reality Coordination System

The relationship between the image plane and the reality coordinate system depends on several parameters of camera. In this case, any coordinate in image plane can be determine by using the image size in pixel (ie. 1280 horizontal pixel * 720 vertical pixel). The coordinate of object from the camera lens that captured in the screen is normally scaled and inverted horizontally and vertically. If the distance of an object in reality which is perpendicular to the lens is *D*, and the distance of image plane from lens is I_{cam}, their relationship is called magnification factor:

$$M = \frac{D}{I_{cam}} \quad (1)$$

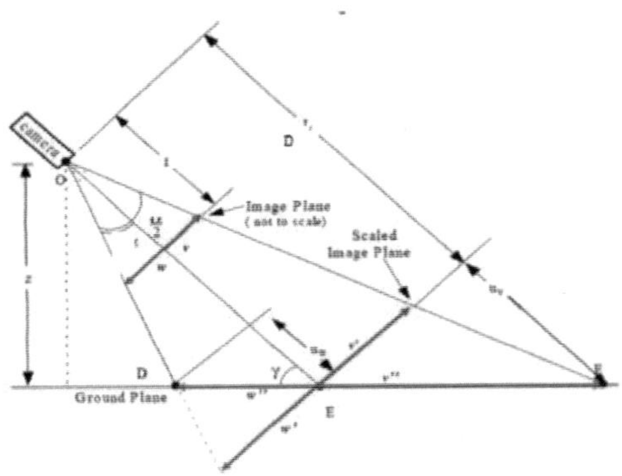

Figure 4. Elevation view of camera

The position of any point on the vertical plane of reality, y_g is as follow:

$$y_g = \frac{z y_i}{\sin(\gamma)(I_{cam}\sin(\gamma) - y_i\cos(\gamma))} \quad (2)$$

IOP Conf. Series: Materials Science and Engineering **319** (2018) 012056 doi:10.1088/1757-899X/319/1/012056

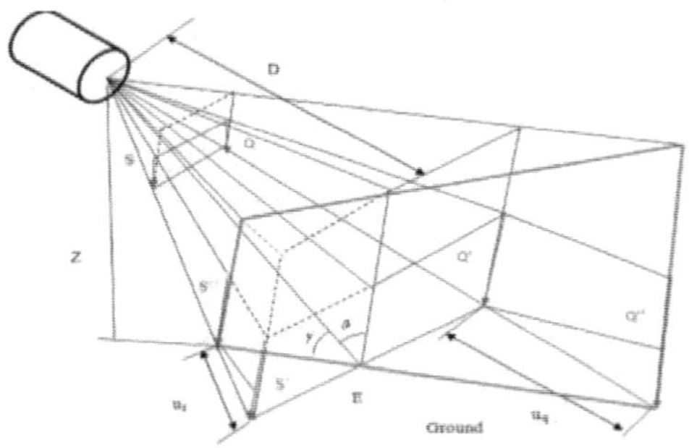

Figure 5. Camera transformations

The position of any point on horizontal plane of reality, x_g is as follow:

$$x_g = \frac{z\cos(\vartheta) - Mx_i\sin(\gamma)}{\sin(\gamma)\,(l_{cam}\cos(\vartheta) - |x_i|)}|x_i| \quad (3)$$

These formulas are base on basic calculations without considering the effect of image distortion. Hence the result is the estimated position based on the image captured.

3.3 Mobile Robot Prototype

The mobile robot is a three wheel robot which has two main wheels with DC gear motor and one free wheel. This mobile robot is fabricated using 40mmx40mm aluminium profile and 10mm thick of aluminium plate. The mobile robot main body has 2 layers, where bottom layer is fixed with two DC gear motor and electrical components while the upper layer is to mount sensors and for other uses purpose.

1) Mobility System

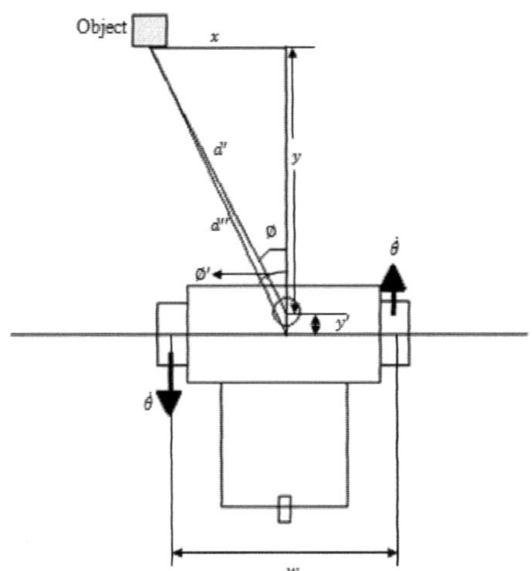

Figure 6. FBD of robot angle to object

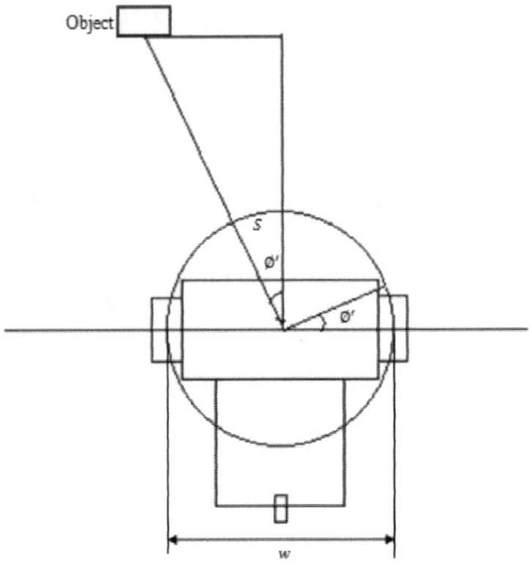

Figure 7. Rotary trajectory of robot

The angle of the object from the center of main body is as follow:

$$\emptyset' = tan(\frac{x}{y+y'}) \tag{4}$$

The distance required to travelled by the wheel to meet the object perpendicularly with the camera at any angle can be simplified as follow:

$$S = \frac{w}{2}\left|tan(\frac{x}{y+y'})\right| \tag{5}$$

2) Error Calculation for Turning Speed of Wheel
For the robot to travel in straight line with least error due to misalignment of the robot body, the angular speed required by both wheel can be acquire by using the following formula:

$$\theta_{error} = r\frac{(\dot{\theta}_L - \dot{\theta}_R)}{w/2} \tag{6}$$

Where $\dot{\theta}_L$ indicates the angular velocity of left wheel and $\dot{\theta}_R$ is the angular velocity of right wheel. The relationship between PWM and speed of wheel can be determined using the following formula:

$$\frac{PWM}{PWM_i} = \frac{v}{r\dot{\theta}_i} \tag{7}$$

4. Results and Discussion

The results are illustrated as follows:

4.1 Detection and Recognition System

Fig. 8(a) shows the image of red drinking can is subtracted out from the surrounding and converted to gray scale. Fig. 8 (b) shows the image is filtered by calculating the large pixel with median. Fig. 8 (c) shows the image is converted to binary scale. Fig. 8 (d) is the result from removing noises. Fig. 8(e) is the result after recognizing the drinking can and bounded with rectangle box and centroid. Fig. 8 (f) shows the result of detection and recognition in real image.

IOP Conf. Series: Materials Science and Engineering **319** (2018) 012056 doi:10.1088/1757-899X/319/1/012056

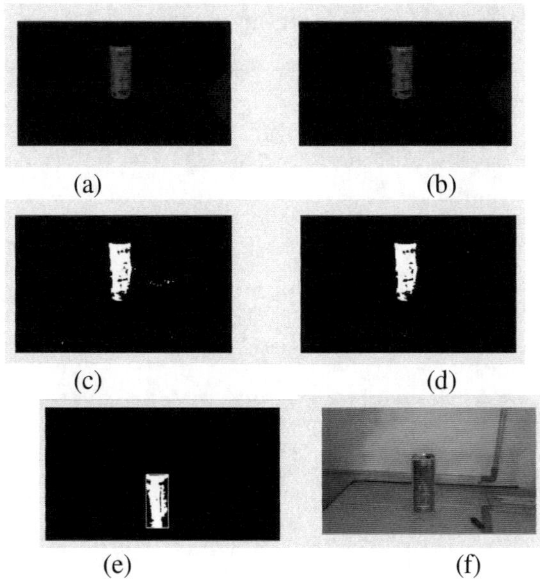

Figure 8. Image Filter Process

4.2 Image to Reality Coordination System (Magnification factor calculation)

A paper is placed in parallel with the webcam lens and scaled in centimetre. The distance of paper from camera is 18cm. The length of the paper captured by image is approximately 20cm as in Fig. 9. By applying magnification formula in section 2 eq. 2, the estimated value of *Icam* is calculated which is 1152cm/pixel.

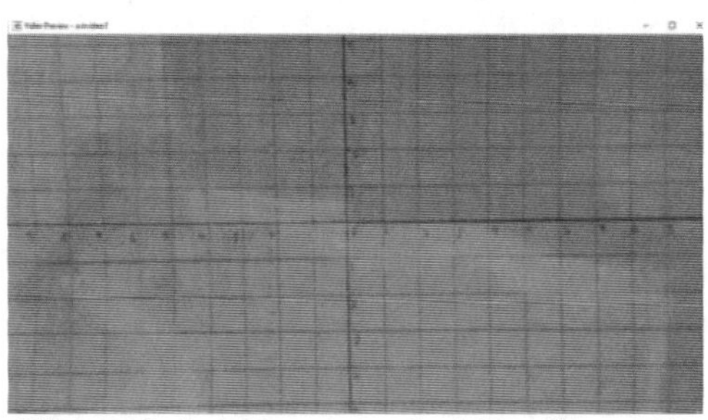

Figure 9. Image captured with reality scale

4.3 Image to reality coordinate comparison

Fig. 10. shows the x and y coordinate calculated by using the image pixel coordinate. The coordinate shows (x,y) distance in cm and the value at bottom is the displacement of the object from camera.

Figure 10. Object coordinate based on image

Table 1. Y-coordinate (vertical) result of reality and result from image

Real Y distance (cm)	Image Y distance (cm)	Difference (cm)	Ratio difference (Y_{image} : 1)
10	8	2	1.250
20	17	3	1.176
30	26	4	1.154
40	35	5	1.143
50	45	5	1.111
60	54	6	1.111

Table 1 shows the result of Y (vertical) distance show in image and also their respective reality distance. The results show that the difference of distance from image and reality distance increases. This shows that as the object is moved away more from the camera, the difference will be larger.

Table 2. X-coordinate (horizontal) result of reality and result from image

Real X distance (cm)	Image X distance (cm)	Difference (cm)	Ratio difference (X_{image} : 1)
10	9	1	1.111
20	18	2	1.111
30	27	3	1.111
40	37	3	1.081
50	45	5	1.111
60	53	7	1.132

Table 2 shows the result of X (horizontal) distance show in image and also their respective reality distance. The results show that the difference of distance from image and reality distance increases. This shows that as the object is moved more to the right or left in the image, the difference will be larger.

4.4 Error of Angle Displaced

The error is tested by giving both DC gear motor the same PWM value and recording the horizontal displacement travelled by the mobile robot. Hence calculate the angle displaced using the equation (6).

Table 3. Result of angle displaced from origin

A (Y travelled /cm)	B (X travelled /cm)	Angle displaced, θ_{error} (°)
180	7	-2.227
181	-1	-0.317
180	5.5	1.750
178	0.7	0.225
180	9.5	3.021
180	-12	-3.814

180	-3	-0.955
180	-1.7	-0.541
180	-5.1	-1.623
180	-2	-0.637
180	0.7	0.223
Average		-0.445

Negative sign indicates direction

Table 3 shows the result of angle displaced by mobile robot when tend to move in straight line. The error of angle displaced is used to calculate the angular velocity of wheel and hence determine the estimated PWM value as input for DC gear motor.

4.5 Wheel Speed

The speed of wheel is determined using the same method as the error of angle displaced. By using the data collected from the experiment, the relationship between PWM and speed of wheel can be related by using formula at Chapter 3 equation (12).

Table 4. Speed of wheel based on time taken and displacement

A (Y travelled /cm)	**B (X travelled /cm)**	**Time taken (s)**	**Speed, v (cm/s)**
180	7	22.47	8.017
181	-1	22.73	7.963
180	5.5	22.91	7.860
178	0.7	22.75	7.824
180	9.5	23.47	7.680
180	-12	24.53	7.354
180	-3	22.85	7.878
180	-1.7	23.22	7.752
180	-5.1	22.90	7.863
180	-2	23.16	7.773
180	0.7	23.54	7.647
	Average		7.783

Negative sign indicates direction

5. Conclusion

The system is able to detect and recognize rubbish which has RGB (red, green, blue) colour components and approach the rubbish for more action. The system is able to work continuously after approaching one object. It will continue to move forward until the detection and recognition system senses another object. The detection of object is dependent on the brightness of the surroundings. The light reflection will alter the colour captured by the camera and causes failure of detection. the errors occur in this system may due to distortion of image where the image is distorted at the border of four sides of the image. The distortion is normally associated with zoom lenses. The types of distortion occurs is differently according to the focal length of the camera. The mobility error is mainly due to alignment of the robot wheel which causes the mobile robot unable to move according to the calculation. The type of free wheel is also a factor to the angle change. The free wheel must able to withstand the pressure of the weight of mobile robot and able to move freely in the meantime.

Acknowledgment

Authors would like to thank the Ministry of High Education and Universiti Malaysia Pahang for supporting this project under research grant RDU160131.

IOP Conf. Series: Materials Science and Engineering **319** (2018) 012056 doi:10.1088/1757-899X/319/1/012056

References

[1] Hani Hunud A. Kadouf, 2013. Colour-based Object Detection and Tracking for Autonomous Quadrotor UAV

[2] M.A. Ali, M. Mailah, W.A. Yussof, Z.B. Hamedon, Z.B. Yussof, A.P. Majeed. Sensors Fusion based Online Mapping and Features Extraction of Mobile Robot in the Road Following and Roundabout. IOP Conference Series: Mat. Sc. and Eng. Vol. 114, no. 2016, pp. 1-12. 2016.

[3] M. Ali, M. Mailah and H.H. Tang, "A novel Approach for Visibility Search Graph Based Path Planning," *13th Int. Conf. Robot., Control Manuf. Syst.*, Kuala-Lumpur, Malaysia, 2-4th April, 2013, pp. 44-49.

[4] M. Ali, M. Mailah and H.H.Tang, "Implementation of Laser Simulator Search Graph for Detection and Path Planning in Roundabout Environments," *WSEAS Trans. Signal Process.*, vol. 10, no. 2014, pp. 118-126, 2014.

[5] M.A. Ali, M. Mailah, T.H. Hing. Path Planning of Mobile Robot for Autonomous Navigation of Road Roundabout Intersection. Inter. Jour. of Mech; vol. 6, no.4. pp 203-11. 2012.

[6] MAH Ali, WABW Yusoff, ZB Hamedon, ZBM Yusssof, M Mailah. Global mobile robot path planning using laser simulator. 2016 2nd IEEE International on Robotics and Manufacturing Automation (ROMA),

[7] M A H Ali, M Mailah, W. A. Yussof, Z. B. Hamedon, Z. M. Yussof. Wheeled Mobile Robot Path Control in a Complex Trajectory using Hybrid Methods. 15th Inter conf. on Robot, Control and Manuf. Systems. Kuala-lumpur, Malaysia, April 2-4 April. 2015, pp 38-43.

[8] Mohammed A. H. Ali, Musa Mailah, Wan Azhar B. Yussof, Zamzuri Bin Hamedon, Zulkili B. M. Yussof. (2017) An Intelligent Robust Control of Wheeled Mobile Robot in Restricted Environment. *International Journal of Control Systems and Robotics*, 2, 6-11

Control of wheeled mobile robot in restricted environment

Mohammed A H Ali and Chang Yong En

Faculty of Manufacturing Engineering, Universiti Malaysia Pahang, 26600 Pekan, Malaysia

E-mail: hashem@ump.edu.my

Abstract. This paper presents a simulation and practical control system for wheeled mobile robot in restricted environment. A wheeled mobile robot with 3 wheels is fabricated and controlled by proportional derivative active force control (PD-AFC) to move in a pre-planned restricted environment to maintain the tracking errors at zero level. A control system with two loops, outer by PD controller and inner loop by Active Force Control, are designed to control the wheeled mobile robot. Fuzzy logic controller is implemented in the Active force Control to estimate the inertia matrix that will be used to calculate the actual torque applied on the wheeled mobile robot. The mobile robot is tested in two different trajectories, namely are circular and straight path. The actual path and desired path are compared.

1. Introduction

Mobile robot plays an important role in industrial society today. There are several types of mobile robot, such as legged, wheeled and tracked mobile robot [1].. Wheeled mobile robots are used in many applications especially in factory and some restricted environments that human cannot perform well [2, 3].

The motion of the wheeled mobile robot is controlled by controller such as Proportional-Integral-Derivative (PID) [4, 5], Fuzzy logic control (FLC) [5-7] and Active force control (AFC) [2,3,8]. By implementing a useful controller, the motion of the wheeled mobile robot will become stable and able to move in different types of trajectories with zero tracking errors [9].

Moreover, tele-operated system which controls the robot remotely, will cause inaccurate control operation of mobile robot. Tele-operated mobile robot control system may need the manpower and some error un-estimated disturbances may occur during the control operation. Therefore, an appropriate controller should be used to increase the predictability of the mobile robot's behaviour, decrease tracking errors and can make sure that the control of mobile robot is robust enough.

In this research, the Active force control (AFC) algorithm will be implemented to a three wheeled mobile robot. The motion of the mobile robot is planned firstly and then the actual of the wheeled mobile robot are compared with desired paths. Moreover, both kinematic and dynamic modelling of the mobile robot are derived and considered. In this project, it is only focused on the three wheels differential drive mobile robot which will be tested in circular and straight path.

2. Modelling of robot

The mobile robot comprises of three wheels driving system, as show in Fig. 1. The motion of the mobile robot can be described in global (x, y, φ) and local coordinate system ($\dot{\theta}_r$, $\dot{\theta}_l$).

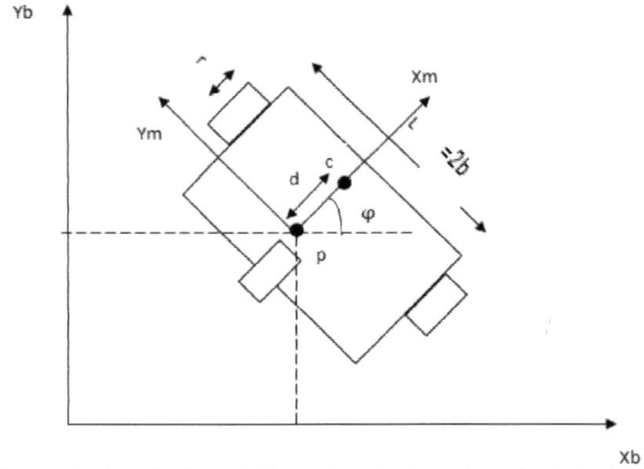

Figure 1. Motion of wheeled mobile robot in local and global coordinate system

Where r is the radius of the wheel and L is the distance between the wheels of mobile robot.

2.1. Kinematic modelling

The mobile robot is moved with two differential drive wheels and a castor wheel, so the velocity is:

$$v = \begin{bmatrix} \frac{r}{2} & \frac{r}{2} \end{bmatrix} \begin{bmatrix} \dot{\theta}_R \\ \dot{\theta}_L \end{bmatrix} \quad (1)$$

The difference between angular velocity of right and left wheels can be illustrated as in Eq. 2.

$$\dot{\varphi} = \begin{bmatrix} \frac{r}{2b} & -\frac{r}{2b} \end{bmatrix} \begin{bmatrix} \dot{\theta}_R \\ \dot{\theta}_L \end{bmatrix} \quad (2)$$

The local coordinate system is show in Eq. 3

$$\begin{bmatrix} \dot{X} \\ \dot{Y} \\ \dot{\varphi} \end{bmatrix} = \begin{bmatrix} \cos\varphi & -d\sin\varphi \\ \sin\varphi & d\cos\varphi \\ 0 & 1 \end{bmatrix} \begin{bmatrix} V \\ \dot{\varphi} \end{bmatrix} \quad (3)$$

The Equation (4) thus can be written in the term of angular velocity $\dot{\theta}_R$ and $\dot{\theta}_L$ as show in Eq. 4

$$\begin{bmatrix} \dot{X} \\ \dot{Y} \\ \dot{\varphi} \end{bmatrix} = \begin{bmatrix} \frac{r}{2}\cos\varphi - \frac{dr}{2b}\sin\varphi & \frac{r}{2}\cos\varphi + \frac{dr}{2b}\sin\varphi \\ \frac{r}{2}\sin\varphi - \frac{dr}{2b}\cos\varphi & \frac{r}{2}\sin\varphi - \frac{dr}{2b}\cos\varphi \\ \frac{r}{2b} & -\frac{r}{2b} \end{bmatrix} \begin{bmatrix} \dot{\theta}_R \\ \dot{\theta}_L \end{bmatrix} \quad (4)$$

2.2 Dynamic modelling

The dynamic of the mobile robot can be derived using Lagrange equation as show in Eq. 5.

$$\frac{d}{dt}\left(\frac{\delta k}{\delta \dot{q}}\right) - \frac{\delta k}{\delta q} = \tau_i - A^T(q)\lambda \quad (5)$$

Where k = kinetic energy, q = coordinate system, τ_i = exerted torque on the robot, $A^T(q)$ = constraints of robot.

The total kinetic energy can be written as in Eq. 6:

$$KE_{total} = KE_{body} + KE_{RW} + KE_{LW} \quad (6)$$

The general form of Lagrange Equation:

$$M(q)\ddot{q} + V(q,\dot{q})\dot{q} = B(q)r - A^T(q)^{\lambda} \quad (7)$$

Where:

$$M(q)=\begin{bmatrix} m & 0 & 2m_w d\sin\varphi & 0 & 0 \\ 0 & m & -2m_w d\cos\varphi & 0 & 0 \\ 2m_w d\sin\varphi & -2m_w d\cos\varphi & I & 0 & 0 \\ 0 & 0 & 0 & I_w & 0 \\ 0 & 0 & 0 & 0 & I_w \end{bmatrix}$$

$$V(q,\dot{q})\begin{bmatrix} 2m_w d\dot{\varphi}^2\cos\varphi \\ 2m_w d\dot{\varphi}^2\sin\varphi \\ 0 \\ 0 \\ 0 \end{bmatrix} \quad \lambda=\begin{bmatrix}\lambda_1\\ \lambda_2\\ \lambda_3\end{bmatrix} \quad B(q)r=\begin{bmatrix}\tau_R & 0\\ 0 & \tau_L\end{bmatrix}$$

3. Robot Design and Fabrication

Fig. 2 shows the 3D view of wheeled mobile robot. The wheeled mobile robot is non-holonomic mobile robot, which consists of two differential wheels and one castor wheel.

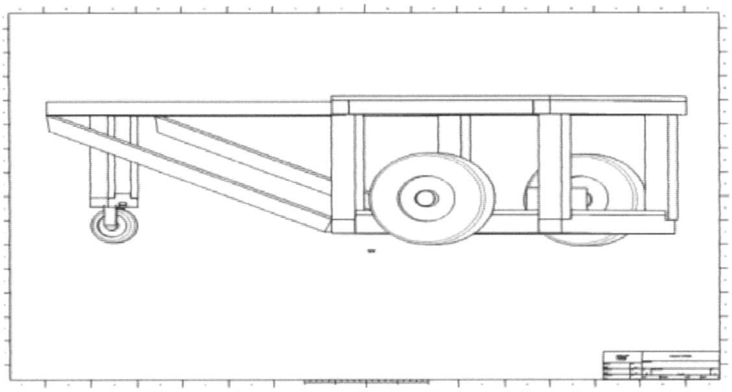

Figure 2. Design of wheeled mobile robot

Fig. 3 shows the fabricated wheeled mobile robot. The body structure of the mobile robot is made of 30x30mm and 40x40 aluminium profile. 10mm thick aluminium plate is used as the cases of the mobile robot. The mechanical components are illustrated in Table 1.

Figure 3. Wheeled mobile robot

Table 1. Mechanical part fabrication

No.	Parts	Material	Process
1	Body support	Aluminium	Sawing & Drilling
2	Body casing	Aluminium	Sawing & Drilling
3	Motor shaft	Mild steel	Turning, Drilling, Electric Discharges & Welding

4. Robot Controller Design

Fig. 4 shows the design of PD-AFC implemented in the control of wheeled mobile robot. The control loop consists of two loops, outer loop and inner loop. PD controller is used in the outer loop of the control system. The inner loop of the control loop is AFC control strategy, which is able to make the dynamic system stable and robust.

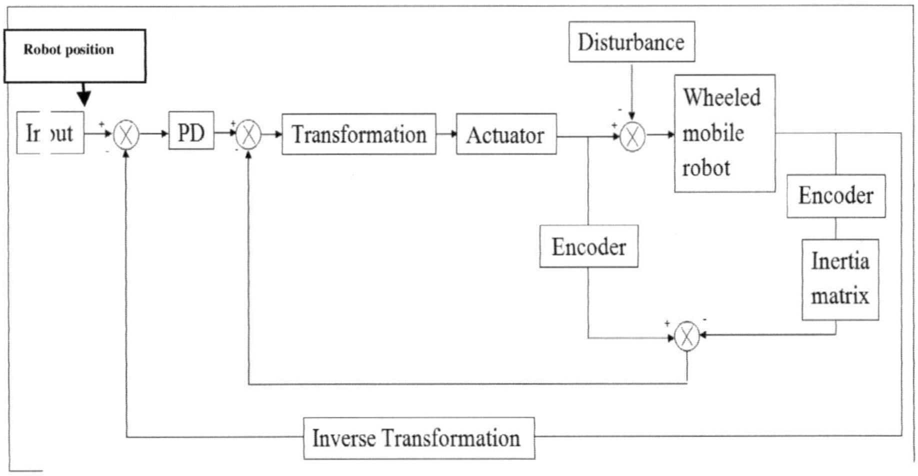

Figure 4. PD-AFC scheme

5. Result and Discussion

5.1 MATLAB simulation

The simulation of the control of the wheeled mobile robot is simulated in MATLAB SIMULINK. To control the wheeled mobile robot in restricted environment, the mobile robot is simulated to move in two different trajectories, which are circular path and straight path. The references path and the actual path are compared in the section below.

In the simulation, the assumptions of the wheel mobile robot parameters are shown in the Table 2. The parameter of the robot included: mass of the robot, mass of the wheel, movement of inertia of the robot and movement of inertia of the wheel, radius of the wheel, length of the mobile robot and distance between the references point and centre point.

Table 2. Wheel mobile robot parameters

Mass of the robot, m	36 kg
Mass of the wheel, m_w	0.8 kg
Moment of inertia of the robot, I	18 kgm^2
Moment of inertia of the wheel, I_w	0.00818kgm^2
Radius of the wheel ,r	0.1m
Length of the mobile robot., b	0.02m
Distance between references point and centre point, d	0.5m

The simulation of wheeled mobile robot in circular path is shown in the Fig. 5 and Fig. 6. The reference path and actual path of the wheeled mobile robot are compared.

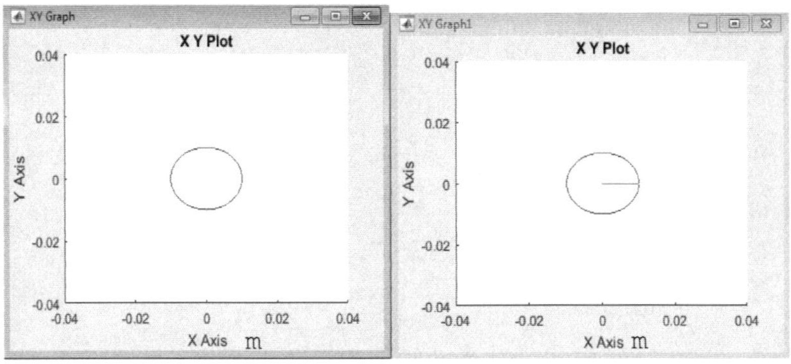

(a) (b)

Figure 5. Straight path control in practical control : (a) reference (b) simulated

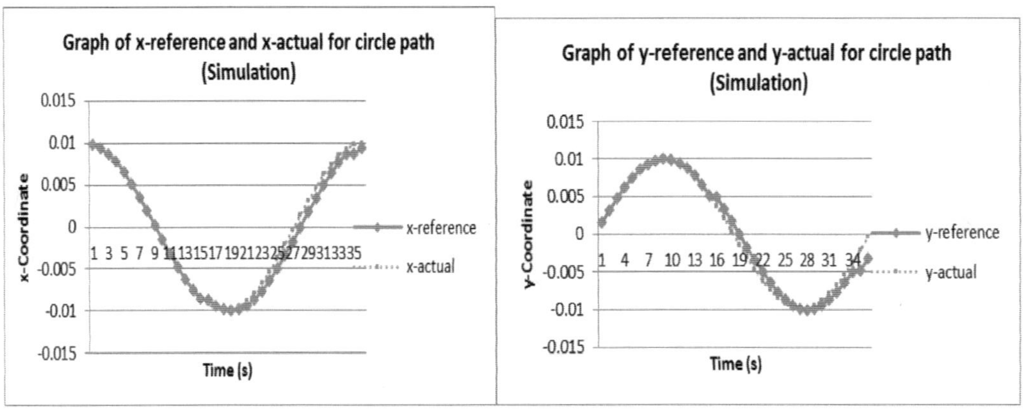

(a) (b)

Figure 6. Tracking error for straight path in practical: (a) x-direction (b) y-direction

Fig. 6 shows the comparison of x-references and x-actual for circular path and the graph of y-references and y-actual for circular path. The x and y references path is compared with the x and y actual value from the simulation for every 10 s. The x and y equation of the graph is given in Eq. 3 and 4. It is noticed, that there is only slightly different between the xy-references values and xy-actual values, which mean that the errors of the y are small. So it can be said that the system is robust.

5.2 Practical part

The wheeled mobile robot is controlled practically in straight line as show in Fig. 7 and 8. The mobile robot is set to move in straight line for 20s before it is automatically stopped. From the Fig. 7 and 8, it can be seen that the mobile robot is starting to run at the origin point until it reach the end point. By comparing both Fig. 7 a and b, the actual path is not same with the reference path. From Fig. 8, it clearly shows that there is a big error occurs between the reference path and actual path.

Fig. 7 shows the tracking errors of x-references and x-actual for straight path and tracking errors of y-references and y-actual for straight path in practical. In the practical part, the mobile robot is set to move in straight line for 20s. The x and y references path are compared with the x and y actual value for 20s. From Fig 8, it shows that the mobile is not able to move in straight line. By comparing the x-actual and x-reference graph, at first 8s, the mobile robot is move in the reference path, after that the mobile robot is move to the right. While for the y-actual and y-reference graph, the mobile robot is stable at first 9s, which means that it is able to move in straight line by following the references coordinate for first few seconds, after that the mobile robot is not moved similar to the references coordinate.

IOP Conf. Series: Materials Science and Engineering **319** (2018) 012057 doi:10.1088/1757-899X/319/1/012057

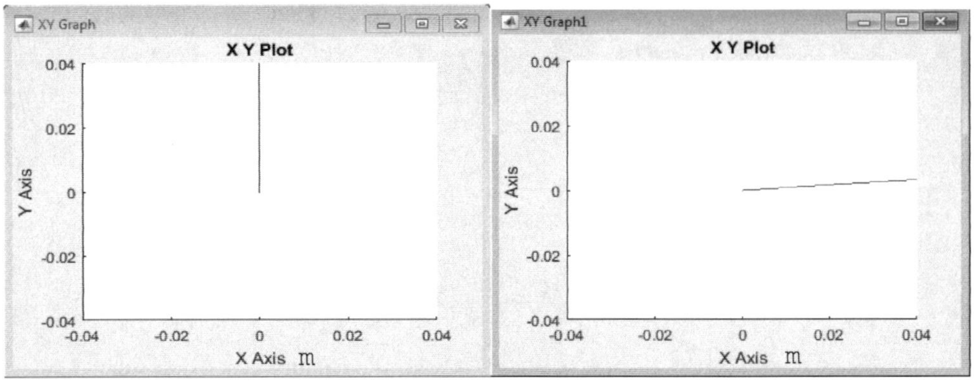

(b) (b)

Figure 7. Straight path control in practical control : (a) reference (b) actual

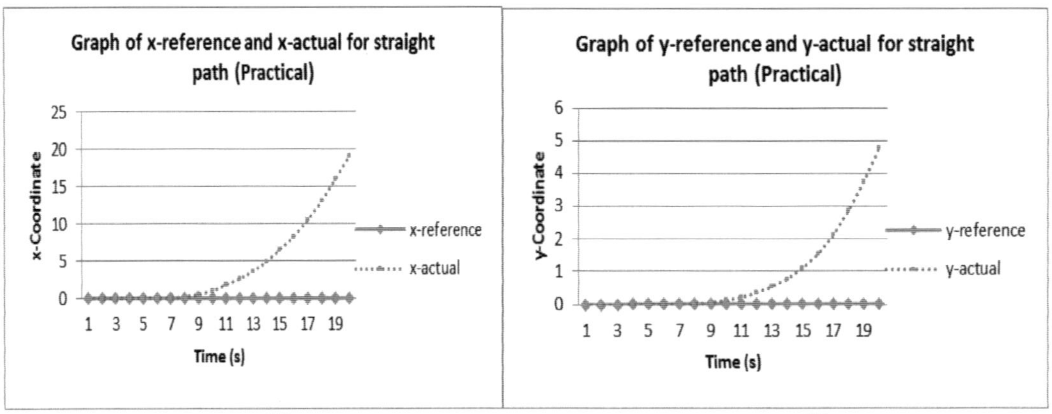

(b) (b)

Figure 8. Tracking error for straight path in practical: (a) x-direction (b) y-direction

The bad performance in the practical part may due to many reasons. In the practical part, in order to make the wheeled mobile robot to move in its reference path, there are many factors need to be considered, such as the weight of the mobile robot, friction of the wheel of the mobile robot with the road, the friction of the motor shaft with the bearing, and the motors characteristic.

From the design, the mobile robot is supported by the bearing and the wheel, in addition, the mobile robot is made by aluminum plate and aluminum profile, so the weight of the mobile robot is around 35kg to 40kg which will cause the friction between the shaft and bearing become high. Moreover, due to the parallax error during assemble the bearing and motor shaft on the body of the mobile robot, it may cause the wheel not align and unbalance friction act on the wheel. This factor may affect the mobile robot unable move in straight line. Besides that, different motor characteristic also will cause the mobile robot unable to move in straight line. During practical, the motor of the mobile robot are not starting at the same time, so in the starting point, the mobile robot already not move in straight line.

6. Conclusion

In a nutshell, the objective of the project is fully achieved, which is fabricated the wheeled mobile robot, apply the Proportional Derivative-Active force control (PD-AFC) algorithm with different types of restricted environments, and control the wheeled mobile robot in restricted environment with zero track error. However, in both straight and circular path, the errors between the references value and actual values are approximately zero. So it is conclude that the PD-AFC algorithm is able to control the wheeled mobile robot in simulation part with zero track error. For the practical part, due to many factors, the mobile robot is not follow exactly the reference path, so it needs further improvement.

Acknowledgment
Authors would like to thank the Ministry of High Education and Universiti Malaysia Pahang for supporting this project under research grant RDU160131.

References

[1] Elyoussef, E. S., De Pieri, E. R., Moreno, U. F., & Jungers, M. (2012). *Super-Twisting Sliding Modes Tracking Control of a Nonholonomic Wheeled Mobile Robot. IFAC Proceedings Volumes* (Vol. 45). IFAC. https://doi.org/10.3182/20120905-3-HR-2030.00124
[2] Ali, M. A. H., Yusoff, W. A. B. W., Hamedon, Z. B., & Yusssof, Z. B. M. (2015). Wheeled Mobile Robot Control in Restricted Environment using PD-Active Force Control Scheme.
[3] Abdalla, T. Y. (2013). Trajectory Tracking Control for Mobile Robot using Wavelet Network, *74*(3), 32–37.
[4] Astrom, K. J., & Hagglund, T. (2011). Auto-tuners for PID Controllers.
[5] Godjevac, J. (2005). Comparison Between Pid & Fuuzy Logic.Pdf. *Comparison Between Pid & Fuzzy Logic*.
[6] Hashemi, E., Ghaffari Jadidi, M., & Ghaffari Jadidi, N. (2011). Model-based PIfuzzy control of four-wheeled omni-directional mobile robots. *Robotics and Autonomous Systems*, *59*(11), 930–942. https://doi.org/10.1016/j.robot.2011.07.002
[7] Leena, N., & Saju, K. K. (2016). Modelling and Trajectory Tracking of Wheeled Mobile Robots. *Procedia Technology*, *24*, 538–545. https://doi.org/10.1016/j.protcy.2016.05.094
[8] Mailah, M., & Priyandoko, G. (2007). Simulation of a suspension system with adaptive fuzzy active force control. *International Journal of Simulation Modelling*, *6*(1), 25–36. https://doi.org/10.2507/IJSIMM06(1)3.079
[9] Mohammed A. H. Ali, Musa Mailah, Wan Azhar B. Yussof, Zamzuri Bin Hamedon, Zulkili B. M. Yussof. (2017) An Intelligent Robust Control of Wheeled Mobile Robot in Restricted Environment. *International Journal of Control Systems and Robotics,* **2**, 6-11

IOP Conf. Series: Materials Science and Engineering **319** (2018) 012058 doi:10.1088/1757-899X/319/1/012058

The Determination of Production and Distribution Policy in Push-Pull Production Chain with Supply Hub as the Junction Point

A T Sinaga and R Wangsaputra

Industrial Engineering and Management, Institut Teknologi Bandung, Ganeca Street 10, Bandung, West Java, Indonesia, 40132

astridtheresia27@gmail.com; rwangsa@ti.itb.ac.id

Abstract. The development of technology causes the needs of products and services become increasingly complex, diverse, and fluctuating. This causes the level of inter-company dependencies within a production chains increased. To be able to compete, efficiency improvements need to be done collaboratively in the production chain network. One of the efforts to increase efficiency is to harmonize production and distribution activities in the production chain network. This paper describes the harmonization of production and distribution activities by applying the use of push-pull system and supply hub in the production chain between two companies. The research methodology begins with conducting empirical and literature studies, formulating research questions, developing mathematical models, conducting trials and analyses, and taking conclusions. The relationship between the two companies is described in the MINLP mathematical model with the total cost of production chain as the objective function. Decisions generated by the mathematical models are the size of production lot, size of delivery lot, number of kanban, frequency of delivery, and the number of understock and overstock lot.

1. Introduction

An efficient production chain can be achieved by harmonizing production and distribution activities [12]. Distribution activities involve more than a company. Therefore, the management of intercompany interactions within the production chain is needed to improve the efficiency of the production chain. Improved efficiency results are generally seen in decreasing cost of production chains and increasing end-customer satisfaction [4]. Improved efficiency is positively correlated with increased company competitiveness. Increased competitiveness of a company causes the company to maintain and improve its position in competition.

The simplest production chain is a production chain that consists of 2 (two) companies. One characteristic of inefficient production chains is when companies in the production chain are still optimizing individually. Individual optimization can be beneficial to a party and bring harm to the other party. In addition, the production chain is said to be inefficient if the distribution activities between two companies are not harmonious. This disharmony can occur because the two companies implement different production systems. One of the symptoms of disharmonization of production and distribution processes are the incidence of excess inventory or shortage of products.

The main objective of this research is to improve the efficiency of the production chain between two companies so that both companies can maintain and improve their competitive ability. The main proposal of this research is to change the contract-based interaction into a partnership based. With partnership interaction, optimization between 2 (two) companies is done together. The optimization goal used in this research is to minimize the cost of production chain.

Harmonization of production and distribution activities between 2 (two) companies is done by applying the push-pull system. The push-pull system is chosen to accommodate the different production systems used by both companies. Push-pull system is a system that aims to combine the advantages possessed by the push system and pull system. The development done in this research is to involve supply hub as junction point between 2 (two) companies in a production chain. The use of a supply hub can help a company to receive just-in-time raw materials, especially if the company is located far away from its suppliers [7]. Under these circumstances, suppliers can deliver large quantities of products to the supply hub to minimize transportation costs, while customers still get the desired product at a smaller and precise batch size and just when the products are needed.

2. Literature Review

The early studies about push-pull systems summarized in **Table 1**.

Table 1. Previous studies about push-pull system

Reference	Year	Research Model Characteristics					Algorithm
		Stages	Type	Number of Product	Junction Point Type	Objective	
[9]	1996	Four	Production line	Single	Work station	Inventory minimization	DES
[3]	1998	Multi	Production line	Single	Work station	Production cost minimization	DES, SA
[1]	2004	Two	Produuction chain	Single	Company	Production chain cost minimization	Dynamic Prog.
[12]	2004	Multi	Production line	Single	Work station	Product variance minimization	DES
[13]	2004	Single	Produuction chain	Single	-	Production chain cost minimization	B&B
[11]	2006	Single	Produuction chain	Single	-	Production chain cost minimization	B&B
[2]	2008	Multi	Production line	Multi	Work station	Production cost minimization	DES, GA
[5]	2009	Multi	Production line	Single	Work station	Production cost minimization	DES, GA
[6]	2012	Multi	Produuction chain	Single	Company	Production chain cost minimization	SA
[8]	2012	Three	Production line	Single	Work station	Inventory minimization	Simulation
[10]	2013	Six	Production line	Single	Work station	Inventory minimization and number of lost customers	DES

From research on push-pull systems that have been done earlier, research is generally done on production lines, using near optimal algorithms, and using work station or company as the junction point. Research on the scope of the production line reviews the performance of the push-pull system on the interaction between machines or between work stations. Previous research has generally shown that push-pull systems are superior to push and pull systems in terms of cost and inventory.

3. Mathematical Model

The mathematical model is the development of Wang and Sarker [13] and Shah and Goh [11]. Implementation of push-pull production chain is done by involving supply hub as junction point, which is where switching of production system from push production system become pull production system. Research is limited to the relationship between one supplier, one supply hub, one customer, and one product. Supplier has real time data inventory in the supply hub to be able to react to the sudden change in customer's demand. The products in the supply hub become the supplier responsibility. After the product is transferred to the customer's location, the ownership of the product is held by the customer. At the supply hub is applied the lower limit and upper limit of inventory. Supplier is charged by penalty fee if the product in the supply hub exceeds a predetermined limit. The assumptions used in this research are:

- Demand is known and deterministic
- Customer production rate is known and bigger than demand rate
- Production facility is not deformed
- Raw material accepted by supplier in time so the fulfilment of customer need is not disturbed
- Supplier production and supply hub replenishment lead time are zero

Variables and parameters notation used in this research listed in **Table 2** and **Table 3.**

Table 2. Decision variables

Notation	Definition	Unit of Measurement
Q_d	Supplier's dispatch lot size	unit
x	Integer to determine the supplier's production lot size	unit
y	Number of understock lot at the supply hub	*unitless*
z	Number of overstock lot at the supply hub	*unitless*
k	Number of kanban	*unitless*
n	Number of shipments placed at customer	*unitless*
Q	Total quantity of finished goods produced over specific period	unit

Table 3. Parameters

Notation	Definition	Unit of Measurement
I_b	Minimum inventory limit at the supply hub	unit
I_a	Maximum inventory limit at the supply hub	unit
D	Demand rate	unit/ month
p	Customer's production rate	unit/ month
A_p	Supplier's production cost	rupiah/ batch
A_d	Supplier's dispatching cost	rupiah/ batch
h_p	Inventory carrying cost at supplier's warehouse	rupiah/unit/ month
h	Inventory carrying cost at supply hub	rupiah/unit/ month
p_s	Penalty cost for under stocking at supply hub	rupiah/unit/ month
p_e	Penalty cost for over stocking at supply hub	rupiah/unit/ month
A_w	Customer's ordering cost	rupiah/ batch
A_s	Customer's production cost	rupiah/ batch
H_f	Inventory carrying cost at customer's warehouse	rupiah/unit/ month
A_f	Customer's dispatching cost	rupiah/ batch

There are 10 cost components, 6 (six) supplier-related costs and 4 (four) customer-related costs. 6 (six) supplier costs consist of 3 (three) costs related to the supplier's plant and 3 (three) costs related to the supply hub.

3.1 Supplier's plant cost

Production lot size at supplier's plant are always integer multiples of the dispatch lot size. Therefore, supplier's production and dispatching costs are shown in **Equation (1)** and **Equation (2)**.

$$Supplier's\ production\ cost = A_p \times \left(\frac{D}{Q_p}\right) = A_p \times \left(\frac{D}{xQ_d}\right) \quad (1)$$

$$Supplier's\ dispatching\ cost = A_d \times \left(\frac{D}{Q_d}\right) \quad (2)$$

IOP Conf. Series: Materials Science and Engineering **319** (2018) 012058 doi:10.1088/1757-899X/319/1/012058

As shown in **Figure 1**, inventory at the supplier's warehouse follows a step function. After one production cycle, supplier would hold $(x\text{-}1)Q_d$ products for the first period T, $(x\text{-}2)Q_d$ products for the second period T, and so on. Supplier's inventory cost shown in **Equation (3)**.

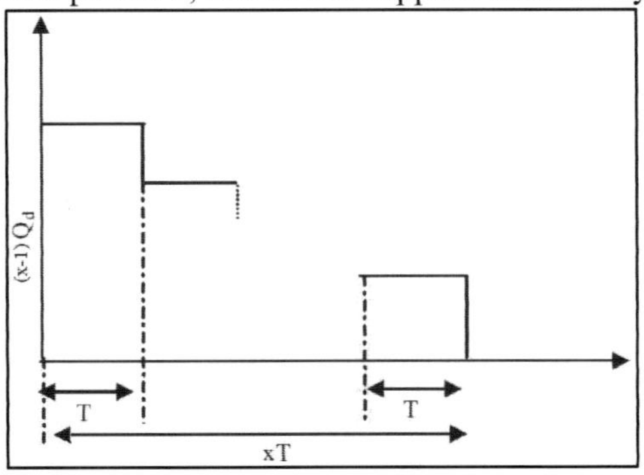

Figure 1. Inventory at supplier's warehouse

$$\begin{aligned} & Supplier's\ inventory\ cost \\ & = \left[\frac{(x-1)Q_d^2}{D} + \frac{(x-2)Q_d^2}{D} + \cdots + \frac{Q_d^2}{D}\right] \times \frac{D}{xQ_d} \times h_p \\ & = \frac{(x-1)Q_d}{2} \times h_p \end{aligned} \tag{3}$$

3.2 Supply hub cost

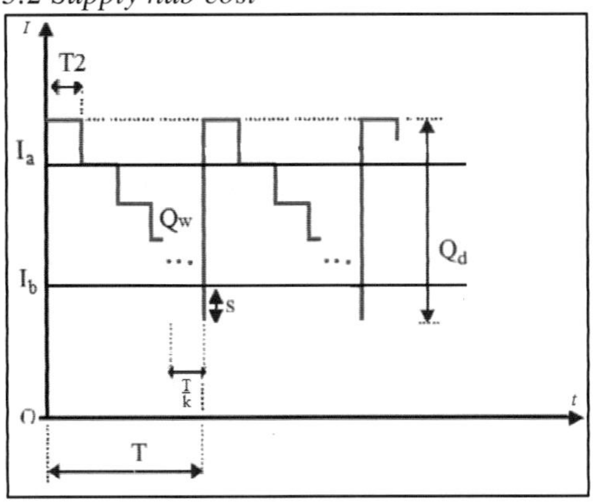

Figure 2. Inventory at supply hub

Products in the supply hub are delivered to the customer with the lot size Q_w. Q_w is defined as the lot size Q_d divided by the k kanban used. Understock products in supply hub represented by s at **Figure 2**. The number of understock product is always integer multiples of the supply hub's dispatch lot size. Therefore, number of understock product follows **Equation (4)**.

$$s = yQ_w = y\frac{Q_d}{k} \tag{4}$$

Supply hub's inventory cost follows **Equation (5)**.

$$\begin{aligned} Supply\ hub'sinventory\ cost &= \left\{\frac{1}{T}\left[Q_w\frac{T}{k}(1+2+\cdots+k)\right] + (I_b - s)\right\} \times \frac{Q_d}{D} \times \frac{D}{Q_d} \times h \\ &= \left[\frac{Q_w(k+1)}{2} + (I_b - s)\right]h \\ &= \left[\frac{Q_d(k+1)}{2k} + (I_b - s)\right]h \\ &= \left[\frac{Q_d(k+1)}{2k} + \left(I_b - \frac{yQ_d}{k}\right)\right]h \end{aligned} \tag{5}$$

Similar to s, number of overstock product is always integer multiples of supply hub's dispatch lot size. Therefore, number of overstock product follows **Equation (6)**.

$$Overstock\ product = zQ_w = z\frac{Q_d}{k} \tag{6}$$

Supply hub's understocking and overstocking penalty costs shown in **Equation (7)** and **Equation (8)**.

$$\begin{aligned} Supply\ hub's\ understocking\ penalty\ cost &= \left\{\left[\frac{1}{2}y(1+y)\right]\left[\frac{Q_d}{k}\right]\right\} \times \frac{T}{k} \times \frac{D}{Q_d} \times p_s \\ &= \left\{\left[\frac{1}{2}y(1+y)\right]\left[\frac{Q_d}{k}\right]\right\} \times \frac{Q_d}{kD} \times \frac{D}{Q_d} \times p_s \end{aligned}$$

$$= \frac{\left\{\left[\frac{1}{2}y(1+y)\right]\left[\frac{Q_d}{k}\right]\right\}}{k} p_s \tag{7}$$

$$Supply\ hub's\ overstocking\ penalty\ cost = \left\{\left[\frac{1}{2}z(1+z)\right]\left[\frac{Q_d}{k}\right]\right\} \times \frac{T}{k} \times \frac{D}{Q_d} \times p_s$$

$$= \left\{\left[\frac{1}{2}z(1+z)\right]\left[\frac{Q_d}{k}\right]\right\} \times \frac{Q_d}{kD} \times \frac{D}{Q_d} \times p_s$$

$$= \frac{\left\{\left[\frac{1}{2}z(1+z)\right]\left[\frac{Q_d}{k}\right]\right\}}{k} p_e \tag{8}$$

3.3 Customer cost

Customer's ordering cost, production cost, and dispatching cost are shown in **Equation (9)**, **Equation (10)**, and **Equation (11)**.

$$Customer's\ ordering\ cost = A_w \times \left(\frac{D}{Q_w}\right) = A_w \times \left(\frac{D}{Q_d/_k}\right) \tag{9}$$

$$Customer's\ production\ cost = A_s \times \left(\frac{D}{Q}\right) \tag{10}$$

$$Customer's\ dispatching\ cost = A_f \times \left(\frac{D}{Q_f}\right) = A_f \times \left(\frac{D}{Q/_n}\right) \tag{11}$$

Inventory at customer's warehouse described through **Figure 3**.

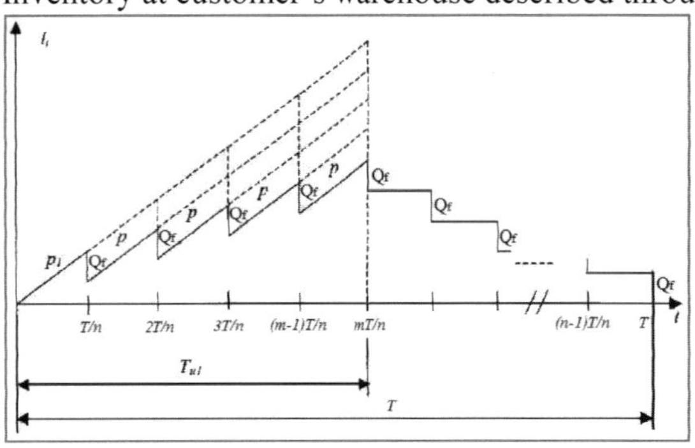

Figure 3. Customer's inventory

Customer's inventory cost follows **Equation (12)**.

$$Customer'sinventory\ cost = H_f\left\{\frac{1}{T}\left[\frac{1}{2}QT_{u1} - Q_f\frac{T}{n}[1+2+\cdots+(m-1)] + Q_f\frac{T}{n}[1+2+\cdots+(n-1)]\right]\right\}$$

$$Customer'sinventory\ cost = H_f\frac{Q}{2n}(n-m+1) = H_f\frac{Q}{2}\left(\frac{1}{n}+\left(1-\frac{D}{p}\right)\right) \tag{12}$$

Therefore, total objective function in this research is shown at **Equation (13)**.

$$= A_p \times \left(\frac{D}{xQ_d}\right) + A_d \times \left(\frac{D}{Q_d}\right) + \frac{(x-1)Q_d h_p}{2} + \left[\frac{Q_d(k+1)}{2k} + (I_b - s)\right]h + \frac{\left\{\left[\frac{1}{2}y(1+y)\right]\left[\frac{Q_d}{k}\right]\right\}}{k} p_s$$
$$+ \begin{cases} 0 & if\ (Q_d - s) \le (I_a - I_b) \\ \frac{\left\{\left[\frac{1}{2}z(1+z)\right]\left[\frac{Q_d}{k}\right]\right\}}{k} p_e & if\ (Q_d - s) > (I_a - I_b) \end{cases} + A_w \times \left(\frac{Dk}{Q_d}\right) + A_s \times \left(\frac{D}{Q}\right) + A_f$$
$$\times \left(\frac{Dn}{Q}\right) + H_f\frac{Q}{2}\left(\frac{1}{n}+\left(1-\frac{D}{p}\right)\right) \tag{13}$$

Subject to: $\boldsymbol{s \le I_b}$; $\boldsymbol{Q_d \ge s}$; $\boldsymbol{y, z < k}$; $\boldsymbol{s, Q \ge 0}$; $\boldsymbol{k, n \ge 1}$; $\boldsymbol{k, n \in integer}$

4. Numerical Example

From **Equation (13)** it is known that the objective function has *if else condition*. Therefore, this research has 2 (two) models, that is model without considering overstocking and model by considering overstocking. The model chosen to apply is a model that has a lower cost. Parameters that used for numerical example listed in **Table 4**.

Table 4. Numerical Example

Demand and production rate (unit/month)	Dispatch and order cost (rupiah/setup)	Production cost (rupiah/setup)	Inventory carrying cost (rupiah/month/unit)	Penalty cost (rupiah/month/unit)	Inventory limit at supply hub (unit)
$D = 3840$ $p = 7000$	$A_d = 13.09$ $A_w = 3.11$ $A_f = 13.38$	$A_p = 96$ $A_s = 129$	$h_p = 0.55$ $h = 10$ $H_f = 35$	$p_s = 5.46$ $p_e = 10.91$	$I_a = 1000$ $I_b = 40$

Solution for **Equation (13)** reported in **Table 5**.

Table 5. Best solution

Dec. variables	Without Overstock	With Overstock
Q_d	573 unit	733 unit
x	3	2
y	0	1
z	-	1
k	73	11
n	5	5
Q	1442 unit	1442 unit
Total Cost	Rp199,253,000.00	Rp214,359,000.00

5. Sensitivity Analysis

Sensitivity analysis is done by changing the model's parameters. The summary of sensitivity analysis shown in **Table 6**.

Table 6. One parameter sensitivity analysis

No	Parameters	Without Overstock			With Overstock		
		Significant Change in Objective Function Value	Sensitive Decision Variable	Robustness Range	Significant Change in Objective Function Value	Sensitive Decision Variable	Robustness Range
1	Demand rate	Lowered over 30%	Q	-5% – 20%	Lowered over 30%	Q	-15% – 0%
2	Minimum inventory at the supply hub	Lowered over 40%	Q_d	0% – 75%	Lowered over 50%	Q_d	-5% – 20%
3	Maximum inventory at the supply hub	Lowered over 70%	Q_d	-75% – 5%	Lowered over 70%	Q_d	-15% – 0%
4	Customer's production rate	Lowered over 30%	Q	-15% – 5%	Lowered over 30%	Q	-15% – 5%
5	Supplier's production cost	Lowered over 70%	Q_d	-20% – 40%	Lowered over 70%	Q_d	-60% – 10%
6	Customer's production cost	Lowered over 40% Lowered over 40%	Q	-5% – 40%	Lowered over 50% Raised over 70%	Q	-5% – 40%

As shown in **Table 6**, demand, maximum inventory at the suppy hub, and customer's production rate are the parameters that causing significant change in objective function value. Therefore, two parameters sensitivity analysis is conducted. It is summarized in **Table 7**.

Table 7. Two parameters sensitivity analysis

<table>
<tr><td>
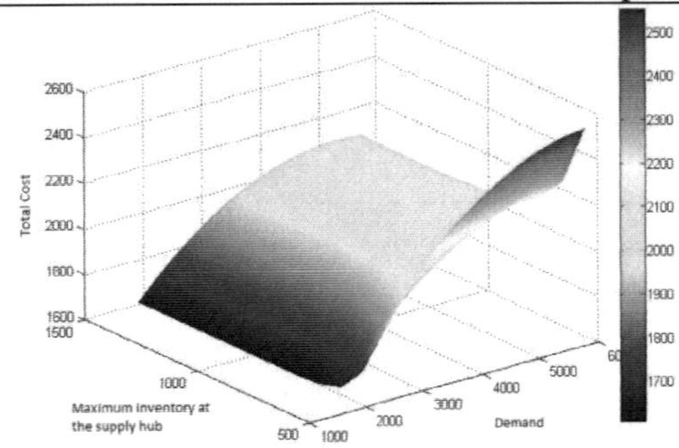

Figure 4. Total cost changes due to changes in demand and maximum inventory parameters
</td><td>As shown in **Figure 4**, it is known that the total cost increase due to the increasing of the demand and decreasing of maximum inventory at the supply hub. The optimum total cost is Rp160,709,300. This condition occurs when the demand decreases by 50% and the maximum inventory limit is in the range -10% to +50%.</td></tr>
<tr><td>
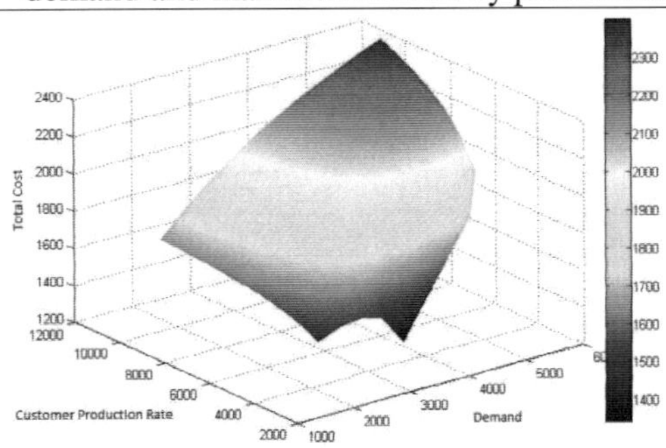

Figure 5. Total cost changes due to changes in demand and customer's production rate parameters
</td><td>As shown in **Figure 5**, it is known that the total cost increase due to the increasing of the demand and customer's production rate. The optimum total cost is Rp134,402,300. This condition occurs when the demand and the customer's production rate decrease by 10%.</td></tr>
<tr><td>
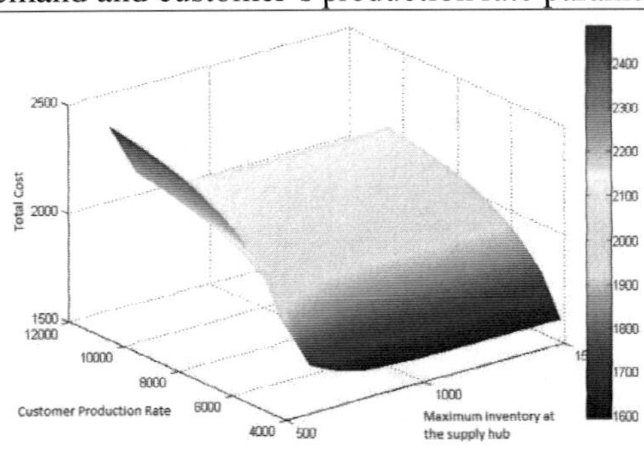

Figure 6. Total cost changes due to changes in maximum inventory and customer's production rate parameters
</td><td>As shown in **Figure 6,** it is known that the total cost increase due to the increasing of customer's production rate and decreasing of maximum inventory at the supply hub. The optimum total cost is Rp159,436,500. This condition occurs when customer's production rate decrease by 40% and maximum inventory at the supply hub increase by 40%.</td></tr>
</table>

6. Conclusion

The conclusions of this research are consolidated below:

- Improved efficiency of the production chain can be done by collaborative optimization and harmonization of production and distribution activities.
- Harmonization of production and distribution activities in this research is done by applying push-pull system and supply hub on production chain relation between 2 companies.
- The purpose of this research is to do supply chain optimization collaboratively by determining the size of production lot, lot size of shipment, number of kanban, and frequency of delivery done by company in production chain, and the number of understock and overstock lot.
- Demand, maximum inventory limit at the suppy hub, and customer's production rate are the parameters that causing significant change in objective function value.

Acknowledgement
We would like to express our gratitude to the ITB internal research grant and Manufacturing Systems Research Group-ITB for the financial support to the research.

References

[1] Ahn, H.-S. & Kaminsky, P., 2004. Push, pull, or hybrid control in supply chain management. *IIE Transactions* , 37(1), pp. 609-621.
[2] Cochran, J. K. & Kaylani, H. A., 2008. Optimal design of a hybrid push/pull serial manufacturing system with multiple part types. *International Journal of Production Research,* 46(4), p. 949–965.
[3] Cochran, J. K. & Kim, S., 1998. Optimum junction point location and inventory levels in serial hybrid push/pull production systems. *International Journal of Production Research,* 36(4), pp. 1141-1155.
[4] Fugate, B., Sahin, F. & Mentzer, J. T., 2006. Supply Chain Management Coordination Mechanisms. *Journal of Business Logistics,* 27(2), pp. 129-161.
[5] Ghrayeb, O., Phojanamongkolkij, N. & Tan, B. A., 2008. A hybrid push/pull system in assemble-to-order manufacturing environment. *J Intell Manuf,* 20(1), pp. 379-387.
[6] Kim, S.-H., Fowler, J., Shunk, D. & Pfund, M., 2012. Improving the push–pull strategy in a serial supply chain by a hybrid push–pull control with multiple pulling points. *International Journal of Production Research,* 50(1), p. 5651–5668.
[7] Liu, M., Li, G. & Gao, T., 2014. Superiority of Collaborative Replenishment with Supply-hub in Assembly System. *Information Technology Journal,* 13(1), pp. 1062-1069.
[8] Mahaprata, S., Yu, D. & Mahmoodi, F., 2012. Impact of the pull and push-pull policies on the performance of a three-stage supply chain. *International Journal of Production Research,* 50(16), pp. 4699-4717.
[9] Pandey, P. C. & Khokhajaikiat, P., 1996. Performance modeling of multistage production systems operating. *Int. J. Production Economics,* 43(1), pp. 17-28.
[10] Phonsuwan, S. & Kachitvichyanukul, V., 2013. Simulation for OTOP supply chain with hybrid push-pull flow control. *Int. J. Services and Operations Management,* 15(2), pp. 196-214.
[11] Shah, J. & Goh, M., 2006. Setting Operating Policies for Supply Hubs. *International Journal Production Economics,* 100(1), pp. 239-252.
[12] Takahashi, K. & Nakamura, N., 2004. Push, pull, or hybrid control in supply chain management. *International Journal of Computer Integrated,* 17(2), pp. 126-140.
[13] Wang, S. & Sarker, B. R., 2004. A Single-stage Supply Chain System Controlled by Kanban Under Just-in-time Philosophy. *Journal of the Operational Research Society,* 55(5), pp. 485-494.

IOP Conf. Series: Materials Science and Engineering **319** (2018) 012059 doi:10.1088/1757-899X/319/1/012059

Ergonomics Risk Assessment among support staff in Universiti Malaysia Pahang

Faisal Jusoh, Muhammed Nafis Osman Zahid[1]

[1]Faculty of Manufacturing Engineering, Universiti Malaysia Pahang 26660, Pekan, Pahang, Malaysia

nafis@ump.edu.my

Abstract. Awareness of ergonomic risk assessment among workers are getting intense in many industries nowadays. It is essential since most of the workers spend 7 to 8 hours of their time in the workplaces. Previous study shown that spending too much time with static posture in sitting at workplace leads to the problem of Musculoskeletal Disorders (MSDs). The implications are not only harmful to human body but also effect the productivity. Currently, there are no scientific study conducted to assess the conditions of workers in Universiti Malaysia Pahang (UMP). Therefore, the problem of MSDs could not be justified clearly and the top management did not acknowledge this issue. This study aims to present current scenario of ergonomic risk level at UMP by using structured model. It focuses on operational staff from faculties and Human Resources Department (HRD). Initially, three types of assessments are executed based on general working condition, Cornell Muscokeletal Discomfort Questionnaire (CMDQ) and Rapid Office Strain Assessment (ROSA). Based on the findings, 90% of the respondents felt discomfort at workplace but prefer to rectify the issues by themselves. Almost 50% of them evaluated themselves in level 4-5 of discomfort level. The CMDQ result shown the discomfort area at faculties and HRD. The workplace at faculties and HRD had been assessed through ROSA and the overall result shown the risk level is medium level respectively. Therefore, further investigation is requires and improvement of workplace need to be proposed to establish good working condition.

1. Introduction

The ergonomics study had played major important role to provide pleasant interaction between people and workplace in many field such as manufacturing, agriculture, mining, services or others industries. The proper coordination of human-workplace had contributed toward organisation productivity and indirectly increased the performance to achieve their goal. The type of workplace is differentiate according to the nature of the industry. For example, the workplace in manufacturing industry usually consist the element of peoples, machinery, material handler, equipment, tool, jig and fixture, workbench, computer, desk, chair and etc. However, in services industry, the workplaces are normally consists the element of peoples using computer, telephone, work document and stationary while dealing with customers.

Previous studies have recognized the contribution of ergonomics in enhancing the performance in service and manufacturing sectors. Ergonomics studies provide sufficient data in designing the organization workplace according to the philosophy of "fit man to job" or "fit job to man". According to Bridger (2008), attempt to "fit man to job (FMJ)" were based on the idea that productivity or efficiency could be improved by selecting workers with the right aptitude for a particular job[1]. Meanwhile, "fit job to man (FJM)" is an attempt to design task to suit the characteristic of the workers.

In addition for services industry, the roles of ergonomics has becomes more essential and significant due to the meaningful information in designing an ideal workplace and human posture interaction. Basically, the natures of work for service industry required them to work in sitting position for a long time per day. This lead to a sedentary behaviours (SB) which lead to low energy expenditure in a reclining position. If this condition prolonged, it can lead to severe injuries caused by static posture, forceful exertion or repetitive movement [2]. Ultimately, this will caused discomfort or unpleasant condition to human body of workers. Most office works today have undergone radical changes over recent years. Technologies changes have led to a situation where users have little reason to leave their workstations. As a consequence, many people start to experience discomfort and pain brought by the repetitive nature of task, lack of break and task variety. In order to sustain the human and pleasant working environment, the evaluation toward risk surround should be done. The risk assessment is a part of a whole health and safety management system [3]. The purpose of this assessment is to identify hazard or potential situations that can cause harm to human. Ultimately, it assists to identify whether the risk control measure is already in place and provide recommendation if it is not yet considered.

2. Ergonomics Assessment

Cornell Muscokeletal Discomfort Questionnaire (CMDQ) and Rapid Office Strain Assessment (ROSA) are among the approaches widely used to investigate the problem of Musculoskeletal Discomforts (MSD). Shariat A. (2016) has implemented these approaches among Telecom Malaysia staffs to investigate MSD problems [4]. They need to get involve in one of the approach which is intervention groups (receiving training exercise, receiving modified ergonomics, receiving a combination of exercise and ergonomics modification) and a control group (receiving none of these interventions). The CMDQ Questionnaire was used to measure musculoskeletal discomforts, with focus on pain severity, before treatment and after 2, 4 and 6 months of the interventions. The range of motion (ROM) of the hip, neck, shoulder and knee were measured by goniometer, and the Borg CR10 scale was used to measure the perceived exertion of training exercises. The ROSA questionnaire was used to assess the strain associated with office work. Height and weight were also measured to calculate the body mass index (BMI).

Meanwhile, Liebregts, (2015) studied on how to get more accurate result of ROSA by examining the validity and reliability through photo-based assessments [5]. Here, twenty-three office work stations were assessed on-site by an ergonomist, and 5 photos were obtained. As a result, the photo-based assessments had identified as most appropriate when applied as a "triage" method for a big group of workstations. Matos. and Arezes, (2015) studied on how the regular use of computer in the office contributed to many risk factors related to Work Musculoskeletal Disorders (WRMSD) such as maintaining static sitting postures for long time and awkward postures of the head, neck and upper limbs [6]. This study evaluates the present of WRMSD in workstation and found that the cause may not be related to the poor availability of equipment but the need to optimize the usage by the workers. It was noticed that the interaction of workers with the tasks and the sitting posture at the workstation along the day have repercussions at a muscular level, essentially for the cervical and shoulders segments.

Poochadaa and Chaiklieng (2015) had studied the issues of Musculoskeletal disorders (MSDs) which complained by call centre workers who involved in static work or tasks required the repetitive motion of the upper limbs and prolonged computer work [7]. The survey was conducted among 216 call centres in Khon Kaen Province, Thailand. The workers are selected from those who had experienced the job more than 6 months, working time at least 32 hours per week and dealed with computer at least 4 hours per days. By using the ROSA, this study showed that most workers were exposed to the high ergonomics risk for MSDs development. In order to prevent MSDs, there should be ergonomics awareness programme implemented among the workers. A part of that, personal working behaviour and the design of the workstations should be improved according to ergonomics principles.

3. Woking Nature in UMP

As a player in servicing a nation through education mainstream, UMP strives to deliver high quality service to the stakeholder. For a past 10 years, UMP has gone through a lot of development phases and these effect the workstation design and equipment. Therefore, most of the offices are equipped with recent design of writing desk and chairs. However, the effectiveness and how it reduce the discomfort for a long sitting work are not proven yet. If the issue is not investigated, there is a high tendency for the worker to face a syndrome called as Work Musculoskeletal Disorder (WMSD's) or Low Back Pain (LBP) injury.

Particularly in UMP, there are no further assessment and analysis to address ergonomics issue among workers. Hence, this study is conducted to assess MSD problem and determine either it is presented or not and how serious the condition. Since support staffs played a major role in university performance, their absenteeism to work may contribute to the low productivity, decrease their availability and performance lost to achieve university yearly target. Basically, the objective of this project is to evaluate the current working condition at UMP which relate to the Work Muscokeletal Decease (WMSD) problem for support staffs at faculties and Human Resources Department. Besides, it aim to study, measure and compare the issue at both area by using Cornell University Discomfort Questionnaire (CMDQ) and Rapid Office Strain Assessment (ROSA) approach. Finally, the study intended to formulate and propose corrective actions to related departments based on the findings.

In order to further assess this issue, a comprehensive study has been conducted by Damanhuri, Z. *et al.* (2014) to evaluate the discomfort or low back pain among the office workers in Malaysia public universities [8]. Based on their findings, LBP is identified as a highest injury for these category of workers; worked for 10 years or less (66.1%), not attended office ergonomics course (91.2%), used non-ergonomic chair (98.2%) and leave their working chair infrequently (62.5%) and in short duration (74.1%).

4. Methodology

There are three type of questionnaire used during this survey. The first questionnaire initially wants to check the general condition of existence problem in UMP. It contained seven (7) questions about discomfort scenario at present workplace, time consuming at workplace, the current symptom, discomfort scale, starting time to occurred and current action take. The second questionnaire collects background data of the staffs like gender, weight, height and embedded the standard form of Cornell Muscokeletal Discomfort Questionaire (CMDQ) which measured the eleven (11) area of human body. The score is calculated based on weightage of frequency, discomfort and interference. The third assessment used standard form of Rapid Office Strain Assessment (ROSA) to check the body contacting at workplace in terms of chair condition, computer accessories (mouse, monitor, and keyboard) and telephone. This questionnaire use to verify the discomfort found in initial survey and CMDQ.

Basically, working conditions of the staffs can be categorized based academic and non-academic staffs. The nature is quite different since academic staffs spend a lot of time in the classroom, laboratories, workshop, meeting room and their own office. Meanwhile, the support staffs work in the nature of less movement, static and spend more time at workstation every day (sitting based). There are 86 respondents participated in the survey and comes various faculties and HRD. The composition of respondents is summarize in Table 1.

Table 1. Respondent composition in the survey.

Gender	Faculty	HRD	Total	Percentage
Male	12	10	22	26%
Female	37	27	64	74%
Total	49	37	86	100

5. Results and Discussion

Most of the staffs are female, young and half of them having an ideal weight or Body Mass Index (BMI) and almost 40% of them are fall in excessive weight and obesity 1 and 2. We found that 36% of staff which have ROSA score 4-5 faced a problem of excessive body weight. Almost 85% of staff are satisfied with facilities provided as it is new but they realized the issue of discomfort in their workplace. 50% rate themselves in between 4~6 discomfort score. Support staff normally spent 6 hours to 8 hours of their time at workplace per day and 44% claimed they have problem of fatigue/illness at shoulders/low back /thigh /buttock/ knee. Meanwhile, 37% reported a problem of eye fatigue while focusing for long hours, 6% have a problem of drizzling after sitting for long hours per day. This figure means that even though they felt convenience on working environment, they also have a problem of sitting with their workplace.

5.1 Cornell Muscokeletal Discomfort Questionnaire (CMDQ)

The result of CMDQ reported total body discomfort and observe the human body area of high impact due to long hours sitting. Based on the result in table 2, the trend shown that every faculty and HRD faced a different problematic area. In general, lower back is the most problematic area and the top (3) three highest is at FKM, FKASA and FKEE. This followed by upper right arm, shoulder and neck. The area of discomfort found in this study are almost similar with the previous studies related to office worker [8, 9]. Therefore, particular attention need to be addressed on these area while developing the workstation for office worker.

Table 2. Discomfort score (Highest score indicates the most affected area)

Discomfort area	**Discomfort score**							
	FTek	FIST	FKASA	FKEE	FKM	FIM	FSKKP	HRD
Neck	7.42	7.83	9.94	6.25	14.56	7.57	8.00	6.83
Shoulders (right)	7.42	5.83	6.78	8.58	14.63	5.00	3.71	7.46
Upper back	3.08	8.08	8.33	12.50	12.63	9.57	5.79	11.13
Upper arm (Right)	3.33	2.75	11.83	1.33	2.00	3.29	1.07	4.32
Lower back	6.33	10.50	19.22	16.33	23.06	8.14	12.21	8.69
Wrist(Right)	1.67	4.08	10.33	5.42	20.56	4.93	5.36	4.76
Hip/buttock	4.00	11.42	5.17	7.9	6.13	4.86	10.71	9.10
Knee(right)	1.92	0.67	12.83	3.42	15.75	4.36	2.21	4.32
Knee (left)	1.92	0.67	8.94	3.42	10.75	4.36	10.79	3.45

FTek-Faculty of Engineering Technology
FIST-Faculty of Industrial Sciences & Technology
FKASA-Faculty of Civil Engineering & Earth Resources
FKEE- Faculty of Electrical & Electronics Engineering
FSKKP- Faculty of Computer Systems & Software Engineering
FKM- Faculty of Mechanical Engineering
FIM- Faculty of Industrial Management

5.2 Rapid Office Strain Assessment (ROSA)

The result of ROSA score was collected through self-assessment at respected area based on Section A (The Chair - Chair Height and Seat Pan Depth), Section B (Telephone and Monitor) and Section C (Keyboard and Mouse). The highest Final ROSA Score for faculty is FKASA (5.00) and followed by

FKM (4.33), FKEE (4.17), FSKKP (3.71), FIM (3.57), FIST (3.17) and FTek (3.00). Referring to table 3, the final ROSA score for faculties and HRD is 3.45 and 3.38, respectively. ROSA standard guidelines stated score 1-2 is consider as low risk, 3-4 is medium risk, 5-7 is high risk and 8-9 is very high risk. The average score of all department in UMP is around 3-4 and therefore fall under medium risk level.

Table 3. Overall score for ROSA analysis based on faculty and HRD

Section	Content	Faculty	HRD	Overall Result
A	Seat Pan Height / Depth Arm Rest/ Back Support	3.04	2.97	3.00
B	Phone / Monitor	2.73	2.51	2.62
C	Mouse / Keyboard	2.96	2.78	2.87
Monitor& Peripheral	Monitor & Phone Mouse& Keyboard	3.18	2.97	3.07
	Final ROSA Score	**3.45**	**3.38**	**3.42**

Based on the finding, there are several actions can be initiated by the UMP staffs which relate to proper arrangement of workplace design, knowledge about body posture at workplace and minor adjustment to the working environment. The summary is provided in table 4 and figure 1. The current action taken such as coaching and frequent monitor at workplace is suitable to improve the working environment but need consistency since it relates to staff working culture. By continuous effort and monitoring, a minor positive result can be seen in short terms. Therefore, additional approaches and educating the staff will minimize the risk level in the future.

Table 4. Corrective actions for identified problem in the workplace

No	Problem Description	Corrective Action
1	Discovered an obstacles or unnecessary items like computer CPU, storage box and unused material that blocked the worker leg from free movement.	The related staffs had been taught on the basics of 5'S and be advised to not placing any items under workstation.
2	Workers not follow standard way of working positions such as did not use chair arm rest for a long hours, improper location of monitor, keyboard and telephone.	Coach the workers on how to adjust the workplace condition and suggest optimum distance to the equipment.
3	Workers have a lack of knowledge on how to organise his/her sitting position, how to sit and adjust the chair correctly.	The hands on training has been given to related staff on how to apply the correct method of sitting at workplace included the function of controller at chair.
4	The light seems too glare due to the bright illumination. This condition is unpleasant which had annoyed staff focus and caused to eye fatigue and stress.	The technical evaluation by expertise in UMP in terms of illumination lux and space measurement is needed to evaluate the condition by using special equipment.

6. Conclusion

The ergonomics assessment conducted can be summed up to several points. Based on current working condition survey, most of the UMP staffs are not aware with the ergonomic problem in their workstations. Problems like eye fatigue, drizzling, illness at certain body parts are stated in the survey but they did not seeking advice from medical doctor or inform their superior. Based on CMDQ survey,

IOP Conf. Series: Materials Science and Engineering **319** (2018) 012059 doi:10.1088/1757-899X/319/1/012059

several discomfort area are identified and it is varies between the faculties and HDR. Lower back pain is one of the critical discomfort area faced by the workers in all department. ROSA analysis has further validated the ergonomics condition in UMP. The score indicates the risk is at medium level which require further actions for the institution to improve their employees working environment. Overall, the study on ergonomics in service sector is very crucial to address the problem of productivity and health issues. Awareness on ergonomics aspect and knowledge about good working condition will assist them to initiate changes within their capacities.

Figure 1. Corrective actions to improve workplace.

References

[1] Bridger, R., *Introduction to ergonomics*. 2008: Crc Press.

[2] Waters, C.N., et al., *Assessing and understanding sedentary behaviour in office-based working adults: a mixed-method approach.* BMC public health, 2016. **16**(1): p. 360.

[3] McKeown, C., *Office ergonomics: practical applications*. 2007: CRC Press.

[4] Shariat, A. *Effective Methods of Reducing Lower Back, Neck and Shoulder Pain Among Office Workers*. U.S. National Institutes of Health, 2016 [cited 2017; Available from: https://clinicaltrials.gov.

[5] Liebregts, J., M. Sonne, and J. Potvin, *Photograph-based ergonomic evaluations using the Rapid Office Strain Assessment (ROSA).* Applied ergonomics, 2016. **52**: p. 317-324.

[6] Matos, M. and P.M. Arezes, *ergonomic evaluation of office workplaces with rapid office strain assessment (ROSA).* Procedia Manufacturing, 2015. **3**: p. 4689-4694.

[7] Poochada, W. and S. Chaiklieng, *Ergonomic risk assessment among call center workers.* Procedia Manufacturing, 2015. **3**: p. 4613-4620.

[8] Damanhuri, Z., et al., *Low back pain among office workers in a public university in Malaysia.* International Journal of Public Health and Clinical Sciences, 2014. **1**(1): p. 99-108.

[9] Sonne, M., D.L. Villalta, and D.M. Andrews, *Development and evaluation of an office ergonomic risk checklist: ROSA–Rapid office strain assessment.* Applied ergonomics, 2012. **43**(1): p. 98-108.

IOP Conf. Series: Materials Science and Engineering **319** (2018) 012060 doi:10.1088/1757-899X/319/1/012060

Modelling Electrical Energy Consumption in Automotive Paint Shop

Muchamad Oktaviandri[1,2], Aidil Shafiza Bin Safiee[1]

[1]Faculty of Manufacturing Engineering, Universiti Malaysia Pahang, 26600, Malaysia.
[2]Fakultas Teknologi Industri, Universitas Bung Hatta, Padang, 25143, Indonesia.

muchamad_oktaviandri@yahoo.com

Abstract. Industry players are seeking ways to reduce operational cost to sustain in a challenging economic trend. One key aspect is an energy cost reduction. However, implementing energy reduction strategy often struggle with obstructions, which slow down their realization and implementation. Discrete event simulation method is an approach actively discussed in current research trend to overcome such obstructions because of its flexibility and comprehensiveness. Meanwhile, in automotive industry, paint shop is considered the most energy consumer area which is reported consuming about 50%-70% of overall automotive plant consumption. Hence, this project aims at providing a tool to model and simulate energy consumption at paint shop area by conducting a case study at XYZ Company, one of the automotive companies located at Pekan, Pahang. The simulation model was developed using Tecnomatix Plant Simulation software version 13. From the simulation result, the model was accurately within ±5% for energy consumption and ±15% for maximum demand after validation with real system. Two different energy saving scenarios were tested. Scenario 1 was based on production scheduling approach under low demand situation which results energy saving up to 30% on the consumption. Meanwhile scenario 2 was based on substituting high power compressor with the lower power compressor. The results were energy consumption saving of approximately 1.42% and maximum demand reduction about 1.27%. This approach would help managers and engineers to justify worthiness of investment for implementing the reduction strategies.

1. Introduction

In a recent economic downtrend environment, the industrial player must be smart enough to control their expenses and utilize their resources optimally. In order to sustain in their business, the company must find a solution to reduce their operational cost so that they can survive and keep maintaining their reputation in front of customer view. One aspect that has big potential for cost reduction is an energy consumption cost.

In general, there are some barriers that limit the industrial player to extensively explore on improving energy efficiency. Such barriers are the limitation of capital investment, technical risks, challenging in collecting energy related data, very little information about efficiency opportunities and the cost of identifying and assessing opportunities. This study tries to analyse energy consumption level in automotive industry by using simulation-based model approach and observe the effectiveness of energy saving methods through virtual analysis. In automotive manufacturing plant, energy is used throughout the plant for many different end-uses. The common types of energy used are electricity, steam, gas and compressed air. About two-thirds of total expenditure on energy is spent on electricity [1]. In automotive assembly process plant, generally, consists of body shop, paint shop, assembly shop [2]. Paint shop is a general name called for area where the painting process is conducted. Inside the paint shops there are several main processes such as pre-treatment, electro-coating, sanding, sealant, primer coating, base coating and top coating [3]. Meanwhile, in some processes like pre-treatment and electro-coating there are several sub-processes.

In regards of portion energy consumption, paint shop is considered the highest consumer of energy which estimated around 50-70% of total energy consumed [4]. Pre-treatment, paint application and curing are the main processes in the area that use major energy, while the detail distribution depends on the system's design and the technology built in [4].

In a relation with the high energy used in the paint shop area, many studies on energy consumption can be found using paint shop as a case study. Guerrero et al. [10] introduced a model to optimize painting process by reducing the number of unnecessary repaints, and reported that this method successfully reduced the energy consumption by 4%. Rohdin et al. [5] studied the effect of controlling air flow in the spray booth line in order to reduce energy caused by large quantities of air used to maintain the air quality. Chen and Zang [6] investigated energy consumption reduction through effective scheduling of machine start up and shutdown. Trakultongchai [11], used simulation model approach to study on process improvement. Effect of car sequencing to the energy consumption was studied by Zhang et al. [7]. Earlier, Cronrath et al. [4] came out with environmental discrete event simulation approach for energy analysis to overcome barriers that facing in implementation of energy analysis study.

Several researchers have conducted energy analysis study using various simulation tools available nowadays. Simulation modelling has gain interest to researcher due to its ability to model real system environment without affecting actual system. Al-Ajmi and Handy [12] used plant simulation as a design tool for to analyse energy consumption performance in proposed building design as well as evaluating the effect of varying certain design parameters. Fysikopoulus, et al. [13] conducted an empirical study of energy consumption at body-in-white subassembly. Meanwhile, thermodynamic simulation model has been used by Feng and Mears [8] for analysis of energy consumption by Heat, Ventilation and Air Conditioning (HVAC) system in automotive paint shop.

In recent development, Tecnomatix Plant Simulation (TPS) is increasingly been used in the energy analysis study. It is due to introduction of energy analyser function inside the Plant Simulation software version 11 [9]. Trebuna and Popovic [14] used Tecnomatix to develop digital factory model and demonstrate the power consumption capability of the software. It is almost similar with the study by Cronrath [4].

This paper aims at reporting a tool to model and simulate energy consumption at paint shop area by conducting a case study at XYZ Company, one of the automotive companies located at Pekan, Pahang. The model is conducted based on discrete event simulation approach by using energy analysis function of Tecnomatix Plant Simulation software version 13. The painting process covered in the study is starting from pre-treatment until final touch up process while the energy analysed is based on electrical control panels directly connected to the process equipment's in paint shop area. Objectives of this study were; (i) to develop energy simulation-based model for paint shop area, (ii) to validate energy consumption at the paint shop area, and (iii) to evaluate energy saving potential through different scenarios using simulation model.

This paper is organized as follows. First Section serve as an essential introduction to research. Second section describes the research methodology and its rationale. Next section provides an overall result and discussion of research finding. Last section concludes the study and suggestion for future work.

2. Methodology

The methodology carried out in order to fulfil the objectives stated in the study. Generally, the methods can be divided into four stages which relate to the particular objective, namely: (i) Data collection, (ii) Initial model development, (iii) Model validation, and (iv) Simulate alternative scenarios, as discusses as follow:

2.1. Data Collection and Initial Model Development

Several information and parameters have been defined as the necessary input data in this study. The data can be classified into 3 type of data; process flow data, energy data and validation data.

The gathered data under process flow category were overall paint shop layout (required to developed the mimic model that almost similar with the actual system), Process flow information (required to know the sequence of processing for the paint shop process), Time study (processing time & cycle time), and Production working hours (required to simulate the system).

The focus of energy to be analyzed in this study is the electrical energy. In line with that, the elements that use electrical power are identified and the power measurements are made on each of the elements. The measurement conducted using AC clamp meter to measure Phase Current (A) and Line Voltage while Power Factor, PF value is taken directly from the electrical element specification. Equation (1) is then been used to calculate Phase power, P_n for each phase where for 3-phases system there will be Red-phase Power, P_{red}, Yellow-phase Power, P_{yellow} and Blue-phase Power, P_{blue}. Summation of each phase power will result the Total Power, P_{total} used by each of the element as specify in equation (2).

$$P_n = \frac{Line\ Voltage}{\sqrt{3}} \times \frac{Phase\ Current\ (A)}{1000} \times PF \tag{1}$$

$$Total\ Power\ (kW), P_{total} = P_{red} + P_{yellow} + P_{blue} \tag{2}$$

The P_{total} value calculated is the main input parameter for simulation model. For validation purpose, the latest 5 months electricity consumption (EC) and maximum demand (MD) data are collected from Tenaga Nasional Berhad (TNB) smart meter. EC is the total amount of energy been used for a month period and it is measured in kWh while MD is the highest level of electrical demand recorded during a 30-minute interval in a month and it is measured in kW. The interval time to determine MD may vary by different energy provider and country. The type of tariff implemented at the company is determined to calculate the electricity cost.

Next step after all the relevant data have been collected is to start building the simulation model of the system. At first, it is defined as initial because adjustment and modification might be needed after run test and validation process. Under this phase, three main activities have been defined which are constructing model conceptual and assumptions, programming the model and validation and analysis of the model.

2.2. Model Validation

To ensure the simulation model is valid, the total EC and MD value from the simulation result are being compared with the actual data recorded from the company. The confidence interval set for the simulation model is 95% for EC and 85% for MD. Lower value set for MD because of the study scope did not cover in details the operation sequence and running time for every single energy consumer element which is needed to accurately simulate MD value. Thus, the error within 5% for EC and 15% for MD are considered valid. After the result validated, the scenario for energy saving measure is being simulated and the amount of energy saving is being presented in term of amount and cost saved.

2.3. Simulate alternative scenarios

In this study, two different scenarios are evaluated in order to analyse the energy consumption reduction by implementation of the alternative scenarios. The scenarios chosen for evaluation were based on current situation at XYZ Company.

The technique applied in scenario 1 is considering low demand situation where the production running below its actual capacity. The processes which recorded the high amount of energy consumption had been chosen which is in this case Primer and Top Coat processes. Both processes are consequent where the part from primer operation will go into top coat operation. In this scenario, the production time for both processes are reduced into half than normal working hours. In the first half, the Primer process will operate first and in the second half, Primer will be shut down and Top Coat will begin operation.

Meanwhile, scenario 2 is using equipment change method. Under simulation approach, it is easy to simulate the equipment change. In this case, the compressors power change has been modelled and the effect on total energy consumption had been simulated to study the amount of reduction. The change is done by reducing the power input for high power compressor to follow the value of lower power input compressor. In reality, the new equipment must be ready first before it can be tested but in virtual, it is simply done by changing the power setting at simulation model. The result could be used to justify whether the changing is worth or not for implementation.

3. Results and Discussion

3.1. Data Collection

3.1.1. Process Flow, Processing Time and Production Working Hour

Paint shop of the company XYZ consists of eight main processes. It starts with Pre-Treatment and Electro-deposition (PTED) section involving several sub-processes such as hot water cleaning, degreasing, rinsing, conditioning, phosphate, electro-deposition, ultra-filter and oven. Then it goes to ED Sanding, sealant, primer, moist sanding, major rework, top coat and touch. Processing time data is directly taken from time study data provided by the company. This scope study is not going detail on processing time distribution pattern. Our general assumption for manual process, the time follow the data provided in the time study table while for automatic process the time set constantly 5.00 minutes based on production cycle time. The company XYZ is running production in normal working hour, 5 days per week. The start working time is at 8:00 AM and end at 5:30 PM.

3.1.2. Energy Data

Current measurement has been conducted for 58 control panels which have been identified having direct relation with the production process. Although it is not completely covers all energy consuming elements for the whole paint shop like administrative room, air conditioning system and a few control panels that difficult to access, the result is still considered valid because all major systems that have significant effect to energy consumption have been covered. For the example, the total power input result calculated using previous mentioned equations for Top Coat process is shown in Table 1. The top coat involves 11 control panels with different power input along its processes.

3.1.3. Validation data

In order to validate the result from simulation model developed, the current electricity usage for the company XYZ is recorded. The available data provided by the company is the latest 5 month. Table 2 summarized the electricity usage and the cost based on TNB tariff. The company XYZ is using Tariff E1- Medium Voltage General Industrial Tariff. Under this tariff, the cost of RM29.60/kW will be charged for each kilowatt of MD per month and 33.70cent/kWh for total EC.

Table 1. Power Calculation for Top Coat process.

Main Process	System Control Panel	Control Panel ID	Line Voltage (V)	Power Factor; PF	Total Power (kW)
Top Coat	ASU Top Coat Blower 2	TopCoat_CP1	430	0.75	129.75
	ASU Top Coat Blower 3	TopCoat_CP2	430	0.75	123.41
	ASU Top Coat Blower 4	TopCoat_CP3	430	0.85	60.78
	ASU Top Coat Humidifier 1	TopCoat_CP4	430	0.85	5.87
	ASU Top Coat Humidifier 2	TopCoat_CP5	430	0.85	5.96
	Top Coat Exhaust Fan No 1	TopCoat_CP6	430	0.85	29.58

Top Coat Exhaust Fan No 2	TopCoat_CP7	430	0.75	98.61
Top Coat Exhaust Fan No 3	TopCoat_CP8	430	0.85	35.07
Top Coat Exhaust Fan No 4	TopCoat_CP9	430	0.75	96.19
Top Coat Oven Zone 1	TopCoat_CP10	430	0.85	32.73
Top Coat Oven Zone 2	TopCoat_CP11	430	0.85	32.60

Table 2. Energy consumption records for Company XYZ.

	November	December	January	February	March	Average
EC (kWh)	779,203	813,545	820,156	846,371	1,112,628	874,380.6
Tariff (RM)	0.337	0.337	0.337	0.337	0.337	0.337
Cost usage (RM)	262,591.4	274,164.7	276,392.6	285,227.0	374,955.6	294,666.3
MD (kW)	3,083	2,763	3,090	3,169	3,236	3,068.2
MD Tariff (RM)	29.60	29.60	29.60	29.60	29.60	29.60
MD cost (RM)	91,256.80	81,784.80	91,464.00	93,802.40	95,785.60	90,818.72
Total Cost (RM)	353,848.2	355,949.5	367,856.6	379,029.4	470,741.2	385,485.0

3.2. Initial Model Development

From the data collection result, it is known that 58 control panels are going to be simulated in the model for energy consumption. The control panel may be connected with more than one process or system. For the purpose of modelling, the energy parameter setting for control panel will be set to either one of related process to avoid the multiplication of the result if each process is setting with the power input parameter. Thus, the sub-processes for every main process are mapped with the related energy system control panel. For each process that was mapped with energy control panel, the power input value will be assigned. In our model, we assume that operating power and working power for all the systems are same but in actual certain systems would have slightly lower operating power than working power.

Several assumptions have been made on the simulation model developed, such as; The power input used for operational, working and setting up state is considering same; assuming each station has buffer for part waiting to be processed; the arrival rate for incoming parts is assumed constant; Part entering the system considered like triggering switch to activate the energy usage. For manual processes; the processing time is considered varies in normal distribution based on data from time study; Upper and lower bound is set at 30% from the mean value. For automatic process; the processing time is considered constant and set same as cycle time which is 5.00 minutes; the operation machine is assumed running at 98% availability rate in order to model unexpected downtime that might occur throughout the operation. The peripherals energy consumers such as light and air-conditioning system along the operation are neglected.

3.3. Model Validation Results

The simulation model has been set to run for 20 days in order to get the total energy consumption in a month. The total EC result from simulation is 916,350 kWh in a month while the MD is 2,741.2 kW. To validate the result, the average data of the latest five months usage had been taken for comparison. Table 3 shows the comparison between simulation result and actual data.

Table 3. Comparison between simulation result and actual data.

	Actual	Simulation	Difference (Simulation -Actual)	Percentage Error
EC (kWh)	874,380.6	916,350	41,969.4	4.8
MD (kW)	3,068.2	2,741.2	-327	-10.7

From Table 3, the percentage error calculated between simulation and actual result is 4.8% for EC and 10.7% for MD. Thus, the result is considered valid because both EC and MD results are within targeted percentage error.

3.4. Simulating Alternative Scenario

3.4.1. Scenario 1: Reschedule Primer and Top Coat Operation

In a low demand situation as facing by company XYZ recently, one possible approach could be consider for energy saving is to reschedule the operation of primer and top coat process. Conducting this operation strategy could also reduce the requirement of workers where the same operators for primer process are then use to execute top coat process. But, need to be reminded this only works in a low demand and loose production situation only. Table 4 shows the operation time for both processes under normal production and rescheduled production. In a process flow, primer comes before top coat. Therefore, primer processes are being executing first in first half operation time from 8:00 AM till 12:30 AM. During that time, all energy resources for Top Coat is still under shutdown or sleep mode. When the primer process finished, the energy resources at Primer will be shut down and Top Coat starts operation for the second half of operation time.

Table 4. Operation time for Primer and Top Coat.

	Normal Operation	Reschedule Under Low Demand
Primer	08:00 AM - 17:30 PM	08:00 AM - 12:30 AM
Top Coat	08:00 AM - 17:30 PM	2:00 PM - 18:30 PM

Total simulated energy consumption for one month under this operation had been reduced to 641,445 kWh while maximum power input is not changed. Calculation for reduction amount is tabled in Table 5. Total energy consumption reduced in Scenario 1 compare to the current simulated system is 274,905 kWh per month which is equal to RM 92,642.99.

Table 5. Reduction amount for Scenario 1.

	Current	Scenario 1	Amount Reduction	
			kWh	RM
Consumption (kWh)	916,350.0	641,445.0	(274,905.0)	(92,642.99)
MD (kW)	2,741.2	2,741.2	-	-
Percentage (%)			30	

3.4.2. Scenario 2: Changing Compressor Power

In scenario 2, the effect of changing the power input of compressor is being studied. We knew from the simulation results for current system, compressors were among top energy consumers. To study this in real system is very difficult because it involves modification on current system. Thus, production has to be interrupted and some investment cost need to be spent.

Meanwhile, conducting such studied in virtual environment could be done easily. Table 6 shows current compressors used in paint shop area and their power input respectively. If Compressor No. 21 and Compressor No. 12 been changed to the same type with compressor No. 18 and 20, the power input is expected be reduced to 107.7 kW.

Table 6. Current compressors in paint shop operation.

No.	System	Control Panel	Power Input (kW)
1	Compressor No.12	Compressor_CP1	117.3
2	Compressor No.18	Compressor_CP2	107.7
3	Compressor No.20	Compressor_CP3	107.7
4	Compressor No.21	Compressor_CP4	133.1
5	Compressor No 22	Compressor_CP5	76.1

To simulate this, we simply change the input power parameter at our simulation model for compressor No. 21 and 12 from 133.1 and 117.3 respectively to 107.7 kW. After running the simulation, the total energy consumption recorded was 903,337.8 kW. The result is tabled as in Table 7. It has been noticed that the MD value is also affected with the reduction amount of 34.90 kW. Summation of reduction both in energy consumption and MD results the total saving of RM 5,418.15 per month under scenario 2 implementation.

Table 7. Reduction of energy consumption under Scenario 2.

	Current	Scenario 2	kWh	Amount Reduction	
				%	RM
Consumption (kWh)	916,350.0	903,337.8	(13,012.20)	1.42	(4,385.11)
MD (kW)	2,741.2	2,706.3	(34.90)	1.27	(1,033.04)
Total (RM)					(5,418.15)

4. Conclusion

4.1. Conclusion

This study has successfully come out with the automotive paint shop simulation model for XYZ Company that could be used for energy consumption analysis. The model has been developed using Tecnomatix Plant Simulation software version 13 which embedded with the energy analysis function. Comparison with actual consumption data recorded by the company, shows that the model was accurate within ±5% with the average recorded EC and ±15% for the MD data.

In this study, two scenarios on energy saving methods had been tested using the model developed. In scenario 1, it can be noticed that it is possible to reduce the EC simply by managing the production schedule and the way production running. By running half working hours for Primer and Top Coat respectively and consequently, the amount of EC could be reduced up to 30% which is about RM 92,642.99 saving in electricity charges.

Moreover, in the scenario 2, the higher power input compressors which are compressor 12 and 21 had virtually changed to the same power type with compressor 18 and 20. The result of the saving amount is approximately 1.42% in EC and 1.27% in MD which is about RM5,418.15 total reduction in electricity charges.

The simulation result for current system and energy saving scenarios had been summarized as shown in Table 8. It can be concluded that large amount of saving could be resulted from implementation of scenario 1. Meanwhile, for scenario 2 further study on amount of investment and return of investment should be conducted to justify whether the implementation is worth or not.

Table 8. Summary of simulation result for all scenarios.

	Current	Scenario 1	Scenario 2
EC (kWh)	916,350.00	641,445.00	903,337.80
Reduction (kWh)		274,905.00	13,012.20
Percentage (%)		30.0	1.42
MD (kW)	2,741.20	2,741.20	2,706.30
Reduction (kW)		0.00	34.90
Percentage (%)		0	1.27

4.2. Future Work

In this study, the model paint shop being studied is an old fashion type paint shop line which most of the processes are manually conducted by human such as sanding, primer and top coat spray. In modern style, mostly the new paint shop line involves robots to execute the processes. However, this type of paints shop line is still unavailable at company being studied. Other than that, the major limitation in this study is the availability of data. Several assumptions have to be made because of lack of data available. This is because company policy state limited access to the outsider in accessing company internal information. Therefore, only data that are provided by the company are allowed to be used in study purpose. Even no single photo is allowed to be captured inside the company.

The same model and approach could be extended further for other type of analysis which has not been covered in this project scope. At least three other topics could be considered for further research as follows:

- Energy efficiency could be related with how well the energy been used to produce one unit of output. Data on processing time, machine availability, reject rate and current production output at least must be well collected to realise this study.
- The capability of Technomatix Plant Simulation is also to analyse different power operation state if details power input for each state is known. By utilizing the operation state, energy optimization study could be conducted. Again, information on failure rate, machine availability, production schedule is crucial in order to conduct energy optimization study.
- An extended study using the same approach could be done on automatic paint shop model that most of processes done by robots. It should be more interesting because this paint shop is expected to consume more energy compare to manual paint shop

Acknowledgement
The authors are grateful to the Jabatan Penyelidikan dan Inovasi, Universiti Malaysia Pahang, for financial support under RDU1703157.

References
[1] Galitsky, C. and Worrell, E. (2008). Energy Efficiency Improvement and Cost Saving Opportunities for the Vehicle Assembly Industry: An ENERGY STAR® Guide for Energy and Plant Managers. Technical Report. USA: LBNL-50939

[2] Jerry, B. 1998. *Handbook of Simulation.* Atlanta: John Wiley& Sons, Inc.

[3] Sutherland, J. W. 2014. Automobile Assembly Painting: Energy and Environmental Impacts (online). http://www.me.mtu.edu/~jwsuther/erdm/lect36a.pdf (15th November 2016)

[4] Cronrath, C., et al. 2016. Energy Reduction in Paint Shops Through Energy-Sensitive On-Off Control. *International Conference on Automation Science and Engineering (CASE).* USA

[5] Rohdin, P. et al. 2012. Energy efficient process ventilation in paint shops in the car industry – experiences and an evaluation of a full scale implementation at Saab Automobile in Sweden. *Paper presented at the International Conference on Industrial Ventilation.* Paris

[6] Chen, G., and Zhang, L. 2013. Energy-Efficient Production Systems Through Schedule-Based Operations. *IEEE Ttransactions on Automation Science and Engineering,* **10**(1)

[7] Zhang, S., et al. 2015. A novel artificial ecological niche optimization algorithm for car sequencing problem considering energy consumption. *Journal of Engineering Manufacture,* **229**(3):546-562

[8] Feng, L. and Mears, L. 2015. Analysis of HVAC energy in automotive paint shop. *International Manufacturing Science and Engineering Conference.* USA

[9] Heinicke, M. 2003. Optimized Energy Efficiency with Tecnomatix (online). https://community.plm.automation.siemens.com/t5/General-Tecnomatix-Forum/Plant-Simulation-Energy-Efficiency-Optimization/td-p/612 (20th November 2016).

[10] Guerrero, C. A., et al. 2010. Energy reduction in automotive paint shops a review of hybrid/electric vehicle battery manufacturing. Master. Thesis. University of Kentucky. USA

[11] Trakultongchai, A., et al. 2013. Process Simulation and Improvement of Automotive Paint Shop. *International Journal of Mining, Metallurgy & Mechanical Engineering (IJMMME),* **1**(1)

[12] Al-Ajami, F.F. and Hanby, V.I. 2008. Simulation of energy consumption for Kuwaiti domestic buildings. *Journal of Energy and Buildings.* **40**(2008): 1101-1109.

[13] Fysikopoulos, A., et al. 2012. An Empirical Study of the Energy Consumption in Automotive Assembly. *Paper presented at the 45th CIRP Conference on Manufacturing Systems 2012.*

[14] Trebuna, P. and Popovic, R. 2015. Power Consumption Management in the Digital Factory. *American Journal of Mechanical Engineering,* **3**(6): 272-274

IOP Conf. Series: Materials Science and Engineering **319** (2018) 012061 doi:10.1088/1757-899X/319/1/012061

Cost viability of 3D printed house in UK

A L Mohd Tobi[1,*], S A Omar[2], Z Yehia[3], S Al-Ojaili[3], A Hashim[3], O Orhan[3]

[1]Faculty of Engineering Technology, Universiti Tun Hussein Onn Malaysia, Batu Pahat, Malaysia,
[2]Faculty of Accountancy and Business, Universiti Selangor, Shah Alam, Malaysia,
[3]Faculty of Engineering, University of Nottingham, Nottingham, United Kingdom.

*abdlatif@uthm.edu.my

Abstract: UK has been facing housing crisis due to the rising price of the property on sale. This paper will look into the viability of 3D printing technology as an alternative way for house construction on UK. The analysis will be carried out based on the data until the year of 2014 due to limited resources availability. Details cost breakdown on average size house construction cost in UK were analysed and relate to the cost viability of 3D printing technology in reducing the house price in UK. It is found that the 3D printing generates saving of up to around 35% out of total house price in UK. This cost saving comes from the 3D printed construction of walls and foundations for material and labour cost.

Keywords: UK, house price, 3D printing

1. Introduction

The UK's housing crisis is increasing in severity, with an ever growing gap between the supply and demand of housing. The average house price for England and Wales in 2015 is £179,492 with areas such as London experiencing average prices of £458,283 [1]. The average 2014 salary for individuals who had been in the same role for more than 12 months is £27,200 [2]. By using average house price data from HM Land Registry [1], this makes the average house price for England and Wales 6.6 times the average annual salary as shown based on data from and of average salaries from Office for National Statistics [2]. **Figure 1** shows a plot of the house price to earnings ratio for various areas in the UK.

House price is growing at an average rate of 6.7% per annum from Jan 2014 to Jan 2015 (**Table 1**) with average salaries only increasing at 0.7% per annum [1]. Homeownership is put further out of reach for many people as time goes by. Whilst many solutions to this issue are posed, one commonly suggested solution is to simply build more houses. If the supply of housing increases, the price should fall and allow easier acquisition of houses.

In 2007, the UK government set a target to build 240,000 new houses per year by 2016 [3]. In 2014, only 118,760 houses were built, an 8% increase from the previous year [4]. The lack of production is largely due to lingering risk aversion from the 2008 recession [5]. Private construction enterprises do not wish to risk their money by investing into new housing, when there are no buyers due to economic downturn. For 3D printing house construction to be a suitable solution to the UK's housing crisis, it needs to perform better than traditional methods. It must reduce the costs of construction significantly.

Reducing the costs of construction reduces the risk to construction companies and encourages them to commit to new projects.

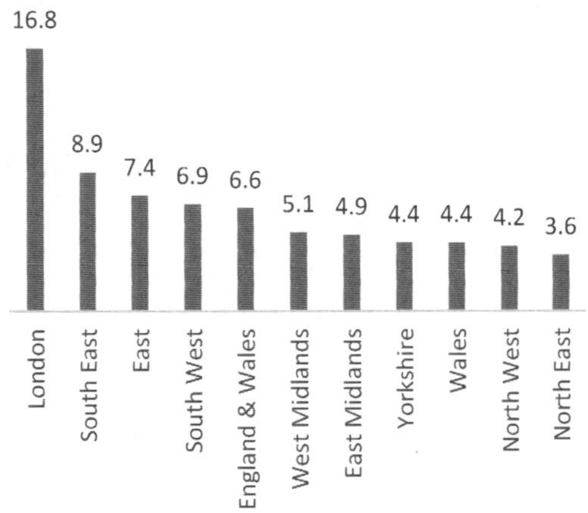

Figure 1. Plot of house price to average earning ratio for various areas in the UK. This plot uses house price data from [1] and earning data from [2].

Table 1. Average house price by property type adopted from [3].

Average price by property type (England & Wales)	January 2014	January 2015	Difference
Detached	£265,814	£282,338	6.2%
Semi-detached	£159,184	£170,926	7.4%
Terraced	£126,931	£134,757	6.2%
Flat/maisonette	£160,016	£171,669	7.3%
All	£168,231	£179,492	6.7%

2. 3D printed house technology

The early technology of 3D printed house is pioneered by a technology called Contour Crafting [6]. It is a technology resulting from the application of the principles of 3D printing to the construction of houses. The technology is a method of automated additive fabrication applied to construction. Single layers of concrete are repeatedly extruded using a three dimensional gantry system. Layers of concrete are built upon each other sequentially until the wall structure is complete (**Figure 2**). The technology applies the modern principles of automation to the traditional principles of construction.

IOP Conf. Series: Materials Science and Engineering **319** (2018) 012061 doi:10.1088/1757-899X/319/1/012061

Figure 2. Illustration of house construction using Contour Crafting 3D printing technology. Image adopted from [6].

Historical construction tools such as trowels, sculpturing knives and putty knives have been used for thousands of years in construction [7 & 8]. The 3D printed house uses the same approach by extruding wet concrete and using the trowel like attachment to shape it, as it would be done with manual tools. The additive approach employed by the technology also allows complex geometries to be created as easily as planar walls, whereas these may take significant expertise to create by manual methods.

Each layer is constructed in two passes. In the forward pass, two parallel surfaces are extruded. These surfaces bound a hollow cavity, and are known as the "filling rims". On the backwards pass, triangulated filler is extruded in the space between the filling rims, in order to provide strength to the wall. This can be seen in **Figure 3**. By only providing a triangulated filling between the rims, a lot of material is saved compared to traditional construction which required filling the entire cavity with solid material. These savings can be rather significant and easily predicted. **Figure 3** shows that curved walls are also possible. These are built as easily as standard rectilinear ones; allowing for geometries not easily acquired by traditional methods.

Figure 3. A section of wall constructed using Contour Crafting. Image adopted from [8].

3. Development in 3D printed house

Winsun, a construction company based in Shanghai, China are well known for 3D printed construction. In March 2014, Winsun announced that they had constructed ten concrete houses, in the space of 24 hours. Each house has 200 m^2 usable floor area and cost only £2,970 (**Figure 4**). The 3D printed parts of the house are mainly the exterior wall structure and base. They were mentioned that the total cost for the house construction to be halved from the traditional one [9].

In January 2015, Winsun completed a 3D printed villa that costs about £105,000 (**Figure 5**). The construction of the villa took eight people in a month to be completed. It is also claimed to have otherwise taken 30 people in three months. The claim indicates that a reduction of 91% in man-hours is possible by using 3D printing house construction. Saving from the labour cost is also contributed by employing less skilled workers since the wall construction is mainly done by the printer. Not only that, the number of workers has been reduced. Less expensive workers can also be employed [10].

Figure 4. A selection of the 10 3D printed houses made in 24 hours for £2,970 each by Winsun construction company in march 2014. Image adopted from [9].

Figure 5. Image of the villa created using 3D printing technology by Suzhou Ltd, a construction company in Shanghai, China. Image adopted from [10].

4. Construction cost analyses

A study done by Cuéllar-Franca & Azapagic [11] looks at the cost of building a house in UK for detached, semi-detached, and terrace house. The cost of the house construction can be divided in terms of labour cost, material cost, energy cost and profit taken by the developer. The construction cost in building a semi-detached house in UK with 90 m^2 usable floor area is about £68,000. While 52% of this contributes to the labour cost in the construction, another 37% for materials, less than 1% on the energy cost and the remaining are the profit taken by the construction company. The average size of the house presented in [5] which is about 91 m^2 usable floor area is similar to the one used in the study.

Based on the information from [11], the construction cost of internal wall, external wall and foundation for a semi-detached house is about £36,000 in material and labour cost (**Table 2**). These constructions are potentially can be built using 3D printing technology. This is about 53% from total construction cost of £68,000 as mentioned above. Assuming of 90% reduction in labour (as claimed by Winsun [10]) and 50% reduction in materials [9], the estimated savings can be gained are £19,000 from labour cost and £7,400 from materials cost. This total up to £26,400 saving in construction cost per house.

Table 2. Construction cost for semi-detached house in UK, adopted from [11].

COMPONENTS	Materials cost (£)	Labour cost (£)	Fuel and machinery cost (£)	Construction company's profit (£)
Construction activities			390	
External wall	9,568	13,733		2,330
Internal wall	3,911	6,041		995
Foundation	1,396	1,433		283
First Floor	2,423	3,880		630
Ground Floor	1,605	3,384		499
Roof	5,077	5,137	81	1,021
Ceiling	663	1,019		168
Windows	359	592		95
Doors	603	377		98
TOTAL	**25,605**	**35,596**	**471**	**6,119**

UK development land price in average for England during 2007 peaked at £4 million per hectare [12]. Combining data form Savills [13] and Land Registry [12] the estimated price for land development in 2013 was about £2.4 million per hectare. By assuming 100 m^2 land being used for building average-sized UK house, the cost of the land equates to £24,000 per house.

As mentioned earlier, the average house price in UK is around £180,000 [1], the construction cost is £68,000 [11] and the land cost is £24,000 [12, 13]. As calculated earlier, the 3D printing house creates saving in construction cost of about £26,400. By assuming the same ratio of profit to cost being taken by the developer with the same land cost, the new 3D printed house can be priced at 30% lower than the traditional construction house (**Figure 6**). However, it is important to note that the cost of the 3D printing technology that is going to be used is not included in this calculation. The investment in the 3D printing technology is a one off setup cost which will be reduced with the quantity of the houses built. The cost of the technology is not publicly available yet.

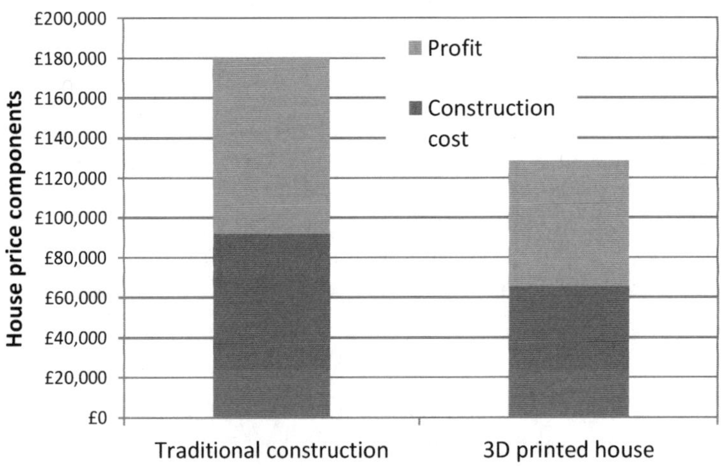

Figure 6. House price comparison for traditional construction and 3D printed house for average sized house in UK.

5. Conclusion

It has been discussed that the technology of 3D printed house must able to construct houses at lower cost than traditional construction methods in order for it to be a viable method. Work presented is a simple analogy on saving estimation for 3D printed house to be implemented in UK. The work will encourage further detail analyses on the cost implication of the technology in the future. This paper investigates the cost viability for 3D printed house technology in UK for average sized house with usable space of 90 m^2. The presented results allow the following conclusions to be made:

- 3D printing technology is able to replace the traditional construction of internal wall, eternal wall and foundation.
- There is possibility of house price reduction up to 30% which can be obtained through the introduction of 3D printed house in UK.
- The cost of the technology is not included in the calculation due to data unavailability.

Acknowledgement

The authors acknowledge the financial support by the Universiti Tun Hussein Onn Malaysia, Short Term Grant Vot. No. U643

References

[1] HM Land Registry. (2015). January 2015 *Market Trend Data: Land Registry's Market Trend Data for England and Wales*. Available: https://www.gov.uk/government/news/january-2015-market-trend-data. Last accessed 20th Aug 2017.
[2] Office for National Statistics. (2014). *Annual Survey of Hours and Earnings: 2014 Provisional Results*. Available: https://www.ons.gov.uk/employmentandlabourmarket/peopleinwork/earningsandworkinghours/bulletins/annualsurveyofhoursandearnings/2014-11-19. Last accessed 20th Aug 2017.
[3] Wilson, W. (2010). *Housing supply and demand. London: House of Commons Library*. Available: https://www.parliament.uk/documents/commons/lib/research/key_issues/Key-Issues-Housing-supply-and-demand.pdf. Last accessed 20th Aug 2017.
[4] Department for Communities and Local Government. (2015). *House building in England: October to December 2014*. Available: https://www.gov.uk/government/statistics/house-building-in-england-october-to-december-2014. Last accessed 20th Aug 2017.
[5] Department for Communities and Local Government. (2010). *English Housing Survey Housing stock report 2008*. Available: https://www.gov.uk/government/uploads/system/uploads/attachment_data/file/6703/1750754.pdf. Last accessed 20th Aug 2017
[6] Contour Crafting Corporation. (2017). *Introducing Contour Crafting Technology*. Available: http://contourcrafting.com/. Last accessed 20th Aug 2017
[7] Zhang, J. & Khoshnevis, B. (2013). *Optimal machine operation planning for construction by Contour Crafting*. Automation in Construction, Volume 29, pp. 50-67.
[8] Hwang, D. & Khoshnevis, B. (2005). *An Innovative Construction Process-Contour Crafting*. 22nd International Symposium on Automation and Robotics in Construction ISARC 2005 - September 11-14, 2005, Ferrara (Italy).
[9] ITN. (2014). *3D printer builds houses in China - video*. Available: https://www.theguardian.com/technology/video/2014/apr/29/3d-printer-builds-houses-china-video. Last accessed 20th Aug 2017.
[10] Davidson, N. (2015). 3D-printed cities: is this the future?. Available: https://www.theguardian.com/cities/2015/feb/26/3d-printed-cities-future-housing-architecture. Last accessed 20th Aug 2017.

[11] Cuéllar-Franca, R. M. & Azapagic, A. (2014). *Life cycle cost analysis of the UK housing stock*. The International Journal of Life Cycle Assessment, 19(1), pp. 174-193.

[12] Department for Communities and Local Government. (2012). *Table 563: average valuations of residential building land with outline planning permission (final version)*. Available: https://www.gov.uk/government/statistical-data-sets/live-tables-on-housing-market-and-house-prices. Last accessed 20th Aug 2017.

[13] Savills. (2014). *Market in Minutes UK Residential Development Land: February 2014*. Available: http://pdf.euro.savills.co.uk/residential---other/mim-resi-development-land.pdf. Last accessed 20th Aug 2017.

Gradient Evolution-based Support Vector Machine Algorithm for Classification

Ferani E. Zulvia[1*] and R. J. Kuo[2]

[1]Department of Logistics Engineering, Universitas Pertamina, Teuku Nyak Arief Road, Simprug, Kebayoran Lama, Jakarta, Indonesia
[2]Department of Industrial Management, National Taiwan University of Science and Technology, No. 43, Section 4, Kee-Lung Road, Taipei, Taiwan

*corresponding author. Email: Feranieva.z@universitaspertamina.ac.id

Abstract. This paper proposes a classification algorithm based on a support vector machine (SVM) and gradient evolution (GE) algorithms. SVM algorithm has been widely used in classification. However, its result is significantly influenced by the parameters. Therefore, this paper aims to propose an improvement of SVM algorithm which can find the best SVMs' parameters automatically. The proposed algorithm employs a GE algorithm to automatically determine the SVMs' parameters. The GE algorithm takes a role as a global optimizer in finding the best parameter which will be used by SVM algorithm. The proposed GE-SVM algorithm is verified using some benchmark datasets and compared with other metaheuristic-based SVM algorithms. The experimental results show that the proposed GE-SVM algorithm obtains better results than other algorithms tested in this paper.

1. Introduction

Classification is one of data mining approach which has been widely applied in many applications. Therefore, many studies have proposed classification methods such as decision tree, support vector machine, k-nearest neighbour, etc. [1]. Among these methods, support vector machine (SVM) has become popular among classification techniques. SVM is a supervised learning model which construct a set of hyperplanes to separate the dataset [2]. Although it has been applied in many applications and shown a good performance, this algorithm highly depends on the parameters [3, 4]. Therefore, determining SVM parameter is a crucial issue before applying SVM algorithm. Many studies have attempted to overcome this problem [5-7]. Some of the proposed methods employ metaheuristic to find the best parameters for the SVM. Lin, Lee, Chen and Tseng [8] use a simulated annealing (SA) algorithm, Huang and Wang [9] apply genetic algorithm (GA) algorithm, Lin, Ying, Chen and Lee [10] use particle swarm optimization (PSO) algorithm, and etc. [11, 12]. This paper also employs a metaheuristic, named gradient evolution (GE) algorithm to automatically tuning the SVM parameters. GE algorithm is a new metaheuristic inspired from Newton-Raphson algorithm [13]. It explores the search space using three operators, namely the vector updating, jumping, and refreshing. According to the previous researches, GE algorithm has promising results [13, 14]. Therefore, this paper attempts to apply GE algorithm to automatically optimize SVMs' parameters.

The reminder of this paper is organized as follows. Section two presents the literature review. Section three discusses the proposed algorithm while the experimental results are discussed in Section four. Finally, the concluding remarks are made in Section five.

2. Literature Review

This section briefly reviews some basic theories applied in this paper.

2.1. Support Vector Machine

SVM has been used for classifying both linear and nonlinear separable data. In order to classify nonlinear data, SVM uses a Kernel function. This paper applies a Gaussian Kernel $k(\bar{x}, \bar{y})$ as given in Eq. (1).

$$k(\bar{x}, \bar{y}) = e^{-(|\bar{x}-\bar{y}|^2/2\sigma^2)}, \tag{1}$$

where $\bar{x}, \bar{y} \in \mathbb{R}^n$ and $\sigma > 0$ is a free parameter. The non-linear dual Lagrangian model of non-linear SVM can be defined in Eq. (2).

$$\hat{f}(\bar{x}) = \operatorname{sgn}\left(\sum_{i=1}^{l} \alpha_i^* y_i k(\bar{x}_i, \bar{x}) + b^*\right), \tag{2}$$

The training algorithm in terms of a kernel function is given by Eq. (3).

$$\bar{\alpha}^* = \underset{\bar{\alpha}}{\operatorname{argmax}}\left(\sum_{i=1}^{l} \alpha_i - \frac{1}{2}\sum_{i=1}^{l}\sum_{j}^{l} \alpha_i \alpha_j y_i y_j k(\bar{x}, \bar{y})\right), \tag{3}$$

Subject to,

$$0 \le \alpha_i \le C, i = 1, \dots, m \text{ and } \sum_{i=1}^{l} \alpha_i y_i = 0, \tag{4}$$

where l is the number of support vectors in the trained model and C is weight parameter.

2.2. Gradient Evolution Algorithm

Gradient Evolution (GE) algorithm is a new metaheuristic developed from Newton-Raphson Optimization algorithm [13]. This algorithm represents the solution as a vector. GE algorithm explores the search space using three operators, vector updating, jumping, and refreshing. Vector updating focuses on deep search while vector jumping focuses on wide search. The last operator, vector refreshing is an operator which manage the diversity in the population.

3. Methodology

This paper applies GE algorithm to fine tune SVMs' parameters. There are two parameters in non-linear SVM using Gaussian, number of support vectors in the trained model and weight parameter. GE algorithm is employed to optimize these parameters. In GE algorithm, they are represented as a vector. Fitness of each vector is defined as the error of the training test. These parameters are improved iteratively. After all iterations, the final model obtained using the best parameter is used for predicting the testing error.

The flowchart of the proposed GESVM is given in Figure 1 while the procedure is as the following.

Step 1: Set parameter for GE: number of vectors, number of iterations, jumping rate, reduction rate, and minimum saturation.

Step 2: Generate initial vectors, $\mathbf{P}^t = \{\mathbf{X}_i^t | i = 1, \dots, L\}$.

Step 3: Calculate fitness of each vector by performing SVM with parameter as given by the vector. Use the average of training error as the fitness.

Step 4: Record the best vector.

Step 5: For each iteration, t, update all vectors.

Step 6: For each vector, p, perform vector updating towards p.

$$u_{ij}^t = x_{ij}^t - \left(r_g \cdot \frac{\Delta x_{ij}^t}{2}\right) \cdot \left(\frac{x_{ij}^W - x_{ij}^B}{x_{ij}^W - 2x_{ij}^t + x_{ij}^B}\right) + r_a \cdot (y_j - x_{ij}^t), \forall j = 1, \dots, D, \tag{5}$$

$$\Delta x_{ij}^t = \frac{|x_{ij}^t - x_{ij}^B| + |x_{ij}^W - x_{ij}^t|}{2}, \forall j = 1, \dots, D \tag{6}$$

where u_{ij}^t is a mutant vector, y_j is the best vector, x_{ij}^B are x_{ij}^W a better and a worse vector than vector x_{ij}^t, respectively. The r_g and r_a are random numbers. For the best vector, x_{ij}^B is replaced by b_j defined in Eq. (7).

$$b_j = x_{ij}^t - \Delta x_{ij}^t, \forall j = 1, \dots, D. \tag{7}$$

Where,

$$\Delta x_{ij}^t = \frac{\gamma + |x_{ij}^W - x_{ij}^t|}{2}, \forall j = 1, \dots, D. \tag{8}$$

On the other hand, x_{ij}^W is replaced by w_j defined in Eq. (9).

$$w_j = x_{ij}^t + \Delta x_{ij}^t, \forall j = 1, \dots, D \tag{9}$$

Where,

$$\Delta x_{ij}^t = \frac{|x_{ij}^t - x_{ij}^B| + \gamma}{2}, \forall j = 1, \dots, D, \tag{10}$$

where γ is a pre-defined parameter, defined as initial step size.

Step 7: Generate a random number r between $[0,1]$. If r less than jumping rate, perform vector jumping towards vector p.

$$u_{ij}^t = -u_{ij}^t + r_m \cdot (u_{ij}^t - x_{kj}^t), \forall j = 1, \dots, D, \tag{11}$$

where $x_{kj}^t \epsilon \mathbf{X}_{kj}^t$ is a random neighbor vector in $\mathbf{P}^t$, $\forall i \neq k$ and r_m is a random number in range $[0,1]$.

Step 8: Calculate fitness

Step 9: If the new fitness is worse than the previous fitness, use the previous vector p as the next vector p and reduce the vector saturation p.

Step 10 : If vector saturation p is less than minimum saturation, regenerate vector p.

Step 11 : Update the best vector.

Step 12 : Back to Step 5 until the end of iteration.

Table 1. Parameter Setting.

Algorithm	Parameter	Value
GESVM	Jumping rate	0.3
	Reduction rate	0.05
	Minimum saturation	0.2
	Initial Delta	0.1
PSOSVM	Inertia weight	0.8
	Learning rate 1	1.5
	Learning rate 2	1.5
DESVM	Crossover rate	0.2
	Alpha	0.5

4. Experimental Results and Analysis

The proposed algorithm is tested using two benchmark datasets, WDBC and Ionoshpere taken from UCI repository. The WDBC dataset comprises of 569 instances with 30 attributes while ionosphere dataset has 351 instances with 34 attributes. For each dataset, 10-fold cross validation is applied. The results are compared with other metaheuristic-based SVM algorithms. They are PSO-SVM and DE-SVM algorithms. Since in this problem the metaheuristic algorithm focuses on performing global exploration, the parameter setting for metaheuristic algorithm is also set to achieve this purpose. Table 1 lists the parameter setting for each algorithm.

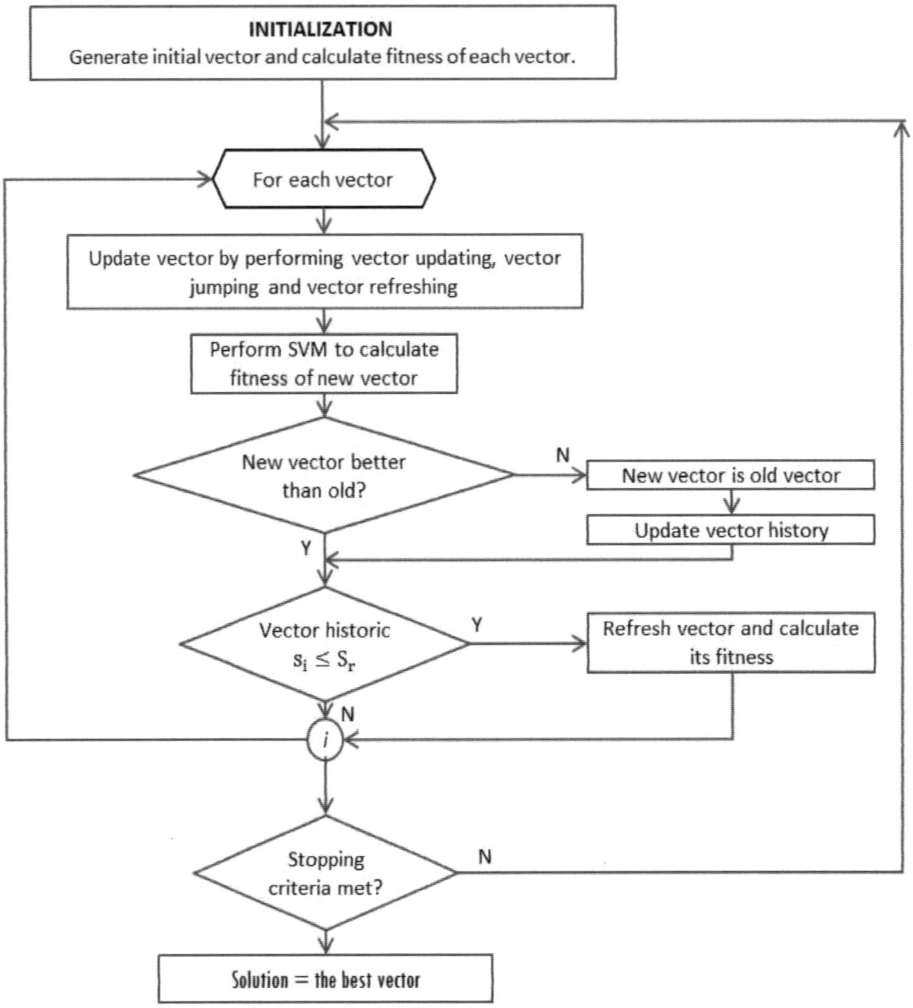

Figure 1. The proposed GESVM Flowchart

Table 2. Experimental Results.

Dataset	comparator	GESVM Training error	GESVM Testing error	PSOSVM Training error	PSOSVM Testing error	DESVM Training error	DESVM Testing error
WDBC	average	0.00E+00	1.22E-01*	0.00E+00	1.28E-01	0.00E+00	2.24E-01
	St. deviation	0.00E+00	1.41E-02	0.00E+00	9.36E-03*	0.00E+00	5.43E-02
	min	0.00E+00	9.83E-02*	0.00E+00	1.11E-01	0.00E+00	1.47E-01
	max	0.00E+00	1.37E-01	0.00E+00	1.37E-01	0.00E+00	3.11E-01
Ionoshpere	average	0.00E+00	2.64E-01*	0.00E+00	2.70E-01	0.00E+00	2.77E-01
	St. deviation	0.00E+00	3.43E-02*	0.00E+00	2.44E-02	0.00E+00	2.26E-02
	min	0.00E+00	1.96E-01*	0.00E+00	2.25E-01	0.00E+00	2.54E-01
	max	0.00E+00	2.91E-01	0.00E+00	2.93E-01	0.00E+00	3.16E-01

**:best result*

Ten independent runs are executed for each algorithm. Table 2 summaries the result. A further statistical test analysis is also conducted to evaluate if the results of these algorithms are significantly different. Table 3 gives the statistic test result. These tables show that combining SVM with metaheuristic helps

SVM to get very low training error. In terms of testing error, GESVM algorithm obtains significantly better results for WDBC datasets. For ionosphere dataset, GESVM algorithm still get better result but not significantly different than PSOSVM and DESVM algorithms. The reason might be because GE algorithm has operator to avoid local optima and maintaining diversity in the population. On the other hand, PSOSVM and DESVM do not have a specific operator focused on avoiding local optimal and maintaining diversity in the population.

Table 3. Statistic Test Results.

Testing	WDBC	Ionosphere
GESVM vs PSOSVM	0.499	0.801
GESVM vs DESVM	0.006*	0.556
PSOSVM vs DESVM	0.008*	0.679

**:significantly different*

5. Conclusion

SVM algorithm is a popular classification algorithm. Many studies have shown good performance of SVM algorithm. However, SVM involves parameters which significantly influence the result. This paper proposes an algorithm which can automatically fine tune SVMs' parameters. The proposed algorithm employs a metaheuristic named GE algorithm as the parameter generator. This algorithm helps to find the best parameter setting for SVM algorithm.

The proposed algorithm is verified using two benchmark datasets and compared with PSOSVM and DESVM algorithms. The results show that the proposed GESVM can provide better results than PSOSVM and DESVM algorithms. It might be happened due to GESVM has two operators focus on wide exploration. They are the vector jumping and vector refreshing.

This study should be extended by testing more datasets and comparing with more algorithms. Other approaches to optimize SVM parameters are also should be further studied.

References

[1] Fayyad U M, Piatetsky-Shapiro G, Smyth P and Uthurusamy R 1996 *Advances in knowledge discovery and data mining* (CA, USA: American Association for Artificial Intelligence Menlo Park)

[2] Kuo R J, Mei C H, Zulvia F E and Tsai C Y 2016 An application of a metaheuristic algorithm-based clustering ensemble method to APP customer segmentation *Neurocomputing* **205** 116-29

[3] Matheny M E, Resnic F S, Arora N and Ohno-Machado L 2007 Effects of SVM parameter optimization on discrimination and calibration for post-procedural PCI mortality *Journal of Biomedical Informatics* **40** 688-97

[4] Cao L J and Tay F E H 2003 Support vector machine with adaptive parameters in financial time series forecasting *IEEE Transactions on Neural Networks* **14** 1506-18

[5] Devos O, Ruckebusch C, Durand A, Duponchel L and Huvenne J-P 2009 Support vector machines (SVM) in near infrared (NIR) spectroscopy: Focus on parameters optimization and model interpretation *Chemometrics and Intelligent Laboratory Systems* **96** 27-33

[6] Min J H and Lee Y-C 2005 Bankruptcy prediction using support vector machine with optimal choice of kernel function parameters *Expert Systems with Applications* **28** 603-14

[7] Guo H, Liu H and Wang L 2006 Method for selecting parameters of least squares support vector machines and application *Journal of System Simulation* **18** 2033-6

[8] Lin S-W, Lee Z-J, Chen S-C and Tseng T-Y 2008 Parameter determination of support vector machine and feature selection using simulated annealing approach *Applied Soft Computing* **8** 1505-12

[9] Huang C-L and Wang C-J 2006 A GA-based feature selection and parameters optimizationfor support vector machines *Expert Systems with Applications* **31** 231-40

[10] Lin S-W, Ying K-C, Chen S-C and Lee Z-J 2008 Particle swarm optimization for parameter determination and feature selection of support vector machines *Expert Systems with Applications* **35** 1817-24
[11] Lorena A C and de Carvalho A C P L F 2008 Evolutionary tuning of SVM parameter values in multiclass problems *Neurocomputing* **71** 3326-34
[12] Huang H-L and Chang F-L 2007 ESVM: Evolutionary support vector machine for automatic feature selection and classification of microarray data *Biosystems* **90** 516-28
[13] Kuo R J and Zulvia F E 2015 The gradient evolution algorithm: A new metaheuristic *Information Sciences* **316** 246-65
[14] Kuo R J and Zulvia F E 2016 Cluster analysis using a gradient evolution-based k-means algorithm. In: *2016 IEEE Congress on Evolutionary Computation (CEC),* pp 5138-45

Improving student satisfaction of Andalas University Dormitory through Service Quality and Importance Performance Analysis

Nilda Tri Putri and Larisa Anggraini

Department of Industrial Engineering, Faculty of Engineering, Andalas University, Padang, West Sumatra Indonesia

nilda@ft.unand.ac.id, larisangrn@gmail.com

Abstract. Residential satisfaction of university dormitories serve as one of the significant aspects in the framework of sustainability in higher education. This research investigated the quality of dormitory services in Andalas University Dormitory based on student's satisfaction. According to management residential, the enrollment of residential student has increased gradually in Andalas University. In 2016, capacity of residential student is 1686, but only 1081 students can stay at dormitory because some rooms in bad condition. There are a lot of problems and complaints regarding dormitory's service quality i.e water problems, leaky rooms and bathrooms, cleanliness and inadequate facilities in residential college. In addition, there are 20% of last year student's residential check out before the time of contract runs out. The aim of this research are understanding the level of GAP exists between expectation and perception students' residential in the content of service quality and evaluating the improvement priority services using Importance Performance Analysis. This study is measuring service quality by using Responsiveness, Assurance, Empathy, Reliability and Tangible dimension. A negative GAP indicates that the actual services are than what was expected and the GAP is highlighted area for improvement. Based on IPA, management should improve this following dimension services : responsiveness, tangible and assurance dimension.
Keywords: Dormitory, GAP Analysis, Importance Performance Analysis, Service Quality

1. Introduction

In accordance with DIKTI and KEMENDIKBUD RI decree (Number 25/DIKTI/Kep/2014) about general guidelines in educational system for new students, Higher Learning Institutions are trying to attract more new students to stay in residential and employing various strategies to maintain them. New students need psychological and social readiness to be able quickly adapt to college life and in particular learning system. Andalas University located in Padang, West Sumatera Regions of Indonesia, provides a low-cost, attractive, safe, and convenient living quarters for undergraduate students. Andalas University obligates new students with BIDIKMISI Grants (National Scholarship) to stay in residential college (dormitory) and advises other new students (especially students come from outside West Sumatera) to live in dormitory. To achieve a good educational system in national education, Andalas University should examine and understand how students receive a well services and best physical environment of higher institution of learning. Service quality is one of the factors that can lead to customer satisfaction [1]. It can be used as a strategy for achieving the residential college satisfaction. Service quality is among factors that determine customer satisfaction and customer satisfaction can lead to customer loyalty. Customer satisfaction and service quality have been for this latest year's important topic both for the researches. Many studies talked about the relationship between service quality and customer satisfaction. [2] revealed that students are satisfied with the student housing facilities. Overall satisfaction is positively related to loyalty behaviour. [3] identified factors contributing to satisfaction consist of the social qualities (e.g., social densities, the kitchenette, bathroom, and storage facilities) and some demographic, characteristics of the students. Hall configuration is a predictor of satisfaction. Another researcher [4] refered using SERVQUAL model and tested the conceptual model of service quality based on five service quality dimensions of the student residential in China. Consequently, the result indicated the important link between service quality and student satisfaction.

According to residential management, the enrolment of residential students has increased gradually in Andalas University. In 2016, capacity of residential student is 1686 students, but only 1081 students can stay at dormitory because some rooms in bad condition. Based on the results of the suggestion box in every building Andalas Dormitory, the complaints from student residential college quite a lot. In addition, there are 20% of last year student's residential leave the dormitory (check out) before the time of contract runs out. The facts above indicates that there are many student residential who are not satisfied with the service provided. So that, it is important to improve the quality standard of service in residential student in Andalas Universities by evaluating the facilities in residential students or measuring user experience of residential colleges.

This research limits on Andalas University Dormitory which has three buildings for female students which are Oren Dormitory, Hijau Dormitory and RPX Dormitory, and two buildings for male students which are Roesma-M.Shaff Dormitory and Menpera Dormitory. The objectives of the current study are to analyze the effect service quality provided by Andalas University Dormitory to student residential satisfaction, to identify the gap between student's perception and expectation on the service quality in Andalas University Dormitory and to evaluate the improvement priority service based on students perception using Importance Performance Analysis.

2. Service Quality Dimensions

The service quality in the field of education and higher learning institution particularly is not essential, but also important parameter of educational excellence [5]. Consequently, the measurement of service quality, including student residential service quality, has to be based on perceived quality rather than objective quality. The service quality instrument is structured in five dimension, namely [6]:

(i) Responsiveness : The focus of this dimension is on promptness and attentiveness to deal the customer's requests, complain, problems and questions.
(ii) Tangibles : Tangibles was defined as physical appearance of facilities, personnel, communication material, equipment etc.
(iii) Empathy : Caring, individualist attention provided to customer
(iv) Assurance : This dimension is refered to employee's courtesy and knowledge, and ability to gain customer's confidence and trust.
(v) Reliability : Reliability is service provider's ability to perform certain service accurately and dependably.

[7] research show important indicator factors of responsiveness dimension in student residential such as perceptive to provide information, fast respond in complaints and respond innovation/suggest from customers. The findings of [8] revealed another important indicator factors of responsiveness are service in student complaints, willingness to provide information and alertness of management in handling students complaints. According to [9] tangible dimension measured by this following attributes: physical appearance (cleanliness of dormitory, view from residential, adequacy of living space, furniture adjustability, building appearance etc), facilities (electricity &water system, secure facilities, sanitary facilities, recreation/sporting facilities, parking facilities etc) and equipment. [9] categorised ventilation, cost of housing, noise level in the housing estate, lighting in residence, accessibility to public transportation as attributes in Tangible Dimension. [10] defined the factors that raise the level of service quality of Empathy Dimension such as attitude and manner staff or management, privacy in the residence, and neighborhood relations. Besides this, the ability

management to communicate with customer, understanding customer needs and wants, sympathy receiving complaints also affected the service quality level of empathy dimension [11]. [12] proposed 5 items of Assurance dimension that improve service quality in residential. The items are security control and patrol in residential, ability of handling emergency situation, crime rate of neighborhood, potential for flood damage and maintenance regularly facilities. [7] also found the following reliability aspect : professional knowledge, good communication, rigor in service and the conformity service with the promise offered. This indicator will be used as literature to determine the questionnaires in survey.

3. Research Methodology

We used the GAP Analysis and IPA for assessing students' perception of Andalas University Dormitory. The methodology of the empirical research was articulated in this main steps :

(i) Selection of Attributes
Based on literature review [13],[8],[9],[10],[11],and [12] about attributes in service quality dimension, researcher validated all atributes to management residential andalas university. Therefore, valid attributes was selected as questionaires in Survey. Selection of Attributes to be included in the Gap Analysis and Importance Performance Analysis was 11 attributes on Responsiveness Dimension, 11 attributes on Empathy Dimension, 10 attributes on Assurance Dimension, 10 attributes on Reliability Dimension, and 46 attributes on Tangible Dimension. Total of attributes are 88 attributes.

(ii) Reliability and Validity
To obtain the reliability of instrument testeed Cronbach's Alpha statistics contained in software SPSS 16.0. Instrument in all dimension was reliable through the result of Cronbach's alpha calculation show values greater than 0.70. The validity will be tested by comparing the R-value and R-Table (0,361). The all item was considered valid through thr R value are bigger than R-table.

(iii) Data collection
In this research data were collected by using questionnaire. The questionnaire is using Likert scale, which range from 1-5 (1=Strongly Disagree, 2=Disagree, 3=Moderate, 4= Agree, 5=Strongly Disagree.) Sample of the study will be determined based on Slovin formula from 1081 population of students' residential year 2016, there will be 292 respondents. The response rate was drawn 68 % in consideration of only 200 students filled the questionnaire.

(iv) Data Analysis
GAP analysis was applied to analyze the gap between perception and expectation from students. Meanwhile Importance Performance Analysis (IPA) was used to evaluate the improvement priority service based on students perception.

4. Result and Discussion

4.1 Gap Analysis

Table 1 represents the summary of gap scores for each dimension of SERVQUAL in Andalas University Dormitory. The study showed (Table 1) there is a huge gap exist between service quality all dimension and expectation from students residential ands also all service quality dimensions have negative gap between expectation and perception. It means the result of the research showed that students residential are not satisfied with dormitory services. The responsiveness dimension has the widest gap among other SERVQUAL dimension. It means that Andalas University Dormitory should prioritize responsiveness improvement to improve its service quality. Empathy dimension has the narrowest gap point among all dimensions. As shown on the table 1, it can be seen that empathy has GAP score above -1 for Hijau, RPX, Roesma and Menpera building. Based on the score, assurance dimension is the best service quality dimension that has been provided by Andalas Univeristy Dormitory among other dimension.

Table 1. Table Gap of Service Quality Dimension in Andalas University Dormitory

Dimension	BUILDINGS				
	Hijau	Orens	RPX	Roesma	Menpera
Responsiveness	-1.37	-1.55	-1.38	-1.25	-1.51
Tangible	-1.09	-1.3	-1.29	-1.01	-1.11
Empaty	-0.8	-1.2	-0.89	-0.85	-0.77
Reliability	-0.94	-1.19	-1.03	-0.91	-0.99
Assurance	-1.21	-1.47	-1.1	-1.14	-1.12

4.2 Importance Performance Analysis (IPA)

Importance – Performance Analysis applied with the calculation of the average value each attributes. The statement from the importance variable (expectation) and performance variable (perception). The higher gap from perception-expectation calculated become a view for the dormitory management to focused on the improvement on the service attributes which need an improvement. The priority mapping analysis in this research used in all of respondents for every buildings. From all of the data calculation with SPSS 16 resulted a IPA cartesian diagrams for all student respondent to the service shown in this below.

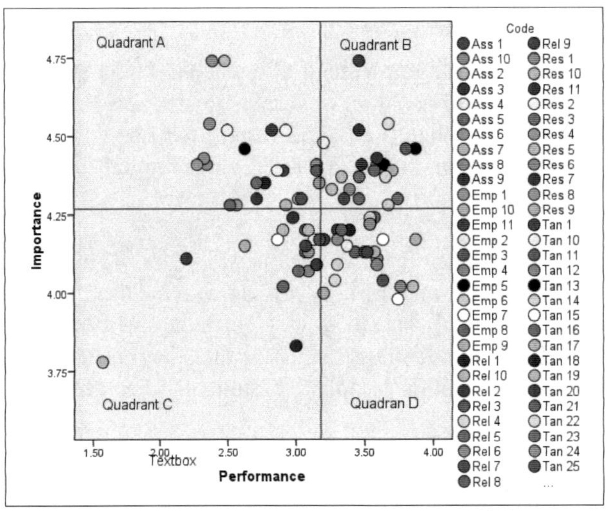

Figure 1. IPA for Hijau Dormitory Building

In Figure 1, the X- axis represents the perception of performance scores relating to students' residential experience of Andalas University Dormitory. The Y-axis represents the relative importance that five Importance factors Service Quality had to the students when choosing a residential. The mean Importance rating for the pooled data was 4,267 while the mean Performance rating was 3,178. The four quadrants are constructed based on the means scores of the Importance and Performances ratings.

From the description of Cartesian diagram Figures indicates that twenty-two (22) attributes are in A quadrant, twenty-two (22) attributes in quadrant B, sixteen (18) attributes in quadrant C and twenty-six (26) attributes D quadrant. So the total attributes seen in Cartesian diagram are 88 attributes.

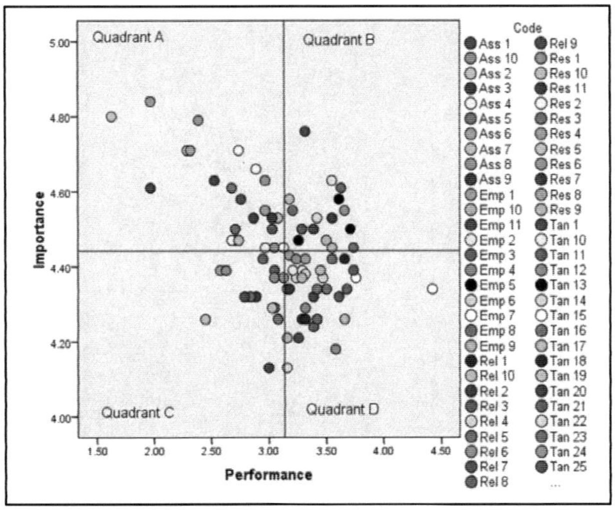

Figure 2. IPA for Orens Dormitory Building

IPA analysis in Figure 2 is conducted through a questionnaire with 88 questions relating to "importance" and "performance" on dormitory. The result are expressed ad a coordinate diagram with "importance" being horizontal x-axis and "performance" being the vertical y-axis. The mean value fott "importance" and "performance" (4,442; 3,1375) is defined as the origin and used to partition the coordinate space into four quadrants. The Quadrant A is "Concentate Here", the Quadrant B is "Keep up the Good Work", the Quadrant C is "Low priority" and the Quadrant D is "Possible Overkill".

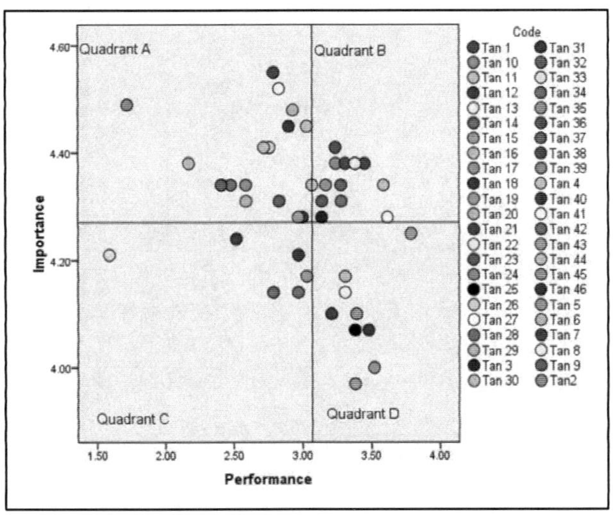

Figure 3. IPA for RPX Dormitory Building

In Figure 3, the X- axis represents the perception of performance scores relating to students' residential experience of Andalas University Dormitory. The Y-axis represents the relative importance that five Importance factors Service Quality had to the students when choosing a residential. The mean Importance rating for the pooled data was 4,271 while the mean Performance rating was 3,0697. The four quadrants are constructed based on the means scores of the Importance and Performances ratings. From the description of Cartesian diagram Figures indicates that twenty-eight (28) attributes are in A quadrant, eighteen (18) attributes in quadrant B, fifteen (15) attributes in quadrant C and twenty-seven (27) attributes D quadrant. So the total attributes seen in Cartesian diagram are 88 attributes.

IOP Conf. Series: Materials Science and Engineering **319** (2018) 012063 doi:10.1088/1757-899X/319/1/012063

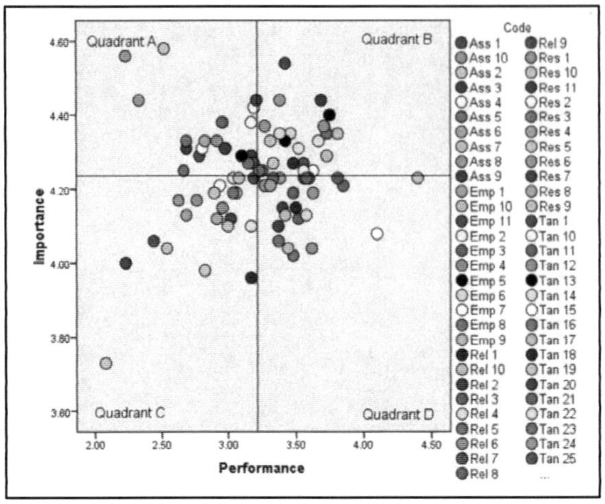

Figure 4. IPA for Roesma Dormitory Building

This study used IPA to establish the variation between "importance" and "performance". IPA analysis in Figure 4 is conducted through a questionnaire with 88 questions. The result are expressed ad a coordinate diagram with "importance" being horizontal x-axis and "performance" being the vertical y-axis. The mean value for "importance" and "performance" (4,236; 3,2147) is defined as the origin and used to partition the coordinate space into four quadrants.

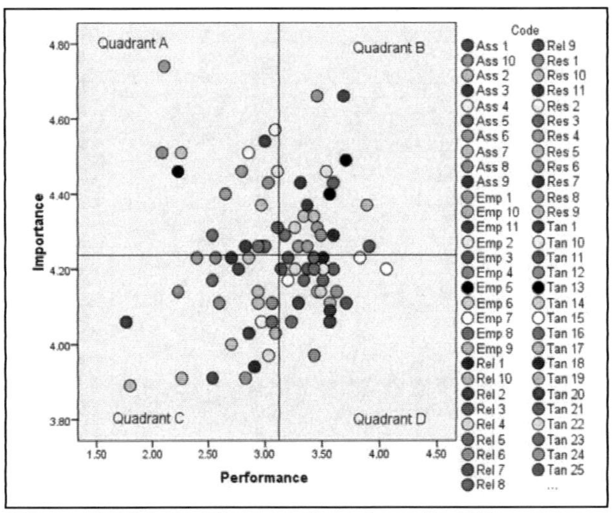

Figure 5. IPA for Menpera Dormitory Building

The X- axis represents the perception of performance scores relating to students' residential experience of Andalas University Dormitory. The Y-axis represents the relative importance that five Importance factors Service Quality had to the students when choosing a residential. The mean Importance rating for the pooled data was 4,2369 while the mean Performance rating was 3,121. The four quadrants are constructed based on the means scores of the Importance and Performances ratings.

From the description of Cartesian diagram Figures indicates that eighteen (18) attributes are in A quadrant, nineteen (19) attributes in quadrant B, twenty-three (23) attributes in quadrant C and twenty-seven (27) attributes D quadrant.

5. Conclusion

Through the GAP Analysis Importance-Performance Analysis we concluded that the management should focus on the Concentrate Here attribute, and allocate resources to improve service quality to :

Fast respond in complaints (Res 4), Check regularly facilities (Res 5), Replacing damaged facilities (Res 6), water supply system (Tan 2), Arrangement of waste collection (Tan 6), availability of health care center (Tan 23), Availability of internet (Tan 30), Speed of internet (Tan 32), Cleanliness of shared bathrooms/toilets (Tan 41), easy check in and check out (Rel 7), Speed of service management (Rel 10), security patrol and control 24 hours (Ass 1), and crime rate of occupants residential/neighborhood (Ass 4). In the possible overkill quadrant, there is an over emphasis in the government's allocation of resources on attributes that are deemed to be unimportant and it is time to consider divesting investment in these area by allocating more resources to concentrate here quadrant.

GAP Analysis and Importance Performance Analysis (IPA) are simple and useful techniques that can help management residential in Andalas University to identify which attributes should be improved to increase overall students residential satisfaction. From the research prospective, this study support the adoption of GAP analysis and IPA as an alternative framework for evaluating students' satisfaction. Such framework can be used in further research on students' residential satisfaction.

References

[1] Kasper, H.; Helsdingen, van P.; Gabbott, M. 2006. Service Marketing Management, A Strategic Perspective. West Sussex: John Wiley & Sons.

[2] Najib, N.; Yusof, N.; Abidin, N. 2011. Student residential satisfaction in research universities. J. Facil. Manag. 2011, 9, 200–212.

[3] Amole, D. 2009. Residential satisfaction and levels of environment in students' residences. Environ. Behav., 41, 866–879.

[4] Chow, I.H., Lau, V.P., Lo, T.W., Sha, Z. & Yun, H. 2015. Service quality in universities operations in China: decision- and experiential-oriented perspectives. International Journal of Hospitality Management, 26(3), 698–710.

[5] Alves, H. and M. Raposo, 2010. The Influence of University Image on Students' Behaviour. International Journal of Educational Management, 24 (1):73-85

[6] Fen, Y. S. and Lian, K. M. 2016. "Service quality and customer satisfaction: Antecedents of customer's repatronage intentions", Sunway Academic Journal, Vol. 4, 59–73.

[7] Freeman, C., Carlson, J., & Sperry, L. 2013. Adlerian marital therapy strategies with middle income couples facing marital stress. The American Journal of Family Therapy, 21(4), 324-332. doi: 10.1080/01926189308251003.

[8] Kim, S., Yang, I., Yeo, M., Kim, K., 2015. Development of a housing performance evaluation model for multi-family residential building in Korea. Building and Environment 40 (2015), 1103–1116.

[9] Palmer, A. 2011. Principles of services marketing. 6th ed. Maidenhead: McGraw-Hill Education

[10] Lai, J. H. K. Yik, F. W. H. 2011. An analytical method to evaluate facility management services for residential buildings. Building and Environment, 46(1), 165-175.

[11] Ning, Yan & Chen, Jiaojiao. 2016. Improving Residential Satisfaction of University Dormitories through Post-Occupancy Evaluation in China: A Socio-Technical System Approach. Department of Construction and Real Estate, Southeast University, Nanjing.

[12] Xueqin, He. (2009). Residential Satisfaction with Home Location: Examination of The Relationship between Location-Embedded Benefits and Risk Perception.

[13] Jiboye, Adesoji David. 2012. Post Occupancy Evaluation, of residential satisfaction in Lagos, Nigeria: Feedback for residential improvement: Department of Architecture, Obafemi Awolowo University, Ile-Ife, Nigeria

IOP Conf. Series: Materials Science and Engineering **319** (2018) 012064 doi:10.1088/1757-899X/319/1/012064

Optimization of diesel engine performance by the Bees Algorithm

Siti Azfanizam Ahmad, Devaraj Sunthiram

Department of Mechanical and Manufacturing Engineering, Faculty of Engineering, Universiti Putra Malaysia, 43400 UPM Serdang, Selangor, Malaysia

s_azfanizam@upm.edu.my

Abstract. Biodiesel recently has been receiving a great attention in the world market due to the depletion of the existing fossil fuels. Biodiesel also becomes an alternative for diesel No. 2 fuel which possesses characteristics such as biodegradable and oxygenated. However, there are facts suggested that biodiesel does not have the equivalent features as diesel No. 2 fuel as it has been claimed that the usage of biodiesel giving increment in the brake specific fuel consumption (BSFC). The objective of this study is to find the maximum brake power and brake torque as well as the minimum BSFC to optimize the condition of diesel engine when using the biodiesel fuel. This optimization was conducted using the Bees Algorithm (BA) under specific biodiesel percentage in fuel mixture, engine speed and engine load. The result showed that 58.33kW of brake power, 310.33 N.m of brake torque and 200.29/(kW.h) of BSFC were the optimum value. Comparing to the ones obtained by other algorithm, the BA produced a fine brake power and a better brake torque and BSFC. This finding proved that the BA can be used to optimize the performance of diesel engine based on the optimum value of the brake power, brake torque and BSFC.

1. Introduction

The BA is one of popular swarm-based algorithms and it mimics the real foraging behavior of honey bees. In natural, a colony of bees extends themselves over a distance in multiple directions to explore a big number of food sources for their survival. Bees normally deployed their foragers to good fields. Theoretically, bees usually go to the flowers which contain more nectar which can be collected in minimum efforts. The foraging process of the bees starts with a colony of bees in the form of scout bees are being sent to look for some exciting flower patches which contain large amount of nectar. Once the scout bees return to the hive, they deposit their nectar and go to the "dance floor" where they perform the "waggle dance" [1].This "waggle dance" is vital in bee's colony because it acts as a language to communicate with other bees. This communication contains three important messages about the flower patch which are the direction of the flower patch found, the distance of the hive and their rating on the nectar found. These three pieces of information help them to return to the flower patch precisely, without any directions or maps. The "waggle dance" is also enabled the bees to spread in different flower patches to make sure the food that they are going to collect is in a quality manner and sufficient amount of energy is invested. After the dance, the scout bees return to the flower patches with other bees which are known as followers. This mechanism makes them gather their food in a quick and efficient way [1].

The BA is a well-researched study that takes place in social insects. Bees exhibit many characteristics which proved their usage as models for optimization problems. Examples of their features are a division of labors, communication in group levels and cooperative behavior. In addition, these insects' algorithm performs a neighborhood search which is combined with random search. This makes the algorithm to perform on optimization problem. This algorithm proved that it can be used to optimize problems as other algorithms do [2], [3], [4], [5].

Meanwhile, diesel fuel in general is any liquid fuel used in diesel engines, whose fuel ignition takes place as a result of compression of the inlet air mixture and then injection of fuel. The diesel engine is an internal combustion engine in which ignition of the fuel that has been injected into the combustion chamber is caused by the high temperature. Diesel engine works by compressing only the air. This increases the air temperature inside the cylinder to such a high degree that it ignites atomised diesel fuel that is injected into the combustion chamber. This contrasts with spark-ignition engines such as a petrol engine or gas engine, which use a spark plug to ignite an air-fuel mixture.

The performance of diesel engine is significantly influenced by the biodiesel percentage in fuel mixture, engine speed and engine load [6]. The fuel consumption increases with higher biodiesel concentration [7] while different vehicle speed and engine load results in different impact on vehicle emissions [8]. The brake power is the power available at the crankshaft and brake torque is the power of the braking system. Higher brake power and brake torque is more preferable [9]. The BSFC is a measure of the fuel efficiency burns and causes rotational power. In other words, it is the rate of fuel consumption divided by the power produced. In contrast to the brake power and brake torque, the minimum BSFC is aimed in a vehicle [9]. The configuration of percentage in fuel mixture, engine speed and engine load affects the performance of the brake power, brake torque and BSFC. The use of diesel fuels blended with certain percentage of biodiesel are expected to bring positive impacts to the society and environments as this mixture fuel is able to reduce emissions along-with reduction of the petroleum fuel consumption.

2. Methodology

The BA is known for optimizing problems which uses the inspiration from natural honey bees in their characteristics which help to find an optimal solution for the problem [1]. The BA has its parameter namely, the number of scout bees (n), number of sites selected out of n visited sites (m), number of elite sites out of m selected sites (e), number of bees recruited for the elite sites (nep), number of bees recruited for other selected sites (nsp), size of patches (ngh) and stopping criterion. Optimization process starts by forming a number of scout bees (n) in their colony. The scout bees associate to a possible solution to an optimization problem. Each solution found by the scout bees is to be arranged in a descending order of the fitness. Then, the best m sites are selected for a neighborhood search. In the neighborhood search, more bees (nep) are sent to elite (e) sites whereas fewer bees (nsp) are sent to non-elite (m-e) sites. The solutions generated by this exploitative search are expected to be better than the one found by a single scout bee.

Unsettled scout bees (n-m) are used to search for random solution in space. The purpose of this explorative search is to prevent the bees from being trapped in local optima. At the end of each iteration, new scout bees are formed and each solution found by the scout bees are sorted out for the next iteration. The combination of exploitative and explorative search helps to find the best solution in an effective manner. The steps are then repeated until the intended solution is found or the stopping criterion is met.

Figure 1 shows a flowchart which illustrates the steps of the study. The research starts with the problem identification, followed by the literature review which briefly explains about the BA and diesel engine. Then, the research continues with a preliminary test to find the best configuration of biodiesel percentage (x_1), engine speed (x_2) and engine load (x_3) that makes up the maximum brake power, maximum brake torque and minimum BSFC. The next step is to find the optimum value of x_1, x_2 and x_3 using the BA, within a guided searching space. The optimization process is carried on until a stopping criterion is met. In this problem, 100 iterations was set as the stopping criterion. The final step is to find the optimum brake power, brake torque and BSFC using the optimum x_1, x_2 and x_3, obtained from the previous step.

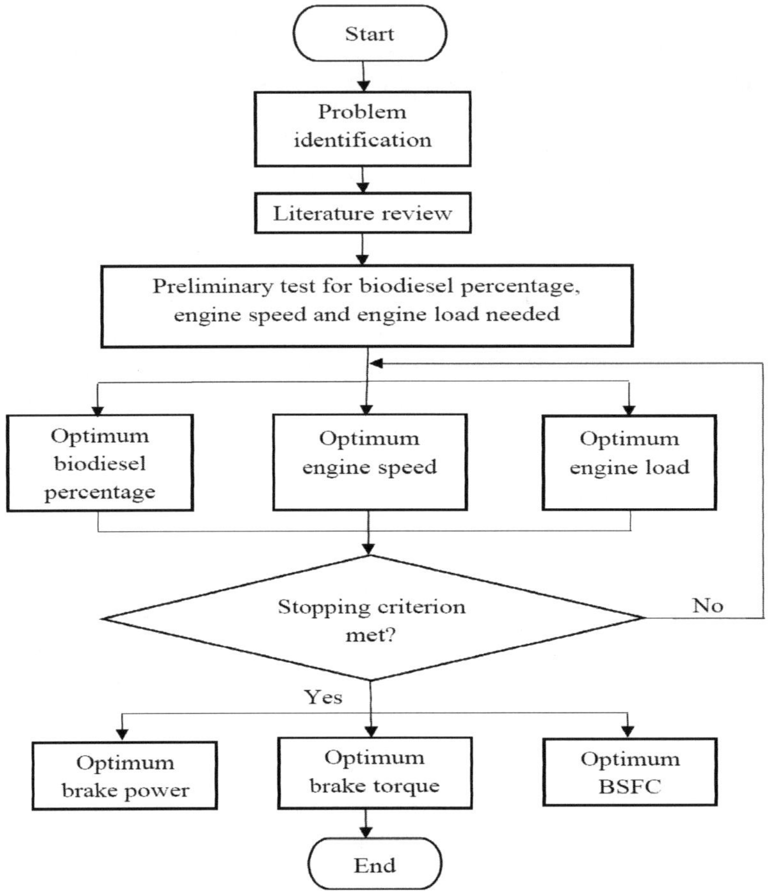

Figure 1. Flowchart of the study.

The equation (1), (2) and (3) is used to find the brake power, brake torque and BSFC respectively [6].

$$\text{Brake Power }(kW) = -47.32 + (-0.08)x_1 + (0.056)x_2 + (0.205)x_3 + (0.0002){x_1}^2 + (-1.4\text{x}10^{-5}){x_2}^2 + (-5.499\text{x}10^{-4}){x_3}^2 + (0.0004)x_1x_3 + (0.0002)x_2x_3 \quad (1)$$

$$\text{Brake Torque }(N.m) = -299.277 + (-0.524)x_1 + (0.302)x_2 + (4.654)x_3 + (0.00362){x_1}^2 + (-7.401\text{x}10^{-5}){x_2}^2 + (-0.00637){x_3}^2 + (-3.771\text{x}10^{-4})x_2x_3 \quad (2)$$

$$\text{BSFC }(g/kW.h)) = 298.74 + (0.5)x_1 + (-0.088)x_2 + (-0.236)x_3 + (0.0014){x_1}^2 + (2.67\text{x}10^{-5}){x_2}^2 + (-0.00018)x_1x_2 + (-0.00336)x_1x_3 \quad (3)$$

3. Result and Discussion

3.1 Preliminary test

A total of 15 preliminary tests were conducted by using multiple combinations of percentage of biodiesel (x_1), revolution per minute (rpm) (x_2) and percentage of engine load (x_3) in order to determine which combinations led to high quality of performance. Table 1 shows the combinations of biodiesel percentage, rpm and engine loads used to determine the amount of brake power, brake torque and BSFC [6]. The value of x_1, x_2 and x_3 was simply substituted into equation (1), (2) and (3) to find the brake power, brake torque and BSFC, respectively. The results showed that the maximum brake power, 60.04 kW, maximum brake torque, 320.25 N.m and minimum BSFC, 198.93 g/(kW.h) all took place when the biodiesel percentage was 50%, engine speed was 1900rpm and 100% engine load.

Table 1. The experimental design [6].

Test number	Biodiesel percentage in fuel mixture (%), x_1	Engine speed (rpm), x_2	Engine load (%), x_3	Brake Power (kW)	Brake Torque (N.m)	BSFC (g/(kW.h)
1	20	1365	40	20.08	121.40	221.89
2	80	1365	40	17.44	111.68	237.48
3	20	2435	40	31.63	127.48	232.44
4	80	2435	40	28.99	117.76	236.47
5	20	1365	80	36.88	256.40	209.76
6	80	1365	80	35.20	246.68	217.29
7	20	2435	80	56.99	246.33	220.31
8	80	2435	80	55.31	236.61	216.28
9	0	1900	62.5	42.95	228.56	213.18
10	100	1900	62.5	39.45	212.36	221.98
11	50	1000	62.5	15.59	153.99	231.69
12	50	2800	62.5	43.13	148.93	239.72
13	50	1900	25	19.82	84.65	229.23
14	50	1900	100	**60.04**	**320.25**	**198.93**
15	50	1900	62.5	40.70	211.41	214.08

3.2. Optimization of diesel engine with the Bees Algorithm

An optimum point of diesel engine was calculated by applying the equation (4) [6].

$$\text{Goal function} = -\frac{\text{equation (1)}}{\text{Power (max)}} - \frac{\text{equation (2)}}{\text{Torque (max)}} + \frac{\text{equation (3)}}{\text{BSFC (min)}} + \frac{0.00012}{0.00014} + \frac{88}{83} + \frac{560}{589} \tag{4}$$

This goal function was used to determine the optimum values of the biodiesel percentage, engine speed and engine load. The denominator of the first three terms respectively is the maximum brake power, maximum brake torque and minimum BSFC that was obtained in preliminary test. The last three terms are the emissions characteristics from the diesel engine, which are the emissions of hydrocarbon (HC), carbon monoxide (CO) and nitrogen oxide (NO_X) respectively [6].

The parameter of the BA is shown in table 2 and the algorithm was run for 100 iterations. The searching space for the BA was set based on the best value of x_1, x_2 and x_3 in table 1 and optimum values gained by ABC [6], which was {50, 1900, 97} for the lower limit and {85.63, 2208, 100} for the upper limit. From this experiment, the goal function or fitness produced was 1.92951. The trend of the fitness value against the number of iteration is illustrated in figure 2. The fitness was improving drastically at the early stage of optimization but then stagnated when iteration reached 82. The fitness remained 1.92951 until 100 iterations and the optimum value gained by the BA is shown in table 3.

Table 2. Parameter setting of the BA.

n, number of scout bees	10
m, number of sites selected out of n visited sites	3
e, number of elite sites out of m selected sites	1
nsp, number of bees sent to non-elite sites	2
nep, number of bees sent to elite sites	3
ngh, size of patches	1

Figure 2. The trend of the fitness against number of iterations.

Table 3. Optimum values gained by the BA.

	Biodiesel Percentage (x_1)	Engine Speed (x_2)	Engine Load (x_3)
Optimum Values	74.67%	1907.81rpm	97.02%

Subsequently, these optimum values were substituted back into equation (1), (2) and (3) to find the optimum condition for the brake power, brake torque and BSFC, respectively. The result is shown in table 4 in which the values of the brake power, brake torque and BSFC was 58.33kW, 310.33N.m and 200.29g/(kW.h) respectively. Comparing to the Artificial Bee Colony (ABC) [6], the BA produced less brake power but outperforming the former on the brake torque and BSFC. Lower engine speed, 1907.81rpm comparing to 2208.00rpm [6] at similar percentage of engine load 97.02% has given an advantage to the BA to produce a higher brake torque and lower BSFC. This finding proved that the BA was able to estimate the optimum point in diesel engine and hence the algorithm has potential to improve the diesel engine performance.

Table 4. Performance of the BA.

	Brake Power (kW)	Brake Torque (N.m)	BSFC (g/(kW.h))
Optimum Value by ABC [6]	63.00	298.00	202.85
Optimum Value by BA	58.33	310.33	200.29

4. Conclusion

In conclusion, the BA can be used to optimize the brake power, brake torque and BSFC which influence the performance of the diesel engine. The objective which was to optimize the diesel engine performance using biodiesel blended fuels was accomplished. The goal function value was obtained by setting a searching space based on a preliminary test and the optimum values from literature. The optimum conditions for biodiesel percentage, engine speed and engine load was 74.67%, 1907.81rpm and 97.02% respectively. The optimum values of the brake power, brake torque and BSFC to optimize diesel engine performance was 58.33 kW, 310.33 N.m and 200.29g/(kW.h) respectively. The brake power was less than the one obtained by the ABC but the brake torque and BSFC was improved with the BA. This finding proved that the BA can be used to optimize the diesel engine performance. For further research, it is necessary to consider the emissions of HC, CO and NO_x in the problem. It is also interesting to study the sensitivity of the parameter setting of the BA in this optimization problem.

References

[1] Pham, D T, Ghanbarzadeh, A, Koç, E, Otri, S, Rahim, S and Zaidi, M 2006 The bees algorithm - a novel tool for complex optimisation problems. 173 Proceedings 2nd International virtual conference on Intelligent Production Machines and Systems (IPROMS), pp. 454-459

[2] Pham, D T, Soroka, A J, Ghanbarzadeh, A, Koc, E, Otri, S and Packianather, M 2006 Optimising neural networks for identification of wood defects using the bees algorithm. IEEE International Conference on Industrial Informatics, (INDIN), pp. 1346-1351

[3] Gupta T and Kumar D 2014 Optimization of clustering problem using population based artificial bee colony algorithm: a review. International Journal of Advanced Research in Computer Science and Software Engineering, 4 (4), 491-502

[4] Pham, D T and Haj Darwish, A 2010 Using the bees algorithm with Kalman filtering to train an artificial neural network for pattern classification. Proceedings of the Institution of Mechanical Engineers. Part I: Journal of Systems and Control Engineering 224(7), 885-892

[5] Pham, D T, Pham, Q T, Ghanbarzadeh, A and Castellani, M 2008 Dynamic optimisation of chemical engineering processes using the Bees Algorithm. Proceedings of the 17th World Congress The International Federation of Automatic Control, pp. 6100-6105

[6] Shirneshan A, Samani B H and Ghobadin B 2016 Optimization of biodiesel percentage in fuel mixture and engine operating conditions for diesel engine performance and emission characteristics by Artificial Bees Colony Algorithm. Fuel, 184, 518-526

[7] Valente, O S, da Silva, M J, Pasa, V M D, Belchior, C R P and Sodré, J R 2010 Fuel consumption and emissions from a diesel power generator fuelled with castor oil and soybean biodiesel. Fuel 89 (12). 3637-3642

[8] Kean, A J, Harley, R A and Kendall, G R 2003 Effects of vehicle speed and engine load on motor vehicle emissions. Environmental Science and Technology.37 (17), 3739–3746

[9] Shirneshan A and Nedayali A 2016 Investigation of the effects of biodiesel-diesel fuel blends on the performance and emission characteristics of a diesel engine. Jurnal Teknologi, 78 (6), 169-177

IOP Conf. Series: Materials Science and Engineering **319** (2018) 012065 doi:10.1088/1757-899X/319/1/012065

Stress analysis on the reinforcement particles of the metal matrix composite by Raman spectroscopy

A K M Asif Iqbal[1], Yoshio Arai[2]

[1]Faculty of Manufacturing Engineering, University Malaysia Pahang, 26600, Pekan, Pahang, Malaysia.
[2]Division of Mechanical Engineering and Science, Graduate School of Science and Engineering, Saitama University, 338-8570, Japan.

asifiqbal@ump.edu.my

Abstract. In this research, the stress state of the reinforcing SiC particles in a hybrid MMC is investigated by micro Raman spectroscopy. The experiment was carried out *in situ* in the Raman spectroscopy. Experimental results show that cracks due to monotonic loading propagates by the debonding of the particle/matrix interface and particle fracture. Moreover, secondary cracks form in front of the main crack tip coalesce with the main crack in subsequent loading and final failure occurs. A high decrease in stress (several hundred in MPa) is observed with the interfacial debonding at the interface and with the particle fracture on the particle. The critical tensile stresses for particle-matrix interface debonding and particle fracture develop in hybrid MMC are also estimated during the crack propagation.

1. Introduction

Nowadays, aluminium based MMCs have made a substantial research interest due to their improved mechanical properties compared to conventional engineering materials [1-3]. The incorporation of the reinforcement particles, such as SiC and Al_2O_3 particles to conventional aluminium alloys exceptionally increases its modulus and fracture resistance. Therefore, these materials are now particularly attractive for structural and wear applications [3]. However, these materials show relatively poor fracture properties as they become often brittle once the higher volume fraction of reinforcement is added to them [4]. Previous studies explained that the key mechanism of the fracture of MMCs is the huge differences in strain carrying capability of reinforcement and the matrix alloy as the reinforcement deforms elastically and the matrix deforms plastically [5–7]. Due to the constrained plastic flow in the matrix that exists between the reinforcement particles in the MMCs, hydrostatic stresses develop which plays a significant role in the failure mechanism during monotonic and cyclic deformations [8]. Among the different constraint level, interfacial bonding between reinforcing particles and matrix alloy becomes a dominating factor in local failure processes and the strengthening of MMCs. A good bonding in the interface results high dislocation density in the matrix which improves the strength of MMC [9]. Besides, the damage initiation mechanism in the different types of MMCs has been broadly investigated and found that the damage often starts due to the interface debonding of the reinforcement and matrix [10, 11]. Iqbal et al. [11] have investigated the fatigue fracture initiation and propagation mechanism in the hybrid MMC and observed that the fracture was initiated by the debonding of particle and matrix located in the hybrid clustering region. The initiated microcracks coalesced to the nearby microcracks in front of the crack tip and propagated through the reinforcement-matrix interface. This phenomenon

Table 1. Chemical composition of AC4CH alloy, (wt%)

Si	Fe	Mg	Ti	Al
7.99	0.2 (max.)	0.57	0.07	Bal.

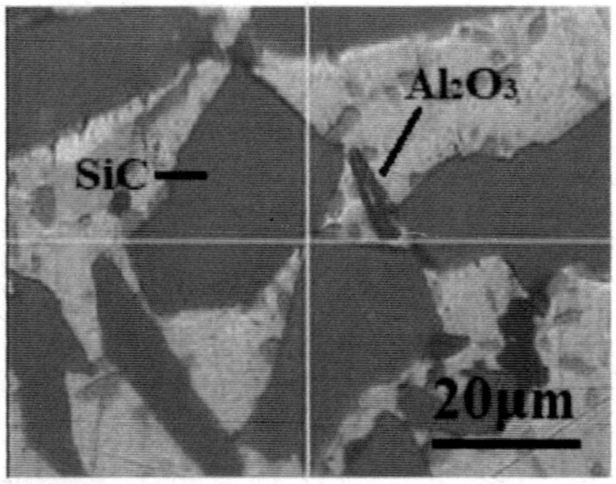

Figure 1. Microstructure of hybrid MMC

was also investigated numerically and found that high level of stress concentration occurred on the particle-matrix interface and the area is highly vulnerable to fracture initiation [12, 13]. Usually, the crack propagation occurs in two ways, partial debonding and reinforcement breakage. In case of reinforcement partial debonding, the crack propagates along the reinforcement–matrix interface while the reinforcement breakage occurs when the reinforcement fails as the load exceeds the reinforcements tensile strength [14]. It is observed that if the interfacial crack spreads far enough, the connecting reinforcement breaks finally, after the reinforcement-matrix interfacial debonding. Therefore, it would be very important to know the stress state of the reinforcing particles during the crack propagation as materials reliability in service depends on it.

Micro-Raman spectroscopy (MRS) is a potential tool in experimental mechanic's field as this technique provides some unique benefits, for example, non-destructive, non-contact, high spatial resolution and depth focus. This technique is also very useful to measure the distribution of the stress or strain on the reinforcement at microscale [15]. MRS has widely been used to study the load transfer behaviour of reinforcement debonding, bridging, pull-out and fracture [16]. In the present study, micro Raman spectroscopy is used to analyze the stress developed in the reinforcement particles during crack propagation in a hybrid MMC. The reinforcement fracture and the interfacial fracture are also observed in this study.

2. Materials and experimental procedures

Hybrid metal matrix composite with 21 Vol% SiC particles and 9 Vol% Al_2O_3 whiskers as reinforcements and the aluminium alloy JIS-AC4CH as matrix is used for this study [17]. The MMC material is fabricated by the squeeze casting process. The maximum casting pressure is used to 100 MPa. The squeeze casting pressure of 100 MPa is adequate to overcome the resistance against the flow and to press the melt into all the open pores of the hybrid preform. Finally, the material is heat treated using the T7 process. The chemical composition of AC4CH alloy is listed in Table 1. The microstructure of the hybrid MMC is shown in figure 1. The SiC particles in the hybrid MMC are mostly rectangular with sharp corners, and the Al_2O_3 whiskers are roller-shaped. The average length of SiC particles and the Al_2O_3 whiskers is 23 μm and 35 μm respectively. The average diameter of the Al_2O_3 whiskers is 2 μm. The mechanical properties of reinforcement materials and hybrid MMC are shown in Table 2.

Table 2. Mechanical properties of reinforcement and tested materials

Parameters	Al_2O_3	SiC	Al alloy	Hybrid MMC
Young's modulus (GPa)	380	450	70.0	142
Poisson's ratio	0.27	0.20	0.33	0.28
Yield strength (MPa)	-	-	131	166
Tensile strength (MPa)	-	-	262	228
Tensile elongation (%)			9.22	2.77

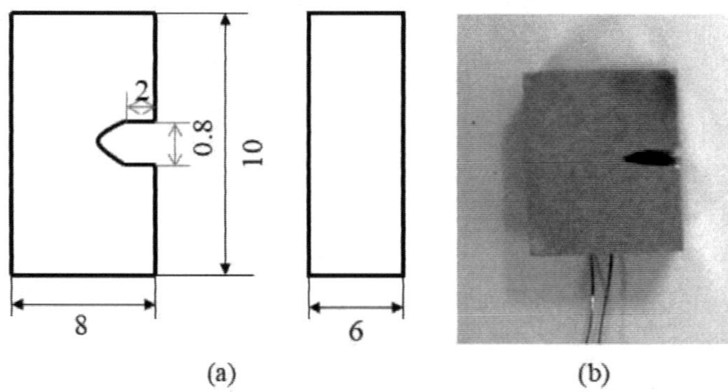

Figure 2. Specimen used in the experiment (a) schematic illustration of dimension, (b) original specimen

A rectangular notched specimen is used for the experiment as shown in figure 2. The machined surfaces of the specimens are polished by using a polishing machine with 15, 3, and 1 μm diamond particles sequentially until all scratches and surface machining marks are removed and become smooth enough to prevent attenuation of the scattered light from the specimen. The experiment is carried out *in situ* in the Raman spectroscopy with special loading fixture consists with a wedge and a mechanical actuator. A schematic illustration of the experimental setup is shown in Fig. 3. By rotating the screw in the fixture, the wedge moves towards the notch which provides crack face opening of the specimen. The wedge angle is 5.4 deg. The rotation of the screw is manually controlled and monitored carefully. The data presents in this article are in relation with the rotation of the screw in the fixture. A strain gauge is attached at the bottom of the specimen to determine the contact of the wedge to the specimen notch. The strain data is recorded by a computer data acquisition system. All tests are carried out at room temperature. The initiation and propagation of micro-cracks are observed by the optical microscope attached in the micro Raman spectroscopy.

2.1. Raman spectroscopy measurement

The Raman microscopic analysis is carried out on a laser micro-Raman spectrometer (Lamda vision LV-RAM 532). A green laser with a wavelength of 532 nm and a power of 70 mW is used. The laser beam is focused to a 2μm diameter spot on the specimen surface using an optical microscope with X50 objective lens. The backscattered light is dispersed by leading it to the spectrometer through the microscope objective. Raman spectra are recorded using a charge coupled device (CCD). The Raman frequencies of the Raman bands are obtained with an accuracy of ±0.5cm^{-1}.

3. Results and Discussion

To analyze the stress concentration on the reinforcing particles, first, the Raman gage factor of SiC particle is determined. Therefore, an axial compressive load of 200 N is applied to a small SiC bar and Raman peaks are observed. The gage factor (GF) is calculated to be 0.0023 cm-1.

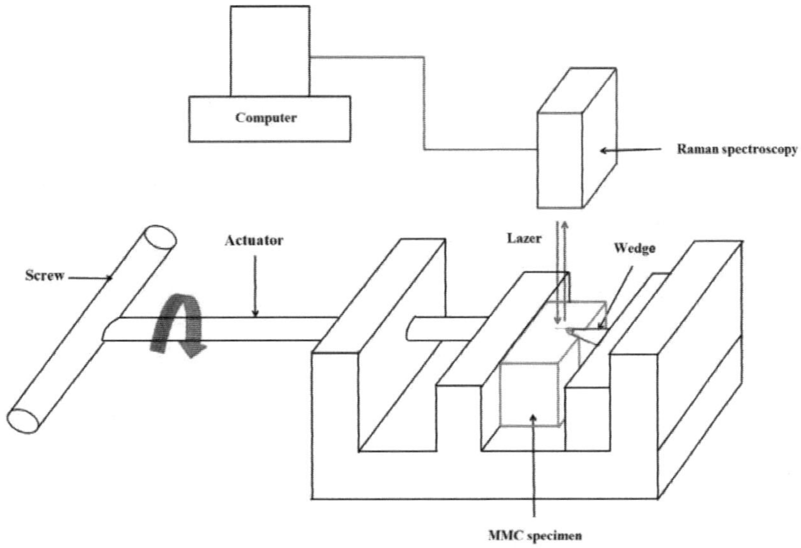

Figure 3. Schematic illustration of experimental setup

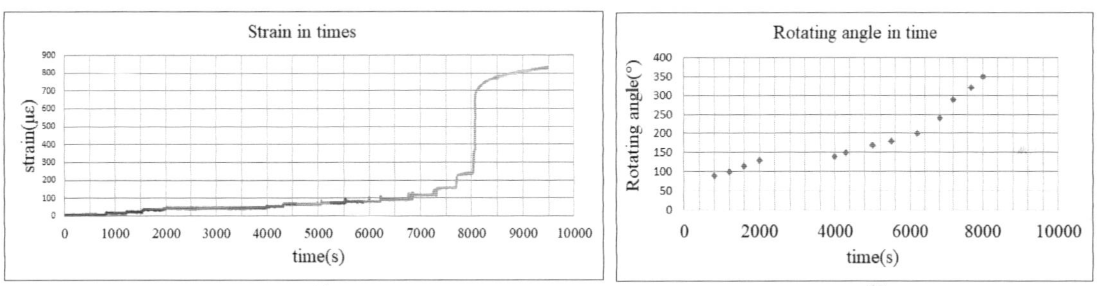

(a) (b)

Figure 4. (a) Change of strain in different times, (b) Rotation of screw in different times.

The change in strain in different times and with different angles of the rotating screw is presented in Figure 4. The crack nucleation at the notch during the application of monotonic load as a sudden increase of the strain is shown in the figure 4. The screw is rotated to 300° from 0 to 7000s (Fig. 4b), however, no substantial change in strain is found. This indicates that the wedge did not touch the notch at this stage. The strain starts increasing at 7700 s and 320° rotation of the screw, indicating the contact of the wedge to the notch. Besides, at 8000 s and 360° rotation of the screw, the strain increases rapidly, indicating the crack initiation at the tip of the notch at this stage. This result clearly demonstrates that the crack initiates at 40° increment in rotation of the screw from the absolute value of rotation angle 320° at which the wedge contacts the notch. In the following discussion, we use the increment value in rotation from the 320° and call it 'rotation angle'. To determine the state of the stress and the state of crack propagation, Raman spectrum is taken on several particles in different loading angle. One of the particles has been chosen for this paper is particle 8. The optical micrograph of the particle 8 in different rotating angle and the behaviour of a propagating crack at the interface between particle 8 and matrix alloy is shown in the figure 5. The red spots on the particle indicate the Raman laser spots. Initially, no cracks are observed at the interface of the particle (Fig. 5a). Besides, at 45° rotation of the screw (Fig. 5b), the main crack propagates along the interface of the particle as indicated in the Fig 5b. At the same time, another crack is observed in front of the main crack tip (as indicated by the arrow in Fig 5b). The presence of arrested crack at this stage indicates the interface debonding seems to be stable. The crack then further propagates by deflecting into SiC particle at the 50° rotation of the screw (Fig. 5c). The particle fracture is observed once the rotation of the screw reached to the 55° angle (Fig. 5d).

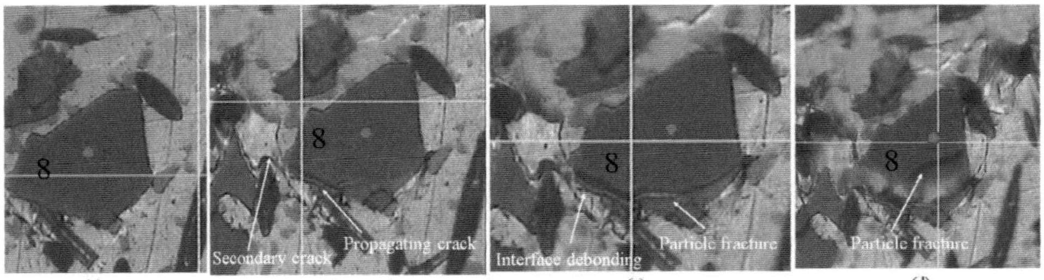

(a) (b) (c) (d)

Figure 5. Interface crack on particle at different rotating angle of the screw.

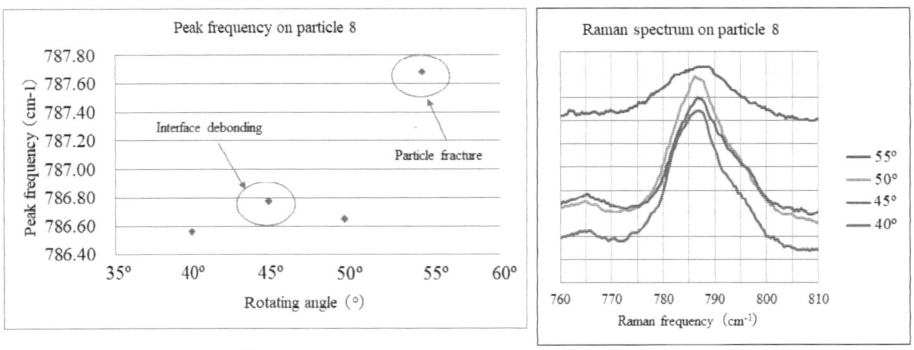

(a) (b)

Figure 6. Raman spectroscopy on particle 8 in different rotating angle (a) peak wave number (b) Raman spectrum

Table 3. Change of accumulated stress on particle 6 in different rotating angle

Rotating angle (°)	Peak wave number (cm^{-1})	Change in stress (MPa)	Accumulated change in stress (MPa)
40	786.56		
		-435	
45	786.77		-435
		-146	
50	786.65		-581
		-245	
55	787.7		-826

The same phenomena are also observed in other particles lies in the crack propagation direction. Figure 6 represents the results of the Raman spectrum taken on the particle 8 in different rotating angle. At the angle of 45°, the peak value of Raman spectrum is found to be 786.77 cm-1. At this stage, interfacial debonding occurs at particle 8. The Raman spectrum also moves to higher peak value when the crack deflects into the particle at 50° rotation. Besides, the highest value of the Raman spectrum is found to be 787.7 cm-1 at 55° rotating angle. At this stage particle 8 becomes fractured. From the obtained peak wave number, the change in stress for debonding to take place in particle 8 is calculated by using the SiC gage factor and the result is presented in Table 3. The stress concentration on the particle 8 is calculated to be -435 MPa (- sign indicates a decrease in stress) once the particle is debonded from the matrix. The decrease in stress compared to the un-cracked state means an unloading of originally tensile stress develops due to the opening of the crack under wedge loading. The change in stress due to crack deflection from the interface to the SiC particle is -146 MPa. Besides, the maximum change in stress measured in the particle 8 after particle fracture is found to be -245 MPa at the angle of 55° of the screw. The accumulated change in stress from the un-cracked state to the particle fracture is -826 MPa which means 826 MPa is the estimated value of tensile stress developed in the particle located just ahead of the crack tip before fracture. The similar high value of change in stress is also measured in other particles in front of the crack tip.

The above results clearly show that the crack due to the waged-type monotonic load propagates along the interface of particle-matrix as well as through the particle in the hybrid MMC. The local effective stress, which is acting on the particle ahead of the crack tip reaches extremely high values (800 – 900 MPa). At the fracture stress under monotonic loading condition, the reinforcing particles deformed elastically within the plastically deforming matrix alloy. Thus, a large strain mismatch occurs between these two materials. For this mismatch of strain, a consequent concentration of stress is generated in the particles and at the interface between the reinforcing particles and matrix alloy. These stresses cause the interface separation of particle-matrix as well as the particle fracture, and the crack propagates till the final failure occur.

4. Conclusions

The stress state of the reinforcing particles in hybrid MMC has been investigated by *in situ* fracture experiment in the micro Raman spectroscopy. The results show that the cracks developed due to the wedge type monotonic load propagates by the debonding of the particle/matrix interface and particle fracture. Secondary cracks also generate in front of the main crack tip, which coalesce with the main crack and the final failure occurs. The Raman spectroscopy confirms that stress drops rapidly (several hundred in MPa) once the interfacial debonding and particle fracture occur.

Acknowledgement

The authors express gratitude to the University Malaysia Pahang for providing financial support to present the paper at the conference under RDU 150350.

References.

[1] Clyne TW, Withers PJ. An introduction to metal matrix composites. Cambridge University Press, Cambridge; 1993.
[2] Chawla KK. Composite materials—science and engineering. Springer-Verlag, New York; 1987.
[3] Suresh S, Mortensen A, Needleman A. Fundamentals of metal matrix composites. Butterworh / Heinemann, London; 1993.
[4] Miserez A, Rossoll A, Mortensen A. Investigation of crack-tip plasticity in high volume fraction ceramic particle reinforced metal matrix composites. Eng Fract Mech 2004;71:2385–406.
[5] Levin M, Karlsson B. Influence of SiC particle distribution and prestraining on fatigue crack growth rates in aluminium AA 6061–SiC composite material. Mater Sci Technol 1991; 7: 596–607.
[6] Davidson DL. Fatigue and fracture toughness of aluminium alloys reinforced with SiC and alumina particles. Compos 1993;24:248–55.
[7] Chen AL, Arai Y, Tsuchida E. A numerical study on the effect of thermal cycling on monotonic response of cast aluminium alloy-SiC particulate composites. Theor Appl Mech 2004;53:63–73.
[8] Chawla N, Andres C, Jones JW, Allison JE. Effect of SiC volume fraction and particle size on the fatigue resistance of a 2080 Al/SiCp composite. Metall Mater Trans A 1998;29:2843–54.
[9] Chen EY, Lawson L, Meshii M. The effect of fatigue microcracks on rapid catastrophic failure in Al–SiC composites. Mater Sci Eng A 1995;200:192–206.

[10] Chen ZZ, Tokaji K. Effects of particle size on fatigue crack initiation and small crack growth in SiC-particulate reinforced aluminium matrix composites. Mater Lett 2004;58:2314–21.
[11] Iqbal AA, Arai Y, Araki W. Effect of Hybrid Reinforcement on Crack Initiation and Early Propagation Mechanisms in Cast Metal Matrix Composites during Low Cycle Fatigue. Mater Des 2013;45:241-52.
[12] Yan YW, Geng L, Li AB. Experimental and Numerical Studies of the Effect of Particle Size on the Deformation Behavior of the Metal Matrix Composites. Mater Sci Eng A 2007;448:315-25.
[13] Iqbal AA, Arai Y, Araki W. Effect of Reinforcement Clustering on Crack Initiation Mechanism in a Cast Hybrid Metal Matrix Composite during Low Cycle Fatigue. Open J compos mater 2013:3:97-106.
[14] Zhenkun L, Wang Quan, Qiu W. Micromechanics of fiber–crack interaction studied by micro-Raman spectroscopy:Broken fiber. Opt Laser Eng 2013;51:1085-91
[15] Lei ZK,QiuW, Kang YL, Liu G,Yun H. Stress transfer of single fiber/micro droplet tensile test studied by micro-Raman spectroscopy. Compos A 2008;39:113–8.
[16] Lei ZK,Wang Q, Kang YL, Qiu W, Pan XM. Stress transfer in microdroplet tensile test:PVC coated and uncoated Kevlar-29 single fiber. Opt Laser Eng 2010;48:1089–95.
[17] "Aluminium Alloy Castings," JIS H5202, Japan Industrial Standard, 2002. (In Japanese).

IOP Conf. Series: Materials Science and Engineering **319** (2018) 012066 doi:10.1088/1757-899X/319/1/012066

Effect of 1.0% Ni on high-temperature impression creep and hardness of recycled aluminium alloy with high Fe content

M Faisal[1], Noor Mazni[1] and A K Prasada Rao[2]

[1] Faculty of Manufacturing Engineering, Universiti Malaysia Pahang, Pekan, Pahang Darul Makmur, Malaysia
[2] Department of Mechanical Engineering, School of Engineering and Technology, BML Munjal University, Gurgaon, Haryana, India

corresponding author's e-mail: drmazni@ump.edu.my

Abstract. Reported work focusses on the effect of 1.0% Ni addition on the microstructure, high-temperature impression creep and thereby the hardness of recycled Al-alloy containing >2wt% Fe, obtained from automotive scrap. Present studies have shown that the addition of 1.0% Ni have supress the formation of β-phase (Al_5FeSi) by supressing the peritectic transformation of α-phase (Al_8Fe_2Si). Such suppression is found to improve the hardness and high-temperature impression creep of the recycled aluminium alloy.

1. Introduction

Recycling of aluminium has been gaining its importance due to the energy concerns in extraction of primary aluminium [1,2]. Fe impurity in recycled aluminium is highly undesirable due to the tendency of the formation of brittle intermetallics such as Al_5FeSi (β-phase); which exists in a needle/plate-like morphology [3,4]. On the other hand, literature suggests that the formation of Al_8Fe_2Si phase, decreases the brittleness caused by the Al_5FeSi (β-phase), owing to its polyhedral and star-like morphology [4]. However, it is known that Al_8Fe_2Si phase undergoes peritectic transformation leading to the formation of Al_5FeSi (β-phase), which is sensitive to the cooling rate [4]. Literature reports that Al_8Fe_2Si (α-phase) is prerequisite to form Al_5FeSi (β-phase). Al_5FeSi (β-phase) has been assumed to act as stress raisers and are points of weak coherence, causing reduction in mechanical properties, particularly reducing the ductility of cast alloys. The reduction of mechanical properties depends on the Fe-rich intermetallics' amount, size and type. Al_5FeSi only dissolves limited amount of Fe-correctors such as Mn and Cu. It appears as faceted platelets (needle-shape structure), extended up to several millimetres long. This phase, therefore, has been identified as the Fe phase that causes the most serious loss of strength and ductility in Al-Si alloy castings [4,5]. Owing to its similarity with Fe in the lattice structure and atomic radius, Ni is expected to intervene the nucleation and growth of the Fe-rich intermetallics. Hence, Ni was added as an iron-corrector, to the recycled Al-alloy investigated in the present study. Influence of Ni on the microstructure and thereby the high-temperature impression creep resistance and hardness of the recycled alloy has been discussed herein.

2. Materials and Methods

About 1000 gm of recycled Al-alloy ingot was cut and melted in a tilting-type resistance furnace. Predetermined amounts of 1.0% Ni were added into the melt at 730±5°C followed by 30 s stirring. The melt was then poured into a pre-heated cast-iron cylindrical mold. The as-cast specimen was cut along their transverse axis and was subjected to standard metallographic preparations. Optical and scanning electron microscopy was performed on the specimen in order to study the microstructure. The high-temperature (250°C) impression creep Studies (at constant indentation stress 360 MPa) were conducted using an impression creep testing equipment, in order to understand the effect of Ni additions on the high-temperature strength of the recycled Al-alloy. The high-temperature impression creep tester consists of Inconel indenter (cylindrical, 0.8 mm diameter) enclosed in an electric resistance furnace capsule. The time-displacement curves obtained from the high-temperature impression creep experiments are discussed herein. Vickers micro hardness test was conducted with 0.98N load and 10s dwell time. The hardness test for each sample was repeated for six times. Average hardness value is presented in this work to examine the relationship between the microstructure and the hardness of the recycled aluminum alloys. The results obtained in the above measurements are closely studied for a detailed understanding of the influence of 1.0% Ni on the high-temperature impression creep and hardness of the recycled alloy.

3. Results and Discussion

3.1. Microstructure

As-received automotive recycled alloy used in the present work has been found to consist Fe in the form of star-like Al_8Fe_2Si phase and long needle-like Al_5FeSi phase. SEM-EDS analysis shown in figure 1 confirms the Al_8Fe_2Si phase found in the recycled alloy without addition of Ni; on the other hand, Al_5FeSi phase is confirmed through the SEM-EDS analysis shown in figure 2. It is well documented by Belov *et al.* [6] that hypo-eutectic Al-Si alloys containing Fe >2 wt.%, Al_8Fe_2Si phase undergoes peritectic transformation which results in Al_5FeSi phase, under ideal equilibrium solidification conditions. It has also been reported that these two phases coexist if the ideal equilibrium conditions are not met during solidification, as the above peritectic transformation is rather sluggish. Figure 3(a) reveals that as-cast microstructure of the recycled alloy without the addition of Ni contains both star-like Al_8Fe_2Si phase and long needle-like Al_5FeSi phase which confirms Belov *et al.* [6] report on non-equilibrium microstructure. The microstructures of the recycled alloy treated with 1.0 wt.% of Ni in figure 3(b) reveal a hexagonal faceted morphology of Al_8Fe_2Si phase, unlike in the case of figure 3(a), indicating an active role of Ni on the morphological transformation of the Al_8Fe_2Si phase. It is can also be seen that Al_8Fe_2Si refine in their size, suggesting that Ni results in nucleation of the Al_8Fe_2Si phase.

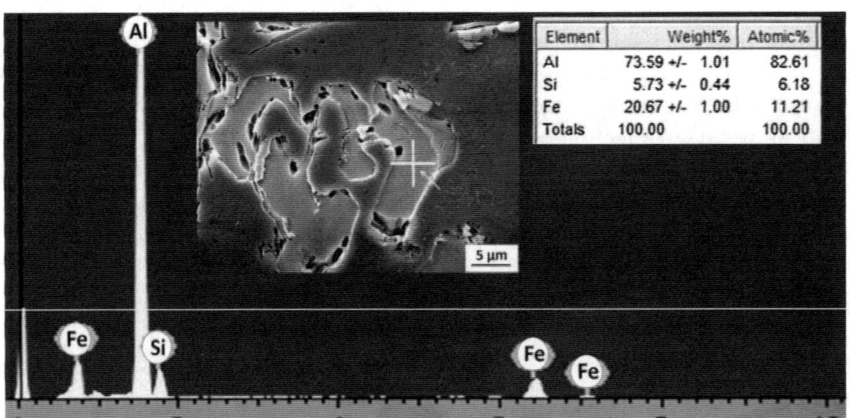

Figure 1. SEM-EDS analysis of the Al_8Fe_2Si phase in the recycled alloy without Ni

IOP Conf. Series: Materials Science and Engineering **319** (2018) 012066 doi:10.1088/1757-899X/319/1/012066

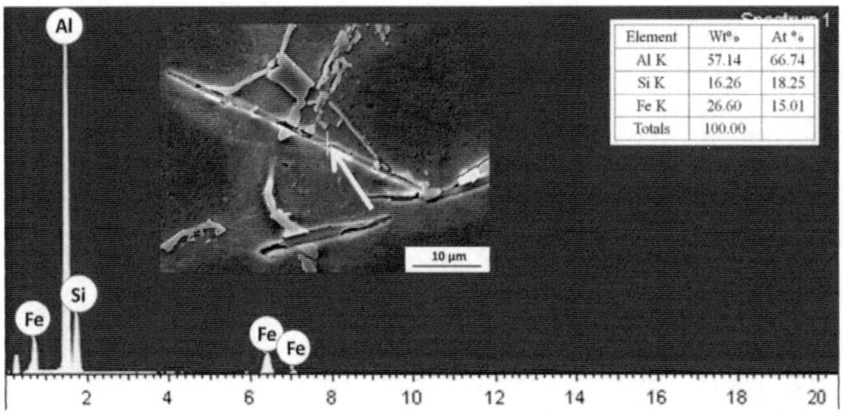

Figure 2. SEM-EDS analysis of the Al_5FeSi phase in the recycled alloy without Ni

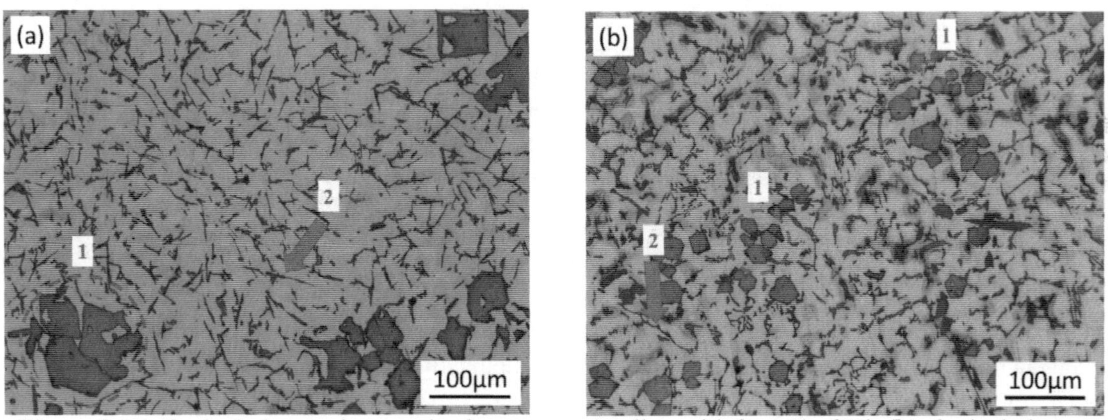

Figure 3 (a). Optical photo-micrograph of as-cast scrap alloy with 0 wt.% of Ni. **(b).** Optical photo-micrograph of as-cast scrap alloy with 1.0 wt.% of Ni.

3.2. High-Temperature Impression Creep

The time-displacement curves obtained from the impression creep experiments conducted on the recycled alloy specimens have been plotted in figure 4. It has been found that the depth of the indentation decreases with the increase in the Ni content, suggesting a significant strengthening effect of Ni. Decrease in the slope of the time-displacement curve with increase in the Ni content, also suggests an increase in the activation energy of creep deformation in the Ni containing alloy.

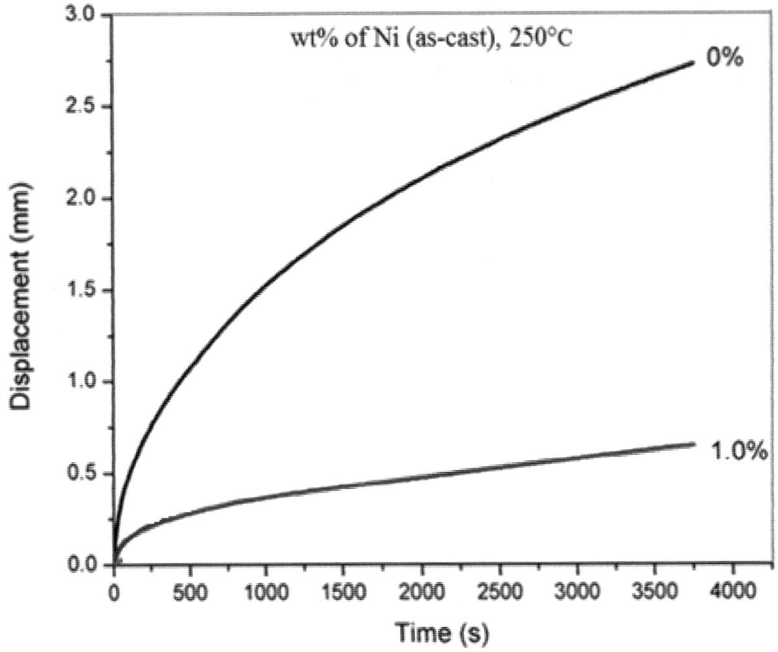

Figure 4. Effect of 1.0% Ni on the impression creep behaviour of the recycled alloy at elevated temperature (250°C) and a constant stress of 360 Mpa

3.3. Vickers Micro Hardness Test

Figure 5 shows the evolution of hardness value with increasing addition of Ni. It is well documented by A. Školáková *et al.* that Al-alloys contained very fine structure and intermetallic phases. As the addition on Ni in recycled Al-alloys increase, the size and morphology of the star-like Al_8Fe_2 refines. These very fine structure and intermetallic phases significantly led to higher hardness as internal stress is well distributed along the marix. Besides refine in size, addition of Ni also suppresses the peritectic transformation of Al_8Fe_2Si. This leads to the depletion of the brittle Al_5FeSi phase, thereby improves the hardness of recycled Al-alloys. In addition, the alloy may contain ultrafine coherent particles of intermetallics, which are not detectable by XRD. Thus, by evaluating the result in figure 6 it can be seen clearly that see that the hardness values of recycled Al-alloys can be improve with increase Ni addition.

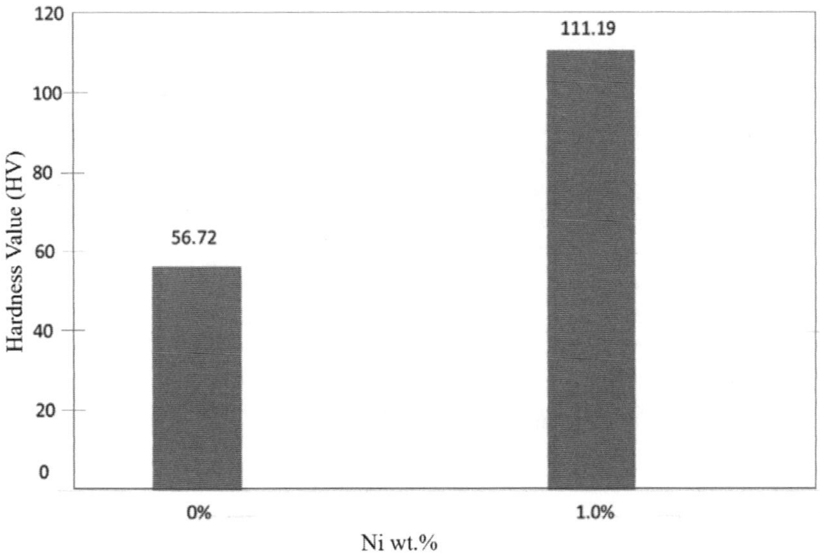

Figure 5. Hardness of recycled aluminium with increase Ni addition

4. Conclusion

• Nickel refines the size and morphology of the star-like Al_8Fe_2Si phase in Al-Si recycled alloy with Fe >2 wt.%.

• It has been found that Ni suppresses the peritectic transformation of Al_8Fe_2Si, which leads to the depletion of the Al_5FeSi and thereby improves the high-temperature strength and mechanical behavior of the Al-recycled alloy with high Fe impurity.

5. Acknowledgement

The authors would like to express gratitude to Universiti Malaysia Pahang for the financial support of this research through RDU150377 grant.

6. References

[1] Gabrielle G, Elsa O and Randolph K 2012 Improving Aluminum Recycling: A Survey of Sorting and Impurity Removal Technologies *Resour. Conser.. Recycl.* **58** 79-87

[2] Subodh K D 2006 Designing Aluminum Alloys for a Recycle-Friendly World *Mater. Sci. Forum* **519–21** 1239–44

[3] Moustafa M A 2009 Effect of Iron Content on the Formation of β-Al_5FeSi and Porosity in Al-Si Eutectic Alloys *J. Mater. Process. Technol.* **209(1)** 605–10

[4] Xinjin C and John C 2006 Morphology of β-Al_5FeSi Phase in Al-Si Cast Alloys *Mater. Trans.* **47(5)** 1303–1312

[5] Backerud L, Chai G and Tamminen J 1990 Solidification Characteristics of Aluminum Alloys *Foundry Alloys* **2** 71-84

[6] Belov N A, Aksenov A A and Eskin D G 2002 Iron in Aluminium Alloys: Impurity and Alloying Element

[7] Školáková A, Novák P, Vojtěch D and František K T 2016 Microstructure and Mechanical Properties of Al-Si-Fe-X alloys *Mater. Design* **107** 491-502

IOP Conf. Series: Materials Science and Engineering **319** (2018) 012067 doi:10.1088/1757-899X/319/1/012067

Laser surface modification of Yttria Stabilized Zirconia (YSZ) thermal barrier coating on AISI H13 tool steel substrate

M S Reza[1], S N Aqida[1], I Ismail[2]

[1] Faculty of Mechanical Engineering, Universiti Malaysia Pahang, 26600 Pekan, Pahang, Malaysia

[2] Faculty of Manufacturing Engineering, Universiti Malaysia Pahang, 26600 Pekan, Pahang, Malaysia

E-mail: [1]mdreza@ump.edu.my

Abstract. This paper presents laser surface modification of plasma sprayed yttria stabilized zirconia (YSZ) coating to seal porosity defect. Laser surface modification on plasma sprayed YSZ was conducted using 300W JK300HPS Nd: YAG laser at different operating parameters. Parameters varied were laser power and pulse frequency with constant residence time. The coating thickness was measured using IM7000 inverted optical microscope and surface roughness was analysed using two-dimensional Mitutoyo Surface Roughness Tester. Surface roughness of laser surface modification of YSZ H-13 tool steel decreased significantly with increasing laser power and decreasing pulse frequency. The re-melted YSZ coating showed higher hardness properties compared to as-sprayed coating surface. These findings were significant to enhance thermal barrier coating surface integrity for dies in semi-solid processing.

1. Introduction

Thermal barrier coatings (TBCs) are frequently used in die casting industry for insulation system which due to repetitive high temperature liquid flow and solidification process [1]. Rapid high temperature liquid flow produces erosion while high operating temperature gasses caused hot corrosion to metallic component which penetrated into ceramic coating. Erosion occurred due to repetitive rapid heating and cooling of die surface [2]. The cyclic process can cause premature failure in dies due to physical impingement of incoming liquid on die surface. Along with corrosion effect, oxygen diffusion occurred and was dominated in high porosity regions. Consequently, bond coat layer oxidized, and formed thermally grown oxide (TGO) layer [3]. Hence, decreasing porosity content in top coat layer is crucial to overcome TGO layer formation, as well as improving the coating surface integrity.

Previous studies on surface processing indicated laser modification is among the methods for improvement of plasma sprayed TBCs [4]. Laser treatment of YSZ TBCs was conducted to reduce surface roughness of top coat, increase surface hardness and seal open pores. Laser power, pulse duration and overlapping pulses were controllable parameters that can affect fracture strength of the ceramic coating layer [5]. In most TBC systems, substrate materials are super alloys or stainless steels coated using atmospheric plasma spraying technique. In this study YSZ coating surface was laser modified using pulse Nd:YAG laser processing to enhance TBC surface integrity and the substrate material was surface treated H13 tool steel. Effects of surface laser modification on surface roughness and hardness properties of plasma sprayed YSZ thermal barrier coatings were investigated.

2. Methodology

Coating layers were deposited by atmospheric plasma spray on cylindrical AISI H13 tool steel substrate of 10 mm (d) x 150 mm (l) dimension. Bond coat material was Praxair NiCrAlY (Ni 164/211 Ni 22% Cr 10% Al 1.0%Y) while top coat material was Praxair Ai-1075 yttria stabilized zirconia, YSZ (ZrO_2 + 8wt.% Y_2O_3). A 300W JK300HPS Nd:YAG laser with TEM_{00} mode was used to modify the YSZ thermal barrier coating surface. In Table 1, laser power and pulse frequency were varied to produce constant residence time of 0.2 ms. Traverse speed of laser and sample rotation speed was constant at 2 m/min and 58 rpm respectively. The resulted laser energy range to melt the surface was between 0.03 and 0.07 J. Surface roughness of the lasered and without lasered YSZ TBC surface was measured using Mitutoyo Surface Roughness Tester. For cross sectional micrograph, metallographic study was conducted using IM7000 inverted optical microscope and cross sectional surface of microhardness were measured using MMT-X7 Matsuzawa Hardness Tester Machine with Vickers hardness scale.

Table 1. Laser surface modification parameter settings

Sample	Pulse frequency (Hz)	Peak power (W)	Energy (J)	Residence time, T (ms)
1	70	2	0.04	0.2
2	73	3	0.05	0.2
3	76	4	0.06	0.2
4	79	5	0.07	0.2
5	82	2	0.03	0.2
6	85	3	0.04	0.2
7	88	4	0.05	0.2
8	91	5	0.06	0.2

3. Results and discussions

Figure 1 shows micrograph of cross-sectional surface of coated laser modified surface H13 tool steel which comprised of laser surface modification of YSZ top coating [A], YSZ top coating [B], NiCrAlY bond coating [C], laser modified surface of substrate H13 [D] and substrate H13 tool steel [E]. Lasered surface modification of top coating layer thickness average from 10 to 30 micron. Whereby, YSZ top coating layer thickness average from 180 to 360 micron, while bond coat NiCrAlY thickness average from 100 to 150 micron. Laser modified surface of substrate H13 tool steel thickness average from 20 to 30 micron.

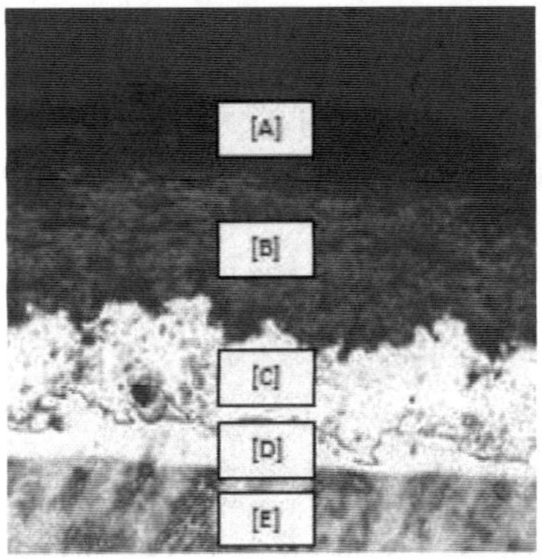

Figure 1. Micrograph of cross-sectional surface of laser modified YSZ H13 tool steel

Figure 2 shows results of surface roughness (SR) measurement for lasered YSZ TBC surface and without lasered YSZ TBC surface. Based on Figure 2, reduction of SR, R_a was measured on all samples which were laser modified after YSZ coating deposition. Sample 2 exhibited the lowest surface roughness with highest surface roughness reduction of 42 % after laser surface modification. While, the lowest SR reduction occurred in sample 8 of 7%. Comparing between sample 2 and sample 8 SR result, sample 2 has lower SR value than sample 6 with similar pattern exhibited for sample 4 and sample 8. For other samples comparison between sample 1 and sample 7 SR result, sample 1 has lower SR value than sample 5. While, similar pattern exhibited too for sample 3 and sample 7.

According to table 1, where laser power was constant by 3W, sample 2 laser energy setting was set higher than sample 6 and as for sample 4 and sample 8 the energy setting was also set higher for sample 4 than sample 8 but with laser power constant by 5W. Similar pattern to sample 1 and sample 5, where higher laser energy was set for sample 1 with constant laser power of 2W and as if for sample 3 and sample 7, higher laser energy for was set for sample 3 with constant laser power of 4W. All the comparison between sample 1 and 5, sample 2 and 6, sample 3 and 7 and sample 4 and 8, generally showed the pattern of lower pulse frequency leads to lower surface roughness.

Effects of higher laser energy setting, explains the phenomenon where laser beam that was radiated on the surface, remelting it by high heat creation. When the laser beam passed through the molten area, this zone was solidified rapidly [4]. These remelting and re-solidification phenomena resulted in the reduction of surface roughness of plasma sprayed coating. Reduction in the surface roughness leads to the improvement of corrosion as it is due to the reduction of specific surface area of the laser layer. The corrosion reaction of the molten salts of semi-solid liquid in die with the top coat will be reduced as compared with the coatings in the as-sprayed condition [5].

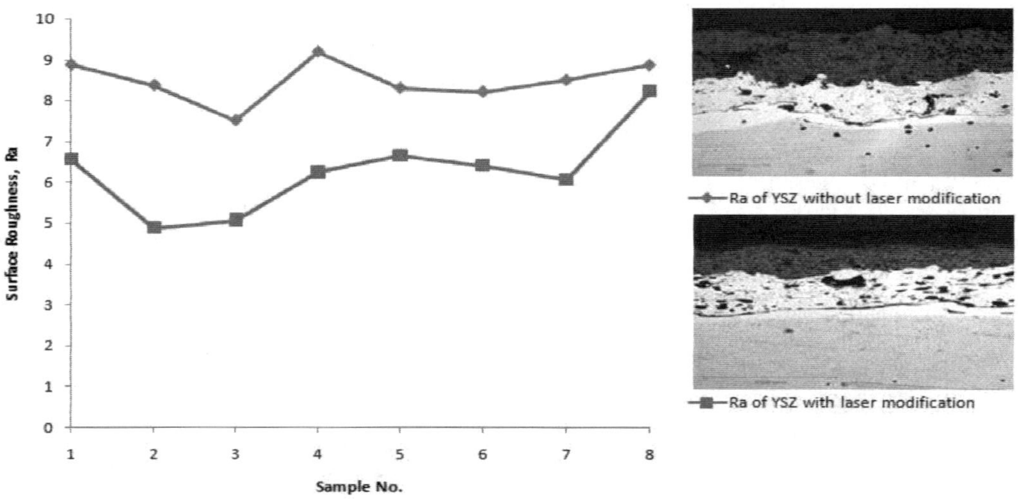

Figure 2. Surface roughness (SR) of laser surface modification of YSZ layer

Figure 3 illustrates the variation of micro-hardness of cross sectional surface of laser modified H13 tool steel as a function of depth from top laser surface modification to the substrate. In general the laser surface modification has improved the hardness of specimen for all incident energies. The micro-hardness initially increases with laser power and pulse energy and begins to decrease at a certain parameters. The micro-hardness of the irradiated area is larger than the non-irradiated area, by reason of that the grains after laser irradiated become smaller and closer compared with raw materials under the action of the temperature gradient [5].

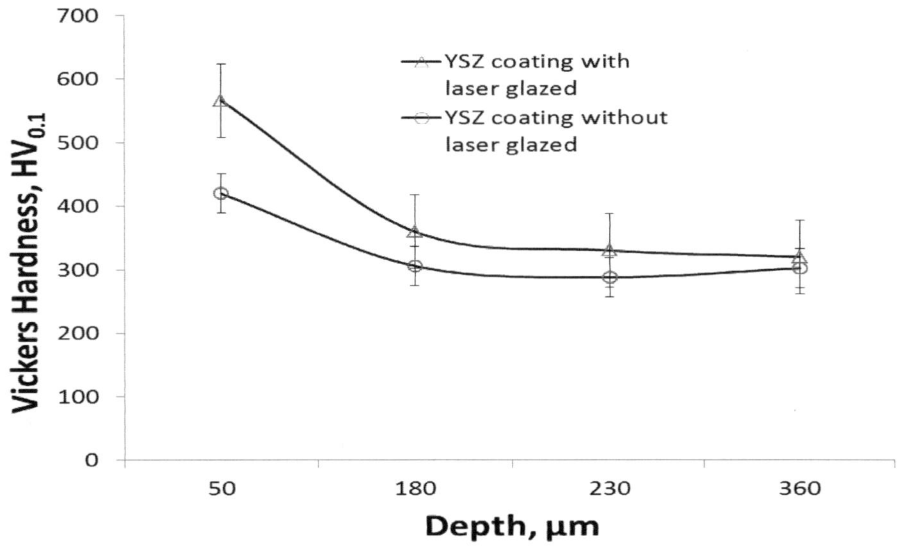

Figure 3. Micro-hardness of cross-sectional surface of laser modified YSZ H13 tool steel

4. Conclusion

Higher laser pulsed energy and lower pulse frequency parameter setting leads to decrease in surface roughness. Increase in micro-hardness of specimen for all incident energies was also obtained. Thus, lead to the improvement of the resistance to corrosion of YSZ thermal barrier coating of H13 tool steel.

Acknowledgements

Authors would like to acknowledge support from the Ministry of Education Malaysia under Fundamental Research Grant Scheme RDU160141 and Universiti Malaysia Pahang for funding this research.

References

[1] Rahelah Ahmadi-Pidani, Reza Shoja-Razavi, Reza Mozafarinia, Hossein Jamali, Laser surface modification of plasma sprayed CYSZ thermal barrier coatings, Ceramics International 39 (2013) 2473-2480.

[2] Hamali,R.Mozafarinia,R.Shoja-Razavi,R.Ahmadi-Pidani,Comparison of thermal shock resistances of plasma-sprayed nanostructured and conventional yttria stabilized zirconia thermal barrier coatings, CeramicsInternational38(2012)6705–6712.

[3] H.Jamali,R.Mozafarinia,R.Shoja-Razavi,R.Ahmadi-Pidani,Investigation of thermal shock behavior of plasma-sprayed NiCo-CrAlY/YSZ thermal barrier coatings, Advanced Materials Research 472–475 (2012)246–250.

[4] M. Bonek, Characterization and performance of laser alloyed commercial tool steels, Division of Materials Processing Technology, Management and Computer Techniques in Materials Science, Institute of Engineering Materials and Biomaterials, Silesian University of Technology, ul. Konarskiego 18a, 44-100 Gliwice, Poland,

[5] C.Batista, A. Portinha, R.M. Ribeiro, V. Teixeira, C.R. Oliveira, Evaluation of laser-glazed plasma-sprayed thermal barrier coatings under high temperature exposure to molten salts, Surface and Coatings Technology 200 (2006) 6783–6791.

Effect of Heating Time on Hardness Properties of Laser Clad Gray Cast Iron Surface

B. Norhafzan[1], S. N. Aqida[2], F. Mifthal[3], A. R. Zulhishamuddin[4], I. Ismail[5]

[1,2,3,4]Faculty of Mechanical Engineering, University Malaysia Pahang, 26600 Pekan, Pahang, Malaysia
[1]Department of Mechanical Engineering, Politeknik Muadzam Shah, 26700 Muadzam Shah, Pahang, Malaysia
[5]Faculty of Manufacturing Engineering, University Malaysia Pahang, 26600 Pekan, Pahang, Malaysia

norhafzan23@gmail.com

Abstract. This paper presents effect of heating time on cladded gray cast iron. In this study, the effect of heating time on cladded gray cast iron and melted gray cast iron were analysed. The gray cast iron sample were added with mixed Mo-Cr powder using laser cladding technique. The mixed Mo and Cr powder was pre-placed on gray cast iron surface. Modified layer were sectioned using diamond blade cutter and polish using SiC abrasive paper before heated. Sample was heated in furnace for 15, 30 and 45 minutes at 650 °C and cool down in room temperature. Metallographic study was conduct using inverted microscope while surface hardness properties were tested using Wilson hardness test with Vickers scale. Results for metallographic study showed graphite flakes within matrix of pearlite. The surface hardness for modified layer decreased when increased heating time process. These findings are significant to structure stability of laser cladded gray cast iron with different heating times.

1. Introduction

Gray cast iron usually used for automotive and most engineering component were operate in high temperature condition such as brake disc, stove part and die application. The ability to withstand in thermal cyclic condition caused gray cast iron suitable use in high working temperatures in manufacturing process. However when the temperature is above 600°C, the grey cast iron are easy to wear and collapse due to brittleness, thermal wear and fatigue, so the grey cast iron life at high temperature is not long enough. Laser cladding was implemented to improve die life due to excellent bonding, improve hardness, absence of flaws, produce new mechanical properties and high temperature stability of surface layer [1]. Laser melting on gray cast iron surface shown hardness increased after laser procesed [2].

Thermal stability is the important factor of material to stabilized its molecule at high temperature. Study of thermal stability for various material shown hardness of the sample decreased when the temperature increased [3]. Thermal stability is a tendency for modified surface to retain it properties at high working temperature. Thermal stability on Ti-6Al-4V improve after cladding process with TiVCrAlSi [4] . Alloying element such as Molybdenum and chromium improve strength, hardness, toughness and hardenibility also improve thermal stability [5]. Another thermal stability study shown, increased annealing temperature caused decreased microhardness and enlarge grain in Ni-based alloy

[6, 7]. The thermal stability of die material at elevated temperature was investigated through micro hardness testing and metallographic study. The combination of wear resistance, high strength and thermal stability can increase significantly the lifetime. This study investigates the hardness properties of laser cladded grey cast iron with Mo-Cr powder addition after heating at 650 °C temperature and different heating time of 15, 30 and 45 minutes.

2. Experimental

Laser cladded surface of gray cast iron with Mo-Cr powder was used as a material in this study. The cladded samples were cut using diamond blade. Samples surface were prepared using abrasive papers of 400, 800 and 1200 grit size, and polish using red felt with and imperial cloth with 6 and 1 µm diamond paste before and after heating. The 4% alcoholic nitric acid (nital) used for 5 seconds at room temperature to etched sample before surface characterization to reveal the ferrite grain boundaries and reveal phases. Samples were then heated for 15, 30 and 45 minutes using verbatim furnace at 650°C. After that, metallographic study was conducted to verify the grain size and phase structure. Metallographic study was conducted using ECON-2010L0 optical microscopes with image analysis software at 100 µm. Surface hardness was analysed using Vickers scale of Wilson Hardness tester with 100 kN load.

3. Result and Discussion

3.1. Metallographic Study

Micrograph in **Figure** 1 indicates cross-sectional surface of laser clad grey cast iron with three layers namely substrate or based materials (BM), heat affected zone (HAZ) and cladding zone (CZ). The heat-affected zone is a transition zone where substrate was not fully melted and the chemical properties were altered by high temperature heat [8]. The cladding zone produced fine grain structure with high metallurgical bond. The cladding zone is important to shown the effect of Mo-Cr powder addition via laser cladding process at heating temperature of 650°C.

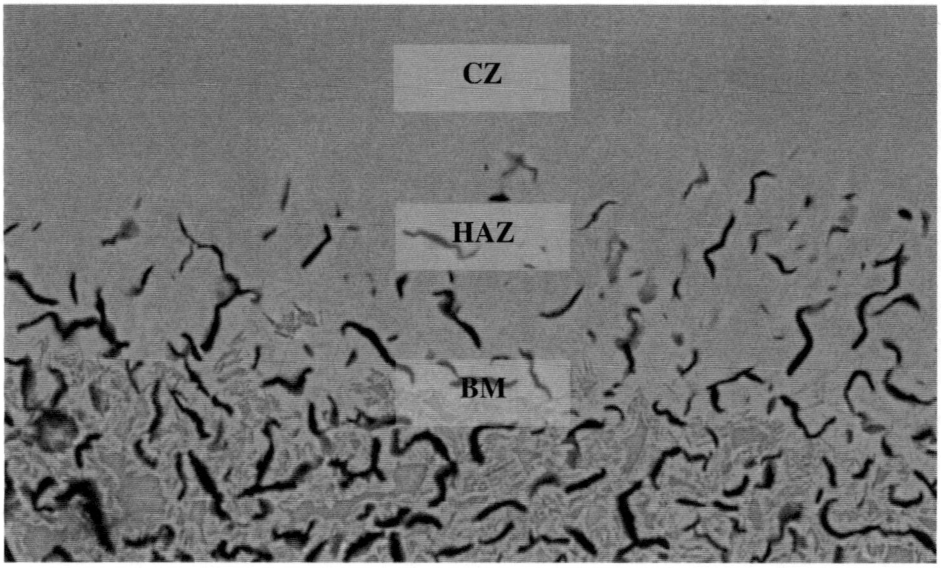

Figure 1 : Cross-sectional of laser clad grey cast iron with three layers namely cladding zone, heat affected zone and based materials (substrate)

Substrate of grey cast iron has an interconnected flake graphite shape. After laser processing, some gray cast iron structure were mixed together with Mo-Cr and produced new microstructure with grain refinement [9]. Grain refinement in the cladded layer is shown as white layer with graphite phase due to rapid heating and cooling during laser processing [10]. The metallographic study was performed to analysed clad surface microstructure after heated at 650°C for 15, 30 and 45 minutes. **Figure 2** shows transformation of grey cast iron microstructure after heated at elevated temperature and cooling down at room temperature.

The microstructure before and after heating of 15, 30 and 45 minutes at 650°C is shown in **Figure 2** (a) to (d). Increasing heating time produced larger ferrite structure. **Figure 2** (d) shows the effect of the longest heating time of 45 minutes on ferrite size of cladded gray cast iron with Mo-Cr powder addition. Decrease of graphite flake in clad layer produced hard phase on surface layer. Graphite flake increased when the duration of heating time increased. The microstructure of grey cast iron is a matrix of pearlite with the graphite flakes. Ferrite can only absorb a very low amount of carbon at room temperature, the un-absorbed carbon separates out of the body centered cubic structure to form carbides which mixed together to create small pockets of an extremely hard crystal structure called cementite [11]. However, when ferrite is heated the body-centered-cubic structure changes to a face-centered-cubic structure, thus allowing for absorption of the carbon into the crystal structure. Decreased graphite phase near the surface contribute to prevent wear and increased lifetime [2].

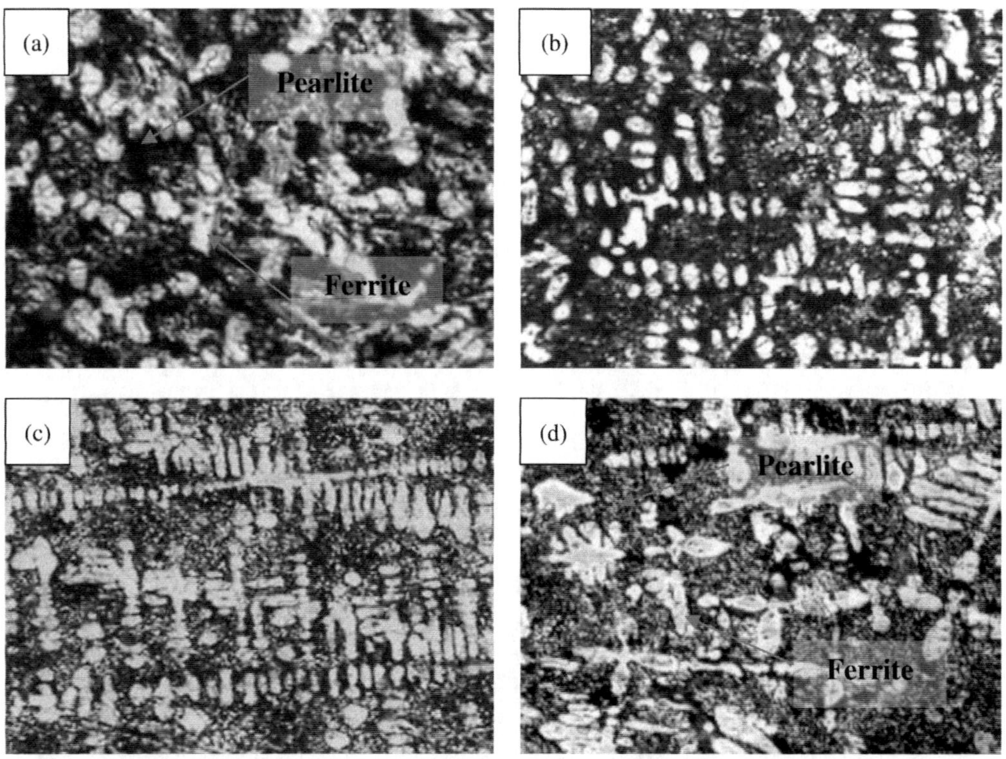

Figure 2: Micrograph of Gray cast iron with Mo-Cr (a) before heated at 650°C, (b) after heated 15 minutes, (c) after heated 30 minutes and (d) after heated 45 minutes

3.2 Hardness properties

The hardness properties of modified layer decreased directly proportional to heating period. Surface hardness in cladded zone increase due to brittle martensite structure formation. Surface hardness properties of gray cast iron before laser surface cladding were approximately 177 $HV_{0.1}$. Surface hardness of cladded layer increased up to 1000 $HV_{0.1}$ after cladding process. The surface hardness of melted layer reported previously was increased up to 737 $HV_{0.1}$ [2]. Hardness of modified layer increased up to four times of the substrate hardness due dissolved carbon content in melted zone. After heated to 650°C for 15, 30 and 45 minutes, the surface hardness of melted layer decreased to 684, 536 and 436 $HV_{0.1}$. The surface hardness of cladded layer after heated for 15, 30 and 40 minutes decreased to 912, 692 and 668 $HV_{0.1}$ respectively.

The hardness of modified surface decreased when heating period increases. **Figure 3** shows the hardness properties across layer depth of melted gray cast iron and cladded grey cast iron before and after heated at 650°C with different heating time. At 650 °C temperature surface hardness of each sample decreased when the duration of heating time increased. The cladded grey cast iron was more stable at 650°C after heated for 45 minutes compare to laser melted grey cast iron. The surface hardness of cladded layer was higher than melted layer before and after heating for 15, 30 and 45 minutes due to effect of adding Mo-Cr on gray cast iron surface. Addition of Mo-Cr powder can simplify crystallization process thus improve thermal stability of cladded layer [12]. Hard phase of clad layer surface due to decreased graphite phase on cladded layer improve hardness properties and wear resistance of gray cast iron [13]. These properties are important to increase thermal stability of gray cast iron for enhanced lifetime when used at cyclic heating and cooling condition.

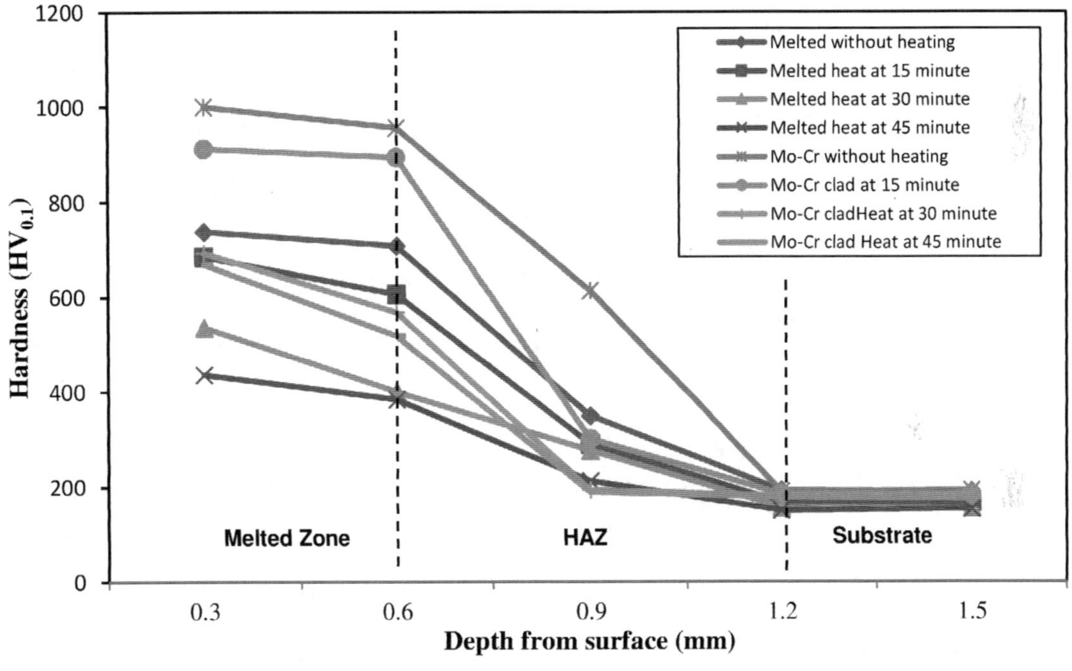

Figure 3: Microhardness across layer depth of melted gray cast iron and cladded gray cast iron before and after heating at 15, 30 and 45 minutes at 650°C

4. Conclusion

The laser cladding process on gray cast iron with Mo-Cr powder addition can improve surface hardness of gray cast iron. The microstructure of cladded gray iron is a matrix of pearlite with the graphite flakes. The graphite flake size of cladded layer become enlarge with increased heating time at 45 minutes. By adding alloying element of Mo-Cr in gray cast iron the hardness of cladding layer increased to 1000 $HV_{0.1}$ and decreased to 668 $HV_{0.1}$ after heating time of 45 minutes. Increase of heating time at high temperature at 650°C decreased the surface hardness properties of clad layer. The clad layer exhibits structure stability with lower hardness reduction after heating. These findings are significant to enhance thermal stability on laser clad gray cast iron with Mo-Cr powder addition.

Acknowledgements
The authors would like to acknowledge the support from University Malaysia Pahang under GRS150306 and Ministry of Higher Education Malaysia.

References

[1] B. Norhafzan, S. Aqida, E. Chikarakara, and D. Brabazon, "Surface modification of AISI H13 tool steel by laser cladding with NiTi powder," *Applied Physics A,* vol. 122, pp. 1-6, 2016.
[2] A. Zulhishamuddin, S. Aqida, and E. Rahim, "Optimization of pulsed Nd: YAG laser melting of gray cast iron at different spot sizes for enhanced surface properties," in *AIP Conference Proceedings*, 2016, p. 030004.
[3] S. Aqida, M. Maurel, D. Brabazon, S. Naher, and M. Rosso, "Thermal stability of laser treated die material for semi-solid metal forming," *International Journal of Material Forming,* vol. 2, pp. 761-764, 2009.
[4] C. Huang, Y. Zhang, J. Shen, and R. Vilar, "Thermal stability and oxidation resistance of laser clad TiVCrAlSi high entropy alloy coatings on Ti–6Al–4V alloy," *Surface and Coatings Technology,* vol. 206, pp. 1389-1395, 2011.
[5] A. Zulhishamuddin and S. Aqida, "An overview of high thermal conductive hot press forming die material development," *Journal of Mechanical Engineering and Sciences,* vol. 9, pp. 1686-1694, 2015.
[6] Y. Sun, L. Fu, Z. Fu, A. Shan, and E. J. Lavernia, "Enhanced thermal stability and ductility in a nanostructured Ni-based alloy," *Scripta Materialia,* vol. 141, pp. 1-5, 2017.
[7] N. Gordillo, C. G. de Castro, E. Tejado, J. Pastor, G. Balabanian, M. Panizo-Laiz, *et al.*, "On the thermal stability of the nanostructured tungsten coatings," *Surface and Coatings Technology,* vol. 325, pp. 588-593, 2017.
[8] S. Aqida, D. Brabazon, and S. Naher, "Atomic diffusion in laser surface modified AISI H13 steel," *Applied Physics A,* vol. 112, pp. 139-142, 2013.
[9] N. Bariman, S. N. Aqida, and F. Fauzun, "Laser Melting of High Thermal Conductivity Steel (HTCS) Surface," in *Materials Science Forum*, 2017.
[10] L. Wang, J. Zhou, Y. Yu, C. Guo, and J. Chen, "Effect of powders refinement on the tribological behavior of Ni-based composite coatings by laser cladding," *Applied Surface Science,* vol. 258, pp. 6697-6704, 2012.
[11] W. Cui, D. San-Martín, and P. E. Rivera-Díaz-del-Castillo, "Stability of retained austenite in martensitic high carbon steels. Part I: Thermal stability," *Materials Science and Engineering: A,* 2017.
[12] W. Yu, H. Zeng, Y. Sun, and Z. Hua, "Effect of Mo addition on the thermal stability, microstructure and magnetic property of FeCoZrBCu alloys," *Vacuum,* vol. 137, pp. 175-182, 2017.
[13] M. Paczkowska, "Analysis of gray iron microstructure influence on potential laser heat treatment effects," *Archives of Mechanical Technology and Automation,* vol. 32, pp. 27-36, 2012.

Competing risk models in reliability systems, an exponential distribution model with Bayesian analysis approach

I Iskandar

Faculty of Manufacturing Engineering, Universiti Malaysia Pahang, Pekan 26600, Pahang, Malaysia

ismed_iskandar@yahoo.com

Abstract. The exponential distribution is the most widely used reliability analysis. This distribution is very suitable for representing the lengths of life of many cases and is available in a simple statistical form. The characteristic of this distribution is a constant hazard rate. The exponential distribution is the lower rank of the Weibull distributions. In this paper our effort is to introduce the basic notions that constitute an exponential competing risks model in reliability analysis using Bayesian analysis approach and presenting their analytic methods. The cases are limited to the models with independent causes of failure. A non-informative prior distribution is used in our analysis. This model describes the likelihood function and follows with the description of the posterior function and the estimations of the point, interval, hazard function, and reliability. The net probability of failure if only one specific risk is present, crude probability of failure due to a specific risk in the presence of other causes, and partial crude probabilities are also included.

1. Introduction

Reliability theory concerns the ability of a component or a system, either a life creature or an equipment, to be functioning during its expected length of life. The studies of reliability theory become more important because they are generally more expensive than measurement data in most quality control processes. The estimation of the parameters of the assumed failure models may be based on data collected over life tests or obtained from past engineering experience or from handbook data, if available. Analysis of data, either historical or from life tests, may use any of the classical parameter estimation methods such as maximum likelihood, matching moments, regression, etc. However, most of recent analysis so far has been limited to single risk-mode models. A system or component only considered as in function or fail only without considering the causes of its failure.

But in the real systems, the failure may be a phenomenon within a single item such as the tread wear, puncture, or defective side walls or from other causes of an automobile tire. Or it can happen in physically distinct components of a system such as the CPU, hard-drive, or monitor, of a computer system. This phenomenon is commonly called as competing risks. The competing risks theory describes how many causes of failure act together to affect the performance of a system.

Another problem in this area: the failure data together with their causes of failure are simply quantitatively inadequate, time consuming and expensive to perform the life tests, especially in engineering areas. For these cases, Bayesian analyses are more beneficial than classical one. The Bayesian estimation analyses allow us to combine past knowledge or experience with less data available (which reflect the competing risks phenomenon) in the form of an apriori distribution to make inferences of the parameter of interest. Although Bayesian estimation has been used widely in reliability analysis, the application so far has been limited to single risk-mode models.

Competing risks models have been studied for many years but not extensively. The estimation methods, however, have been classical. By considering all possible causes of failure for our estimations will lead us to the possibility of data scarcity for each cause and make the classical estimation methods difficult to apply. As the result it will also lead us to the wrong conclusion about our system reliability. To address this problem, the author propose to develop the robust and more precised competing risk models of a complex reliability system using the Bayesian analyses approach that can address the problems and provide the scientists and engineers with a better reliability design models of products and maintenance systems. Three models have been developed as mentioned above. More models are still available to be explored and developed.

This paper is an extended works of the author on developing some competing risks model in reliability engineering. Three models by using Weibull and Gamma distributions as well as for attribute type of data have been published previously [1][2][3].

2. Literature Review

The theory of competing risks has been in use in bio-medicine for a long time. It began in 1760 with Daniel Bernoulli when he tried to determine mathematically what would happen to a population mortality structure at different ages if smallpox were eliminated from that population. Most of the currently accepted techniques were developed during the 19th century when the problem had been of great interest and importance to actuaries for over 100 years. Since that time scientists, mostly in Bio-medical areas but a few in engineering , have put their contributions for the basic understanding and theory of the competing risks and its application.

The methods used in analyzing survival data was discussed with competing risks by using SAS software [4]. A software was used to implement the appropriate nonparametric methods for estimating cumulative incidence functions and for testing differences between these functions in multiple groups. A model was proposed in which it was applied to accelerated degradation data from plastic substrate active matrix light-emitting diodes (AMOLEDs), along with sensitivity analysis [5]. The reliability of the model was estimated with competing risks analyses from linear degradation and failure time data. Competing risk data were analyzed in the presence of complete information of failure cause [6]. In their paper, the authors interest was to consider the occurrence of missing causes as well as interval censored failure time. The competing risk of death and premature assessment of neurological prognosis pose significant challenges was studied to measuring these surrogate endpoints after cardiac arrest in order a need for new therapies to improve outcomes after cardiac arrest [7]. Competing risk analysis was also applied for patients who die early are precluded from developing on the incidence of ventilator-associated pneumonia (VAP) [8]. Interval censored failure time data with competing risks and a new estimator was analyzed for the cumulative incidence function [9]. It was derived using an approximate likelihood and a test statistic to compare two samples and then obtained by extending Sun's test statistic. Small sample properties of the proposed methods are examined by conducting simulations and a cohort dataset from AIDS patients is analyzed as a real example.

Numerous scientists have contributed and provided the philosophical basis for Bayesian analysis in reliability systems and use it in many areas of application and research. They are not to be mentioned here because so many of them doing research in this area and because their Bayesian method of analysis were for single risk-mode model only. There were a limited number of scientists who studied and analysed the competing risks problems by using a Bayesian analyses approach in order to estimate the parameters of interest. Their models were mostly for applications in a specific condition but not for the basic understanding of general model applications. Bayesian analysis of system failure data was used under a competing-failure framework when the failure causes have not been exactly identified but narrowed down to a subset of all potential causes of failure [10]. The model of recurrent events used Bayesian reference-analysis with competing risks, where the sampling distribution of the observations due to some failure types is modeled through a proportional intensity homogeneous Poisson process [11]. Individuals are expected to experience repeated events along with the fixed covariates. The goal was to estimate the cause-specific intensity functions.

For competing risks models where the causes of failures are being considered, the life test data scarcity for each cause is inevitable. In many situations, especially in engineering areas, these life test data are time consuming to obtain and expensive to perform the life tests. For these cases, Bayesian analyses are more beneficial than classical one. To deal with this situation, some basic notions have been developed on a general competing risks model in reliability system using a Bayesian analysis approach [12]. Also three papers that deal with this matter using Weibull and Gamma as posterior functions and Uniform distribution as prior distribution for both posteriors have been published [2][3]. Another paper was for attribute type of life test data with Multi-nomial distribution model as the posterior and Beta distribution function as its prior [1].

3. Exponential Distribution Model

This section describes the likelihood function and follows with the description of the posterior function. A non-informative prior is used in our analysis. It is followed by the estimation of the failure rate and hazard function. The net, crude, and partial crude probabilities are also included.
The individual p.d.f. of the length of life t_i is given by

$$g_i(t_i|\lambda_i) = \lambda_i e^{-\lambda_i t_i} , \tag{1}$$

for $i = 1,\dots,k$, $\lambda_i > 0$, and $t_i \geq 0$. The parameter λ_i is known as the failure rate.

3.1.The Likelihood Function

Let us consider the case in which all lengths of life and associated causes of failure are known. The individual likelihood function is given by

$$L_i(\lambda_i) \propto \lambda_i^{n_i} \exp(-\lambda_i T) \tag{2}$$

and the statistic

$$T = \sum_{i=1}^{k} \sum_{j=1}^{n_i} t_{ij} \tag{3}$$

is the suffient for the estimation of λ_i. The t_{ij} are the failure times for` $i = 1,2,\dots,k$ and $j = 1,2,\dots,n_i$.

3.2. The Posterior Function

The individual posterior function is given by

$$g_i\left(\lambda_i \middle| t_{ij}\right) = \frac{\lambda_i^{n_i} \exp(-\lambda_i T) f_i(\lambda_i)}{\int_0^\infty {}_i^{n_i} \exp(-\lambda_i T) f_i(\lambda_i) d\lambda_i} , \tag{4}$$

where $f_i(\lambda_i)$ is the individual prior distribution.

3.3. Non-informative Prior Distribution

Taking the logarithm of the individual likelihood function in equation (4.33) above, we will obtain

$$\ln L_i(\lambda_i) = s \ln \lambda_i - \lambda_i T + c , \tag{5}$$

where c is a constant.

Since the derivation of non-informative prior is mathematically very close associated with the Fisher's information [13] that is given by

$$J\left(\hat{\lambda}_i\right)=\left(\frac{n_i}{\hat{\lambda}_i^2}\right)_{\lambda_i=\hat{\lambda}_i}, \tag{6}$$

a non-informative prior distribution for λ_i can be obtained as follows

$$\begin{aligned} f_i(\lambda_i) &\propto J^{1/2}(\lambda_i) \\ &\propto \frac{1}{\lambda_i}, \qquad \lambda_i > 0. \end{aligned} \tag{7}$$

3.4. Failure Rate Estimation

The individual posterior function in equation (4.35) becomes

$$g_i\left(\lambda_i \middle| t_{ij}\right)=\frac{T^{n_i}}{\Gamma(n_i)}\lambda_i^{n_i-1}\exp\left(-\lambda_i T\right), \qquad 0<\lambda_i<\infty. \tag{8}$$

This is *a gamma distribution* with the shape parameter n_i and the scale parameter T.

3.5. Point Estimation

The individual posterior mean is given by

$$E_i\left(\lambda_i \middle| t_{ij}\right)=h_i(t)=\frac{n_i}{T}, \tag{9}$$

and the individual posterior variance is as follows

$$Var_i\left(\lambda_i \middle| t_{ij}\right)=\frac{n_i}{T^2}. \tag{10}$$

This posterior mean is also called as the hazard function or the failure rate.
The hazard function for the system is as follows

$$h(t)=\sum_{i=1}^{k}\frac{n_i}{T}, \tag{11}$$

We know that all functions above are constant over time.

3.6. Interval Estimation

The $100(1-\gamma)\%$ *two-sided Bayes confidence interval (TBCI)* equations for each individual posterior mean are given by

$$\lambda_{i,L}=\frac{\chi^2_{\gamma_i/2}(2n_i)}{2T}, \tag{12}$$

and

$$\lambda_{i,U} = \frac{\chi^2_{1-\gamma_i/2}(2n_i)}{2T} , \tag{13}$$

where $\lambda_{i,L}$ and $\lambda_{i,U}$ are the lower and the upper bounds of the interval, respectively.

3.7. Reliability Estimation

The reliability of the system is given by

$$R_{sys}(t|\lambda_i) = \exp\left\{-(t)\sum_{i=1}^{k}\frac{n_i}{T}\right\}. \tag{14}$$

3.8. The Net, Crude, and Partial Crude Probabilities

The net probability of failure, i.e. probability of failure within a particular time period if the specific risk is the only risk present, is given by

$$q_{i,net} = 1 - \exp\left\{-(t)\sum_{i=1}^{k}\frac{n_i}{T}\right\}. \tag{15}$$

The crude probability of failure, i.e. the probability refers to the failure within a particular time interval from a specific cause in the presence of all other risks, is given by.

$$Q_{i,crude} = \frac{n_i}{\sum_{i=1}^{k} n_i}\left\{1 - \exp\left(-(t)\sum_{i=1}^{k}\frac{n_i}{T}\right)\right\} . \tag{16}$$

and the partial crude probability of failure, i.e. the probability of failure from a specific cause when another risk (or risks) is eliminated from the population, is given by

$$Q_{i.1,prt.crude} = \frac{n_i}{\sum_{i=1}^{k} n_i - n_1}\exp\left\{-\frac{t}{T}\left(\sum_{i=1}^{k} n_i - n_1\right)\right\}. \tag{17}$$

4. Conclusions

There are some models that are open to be developed in this area, depending on the problems of interest. In this paper we select our investigation by using the exponential distribution as a posterior function and non-informative distribution as its prior. This distribution is widely used in reliability analysis for a constant hazard rate and is one of some other common failure models as well as Weibull, exponential, Normal, Log-normal, and Inverse Gaussian models. It should be noted that the choice of the time to failure distribution functions depend on the choice of its reliability measures that is meaningful and useful to the problem of interest. Different pattern of operating life of a device may required to have a specific probability of successfully performing its required function.

The estimations of this exponential model for its individual posterior function (the individual mean time to failure / MTTF) and variance are developed and followed by the estimations of the hazard and the reliability functions. Based on these estimations, others such as the estimations for reliable or design life for a specific reliability and the probability of survival between a specific interval of time can also be calculated. One important part of competing risks analysis is its ability to elaborate the net probability if only one risk is present and the crude / partial crude probablitiies if one or more risk is eliminated from the system while other risks remain.

The author realizes that for all descriptions above of this model need a real or simulated data to test and prove the validity of the model. But the limitation of time to obtain or to conduct a simulation for such a data has put this matter for further investigation without eliminating the importance of this model. Another important measures of this analysis is to include the dependent risks in our model for further research.

Acknowledgments

Authors wishing to acknowledge assistance or encouragement from Universiti Malaysia Pahang through RDU130341.

References

[1] Iskandar I, Mohd Razali N 2014 *Advanced Material Research* Multi-mode Failure Models for Attribute Test Data in Reliability Systems, a Bayesian Analysis Approach Using Multi-nomial Distribution Model **903** 419-24
[2] Iskandar I, Gondokaryono Y S 2016 *IOP Conf. Series: Materials Science and Engineering* Competing Risk models in Reliability Systems, a Weibull Distribution Model with Bayesian Analysis Approach 114
[3] Iskandar I 2016 *IOP Conf. Series: Materials Science and Engineering* Competing Risk Models in Reliability Systems, a Gamma Distribution Model with Bayesian Analyses Approach 114
[4] Lin G, So Y, Johnston G 2012 *SAS Global Forum 2012 Statistics and Data Analysis* Analyzing Survival Data with Competing Risks Using SAS® Software SAS Institute Inc Cary NC 344-2012
[5] Haghighi F, Bae S J 2015, *IEEE Trans. on Reliability* Reliability Estimation from Linear Degradation and Failure Time Data With Competing Risks Under a Step-Stress Accelerated Degradation Test **64** N3
[6] Do G, and Kim Y J 2016 *J. Statistical Computation and Simulation* Analysis of interval censored competing risk data with missing causes of failure using pseudo values approach **87** 631-9 Issue 4
[7] McCredie V A, Scales D C 2016 *Critical Care* Surrogate endpoints and competing risk of death in cardiac arrest research Online ISSN 1364-8535 **20**:189
[8] Ranzani O T, Li Bassi G, Torres A 2016 *Intensive Care Medicine* Prone position and VAP incidence in the PROSEVA trial: attention to the causal question when interpreting competing risk analysis **42** Issue 12 2119–20
[9] Kim Y J 2016 *Journal of Applied Statistics* Analysis of interval censored failure time data with competing risk 1-13
[10] Xu A, Tang Y 2011, *Communications in Statistics - Theory and Methods* Nonparametric Bayesian Analysis of Competing Risks Problem with Masked Data **40** Issue 13 2326-36.
[11] Fu J, Tang Y, Guan Q 2014 *Quality Technology & Quantitative Management* Objective Bayesian Analysis for Recurrent Events in Presence of Competing Risks **11** Issue 3 265-79
[12] Iskandar I 1996 *Multi-mode Failure Models in Reliability Systems: a Bayesian analysis approach* (a Ph.D dissertation in Industrial Eng. Dept. New Mexico State Univ. NM USA)
[13] Martz H F, Waller R A 1991 *Bayesian Reliability Analysis* Krieger Publishing Company Malabar Florida USA 336

Optimal Lease Contract for Remanufactured Equipment

B P Iskandar[1], R Wangsaputra[1], U S Pasaribu[2] and and H Husniah[3]

[1]Department of Industrial Engineering, Bandung Institute of Technology, Ganesha 10,Bandung, 40132, Indonesia
[2]Department of Mathematics, Bandung Institute of Technology, Ganesha 10,Bandung, 40132, Indonesia
[3]Department of Industrial Engineering, Langlangbuana University,Karapitan 116,Bandung, 40261, Indonesia

Email: bermawi@mail.ti.itb.ac.id, hennie.husniah@gmail.com

Abstract. In the last two decades, the business of lease products (or equipment) has grown significantly, and many companies acquire equipment through leasing. In this paper, we propose a new lease contract under which a product (or equipment) is leased for a period of time with maximum usage per period (e.g. 1 year). This lease contract has only a time limit but no usage limit. If the total usage per period exceeds the maximum usage allowed in the contract, then the customer (as a lessee) will be charged an additional cost. In general, the lessor (OEM) provides a full coverage of maintenance, which includes PM and CM under the lease contract. It is considered that the lessor offers the lease contract for a remanufactured product. We presume that the price of the lease contract for the remanufactured product is much lower than that of a new one, and hence it would be a more attractive option to the customer. The decision problem for the lessee is to select the best option offered that fits to its requirement, and the decision problem for the lessor is find the optimal maintenance efforts for a given price of the lease option offered. We first find the optimal decisions independently for each party, and then the joint optimal decisions for both parties.

1. Introduction

Recently, leasing an equipment is increasingly common for companies rather than to own it [1]. Over 80% of American businesses lease at least one of their equipment acquisitions, and nearly 90% say they would choose to lease again [2]. Some motivations of leasing are saving on initial investment, flexibility on equipment upgrading, and cost reduction in maintenance and inventory. Lease contracts have been studied by many researchers [3] provides a comprehensive review and the study can be done from the lessor's or lessee's perspective. From the lessor's perspective, one can study from two levels of decision problems i.e. the strategic level (issues relevant such as type and number of equipment lease, upgrade options to compensate with technological obsolescence, etc.) and the operational level (dealing with issues consisting of maintenance servicing, spare part stock, crew size, etc.).

We can view the lessee as individual households (for consumer equipments such as electronic appliances, automobiles, etc.) or businesses (for commercial or industrial equipments). The lessee wants to decide which equipment to lease for the case where several brands offered, and the best lease option from the options available for a given equipment. For the case where the study is done from the lessor and lessee point of views then a game theory formulation is needed to modelling the decision problems (See [4]).

Furthermore, the lease contracts can be grouped into two categories – one dimensional and two dimensional lease contracts. The one dimensional lease contract has only time/age limit whereas the two dimensional one has both age and usage limits. For the one-dimensional lease contract, the study have been done by many researchers but only a few works for the two dimensional lease contract.

For the one-dimensional case, the study focuses on finding an optimal maintenance policy which minimizes the total maintenance cost for the lessor, and the leased equipment considered can be a new, used or remanufactured item. The lease contract for the new lease item, can be found in [1], [5] to name a few, whilst [6], [7] examined the lease contract for used items. Finally, [8] considered lease options which include a remanufactured equipment. When the equipment is used intensively (or with high usage) per unit of time, the usage experienced affects significantly the deterioration of the equipment. This indicates the need to consider age and usage in modelling the failure and also defining the lease contract which involves two parameters –i.e. age and usage limits (called a two dimensional lease contract). We are aware only the works by [9] and Iskandar and Husniah [10] belong to this group. In [9] the period of the contract is always the same (i.e. Γ_0) with maximum usage rate r_m whilst [10] consider a two dimensional lease contract for maximum Γ_0 (age) or U_0 (usage).

In this paper, we propose a two dimensional lease contract which has a time limit but no usage limit. However, if the usage exceeds the maximum usage allowed in the contract, then the lessee has to pay some additional cost. This lease contract can be view as the extension of the lease contract studied by [9] which confine the usage rate to r_m, whilst in this paper there is no upper limit of usage rate. However, if the usage exceeding the maximum value results in some additional cost to the lessee due to more failures would likely to occur (and hence a higher maintenance cost for servicing the equipment) when the usage is beyond the maximum value. In general, OEM (as a lessor) offers not only a lease contract for a brand new equipment but also a lease contract for a remanufactured one. As the price of the lease contract for the remanufactured equipment is much lower than the price of a new one, and hence it would be a more attractive option to the companies.

The paper is organised as follows. In section 2 we give model formulation for the two dimensional lease contract studied. Sections 3 and 4 deal with model analysis and the optimal decisions for the lessor and the lessee. In section 5, we provide with a numerical example. Finally, we conclude with topics for further research in Section 6.

2. Model Formulation

In this section, we first define a new lease contract, describe failure model, formulate a PM policy and its effect on reliability, and then obtain the expected profit for a lessor and a lessee.

Notations:

$\Omega = [0,\Gamma) \times [0,\infty)$	: lease coverage region
δ_y	:Preventive maintenance level
ω	:Down time target
U_y	:Total usage in $[0, m\Gamma)$
$G(x)$	:Distribution function of downtime X_i
$m\Gamma$	:Lease contract periods, $\Gamma > 0$ and $m = 1,2,...$
y	:Usage rate
N	:Number of PM during lease contract
T_r	:The time to the first failure of the remanufactured equipment
P	:Lease contract price
C_b	:Annual cost
C_p	:Preventive maintenance cost
C_r	:Average repair cost
$F(t,\alpha_r)$ $f(t,\alpha_r)$	:Distribution function and density function for T_r

$\lambda(t,\alpha_r)$, $\Lambda(t,\alpha_r)$	:Hazard function and cumulative hazard function associated with $F(t,\alpha_r)$
$\Pi_r(m,y),\Pi_s(m,y)$	:Expected profit for a lessor, and for a lessee
$J^1(K,\delta_y),J^2(K,\delta_y)$	: expected preventive maintenance cost, and minimal repair cost

2.1 New Lease Contract

We consider that a lessee uses an equipment with a certain usage rate. A different lessee has a different usage rate. As a result, the usage rate varies from lessee to lessee. A lease contract studied is as follows. The the remanufactured equipment is leased for period of $m\Gamma$ ($\Gamma>0$ and $m=1,2,...$) with a maximum usage ($U_{\max}$) (e.g. number of copies/time period, km travelled/ time period or machine-hours / time period).For a given lessee (or usage rate y), if the total usage at the end of a lease period, U_y exceeds $U_{\max}$ (See Fig.1), then the lessee (or customer) will be charged an additional cost which is proportional to $\varphi=Max\{0,U_y-U_{\max}\}=Max\{0,y\Gamma-U_{\max}\}$ for one period of a lease contract. This additional cost is viewed as a compensation to the lessor as larger usage rate $(U_y>U_{\max})$ will result in more failures under lease contract and hence higher maintenance cost.

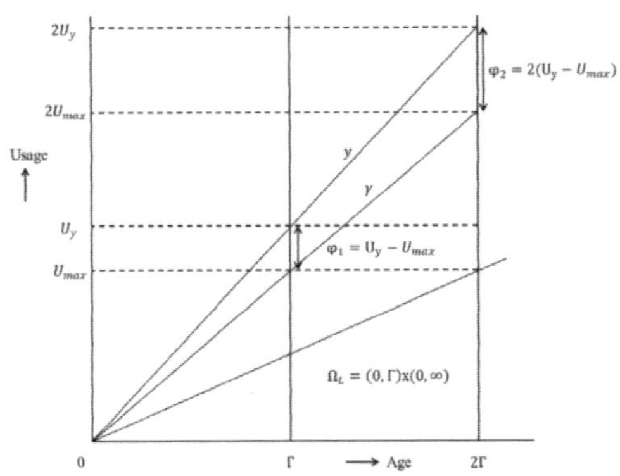

Figure 1. A lease contract with m periods ($m\Gamma, m=2$) and the maximum usage ($U_{\max}$)

2.2. Modelling Failure

In general, most products at the end of the first life or EOU (end-of-use) have a low reliability (or their reliability is below the threshold value of reliability R^*). Remanufacturing efforts improve the reliability of the product to at least the same level of the threshold reliability as all parts have been undergone inspection to ensure that each parts meets the quality requirements. Let T_r be the time of the first failure the remanufactured product. $F(t)$ is the distribution function for T_r . If $F(t)$ is given by Weibull distribution function with $F(t,\alpha_r)=1-e^{-(t/\alpha_r)^\beta}$, then the reliability of the remanufactured product is $R(t,\alpha_r)=1-F(t,\alpha_r)=e^{-(t/\alpha_r)^\beta}$. As $R(t,\alpha_r)\geq R^*$, then we have $\alpha_r\geq t/\left\{(\sqrt[\beta]{-\ln R^*}\right\}$. The hazard and the cumulative hazard functions associated with $F(t,\alpha_r)$ are given by $\lambda(t,\alpha_r)$ and $\Lambda(t,\alpha_r)=\int_0^t \lambda(x,\alpha_r)dx$ respectively.

Now, we consider that failure is not only influenced by age but also usage. As the remanufactured product is a repairable item then it is appropriate to modelling failures using a failure rate function. For a given customer (y), let $\lambda_y(t)$ be the conditional hazard (failure rate) function. We will use PHM (Proportional Hazard Model) whichallows to incorporate the age and usage for modelling the degradation the equipment. Using PHM, the conditional hazard function, $\lambda_y(t)$ is given by

$$\lambda_y(t) = \phi(y, y_0, \rho)\lambda_0(t, \alpha_0) \tag{1}$$

where $\phi(y, y_0, \rho) = e^{\rho(y-y_0)/y_0))}$, $\rho \geq 1$ and $\lambda_0(t, \alpha_0)$ is the base line failure rate function when the usage rate, y is equal to y_0 (a nominal usage of the equipment).

The reasons to support the form of $\phi(y, y_0, \rho)$ is as follows. Note that every equipment is designed with a nominal usage. If the lessee uses the equipment with the usage rate exceeding the nominal value, $y > y_0$, then the equipment will deteriorate faster, otherwise when $y < y_0$ then it goes slower. If the lessee uses the equipment with the usage rate exceeding the nominal value (or the maximum usage allowed in the contract), then the equipment will deteriorate faster. This in turn will result in more failures and hence repair cost under the lease contract period. As a result, the lessee will be charged an additional cost. This cost is viewed as a compensation to the lessor due to the increase in repair cost. It is considered that $\lambda_0(t, \alpha_0) = \lambda(t, \alpha_r)$ with $\alpha_0 = \alpha_r$ (or the base line failure rate function is the failure rate function influenced by age only). For the case where $y = y_o$ then $\phi(y, y_0, \rho) = 1$ and hence $\lambda_y(t) = \lambda_r(t, \alpha_r)$. If $F(t, \alpha_r)$ is a two parameter Weibull distribution function, then $\lambda(t, \alpha_r) = \beta/\alpha_r (t/\alpha_r)^{\beta-1}$. Let $T_{r|y}$ be the time to first failure of the remanufactured product a given lessee (y). The distribution function for $T_{r|y}$ is given by $F_y(t)$. Let $N_y(t)$ be the number of failures in $(0, t]$ for a given y. If all failures under the lease contract are minimally repaired and repair times are very small relativeto the mean time between failures, then $N_y(t)$ is a non-homogeneous Poisson process (NHPP) with intensity function $r_y(t) = \lambda_y(t)$. The cumulative hazard functions associated with $r_y(t)$ is given by $R_y(t) = \int_0^t r_y(x)dx$.

2.3. Modeling the PM effect

The preventive maintenance (PM) policy is defined as follows. For a truck with usage rate y, the equipment is periodically maintained at $k\tau_y$, $k = 1, 2, ..., K$. This involves k disjoint intervals - $[0, \tau_y)$, $[\tau_y, 2\tau_y)$...$[(K-1)\tau_y, K\tau_y)]$ in which all failures within PM period are minimally repaired, where k is an integer value. The effect of PM actions are modelled through either the reduction in (i) the intensity function or (ii) the age. Here, we model using (ii) and the virtual age of the equipment after PM at $k\tau_y$, $k = 1, 2, ..., K$ is given by $v_k = \delta k\tau_y$, $k = 1, 2, ..., K$ (or the reduction in age is $k\tau_y - \delta k\tau_y = k(1-\delta)\tau_y$). As a result, the virtual age at time t, $k\tau_y < t < (k+1)\tau_y$ is given by $v_y(t) = k\delta\tau_y + (t - \tau_y)$. The conditional hazard function at time t, $k\tau_y < t < (k+1)\tau_y$ is given by

$$r_y(v_y(t)) = \phi(y, y_0, \rho)\lambda(v_y(t), \alpha_r) \tag{2}$$

Since any failure occurring between PM is minimally repaired, then the expected total number of minimal repairs in $\left(k\tau_y, (k+1)\tau_y\right), 0 \leq k \leq N$ is given by

$$E\left[N(k\tau_y, (k+1)\tau_y\right] = \int_{k\tau_y}^{(k+1)\tau_y} r_y\{v_y(t)\}dt \tag{3}$$

where $r_y(v_y(t))$ is given in (2). As a result, we have

$$M_y(m\Gamma) = \sum_{k=0}^{K} E\left[N(k\tau_y, (k+1)\tau_y\right] \tag{4}$$

where $m\Gamma = (K+1)\tau_y$, m is the number of lease periods.

3. Simulation of The System

Under the proposed lease contract, we first obtain the expected profit (i.e. the expected total revenue - the total expected cost) for the lessor and then for the lessee.

3.1. Lessor's expected profit

The lessor's expected total cost consists of PM cost, CM cost and the penalty cost paid by the lessor due to the down time of the equipment exceeds the predetermined target ω. These costs are obtained as follows.

Preventive maintenance (PM) cost:

Let $C_p(\delta_y)$ and C_r be the cost of the k-*th* PM and the cost of each minimal repair. If $C_p(\delta_y) = C_0 + C_1(1-\delta_y)\tau_y$ as in [6] and [7] then the expected total PM cost over the lease period $(0, m\Gamma), m \geq 1$ is given by

$$J^1(K,\delta_y) = \sum_{k=1}^{N} C_p(\delta_y) = KC_0 + KC_1(1-\delta_y)\tau_y = KC_0 + KC_1(1-\delta_y)\frac{m\Gamma}{(K+1)} \tag{5}$$

where $m\Gamma = (K+1)\tau_y$.

Corrective maintenance (CM) cost:

Let C_r be the cost of each minimal repair, then the expected minimal repair cost is given by

$$J^2(K,\delta_y) = C_r\left(\sum_{j=0}^{K}\int_{k\tau_y}^{(k+1)\tau_y} r_y\{v_y(t)\}dt\right) \tag{6}$$

As a result, the lessor's expected total cost is given by $J^1(K,\delta_y) + J^2(K,\delta_y)$.

The penalty Cost:

Under the lease contract, a penalty occurs if the down time of the equipment exceeds the predetermine d target ω. Let X_i be a down time caused by the i-th failure. X_i is i.i.d with a common distribution function $G(x)$. The expected value of X_i is μ. The penalty occurswhen $X_i > \omega$ and hence, the expected penalty over $m\Gamma$ is given by

$$EP_y(m,y) = q \times y \times M_y(m\Gamma)E\left[Max\{0, X_i - \omega\}\right] = q \times y \times M_y(m\Gamma)\int_{\omega}^{\infty}[1-G(x)]dx \tag{7}$$

where q is a penalty cost per unit usage. Note that $q \leq \upsilon$ (υ is revenue per unit usage).

Expected total revenue:

The expected total revenue is the sum of the price of lease contract and some additional revenues due to the monthly usage of the equipment is greater than $U_{\max}$ (the maximum usage). The price of lease contract is dependent on the usage rate and the number of lease contract periods. The price of the lease contract for the first period is P. The lessor will give an incentive if the equipment is leased for more than one periods (or $m\Gamma, m \geq 2$). If the equipment is leased for more than one periods, then the lessee will get a discount price given by

$$P(m) = P\left\{e^{-\mu m}/e^{-\mu}\right\} = P\left\{e^{-\mu(m-1)}\right\} \tag{8}$$

where $0 \leq \varphi < 1$ and m represents a discount parameter and the number of lease periods, respectively. Note: We use a different form of $P(m)$ as in [8]. For a lessee's usage rate y, the additional revenues earned by the lessor at the end of period $m\Gamma, m = 1,2...$ given by

$$\Phi(m,y) = C_d\sum_{j=1}^{m} Max\{0, (y \times j\Gamma - jU_{\max})\}, \tag{9}$$

where C_d is the additional cost charged (e.g. $/km or $/page copied). $\Phi(m,y)$ is viewed as some additional revenues for the lessor. As a result, we have the expected profit given by

$$\Pi_r(K,\delta_y) = P(m) + \Phi(m,y) - EP_y(m,y) - J^1(K,\delta_y) - J^2(K,\delta_y) - mC_b^r \tag{10}$$

3.2. Lessee's expected profit

The lessee's expected profit is equal to the expected total revenue minus the total expectedcost, that will be obtained as follows. The total expected cost consists of (i) the price of lease contract paid and (ii) some additional charges due to total usage exceeds the maximum usage allowed in the contract.

Lease contract price:

For a given U_{max} (maximum of usage rate allowed), the price of a lease contract depends on the number of lease periods (m) chosen by the lessee, and it is given by $P(m)$. A more periods leased results in a lower price of the lease contract as in eq. (8). This in turn gives an incentive for the lessee to lease more than 1 period.

Expected total revenue:

Let υ be the revenue per unit of usage (e.g. \$/km, \$/page copied). The total revenue earned by the lessee, R_y consists of the revenue generated by the equipment and some compensation, $EP_y(m)$ paid by the lessor due to the down time of the equipment exceeds the predetermined target ω .It is given by

$$R_y(m,y) = y \times \upsilon \times \left(m\Gamma - E\left[D\right]\right) + EP_y(m) \quad (11)$$

where $\Gamma - E[D]$ is total time available, $E[D]$ is the expected total downtimes over the contract periods ($m\Gamma$), given by $E[D] = M_y \times \mu$, M_y is the expected number of failures over $m\Gamma$ and μ is the expected downtime caused by each failure.

Expected additional charges:

The lessee incurs some additional costs when a monthly usage is greater than U_{max}. The additional cost, $\Phi(m,y)$ is given by (9). As a result, the expected profit is given by

$$\Pi_s(m,y) = y \times v \times \left(m\Gamma - E\left[D\right]\right) + EP_y(m) - P(m) - \Phi(m) \quad (12)$$

4. Optimization

We first consider the case where the optimization is carried out independently from the viewpoint of each party, and then the situation where the lessor and the lessee would like to achieve a win-win solution (or the optimization is done simultaneously from both parties).

Case 1a: **[Lessor's viewpoint]**

The lessor offers lease contracts for a fixed price which covers all maintenance costs, administration cost and a profit margin. As mentioned previously, the lease contract provides a full coverage of maintenance (including PM and CM). In order to maximise its profit, the lessor tries to improve the maintenance performance such that to minimize the maintenance costs, which is conditional on the lessee's usage rate. The nominal usage rate can be used to estimate the lessee's usage rate. Hence, the optimization problem is given as follows.

$$Max\left\{\Pi_r(K,\delta_y)\right\} = Max\left\{P(m) + \Phi(m) - EP_y(m) - J^1(K,\delta_y) - J^2(K,\delta_y) - mC_b^r\right\}$$
$$s/t: 0 \le \delta_y < 1; K = 1,2,... \quad (13)$$

Case 1b: **[Lessee's viewpoint]**

The equipment is used to generate revenues or to support a lessee's business process. In both cases, the equipment plays a key role to achieve the lessee's revenue target. Of interest to the lessee is to find the optimal usage rate and the number of lease periods maximising its profit. We view that each lessee has a different optimal usage rate. Hence, the optimization is done from a given lessee seeking for y and m that maximize its profit. As a result, the optimization problem is given by

$$Max\left\{\Pi_s(m,y)\right\} = Max\left\{y \times v \times \left(m\Gamma - E\left[D\right]\right) + EP_y(m) - P(m) - \Phi(m)\right\}$$
$$s/t: y_{min} \le y < y_{max}; m = 1,2,...L/\Gamma \quad (14)$$

where L is a life cycle of the equipment.

Case 2: **[Joint Optimization]**

Here, we assume that both parties (the lessor and the lessee)would like to cooperate or to achieve a win-win solution. In other words, the optimization is done jointly from both parties. This joint optimization is expected to give more profit for each party, which is at least the same as the profit earned if the optimization is done independently (Case 1). This incentive would give a strong motivation for both parties to cooperate. As a result, the optimization problem is given by

$$Max\{\Pi_r(K,\delta_y)+\Pi_s(m,y)\} \quad (15)$$

$$s/t:\Pi_r(N,\delta_y)\geq \pi_r,\ \Pi_s(m,y)\geq \pi_s\ y_{\min}\leq y<y_{\max};m=1,2,...L/\Gamma,0\leq\delta_y<1;K=1,2,...$$

Where π_s, π_r represent the profit obtained in Case 1 (when the optimization is done separately from each viewpoint). If $\tilde{\pi}_s$ and $\tilde{\pi}_r$ represent the optimal profit resulting from (15) for the lessor and the lessee, respectively, then $\tilde{\pi}_s\geq\pi_s$ and $\tilde{\pi}_r\geq\pi_r$.

5. Numerical Example

We consider the following parameter values as shown in Table 1.

Table 1. Parameter values used

Parameter	Value	Parameter	Value
p	1	y	1.5x 60000
m	1	y_0	1 x 60000
β	2.5	C_r	1
β_2	0.5	C_0	1.5
α_r	6	C_1	0.25
α_2	0.25	v	0.8
R	0.8	q	$0.8v$
C_b	10	C_d	$0.7v$
δ	0.027	K	1
U_{max}	60000	Γ	12
ω	4	τ	$\frac{\Gamma}{(K+1)}$

Table 2. Results for Case 1 and Case 2

Case 1: Individual Optimization			
Lessor		Lessee	
δ^*	0.668	y	1.5
$y=y_0$	1		
Profit	354.705	Profit	308.614
Case 2: Joint Optimization			
Lessor		Lessee	
δ^*	0.452	y	1.5
Profit	527.560	Profit	357.632

Table 2 shows optimal solutions for cases 1 and 2. In Case 1, the optimization is carried out independently for each party. Profit obtained is 354.705 for the Lessor, and 308.614 for the Lessee, whilst the profit resulting from the joint optimization (Case 2) is 527.560 for the Lessor, and 357.632 for the Lessee. This is as expected since (i) the Lessor performs a more effective PM (i.e. with lower

$\boldsymbol{\delta}^*$ (=0.452)) and the PM in turn will decrease the number of failures and hence the total maintenance cost and downtime, and (ii) the usage increases from 60000 to 90000 (this causes the profit for the Lessor increases significantly).

6. Conclusion

In this paper we have studied a two-dimensional lease contract for equipment in which the lessee will be charged some additional cost if the usage at the end of the contract exceeds the maximum usage allowed. We find the optimal solution independently for each party and then seek for the joint optimization for both parties. The profit obtained from joint optimization is at least the same as the optimal profit obtained independently. One can model the decision problems for the lessor and the lessee using a Stakelberg game theory formulation the lessor as leader and the lessee as follower, or a Nash game theory formulation. Another topic is to modelling failures using the twodimensional approach (or using a bivariate distribution function). These two topics are currently under investigation.

Acknowledgements

This work is funded by the Ministry of Research, Technology, and Higher Education of the Republic of Indonesia through the scheme of "PUPT 2017" with contract number SP DIPA-042.06.1.401516/ 2017.

References

[1] Jaturonnatee J, Murthy D N P, and Boondiskulchok R 2006 *European Journal of Operational Research* Optimal preventive maintenance of leased equipment with corrective minimal repairs **174** 201–215.

[2] Giddy 2004 *Giddy .org* Equipment Leasing.

[3] Murthy D N P and Jack N 2014 *Springer Series in Reliability Engineering Springer-Verlag* Extended warranties, maintenance service and lease contracts London.

[4] Bhaskaran, Sreekumar R and Gilbert S 2007 *McCombs Research Paper Series* Implications of channel structure for leasing or selling durable goods No. IROM-01-07.

[5] Pongpech J and Murthy D N P 2006 *Reliability Engineering & System Safety* Optimal periodic preventive maintenance policy for leased equipment **91** 772-777.

[6] Pongpech J, Murthy D N P and Boondiskulchok R 2006 *J. Qual Maintenance Eng* Maintenance strategies for used equipment under lease **12** 52-67.

[7] Yeh R H, Kao K C and Chang W I 2011 *Expert Systems with Applications* Preventive-maintenance policy for leased products under various maintenance costs **38**(4) 3558-3562

[8] Aras F, Gullu R and Yurulmez S 2011 *Int. J . Production Economics* Optimal inventory and pricing policies for remanufacturable leased products **133** 262-271

[9] Hamidi M, Liao H and Szidarovszky F 2016 *European Journal of Operational Research* Non-cooperative and cooperative game-theoretic models for usage-based lease contracts **255** 163–174

[10] Iskandar B P and Husniah H 2017 Computers & Industrial Engineering Optimal preventive maintenance for a two dimensional lease contract **113** 693-703.

[11] Yeh R H, Kao K C and Chang W I 2009 *Computers & Industrial Engineering* Optimal maintenance policy for leased equipment using failure rate reduction **57**(1) 304-309.

IOP Conf. Series: Materials Science and Engineering **319** (2018) 012071 doi:10.1088/1757-899X/319/1/012071

Multi Objective Optimization Using Genetic Algorithm of a Pneumatic Connector

HA Salaam [1], Zahari Taha [2], TMYS Tuan Ya [3]

[1] Faculty of Mechanical Engineering
[2] Faculty of Manufacturing Engineering,
Universiti Malaysia Pahang, 26600, Pekan, Pahang Darul Makmur, Malaysia.
[3] Department of Mechanical Engineering, Faculty of Engineering, Universiti Teknologi PETRONAS, Tronoh, Perak Darul Ridzuan, Malaysia.

Corresponding author's e-mail address: hadisalaam@ump.edu.my

Abstract. The concept of sustainability was first introduced by Dr Harlem Brutland in the 1980's promoting the need to preserve today's natural environment for the sake of future generations. Based on this concept, John Elkington proposed an approach to measure sustainability known as Triple Bottom Line (TBL). There are three evaluation criteria's involved in the TBL approach; namely economics, environmental integrity and social equity. In manufacturing industry the manufacturing costs measure the economic sustainability of a company in a long term. Environmental integrity is a measure of the impact of manufacturing activities on the environment. Social equity is complicated to evaluate; but when the focus is at the production floor level, the production operator health can be considered. In this paper, the TBL approach is applied in the manufacturing of a pneumatic nipple hose. The evaluation criteria used are manufacturing costs, environmental impact, ergonomics impact and also energy used for manufacturing. This study involves multi objective optimization by using genetic algorithm of several possible alternatives for material used in the manufacturing of the pneumatic nipple.

1. Introduction

Manufacturing cost to produce a product increased yearly because of increasing material, tool, energy, man power, energy, coolant and lubricant cost due to un-favourable currency exchange rate, inflation rate and government taxes [1]. The costs associated with environmental impact and worker health need to be managed properly because it can be an extra burden. In manufacturing there are a few stakeholders interested in the economic profitability, environmental integrity, energy integrity and social equity. Consumer and end users are those who are interested in the economics aspects because they want to get the best value for money products. The trade associations are interested in the social aspects of the workers such as the quality of life of a worker while the energy and environmental aspects are interested by national environmental agencies, public community and private companies particularly in fulfilling environmental regulations set by the governments. Thus making decision on a product design is difficult since it involves multiple objectives. One approach is to use multi-objective optimization using genetic algorithm. Here, all alternatives are modelled and optimized based on the four objectives by using pneumatic connector as a case study as shown in Figure 1.

2. Triple Bottom Line (TBL) Concept

The Triple bottom line concept was introduced by John Elkington in the 1990's based on the sustainability concept proposed by Dr Harlem Brutland in the 1980's. It encompasses a framework to measure corporate organization performance which went beyond the traditional way of measuring performance where it measures profits, share holder value and return on investment used by company previously. This approach also included environmental and social dimensions assessment. By focusing on financial profitability, social equity, energy integrity and environmental integrity criteria; TBL reporting can be a useful tool to sustain sustainability goals.

The first criterion in TBL is financial profitability which can be measured by using the cost of pollution, cost of products fabrication and other factors in profit calculations. The second criterion is social equity and ergonomics assessment can be used to evaluate social equity [2]. The third criterion is energy integrity which can be measured by using the energy consumed during the product fabrication process [1]. The fourth criterion is environmental integrity where it is a company's commitment towards reduction of environmental impact during product fabrication process. There are no universal indicators that can be used in accessing triple bottom line approach and each problem can be solved by using different indicators which suit the case study [3]. The indicators that can be used to measure financial profitability are personal income, cost of underemployment and total manufacturing cost; while environmental integrity can be accessed by measuring impact to the environment when fabricating products. The suitable assessment methods are air and water quality, energy consume during producing products and disposal of toxic waste [1] and social equity represents the measurement of human happiness in life such as median household income or ergonomics assessment at work such as weight lifting index.

3. Multi-Objective Optimization using Genetic Algorithm

In engineering optimization problems, the objectives are conflicting of each other and it prevents simultaneous optimization of each of the objective [4]. There are 2 approaches to do multi-objective optimization. The first one is to combine the individual objective functions into a single composite function. Here, the determination of a single objective is possible by using the weighted sum method and utility theory but there is a problem when to select a proper solution because it applied decision maker's preferences. The second approach is usually used in real-life case study by determining an entire Pareto optimal solution set or a representative subset. Here, solutions that are non-dominated with respect to each of the criteria are called Pareto optimal solution. When the evaluation is moving from one Pareto to another one, a certain amount of sacrifice in objective(s) is needed to achieve a certain amount of gain in the others. Pareto optimal solutions sometimes referred to a single solution because they can be practical when real-life problem are being considered. Both approaches will give either a single solution or a set of optimized solutions after some trade-offs have been considered. The advantages of using genetic algorithm measurement are that it can handle nonlinear problems and its implicit parallelism where the solution space is explored in multiple directions [5]. Most of the application area of multi-objective optimization using genetic algorithm is in engineering field such as in product design and sustainable product development area [6].

4. Methodology

The pneumatic connector used in this study is shown in Figure 1. This product was selected because the demand for it is high since it has been used in many industries to connect high compressed air hose for multi-purpose usage.

Figure 1. Pneumatic Connector.

In financial profitability criteria, the total manufacturing cost approach is adopted because it represents the cost needed to produce pneumatic connector. The total manufacturing cost calculation was adopted from [7] with some modifications as shown in Equation (1).

Total cost=Material cost+ Tool cost+ Coolant cost+ Lubricant cost+ Energy cost+ Labor cost (1)

Material Cost = Standard size price (RM/gram) x Required raw material weight (gram) (2)

Tool cost = tool cost per unit (RM) / number of product produced (n) (3)

This method has been adopted because the determination of tool life based on experimentally is totally different compared to reality. Researchers used new cutting tool in the experiment to get the results but in reality, the cutting tool being used until it wear which can produced many products using one cutting tool [8]. Hence Equation (3) is proposed to measure tool cost in this study.

Energy cost = Energy used (kWh) x electrical tariff (RM/kWh) (4)

Labor cost = Monthly Salary (RM) / average monthly output (5)

In this paper, machining time did not being considered in the labor cost since the worker perform multi tasking work such as control the machine, record the waste product amount and do product quality checking. For coolant and lubricant cost, equation (6) – (8) were used.

Coolant Cost = Coolant or lubricant volume x Coolant or lubricant cost rate (6)

Coolant volume = (tank capacity + makeup volume) / (month used x actual output) (7)

Makeup volume = (tank capacity x coolant loss rate) / (1 – coolant loss rate) (8)

The environmental integrity assessment in a production line consists of cutting tool impact, chip recycling impact, disposal of coolant and lubrication impact; and energy impact (Narita et al., 2012). This study only considered chip re-cycling impact and energy impact because the number of product being produce when using the same cutting tool is high compared to the weight of a cutting tool, coolant and lubricant used; hence it can be neglected [8]. The chip recycling impact is assessment adopt [9] method which considers the amount of carbon weight released into the air by the scrap material produced from the fabrication process such as shown in equation (9).

$$Ch_e = (raw\ material\ weight(kg) - finish\ product\ weight(kg)) \times LCI(M) \quad (9)$$

Where Che is chip re-cycling impact; LCI (M) is metal chip recycling emission intensity. The energy impact for CNC turning process is given by Equation (10) - (13) adopted from [10]:

$$P_{c\_turn} = \left(\frac{V_c \times a_p \times f_n \times K_c}{60000}\right) \quad (10)$$

$$P_{c\_drill} = \left(\frac{V_c \times a_p \times f_n \times K_c}{240000}\right) \quad (11)$$

$$P_{c\_boring} = \left(\frac{V_c \times a_p \times f_n \times K_c}{60000}\right) \times \left(1 - \frac{a_p}{D_c}\right) \quad (12)$$

$$\sum P_{c\_total} = \sum P_{c\_turn} + \sum P_{c\_drill} + \sum P_{c\_boring} \quad (13)$$

Where P_{c_turn} is power required to perform turning; P_{c_drill} is power required to perform drilling; P_{c_boring} is power required to perform boring; V_c is cutting speed (m/min); a_p is depth of cut (mm); f_n is federate (mm/min); K_c is Specific cutting force (N/mm^2) for Aluminum 6061 is 400 and 550 for Brass C3604 and Dc is drill diameter. In this paper the workers life quality is adopted for ergonomic assessment because it reflects the immediate impact on labor of the machining production floor [1]. The assessment is based on the revised Lifting Equation with some modification as proposed by [11] specific for South East Asia male, where the evaluation method is based on Equation (14) and (15).

$$LI = \frac{Load\ Weight}{Recommended\ Weight\ Limit} = \frac{L}{RWL} \quad (14)$$

$$RWL = LC \times HM \times VM \times DM \times AM \times FM \times CM \quad (15)$$

Where LC is load constant = 23kg; HM is Horizontal Multiplier; VM is Vertical Multiplier; DM is Distance Multiplier; AM is Asymmetric Multiplier; FM is Frequency Multiplier, and CM is Coupling Multiplier which can be referred in tables provided by the developer. In this study, Multi-objective optimization using genetic algorithm (GA) is used to optimize the results by using Matlab Software such as proposed by [4]. In GA, a solution vector x∈X is called a chromosome. Chromosome are made from discrete units called genes, while each genes controls one or more features of the chromosome. In this paper, the genes are represented by 14 variables used in the study such as shown in Figure 2. Normally a chromosome correspond to a unique solution x in the solution space. This requires a mapping mechanism (encoding) between the solution space and the chromosome.

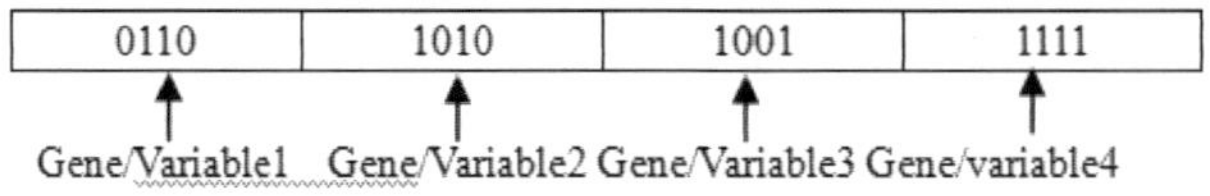

Population	Chromosome1/ Solution A1/B1	01101010
	Chromosome2/ Solution A2/B2	10101001
	Chromosome3/ Solution A3/B3	10011111
	Chromosome4/ Solution A4/B4	11110110

Figure 2. The example of case study to understand the gene, chromosome and population in GA.

The operation of GA starts with a collection of chromosome called population. The population in GA is randomly initialized. As the search for optimization evolves, the population includes fitter and fitter solutions until it eventually converges or dominated by a single solution. GA use crossover and mutation to generate new solutions. By iteratively applying the crossover operator, the good chromosome are expected to appear more frequently in the population which leads to producing a good solution [4].There are 5 steps involved where it starts with a randomly initial population, P_0. Set *t=0*. The A1, A2, A3, A4, and B1, B2, B3, B4 are two different initial populations based on material type. Then, if the stopping criterion is satisfied, return toP_t. Next, evaluate the fitness population by assigning a rank $r(x,t)$to each solution x ϵ $P_t$using the ranking scheme$r_2(x,t) = 1 + nq(x,t)$; assigning a fitness values to each solution based on the solution's rank as: $f(x,t) = N \sum_{k=1}^{r(x,t)=1} n_k - 0.5 \times \left(n_{r(x,t)} - 1\right)$ where n_k is the number of solutions with rank k. Then,calculate the niche count nc(x, t) of each solution x ϵ $P_t$using $nc(x,t) = N \sum_{y \epsilon P, r(x,t)} max \left\{\frac{\sigma_{share} - dz(x,y)}{\sigma_{share}}, 0\right\}$. Next, calculate the shared fitness value of each solution x ϵ P_t by using *f'(x, t) =f(x, t)/nc(x, t)* and then, normalized the fitness values by using the shared fitness values. To select parents for mating pool a stochastic selection is used based on *f''*. Crossover and mutation on the mating pool is applied until the offspring population Qt of size N is filled; Set $P_{t+1} = Q_t$. In step 5; Set *t=t+1*. Lastly, Step 2 will start back the step until exits.

5. Results and Discussions

Two types of materials involved in this study are Aluminum 6061 steel and Brass C3604. Both materials were chosen because they are commonly used to produce pneumatic nipple hose connector. The design of experiment used in this study is full factorial with three cutting parameters; hence the number of experiment needed is 2^3=8 experiments. The machining process involved are rough turning, fine turning, center drill, drilling three different size of drill bit and thread cutting using OKUMA LB15-II CNC lathe machine. The energy consumed being measured by power harmonic analyzer during the machining process. Machining parameters used in this study follows [12] as shown in Table 1.

Table 1. Machining parameters used in the case study

SET	Description
1	Cutting Speed: 42m/min; Feedrate: 0.1mm/rev; Depth of Cut: 0.50, 0.25mm;
2	Cutting Speed: 42m/min; Feedrate: 0.2 mm/rev; Depth of Cut: 0.50, 0.25mm
3	Cutting Speed: 83m/min; Feedrate: 0.1mm/rev; Depth of Cut: 0.5, 0.25mm
4	Cutting Speed: 83m/min; Feedrate: 0.2 mm/rev; Depth of Cut: 0.50, 0.25mm

Drilling process used center drill, diameter 10.0 mm, 13.0 and 14.5 mm drill tools with feed rate of 0.1 mm/rev and cutting speed of 30 m/min for both materials. Lastly for threading process, the cutting depth is 0.25mm and the cutting speed is 30 m/min. Both materials were sent to the laboratory to confirm the material grades. Figure 3 shows the comparison chart summary for theoretical and experimental results for all four criteria where A stands for Aluminum and B stands for Brass.

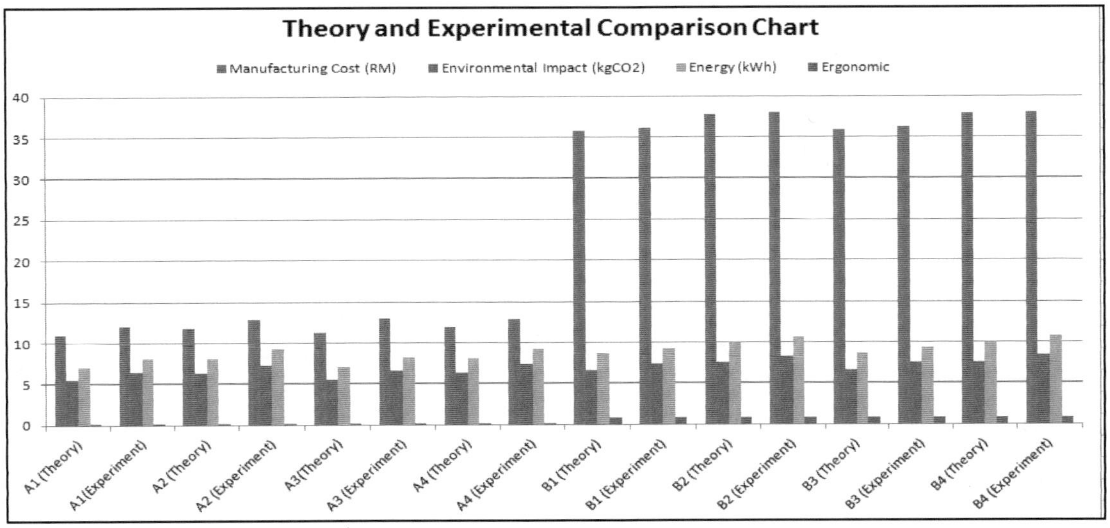

Figure 3. Comparison chart summary for theoretical and experimental results for all criteria

Manufacturing cost, environmental impact and energy increases when the cutting parameters such as cutting parameter for turning process increases such shown in figure 3. Manufacturing cost to produce brass pneumatic nipple hose is higher compared to Aluminum because of brass is more expensive. Increasing cutting speed and feed rate increases the energy consumption during the machining process. For the ergonomics criterion, the results are the same since for all solutions the location of raw material pallet is the same. When comparing both theoretical and experimental data, it shows that experimental result recorded a higher value compared to theoretical results. The difference is less than 10% for Brass material and less than 20% for the Aluminum material for energy criteria and in the acceptable range such as stated by [13]. Few factors that contribute to this results such as the chips stuck at the cutting tool during machining process, wire and spare part components being replace supplied by different manufacturers have been identified in this study. Lastly, the machining process is assumed to run smoothly but in reality it need to be stopped for trouble shooting such as chip stuck at the cutting tool during machining process. In calculating environmental impact and energy impact criteria, the raw material weight and finish product weight used is assume the same value theoretically but in reality they are different. Since the energy consume during the machining is different, the energy impact to the environment is also different because they are inter-related to each other. The same thing goes for the manufacturing cost. That is why it is higher compared to the theoretical results. There are two mathematical functions used in this study; one for Aluminum and the other one for Brass. Pareto front plot function is used in this study since it plots the first objective function against the second objective function. There are 14 variables involved in this study which can be manipulated to get the optimum results based on situation that contribute to the 4 criterion results. The 14 variables are seven types of tool life, feed rate, cutting speed, coolant and lubricant loss rate; average monthly output, raw material length and finished product weight. The initial population used in this study is 200 and the stopping rule used is the default setting. Optimization result shows that for each different case, the optimization result is different; such as shown in Table 2.

Table 2. Optimization results for each criteria.

Material	Manufacturing Cost	Environmental Impact	Energy Impact	Ergonomics
Aluminum	Raw material Length: 54.58mm Cutting Speed:63.08 mm/min Feedrate:0.1975 mm/rev	Raw material Length: 54.50 mm Cutting Speed: 42.06m/min Feedrate: 0.1001mm/rev	Raw material Length: 54.50mm Cutting Speed: 42.47 m/min Feedrate: 0.1000mm/rev	Raw material Length: 54.50mm Cutting Speed: 42.00 m/min Feedrate: 0.1000 mm/rev
Brass	Raw material Length: 54.70 mm Cutting Speed: 83.78m/min Feedrate: 0.1044 mm/rev	Raw material Length: 54.50mm Cutting Speed: 42 m/min Feedrate: 0.1000 mm/rev	Raw material Length: 54.50mm Cutting Speed: 42 m/min Feedrate: 0.1000 mm/rev	Raw material Length: 54.50mm Cutting Speed: 42.00 m/min Feedrate: 0.1000 mm/rev

Based on the results in Table 2, the results are different for each criteria. For manufacturing cost criteria different cutting speed, federate and raw material length recorded for both aluminum 6061 and Brass C3604. For environmental impact criteria, cutting speed and federate is different while raw material length is the same for both material. For energy criteria the results for both materials are almost the same for both material while for ergonomics criteria the optimization results is the same for both materials. From the optimization that has been done, optimization evaluation for manufacturing cost criteria is the most comprehensive evaluation because 11 out of 14 parameters were varied.

6. Conclusions

As a conclusion, triple bottom line concept can be implemented at the product design stage and the machining parameters can be optimized when evaluating its impact on the environment, energy, economics and social dimension. Multi-objective optimization by using genetic algorithm method can be used to optimize the machining parameters and material for the product based on the four dimensions optimization evaluation.

Acknowledgement

The authors like to acknowledge IMAMS Lab, Faculty of Manufacturing Engineering, Faculty of Mechanical Engineering; and Research and Innovation Department Universiti Malaysia Pahang, for their assistance in completing this study.

References

[1] Zahari Taha, H. A. Salaam and T.M.Y.S Tuan Ya. "Triple Bottom Line Optimization for Sustainable Product Design" *Proceedings of APIEMS 2015.*

[2] H. A. Salaam, Zahari Taha and T.M.Y.S Tuan Ya. "Triple Bottom Line Optimization for Sustainable Product Design" *Proceedings of APIEMS 2016.*

[3] Slaper, T. F., and Hall, T.J. 2011. Triple bottom Line: what is it and how does it works?. *Indiana Business Review Spring 2011, Volume 86*, No 1

[4] Konak Abdullah, David W. Coit and Alice E, Smith (2006). Multi-Objective Optimization Using Genetic Algorithm: A Tutorial, *Realibility Engineering and System Safety 91 (2006)* pp 992-1007

[5] Hasan, S., Chordia, S., Varshneya, R. 2012. *Lecture Notes: Genetic Algorithm*

[6] Heintz, J., Belaud J.P., Pandya, N., Santos, M. T. D. and Gerbaud, V., 2014. Computer aided product design tool for sustainable product development. *Computers and Chemical Engineering 71 (2014)* 362–376

[7] Hao, Z. (2012). Integrating Sustainable Manufacturing Assessment into Decision Making for a Production Work Cell. *Master Thessis: Oregon State University*

[8] Dahmus, J., & Gutowski, T. (2004). An Environmental Analysis of Machining. *ASME International Mechanical Engineering Congress and RD&D Expo*, Anaheim, CA, USA

[9] Narita, H. 2012. "*Environmental Burden Analyzer for Machine Tool Operations and Its Application. Manufacturing System.* ISBN 978-953-51-0530-5.

[10] Sandvik Coromant. Retrieve at 11 May 2017. Website address:http://www.sandvik.coromant.com/.../.../formulas-and-definitions.

[11] Erlinda Muslim, Aisyah Ladha Nurani, Maya Arlini Puspasari (2013). Analysis of Vertical Multiplier on Revised NIOSH Lifting Equation for Male Workers in Indonesia Industry. *Advanced Engineering Forum Vol.10 (2013)* 236-242

[12] Kalpakjian, S. and Schmid, S.R., *Manufacturing Engineering and Technology, Sixth Edition*, Pearson Prentice Hall, New Jersey, 2010.

[13] S. Pervaiz, I. Deiab, A. Rashid, M. Necolescu and H. Kishway (2013). Energy consumption and surface finish analysis of machining Ti6Al4V. *World Academy of Science, Engineering and Technology Vol 7.*

IOP Conf. Series: Materials Science and Engineering **319** (2018) 012072 doi:10.1088/1757-899X/319/1/012072

Capacitated set-covering model considering the distance objective and dependency of alternative facilities

I Wayan Suletra*, Yusuf Priyandari, Wakhid A Jauhari

Industrial Engineering, Faculty of Engineering, Universitas Sebelas Maret

*suletra@staff.uns.ac.id

Abstract. We propose a new model of facility location to solve a kind of problem that belong to a class of set-covering problem using an integer programming formulation. Our model contains a single objective function, but it represents two goals. The first is to minimize the number of facilities, and the other is to minimize the total distance of customers to facilities. The first goal is a mandatory goal, and the second is an improvement goal that is very useful when alternate optimum solutions for the first goal exist. We use a big number as a weight on the first goal to force the solution algorithm to give first priority to the first goal. Besides considering capacity constraints, our model accommodates a kind of either-or constraints representing facilities dependency. The either-or constraints will prevent the solution algorithm to select two or more facilities from the same set of facility with mutually exclusive properties. A real location selection problem to locate a set of wastewater treatment facility (IPAL) in Surakarta city, Indonesia, will describe the implementation of our model. A numerical example is given using the data of that real problem.

Keywords: set-covering, integer programming, either-or constraint, wastewater treatment facility, distance goal

1. Introduction

The location model basically is a model of the relationship between the point of demand and the location point of the service facility. The decision variable in the location model is generally to determine where the optimal locations for service facilities are built. The assumptions and objective functions of the location model vary according to the characteristics of the related problem. Daskin [1] classified the location models based on mathematical modelling (gravity model) into 4 large groups, that is, analytic, continuous, network, and discrete models.

The network models and the discrete models, both assume that the facility locations and demand points are both discrete, that is, they are only present at certain points within the area. The network model assumes a network or path connecting the demand points with facility location points while the discrete model does not require such assumptions.

One kind of the most widely known discrete location models is the set-covering model. The decision variable for the set-covering model is where the optimal locations for service facilities are built so that the objective function is achieved. The objective function of the set covering model is to minimize the cost of the facility location such that a specified level of coverage is obtained. Popular special case of set-covering models is to minimize the number of service facility location points but all demand points are served. In other words, where to build the facilities such that all customers are served with minimum cost is the common goal of the set covering model. Because of its discrete nature, most set-covering models are formulated using integer programming [2, 3].

The following is the integer linear programming model of a basic set-covering problem [4]:

$$\text{Minimize} \quad \sum_{j} c_j X_j \tag{1}$$

$$\text{Subject to:} \quad \sum_{j \in N_i} X_j \geq 1 \qquad \forall i, \tag{2}$$

$$X_j \in \{0,1\} \qquad \forall j. \tag{3}$$

Where,

c_j = construction cost of facility at point j

S = coverage radius of the facility

N_i = a set of facility location j that can serve demand point i,

$$N_i = \left\{ j \middle| d_{ij} \leq S \right\} \ \forall i, \tag{4}$$

d_{ij} = the distance from demand point i to facility location j

$$X_j = \begin{cases} 1, \textit{if point } j \textit{ is selected as a facility location} \\ 0, \textit{otherwise} \end{cases}$$

The above model, equation (1)-(4), explains that there is only one type of demand spread across multiple locations of point and also only one type of facility that can be placed at some points j to serve the demand. Equation (1) is the goal of the model, which is to minimize the facility location cost. Equation (2) assures that all demand points are served by at least one facility. Equation (3) is a logical binary constraint. Equation (4) explains the definition of coverage. However, the above standard model is not suitable to be applied in the case of determining the location of IPAL facilities. The model does not include the capacity and facilities dependency constraints that our model considers. In addition, this standard model does not consider the distance objective in its objective function.

A number of related works have been done to develop or modify the standard set-covering problem (SCP) model so that it's suitable for solving different cases. Rajagopalan et al. [5] developed an SCP model applied to emergency medical service (EMS) by considering demand fluctuations at certain time intervals. Suletra et al. [6] proposed a new constraint in the SCP model to represent the Indonesian government rule that each facility is used by at least two providers in the case of cellular telecommunication. The model is used to optimize the location of the joint base station. Karimi and Bashiri [7] presented a numerical example of a hub of airport location based on the hub covering location model developed in the study. Yaghini et al. [8] proposed a heuristic algorithm to solve an SCP model applied to train driver scheduling problems. Furthermore, Lutter et al. [9] and Zhang et al. [10] developed a new constraint in the SCP model to account an uncertain availability of facility applied to the case of emergency service facilities.

Although the research on SCP has been extensively discussed in the literature, there is no model suitable for solving the problem discussed in this study. The goal of the problem is to minimize the number of facilities and the total travel distance in the condition that the capacity of each facility is limited and there is the dependency property among facilities.

2. Proposed set covering model

The model proposed in this study is an integer programming model that refers to SCP. The basic model of SCP requires that each demand point is served by at least one facility (supply point). In this study, the requirement is rather different, each IKM (Industri Kecil dan Menengah) or SME's (Small and Medium Enterprises) represented by a demand point must be served by only one IPAL facility. The goal is to minimize the number of IPAL facility that can serve all demand points. Besides minimizing the number of IPAL facility, the objective function is formulated to minimize the total distance between IKM and IPAL facility. A big number is used as a weight on the number of facility minimization goal to force the solution algorithm to give first priority to this goal. Therefore, minimization of the total distance is a second priority goal that is very useful when alternate optimum solutions for the main goal exist.

$$\text{Minimize} \quad \sum_i cX_i + \sum_i \sum_j d_{ij} Y_{ij} \tag{5}$$

$$\text{Subject to:} \quad \sum_i Y_{ij} = 1 \quad \forall j, \tag{6}$$

$$Y_{ij} - X_i \leq 0 \ \forall i, j, \tag{7}$$

$$\sum_j l_j Y_{ij} \leq p_i \quad \forall i, \tag{8}$$

$$\sum_i X_i = 1 \quad \forall i \in M, \tag{9}$$

$$X_i \in \{0,1\} \quad \forall i, \tag{10}$$

$$Y_{ij} \in \{0,1\} \quad \forall i, j. \tag{11}$$

where,
c = a big number to prioritize the first goal
d_{ij} = the distance of IPAL *i* to IKM *j*
l_j = the volume of wastewater produced by IKM *j*
p_i = the capacity of IPAL *i*
M = a set of IPAL with mutually exclusive selection (dependency property)
X_i = biner number, 1 if alternative IPAL *i* is selected to build, and 0 is otherwise
Y_{ij} = biner number, 1 if alternative IPAL *i* to serve IKM *j*, and 0 is otherwise

The first term of the equation (5) is the first goal of the proposed model which minimizes the number of facilities, and the second term in the same equation is the second goal which minimize the total distance from IPAL *i* to IKM *j* in the condition that the first goal must be achieved first. A big number, *c*, is used to force the model to assign the priority to the first goal. The second goal is an improvement goal. It will work when the alternate optimum solutions for the first goal exist. Equation (6) is used to model the requirement that each IKM must be served by only one IPAL facility. Equation (7) is a logical constraint to ensure that if the IPAL *i* is allocated to serve the IKM *j* then the IPAL *i* must be selected to build. Equation (8) explains that the capacity of each IPAL cannot be violated, i.e. the total waste distributed to the IPAL *i* from all IKM is less than the capacity of IPAL *i*. Equation (9) describes the mutually exclusive selection representing the dependency property among IPAL. Equation (10) and (11) are logical binary constraints.

3. Numerical implementation

To describe the computation and the applicability of the proposed model, we use the data from a real problem, that is, location selection of IPAL facilities in Mojosongo, Surakarta, Indonesia (figure 1). The facilities serve the surrounding SME's tahu tempe. An IKM is a demand point and the IPAL is the facility that serves the demand points. Having observed in site, there are 41 demand points (j=1,2,…,41) and 7 alternative IPAL locations (i=1,2,…,7). The map of IKM points and IPAL points are depicted in figure 1.

Figure 1. Dispersion of 41 IKM locations and 7 IPAL alternatives.

The first goal is to minimize the number of IPAL facilities and the second is to minimize the total distance of IKM to IPAL facilities. The constraints are capacity of IPAL, each IKM must be served only by one IPAL facility (and all IKM must be served), and facilities dependency constraints. Appendix A shows the distance measure from IKM to IPAL alternatives and appendix B shows the volume of wastewater produced by each IKM.

```
model:
sets:
     ipal/S1..S7/:kapasitas,dipilihtidak;
     ikm/C1..C41/:volumelimbah;
     ipalikmcombination(ipal,ikm):jaraktempuh,alokasi;
endsets
min=@sum(ipal(I):10000*dipilihtidak(I))+@sum(ipalikmcombination(I,J):
jaraktempuh(I,J)*alokasi(I,J));
@for(ikm(J):
      @sum(ipal(I):alokasi(I,J))=1);
@for(ipal(I):
      @sum(ikm(J):alokasi(I,J)*volumelimbah(J))<kapasitas(I));
@for(ipalikmcombination(I,J):
      alokasi(I,J)-dipilihtidak(I)<=0);
      dipilihtidak(1)+dipilihtidak(2)<=1;
      dipilihtidak(6)+dipilihtidak(7)<=1;
@for(ipalikmcombination(I,J):@bin(alokasi(I,J)));
@for(ipal(I):@bin(dipilihtidak(I)));
data:
kapasitas,volumelimbah,jaraktempuh=@OLE('E:\data_lingo\data_ikm_tahu
tempe.xls','kapasitas','volumelimbah','jaraktempuh');
enddata
end
```

Figure 2. Lingo code implemented in the problem

Branch-and-Bound algorithm is used to implement the proposed model to the real problem of IPAL at Mojosongo, Surakarta. Lingo software on a notebook i5 2.6GHz with 8Gb Ram and SSD 256Gb run the model in 5 second to get the global optimum solution. The code listed in figure 2 is the lingo code implemented to solve this problem.

The link between the proposed model and the lingo code can be explained as follows. Y_{ij} is represented by the code 'alokasi', X_i by the code '`dipilihtidak`', c by the number 10000 (big number), d_{ij} by the code '`jaraktempuh`', p_i by the code '`kapasitas`', l_j by the code '`volumelimbah`'. The objective function of the proposed model, equation (5), is represented by the row [7] and [8] of the lingo code list. The first constraint of the proposed model, equation (6), is represented by the row [9] and [10], the second constraint or equation (7) by the row [13] and [14], the third constraint or equation (8) by the row [11] and [12], the fourth constraint or equation (9) by the row [15] and [16], the fifth constraint or equation (10) by the row [18], and the sixth constraint or equation (11) by the row [17]. The row [20] and [21] show that the data of p_i, l_j, and d_{ij}, which are imported from MS Excel file using the Lingo function @OLE.

The optimum solution is depicted by figure 3 and the size of the problem is shown by the figure 3. Branch-and-Bound algorithm needs 169,273 iterations to get the global optimal solution in 5 seconds. Four IPAL facilities are selected to serve 41 IKM, i.e. IPAL 2, IPAL 3, IPAL 5, and IPAL 7. The optimum allocation is as follows. IPAL 2 serves 11 IKM, i.e. IKM 1, IKM 2, IKM 3, IKM 4, IKM 5, IKM 6, IKM 8, IKM 9, IKM 10, IKM 11, and IKM 12. IPAL 3 serves 11 IKM, i.e. IKM 7, IKM 14, IKM 17, IKM 20, IKM 22, IKM 23, IKM 24, IKM 25, IKM 28, IKM 34, and IKM 35. IPAL 5 serves 13 IKM, i.e. IKM 13, IKM 16, IKM 18, IKM 19, IKM 21, IKM 26, IKM 27, IKM 29, IKM 30, IKM 31, IKM 32, IKM 33, and IKM 36. IPAL 7 serves 6 IKM, i.e. IKM 15, IKM 37, IKM 38, IKM 39, IKM 40, and IKM 41.

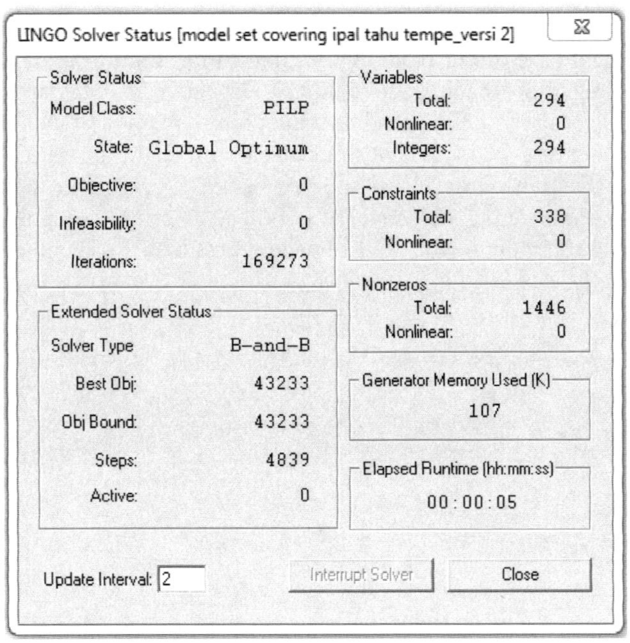

Figure 3. Summary of problem and Lingo output.

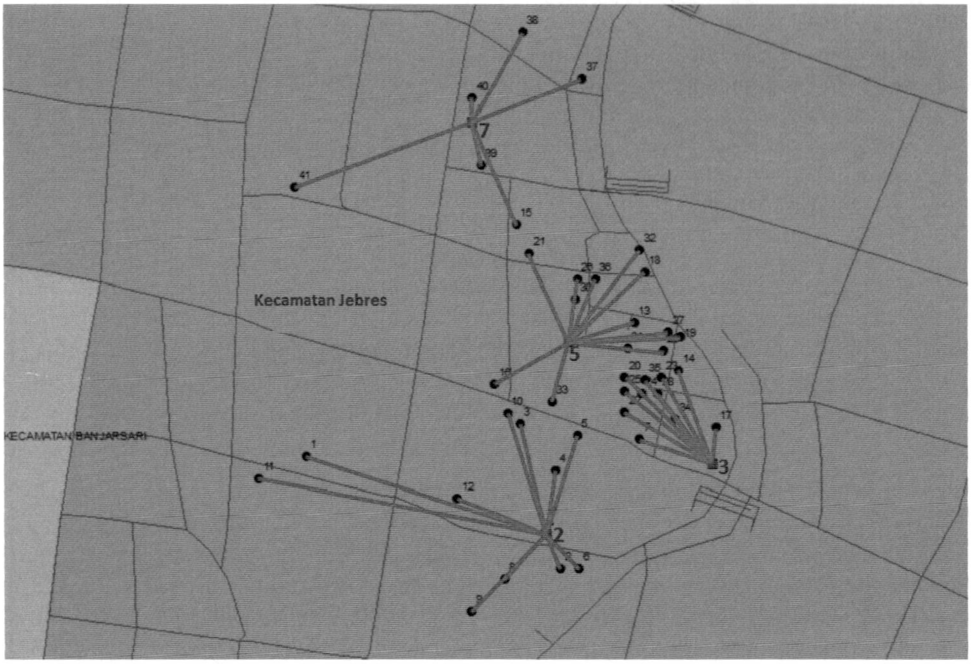

Figure 4. Optimum allocation of 41 IKMs to 4 selected IPAL alternatives.

4. Concluding remark

We propose a kind of set-covering model to minimize two different goals using the single objective function. The problem is formulated as a single objective linear integer programming. The standard set covering model is a special case of our model by setting it to a single goal of minimizing the number of facilities and removing the capacity and facilities dependency constraints. We impose several assumptions about the alternative facilities and demand points. We use Branch-and-Bound Algorithms to solve the integer model and it needs just 5 second to solve the numerical example. The future research can look into accommodating a multiple objective approach to represent a more realistic problem, i.e. a problem considering objective and subjective criteria simultaneously. Further, the relaxation of assumptions about the facility and demand points may also be interesting to be studied.

IOP Conf. Series: Materials Science and Engineering **319** (2018) 012072 doi:10.1088/1757-899X/319/1/012072

Appendix A. Distance between IKM and IPAL alternatives (in meters).

	IPAL1	IPAL2	IPAL3	IPAL4	IPAL5	IPAL6	IPAL7
IKM1	281	233	651	820	652	950	956
IKM2	79	31	989	455	659	555	899
IKM3	233	156	570	453	705	553	763
IKM4	218	141	697	922	766	684	591
IKM5	197	120	681	439	490	786	864
IKM6	65	14	699	605	784	785	697
IKM7	553	601	50	479	667	610	607
IKM8	141	93	707	879	830	680	978
IKM9	5	79	434	537	502	514	415
IKM10	241	164	418	906	655	691	682
IKM11	308	260	435	611	430	644	463
IKM12	176	128	763	704	817	795	579
IKM13	610	926	491	63	30	555	446
IKM14	960	966	61	611	491	656	879
IKM15	647	418	851	897	849	15	70
IKM16	514	682	572	130	92	717	448
IKM17	748	887	37	409	499	958	665
IKM18	665	559	817	75	57	992	659
IKM19	536	490	721	45	75	599	483
IKM20	654	758	68	623	942	646	804
IKM21	489	914	754	60	76	630	762
IKM22	577	894	105	516	792	522	867
IKM23	964	411	52	747	700	454	766
IKM24	666	493	54	890	512	827	502
IKM25	890	428	61	957	741	641	909
IKM26	647	891	898	50	56	828	705
IKM27	660	923	506	40	70	979	629
IKM28	896	420	45	671	420	533	852
IKM29	830	626	449	47	77	895	578
IKM30	430	861	872	28	35	614	822
IKM31	634	411	669	34	10	620	470
IKM32	810	760	536	103	85	975	604
IKM33	893	611	414	55	15	657	818
IKM34	494	969	36	621	543	944	880
IKM35	419	427	61	731	495	545	597
IKM36	622	751	624	51	33	483	572
IKM37	913	609	429	594	784	305	120
IKM38	982	552	670	811	615	250	65
IKM39	965	420	651	533	412	80	20
IKM40	936	904	845	529	742	203	10
IKM41	575	478	994	876	756	219	188

IOP Conf. Series: Materials Science and Engineering **319** (2018) 012072 doi:10.1088/1757-899X/319/1/012072

Appendix B. Wastewater produced by IKM

	Processed soy bean (Kg/day)	Daily Waste (litres)	Waste for 3 days dwell time (litres)
IKM1	100	950	2,850
IKM2	200	1,900	5,700
IKM3	250	2,375	7,125
IKM4	40	380	1,140
IKM5	300	2,850	8,550
IKM6	100	950	2,850
IKM7	200	1,900	5,700
IKM8	25	238	713
IKM9	100	950	2,850
IKM10	100	950	2,850
IKM11	50	475	1,425
IKM12	60	570	1,710
IKM13	150	1,425	4,275
IKM14	150	1,425	4,275
IKM15	90	855	2,565
IKM16	70	665	1,995
IKM17	50	475	1,425
IKM18	50	475	1,425
IKM19	150	1,425	4,275
IKM20	50	475	1,425
IKM21	125	1,188	3,563
IKM22	40	380	1,140
IKM23	125	1188	3,563
IKM24	50	475	1,425
IKM25	200	1,900	5,700
IKM26	100	950	2,850
IKM27	400	3,800	11,400
IKM28	100	950	2,850
IKM29	150	1,425	4,275
IKM30	200	1,900	5,700
IKM31	500	4,750	14,250
IKM32	100	950	2,850
IKM33	50	475	1,425
IKM34	100	950	2,850
IKM35	500	4,750	14,250
IKM36	100	950	2,850
IKM37	250	2,375	7,125
IKM38	100	950	2,850
IKM39	80	760	2,280
IKM40	50	475	1,425
IKM41	50	475	1,425

References

[1] Daskin M S 2008 What You Should Know About Location Modeling, *Naval Research Logistics* **55** 283-294.

[2] Schilling D A, Jayaraman V and Barkhi R 1993 A review of covering problems in facility location *Location Science* **1** 25–55.

[3] Current J, Daskin M and Schilling D 2002 Discrete network location models *Facility Location Theory: Applications and Methods* ed Drezner and Hamacher H (Berlin: Springer-Verlag) Chapter 3, pp. 81–118.

[4] Owen S H and Daskin M S 1998 Strategic facility location: a review *European Journal of Operational Research* **111** 423-447.

[5] Rajagopalan H K, Saydam C and Xiao J 2011 A multiperiod set covering location model for dynamic redeployment of ambulances *Computers & Operations Research* **35** 814–826.

[6] Suletra I W, Widodo and Subanar 2011 Optimum locations of multi-providers joint base station by using set-covering integer programming: modelling & simulation *The 6th SEAMS-UGM Conference 2011*.

[7] Karimi H and Bashiri M 2011 Hub covering location problems with different coverage types *Scientia Iranica, Transactions E: Industrial Engineering* **18** 1571–1578.

[8] Yaghini M, Karimi M and Rahbar M 2015 A set covering approach for multi-depot train driver scheduling *Journal of Combinatorial Optimization* **29(3)** 636–654.

[9] Lutter P, Degel D, Busing C, Koster A M C A and Werners B 2017 Improved handling of uncertainty and robustness in set covering problems *European Journal of Operational Research* **263(1)** 35-49.

[10] Zhang B, Peng J and Li S 2017 Covering location problem of emergency service facilities in an uncertain environment *Applied Mathematical Modelling* **51** 429–447.

IOP Conf. Series: Materials Science and Engineering **319** (2018) 012073 doi:10.1088/1757-899X/319/1/012073

Development of an Assessment Model for Sustainable Supply Chain Management in Batik Industry

G F Mubiena and A Ma'ruf

Institut Teknologi Bandung, Bandung, Indonesia

E-mail: ghaidafatcham@gmail.com, maruf@itb.ac.id

Abstract. This research proposes a dynamic assessment model for sustainable supply chain management in batik industry. The proposed model identifies the dynamic relationship between economic aspect, environment aspect and social aspect. The economic aspect refers to the supply chain operation reference model. The environment aspect uses carbon emissions and liquid waste as the attribute assessment, while the social aspect focus on employee's welfare. Lean manufacturing concept was implemented as an alternative approach to sustainability. The simulation result shows that the average of sustainability score for 5 years increased from 65,3% to 70%. Future experiments will be conducted on design improvements to reach the company target on sustainability score.
Keywords: Sustainable Supply Chain Management, System Dynamics, Lean Manufacturing

1. Introduction

Batik demand in Indonesia increased significantly since UNESCO inaugurated batik Indonesia as the masterpiece of intangible heritage in 2009 [1]. More than 200% of sales volume had achieved in 2011 but for export value decreased 14,71% in 2015 [2]. On the other hand, fierce competition has increased among local and international batik industry such as from China and Malaysia [1]. Sustainability is a necessity for the batik industry [3]. The association between sustainable for internal and external organization is recognized as the concept of Sustainable Supply Chain Management (SSCM) [4][5]. There are three aspects of SSCM, namely economic aspect, environment aspect and social aspect. Several assessment models of SSCM has been proposed. An assessment of SSCM using SCOR 9.0 and Global Supply Chain Forum framework had been conducted [3]. An assessment of SSCM has been conducted using collaboration of SCOR and Base of the Pyramid (BOP) strategies [6]. For Batik Industry in Solo an assessment of SSCM also has been conducted using SCOR method [7]. Among these previous research, the assessment model of SSCM assume a linier correlation between the three aspects which do not consider the dynamic aspect of the system. The objective of this research is to propose a dynamic assessment model for SSCM in Batik Industry. Initial research based on hybrid SCOR 11.0 and dynamic evaluation on economic aspect has been conducted [8]. This paper will discuss not only the economic aspect but also the environment aspect and social aspect. The following chapter will review the reference model. Chapter three describes the proposed model and followed by the result and the discussion of sustainability score. This paper concluded with finding remarks based on the dynamic simulation from 2017 to 2021.

2. Reference Model

2.1. Supply Chain Operations Reference (SCOR)
The SCOR model is a process reference model with standardised terminology developed and endorsed by the Supply Chain Council [9]. The SCOR model has five attributes such as reliability, responsiveness, cost, assets management and agility. The SCOR model version 11 is used throughout this paper. The SCOR model defines two types of attributes performance. The external attributes are divided into reliability, responsiveness, and flexibility, while the internal attributes are cost and assets.

2.2. CO2 Emission and Liquid Waste Management
In case of batik industry, the assessment of environment aspect is conducted through measuring carbon emission and liquid waste. The attribute for measuring liquid waste such as chloride, sulfate, nitrite, nitrate and alkalinity were not filtered and kept in 50 ml PE bottles. These attributes were analyzed according to standard methods APHA, 1998 [10]. The regulation of the local government of Central Java Province also arranges about the standard attributes of liquid waste by using Chemical Oxygen Demand (COD), Biological Oxygen Demand (BOD), Total Suspend Solid (TSS), power of Hydrogen (pH), and temperature that can be delivered to the surrounding environment [11]. The assessment of liquid waste in Batik Industry was conducted by Immawan [1] and then for carbon emission was conducted by Saleh [12]. Carbon emissions is measured from consumption of electricity, kerosene and solar refer to IPCC.

2.3. Employee's Welfare Management
Social aspect is one of the main aspects of SSCM, as well with the economic aspect and environment aspect. Therefore, sustainability has become an important issue in both managers and researches, since managers will increasingly play a major role in writing and instituting CSR policies and code of conduct [13]. Stakeholder perspective suggests that firms need to meet the needs of the stakeholders in addition to the shareholders [14]. According to this perspective, stakeholders are critical for the existence of the firm. The basic idea of the stakeholder perspective is that a firm's success depends on how it is able to manage relationships with key groups, such as customers, employees, suppliers, communities, politicians, and owners, each of which can affect its ability to reach its goals [15]. The previous research has been elaborated a research about an attempt to identify welfare facilities and employee's satisfaction level about welfare facilities [16]. Employee welfare facilities in the organization effect on the behaviour of the employees as well as on the productivity of the organization.

2.4. Lean Manufacturing
Lean Manufacturing is a popular paradigm in the US to identify value and reduce waste begun with the Toyota Production System (TPS) in Japan [17]. To remove waste from plan activity to return of finish good from the customers, Value Stream Mapping (VSM) is used to map the activity at the manufacture and the result is ratio of value added and non-value added time [18].

3. Model Development
Previous research was conducted by developing a model to just identify the relationship among the five attributes assessment on economic criteria namely reliability, responsiveness, cost, assest management and agility [8]. This research propose an additional two assessment attributes for environment, namely carbon emissions and liquid waste and one attribute assessment for social aspect namely employee's welfare. The development of conceptualization model in form of causal loop diagram is illustrated in Figure 1.

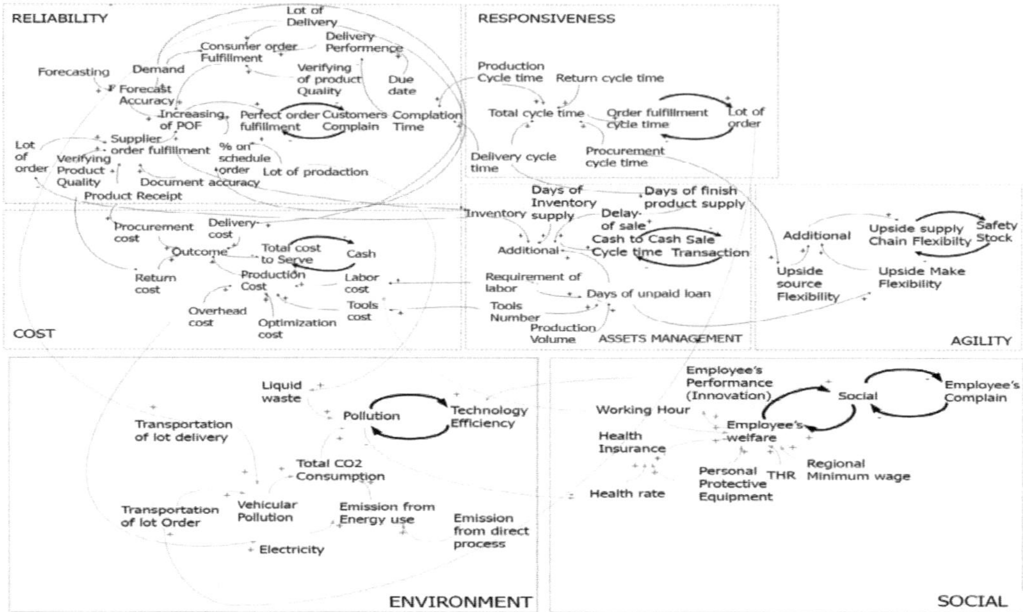

Figure 1. Causal Loop Diagram.

After causal loop diagram created, the relation of variables in the system will be converted into the flow diagram as illustrated in Figure 2. The propose model is validated using U-Theil statistics by comparing the simulation result to the actual data. The U-Theil test is 0.18, which indicates a good forecasting accuracy [19]. In this case, the model is valid.

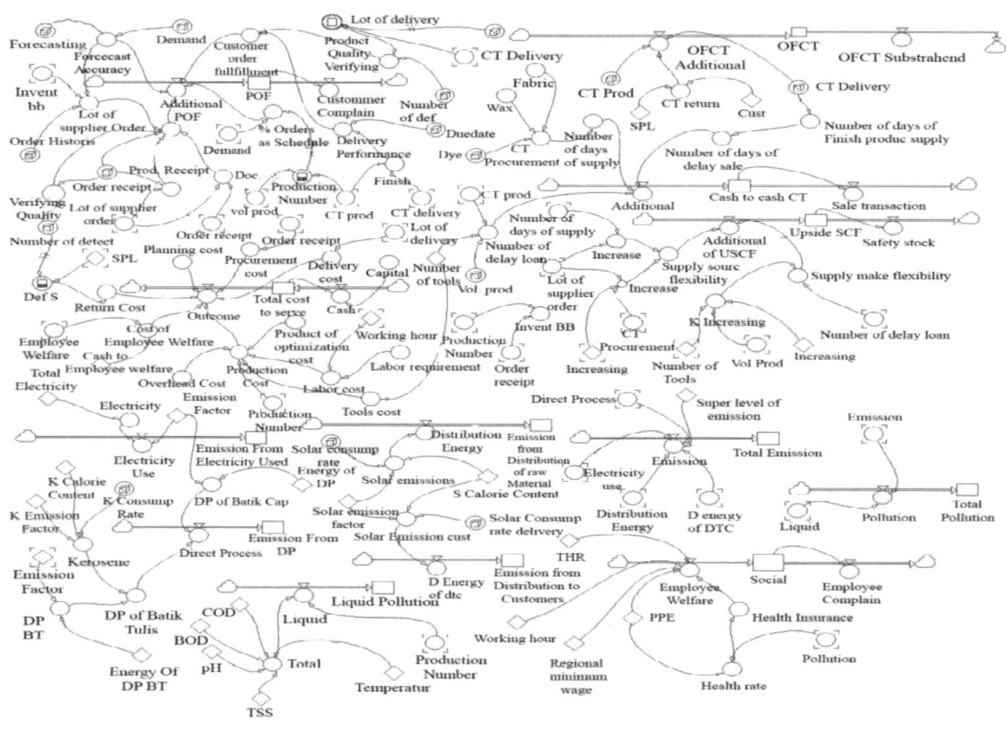

Figure 2. Flow Diagram.

4. Result and Discussion

Assessment of SSCM has been conducted using the proposed model. Assessment of the five attributes SCOR for economic aspect, two attributes (liquid and carbon emissions) for environment aspect and employee's welfare for social aspect for 5 years started from 2017-2021 shows in Table 1.

Table 1. Simulation Result during 5 Years (2017-2021)

No		Years					Average
		1	2	3	4	5	
1	Economic Aspect	52,1%	56,7%	55,8%	62,7%	58,2%	57,1%
2	Environment Aspect	2,0%	2,0%	2,0%	2,0%	2,0%	2,0%
3	Social Aspect	5,9%	6,0%	6,1%	6,1%	6,0%	6,0%
	SUSTAINABILITY SCORE	60,1%	64,8%	64,0%	70,9%	66,3%	65,3%

The average sustainability score is 65,3% which is below the company target. Thus, improvement is required in improving the sustainability score. In this research improvement is conducted through lean manufacturing. Waste identification for lean manufacturing is conducted in the chain of planning, source, manufacture, delivery until return activity. Value Stream Mapping (VSM) is used to map the operating conditions at the company. The mapping process was carried out by interview and observation. Batik Industry has two types product, "batik tulis" and "batik cap". Figure 3 and Figure 4 illustrates VSM for "batik tulis" and "batik cap" in the state before and after implementing lean manufacturing concept.

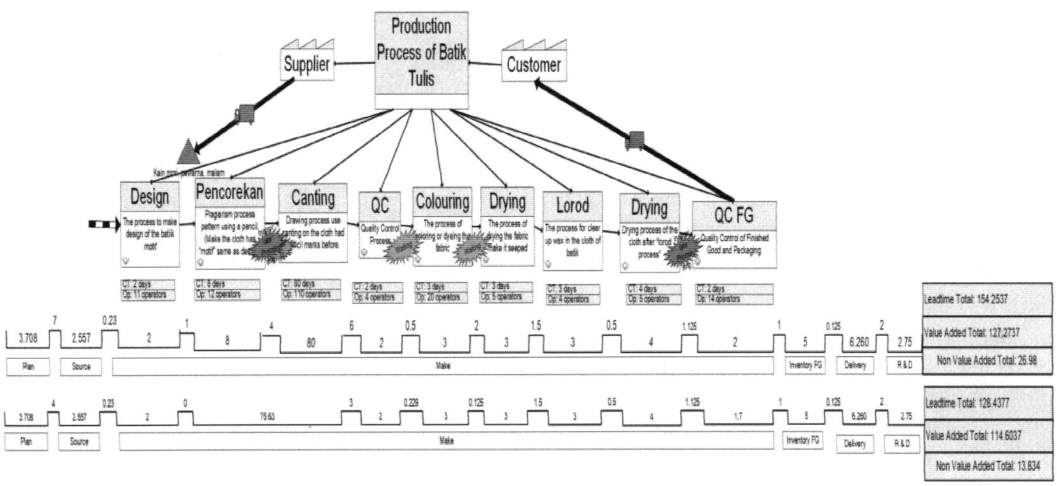

Figure 3. VSM for "Batik Tulis".

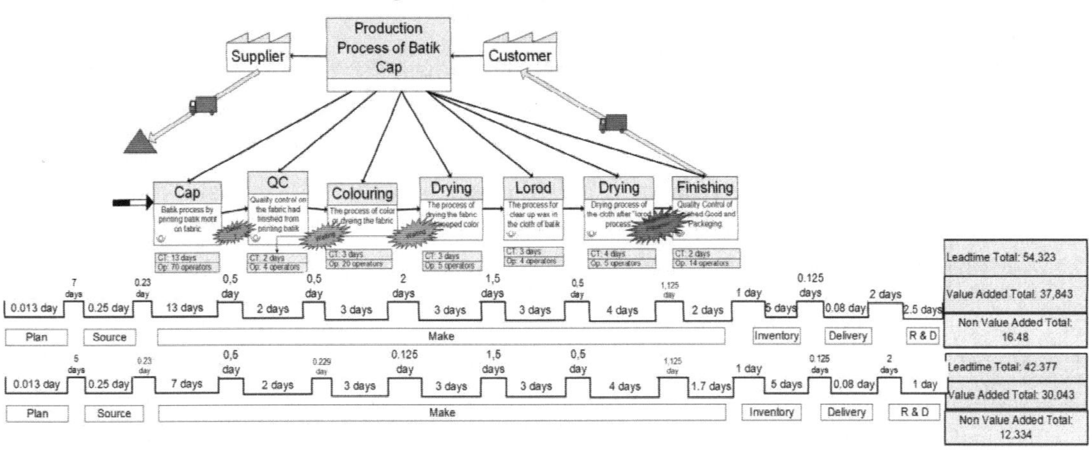

Figure 4. VSM for "Batik Cap".

The VSM in Figure 3 and Figure 4 describe all activities in manufacturing system from planning activity to return activity in Batik Industry. The wastes in Figure 3 occurred in production activity such as waiting, inappropriate processing and defect. Waiting occurred in QC process and drying process. Inappropriate processing occurred in colouring process by the operator with imbalance skill. Defect occurred in canting process. The desired value of the company is a smooth flow of production. To reach the value, the company must eliminate the waste. The proposed elimination of waste that can be done with several actions. For example, using information technology to select the colour in colouring process which is previously used manually by visual. Previously, in colour selection requires 4 hours, if using information technology to select the colour where one colour as one minute for one product, then for 1 batch as 110 products, require only 110 minutes from the previous one needs 4 hours to select the colours. This proposed action to eliminate waste can reduce the production cycle time. The cycle time of "Batik Tulis" for initial production activity is 123,6 days and reduced to 100,8 days after lean manufacturing implementation by eliminating the waste. Figure 4 illustrates cycle time of "Batik Cap", the initial production activity is 36,1 days and reduced to 27,6 days after lean manufacturing implementation. Waste removal occurred in the return activity for "batik cap" from initial 2.5 days reduced to 1 day after lean manufacturing. Figure 3 and Figure 4 also illustrates the different cycle time between "Batik Tulis" and "Batik Cap".

Based on the lean manufacturing activity mentioned above, the simulation model was run to simulate the effect of lean manufacturing. The result reveals an increase average sustainability score from 65,3% to 70,0% as shown in Table 2.

Table 2. Simulation lean manufacturing during 5 years (2017-2021)

		Years					**Average**
No		**1**	**2**	**3**	**4**	**5**	
1	Economic Aspect	57,1%	53,2%	73,3%	67,1%	58,7%	61,9%
2	Environment Aspect	2,06%	2,06%	2,0%	2,0%	2,0%	2,0%
3	Social Aspect	6,0%	6,1%	5,9%	5,9%	6,0%	6,0%
	SUSTAINABILITY SCORE	65,2%	61,4%	81,3%	75,1%	66,8%	70,0%

5. Conclusion

This paper proposed a dynamic assessment model for sustainable supply chain management in Batik Industry. The model covers three aspects: economic, environment and social. Initial assessment reveals that the sustainability score of batik industry is below expectation. Lean manufacturing concept was stimulated in the batik industry as an approach to achieve sustainability. The assessment reveals an increase of the sustainability score. Future experiments will be conducted to design improvements for the company to reach their target of sustainability score.

Acknowledgement

This research was conducted at the Manufacturing System Research Group, funded by the P3MI research grant of ITB

6. References

[1] Immawan, T., Marimin., Arkeman, Y., Maulana, A., 2015 Sustainable Supply Chain Management for Make To Stock-Make To Order To Order Production Typology Case Study: Batik Industry in Solo Indonesia. *Europeaan Journal of Business and Management* Vol. 7, No. 11, ISSN 2222-1905 (Paper) ISSN 2222-2839 (Online)

[2] Disperindag Solo 2012 Data jumlah pengusaha batik

[3] Badurdeen, F., Lyengar, D., Goldsby, T J., Metta, H., Gupta, S., Jahawir, I S. 2009 Extending total life-cycle thinking to sustainable supply chain design. *Int. J. Product Lifecycle Management,* Vol. 4, Nos. 1/2/3

[4] Gimenez, C. and Tachizawa, E.M. 2012, "Extending sustainability to suppliers: a systematic literature review", Supply Chain Management: An International Journal, Vol. 17 No. 5, pp. 531-543.

[5] Carter, C.R. and Rogers, D.S. 2008, "A framework of sustainable supply chain management: moving toward new theory", International Journal of Physical Distribution and Logistics Management, Vol. 38 No. 5, pp. 360-387

[6] Bendul, J C., Rosca, & E., Pivovarova, D. 2016 Sustainable Supply Chain Models For of The Pyramid. *Journal of Cleaner Production.* 1-14

[7] Saleh, C., Agitya, R., Badri, H., & Deros, B., (2015 Operation Overlapping Approach In Mts Production Typology To Assist The Accomplishment Of Sustainable Supply Chain

[8] Saleh C., Mubiena G F., Immawan T., & Hassan A., 2015. Lean Production in Improving Supply Chain Performance Through Hybrid Model SCOR 11.0-System Dynamics. *IOP Conf. Series: Materials Science and Engineering* 114 012069

[9] SCC 2012. Supply chain operations reference model, Cypress, Texas, USA: Supply Chain Council.

[10] APHA, 1998. Standard Methods for Examination of and Wastewater, 19th ed. American Public Health Association, Washington, DC

[11] Mangimbulude, J. C., Breukelen, B. M. Van, Krave, A. S., & Straalen, N. M. Van. (n.d.). Author ' s personal copy Seasonal dynamics in leachate hydrochemistry and natural attenuation in surface run-off water from a tropical landfill. doi:10.1016/j.wasman.2008.06.020

[12] Sukirman · Saleh. C · Dzakiyullah N.R. · Mubiena G.F.· Ismayanda J.A.,2016 "Prediction of Carbon Emissions Footprint in Batik Industry toward Green Manufacturing,"

[13] Hsueh, C. 2014. Improving corporate social responsibility in a supply chain through a new revenue sharing contract. Int. J. Production Economics. 151, 214-222

[14] Freeman, R.E., 1984. Strategic management: a stakeholder approach. Pitman, Boston

[15] Ihlen, O., 2008. Mapping the environment for corporate social responsibility: stakeholdrs, publics and the public sphere. Corp. Commun. An Int. J. 13, 35-146.

[16] George, Z., Kim, I., Springer, M., George, G., & Yu, Y. (2013). Int. J. Production Economics Dynamic pooling of make to stock and make to order operations. *Intern. Journal of Production Economics,* 144(1), 44-56. doi:10.1016/j.ijpe.2013.01.012

[17] K. D. Zylstra, "Lean distribution," Appl. lean Manuf. to Distrib. Logist. supply Chain. Hoboken, NJ, 2006.

[18] D. Gracanin, B. Buchmeister, and B. Lalic, "Using cost-time profile for value stream optimization," Procedia Eng., vol. 69, pp. 1225–1231, 2014.

[19] Department of Treasury, 2008 "Forecasting Accuracy of the Act Budget Estimates", May,

IOP Conf. Series: Materials Science and Engineering **319** (2018) 012074 doi:10.1088/1757-899X/319/1/012074

A Modified Branch and Bound Algorithm for Batch Scheduling in Discrete Manufacture System

A S Indrapriyatna and H Triha

Industrial Engineering Department, Faculty of Engineering, Universitas Andalas, Padang, West Sumatera, Indonesia

ahmadsyaf@ft.unand.ac.id, hadi.06173058@gmail.com

Abstract. Indrapriyatna et al. [1,2] have developed a batch scheduling model. In a model with a discrete batch size, three methods are used to convert batch size to integer. Based on numerical calculations it is found that there is no method that always result in the smallest total cost in every data set. This research is done to try other methods to convert batch size to integer. The approach done is by using the Branch and Bound Algorithm Modification that is adjusted to get scheduling for each batch without changing the number of batch (N) and the number of quantities (q) that are going to be produced. The result is the CSA_Dis_BB Algorithm that gets a discrete (integer) batch size. The implementationed of CSA_Dis_BB Algorithm is done on 7 sets of data from [1,2]. This algorithm does work to produce the smallest total cost.

1. Introduction

Scheduling is essential to produce the order with exact allocations of resources, like the machines used, number of working operators, sequencing the work on parts, and the materials needed. [7] defines scheduling as the process of allocating resources over certain time frame in order to finish certain tasks. Problems of scheduling comes up whenever the work need to be done exceeds the resources.

A good scheduling maximizes the effectivity and usability of the available resources, hence scheduling is a critical activity in each planning, in production, and in production supervision. An effective and efficient scheduling reduces waiting time and idle time, which therefore cuts producton costs.

[3,4,5,6] discuss the problem of sequencing and deciding on the batch size which means the job size. Thus processing time is a variable of decision. The problem of this is how to decide on the batch size (batching) and the sequencing of the processes on those batches; this is called the batch sequencing. [5] have proposed a batch scheduling model that take into account the storage costs and quality costs in a single machine production system.

[1,2] developed a batch scheduling model that considers the storage costs and quality costs and take into account both storage costs of work-in-process batch and finished-batch. This model (1) differentiates storage costs of work-in-process part and finished part in the in-process batch, (2) varies the sample sizes in proportion to the batch size, and (3) has the batch and sample size in integer.

The research that was done resulted in a batch scheduling model that considered storage costs and quality costs and also calculated storage costs of work-in-process batch and finished-batch, had sample size that was proportional to the batch size, for a discrete manufacture system, and applied to a flowshop system made up of one machine. But in this model with the discrete batch size there are 3 methods used to change the batch size into integer, and those methods are: (1) Top Decimal Sum (TDS), (2) Bottom Decimal Sum (BDS), and (3) Rounding-Off Method.

Based on numerical analysis, it can be concluded that from the above three conversion methods into integer, there is not a single method that always results in a best solution. In other words, the method of conversion into integer that is being used does not always give dominant results for each set of data being tested. For that reason, a different conversion method is needed to convert batch size into integer. The approach taken here is by using Branch and Bound Algorithm (because it has been proven to solve integer problems well).

Land and Doig's (1960) mentioned that branch and bound algorithm was developed in the context of finding a solution for linear programs with integer decision variable. The components of this algorithm are [8]: *Branching*, *Bounding*, *Pruning*, and *Retracting*.

In practice, almost all integer problems (IP) is solved by the branch and bound technic. This technic finds optimal solution of IP by enumerating points within feasible area of the subproblem [9]. Each living node is associated with a cost that sets the bound.

2. Research Methodology

Research Methodology is an essential part of a research. This will describe the steps of the research.

Preliminary Study

The objective of preliminary study is to compile the theories that will be the base of finding good solutions.

Problem Identification and Formulation

This is to explain what need to be solved, and formulating it, and explaining and indentifying the problems within certain limits.

Data Collection

The data that is used in this research is theoretical data from the previous researcher [1,2].

Algorithm Design

The process is by designing Branch and Bound Algorithm to get integer Q[i] and algorithm logic test. Next total cost and schedule of each batch is set based on the existing formulas.

Software Design and Analysis

Based on the algorithm, a software is then designed, and the algorithm is analysed.

Closing

Design and results is then concluded and given suggestions for improvement.

3. Research Results

Data Collection

The data used in this research is the 7 sets of data taken from Indrapriyatna et al. [1,2]. Here the units of time and costs are not specified, because any unit can be used as long as it is appropriate, for example minutes for time unit, and monetary currency for the cost. This is done to show that this model is applicable in general.

Branch and Bound Algorithm Modification Design

This algorithm is called a modification because it does not completely follow the BB algorithm. Optimal tests are only made on combinations that give the smallest total cost to the previous batch size. While the original BB algorithm should try also compared with other possible combinations. CSA Algorithm is the basic reference in implementing Branch and Bound Algorithm in the CSA model. The initial phase of designing is by taking the value Q[i] and the optimal number of batches (N), results of CSA Algorithm where i = 1,2,3,...,N. In applying the Branch and Bound Algorithm, some additional variables are needed:

a	= Number of iterations done, where a = 1,2,3,…,N-1.
Qup[a]	= Value of Q[a] rounded up.
Qdown[a]	= Value of Q[a] rounded down.
TC*up*	= TC[N,Q] when Q[a] is rounded up.
TC*down*	= TC[N,Q] when Q[a] is rounded down.
q_awal	= Number of queries (q) in CSA Algorithm.
Sisa_*up*	= result q from q_awal – $\sum_{i=1}^{a} \mathbf{Q[i]}$, when Q[a] is rounded up.
Sisa_*down*	= result q from q_awal – $\sum_{i=1}^{a} \mathbf{Q[i]}$, when Q[a] is rounded down.
Qbaru[N, a]	= Values of Q[i] after converted to integer (Q[i]').
TC_BB[N]	= Total costs of *Branch and Bound.*

The Figure 1 is the Branch and Bound Algorithm to convert Q[i] to its integer form (Q[i]'), called the CSA_Dis_BB Algorithm.

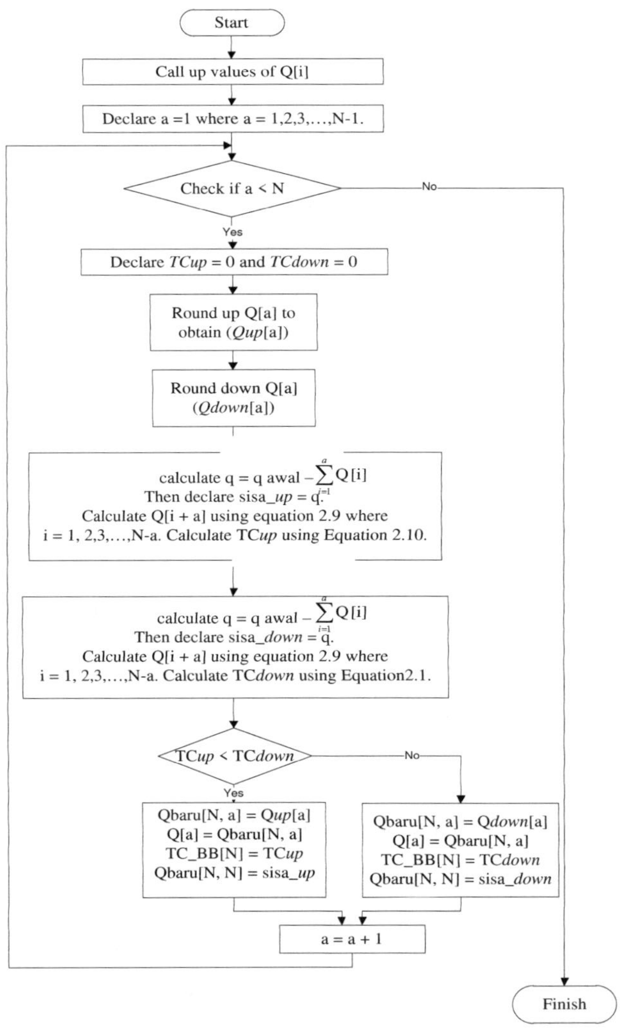

Figure 1. Flowchart of CSA_Dis_BB Algorithm

Testing the CSA_Dis_BB Algorithm
An algorithm is considered correct if it produces the correct outputs for all possibilities of input. Therefore, used to test the algorithm is a set of continuous data with three continuous-sized batches. The test intends to check the logic coherency of the algoritm.

Software Design
After the algorithm is created, the application or software is designed. The programming language used is C# (C Sharp) and the editor used is Visual Studio 2005 and 2008. After obtaining the integer value of batch sizes (Q'[i]), Total Cost (TC[N,Q']) is calculated based on the CSA_Dis Model for Q'[i]. After the integer value of sample size is obtained, Total Cost TC([N.Q']) can be calculated. The following is the total cost of CSA_Dis Method using the integer sample size from CSA_Dis_BB, Rounding Up, TDS, BDS, and rounding-off methods. The recapitulation of total cost is shown in Table 1.

Table 1. Total Cost Recapitulation

No.	Method	Total Cost
1	CSA	79,946,295.95
2	CSA_Dis_BB	79,946,302.67
3	CSA_Dis_RoundingUp	79,995,360.68
4	CSA_Dis_TDS	80,009,882.97
5	CSA_Dis_BDS	80,011,304.63
6	CSA_Dis_Rounding	80,000,334.60

Analysis of CSA_Dis_BB Algorithm Design Result
The CSA_Dis_BB Algorithm design aims to obtain the Q[i] in integer (Q[i]'), Schedule for each batch, and due date for each setup and new parts processing activities (d') and mínimum Total Cost (TC[N,Q']). The Schedule of each batch and d' is shown in Figure 2.

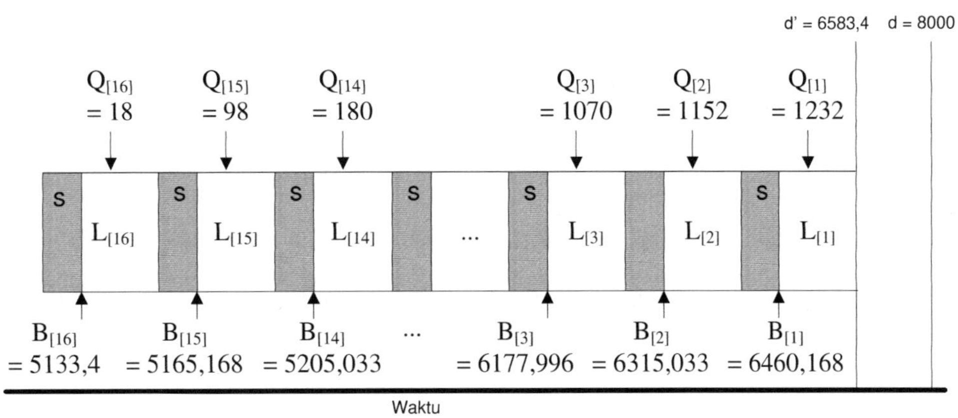

Figure 2. Schedule for Every Batch

Based on Figure 2, the due date for set up and new parts processing activities (d') is smaller than the initial due date. This shows that the resulting schedule and new due date (d') are within the set parameters.

On the other hand, the Total Costs obtained from the CSA_Dis_BB Algorithm (TC[N,Q']) are samallest than the Total Cost from the CSA Algorithm TC[N,Q]. This is affected by the integer batch sizes. TC[N,Q'] from the CSA_Dis_BB Algorithm are all smaller than the integer conversion methods (TDS, BDS, Rounding-Off, Rounding-Up) used by [2].

Table 2 and 3 show recapitulation of total cost for 7 data sets.

Table 2. Calculation Result of CSA, CSA_Dis_BB, and CSA_Dis Models with u = 10%

Set Data	Model CSA (Rp.)	Model CSA_Dis (Rp.)				
		CSA_Dis_BB	Top Decimal Sum	Bottom Decimal Sum	Rounding-off	Rounding-up
Set 1	79,946,295.95	**79,946,302.67**	80,009,882.97	80,011,304.63	80,000,334.60	79,995,360.68
Set 2	6,036,997,149.52	**6,037,012,190.60**	6,037,539,487.26	6,037,543,861.93	6,037,542,765.08	6,037,542,765.08
Set 3	4,683.16	**4,683.80**	4,844.31	4,851.55	4,846.88	4,846.88
Set 4	4,024,001,544.39	**4,024,004.959.71**	4,025,698,186.67	4,025,700,717.75	4,025,662,344.82	4,025,660,400.21
Set 5	7,945,875,401.08	**7,945,878,610.37**	7,946,387,171.50	7,946,389,160.92	NA	7,946,370,307.07
Set 6	1,385.189,731,996.40	**1,385.189,782,217.74**	1,385.193,097,488.45	1,385.193,107,318.21	1,385.193,419,190.73	1,385.193,408,802.91
Set 7	2,171,126,128,412.81	**2,171,126,207,002.90**	2,171,130,933,878.69	2,171,130,947,604.92	2,171,130,805,941.50	2,171,130,804,806.65

Table 3. Calculation Result of CSA , CSA_Dis_BB, and CSA_Dis Models with u = 20%

Set Data	Model CSA (Rp.)	Model CSA_Dis (Rp.)				
		CSA_Dis_BB	Top Decimal Sum	Bottom Decimal Sum	Rounding-off	Rounding-up
Set 1	88,701,045.51	**88,701,051.16**	88,760,219.44	88,751,015.46	88,761,962.08	88,761,962.08
Set 2	6,299,363,829.06	**6,299,365,042.70**	6,037,539,487.26	6,299,849,137.30	6,299,823,235.01	6,299,823,235.01
Set 3	5,164.72	**5,167.53**	5,314.04	4,851.55	NA	5,301.45
Set 4	Not feasible	Not feasible	Not feasible	Not feasible	Not feasible	Not feasible
Set 5	8,008,561,073.33	**8,008,564,331.18**	8,009,020,976.71	8,009,024,453.28	8,009,018,134.11	8,009,018,847.50
Set 6	Not feasible	Not feasible	Not feasible	Not feasible	Not feasible	Not feasible
Set 7	2,176,008,487,542.06	**2,176,008,487,698.73**	2,176,012,679,086.23	2,176,012,705,086.14	2,176,012,555,358.62	2,176,012,540,375.54

Based on Table 2 and 3 (Rounding-Off column), there are cells highlighted in grey that says NA. This means that for the Total Cost (TC[N,Q']) calculations for data set 5 with sample proportion (u) = 10% and data set 3 with sample proportion (u) = 20% in the Rounding-Off Method were not done because the number of query resulted by the integer conversion did not fulfill the parameter of CSA, where the batch sizes were not equal to the initial query.

The CSA_Dis_BB Algorithm obtained discrete batch sizes (integer) without changing the number of batches and queries to be produced. The CSA_Dis_BB Algorithm was implemented on 7 data sets from [1,2]. For all 7 data sets and integer conversion methods, the number of queries after the integer conversion process was equal to the initial number of queries, except for the Rounding-Off Method. The Rounding-Off Method resulted in different number of queries, which were data set 5 with sample proportion (u) = 10% and data set 3 with sample proportion (u) = 20%. The numbers of queries resulted were 1 unit smaller than the intial numbers of queries. This shows that, in those data sets, that the Rounding-Off Method failed to fulfill one parameter of CSA model, that number of discrete batch sizes must be equal to the number of initial queries.

4. Conclusion

We have proposed a modified Branch and Bound algorithm, named The CSA_Dis_BB, to obtain an integer batch size. This algorithm does work to produce the smallest total cost.

5. Recommendations

In order to improve the research on CSA_Dis_BB and CSA_Dis_BB Algorithm design, it is recommended to:

1. Implement other models, as this research only referred to one machine model (CSA), while [2] developed scheduling models for 2, 3, and m machines.
2. Use real value data or field observation data with conditions appropriate for the existing model, in order to test the CSA_Dis_BB method compatibility with all types of data.

Acknowledgment
The authors would like to thank Prof. Dr.-Ing. Hairul Abral, Jonrinaldi, PhD., and Dr. Rika A. Hadiguna for they help.

References

[1] Indrapriyatna A S, Suprayogi, Iskandar B P, and Halim A H 2007a A Batch Scheduling Model for a Single Machine Processing Discrete Parts to Minimize Total Inventory and Quality Costs. *Proceedings of The 1st Asia Pacific Conference on Manufacturing Sistems*. Bali, Indonesia. 5 – 6 September
[2] Indrapriyatna A S, Suprayogi, Iskandar B P, and Halim A H, 2007b Model Penjadwalan *Batch* pada *Flowshop* untuk Minimasi Biaya Simpan dan Kualitas *Jurnal Teknik dan Manajemen Industri ITB* **27** 142 – 163
[3] Halim A H and Ohta H 1993 Batch Scheduling Problem through The Flow Shop with Both Receiving and Delivery Just In Time, *Int. J.l of Production Research* **31** 1943-1955
[4] Halim A H and Ohta H 1994 Batch Scheduling Problems to Minimize Inventory Cost in the Shop with Both Receiving and Delivery Just in Times, *Int. J. of Production Eco.* **33** 185-195
[5] Halim A H, Silalahi J, and Ohta H 2001 A Batch Scheduling Model Considering Quality Costs for the Shop with Receiving and Delivery Just in Time *Proceeding of the 2001 International Conference on Production Research* Prague Czech Republic July 29- August 3
[6] Buckchin J, Tzur M, and Jaffe M 2002 Lot Splitting to Minimize Average Flow-Time in a Two-Machine Flow-shop IIE Transactions **34**, 953 - 970
[7] Baker K R 1974 *Introduction to Sequencing and Scheduling* New York John Wiley & Sons Inc.
[8] Brusco M J and Stephanie S 2005 *Statistics and Computing : Branch and Bound Applications in Combinatorial Data Analysis* New York Springer Science+Business Media Inc.
[9] Dimyati T T and Dimyati A 2006 *Operations Research: Model-model Pengambilan Keputusan*. Sinar Baru Algensindo. Bandung

The Integration of Production-Distribution on Newspapers Supply Chain for Cost Minimization using Analytic Models: Case Study

Era Febriana Aqidawati, Wahyudi Sutopo, Muh. Hisjam
Department of Industrial Engineering, Sebelas Maret University, Surakarta, Indonesia

erafebriana1@gmail.com, wahyudisutopo@staff.uns.ac.id, hisjam@gmail.com

Abstract. Newspapers are products with special characteristics which are perishable, have a shorter range of time between the production and distribution, zero inventory, and decreasing sales value along with increasing in time. Generally, the problem of production and distribution in the paper supply chain is the integration of production planning and distribution to minimize the total cost. The approach used in this article to solve the problem is using an analytical model. In this article, several parameters and constraints have been considered in the calculation of the total cost of the integration of production and distribution of newspapers during the determined time horizon. This model can be used by production and marketing managers as decision support in determining the optimal quantity of production and distribution in order to obtain minimum cost so that company's competitiveness level can be increased.

Keywords: Analytical Model, Integration of Production and Distribution, Newspaper Supply Chain.

1. Introduction

Due to rapid development in industries, the dependency among the supply chain members has been increasing and brings some extent of risk and uncertainty along with benefits [1]. To meet these challenges, a coordinated mechanism among the members is needed in decision making in order to mitigate the current challenges and to increase the system performance and individual profitability of the supply chain [2]. By developing this coordination, manufacturer can minimize total cost while maintaining benefits such as favorable solution considering economical, environmental, and social aspect among the supply chain members [3,4,5].

Production and distribution planning are the main activities in the supply chain [6]. Therefore, an efficient integration of production and distribution plans is critical to achieving competitive advantage [7].

Newspapers are commodities with special characteristics which are perishable, have a shorter range of time between the production and distribution, zero inventory, and decreasing sales value along with increasing in time. Due to the perishable nature of the product, the time from the start of production to the delivery of the finished product to the end-customer is shorter than for most other industries [8].

Thus, as a commodity that has been declining and in a crisis, newspapers need a coordinated production and distribution planning in their framework that can be used to meet the competitive potentials in the newspaper industry [9,10].

Several researchers have attempted to improve the coordination of production and distribution in the supply chain. These researches attempted in developing models and simulations. Different parameters and considerations were used in the models, such as production capacity constraint [11]. The later research proposed a model that considers the problem with one facility, multiple destinations and an infinite vehicle fleet [12]. Another research even considered many uncertainties along supply chain line such as delays, queues, machine breakdown, vehicle malfunction, and environmental factor [13]. The researchers also attempted to solve the problems in production-distribution planning by different objectives, such as minimizing consideration of many uncertainties along supply chain line such as minimizing total tardiness [14], maximizing total revenue [15] and minimizing total global cost [16].

The purpose of this article is to apply development of previous research method to solve newspapers production and distribution integration problems, taking a case study in a newspaper printing company in Surakarta. The company receives orders to print newspapers from three companies and there are four types of printed newspapers. The main raw materials required in the printing process is paper and ink. The company has 2 plate maker machine units and 3 units of printing machines. There are 7 units of distribution vehicles used to deliver newspapers to 11 cities. This article aims to solve the minimization of the cost of production and distribution so that the company has a higher competitiveness.

This article consists of 4 parts, namely introduction, method, result and discussion, and conclusion. The introduction shows the reasons and importances to integrate production and distribution of newspapers. The method section discusses the steps and model formulation to solve the problem. The results of data processing using related methods are discussed and analyzed in results and discussion section. Furthermore the essence of the entire article is explained in the conclusion section.

2. Method

Garside developed a production-distribution integration model. In this model, it is assumed that the relationships between the variables in the objective and limiting functions are linear. In addition, some parameters in this model considered stochastic due to its accommodation of ongoing uncertainty in the supply chain. Uncertainty in the form of delay, variation of vehicle speed, machine and vehicle break down will affect the value of the parameter of production time, travel time, production time capacity and time capacity when the vehicle becomes random or certain distributed probability [13].

This article uses development from existing model. The difference with existing model is that in the developed model, it is assumed that there is no delay in the production floor, the speed is same for all vehicle, the printing machines speed in producing each newspaper is identified, production time capacity and time capacity is restricted to certain time window. This model based on deterministic data which aims to determine the types of products that must be produced, the amount of production, the quantity and the number of shipments of each plant to each distribution center (DC), as well as inventory at the plant and DC. Since the modeled products are newspapers where no inventory stored in warehouses and DC, variables related to stock up in the decision variables, parameters, the objective function, and the constraints are removed from the model and some new relevant constraints are added.

Index

I = factory sets
J = set of distribution destinations
T = time period sets
K = product sets
V = sets of vehicles owned by the factory *i*

Parameter

$Cp_{ik}(t)$ = Production cost of product *k* in the factory *i* in period *t*
$Cs_{ikt}(t)$ = Set up cost of product *k* in the factory *i* in period *t*
$B_i(t)$ = Production capacity in the factory *i* during *t* period (minute)
Qp_{ik}^{max} = Maximum production capacity to make product *k* in factory *i*
tp_{ik} = Required time to make one unit of product *k* in factory *i*
C_v = Delivery cost for vehicle *v* (per minute)
Q_v = Maximum transport capacity of the vehicle *v* (exemplar newspaper)
$T_v(t)$ = Time capacity vehicle v during the time period *t*
td_{ij}^v = Delivery time from factory *i* to distribution center *j* and back again to the factory *i* by using the vehicle *v* (minutes)
$D_{jk}(t)$ = Demand for product *k* in distribution center *j* in period *t*

Decision Variable

$P_{ik}(t)$ = Quantity of product *k* produced in factory *i* in period *t* (exemplar)
$x_{ik}(t)$ = Decision whether factory *i* produce product *k* in period *t*
$y_{ij}^v(t)$ = The number of direct delivery from factory *i* to distribution center *j* by vehicle *v* in period *t*
$q_{ijk}^v(t)$ = Quantity of product *k* delivered from factory *i* to distribution center *j* by vehicle *v* in period *t* (unit)

Mathematical Model
Objective function

$$\textbf{Minimize}: \sum_{t\in T}\sum_{i\in I}\sum_{k\in K} Cs_{ikt}(t).x_{ik}(t) + \sum_{t\in T}\sum_{i\in I}\sum_{k\in K} Cp_{ikt}(t).P_{ik}(t) + \sum_{t\in T}\sum_{i\in I}\sum_{k\in K} C_v . td_{ij}^v . y_{ij}^v(t) \quad (1)$$

Constraints

$$\sum_{k\in K} tp_{ik} . P_{ik}(t) \le B_i(t), \forall\, i \in I, t \in T \quad (2)$$
$$P_{ik}(t) \le Qp_{ik}^{max}.x_{ik}(t), \forall\, i \in I, k \in K, t \in T \quad (3)$$
$$P_{ik}(t) - \sum_{j\in J}\sum_{v\in V} q_{ijk}^v(t) = 0, \forall\, i \in I, k \in K, t \in T \quad (4)$$
$$\sum_{k\in K} q_{ijk}^v(t) \le Q_v . y_{ij}^v(t), \forall\, i \in I, v \in V, j \in J, t \in T \quad (5)$$
$$\sum_{j\in J} td_{ij}^v . y_{ij}^v(t) \le T_v(t), \forall\, i \in I, v \in V, t \in T \quad (6)$$
$$\sum_{i\in I}\sum_{t\in T} q_{ijk}^v(t) - D_{jk}(t) \ge 0 \quad, \forall\, j \in J, v \in V, k \in K, t \in T \quad (7)$$
$$Q_v - D_{jk}(t) \ge 0, \forall\, j \in J, v \in V, k \in K, t \in T \quad (8)$$
$$P_{ik}(t).x_{ik}(t) \ge 0, \forall\, i \in I, k \in K, t \in T \quad (9)$$
$$x_{ik}(t) = 1, \forall\, i \in I, k \in K, t \in T \quad (10)$$
$$y_{ij}^v(t) \ge 0 \text{ and integer number} \quad (11)$$
$$x_{ik}(t) \in \{0,1\} \quad (12)$$
$$P_{ik}(t) \ge 0\, , Cs_{ikt}(t) \ge 0, q_{ijk}^v(t) \ge 0 \quad (13)$$

This model will minimize the total cost of production set up cost, variable production cost and regular shipping cost. Constraints 2 and 3 guarantee that the production amount of each product does not exceed the capacity of production time and production capacity in each plant. Constraint 4 is the equilibrium equation of the number of each product in each plant and DC. Constraint 5 guarantees the quantity of shipments of all products not to exceed the vehicle carrying capacity of each plant. While the constraint 6 guarantees the number of direct shipments from the factory to the DC does not exceed the regular delivery time capacity of each vehicle. Constraint 11 imposes a non-negative and integer number for the decision variable $y_{ij}^v(t)$. Constraint 12 implies a binary number for the decision variable $x_{ik}(t)$. Constraint 13 implies a non-negative number of decision variables $P_{ik}(t)$, $Cs_{ikt}(t)$, and $q_{ijk}^v(t)$. Next, we conduct data collecting to provide input for parameters in the model. Then this model is calculated by using IBM ILOG CPLEX 12.7.1 software to get optimal result.

3. Result and Discussion

The proposed model was applied to a company that produces 4 types of newspaper and distributes to 11 cities using 7 vehicles during the planning horizon of 3 days. Here, we consider parameters such as production costs, set up cost, delivery time and demand for products for each distribution center. Data related to those parameters are presented in Table 1-4. It is important to note that the production time capacity for each period is 300 minutes, maximum transport capacity for each vehicle is 3550 exemplar and duration provided to deliver the newspapers from the factory is 240 minutes for each vehicle. Furthermore, the production capacities in the factory by coordinating 3 provided printing machines for newspaper 1, 2, 3, 4 are 24000, 28200, 12600 and 6600 exemplar respectively. Meanwhile, required time to produce each newspaper type is 217, 49, 38 and 130 minute respectively. Whereas the delivery cost of each vehicle in IDR/minute is 59, 78, 76,137, 237, 440 and 154 respectively.

Table 1. Production cost per unit (IDR)

Newspaper	Period 1	Period 2	Period 3
1	719	587	719
2	864	864	864
3	1390	1653	1653
4	1954	1954	2100

Table 2. Set up cost per unit (IDR)

Newspaper	Period 1	Period 2	Period 3
1	1491455	1267571	1542081
2	561356	559628	561615
3	471905	334402	337047
4	660182	695741	692931

Table 3. Delivery time from factory to distribution center by each vehicle

Distribution center	Vehicle						
	1	2	3	4	5	6	7
Solo	19						
Sukoharjo	43						
Sragen		57					
Wonogiri				113			
Karanganyar			57				
Klaten					57		
Boyolali						50	
Yogyakarta					120		
Salatiga						91	
Semarang						152	
Purwodadi							125

Table 4. Demand in each distribution center.

Distribution center	Newspaper	Demand (exemplar)		
		Per. 1	Per. 2	Per. 3
Solo	**1**	3244	3500	3447
	2	4000	4215	4112
	3	0	0	0
	4	0	0	0
Sukoharjo	**1**	2000	2674	3082
	2	600	642	517
	3	0	0	0
	4	0	0	0
Sragen	**1**	1500	1632	1530
	2	0	0	0
	3	0	0	0
	4	0	0	0
Wonogiri	**1**	1500	1555	1208
	2	0	0	0
	3	0	0	0
	4	0	0	0
Karanganyar	**1**	1500	1269	1443
	2	0	0	0
	3	0	0	0
	4	0	0	0
Klaten	**1**	1500	1468	1520
	2	0	0	0
	3	0	0	0
	4	0	0	0
Boyolali	**1**	1500	1324	1506
	2	0	0	0
	3	0	0	0
	4	0	0	0
Yogyakarta	**1**	1500	1123	1168
	2	0	0	0
	3	0	0	0
	4	2850	2675	2811
Salatiga	**1**	1000	978	1121
	2	0	0	0
	3	0	0	0
	4	0	0	0

Table 4. (cont'd.)

Semarang	**1**	1000	1002	879
	2	0	0	0
	3	1585	1453	1501
	4	0	0	0
Purwodadi	**1**	1000	992	1023
	2	0	0	0
	3	0	0	0
	4	0	0	0

Table 5. Production plan for each newspaper (exemplar)

Newspaper	Period 1	Period 2	Period 3
1	24000	23870	23965
2	28200	27650	26510
3	12600	12543	12430
4	6600	6543	6578

Table 6. Comparison between previous and proposed system

Aspect	**Previous System**	**Proposed System**
Total cost (production cost + set up cost + delivery cost)	Total cost during one period reaches IDR 15-20 million.	Total cost can be minimized during the planning horizon within 3 periods with optimal result of Rp. 10,181,000
Production quantity plan	Production quantity plans for each product are based on the forecast and has not considered the integration of production-distribution optimization.	Production quantity plans during the planning horizon, consider the constraints and parameters in the production-distribution integration model with the aim of minimizing costs.
Quantity of products delivered	Not all vehicles distribute products based on demand to specific destinations. Only one to two vehicles per destination are available.	Each vehicle has its own portion that has been plotted according to the distribution objectives for a certain period.

4. Conclusion

Analytic model of integration of production and distribution in the supply chain has been developed as a reference for newspaper production and distribution planning with the decision to minimize total cost during the planning horizon. Production and distribution planning can be applied optimally in the newspaper supply chain by integrating the production and distribution planning of four newspaper products, which are distributed to 11 cities in a daily period. Several parameters and constraints have been considered in the calculation of total cost of the integration of newspaper production and distribution during the determined planning horizon. The proposed model results IDR 10,181,000 in total cost, which is lower than the previous total cost before model is applied. The proposed model has the ability-applied to be used by the production and marketing managers as decision support in determining the optimal quantity of production and distribution in order to minimize production and distribution cost so that the level of competitiveness can be improved.

References

[1] Arshinder K, Kanda A and Deshmukh S G 2011 A Review on Supply Chain Coordination: Coordination Mechanisms, Managing Uncertainty and Research Directions *International Handbooks on Information Systems (Supply Chain Coordination under Uncertainty)* ed T M Choi and T Cheng (Berlin: Springer) pp 39-82

[2] Hossain M S, Uddin M F and Mondal M 2017 The Effects of Coordination Mechanism on Vendor Buyer System of a Supply Chain *Procedia Engineering* vol 194 (Elsevier) pp 529-536

[3] Kurniawan B, Hisjam M and Sutopo W 2011 Integration of production and supply chain strategic planning for renewable resources under sustainability considerations: Teakwood case study *Int.Conf. on Industrial Engineering and Engineering Management* (Singapore: IEEE) pp 433-437

[4] Hisjam M, Habibie A, Sutopo W and Widodo K H 2013 A Supplier-Manufacturer Model for Procurement Plan in Export Oriented Furniture Industry with Sustainability Considerations

Lecture Notes in Electrical Engineering (IAENG Transactions on Engineering Technologies vol 186*)* ed G C Yang et al (Dordrecht: Springer) pp 247-260

[5] Sutopo W, Devi A O T, Hisjam M and Yuniaristanto 2012 A model for procurement and inventory planning for export-oriented furniture industry in Indonesia: A case study *Lecture Notes in Engineering and Computer Science (The 2012 IAENG Int. Conf. on Artificial Intelligence and Applications* vol 2*)* ed G C Yang et al (Hong Kong: Newswood Limited) pp 1214-1217

[6] Lee Y H and Sook H K Optimal Production-Distribution Planning in Supply Chain Management Using a Hybrid Simulation-Analytic Approach *Proc. of the 2000 Winter Simulation Conf.* pp 1252-1259

[7] Bilgen B 2010 Application of fuzzy mathematical programming approach to the production allocation and distribution supply chain network problem *Expert Systems with Applications* vol 37 (Elsevier) pp 4488-4495

[8] Golden B L, Assad A A and Wasil E A 2002 Routing vehicles in the real world: Applications in the solid waste, beverage, food, dairy and newspaper industries *The Vehicle Routing Problem* (Philadelphia: Society for Industrial and Applied Mathematics) chapter 10 pp 245-276

[9] Papathanassopoulos S 2001 The Decline of Newspapers: the case of the Greek press *Journalism Studies* vol 2 no 1 (London: Routledge, Taylor & Francis Ltd) pp 109-123

[10] Kaia K F and Ravlo E H 2014 Supply chain optimization in the newspaper industry: Integrated production allocation, scheduling and distribution *Thesis on Industrial Economics and Technology Management* (Norwegia: Norwegian University of Science and Technology) p 1

[11] Lee Y H and Kim S H 2002 Production-distribution planning in supply chain considering capacity constraints *Computers & Industrial Engineering* vol 43 (Elsevier) pp 169-190

[12] Chen Z L and Vairaktarakis G L 2005 Integrated scheduling of production and distribution operations. *Management Science* vol 51 pp 614-628

[13] Garside A K, Wahyuono R H and Widyarini T 2010 Integrasi Produksi - Distribusi pada Supply Chain dengan Pendekatan Hybrid Analitik – Simulasi (Trans: Production Integration - Distribution on Supply Chain with Analytic Hybrid Approach – Simulation) *Jurnal Teknik Industri* **13** pp 27-36

[14] Ullrich C A 2013 *Integrated machine scheduling and vehicle routing with time windows* (European Journal of Operational Research) pp 152–165

[15] Suswaini E 2015 Strategi Mengoptimalkan Perencanaan Produksi dan Distribusi dengan Metode Interger Linier Programming Branch and Bound di Perusahaan Manufakturing Usaha Maju di Kota Jogjakarta (Trans: Strategy to Optimize Production and Distribution Planning with Linear Integer Programming Branch and Bound Method in Manufacturing Company Usaha Maju in Jogjakarta)

[16] Ma Y, Yan F, Kang K and Wei X 2016 A novel integrated production-distribution planning model with conflict and coordination in a supply chain network *Knowledge-Based Systems* vol 105 (Elsevier) pp 119-133

An integrative multi-criteria decision making techniques for supplier evaluation problem with its application

D Fatrias*, I Kamil and D Meilani

Dept. of Industrial Engineering, Universitas Andalas, Padang (25163), INDONESIA

*dickyf@eng.unand.ac.id

Abstract. Coordinating business operation with suppliers becomes increasingly important to survive and prosper under the dynamic business environment. A good partnership with suppliers not only increase efficiency, but also strengthen corporate competitiveness. Associated with such concern, this study aims to develop a practical approach of multi-criteria supplier evaluation using combined methods of Taguchi loss function (TLF), best-worst method (BWM) and VIse Kriterijumska Optimizacija kompromisno Resenje (VIKOR). A new framework of integrative approach adopting these methods is our main contribution for supplier evaluation in literature. In this integrated approach, a compromised supplier ranking list based on the loss score of suppliers is obtained using efficient steps of a pairwise comparison based decision making process. Implemetation to the case problem with real data from crumb rubber industry shows the usefulness of the proposed approach. Finally, a suitable managerial implication is presented.

1. Introduction

Among all strategic decision involved along enterprises' supply chain system, purchasing decision has a great impact on the overall system due to the fact that the cost of raw materials and component parts represents the largest percentage of the total product cost [1]. Establishing strong relationship with selected suppliers not only substantially reduces purchasing cost, but also significantly improves corporate competitiveness [2]. In today's business, however, many enterprises are still dealing with unreliable suppliers. Working with such suppliers may interrupt production, lower product availability and quality, and at the end, decrease enterprise competitiveness. Therefore, enterprises may initiate to develop their suppliers and try to actively improve the performance of their supplier base [3].

Supplier evaluation and selection are a crucial phase before supplier development. With an increasing occurrence of risk events, supplier evaluation and selection process become more and more important for the business. However, in practical decision making it is often found difficult to make a proper decision due to a complex and unstructured problem of supplier evaluation in nature involving quantitative and qualitative criteria simultaneously. To cope with such problem, enterprises can find utility in multi-attribute decision making (MADM) technique that can assist with evaluating and selecting suppliers. In this regard, several studies adopting MADM techniques have been proposed in literature. Jain et al. [4] proposed fuzzy multi-criteria decision-making model of supplier evaluation and selection where Analytical Hierarchy Process (AHP) is used to assign criteria weight and Technique for Order of Preference by Similarity to Ideal Solution (TOPSIS) is used to rank suppliers. Pitchipoo et al. [5] developed a decision support model for supplier evaluation and selection in the process industry by integrating AHP and Grey Relational Analysis (GRA). Yadav & Sharma [6] integrated Data Envelopment Analysis (DEA) into AHP procedures. Liu & Zang [7] proposed a novel integrated method of entropy weight and an improved ELECTRE-III to deal with supplier selection. Hsu et al. [8] developed a model to evaluate suppliers based on their carbon performance utilizing Analytical Network Process (ANP) to determine relative weights of each criterion and VIKOR to derive a compromised supplier ranking list. Chang et al. [9] proposed a novel approach of decision-making of supplier selection by applying fuzzy DEcision-MAking Trial and Evaluation Laboratory (DEMATEL) method to find influential factors in selecting suppliers. Parkouhi & Ghadikolaei [10] developed resilient approach to evaluate supplier using fuzzy ANP and grey-VIKOR. Chen et al. [11] investigated the use of The Preference Ranking Organization METHod for Enrichment of Evaluations (PROMETHEE) to evaluate suppliers in group decision making and fuzzy environment. Adali et al. [12] proposed an alternative version of fuzzy-PROMETHEE to solve supplier eavaluation problem in which preference functions are handled in terms of fuzzy distances between alternatives with respect to each criterion. Kuo et al. [13] developed integrated methods composed of DEMATEL and ANP to determine the relative importance of criteria and applied VIKOR to assess green suppliers. Gupta & Barua [14] proposed evaluate suppliers among SMEs (Small and Medium Enterprises) on the basis of their green innovation ability in which best-worst method is utilized to obtain criteria weights and fuzzy TOPSIS is used to rank the suppliers. Zeydan et al. [15] combined fuzzy-AHP, fuzzy TOPSIS and DEA to evaluate suppliers for automotive industry. Büyüközkan & Çifçi [16] proposed evaluation framework using hybrid approach of fuzzy DEMATEL, fuzzy ANP and fuzzy TOPSIS to evaluate green suppliers.

This study relies on the theory of MADM from which integrated methods of TLF, BWM, and VIKOR technique are proposed to evaluate a set of suppliers. The TLF allows the DM to set performance target and its tolerance limit of suppliers for which the loss score of suppliers is calculated. The weight of criteria is computed using the most recent and efficient pairwise comparison – based method named BWM. Finally, a VIKOR's procedures are executed to rank the supplier. To examine its usefulness, the proposed research is implemented to solve the case from crumb rubber industry.

2. The proposed integrated approach

To structure appropriate steps of the proposed approach, each of the method used in this research is briefly discussed.

2.1. Taguchi loss function (TLF)

The original concept of TLF is based on quantifying the loss (in term of cost) incurs at customers if the quality characteristic of product deviates from the target value. Particularly, Taguchi proposes an acceptable quality definition that any deviation from quality's target value results in losses. The loss is zero when quality measurement is the same as the target value. Otherwise, such loss incurs whenever quality measurement fall within specification limit.

When $L(x)$ denotes the loss for quality characteristic x, and k is a loss constant factor, then the formulation for "larger is better" loss function is given in Eq. (1).

$$L(x)=\begin{cases} k\times(\frac{1}{x^2})^2 & \text{for single data} \\ k\times\left(\text{MSD}=\frac{1}{n}\sum_{i=1}^{n}(\frac{1}{x_i^2})^2\right) & \text{for } n-\text{data} \end{cases} \quad ; \; k=100\%\times(T)^2 \tag{1}$$

2.2. Best-worst method (BWM)

Best-worst method (BWM) is a novel pairwise comparison based method proposed by Rezaei [17]. The steps of BWM start by first identifying a set of decision n criteria $\{c_1, c_2, ..., c_n\}$. Then the best (e.g. most desirable, most important) and the worst (e.g. least desirable, least important) criteria are determined. Two vector named as best-to-others (BO) and others-to-worst (OW) vector are determined where BO and OW vector are formulated as $A_B = \{a_{B1}, a_{B1},..., a_{Bn}\}$ and $A_W = \{a_{1W}, a_{1W},..., a_{nW}\}^T$, respectively. Finally, the optimal weights (w_1^*, w_2^*, ..., w_n^*) are obtained using Eq. (2).

$$\text{Min } \varphi \tag{2}$$

s.t.:

$$\left|\omega_B - a_{Bj}\omega_j\right| \le \varphi \quad \text{for all } j$$

$$\left|\omega_j - a_{jW}\omega_W\right| \le \varphi \quad \text{for all } j$$

$$\sum_j \omega_j = 1, \quad \omega_j \ge 0 \quad \text{for all } j$$

2.3. VIKOR technique

VIKOR was first proposed by Opricovic [18, 19] which is utilized to handles the situation of conflicting criteria in decision-making and there might be no solution satisfying all criteria simultaneously. The core feature of VIKOR is to evaluate a set of alternatives and propose a ranking that resolves disagreements by providing compromised solutions based on its closeness to the ideal solution.

The VIKOR procedures is composed of several steps. First, the value of f_{ij}, which represents the value of criterion *i* (*i* =1, 2, 3,…, *n*) for each of alternative *j* (*j* = 1, 2, 3, ..., *m*) are determined. For cost typed criteria, the value of $f_i^+ = \min_j f_{ij}$ and $f_i^- = \max_j f_{ij}$ are then selected. After that, ranking measures are calculated using two formulations, i.e.,

$$S_j = \sum_{i=1}^{n} \frac{\omega_i(|f_i^+ - f_{ij}|)}{f_i^+ - f_i^-} \tag{3}$$

$$R_j = \max_i \left(\frac{(|f_i^+ - f_{ij}|)}{f_i^+ - f_i^-}\right) \tag{4}$$

where S_j and R_j represents the distance of *j*-th alternative from positive ideal solution and from negative ideal solution, respectively, and ω_i denotes criteria weight calculated from measuring criteria relative importance. Next, the value of Q_j are computed where:

$$Q_j = \frac{(S_j - S^+)}{S^- - S^+} + \frac{(1-v)(R_j - R^+)}{R^- - R^+} \tag{5}$$

where $S^+ = \min_j S_j$, $S^- = \max_j S_j$, $R^+ = \min_j R_j$, $R^- = \max_j R_j$, and *v* indicates the weight where the value is assumed to be 0.5 [20]. Finally, ranking of alternatives is obtained by sorting *S*, *R*, and *Q* in increasing order. Assuming $a^{(1)}$ and $a^{(2)}$ are the first and second ranked alternative in *Q* list, respectively, then alternative $a^{(1)}$ is the best compromise solution if the two conditions are satisfied: "Acceptable advantage" and "Acceptable stability in decision-making". First condition satisfies $Q(a^{(2)}) - Q(a^{(1)}) \ge DQ$ where $DQ = 1/(m-1)$, and second condition requires that $a^{(1)}$ should also be ranked first according to *S* and/or *R*. If one of the above conditions is not satisfied then a set of compromised solutions are obtained where the following rules are applied: 1) If only the condition 2 is not satisfied, $a^{(1)}$ and $a^{(2)}$ are both the compromised solutions; 2) If the condition 1 is not satisfied, $a^{(1)}, a^{(2)},..., a^{(n)}$ becomes compromised solutions where $a^{(n)}$ is determined by the relation $Q(a^{(n)}) - Q(a^{(1)}) < DQ$ for maximum *n* (the positions of these alternatives are "in closeness").

3. Industrial case problem

3.1. Result

To illustrate the usefulness of our proposed methodology, a case study was carried out in a crumb rubber industry named as XYZ Co. here (due to confidentiality), located in Padang city, Indonesia. The raw material (raw rubber) is mostly supplied by rubber farmers, classified as smallholders localized in West Sumatera and other surrounding regions. In particular, the supply system of XYZ Co. involves suppliers characterized by limited capacity, uncertain delivery lead time and low raw rubber quality. To anticipate

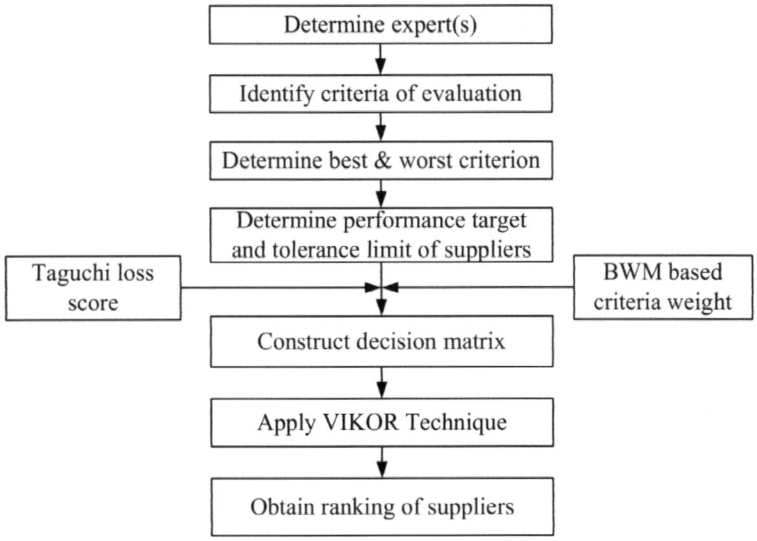

Figure 1. The proposed approach

material shortages, the XYZ Co. adopts a daily-basis supply from suppliers with fixed predetermined price. From current 18 suppliers only five of them constantly provides regular supply to which the proposed approach is applied. The proposed methodology is depicted in Figure 1.

Step 1: Determine the candidate of expert for decision making. To conduct evaluation process, individual opinion by the Chief, as a senior DM, and group opinion by a panel consisting of supporting experts within the company, including the Chief himself, was adopted depending on data requirement. These experts include purchasing, production, quality control, and warehousing supervisor.

Step 2: Identify criteria of evaluation. After presenting some criteria gathered from literature review to the panel, 5 (five) criteria were finally chosen as the most relevant criteria, i.e., quality (C1), quantity (C2), continuity (C3), responsiveness (C4), and reputation (C5).

Step 3: Compute the loss score of supplier using TLF. Due to criteria nature, a larger-is-better type loss function was chosen to specify the performance characteristic of all suppliers. Performance target values and tolerances limit (*T*) of suppliers are then determined through panel discussion (Table 1).

Eight-month performance data of suppliers are gathered for the first three criteria. A qualitative assessment for the last two criteria is performed based on aggregate measures for the entire periods as period-by-period performance does not fluctuate significantly. Due to restricted number of pages, the

Table 1. Target Quality characteristics of criteria

Criteria	Target value	(*T*, %)	Loss at T (%)	*k*	Loss Function
C1	100%	46	100	0.212	$L(x) = 0.212$ MSD
C2	The greatest	30% lower	100	0.090	$L(x) = 0.090$ MSD
C3	The most frequent	20% lower	100	0.040	$L(x) = 0.040$ MSD
C4	Score of 100	70	100	4900	$L(x) = 4900\ (1/x^2)$
C5	Score of 100	75	100	5625	$L(x) = 5625\ (1/x^2)$

Table 2. Decision matrix of loss scores

Supplier *n*	C1	C2	C3	C4	C5
S1	66.59	292.43	10.07	60.49	114.80
S2	61.33	17.02	5.30	100.00	87.89
S3	66.02	2052.68	89.57	87.11	100.00
S4	58.54	1414.49	54.16	76.56	69.44
S5	55.06	91285.71	907.80	87.11	100.00

Table 3. Pairwise comparison of criteria

Best criterion:	The DMs	C1	C2	C3	C4	C5
C1	The chief	1	2	7	4	5
	Pc. M	1	2	5	4	4
	Pd. M	1	2	5	4	2
	Qc. M	1	4	7	5	2
	Wh. M	1	2	5	4	2

Worst criterion: C3	The DMs				
	The chief	Pc. M	Pd. M	Qc. M	Wh. M
C1	7	5	5	7	5
C2	5	4	4	4	4
C3	1	1	1	1	1
C4	3	2	2	2	2
C5	2	2	4	5	4

Pc. M = Purchasing Manager; Pd. M = Production Manager; Qc. M = QC Manager; Wh. M = Warehousing Manager

performance data are not presented in this paper. The value of k and loss functions $L(x)$ are formulated using the data on Table 1. For example, k for quantity criterion is calculated as $k = 100\% \times (30\%)^2 = 0.090$. Since eight performance data are collected, mean squared deviation (MSD) of data is used to formulate the loss function which is $L(x) = 0.090 \times \text{MSD}$.

Step 4: Obtain the weight of criteria using fuzzy-BWM. Best and worst criterion were selected by the panel through mutual consensus, and after that all preference rating used for BO and OW vectors are determined individually by each panel member (Table 3). By the above process, C1 and C3 are chosen as the best and the worst criterion, respectively. Using Eq. (2), the criteria weights along with the values of φ are calculated for five panel members, and by averaging these value the mean weights are obtained (Table 4). Finally, consistency check was carried out for all φ. For example, from comparison performed by the Chief, φ is 0.6834 whose value of $a_{BW} = a_{13} = 7$ with the consistency index of 3.73 (see [18] for details). So, consistency ratio is become 0.6834 / 3.73 = 0.1832, which indicates a good consistency because this value is close to zero.

Step 5: Obtain the rank of supplier using VIKOR. The decision matrix and criteria weights are used to rank suppliers based on VIKOR technique. Noted that the loss scores with respect to each criterion

Table 4. Weights of criteria

Criteria	Mean weights	Individual weights				
		The chief	Pc. M	Pd. M	Qc. M	Wh. M
C1	0,4129	0,4439	0,4362	0,3619	0,4610	0,3619
C2	0,2263	0,2686	0,2545	0,2348	0,1390	0,2348
C3	0,0654	0,0578	0,0767	0,0663	0,0600	0,0663
C4	0,1102	0,1338	0,1315	0,1022	0,0811	0,1022
C5	0,1851	0,0958	0,1010	0,2348	0,2590	0,2348
φ		0,6834	0,6834	0,4586	0,6834	0,4586

Pc. M = Purchasing Manager; Pd. M = Production Manager; Qc. M = QC Manager; Wh. M = Warehousing Manager

Table 5. The value of S, R and Q of suppliers

Supplier n	S	S-rank	R	R-rank	Q	Q- rank
S1	0,8423	4	0,4129	5	0,9255	4
S2	0,4522	1	0,2437	3	0,0531	2
S3	0,9106	5	0,3959	4	0,9550	5
S4	0,4833	2	0,2236	1	0,0338	1
S5	0,5201	3	0,2263	2	0,0811	3
	$S^* = 0.4522$		$R^* = 0.2236$		$j = 5$	
	$S^- = 0.9106$		$R^- = 0.4192$		DQ = 0.25	

uses different scale, making it less comparable with each other. Therefore, those values are first normalized by a linear scale transformation. The normalized value, r_{ij}, is computed as $r_{ij} = x_j^{min} / x_{ij}$, where x_{ij} is the the loss score of supplier i for criteria j and $x_j^{min} = \min_i x_j$. After identifying a maximum and minimum values of r_{ij}, the value of S_i, R_i and Q_i are calculated using Eq. (3)-(5) where v is set to 0.5. Table 5 shows that the supplier with the lowest and the second lowest Q are identified as supplier 4 (S4) and supplier 2 (S2), respectively. Examining the condition of VIKOR, the first conditions of "Acceptable advantage" is not satisfied since the value $Q(S_2) - Q(S_4) = 0.0193$, which is less than $DQ = 1/(5\text{-}1) = 0.25$. However, for the second condition it is found that supplier 4 is obtain the minimum value of R, hence, it is a stable alternative which satisfies the second condition. Therefore, a set of compromised ranking is obtained as S4-S2-S5 >S1>S3 where S4, S2 and S5 are top suppliers, followed by S1 and S3 as the second and the third ranked suppliers, respectively.

3.2. Implications

The proposed approach provides some practical implications for decision makers. This research adopts the value of quality loss by Taguchi as a loss score of suppliers in performance evaluation. This common value provides understandable language in decision making, and utilizing this value to evaluate suppliers makes evaluation process will be much easier and meaningful. More specifically, Taguchi allows decision makers to set performance target values and tolerance limit that is very important in the context of supplier evaluation and selection problem since each enterprise with different organizational goals might use different performance criteria and measurement with different acceptable tolerance limits. Thus, this feature enables decision makers to perform the most precise and comprehensive supplier evaluation with respect to their specific goals.

Using BWM brings a significant benefit for decision making process because aside of improving computational load due to its compact yet efficient steps, this method is capable of producing good consistency rate in the presence of conflicting criteria. By incorporating the BWM's results into VIKOR technique, the DM are able to obtain a compromised supplier ranking lists, providing the closest solution to the ideal. Hence, utilizing it in the proposed approach increases the efficiency of the decision making process.

4. Conclusion

This research proposed a comprehensive approach of multi-criteria decision making that integrates Taguchi loss function (TLF), fuzzy BWM and VIKOR in performance evaluation of suppliers. The merit of using the proposed methodology is that it enables the DM to take the benefit from each method's individual advantage simultaneously in making an appropriate decision. The use of the proposed approach was applied in the industrial case to examine its feasibility in evaluation and selecting the best supplier of crumb rubber industry. Based on the case study, it is convincingly useful for solving an empirical case problem with actual data and feedback from experts both in either single or group opinion.

References

[1] Pazhani S, Ventura J A and Mendoza A 2015 *Appl. Math. Model* **40** 612-34
[2] Xia W and Wu Z 2007 *Omega* **35** 494-504
[3] Glock C H, Grosse E H and Ries J M 2017 *Int'l. J. Prod. Econ* Article in Press.
[4] Jain V, Sangaiah A K, Sakhuja S, Thoduka N and Aggarwal R 2016 *Neur. Comput. & Appl.* 1-10
[5] Pitchipoo P, Venkumar P and Rajakarunakaran S 2013 *J. Ind. Eng. Int'l.* **9** 23 1-15
[6] Yadav V and Sharma M 2015 *J. Enterp. Inform. Manag.* **28** 2 218-42
[7] Liu P and Zhang X 2011 *Int'l. J. Prod. Res.* **49** 3 637–46
[8] Hsu C W, Kuo R J and Chiou CY 2014 *Int'l. J. Env. Sci. Tech.* **11** 3 775–84
[9] Chang B, Chang C and Wu C 2011 *Exp. Syst. App.* **38** 3 1850–58
[10] Parkouhi S V and Ghadikolaei A S 2017 *J. Clean. Prod.* **161** 431-51
[11] Chen Y-J 2011 *Inform. Sci.* **181** 9 1651-70
[12] Adali E, Işık A and Kundakcı K 2016 *Uncert. Supp. Chain Manag.* **4** 3 183-94

[13] Kuo T C, Hsu C W and Li J Y 2015 *Sustainability* **7** 2 1661-1689
[14] Gupta H and Barua M K 2017 *J. Clean. Prod.* **152** 242-58
[15] Zeydan M, Çolpan C and Çobanoğlu C 2011 *Exp. Syst. App.* **38** 2741-51
[16] Büyüközkan G and Çifçi G 2012 *Exp. Syst. App.* **39** 3 3000-11
[17] Rezaei J, Fahim P B M and Tavasszy L 2014 *Exp. Syst. App.* **41**18 8165-79
[18] Opricovic S 1998 *Multicriteria Optimization of Civil Engineering Systems* (Belgrade: Faculty of Civil Engineering) **2** 1 5-21
[19] Opricovic S and Tzeng G H 2007 *Eur. J. Oper. Res.* **178** 2 514-29
[20] Kackar R N 1985 J. Qual. Tech. **17** 4 176–88

IOP Conf. Series: Materials Science and Engineering **319** (2018) 012077 doi:10.1088/1757-899X/319/1/012077

Process Improvement in Outpatient Installation RSUD dr. Soediran Mangun Sumarso Using Lean Hospital Approach

Ghany Sayyida[1], Fakhrina Fahma[1] and Irwan Iftadi[2]

[1]Industrial Engineering Department, Faculty of Engineering, Sebelas Maret University, Jln. Ir. Sutami 36A, Surakarta 57126
[2]Electrical Engineering Department, Faculty of Engineering, Sebelas Maret University, Jln. Ir. Sutami 36A, Surakarta 57126

arifianaghany@gmail.com

Abstract. RSUD dr. Soediran Mangun Sumarso is a public hospital in Wonogiri district which has an outpatient installation service. However, the waiting time of some services in outpatient installations exceeds the standard time set by the health minister of the Republic of Indonesia. It is known from the data waiting time in the outpatient installation. The purpose of this study is to provide improvements using lean hospital approach. Proposed improvement is done by eliminating waste that occurs in outpatient installation service. The methodology used in this study consists of four stages. The first stage is describing the service system using a cross-functional flowchart. The second stage is identifying waste using value stream mapping, observation and interview. The third stage is to determine critical waste by borda method and pareto diagram. The last stage is to provide recommendation improvement using fishbone diagram and FMEA. The result of this research is proposed improvements. The proposed improvements are adding special register counters, implementing an online reservation system, doctors schedule synchronization, adding doctors in polyclinics, fixing queue numbers, applying visual management concepts, making connecting glass in pharmacies and adding multifunction shelves in polyclinics.

1. Introduction

1.1. Background of the study

Hospitals as one of the health services industry are required to maintain the quality of service. There are several methods that can be used to improve the quality of services, especially in the health sector, one of them is lean. Lean thinking has been introduced in healthcare during the latest decades as a quality-improvement method [9]. Lean aim to improve quality by reducing waste and facilitate flow in care processes [11]. Waste can be defined as something that does not add any value to the final product or is not required [3][16]. The concept of lean thinking has rapidly gained followers and has been adopted widely and successfully in the healthcare arena [7][13][15].

Patient waiting time is one of the potential components causing dissatisfaction[14]. Regulation of the health minister of Republic Indonesia Number: 129 / Menkes / SK / II / 2008 About Minimum Hospital Service Standard (SPM) states that the length of waiting time of the Outpatient Installation is less than 60 minutes. However, the waiting time of some services in outpatient installations exceeds the standard time set by the health minister of the Republic of Indonesia. The Percentage of waiting time over 60 minutes and less than 60 minutes was 55.25% and 44.75%. Based on that background, this research will identify the waste that causes long service waiting time in outpatient installation RSUD dr. Soediran Mangun Sumarso using Lean Hospital approach. The purpose of this study is to provide improvements using lean hospital approach. The purpose of the improvement is to reduce the waiting time in the outpatient installation.

1.2. Previous research studies

There have been several studies that have discussed lean hospital as an approach to improve the patient services in hospitals. But in previous research there are several different methods that have been used. The research that discusses the elimination of waste in Brawijaya polyclinic with lean hospital approach [8], using pareto diagram to identify the critical waste. While on lean hospital research at Bethesda Yogyakarta hospital [10], using the Borda method to identify the critical waste.

However, in this study only analyzed the 7 waste that occurred. The research using lean hospital approach in Unisma Hospital [2] has analyzed 8 waste that occured at patient service in hospital. However, the study has not used any method to identify the critical waste. Based on the explanation, then in this study will integrate the tools that have been used in previous research. In this research will analyze the 8 waste that occurred at patient service in outpatient installation and identify the critical waste by using Borda method and Pareto diagram.

2. Methods

The methodology used in this study consists of four stages. The first stage is describing the service system using a cross-functional flowchart. The second stage is identifying waste using value stream mapping, observation, and interview. A value stream map (VSM) is a structured diagram that originated with Toyota, as a tool called material and information flow mapping that used to established a consistent view of the workflow in the same manner the patient experiences it [1][6].

The third stage is to determine critical waste by Borda method and Pareto diagram. The last stage is to provide recommendation improvement using FMEA. The FMEA method is a methodology used to identify and evaluate potential failures, determine the level of risk from failure and priority scale to take the required action [4]. The result of this stage is the recommendation for improvement. Briefly, the methodology used in this research can be seen in table 1.

Table 1. Methodology

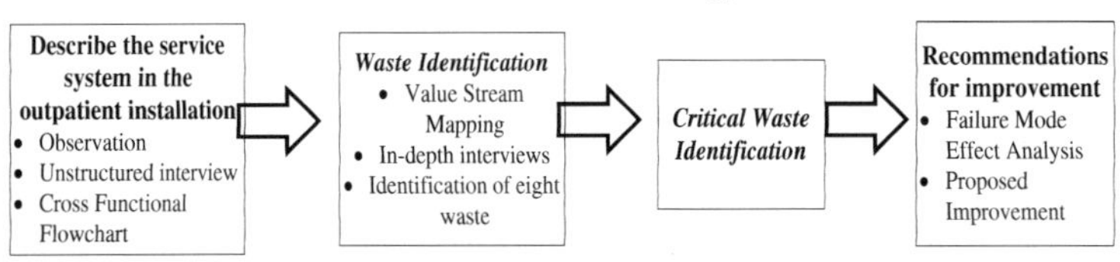

3. Result

3.1. Describe the service system in the outpatient installation

To get a complete picture of the service system in the outpatient installation is done by several steps such as observation, unstructured interview with related parties and cross-functional flowchart depiction. At this stage, the cross-functional flowchart is generated. The cross-functional flowchart describes service activities as well as those responsible for service in the outpatient installations. Cross-functional Flowchart can be seen in figure 1.

IOP Conf. Series: Materials Science and Engineering **319** (2018) 012077 doi:10.1088/1757-899X/319/1/012077

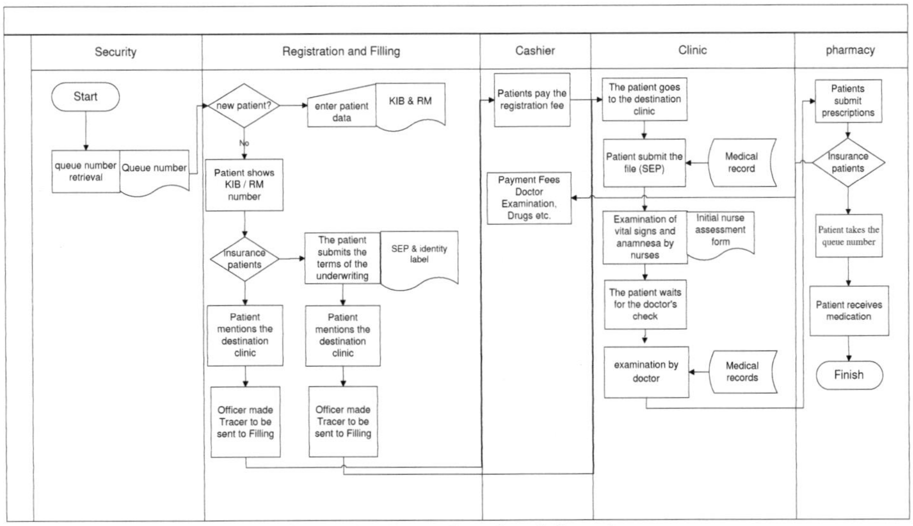

Figure 1. Cross-functional flowchart

3.2. Identification of waste

Value Stream Mapping is used as a tool to identify waste in the service process. Value stream mapping will show the value added and non-value added activity with the total time of each activity. Value Stream is obtained by calculating the processing time and waiting time of each patient service section in the outpatient installation. Furthermore, waiting time and processing time are classified as value added time and non value added time based on activities occurring at outpatient installation services. Based on the data processing time and waiting time obtained, known value added and non value added ratios. Value added ratio is obtained by dividing value added time with lead time. Non-value added ratio is obtained by dividing non-value added time by lead time. Briefly, value stream mapping can be seen in figure 2.

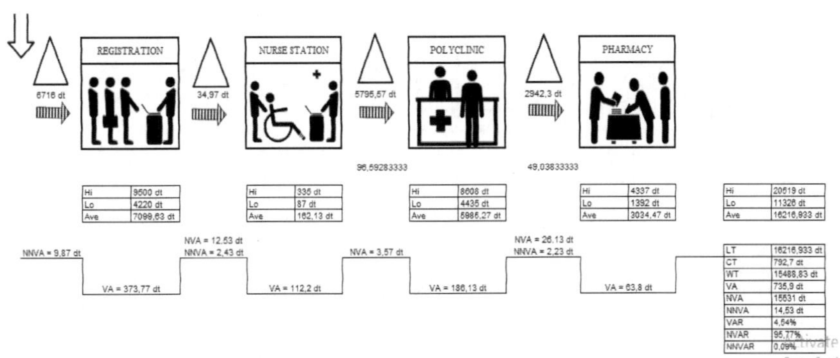

Figure 2. Value stream mapping

Based on Value Stream Mapping can be seen the ratio of value-added, non value-added and necessary non value-added. Value-added ratio is 4.54%, non value-added ratio is 95.77% and the necessary non value-added ratio is 0.09%. Gaspersz (2007) states that if the value added value is still below 30% then the process is "unlean" and still needs improvement.

In addition to using value stream mapping, waste can be identified using in-depth interviews and observations. Based on the results of in-depth interviews and observations of the service process in the outpatient installation, obtained some waste which is the category of 8 waste. The waste of service in the outpatient installation can be seen in Table 2.

Table 2. 8 Waste Activities

No	*Waste*	Activities
1	*Defect*	- The patient does not know where the polyclinic is located - Officer wrongly gives medical record - Medical record is missing / has not been returned
2	*Over Production*	- Overproduction of tracer numbers
3	*Human Potential*	- There is no customer care that serves the complaints/questions of the consumer directly
4	*Waiting*	- Patient waiting at registration - Patient waiting at polyclinic - Patients waiting for the drug to be formulated Patient waiting for Medical record
5	*Transportation*	- The pharmacist walks back and forth from the drug delivery section to the compounding section to take the medicine - filling officers deliver medical record documents find medical record documents that are missing or not returned
6	*Inventory*	- the buildup of medical record documents - certain medicinal stock is empty (unavailable)
7	*Motion*	- looking for equipment - looking for drugs Search for Medical Record documents on storage shelves
8	*Over Processing*	- Inform the same information to the patient repeatedly - checking the vital signs repeatedly - drug inspection repeatedly

3.3. Critical waste identification

Critical waste identification is done by Borda method. The result of critical waste identification with Borda method can be seen in Table 3.

Table 3. Identification of critical waste using Borda method

No	Waste	Rank								Total	Percentage	Rank
		1	2	3	4	5	6	7	8			
1	Defect	0	0	0	0	0	2	4	2	8	4%	8
2	Over Production	0	0	0	0	1	3	4	0	13	6%	6
3	Waiting	4	2	0	0	1	1	0	0	45	21%	1
4	Human Potential	0	0	1	0	1	0	3	3	11	5%	7
5	Transportation	3	0	2	2	1	0	0	0	42	20%	2
6	Inventory	1	2	1	2	2	0	0	0	38	18%	4
7	Motion	0	0	0	0	3	1	3	1	14	7%	5
8	Over Processing	1	3	2	2	0	0	0	0	43	20%	3
		7	6	5	4	3	2	1	0	214	100%	
	Total											

The waste rank that has been obtained through the Borda method is then identified using Pareto diagram. Pareto diagrams are used to determine the critical waste that becomes a priority improvement on the service in the outpatient installation. The following is a Pareto diagram of the occurrence of waste:

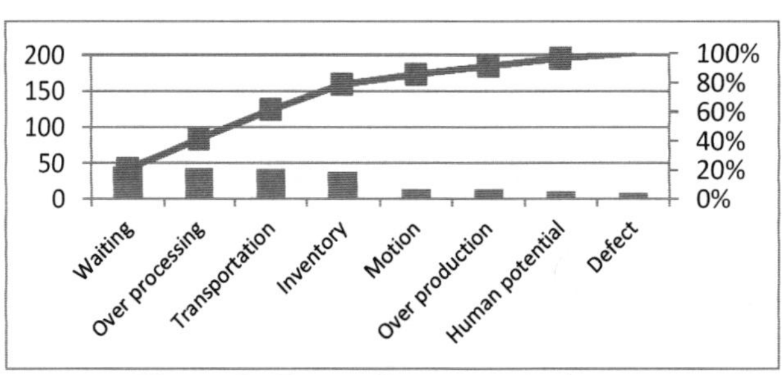

Figure 3. Pareto diagram *critical waste*

In accordance with the principle of Pareto 80/20, then the critical waste selected is waste over processing, waste waiting, waste transportation, and waste inventory.

3.4. Recommendation for improvement

Before giving recommendation improvement, firstly weighted failure mode using FMEA. Through FMEA will be known the cause of the problem with the highest risk so that it can be immediately made suggestions for improvement. By multiplying the ranking for the three factors (severity × occurrence × detection), a risk priority number (RPN) will be determined for each potential failure mode and effect. Severity is the consequence of the failure should it occur. Occurrence is the probability or frequency of the failure occurring. Detection is the probability of the failure being detected before the impact of the effect is realized. Table 4 shows the FMEA identification with the highest RPN value for each waste.

Table 4. FMEA with the highest RPN value

Waste	Service Section	Failure Mode	Severity	Causes of Failure	Occurrence	Control	Detection	RPN
Waiting	Registration	The queue of patients in the morning	4	The distance of the opening hours of the queue number and the registration counter is too far away	10	Provides 7 registration booths	4	160
	Polyclinic	The doctor came late	2	Doctor visits to hospitalization	8	Not available	10	160
Over Processing	Registration	Notify the same information repeatedly	2	Patients arrived late	9	Not available	10	180
				There is no direction to polyclinic	9	Not available	10	180
	Polyclinic			There is no doctor information	9	Not available	10	180
Transportation	Pharmacy	The pharmacist walks from the drug delivery section to the compounding section	2	There is no part connecting the compounding section and the delivery of the drug	10	Not available	10	200
Inventory	Polyclinic	Build up of medical records	2	There is no special place of the patient's Medical Record document	10	The documents placed on the desk	4	80

After obtaining the failure mode with the highest RPN value in each waste, the next step is to give the proposed improvement in accordance with the failure mode that occurred. Here is a suggested improvement for each waste.

- The proposed improvements to reduce waste waiting are to open a special registration counter for infants and toddlers, elderly aged 60 and above, pregnant and lactating mothers. To reduce the queues that accumulate during peak hours, an online reservation system is applied. Management of RSUD dr. SoediranMangunSumarso also needs to synchronize the schedule with the doctor in the clinic so that the accuracy of the service schedule is obtained. The last proposed improvement to reduce waste waiting is by adding a doctor in the polyclinic.
- The proposed improvements to reduce waste overprocessing are to provide information on the estimated time of queue so that the patient can come before the queue number is called. Another improvement recommendation is to add a polyclinic signpost at the outpatient installation and add a doctor's attendance board.
- The proposed improvement to reduce the waste of transportation is to provide a connecting glass that will connect the drug delivery section with the compounding part of the drug so as to minimize the transport carried out by the pharmacist.
- The proposed improvements to reduce waste inventory is to add a multifunctional shelf that can be used as a rack and also a trolley to return medical record documents from polyclinic to filling section.

4. Conclusions

Critical Waste that occurs in outpatient installation services are waste waiting, waste over processing, transportation, inventory. Failure mode with the highest RPN value that occurs in waste waiting is the queue of the patient in the morning and the doctor who arrived late. The failure mode with the highest RPN value in waste over processing is the officer notifying the same information repeatedly. The failure mode with the highest RPN value in waste transportation is the pharmacist's transportation from the drug delivery section to the compounding section. The failure mode with the highest RPN value in waste inventory is the buildup of medical record documents that occurred in the polyclinic. Proposed improvements that can be made to reduce waste waiting, over-processing, inventory, and transportation are open special registration counters for patients with special circumstances, implement an online reservation system, synchronize the schedule of inpatient visits with the physician and add doctor in the polyclinic, provide information on the estimated time of queue number, add a polyclinic signpost at the outpatient installation, add a doctor's attendance board, provide a connecting glass that will connect the drug delivery section with the compounding part of the drug, add a multifunctional shelf.

References

[1] Agarwal S, Gallo J J, Parashar A, Agarwal K K, Ellis S G, Khot U N, Spooner R, Tuzcu E M, Kapadia S R 2016 Impact of lean six sigma process improvement methodology on cardiac catheterization laboratory efficiency *Cardiovascular Revascularization Medicine* **17** 95-101.

[2] Adellia Y, Setyanto N W and Tantrika C F M 2014 Pendekatan lean healthcare untuk meminimasi waste di rumah sakit islam unisma Malang *Jurnal Rekayasa dan Manajemen Sistem Industri* **2** 292-301

[3] Bush R W 2007 Reducing Waste in US Healthcare System *The Journal of The American Medical Association* **297** 871-874

[4] Desy I, Hidayanto B C and Astuti H M 2014 Penilaian risiko keamanan informasi menggunakan metode failure mode and effect analysis di divisi TI PT Bank XYZ Surabaya *SESINDO* 467-472

[5] Gaspersz V 2007 *Six Sigma for Manufacturing and service Industries Strategi Dramatik Reduksi Cacat/Kesalahan, Biaya, Inventori, dan Lead Time dalam Waktu Kurang dari 6 Bulan* (Jakarta : Gramedia pustaka Utama)

[6] Graban M 2012 *2nd Edition Lean Hospital : Improving Quality, Patient Safety, and Employee Engagement* (Boca Raton : RC Press Taylor & Francis Group)

[7] Hicks C, McGovern T, Prior G and Smith I 2015 Applying lean principles to the design of healthcare facilities *International Journal of Production Economics* **170** 677-686.

[8] Himawan R and Arifianto E Y 2017 Minimasi pemborosan pada unit pelayanan kesehatan menggunakan pendekatan lean healthcare *Annual Conference in Industrial and System Engineering 10-20*

[9] Joosten T, Bongers I and Janssen R 2009 Application of lean thinking to health care: issues and observations *Int J Qual Health Care* **21** 341–7.

[10] Kharismawati A and Herliansyah M K 2016 Implementasi lean healthcare pada pasien bpjs rawat jalan di rumah sakit bethesda yogyakarta *Seminar Nasional Teknik Industri Universitas Gajah Mada* 408-417

[11] Liker JK and Kaisha TJKK 2004 *The Toyota way: 14 management principles from the world's greatest manufacturer* (McGraw-Hill : New York)

[12] Menteri Kesehatan Republic Indonesia. (2008). Standar pelayanan minimal rumah sakit (Nomor : 129/Menkes/SK/II/2008).

[13] Mazzocato P, Savage C, Brommels M, et al. 2010 Lean thinking in healthcare: a realist review of the literature *Qual Saf Health Care* **19** 376–82.

[14] Wijono D 2008 *Manajemen Mutu Rumah Sakit dan Kepuasan Pasien Prinsip dan Praktik* (Surabaya: CV. Duta Prima Airlangga)

[15] Womack J P and Miller D 2005 *Going lean in health care* (Cambridge, MA: Institute for Healthcare Improvement).

[16] Womack, J P and Jones D T 2003 *Lean Thinking: Banish Waste and Create Wealth in Your Corporation* (Touchstone Books : London)

IOP Conf. Series: Materials Science and Engineering **319** (2018) 012078 doi:10.1088/1757-899X/319/1/012078

Framework for Designing The Assessment Models of Readiness SMEs to Adopt Indonesian National Standard (SNI), Case Study: SMEs Batik in Surakarta

Fakhrina Fahma, Roni Zakaria, Royan Fajar Gumilang

Department of Industrial Engineering, Faculty of Engineering, Sebelas Maret University, Surakarta, Indonesia

fakhrina09@gmail.com, 2012.ibnu@gmail.com, royanfajargumilang@student.uns.ac.id

Abstract. Since the ASEAN Economic Community (AEC) is released, the opportunity to expand market share has become very open, but the level of competition is also very high. Standardization is believed to be an important factor in seizing opportunities in the AEC's era and other free trade agreements in the future. Standardization activities in industry can be proven by obtaining certification of SNI (Indonesian National Standard). This is a challenge for SMEs, considering that currently only 20% of SMEs had SNI certification both product and process. This research will be designed a model of readiness assessment to obtain SNI certification for SMEs. The stages of model development used an innovative approach by Roger (2003). Variables that affect the readiness of SMEs are obtained from product certification requirements established by BSN (National Standardization Agency) and LSPro (Certification body). This model will be used for mapping the readiness for SNI certification of SMEs'product. The level of readiness of SMEs is determined by the percentage of compliance with those requirements. Based on the result of this study, the five variables are determined as main aspects to assess SME readiness. For model validation, trials were conducted on Batik SMEs in Laweyan Surakarta.

1. Introduction

Indonesia is one of the countries that has high potential in entrepreneurship. According to data from the Asian Development Bank (ADB) Institute in 2015, Indonesia is a country that has the largest Small and Medium Enterprises (SMEs). SMEs has contributed more than 50% to Gross Domestic Product (GDP), and 97% of the employment [3,6]. The contribution of SMEs is vital to Indonesia's economic growth. This is shown by many SMEs that have could survive in the monetary crisis in 1997 and increasing in number each year. So the existence of SMEs is expected to increase Indonesian economic growth [3,6,12].

Since the ASEAN Economic Community (AEC) established on 1st January 2016, Indonesian SMEs have a great opportunity to expand their market share. SMEs's products can expand into the ASEAN market without any barriers either tariff or non-tariff as long as it can be competitive. Products that have a good competitiveness will be able to adapt to the market. The competitiveness indicators are related to the quality assurance of the products. Therefore, the fulfillment of products to standard is an important factor in the expansion of the product market in ASEAN. So only SME products that have a quality assurance certificate will be absorbed into the ASEAN market quickly.

Batik SMEs is chosen as a case study because batik is a superior product in Surakarta. According to data from the Department of Industry and Commerce Surakarta, export volume in 2016 was 579,073.49 kg rose from 488,038.49 kg (in 2015) while the value rose from US $ 9.45 million to US $ 10.88 million. It's made batik as one of the three largest export commodities of Surakarta along with textile products and plastic [17]. Nationally, weaving and batik industries are able to contribute significantly to the national economy with an export value of 151.7 million US dollars in 2016 [8].

Standardization is believed to be an important factor for seizing opportunities in the AEC era and other free trade agreements in the future. Implementation of standardization in the industry is proven by SNI (Indonesian National Standard) certification. This is a challenge for SMEs, because currently there are many SMEs that do not have SNI certification either product or process. The total number of SMEs products in Indonesia is 55 million, but not more than 20% that have SNI certificate [4,6]. This reality is very worrying for the existence of SMEs. The positive impact of AEC about market expansion becomes difficult to realize.

This research will design a model of readiness assessment to obtain SNI certification for SMEs. The model can be used to map how the readiness level of SMEs to obtain SNI certification. It also can be identified stages of the process that the requirements are difficult to be met by SMEs. So that the government can make good program/policy in coaching and empowering SMEs. The model design uses an innovative approach that developed by [11]. This approach provides a framework for applying innovations better. The advantages of this approach are integrated systems ranging from the introduction of needs/problems, required research, model development, the process of commercialization (a publication for the user), diffusion and adoption (through the pilot project) to the final implementation to the wider user. So hopefully this model is easy to apply for assessment in SMEs. This approach model had been used by ref. [9] in the implementation and development of the International Software Process Lifecycle standard in SMEs.

Previous research about assessing the readiness of SNI certification had been done by [1] and [10]. That researchs used a different approach from this study. The framework for assessing the readiness (that done before) consist four critical factors, i.e. perceived national readiness (macro level), perceived industry readiness (meso level), organizational readiness (micro level), and perceived environmental pressure. The variables in the critical factors are derived from the development of 10 CSFs (Critical Success Factors) for quality initiative implementation by Yusof and Aspinwall (1999). But in this study the variables derived from the stages of the certification process that has been determined by BSN.

2. Research Methodology

This research has six stages. Early stage of this research is started from collecting data, i.e. identification certification requirement, identification proses SNI certification, and observe the condition of SMEs in SNI application. The next stage is developing assessment models of readiness SMEs to adopt SNI. In developing the model, there are four steps such as system characterization, develop an influence diagram, determine variables and attribute, and design framework of readiness model. The fourth stage is commercialization. At this stage, all attributes are translated into questions to assess the readiness of SMEs in the questionnaire. The fifth stage is diffusion and adoption. At this stage SMEs were selected as pilot projects to be assessed using a questionnaire. The level of readiness of SNI certification is measured by calculating the number of attributes that SMEs can be met relative to the total number of attributes. Its value is expressed as a percentage. The final step is consequences. This stage contains recommendations that need to be made based on the assessment results on SMEs. The fifth and sixth stage cannot be done because it is still in the research stage. Briefly, the methodology is presented in the Figure 1.

3. Result And Discussion

3.1. *SNI Certification Requirement*

Product certification is a conformity assessment activity to provide confidence that the product meets the established requirements, including performance, security, interoperability, and sustainable products for consumers, regulators, industry and other stakeholder. Certification approval is conducted by a third party called certification body (LSpro) [5,13]. The Batik LSpro is a Center for Crafts and Batik Yogyakarta (known as TOEGOE LSpro). The requirement of batik refers to SNI batik, that is [14,15, 16].

BSN establishes certification requirements that consisting of administrative requirements (i.e. company deed, business legality, brand permit, etc.) and technical requirements (everything related to the implementation of the quality management system, including the document) [1,5,13]. Quality management system covers all processes within the organization, and that organization will operate according to procedures established by organization [7,18].

The certification requirements also depend on the product certification scheme. Product certification scheme are the rules, procedures and management of product requirements. Product certification schemes have seven types, that is 1a, 1b, 2, 3, 4, 5, and 6. The type of certification scheme is differentiated by the type of surveillance activity done [13]. The determination of the batik certification scheme and requirements are determined through FGD with LSpro. The batik certification scheme is type 3. This scheme includes selection, determination (product testing), field evaluation related to production line, review and decission of the certification. Type 3 scheme is followed by surveillance to test and evaluate production line.

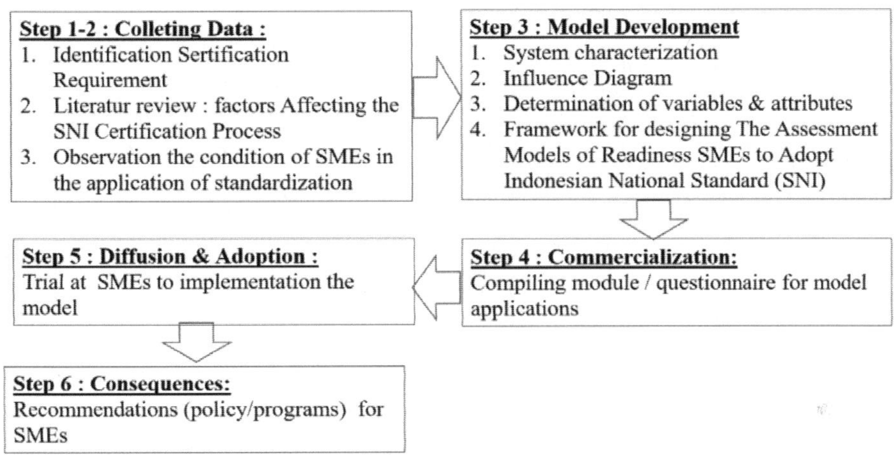

Figure 1. Research methodology

3.2. *Model Development*

Characterization of the system is done by describing the phases of the certification process in detail. All requirements, both administrative and technical are breakdown until the basic requirements are obtained. These requirements are re-identified whether there is a derivative requirement. The process ends when the final requirement can not be broken down again. For example: The business permit, this requirement has requirements such as ID cards, land deed, letter of property tax paid off, notarial deeds, and building permit (IMB). ID cards and letter of property tax paid off are basic requirement, but the notarial deed and IMB aren't. They have derivative requirements so that the breakdown needs to be done. All the stages of the process are presented in the workflow diagram in Figure 2.

All the basic requirements of breakdown results in the final stages are grouped by their character. The results are obtained 25 major attributes, and grouped into five variables (personal, corporate, managerial, technical and economic). Personal variables are attributes that are personal inherent with the personal owners of SMEs. Company variables are attributes that are attached to the company. This means that every company has the right to get this document if they fulfill the requirements (especially the cost). Managerial variables contain attributes that a company can have with a heavier effort because it has to prepare many supporting documents. The technical variables contain the attributes associated with the implementation of the quality management system. The economic variables contain attributes

IOP Conf. Series: Materials Science and Engineering **319** (2018) 012078 doi:10.1088/1757-899X/319/1/012078

associated with the costs that must be incurred to meet the requirements. The framework of SNI certification readiness assessment model in SME is presented in Figure 3 below all the variables and their attributes.

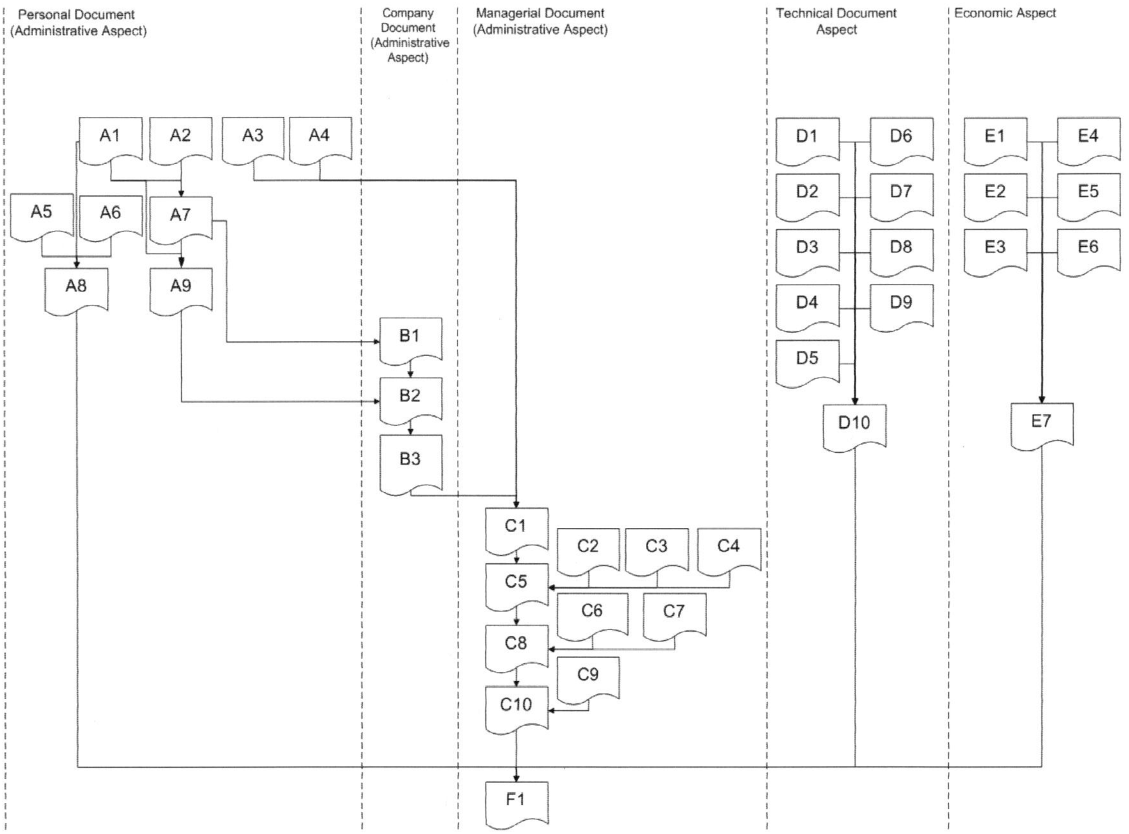

Where:

A1 :	Identity Card	C2 :	Application letter of SPPL	D6 :	Task Allocation
A2 :	Covering Letter from RT/RW	C3 :	Willingness letter of SPPL	D7 :	Specification of Prodution Equipement
A3 :	Letter of Property tax paid off	C4:	Drawing of Business Activity	D8 :	Specification of Production Materials
A4 :	Land Deed	C5 :	SPPL / UKL-UPL / AMDAL	D9 :	Inspection of Product Quality
A5 :	Brand Ettiquette	C6 :	Figure and Calculation of Concrete Construction and Steel Structure	D10 :	Approved Quality Document
A6 :	Brand Submission Letter	C7 :	Architectural Drawings	E1 :	Cost of Notary Deed Establishment
A7 :	Business Domicile Certificate	C8 :	Building Permit Certificate for Business Place	E2 :	Replacement Cost for Printing a Map
A8 :	Brand Certificate	C9 :	Signature and Permission from Neighbor	E3 :	Building Permit Certificate Cost
A9 :	Personal Taxpayer ID Number	C10 :	Business Permit (HO), SIUP, TDP, TDI / IUI	E4 :	Disturbance Permit Cost
B1 :	Notary Deed	D1 :	Policy of Quality	E5 :	Brand Registration Cost
B2 :	Corporate Taxpayer ID Number	D2 :	Structure of Organization	E6 :	Certification and Surveillance Cost
B3 :	Deed of Ratification from District Court or Ministry of Law	D3 :	Production Process Flowchart	E7 :	Total Cost of SNI Certification
C1 :	KRK / IPR	D4 :	Quality Objectives	F1 :	SPPT SNI
C2 :	Application letter of SPPL	D5 :	Procedure of Quality		

Figure 2. Workflow diagram SNI Certification Prosess

3.3. *Commercialization*

At this stage the questionnaires are made for the implementation of the assessment model. The questionnaires will be a guide in the assessment in SMEs. In this section, variables and attributes are translated into questions, as presented in Table 1. The results of the questionnaires will be processed by descriptive analysis to do mapping about the readiness level of SMEs as input materials in focus group discussion with relevant stakeholders. FGDs are conducted to determine critical attributes that were difficult to meet by SMEs and design appropriate policy programs.

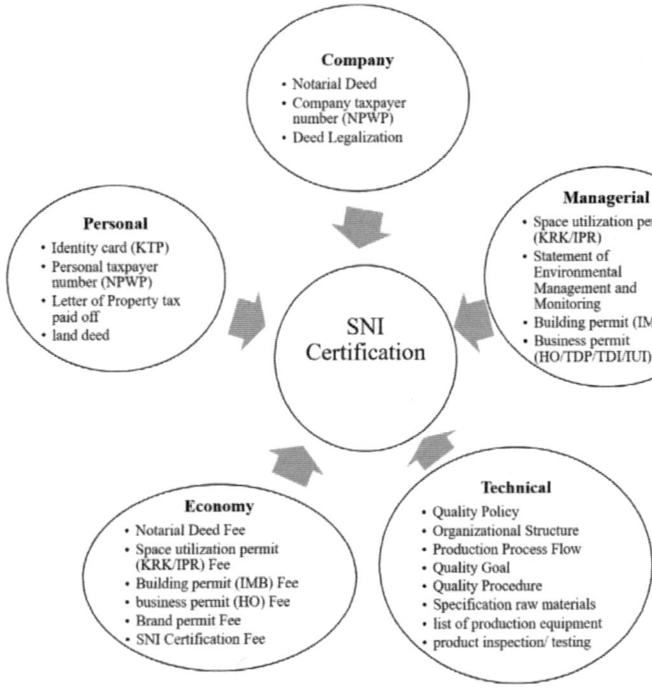

Figure 3. A framework for assessing SMEs Batik readiness on SNI Certification

Table 1. Questionnaire Design

Variable	Item Question
Personal	Checklist of documents that SMEs owner have [v] :
	☐ Identity card (KTP)
	☐ Personal taxpayer number (NPWP)
	☐ Letter of Property tax paid off
	☐ land Deed
Company	Checklist of documents that SMEs/Companies have [v] :
	☐ Notarial Deed
	☐ Company taxpayer number (NPWP)
	☐ Deed Legalization
Managerial	Checklist of documents that SMEs/Company have [v]
	☐ Space utilization permit (KRK/IPR)
	☐ Statement of Environmental Management and Monitoring
	☐ Building permit (IMB)
	☐ Business permit (HO/TDP/TDI/IUI)
Technical	Checklist of documents that SMEs/Company have [v]
	☐ Quality Policy Document
	☐ Organizational Structure document

	☐ Production Process Flow Document
	☐ Quality Goal document
	☐ Quality Procedure Document
	☐ Specification of main and supporting raw materials Document
	☐ list of main and supporting production equipment Document
	☐ product quality inspection / testing Activities Document
Economy	Checklist financial capability of SMEs/Company to fulfill the payment [v]
	☐ Notarial Deed Fee
	☐ Space utilization permit (KRK/IPR) Fee
	☐ Building permit (IMB) Fee
	☐ Business permit (HO) Fee
	☐ Brand permits Fee
	☐ SNI Certification Fee

4. Conclusion

This research has been provided framework for assessing SMEs Batik readiness on SNI certification that can be used to map readiness of SMEs'product. This research has also succeeded in compiling a tool that has been tested for its validity and reliability. Further research can be done by diffusion and adoption stage to implement the assessment models of readiness SMEs complete and comprehensive.

Acknowledgement

The research is supported by The Ministries of Research, Technology, and Higher Education with HIBAH RISET FUNDAMENTAL Research Program (Contract No. 1073/UN27.21/PP/2017).

References

[1] A.Susanty, D. I. Rinawati, B. Purwanggono, D. Puspitasari, Meylani. Assessing SMEs Batik Readiness for SNI Adoption (Case Study SMEs Solo and Yogyakarta). Proceedings of the 2013 IEEE IEEM. 978-1-4799-0986-5/13/$31.00

[2] Badan Standarisasi Nasional., 2014. *Pengantar Standarisasi* (Introduction to standardization), 2nd Edition. BSN, Jakarta.

[3] Badan Standarisasi Nasional. 2013. *Sistem Manajemen Mutu SNI ISO 9001:2008. Penerapan pada Usaha kecil Menengah* (Quality Management System of SNI ISO 9001: 2008. Implementation on Small and Medium Enterprises). BSN, Jakarta

[4] BSN (2016) *Banyak Produk UMKM Belum Memiliki SNI atau Sertifikasi.* (Many SMEs products don't have SNI or certification). [online]. Availlable : http://www.bsn.go.id/main/berita/berita_det/6842/Banyak-Produk-UMKM-Belum-memiliki-SNI-atau-Sertifikasi#.WOYtKNJ942w

[5] BSN. *Bagaimana Proses Mendapatkan Sertifikasi SNI untuk Produk*. (How the Process to Obtain SNI Certification) [online]. Availlable : http://bsn.go.id/main/berita/berita_det/7008/Infografis---Alur-Proses-Sertifikasi-SNI-pada-Produk#

[6] Irjayanti, M., Azis, A.M. Barrier Factors and Potensial Solution for Indonesian SMEs. Procedia Economic & Finance 4 (2012) p : 3-14.

[7] Mahmud, N., and Hilmi, M.F. TQM and Malaysian SMEs Performance : The Mediating Roles of Organization Learning. Precedia Social and Behavioral Sciences 130 (2014) p : 216-225

[8] Ministry of Industry. (2017). *Nilai Ekspor Industri Tenun dan Batik Lampaui USD 151 Juta.* (The Export Value of Weaving and Batik Industry is USD 151 Million). [online]. Availlable : http://www.kemenperin.go.id/artikel/17377/Nilai-Ekspor-Industri-Tenun-dan-Batik-Lampaui-USD-151-Juta

[9] O'Connor, R. V., dan Laporte, Claude. Y., 2013. An Innovative Approach to the Development of an International Software Process Lifecycle Standard for Very Small Entities. International Journal of Information Technologies and Systems Approach.
[10]Rinawati, D. I., Susanty, A., Sari, D. P. dan Meylani. *Penilaian Kesiapan Penerapan Standar Nasional Indonesia (Sni) Pada Ukm Batik Di Solo, Yogyakarta Dan Pekalongan*. (Assessment Readiness of Application Indonesian National Standard (SNI) on Batik SMEs in Solo, Yogyakarta and Pekalongan). Jurnal Standardisasi Volume 17 Nomor 1, Maret 2015: Hal 1 – 10.
[11] Roger, Everett M. 2003. Diffusion Of Innovations Third Edition. The Free Press Collier Macmillan Publisher, London.
[12] Slamet,R., Nainggolan, B., Roessobiyatno, Ramdani, H., Hendriyanto, A, Ilma , L. *Strategi Pengembangan UKM Digital Dalam Menghadapi Era Pasar Bebas*. (Digital SME Development Strategy In Facing Free Market Era). Jurnal Manajemen Indonesia Volume 16- No.2 April 2016 page : 136-147.
[13] SNI ISO/IEC 17067:2013 : *Penilaian kesesuaian - Fundamental sertifikasi produk dan panduan skema sertifikasi produk* (Assessment of conformity - Fundamentals of product certification and product certification scheme guidelines). (ISO/IEC 17067:2013, IDT).
[14] SNI 8302:2016 : *Batik tulis – Kain – Ciri, syarat mutu dan metode uji* (Hand drawn Batik- Fabrics - Characteristics, quality requirements and test methods)
[15] SNI 8303:2016: *Batik cap – Kain – Ciri, syarat mutu dan metode uji* (Hand Stamp Batik - Fabrics - Characteristics, quality requirements and test methods)
[16] SNI 8304:2016: *Batik Kombinasi – Kain – Ciri, syarat mutu dan metode uji* (Mixing between Batik - Fabrics - Characteristics, quality requirements and test methods)
[17] Solopos (2015). 2015, *Realisasi Nilai Ekspor Produk Solo Turun* (The realization of Export Value of Solo Products got down) [online]. Availlable : http://www.solopos.com/2016/01/20/ekspo-solo-2015-realisasi-nilai-ekspor-produk-solo-turun-682633
[18] Tisca, I.A., Cornu, G., Diaconu, N., Dumitrescu, C.D. Diagnosis, risk and efficiency in the implementation of TQM in smal and medium enterprises. Procedia Economic and Finance 26 (2015) p : 215-218.

IOP Conf. Series: Materials Science and Engineering **319** (2018) 012079 doi:10.1088/1757-899X/319/1/012079

Design of conveyor utilization monitoring system: a case study of powder coating line in sheet metal fabrication

Hoedi Prasetyo[1], Yohanes Sugiarto[2] and Cucuk Nur Rosyidi[3]

[1,3] Master Study Program of Industrial Engineering, Faculty of Engineering, Universitas Sebelas Maret, Surakarta 57126, Indonesia
[1,2] Mechatronics Department, Politeknik ATMI Surakarta, Surakarta 57145, Indonesia

E-mail: [1]hoedi@student.uns.ac.id, [2]ysugiarto@atmi.ac.id, [3]cucuk@uns.ac.id

Abstract. Conveyor is a very useful equipment to replace manpower in transporting the goods. It highly influences the productivity, production capacity utilization and eventually the production cost. This paper proposes a system to monitor the utilization of conveyor at a low cost through a case study at powder coating process line in a sheet metal fabrication. Preliminary observation was conducted to identify the problems. The monitoring system was then built and executed. The system consists of two sub systems. First is sub system for collecting and transmitting the required data and the second is sub system for displaying the data. The system utilizes sensors, wireless data transfer and windows-based application. The test results showed that the whole system works properly. By this system, the productivity and status of the conveyor can be monitored in real time. This research enriches the development of conveyor monitoring system especially for implementation in small and medium enterprises.

1. Introduction

Conveyor is a technology which commonly used in industries with high speed and large volume of production. Conveyor has a function as material handling equipment to transport raw materials, work in process (WIP) materials or finished products from one station to another. It is very useful to replace manpower in transporting the goods. This technology was first popularized by Henry Ford in the era of second industrial revolution (beginning of 20th century) to support his automobile mass production. Basically, a conveyor system consists of belt or chain which is driven by rotating electric motors. The goods are put on the belt and move to particular location with a certain speed. Nowadays there are many conveyor systems available which are intended to meet various requirements of industrial process [1].

Conveyor systems are usually used in mass or continuous production systems. As explained in [2], those kind of production systems have several characteristics such as large volume of products, predetermined sequence of operations and higher rate of production with reduced cycle time. Those kind of production systems have risks related to the line balancing and the availability of the machines. If there is a problem in a certain workstation, it can immediately slow down the production speed or even stops the entire production line if a machine's breakdown is occurred. As material handling equipment, conveyor can be considered as a machine also. Conveyor system will highly influences the productivity, production capacity utilization and eventually the production cost. That is the reason why conveyor system plays very important role in production performance.

In order to maintain or to improve the production performance, an accurate production monitoring system is required [3]. This prevails to any kind of production system including productions which utilize conveyor systems. There are some research which have been carried out to create conveyor monitoring systems. Design proposed by [4] aimed to monitor and detect the fault occurring in the coal conveyor using PLC (Programmable Logic Controller) and SCADA (Supervisory Control and Data Acquisition). It was intended to replace manual control which has many disadvantages such as being the major cause of frequent accidents. Another similar research was done by [5] using multivariate data analysis application for fault detection in belt conveyor drive unit. Those previously mentioned research focused only on the condition of the conveyor systems to avoid unexpected breaks in production but did not take into consideration the utilization of the conveyor system. Research by [6] proposed a model to predict and to evaluate the performance of conveyor system. It had a purpose to improve the flows of WIP materials and information with real time traceability and visibility. However, RFID (Radio Frequency Identification) technology has to be adopted to implement this model. Research by [7] resulted a system which was not only able to monitor but also to control the conveyor system. But the system was still in the prototype form and wasn't yet proven to be implemented in real industry.

In order to enrich the development of conveyor monitoring system, this paper proposes a system to monitor the utilization of conveyor at a low cost especially for the conveyor which already existed in a production line. The components used were chosen in such a way as to reduce the investment cost but not eliminate the main function. Research was conducted through a case study at powder coating process line in a sheet metal fabrication owned by PT. ATMI Solo.

2. Research Method

Preliminary observation was conducted to identify the problems exist in the company by interviewing the related staffs. Important information such as production layout, input and output of the process, type of the conveyor system, mechanical construction, electrical control of the conveyor system and other related information were collected. All information gathered were then used to determine the system's specification. The specification is taken from [8] which recommends guiding questions to build production monitoring system. Basically, those guiding questions are explained as follows: what data should be collected and how to collect them, how to save the data, how to connect and communicate the data and how to visualize the data. For the measurement function, direct method was used [9]. The method utilizes laser and proximity sensors which are capable of directly measuring the quantity of products. The signal processing method is taken from [10]. The signal generated by the sensor needs to be processed to produce a good quality of signal. The signal is then reprocessed with a specific descriptor feature so that the information contained can be understood by the user as intended. Examples of feature descriptor are root mean square, peak, mean, standard deviation, etc. The result of signal processing by descriptor is then displayed so that users can access and view information easily for analysis purposes.

There are some works that need to be considered when building the monitoring system, which are mechanical design, electrical wiring and software programming. Afterwards, the system is deployed and tested. The tests consist of verification and validation. Verification means that all functions of the system are tested one by one. If the verification demonstrates proper results, then the system needs to be validated by the associated user. If there were any unsatisfied result related to the specified function, the system is then reviewed to find the solution. The validation results will determine whether the system solves the problem. If the overall test result is feasible then the proposed system can be recommended for implementation.

IOP Conf. Series: Materials Science and Engineering **319** (2018) 012079 doi:10.1088/1757-899X/319/1/012079

3. Results

3.1. Observation result and problems identification

Sheet metal fabrication at PT. ATMI Solo produces various type of office equipments such as chair, desk, drawer, filling cabinet and other similar products. The production begins from material cutting, punching, and bending process. The work in process parts are then hanged on the conveyor as shown in figure 1(a). The existing conveyor system is classified as open track-overhead chain conveyor system [1]. The conveyor has length of 331,2 meters with maximum load capacity of 5 tons. The conveyor runs continuously at a certain low speed. The parts are then moved to the washing and drying facility. The parts need to be painted with specific colour according to the customer order. This process involves powder coating process which is done manually by the operator as shown in figure 1(b). Subsequently, the coated parts need to be heated by the oven facility in order to make the colour permanent. Finally the parts are ready to be assembled into finished products. There is a control panel which is used to turn on and to adjust the speed of the conveyor system. Related operators have an authority to control the conveyor through the control panel in accordance with the situation occurring in production. The productivity is monitored manually, in which an operator writes the number of processed parts on a paper. In the end of workday, the paper is delivered to the supervisor for data checking and validation.

There are three problems identified during the observation. First problem deals with the implementation of manual production monitoring system. The supervisor can only evaluate the productivity data at the end of workday so there is a delay for data collection and analysis. The conveyor system frequently stops because of several causes which are intentional or unintentional. The intentional causes usually come from the unavailability of the operator and the problems occurring on a certain production process. The related operator will deliberately stop the conveyor for a while. This intentional causes have a significant effect on the conveyor utilization.

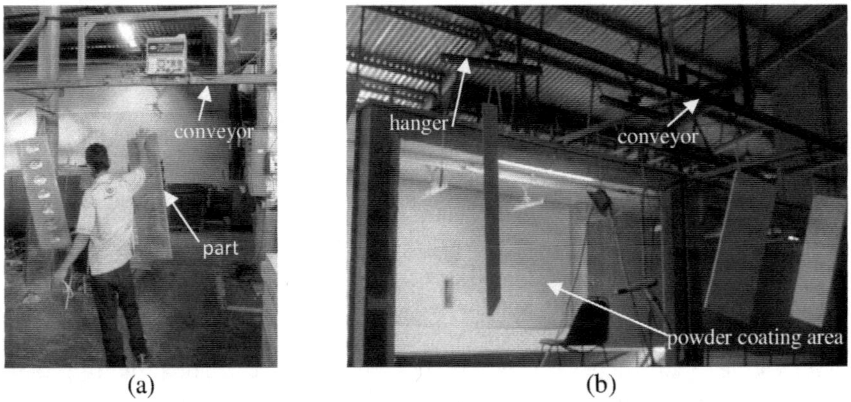

Figure 1. The conveyor system: (a) parts are hanged on the chain, (b) work area for powder coating process.

Longer idle time of the conveyor means lower utilization which results to lower level of productivity. Unfortunately the supervisor is unable to always keep on eye to detect this condition and this is the second problem. The unintentional causes can emerge from the overloading or problems inside the conveyor system which force the system to shutdown by itself. If this happens, the operator should inform the supervisor who then call the maintenance unit to fix the problems. According to the maintenance unit, this condition can be prevented if there is information about the operating hours of the conveyor system. All this time, the operating hours data of the conveyor system are never recorded and that is the third problem.

3.2. System design

Considering the above conditions, it is necessary to design a system capable of monitoring the conveyor utilization automatically. First, we need to determine the system specification using the method proposed by [8]. Table 1 shows the specification. The data need to be collected are the number of processed parts, conveyor status, duration of status and total operating hours. Number of processed parts represents the level of productivity and are detected using proximity and laser sensor. Proximity sensor detects the hangers and laser sensor detects the parts. Conveyor status provides information whether the conveyor is currently at the status of running, idle or down. Duration of status informs the user the duration time of each status in a day. Total operating hours shows the accumulation of conveyor operating hours since the monitoring system is installed.

Table 1. Monitoring system specification.

Guidelines questions	Results	Technology
1.Which data should be collected and how to collect the data?	a. Number of processed parts by detecting the existence of part. b. Conveyor status by detecting the electricity of conveyor control panel. c. Duration of stand by and running by measuring the conveyor motor's rpm (rotation per minute). d. Duration of shut down by detecting the electricity of conveyor control panel. e. Total operating hours	a. Proximity and laser sensor. b. Relay. c. Tachometer measuring. d. Relay. e. Tachometer measuring.
2. How to save the data?	a. Temporary at the work area b. Permanently, PC workstation at the management office	a. SD card module b. MySQL
3. How to connect and to communicate the data?	a. Sensor node to transmitter b. Transmitter to receiver wirelessly c. Receiver to PC workstation	a. Cable wire b. ZigBee protocol c. USB 2.0 cable
4. How to visualize the data?	Windows based application	Visual Studio 2013 and MySQL

The status and its duration data are measured by relay and tachometer circuits. If the conveyor was turned off, then the relay will give "OFF" signal and it will show the shut down status. Motor conveyor rotation is measured by tachometer and indicates that the conveyor is in running state. If the motor is not rotating but the conveyor system is still turned on, the tachometer will give no signal and the conveyor is considered in the idle state.

The data from the measurement instruments are stored temporary in a SD (Secure Digital) card module which then are transmitted wirelessly to a PC workstation and being saved permanently. The wireless communication utilizes ZigBee technology. It is an IEEE 802.15.4-standard for a communication protocols used to create personal area networks with small, low-power digital radios [11]. There is a windows-based application which is dedicated to visualize the data so that the user can make use of the data easily.

IOP Conf. Series: Materials Science and Engineering **319** (2018) 012079 doi:10.1088/1757-899X/319/1/012079

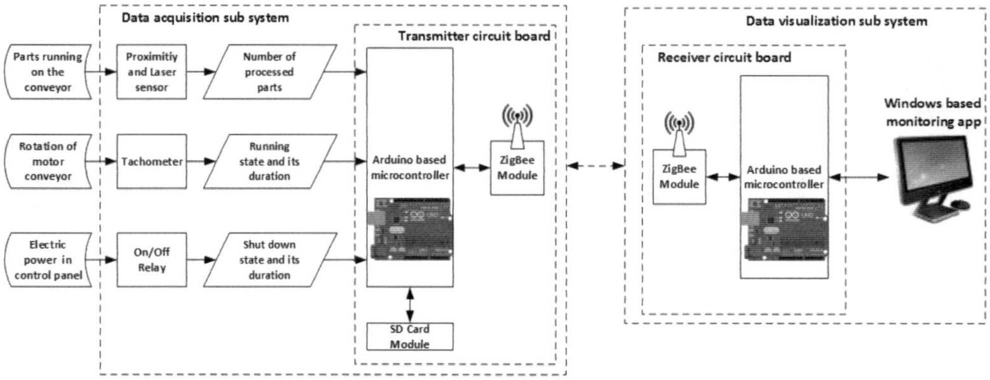

Figure 2. The proposed conveyor utilization monitoring system.

The proposed monitoring system is depicted in figure 2. It consists of two sub systems, data acquisition and data visualization. The data acquisition sub system has a function to acquire data by means of sensors and then transmit them to data visualization sub system. Relay and tachometer sensor are installed on the conveyor's control panel. In particular, the proximity and laser sensor need to be mounted on a mechanical frame which is located next to the conveyor chain. The design of the sensor-mounting frame and its installation are shown in figure 3(a) and 3(b) respectively. All of the signals measured by the sensors are being the input for the transmitter circuit board which is an electronic circuit based on Arduino microcontroller.

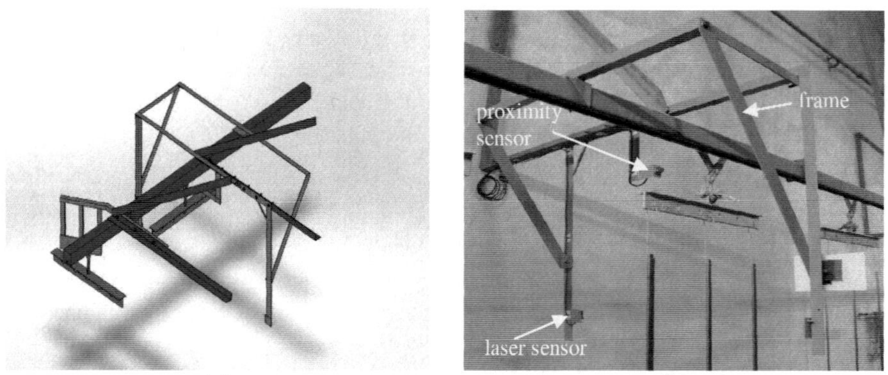

(a) (b)

Figure 3. Sensors mounting frame: (a) 3D design, (b) installation on the conveyor.

The data visualization sub system has a function to receive the data transmitted by data acquisition sub system and then visualize them to the user. This sub system is located at the fabrication management office which is approximately 30 meters from the transmitter. The data are processed by means of a microcontroller circuit and a windows-based application. Figure 4(a) shows the screenshot of the monitoring system application's main view. The application displays the information about recent date and time, total operating hours, recent status of the conveyor, number of processed products and duration of each status for a particular day. Through this monitoring system, the supervisor is now able to monitor the productivity in real time. He can easily supervise the status of the conveyor system remotely from his office without having to visit the shop floor. Information about total operating hours is useful for scheduling the maintenance activity. It can prevent any failure of the conveyor operation. The user can also retrace the data to any date and any time because the system records all status since it was firstly installed. There is an additional feature for the user to easily save a range of data at a certain date and time into spreadsheet form (MS Excel) as shown in figure 4(b).

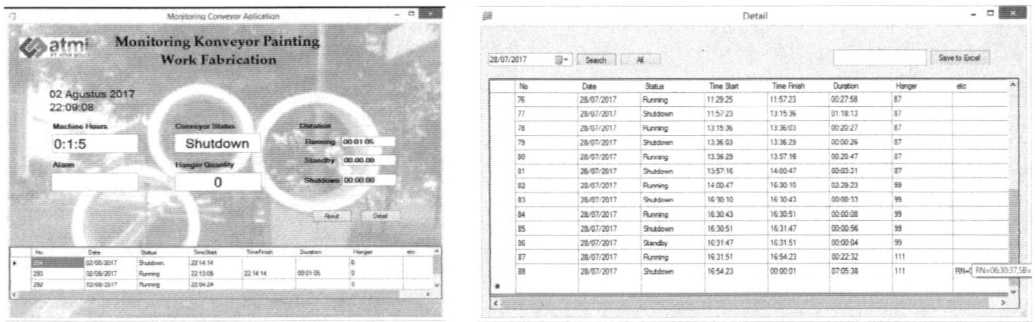

(a) (b)

Figure 4. Monitoring system application: (a) main view, (b) export to spreadsheet feature.

4. Conclusion

In this paper, we developed a conveyor utilization monitoring system for powder coating line in a sheet metal fabrication. The monitoring system consists of data acquisition sub system for collecting and transmitting the required data and data visualization sub system for displaying the data. The test results showed that the whole system works properly. Due to the implementation of the system, the productivity and status of the conveyor can be monitored in real time. Thus, conveyor utilization can be improved and actions can be taken immediately in response to any status of the production line. Information about total operating hours allows maintenance unit to reduce the level of conveyor's down time. The data recording feature also enables user to retrace any data in the past period for any evaluation purposes. The hardware for this proposed system needed relatively low cost investment which was about $ 350. This research enriches the development of conveyor monitoring system especially for implementation in small and medium enterprises. By using more sophisticated sensors and signal processing algorithms, this system can be upgraded to detect and to prevent any fault of the conveyor system.

References

[1] McGuire and Patrick M 2009 *Conveyors: Application, Selection, and Integration* (New York: CRC Press)
[2] Anil Kumar S and Suresh N 2008 *Production and Operations Management* (New Delhi: New Age International (P) Ltd
[3] Subramaniam S K, Husin S H, Singh S S and Hamidon A H 2009 *Int. Journal of Systems Applications, Engineering & Development* **3** 28-35
[4] Kanmani M, Nivedha J and Sundar G 2014 *IJAREEIE* **3** 243-248
[5] Piotr K, Jakub S, Jakub O, Mateusz S, Agnieszka W and Radosław Z 2017 *Cyclostationarity: Theory and Methods III* 173-186
[6] Singh V , Sarwar F, Chan F T S and Tiwari M K 2012 *Int. Journal of Computer Integrated Manufacturing* **25** 551-564
[7] Vardhan D S and Narayan Y S 2015 *Proc. Int. Conf. on Man and Machine Interfacing (Bhubaneswar)* vol 1 (IEEE) p 1-6
[8] Snatkin A, Karjust K, Majak J, Aruvali T and Eiskop T 2013 *Estonian Journal of Engineering* **19** 62-75
[9] Li H and Chen Y 2015 Machining process monitoring *Handbook of Manufacturing Engineering and Technology* ed Andrew Y C N (London: Springer) pp 941-979
[10] Nebot A, Vicente J and Subiron F R 2010 *The Int. Journal of Advanced Manufacturing Technology* **47** 237-257
[11] Kinney P 2003 ZigBee technology: wireless control that simply works *Communications design conference (San Jose)* **2** 1-7

IOP Conf. Series: Materials Science and Engineering **319** (2018) 012080 doi:10.1088/1757-899X/319/1/012080

Profit Analysis Model of Smart Item Implementation in Integrated Supply Chain Process

Yustina Tritularsih[1,2,*], Andhy Rinanto[1], Hoedi Prasetyo[1], Cucuk Nur Rosyidi[3]

[1] Student of Industrial Engineering Program, Faculty of Engineering, Sebelas Maret University, Jl. Ir. Sutami No. 36A, Kentingan, Surakarta, 57126, Indonesia
[2] Staff Lecturer of Mechatronic Departement, Politechnic ATMI Surakarta, Jl. Adisucipto/ Jl. Mojo No.1, Laweyan Surakarta, 57145, Indonesia
[3] Staff Lecturer of Industrial Engineering Program, Sebelas Maret University, Jl. Ir. Sutami No. 36A, Kentingan, Surakarta, 57126, Indonesia

*Email: yustina_tritularsih@atmi.ac.id

Abstract. Nowadays all links of the entire supply chain need to integrate their different infrastructures and they have better control of them to drive better profits. This integration should offer the ability for companies in order to have an overall and transparent insight to its supply chain activities. An intelligent supply chain which is mainly supported by Smart Items technology can satisfy the need of those integration. By means of Smart Items, a company can benefit some advantages. Those are cost reduction and value creation. However, currently there is no comprehensive Smart Item infrastructure exists yet so it is difficult to calculate the true benefit information. This paper attempts to recommend a model for estimating the benefits of implementing Smart Items in a company which has an integrated supply chain process. The integrated supply chain means that three echelons (supplier, shipper and retailer) of supply chain are belonged to a company. The proposed model was used to determine the shrinkage value and RFID tag price which can give the maximum benefit of Smart Items implementation. A numerical example is also provided to give a better comprehension on model calculation.

1. Introduction

Technology advances quite rapidly at this time in which industry needs to adapt with this condition in order to stay competitive. Smart Item technology also grows with a high capability to track goods in more detail and easier to be integrated into the whole business process. Through automatic real-time data acquisition, this technology can improve business operations efficiently. A Smart Item is a device capable of independently collecting information from its environment, processing, and communicating data. Barcode is the first and still largely used technology to recognize items electronically. The barcode label is attached to an item and then optically the item can be detected by the barcode reader. The reader is a printed identification and sends the data to the information system, which updates the data received. Since barcodes can only support tracking, consequently, they can only be used in loading and unloading processes on roads with very rough scales [3]. Radio Frequency Identification (RFID) has a higher level of technology than barcode. It is a method of remotely storing and retrieving data using devices called RFID tags equipped with small antennas. Basically, an RFID infrastructure comprises of RFID tags, readers either fixed or mobile, and device controllers. The information stored includes object location, aggregate information, and information about the environment of a tagged object [5]. Nowadays, sensor nodes are the latest state-of-the-art technology of Smart Item which can detect phenomena at different

IOP Conf. Series: Materials Science and Engineering **319** (2018) 012080 doi:10.1088/1757-899X/319/1/012080

locations, such as temperature changes, vibration, pressure, movement, pollutant levels, etc. The technology doesn't need human beings to serve as intermediaries between the real and virtual worlds. An example of a simple smart item is a temperature sensor that can capture the current temperature, compare it to a certain threshold and decide to send an alert message to an enterprise system if the value of the measurement exceeds the threshold [2]. Smart Item has been integrated to certain supply chain system. Some applications of the smart item in supply chain involved almost all the processes from manufacturing, warehousing, and distributing from the distributor, retailer and transportation agent [6]. The purpose of this paper is to propose a model which can used to analyze the optimum benefits of implementing smart items in an integrated supply chain process consisting of suppliers, shippers and retailers.

2. Notations and Proposed Model

2.1 Notations

The notations used in this research are shown in Table 1.

Table 1. Variable and cost parameters of Smart Item model.

Symbol	Parameter
p_{good}	Price charged for good
$C_{production}$	Variable costs of production per good
q_{sales}	Amount of sold/distributed good
C_{retour}	Cost of manual processing of returned goods (defective or perished)
C_{fix}	Fixed costs
$C_{fix,\,SI}$	Additional fixed costs using Smart Items (infrastructure)
$C_{operation}$	Variable operational costs per Smart Item and shipment (e.g. recharge battery, programming)
C_{SI}	Acquisition costs of Smart Item
s	Penalty depending on cost of goods (shipper → supplier)
$P_{transport}$	Price of shipping per good (to be paid by the customer)
$C_{transport}$	Variable transportation costs per good (for shipper)
$P_{special}$	Additional shipping charge for usage of Smart Item per good
$C_{capacity}$	Costs of capacity loss for reshipping
C_{GSM}	Costs of message sent over GSM to ERP-System
F	Fleet size of shipper
W	Non quantitable advantage through usage of Smart Items (consumer satisfaction, etc.)
$\rho \in (0,1)$	Factor of density, ratio of Smart Item quantity to quantity of goods
$v \in (0,1)$	Factor of maintenance: $v = 0$ all Smart Items get shipped back (reusable); $v = 1$ no Smart Item is returned
$\omega \in (0,1)$	Ratio of defective goods delivered to customer
$\phi \in (0,1)$	Ratio of triggered Smart Items, $0 \le \phi \le \omega \le 1$
$\psi \in (0,1)$	Ratio of searched (potentially lost) goods during shipping
$K \in (0,1)$	Ratio of recovered goods (previously lost)
$\Pi(Qo)$	The expected profit of the news vendor problem
$\Pi(Qi)$	The expected profit of scenario i(i=1,2)
$F(x)$	The cumulative distribution function of x
Qo	The ordering quantity of the newsvendor problem
Qi	The ordering quantity of scenario i(i= 1,2)
x	The random variable representing demand
a	The available rate of ordering quantity
b	The improvement rate of RFID
c	The unit product purchase cost

p	The unit product selling price
g	The shortage cost
s	The unit product salvage price
t	The unit RFID tag price

2.2 Proposed Model

Decker (2008) has developed a model to perform cost benefit analysis to the application of smart item in supply chain involving suppliers, shipper, and retailers. Equation (1) shows the model to determine the profit of a supplier per shipment in Decker (2008).

$$\Pi_1 = \Big((1-\omega)P_{good} - C_{production} - \rho.\big(C_{SI}.v + C_{operation}\big)\Big).-(\omega-\varphi).\ Q_{sales}.C_{retour} + \varphi).Q_{sales}.s + (1-k).\Psi.s.\ Q_{sales} + W - (C_{fix} + C_{fix,SI}) \quad (1)$$

From the equation we can see that the model considered several costs including processing cost, which consists of $C_{production}$, $C_{operation}$, C_{retour}, C_{SI}, C_{fix} and $C_{fix,\ SI}$, one of important parameter in equation (1) is the density factor (ρ) which measure the matter of smart item needed in one pallet. For example in a pallet consisting of 20 boxes and in 1 box consisting of 16 pcs of goods, then ρ = 1/320 if only 1 smart item on the pallet. If 1 smart item per box, then ρ = 1/16. If the smart item is not used anymore, then the maintenance factor *(v)* = 1. In the ideal case, smart item is reused as long as it is not damaged or lost. Ratio of damaged products in shipment process is denoted by φ, supplier will get penalty payment *(s)* from shipper if any goods are damaged during shipment and no need to pay back cost (c_{retour}). With smart items, the location of the goods can be known and the risk of loss of goods can be reduced by the parameter *K*. Variable *W* shows the advantages of Smart Item usage such as customer satisfaction, good reputation due to fast delivery, delivery optimization, and so forth. Equation (2) calculation the profit of the shipper per Shipment [3].

$$\Pi_2 = \Big((1-\phi)(P_{special} + P_{transport}\Big) - C_{transport}).Q_{sales} - C_{GSM}.(\phi + \psi.2.F).Q_{sales} - \phi.Q_{sales}.s - (1-k).\Psi(s + C_{capasity}).Q_{sales} + W - (C_{fix} + C_{fix,SI}) \quad (2)$$

With the use of Smart Item on the shipper side, the shipper service will be better and in the same time requires a harder effort. So the price will be more expensive ($p_{transport} + p_{special}$), but the risk of losing the product becomes smaller and the cost of the shipper will also decrease.

[4] explained the effect of shrinkage on inventory control inaccuracies. The application of RFID technology is considered as a promising solution to solve such problem. The paper proposed a mathematical model that can be used to analyze the effect of RFID application in reducing shrinkage problem on retailer inventory. The relation between availability rate, RFID tag price and other parameters can be described as follows [4] :

$$s\left(a^* + b\left(1-a^*\right)\right) + \frac{1}{a^*}\left(c - a^* s\right)\left(a^* + b\left(1-a^*\right)\right) = c + 1 \quad (3)$$

The determination of optimal order quantitywith RFID implementation has been developed as follows [4] :

$$Q_2^* = \frac{\beta}{\alpha + b(1-\alpha)}\left[1 - \frac{c+t-(\alpha+b(1-\alpha))s}{(p+g-s)(\alpha+b(1-\alpha))}\right] \quad (4)$$

The expression for maximum profit with the implementation of RFID is shown as follows [5] :

$$\Pi(Q_2^*) = \frac{\beta}{2}(p+g-s)\left[1 - \frac{c+t-(\alpha+b(1-\alpha))s}{(p+g-s)(\alpha+b(1-\alpha))}\frac{\beta}{2}g\right]^2 \quad (5)$$

3. Results and Discussions

Based on [3], the proposed model in equation (1) shows that variable W indicates the profits. There are fixed costs(C_{fix} and $C_{fix,SI}$) incurred by the company which previously mentioned the profit of supplier per shipment were eliminated because this paper aims to calculate the optimum profit of Smart Item implementation. So the intangible profit and fixed cost are not considered. Profits are generally came from revenue less the cost of production. In this paper, suppliers and shipper are considered to have their own profit even though they are still the part of supply chain system. Supplier's profit is determined as follows [3] :

$$\boldsymbol{Revenue}_{suplier} = \big((1-\omega).P_s + \emptyset.C_d + (1-k).\psi.C_d\big).Q_{sales} \tag{6}$$

$$\boldsymbol{Cost}_{supplier} = \big(\omega.c_{retour} + c_s + \rho(c_{SI}.v + c_{operation})\big).Q_{sales} \tag{7}$$

$$\boldsymbol{Profit}_{suplier} = \Big(\big((1-\omega).P_s + \phi.C_d + (1-k).\psi.C_d\big) - \big(\omega.c_{retour} + c_s + \rho(c_{SI}.v + c_{operation})\big)\Big) \tag{8}$$

Shipper's profit is modeled as follows [3] :

$$\boldsymbol{Revenue}_{shipper} = (1-\omega).P_t.Q_{sales} \tag{9}$$

$$\boldsymbol{Cost}_{shipper} = \left(C_t + c_{GSM}.\left(\omega + 2.\frac{Q_{sales}}{T_c}.\psi\right) + \phi.C_d + (1-k).\psi.(c_{capacity} + C_d)\right).Q_{sales} \tag{10}$$

$$\boldsymbol{Profit}_{shipper} = \left(((1-\omega).P_t) - \left(C_t + c_{GSM}.\left(\omega + 2.\frac{Q_{sales}}{T_c}.\psi\right) + \phi.C_d + (1-k).\psi.(c_{capacity} + C_d)\right)\right).Q_{sales} \tag{11}$$

The determination of the optimum value of order quantity by implementing RFID (Q_2^*)is based on [4], which referred to as Q_{sales}:

$$\boldsymbol{Q}_{sales} = \frac{\beta}{r}\left(1 - \frac{c+t-(\alpha+b.h).s}{(p+g-s)(\alpha+b.h)}\right) \tag{12}$$

The change of α affects the value of *t* (price tag RFID) so greater shrinkage *(1-α)* or *h* requires higher RFID tag price. The same condition also applies to improvement rate *(b)* component. The value of *t* that is affected by the change of α is calculated as follows:

$$\boldsymbol{t} = \left(s(\alpha + b.h) + \frac{1}{\alpha}(c - \alpha.s)(\alpha + b.h)\right) - c \tag{13}$$

The profit model on the retailer side at in equation (6) will be affected by *t*. A retailer profit model (15) with respect to changes of α can be compiled by substituting *t* element with equation (14) to get equation (15) :

$$\boldsymbol{Profit}_{retailer} = \frac{\beta}{2}(p+g-s).\left(1 - \frac{\frac{1}{\alpha}(c-\alpha.s)}{(p+g-s)}\right)^2 - \frac{\beta}{2}g \tag{14}$$

The total benefits in the supply chain consist of suppliers, shipper, and retailers are obtained by summing up the profits of each level. Any changes of α is followed by RFID tag price (t), and consequently changes of *t* resulted in changes of the optimum order quantity. Equation (16) expresses the total profit of the entire part in supply chain.

$$\boldsymbol{Total\,profit} = \frac{\beta}{r}\left(-\frac{\frac{1}{\alpha}(c-\alpha.s)}{(p+g-s)}\right).\Big(\{[(1-\omega).P_s + \phi.C_d + (1-k).\psi.C_d] - [\omega.c_{retour} + c_s + \rho(c_{SI}.v + c_{operation})]\} + \Big\{(1-\omega).P_t - \Big[C_t + c_{GSM}.\left(\omega + 2.\frac{Q_{sales}}{T_c}.\psi\right) + \phi.C_d + (1-k).\psi.(c_{capacity} + C_d)\Big]\Big\}\Big) + \left(\frac{\beta}{2}(p+g-s).\left(1 - \frac{\frac{1}{\alpha}(c-\alpha.s)}{(p+g-s)}\right)^2 - \frac{\beta}{2}g\right) \tag{15}$$

4. Numerical Example

To get an insight of the model, a numerical example is given, we assume that a company is engaged in the distribution and sale of high quality (premium) fruit products. The company has a delivery service facility and a number of retail stores owned by the company itself. The company's supply chain product delivery line has been equipped with Smart Item's advanced technology of sensor along with its IT system infrastructure. Table 2 shows the data parameters used in the model. The parameters (e.g. costs) from the model are derived from real world data.

Table 2. Data Parameters and Smart Item Impact Parameters

Symbol	Parameter	Value	Symbol	Parameter	Value
v	Maintenance factor	1	p	Retail price/unit	$9.00
c_{retour}	Cost of retour	$2.00	g	Shortage cost	$0.50
$c_{operation}$	Operation/unit	$0.5	s	Salvage price	$3.5
c_{SI}	SI(smart item)/unit	$50	β	Demand	20.000
Cd	Loss/damage cost-shipment	$1.50	b	Improvement rate	0.3
c_{GSM}	GSM	$0.01	Tc	Truck capacity	10000
$c_{capacity}$	capacity	$0.25	ρ	Density factor	0,002
Cs	Production cost	$1.00	ω	Defect rate	5%
Ps	Supplier product price	$2.00	Ψ	Lost product	5%
Ct	Transport cost/unit	$0.05	k	Rate of return of lost product	90%
Pt	Transport price/unit	$0.05	φ	Percentage of triggred items due to defective product	1%
c	Retail purchase/unit	$2.50			

In every shipment, the company uses a truck with a maximum capacity of 20 pallets. Each pallet contains of 10 boxes, and 1 box contains 50 goods. One smart item is installed on each pallet for easy monitoring of goods. The impact of mounting smart items on the pallet is assumed as shown in table 2. We propose an analytical critical tag cost which makes the deployment of RFID cost-effective, and we can derive a threshold value of the RFID tag cost. We illustrate the evolutions of the threshold value of RFID tag cost with available rate (α) of ordering quantity. If the RFID is subject to more shrinkage errors. The results show that the threshold value is decreasing with the available rate of ordering quantity and is increasing with the RFID read rate improvement. Such a result is intuitively expected, since shrinkage errors are not important, the RFID tag costs would be small to be adopted by the retail store. Therefore, for a retailer who suffers from serious inventory inaccuracies caused by shrinkage, it is still worthy to apply RFID technology even if the tag price is at a comparatively high level. Additionally, if RFID technology performs quite well (i.e, improvement rate t), the benefits from RFID technology will exceed the cost induced by the high tag price, which is definitely beneficial to the retailer. To determine the optimal total benefit, the range value of α is set between 0.1 to 0.975 by increasing the interval of 0.025. Based on parameter data in table 2, with α values less than or equal to 0.25, the company does not get a profit, yet but get some losses. This is due to the high price of RFID tags at $\alpha = 0.25$ which is $ 2.25. From figure 1 and 2, we can see that the optimum benefits of each level of supply chain are found at different points. Retailers have an optimum profit at $\alpha = 0.725$ or at the RFID price tag (t) = $ 0.28. Unlike retailer, the supplier and shipper levels have the optimum profit at $\alpha = 0.7$ or t = $ 0.32. The total profits of the company are at the same position as the supplier and shipper. The company's profit graph in figure 1 shows an almost stable value to the shrinkage value of 0.5 and begins to decline afterward the high shrinkage of 0.7, it is not profitable for the company. The use of RFID tags can not be a solution. From the RFID tag price point of view, the graph in figure 2 shows a sharp decline in profit when using RFID tags price more than $ 0.5. It also appears that RFID price tags above $ 2 are even detrimental to the company.

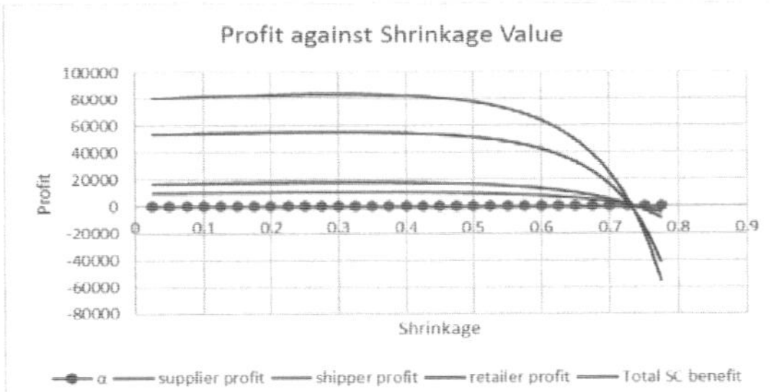

Figure 1. Optimal profit graph based on changes of shrinkage

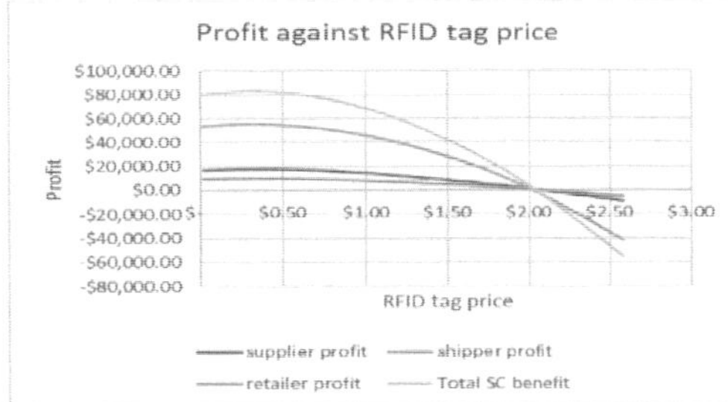

Figure 2. Optimal profit graph based on RFID tag price

5. Conclusion

The proposed model in this paper showed the shrinkage value and RFID tag price that can give company an optimum profit if the company wants to implement Smart Item to its entire supply chain level. As the result of this research, it can be gained that optimum profit value for retailer at α = 0,725 or t = $ 0,28 and for supplier and shipper the optimum profit value at α = 0,7 or t = $ 0,32. Because supplier and shipper have same optimum profit value, it is considered that shrinkage factor is not influential for profit company. The company can be called lose out if the RFID tags price at $ 0,5 until $ 2. There should be other research data that has been empirically proven to be included in the model especially for parameter values.

References

[1] Bhuptani M and Moradpour S 2005 *RFID field guide: deploying radio frequency identification systems* Sun/Prentice-Hall New Jersey.

[2] Chee-Yee Chong and Strkanta P. Kumar 2003 *Sensor Networks: evolution, opportunities and challenges* IEEE Proceeding of the IEEE Vol 91 No.8.

[3] Decker, C., Berchtold, M., Weiss F. Chaves, L., Beigl, M., Roehr, D., Riedel, T and Herzig, D. 2008 *Cost-Benefit Model for Smart Items in the Supply Chain* The Internet of Things pp 155-172 University of Karlsruhe Jerman.

[4] Fan, T. J., Chang, X. Y., Gu, C. H., Yi, J. J., and Deng, S 2014 *Benefits of RFID technology for reducing inventory shrinkage* International Journal of Production Economics 147 pp 659-665 East China University of Science and Technology Shanghai China.

[5] Frank Golatowski, Jan Blumenthal, Matthias Handy, Marc Haase, Hagen Burchardt, and Dirk Timmermann, 2004 *Service-Oriented Software Architecture for Sensor Networks* 2nd IEEE Conference on Industrial Informatics INDIN'04 Berlin.

[6] Zaheeruddin Asif 2005 *Integrating the Supply Chain with RFID: A Technical and Business Analysis* Volume 15 Article 24 Communications of the Association for Information Systems is produced by The Berkeley Electronic Press.

[7] Lee I and Lee B 2010 .*An investment evaluation of supply chain RFID technologies: a normative modeling approach* Int J Prod Econ 125 pp 313–323.

[8] PatrikSpie, ChristofBornho¨vd and Tao Lin,Stephan Haller, and Joachim Schaper 2007 *Going beyond auto-ID: a service-oriented smart items infrastructure* Journal of Enterprise Information Management Vol 20 No 3 pp 356-370 Emerald Group Publishing Limited 1741-0398 DOI 10.1108/17410390710740781.

IOP Conf. Series: Materials Science and Engineering **319** (2018) 012081 doi:10.1088/1757-899X/319/1/012081

Analysis Influence of Managerial Competence, Technical Competence, and Strategic Competence on Firm Performance in Electrical Engineering Company in Bandung

E R Wijaya[1,2], D Irianto[2]

[1]Politeknik ATI Padang, Bungo Pasang Street No. 1, Padang, 25171, Indonesia
[2]Bandung Institute of Technology, Ganesha Street No. 10, Bandung, 40132, Indonesia

Abstract. The industry sectors that have an important role in the era of globalization is the electro engineering sector. The era of globalization led to intense competition. One of the negative effects of the intense competition is declining profits. Drop in profits caused many firms reduces their employees without seeking the root cause of declining profits in detail. Whereas, employee is the important resources to maintain competitive advantage. Competitive advantage can be measured by the performance of which is owned by the firm. The firm's performance can be formed of competencies that is unique, rare, irreplaceable, and difficult to imitate within the firm, one of them is the competence of the individual. According to a competency-based approach and the resource-based approach, individual competence that affect the performance of the firm is managerial competence, technical competence, and strategic competence. Questionnaire is built based on the dimensions of the firm's performance, managerial competence, technical competence, and strategic competence, are processed using partial least squares application. The results indicate that managerial competence negatively impact firm's performance with weak ties. The technical competence and strategic competence positively affect firm's performance with moderate ties.

1. Introduction

The industry sectors that have an important role in the era of globalization is the electro technical sector [14]. The era of globalization led to intense competition. One of the negative effects of the intense competition is declining corporate profits. Drop in profits caused many firms reduces their employees without seeking the root cause of the decline in profit factors in detail. Whereas, the employee is one of the necessary resources to maintain a competitive advantage. According to a competency-based approach [6] [11] and the resource-based approach [6] [11], individual competencies that affect the performance of the firm is managerial competence [3] [15], technical competence [2] [4] [8], and strategic competence [13]. Individuals who have a vital role is at the level of middle management employees such as managers, senior managers, and general managers [8] [11]. The firm which is used as a research study is state-owned enterprises in the field of electrical engineering and its subsidiaries in Bandung. Models built in this study consists of the dimensions of firm performance, managerial competence, technical competence and strategic competence to act as second order latent variables. The dimensions of firm performance are consisted of three indicators, namely market share, sales growth, and customer satisfaction that were acted as first order latent variables [4] [9]. While the managerial competence is composed of 11 dimensions, technical competence is composed of two dimensions, and strategic competence is composed of two dimensions. All dimensions of the three competencies are acts as second order latent variables. The purpose of this study was to confirm the effect of the managerial competence, technical competence, and strategic competence of the firm's performance in state-owned enterprises in Bandung, Indonesia.

2. Material and Methods

2.1. Literature Review on Competencies and Firm's Performance

Resource-based view emphasizes the importance of internal resources to maintain a competitive advantage. This view proved that firm performance is a function of how well the manager build the firm by using the resources of high value, rare, not easily imitated, and difficult to replace. The ability of the manager is a resource that is not easily replicable and related to company performance [10].

The competence became general discussion on human resources systems to compare humans with other resources you need. This view helps firms to identify the knowledge, skills, attitudes, and capabilities that are required in certain occupations to align corporate strategies and priorities [13]. Competence can act as an input or output to human behaviour [13]. Competence is defined as the ability to maintain the coordination of the deployment of assets as the basis for achieving the company's goals. Based on the cognitive aspect and the aspect of action, competence is defined as the ability and the specific skills of the company in the development of their resources by considering the cognitive characteristics to achieve the desired objectives of the company [5]. Personal competence is a combination of experience, knowledge, and skills possessed by an individual or group of individuals [6].

High performance managerial competency is a set of interrelated behaviours to distinguish high performance with average performance [3]. Managerial competence used in this study is high performance managerial competency that has been tested empirically and has 11 dimensions. Dimensions that form the high performance managerial competency are achievement orientation, proactive orientation, presentation, self-confidence, impact, development orientation, managing interaction, interpersonal search, conceptual flexibility, concept formation, and information search.

Technical competence is defined as understanding in the field of Technology, technical equipment and engineering work, product applications, the evolution and direction of technology, and relationships with assistive technologies [8]. Technical competence is important for managers to be able to work effectively. Literature which describes the dimensions that measure the technical competence is not found then the dimensions of technical competence derived from research [8] and a study that used a view based on competence. From the translation of that study concluded that technical competence is formed of two dimensions, technical skills and knowledge that is unique, irreplaceable, rare, and limited which is owned by a manager. Two dimensions is measured using indicators technical skills of research [5] and knowledge of research [2].

The strategic competence is reflection of knowledge and wisdom, meaningful action in deeds, and require a high competence and quality [13]. Rong-rong et al [13] suggest that the strategic competencies that senior managers must have, can be formed from three capabilities. These capabilities include the ability to analyse the strategy, the ability to decide the strategy, and the ability of implementation and control strategy. Two of three capabilities are used in this study whereas the capability of implementation and strategy control is not, because this capability is an element of managerial competence.

A multidimensional approach is used to determine the dimensions that make up the firm's performance [9]. The dimensions consist of market-based performance, financial performance, and customer satisfaction. Market-based performance is measured using indicators of sales growth and market share growth. Financial performance indicators are measured using the total cost reduction, return on investment, return on assets, financial liquidity, and net profit. Customer satisfaction is measured using indicators of the reduction of response time for product design changes, the reduction of response time for product for product volume changes, the accuracy of order processing for customers, the reduction of product return ratio, the speed of order handling, and the reduction of response time for product return. However, the financial dimension is not used in this study by considering the difficulty of getting financial data from the company.

2.2. Partial Least Square

Structured equation modelling (SEM) is one of the multivariate statistical methods to explain the relationship between latent variables based on the variable observed [12]. SEM method has a high degree of flexibility for researchers to connect between theory and data. SEM consists of two types of covariance-based SEM and variance-based SEM (Partial Least Square). Comparison of the two types of SEM methods are described in Table 1.

Table 1. Comparison of the two types of SEM methods

Criteria	CB SEM	Partial Leas Square
Objective	Parameter Oriented	Prediction Oriented
Approach	Covariance	Variance
Assumption	Parametric	Non-parametric
Parameter estimates	Consistent	Consistent
Implication	Optimal for parameter precision	Optimal for prediction precision
Model complexity	small-medium	Large
Sample size	200-800	100-1000
Latent variable score	Indeterminate	Explicitly estimated

PLS is an analysis method that can be relied upon because it does not assume the data must use a scale of specific measurement and small sample size. Determination of the number of samples in the method PLS is (1) ten times the number of the biggest indicators of reflexive or (2) ten times the amount of the biggest indicators of reflexive or (3) to ten times the largest number of lanes structural directed to construct certain structural model [6]. The PLS model evaluation can be done in two ways, namely the evaluation of the measurement model and structural model evaluation.

2.3. Conceptual Framework and Hypothesis

Based on figure 1, which has been through elaboration stages of literature study, hence determined five hypothesis that is:

H1: there is a positive and significant influence given by managerial competence on firm performance
H2: there is a positive and significant influence given by strategic competence on firm performance
H3: there is a positive and significant influence given by technical competence on firm performance
H4: there is a positive and significant influence given by strategic competence on managerial competence
H5: there is a positive and significant influence given by strategic competence on technical competence

Based on figure 2, managerial competence, technical competence, and strategic competence are acted as second order variables whilst firm performance is acted as first order variable. The 11 dimensions of managerial competence, two dimensions of technical competence, and two dimensions of strategic competence are acted as first order variables. Each of first order variables have two or more indicators or item questionnaire.

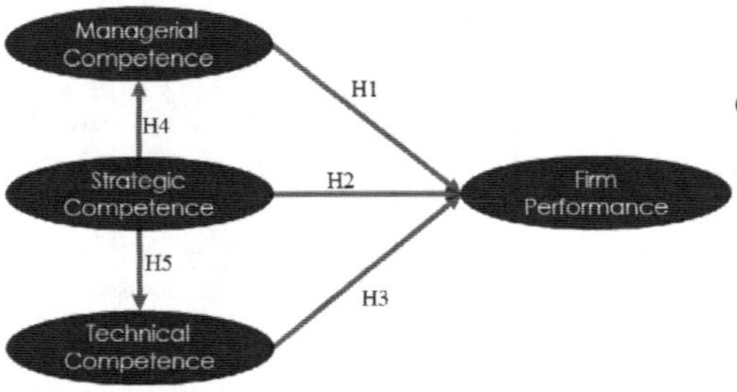

Figure 1. Structural Model

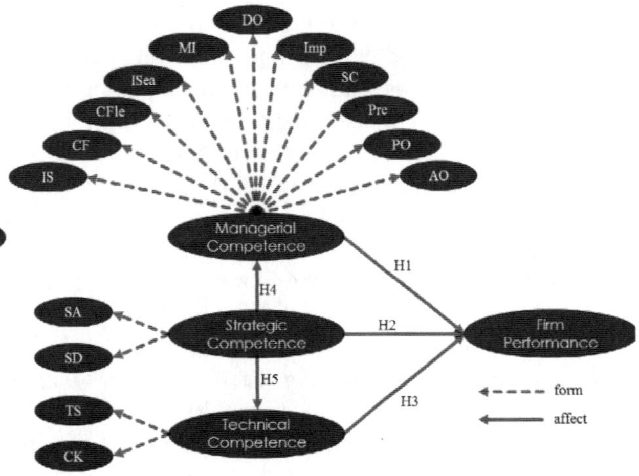

Figure 2. Measurement Model

2.4. Research Methodology

2.4.1. Preparation Research Instrument

The research instrument used in this study was a questionnaire enclosed. Questionnaire established based on the operationalization of latent variables, dimensions / variables manifest and latent variables statement items of managerial competence, technical competence, strategic competence and firm performance. Scale used in the questionnaire is Likert scale consisting of six ratings points. Scoring points for the variable performance of the company is very low, low, low enough, high enough, high, and very high. Scoring points for managerial competence, technical competence, and strategic competence variables are strongly disagree, disagree, somewhat agree, quite agree, agree, and strongly agree.

2.4.2. Identification of the Population and the Respondent

The population of this study is the middle level managers and senior managers of the company from state-owned enterprises in the field of electrical engineering and infrastructure. An overview of the study population is shown in table 2 and table 3.

Table 2. The Result of Questionnaire Distribution

Firms	Manager	Senior Manager	distributed	returned	valid
PT. LEN Industri	17	3	20	18	16
PT. SEI	3	-	3	3	3
PT. ELTRAN	4	-	4	4	4
Total			27	25	23

Table 3. Population of Respondent

Data		Frequency
Genders	Male	22
	Female	1
Level of Educations	S1	21
	S2	2
	S3	6
Managerial Level	Manager	18
	Senior Manager	5
Years of Service (L)	L ≤ 10 years	1
	10 years < L ≤ 20 years	7
	20 years < L ≤ 30 years	15
	L > 30 years	0

IOP Conf. Series: Materials Science and Engineering **319** (2018) 012081 doi:10.1088/1757-899X/319/1/012081

2.4.3.Determining the Sampling Method

The sampling method used is non-probability sampling, by using purposive judgment sampling. This method is used because samplers provide equal opportunities for every member of the target population to be a respondent. Purposive technique used for samples taken pre-defined criteria with the best position to provide the required information. Samples in this study were managers and senior managers at companies in the electrical engineering industry which is a state-owned enterprise. The minimum sample used in this study followed the equations raised by Ghozali [7], which is ten times the amount of structural path. The number of structural path in this study is three that is the path that leads to the variable firm performance. Therefore, the minimum number of samples used is 30 respondents.

2.4.4. Data Collection and Data Proccessing

Data collected in this study consisted of primary data and secondary data. The primary data obtained from questionnaires while the secondary data obtained from historical data of the company. Raw data from the results of questionnaires was processed using SPSS 13 and SmartPLS 2.0 to determine the validity and reliability of the questionnaire and the statement of the questionnaire. Questionnaire statement declared valid and reliable then be processed using SmartPLS 2.0 to perform path analysis, analysis of measurement model and structural model.

2.4.5. Analysis

Analysis used in this study is a qualitative and quantitative analysis. Quantitative analysis based on the analysis of validity, reliability, and path analysis. Qualitative analysis is based on a comparison between the test results with theoretical models of the research with previous research.

3. Result and Discussion

3.1. Reliability and Validity Test of Questionnaire

First, tested the validity and the reliability of all items of the questionnaire statement that compose the first order factor variables. Total items statement is 64 items, 42 items compose 11 first order factor of managerial competence, 10 items compile two first order factor of technical competence, nine items compile two first order factor of strategic competence, and three items compile the firm performance. Reliability and validity test of items questionnaire statements are processed using SPSS with respect to value of Cronbach's Alpha and Spearman Rho. The statement items are declared reliable if it has Cronbach's Alpha greater than 0,6 [1] whilst the item statements are declared valid if the Spearman Rho is greater than 0,5 at the 5% significance level. Based on data processing using SPSS, the 64 items have Cronbach Alpha greater than 0,6, it's indicate that all item is consistent to measure the first order factor. SPSS processed results show that there is a Spearman Rho smaller than 0.5 at a significance level 5%. The item consists of each one item from first order factor of information search, managing interaction, conceptual flexibility, presentation, and firm performance customer satisfaction item, and two items are from the first order factor achievement orientation.

3.2. Reliability and Validity Test of Measurement Model

The reliability and validity test of structural model is processed using SmartPLS by considering the value of Cronbach's Alpha and Composite Reliability as well as loading factor. Reliability test is performed on first order factor variables whilst validity test is performed to item statement of questionnaire. The first order factor variable is declared reliable if it has Cronbach's alpha and composite reliability value greater than 0.6 and the statement item is valid if it has a loading factor greater than 0.5 [6].

Table 4. The value of Cronbach's alpha and Composite Reliability of First Order Factor variables

First Order Factor Variables	Cronbach's Alpha	Composite Reliability
Achievement Orientation	0.642	0.699
Concept Formation	0.779	0.874
Conceptual Flexibility	-0.029	0.296

Table 4. Continued

First Order Factor Variables	Cronbach's Alpha	Composite Reliability
Contextual Knowledge	0.86	0.90
Development Orientation	0.487	0.742
Impact	0.658	0.748
Information Search	0.43	0.18
Interpersonal Search	0.796	0.81
Managing Interaction	0.16	0.32
Presentation	0.32	0.66
Proactive orientation	0.62	0.80
Strategic analysis ability	0.85	0.89
Self-confidence	0.77	0.82
Strategic decision-making ability	0.74	0.78
Technical Skill	0.79	0.86

Based on table 4, there are five first order factor variables that are not reliable because the Cronbach's alpha and Composite Reliability value less than 0.6. These variables are conceptual flexibility (CFle), development orientation (DO), information search (IS), managing interaction (MI), and presentation (P). therefore, the five variables are excluded from measurement model. The measurement model that has experienced the reduction of first order factor variable. Then tested the validity by considering the loading factor value of all statement items. Currently the tested statement item leaves only 49 items, of that amount having a loading factor value greater than 0.5 is 39 items after three validity tests using SmartPLS.

3.3. Reliability and Validity Test of Structural Model

The reliability and validity of structural model is processed using SmartPLS by considering the value of path coefficient and t-statistics between first order factor variable with second order factor variable. Second order factor in this study is managerial competence, technical competence, and strategic competence. Each variable second order factor has its own first order factor. Managerial competence leaves only six first order factor of 11, whilst technical competence and strategic competence still has two first order factor. The relationship between first order factor and second order factor is said to be reliably and valid if it has path coefficient greater than 0.5 and t-statistics value is greater than t-tables with n-1 degrees of freedom at 5% level of significance. The following table provide an overview of the value of path coefficient and t-statistics from the result of the data.

Table 5. The Relationship between first order factor and second order factor

Second order factor	First order factor	PC	t-statistics
Managerial competence	AO	0.92	77.87
	CF	0.74	0.92
	I	0.40	5.21
	ISea	0.45	5.69
	PO	0.80	23.28
	SC	0.73	9.87
Technical competence	CK	0.93	83.73
	TS	0.79	27.22
Strategic competence	SA	0.72	10.69
	SD	0.70	70.99

The value of t-table with degrees of freedom 22 at 5% significance level is 2.074 (two-tailed), according to table 5, it is known that all t-statistics is greater than 2.074. It's indicates that all significant first order factors are used as measure for every second order factor. A first order factor is said to be worthy of use as measure for second order factor if it has a coefficient path value greater than 0.5. Based on table 5, there is still a first order factor which has a coefficient path value less than 0.5, these are I and ISea. It's indicates those two first order factors are less precise used as measure for second order factor managerial competence in stated-owned enterprises. Further, testing of the hypothesis used in this study, hypothesis accepted if it has a value of t-statistics greater than 2.074 and gives positive influence on the dependent variable. Table 6. below provide an overview of the path coefficient and t-statistics values of the hypotheses tested.

Table 6. The Hypothesized construct variable relationship

Hypotheses	Relationship	PC	t-stats
H1	Managerial competence to firm performance	-0.38	4.29
H2	Strategic competence to firm performance	0.40	7.87
H3	Technical competence to firm performance	0.49	6.02
H4	Strategic competence to managerial competence	0.48	5.51
H5	Strategic competence to technical competence	0.55	7.81

Based on table 6, it is known that all hypotheses are accepted because they have t-statistics value greater than 2.074. However, the influence given by the independent variable to dependent variable is not big enough. Furthermore, there is independent variable giving negative effect to firm performance. The relationship is between managerial competence on firm performance, this indicates that in stated-owned enterprises, managerial competence is considered to have negative impact on firm performance. This phenomenon can occur because of shortcomings that exist in the research and environment of stated-owned enterprises are not dynamics.

4. Conclusion

Conclusions of this study is the technical competence and strategic competence have a significant influence and positive impact on the firm performance in the state-owned enterprises. However, managerial competence significantly affects firm performance but have a negative impact on firm performance on stated-owned enterprises. It indicates a model built to confirm the relationship between the managerial competence, technical competence, and strategic competence on firm performance in the state-owned enterprises have differentiation with the model of the referenced based on previous research. In the previous study, population and sample used were functional manager in private companies with sample size greater than 100. In this study, population and sample used are functional managers in stated-owned enterprises with sample size less than 30. Further research is expected to correct the deficiencies that exist in this study by increasing the number of population and sample, not only on one type stated-owned industry but in various types of stated-owned enterprises.

Acknowledgements

We are grateful to Prof. Dr. Dradjad Irianto, M.Eng, Dr. Indriati Sunaryo, Hidea Adinugraha, S.Si, MT, R. Bachtiar Rahman Hakim, S.Si, MT, and Iqbal Yulizar Mukti, S.Si, MT., for insightful discussion and support. We would thanks to Ayu Ekavienamukti Restiwulan, MT. for introducing me to Ms. Nia Herminanti as Human Resources Assistant at PT. LEN Industry. We would also thanks to all staff, manager and senior manager in PT. LEN Industry, PT. SEI, and PT. Eltran Especially Ms. Nia Herminanti for supporting this research and help to spreading the questionnaire. We are grateful to Industrial and Management Engineering Programs of Bandung Institute of Technology for the financial support.

References

[1] Astuti, Y (2010). *Analysis of the Influence of Superior Leadership Style and Subordinate Readiness to Subordinate performance.* Thesis of Industrial Management Engineering Program, ITB, Bandung.

[2] Brill, J.M., Bishop, M.J., and Walker, A.E. (2006). The Competencies and Characteristics Required of An Effective Project Manager: A Web-Based Delphi Study, *Association for Educational Communication and Technology.*

[3] Cockerill, T., Hunt, J., and Schroder, H. (1995). Managerial Competencies: Fact or Fiction, *Business Strategy Review*
[4] Coombs, J.E., Bierly III, P.E. (2006). Measuring Technological Capability and Performance, *R&D Management Journal*
[5] El-Sabaa, S. (2001). The Skills and Career Path of An Effective Project Manager, *International Journal of Porject Management.*
[6] Esrig-Tena, A.B., Bou-Llusar, J.C. (2005). A Model for Evaluating Organizational Competencies: An Application in The Context of Quality Management Initiative, *Journal of Decision Sciences.*
[7] Ghozali, I. 2008. *Structural Equation Modellin: Metode Alternatif dengan Partial Least Square.* Semarang, Badan Penerbit UNDIP.
[8] Grant, K.P., Baumgardner, C.R., Shane, G.S. (1997). The Percieved Important of Technical Competence to Project Manager in The Defence Acquisition Community, *IEEE Transaction on Engineering Management.*
[9] Kim, S.W. (2009): Investigation on Direct and Indirect Effect of Supply Chain Integration on Firm Performance, *International Journal of Production Economics.*
[10] King, A.W., and Zeithaml, C.P. (2001). Competencies and Firm Performance: Examining the Causal Ambiguity Paradox, *Strategic Management Journal.*
[11] Lado, A.A., Boyd, N.G., and Wright, P. (1992). A Competency-Based Model of Sustainable Competitive Advantage: Toward A Conceptual Integration, *Journal of Management.*
[12] Mukti, I.Y. (2010). *ERP management in the post-phase project and its impact on the realization of the benefits of ERP.* Thesis of Industrial Management Engineering Program, ITB, Bandung.
[13] Rong-rong, R., Minhua, W., Xihuai, Y. (2007). Evaluation Model for Strategic Competence of Senior Leaders of Enterprises, IEEE.
[14] Tatsiopoulos, I.P., Avramopoulos, I., Theoharis, I. (1997): A Reference Data Model for Production Control in the Electronics Industry, *Journal of Computers in Industry*
[15] Wickramasinghe, V., and De Zoyza, N. (2009). An Assessment of Managerial Competency Need: Empirical Evidence from Sri Lankan Telecommunication Service Provider, *International Journal of Human Resources Management.*

A Model of Batch Scheduling for a Single Batch Processor with Additional Setups to Minimize Total Inventory Holding Cost of Parts of a Single Item Requested at Multi-due-date

Abdul Hakim Halim[1], Ernawati[2], Nita P A Hidayat[3]

[1]Manufacturing Systems Research Group, Department of Industrial Engineering, Bandung Institute of Technology, Bandung, INDONESIA
[2]Department of Industrial Engineering, Sekolah Tinggi Teknik Cipasung, Tasikmalaya, INDONESIA
[3]Department of Industrial Engineering, Jenderal Achmad Yani University, Cimahi, INDONESIA

Email: ahakimhalim@ti.itb.ac.id

Abstract. This paper deals with a model of batch scheduling for a single batch processor on which a number of parts of a single items are to be processed. The process needs two kinds of setups, i. e., main setups required before processing any batches, and additional setups required repeatedly after the batch processor completes a certain number of batches. The parts to be processed arrive at the shop floor at the times coinciding with their respective starting times of processing, and the completed parts are to be delivered at multiple due dates. The objective adopted for the model is that of minimizing total inventory holding cost consisting of holding cost per unit time for a part in completed batches, and that in in-process batches. The formulation of total inventory holding cost is derived from the so-called actual flow time defined as the interval between arrival times of parts at the production line and delivery times of the completed parts. The actual flow time satisfies not only minimum inventory but also arrival and delivery just in times. An algorithm to solve the model is proposed and a numerical example is shown.
Keywords: batch scheduling, batch processor, actual flow time, inventory holding cost, additional setups,

1. Introduction

Batch scheduling problems to minimize total actual flow time have extensively been discussed in the literature and among the many papers are those written by [1–7] . In the literature, the actual flow time of a part is defined as an interval between arrival time of the part at the production line and delivery time of the completed part. [1] explains the definition of a batch as follows. Let there be a number of parts to be processed on a single machine and to be delivered at a common due date. If all parts arrive at once at a particular point of time, the total actual flow time of all the parts in the shop floor is measured from the arrival time to the common due date. This can be considered that the parts are processed in a single large batch. As it is assumed that parts can be controlled to be available at their respective starting times of processing, the parts can be split into several smaller batches. If this is the case, questions arising are how many batches that the parts should be divided, how many parts that each batch should contain, and how to schedule the resulting batches. The answers to the questions should be directed to achieve the objective of minimizing the total actual flow time. Accordingly, a batch is understood as a number of parts processed in one setup, and each part in the batch waits (on the production facility) until the processing of all the parts in the batch is completed.

[5] show that batch scheduling problems deal not only with the case of job processors where all parts in a batch are processed sequentially part by part (meaning that batch processing time is the multiplication of the part processing time by the number of parts in the batch), but also with the case of batch processors where all parts in a batch are processed at once (meaning that batch processing time equals the part processing time, no matter how many parts in the batch are). It can also be observed from all the papers mentioned above that the setup required before processing each batch is anticipatory (meaning that the corresponding batch is not necessarily available on the machine when a setup is started) and given. It can be comprehended that the existence of this setup is because parts are processed in bathes. In practical production lines, however, there are cases that another kind of setups is also required, that is, when scheduled events happen. For example, in an etching process, a setup is required not only before processing each batch but also every time when the acid used for the etching process should be replaced or added. The replacement or addition of the acid may be done after the batch processor completes a certain number of batches.

The problems that will be encountered in this paper is a model of batch scheduling for a single batch processor processing a number of parts of a single items with additional setups, where the parts arrive at the shop at the times coinciding with their starting times of processing, and the completed parts are to be delivered at multiple due dates. The objective adopted for the model is that of minimizing total inventory holding cost derived from the so-called actual flow time.

2. Total inventory holding cost

Suppose there are a number of parts, n, of a single item to be processed in batches on a single batch processor. The parts can be controlled to arrive at the production line at their respective starting times of processing, and the completed parts are to be delivered at a common due date. The processing time of each part is t, and the setup time required before processing any batch is s. If all parts to be processed have been partitioned into N batches and the batches are backwardly sequenced started from the common due date, moving to time zero on a time scale, the actual flow time for the batches, $F^b_{[i]}$, can be formulated as follows:

$$F^b_{[i]} = d - B_{[i]}, \quad i = 1, 2, 3, \ldots N, \tag{1}$$

where d represents a common due date, and $B_{[i]}$ is starting time for processing the batch scheduled at the ith position counted from d. The actual flow time of parts in the batch, $F_{[i]}$, can be formulated as follows:

$$F_{[i]} = (d - B_{[i]})Q_{[i]}, \quad i = 1, 2, 3, \ldots N, \tag{2}$$

or

$$F_{[i]} = \left\{\sum_{j=1}^{i}(t + s) - s\right\}Q_{[i]}, \quad i = 1, 2, 3, \ldots N, \tag{3}$$

where $Q_{[i]}$ is the number of parts in the batch scheduled at the ith position counted from the common due date. [5] have formulated the total actual flow time, F, of all parts through the shop with a batch processor as follows:

$$F = \sum_{i=1}^{N}\left\{\sum_{j=1}^{i}(t + s) - s\right\}Q_{[i]} \tag{4}$$

where N is the number of batches processed in the shop floor. It should be emphasized here that the actual flow time adopts the so-called backward scheduling, instead of the forward scheduling. The first mentioned approach sequences batches backwardly from the end position on a time scale, i. e., from the due date, moving to time zero, while the later approach sequences batches forwardly from time zero, appearing a mirror image of the backward schedule.

Now, if the process requires additional setups, a (assumed to be fixed and given), performed when it completes a certain number of batches, in addition to the main setups, s, Equation (4) can be rewritten to be the followings:

$$F = \sum_{i=1}^{N}\{\sum_{j=1}^{i}(t + s + a_{[j]}) - (s + a_{[i]})\}Q_{[i]} \quad (5)$$

with

$$a_{[i]} = \begin{cases} a & \forall\, i = N, (N-x), (N-2x), \ldots, (N-(m-1)x) \\ 0 & otherwise \end{cases}$$

where $a_{[i]}$, x and m respectively denote an additional setup required before processing the batch at the ith position on a schedule, the number of batches between two consecutive additional setups, and the ratio of N to x ($m = N/x$). If m is not an integer, m is set to be the minimum integer greater than or equal to N/x (a rounding up of N/x). Note that as both main setups and additional setups are assumed anticipatory, any setup for a batch is not included in the actual flow time of the batch but is included in the actual flow time of the batch processed previously.

If some batches completed before the common due date, the parts in such batches will incur inventory cost. If we follow the condition discussed in [1], distinguishing holding cost per unit time for a part in completed batches, denoted as c_1, from that in process batches, denotes as c_2, the total inventory holding cost, T, can be formulated as follows:

$$T = c_1\left(\sum_{i}^{N}\{\sum_{j=1}^{i}(t + s + a_{[j]}) - (s + a_{[i]})\}Q_{[i]}\right) - (c_1 - c_2)\left(\sum_{i}^{N} tQ_{[i]}\right) \quad (6)$$

Now, if there are r numbers, n_h of parts demanded at respective due dates, d_h, $h = 1, 2, 3, \ldots, r$ (note that d_r is the closest due date to time zero), let divide the whole scheduling period into r partitions of periods each of which constitutes an interval, H_h, $h = 1, 2, 3, \ldots r$, between two consecutive due dates. It can be seen that the batch scheduling problem becomes, firstly, how to determine the numbers, D_h, of parts processed in period H_h, $h = 1, 2, 3, \ldots r$, secondly, how to batch D_h within H_h, and thirdly, how to schedule the resulting batches in periods H_h. The total inventory holding cost, T, can be formulated as follows:

$$T = c_1\left[\sum_{h=1}^{r}\left(\sum_{i=1}^{N_h}\{\sum_{=1}^{i}(t + s + a_{h[j]}) - (s + a_{h[i]})\}Q_{h[i]}\right)\right] + c_1\left(\sum_{g=1}^{r-1}\{\sum_{h=g+1}^{r}(D_h - n_h\}(d_g - d_{(g+1)})\right) - (c_1 - c_2)\sum_{h=1}^{r}\sum_{i=1}^{N_h} tQ_{h[i]} \quad (7)$$

where N_h, $a_{h[i]}$, and $Q_{h[i]}$ respectively represent the number of batches processed in period H_h, an additional setup performed before the batch scheduled in the ith position in period H_h, and the number of parts in the batches scheduled in period H_h, $h = 1, 2, 3, \ldots r$.

3. Model formulation

The model of batch scheduling for a single batch processor with additional setups (BPAS) can be formulated as follows.

Minimize

$$T = c_1\left[\sum_{h=1}^{r}\left(\sum_{i=1}^{N_h}\{\sum_{=1}^{i}(t + s + a_{h[j]}) - (s + a_{h[i]})\}Q_{h[i]}\right)\right] + c_1\left(\sum_{g=1}^{r-1}\{\sum_{h=g+1}^{r}(D_h - n_h\}(d_g - d_{(g+1)})\right) - (c_1 - c_2)\sum_{h=1}^{r}\sum_{i=1}^{N_h} tQ_{h[i]}$$

IOP Conf. Series: Materials Science and Engineering **319** (2018) 012082 doi:10.1088/1757-899X/319/1/012082

Subject to

$$a_{h[i]} = \begin{cases} a & \forall\, i = N_h, (N_h - x), (N_h - 2x), \dots, (N_h - (m_h - 1)x) \\ 0 & otherwise \end{cases} \quad \forall h = 1,2, \dots, r \tag{8}$$

$$N_h t + (N_h - 1)s + \sum_{i=1}^{N_h - 1} a_{h[i]} \le H_h \qquad \forall h = 1,2, \dots, r \tag{9}$$

$$\sum_{h=g+1}^{r} (D_h - n_h) \ge 0 \qquad \forall g = 1, 2, .., (r-1) \tag{10}$$

$$\sum_{h=1}^{r} (D_h - n_h) = 0 \tag{11}$$

$$\sum_{i=1}^{N_h} Q_{h[i]} = D_h \qquad \forall h = 1, 2, \dots, r \tag{12}$$

$$B_{h[1]} + t = d_h \qquad \forall h = 1, 2, \dots, r \tag{13}$$

$$0 \le Q_{h[i]} \le k \,;\; D_h > 0 \,;\; N_h \ge 1 \qquad \forall h = 1, 2, \dots, r; \forall\, i = 1, \dots, N_h \tag{14}$$

Constraint (8) shows that additional setups are conducted repeatedly after the batch processor completes x batches, where m_h is a rounding up of N_h/x. Constraint (9) shows that the total processing time of parts distributed into period $H_h$must be no longer than the length of the period. Constraint (10) describes that the total number of parts processed within a time interval from periods H_r to H_h must be greater than or equal to the demanded parts, n_h, at d_h. However, Constraint (11) describing that the total number of parts processed within the scheduling period must be the same as the total number of parts distributed to all periods, $H_h, h = 1,2, \dots, r$, in the scheduling period must be satisfied. Constrain (12) accomplishes material balances in period H_h respectively. Constrain (13) shows the batches processed last in respective periods H_h, $h = 1,2, \dots, r$, must be completed at d_h. Constrain (14) indicates that all batch sizes ($\forall\, i, h$) must be greater than or equal to zero and less than or equal to the capacity, k, of the batch processor, and that the number of parts distributed to period H_h must be positive and the number of batches in respective periods H_h must be greater than or equal to 1.

4. Algorithm

The following is a proposed algorithm for solving Model [BPAS], developed from the algorithm discussed in [5].

Step 0. Set values of parameters t, s, a, x, k, r, c_1, c_2, n_h and d_h for h = 1,2,..., r.
Step 1. Set h=1, $K_0 = 0$ and $d_{r+1} = 0$, and go to Step 2.
Step 2. Calculate $Y_h = n_h + K_{h-1}$ and $N'_h = Y_h/k$, and go to Step 3.
Step 3. If N'_h is an integer, then set $N_h = N'_h$ and $Q_{h[i]} = k$ for i=1,2,...., N_h. Go Step 4. Otherwise, set N_h = rounding up of N'_h; then set $Q_{h[i]} = k$ for i=1,2,...., $(N_h - 1)$ and $Q_{h[N_h]} = Y_h - (N_h - 1)k$. Go to Step 4.
Step 4. Backwardly sequence the resulting batches in order of non-increasing batch sizes: $Q_{h[1]} \ge Q_{h[2]} \ge \cdots \ge Q_{h[N_h]}$. Go to Step 5.
Step 5. Compute $Z'_h = \frac{d_h - d_{h-1}}{(t+s+a)}$. Go to Step 6.
Step 6. If Z'_h is an integer, then set $Z_h = Z'_h$. Go to Step 7. Otherwise, set Z_h = rounding up of Z'_h. Go to Step 7.
Step 7. If $N_h > Z_h$ then set $N_h = Z_h$ and sequence the bathes in order of non-increasing batch sizes: $Q_{h[1]} \ge Q_{h[2]} \ge \cdots \ge Q_{h[N_h]}$, and go to Step 8. Otherwise go to Step 8.
Step 8. Compute $D_h = \sum_{i=1}^{N_h} Q_{h[i]}$; $Y_h = n_h - D_h$ and $m'_h = N_h/x$. Go to Step 9.
Step 9. If m'_h is an integer, then set $m_h = m'_h$, and go to Step 10. Otherwise, set m_h = rounding up of m'_h, and go to Step 10.
Step 10. Set $a_{h[i]} = a$ for $\forall i = N_h, (N_h - x), (N_h - 2x), \dots, (N_h - (m_h - 1)x)$, and otherwise set $a_{h[i]} = 0$. Go to Step 11

Step 11. For *i*=1,2,..., N_h, compute $FL^a_{h[i]} = \sum_{j=1}^{i}(t+s+a_{h[j]}) - (s+a_{h[i]})$; $B_{h[i]} = d_h - FL^a_{h[i]}$ and set the time interval between d_{h+1} and the starting setup time before processing $b_{h[N_h]}$. IT_h, as the following $IT_h = B_{h[N_h]} - (t+s+a_{h[N_h]}) - d_{h+1}$. Go to Step 12

Step 12. If $IT_h \geq 0$, then compute the objective value (T), and stop. Otherwise, set $N_h = N_h - 1$, and go to Step 11.

5. Numerical experience and analysis

Suppose there is a shop floor with a batch processor producing parts of a single item requested at 3 due dates where the demands at respective due dates are shown in Table 1. The processing time per part is 100, setup time required before processing each batch is 10, and additional setup times required after the processor completes 2 bathes is 15, and the capacity of the batch processor is 20.

Table 1. The numbers of parts demanded at respective due dates

h	1	2	3
n_h	95	100	95
d_h	2000	1525	920

The resulting solution obtained from applying the proposed algorithm is shown in Table 2, and the Gantt chart for the solution is shown in Figure 1. Table 2 shows that the scheduling period is divided into 3 periods, H_1, H_2 and H_3, and the numbers of batches in respective periods are 4, 5, and 6. Table 2 also shows all batch sizes and the positions where the additional setups take places. Other values of decision variables shown in Table 2 are those of actual flow times of respective batches, $F^b_{h[i]}$, the starting times of processing batches, $B_{h[i]}$ and total flow time of parts processed in each period, F_h. The total actual flow time, F, of parts processed in the shop is 95,600 with the total holding cost, *T*, for all parts through the shop is 181200.

The algorithm applies that if the numbers, D_h, of parts distributed into periods H_h, h = 1, 2, 3, ..., *r*, have been decided, the problems in respective periods are to batch D_h and to schedule the resulting batches. The problem in each period constitutes a problem of batch scheduling in the condition of a common due date, is solved separately.

Table 2. The solution for the generated problem

H	1				2					3					
N_h	4				5					6					
i	1	2	3	4	1	2	3	4	5	1	2	3	4	5	6
$Q_{h[i]}$	20	20	20	20	20	20	20	20	20	20	20	20	20	20	10
t	100	100	100	100	100	100	100	100	100	100	100	100	100	100	100
s	10	10	10	10	10	10	10	10	10	10	10	10	10	10	10
$a_{h[i]}$	0	15	0	15	15	0	15	0	15	0	15	0	15	0	15
$F^b_{h[i]}$	100	210	335	445	100	225	335	460	570	100	210	335	445	570	680
$B_{h[i]}$	1900	1790	1665	1555	1425	1300	1190	1065	955	820	710	585	475	350	240
$F_{h[i]}$	2000	4200	6700	8900	2000	4500	6700	9200	11400	2000	4200	6700	8900	11400	6800
IT_h	5				10					215					
F_h	21800				33800					40000					
F	95600														
T	181200														

IOP Conf. Series: Materials Science and Engineering **319** (2018) 012082 doi:10.1088/1757-899X/319/1/012082

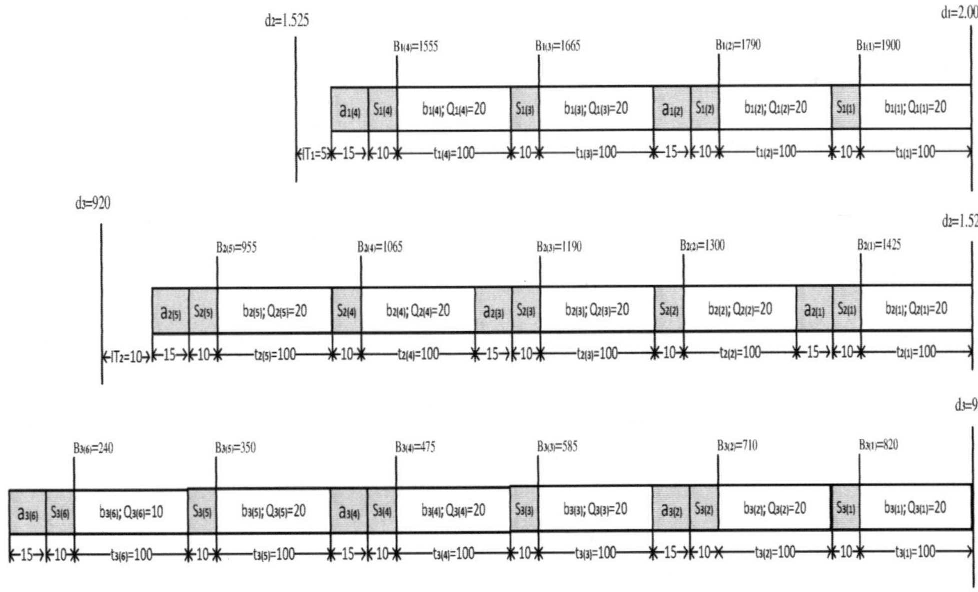

Figure 1. Gantt chart for the resulting solution

From Table 2 and Figure 1, it can be observed that additional setups are required before processing the batch scheduled at the first position in period H_2, b_{21}, and the batch scheduled at the forth position in period H_1, b_{14}, even though both batches are sequenced consecutively. When an additional setup is conducted, the chemicals (acid for an etching process) used in current processing will be replaced or added even though this action has not been necessary, as in this numerical example, the additional setup is required if the batch processor has completed 2 batches. Accordingly, the solution shown in Table 2 may be modified to accommodate this condition.

The modified solution is shown in Table 3 and Figure 2. Comparing the two solutions, the modified solution has made additional setups be conducted every time after the batch processor completes 2 batches. However, the modified solution leads to higher inventory holding cost. As the objective is to minimize holding cost, the solution for the problem is the original one, even though the chemicals currently used should be replaced or added.

Table 3. The modified solution

H	1				2					3					
N_h	4				5					6					
i^*	1	2	3	4	5	6	7	8	9	10	11	12	13	14	15
i	1	2	3	4	1	2	3	4	5	1	2	3	4	5	6
$Q_{h[i]}$	20	20	20	20	20	20	20	20	20	20	20	20	20	20	10
t	100	100	100	100	100	100	100	100	100	100	100	100	100	100	100
s	10	10	10	10	10	10	10	10	10	10	10	10	10	10	10
$a_{h[i]}$	15	0	15	0	15	0	15	0	15	0	15	0	15	0	15
$F^{p}_{h[i]}$	100	225	335	460	100	225	335	460	570	100	210	335	445	570	680
$B_{h[i]}$	1900	1775	1665	1540	1425	1300	1190	1065	955	820	710	585	475	350	240
$F_{h[i]}$	2000	4500	6700	9200	2000	4500	6700	9200	11400	2000	4200	6700	8900	11400	6800
IT_h	5				10					215					
F_h	22400				33800					40000					
F	96200														
T	181500														

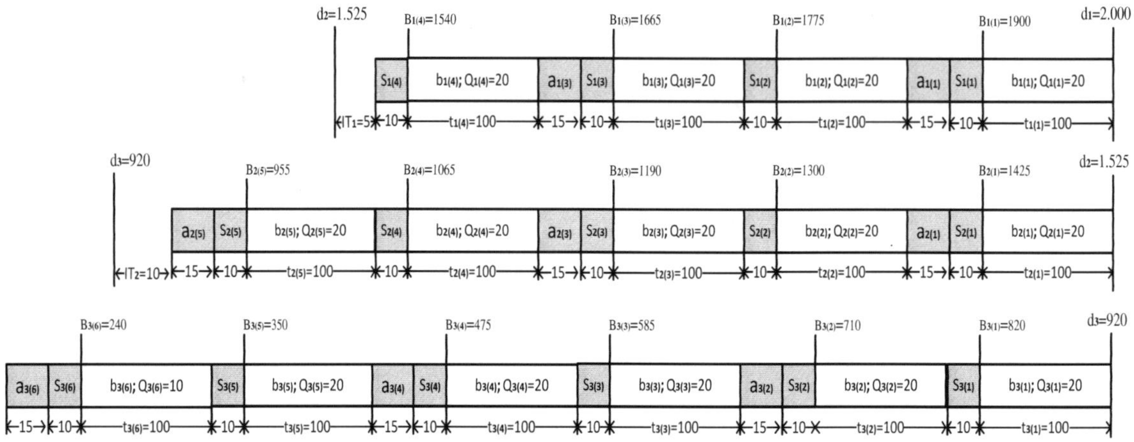

Figure 2. The Gantt chart for modified solution

6. Concluding remarks

This paper has dealt with a problem of batch scheduling for a batch processor on which a number of parts are processed. The processor requires not only main setups conducted before processing any batches but also additional setups conducted after the batch processor completes a number of batches. The objective is to minimize total holding cost derived from the so-called actual flow time. To formulate the problem, the whole scheduling period is partitioned into periods each of which constitutes a time interval between two consecutive due dates. The proposed algorithm solves the problem period by period where in each period the problem is considered as a problem with a common due date, and the algorithm is effective to solve the problem.

References

[1] Halim A H 1993 *Batch scheduling for production systems under just in time environment* Doctorate Dissertation (Japan: University of Osaka Prefecture)

[2] Halim A H and Ohta H 1993 Batch-scheduling problems through the flowshop with both receiving and delivery just in time *Int. J. Prod. Res.* **31** 1943–55

[3] Halim A H, Miyazaki S and Ohta H 1994 Batch-scheduling problems to minimize actual flow times of parts through the shop under JIT environment *Eur. J. Oper. Res.* **72** 529–44

[4] Halim A H, Miyazaki S and Ohta H 1994 Lot scheduling problems of multiple items in the shop with both receiving and delivery just in time *Prod. Plan. Control* **5** 175–84

[5] Hidayat N P A, Cakravastia A, Samadhi T M A A and Halim A H 2016 A batch scheduling model for m heterogeneous batch processor *Int. J. Prod. Res.* **54** 1170–85

[6] Yusriski R, Sukoyo S, Samadhi T and Halim A 2015 Integer batch scheduling problems for a single-machine with simultaneous effect of learning and forgetting to minimize total actual flow time *Int. J. Ind. Eng. Comput.* **6** 365–378

[7] Zahedi Z, Samadhi Tmaa, Suprayogi S and Halim A 2016 Integrated batch production and maintenance scheduling for multiple items processed on a deteriorating machine to minimize total production and maintenance costs with due date constraint *Int. J. Ind. Eng. Comput.* **7** 229–244

IOP Conf. Series: Materials Science and Engineering **319** (2018) 012083 doi:10.1088/1757-899X/319/1/012083

A Framework for the Development of Automatic DFA Method to Minimize the Number of Components and Assembly Reorientations

Alfadhlani[1], T M A Ari Samadhi[2], Anas Ma'ruf[2], Isa Setiasyah Toha[2]

[1]Industrial Engineering Department, Andalas University, Limau Manis, Padang, 25163, Indonesia
[2]Industrial Engineering Department, Bandung Institute of Technology, Jl. Ganesha 10 Bandung, 40132, Indonesia

alfadhlani@ft.unand.ac.id

Abstract. Assembly is a part of manufacturing processes that must be considered at the product design stage. Design for Assembly (DFA) is a method to evaluate product design in order to make it simpler, easier and quicker to assemble, so that assembly cost is reduced. This article discusses a framework for developing a computer-based DFA method. The method is expected to aid product designer to extract data, evaluate assembly process, and provide recommendation for the product design improvement. These three things are desirable to be performed without interactive process or user intervention, so product design evaluation process could be done automatically. Input for the proposed framework is a 3D solid engineering drawing. Product design evaluation is performed by: minimizing the number of components; generating assembly sequence alternatives; selecting the best assembly sequence based on the minimum number of assembly reorientations; and providing suggestion for design improvement.

1. Introduction

Design for Assembly (DFA) is a method to simplify the product structure, so it will reduce the assembly cost. Assembly cost reduction is achieved through reducing the assembly time of product. There are three DFA methods so far: AEM (*Assembly Evaluation Method*) of Hitachi, the Lucas-DFA, and the Boothroyd-Dewhurst DFA [1]. AEM and the Lucas-DFA proposed an assembly evaluation method based on values scale, the values are given as relative measure of assembly difficulties of each component. Assembly evaluation with AEM and the Lucas-DFA tends to be subjective, and these methods do not provide guidelines for design improvements.

DFA method that provides guidelines for design improvements is the DFA developed by Boothroyd and Dewhurst (BD-DFA). In this method, design efficiency is determined by considering the number of components in a product, the ease of handling, insertion and fastening [2]. This method has been widely used in manufacturing industries because it is more applicable [1,3,4,5].

IOP Conf. Series: Materials Science and Engineering **319** (2018) 012083 doi:10.1088/1757-899X/319/1/012083

Inputs for the BD-DFA are product's engineering drawing, exploded views, and assembly sequence. These inputs are used to recognize a number of assembly features in each component of the product design, which consist of: (1) geometric assembly features, such as: basic shapes, rotational symmetry, and feature location; (2) non-geometric assembly features, usually related to the component characteristics, such as: slippery or not, flexible or not, abrasive or not, and their size (small, medium, or large); (3) operational assembly features, related to the insertion problem of component when assembled, such as: a component requires gripping or holding down before it is finally secured or not, easy to align during assembly or not, requires vertical insertion or not, and what fastener type is used.

Outputs of the BD-DFA are assembly time estimation, assembly cost, and design improvement suggestions. Based on assembly features of component, determine the assembly code to find assembly time estimation from handling and insertion time data table. Assembly cost is calculated based on this assembly time estimation. The assembly time is reduced by [2]:

1. Reducing the number of components.
2. Improving the design of components so then:
   a. Easy to grasp and manipulate.
   b. Easy to align.
   c. Easy to insertion.

The first BD-DFA was implemented manually (paper based) using given assembly sequence. Too much effort and time consuming to evaluate assembly design manually, there are some error possibilities in recognizing the assembly features and predicting the assembly time [6]. To overcome these issues, there are several software have been developed, however the developed software are still a manual technique supported by computer [6,7], because the software only used for data manipulation and data calculation, while the required data are still input by the user.

Nowadays the CAD (*Computer Aided Design*) system has been more sophisticated, so that there is a larger opportunity to develop an automatic DFA method, since data extraction, data manipulation and analysis, and re-design suggestions could be performed in computer based. Therefore product evaluation process and design improvement could be done easier.

2. Automatic DFA Method

Li and Hwang [8] proposed a framework for the development of automatic DFA method in CAD system base, the framework consisted of five modules, namely: assembly sequence generation, assembly features extraction, assembly codes determination, data manipulation and computation, and re-design suggestion. Each module was an evaluation step constructed based on experiences, rules, or theory in DFA methodology and it was developed as automatic DFA method. But the effort met some difficulties in developing re-design suggestion module. Soh et al [9] proposed a systematic methodology of DFA and DFD (design for disassembly) for concurrent consideration on remanufacturing perspective.

The automatic DFA method is usually manifested in software which is integrated with CAD system. Refer to the framework proposed by Li and Hwang [8], it is necessary to develop some automatic methods, namely: automatic method for assembly features extraction and assembly codes generation; automatic method for re-design suggestion, and; automatic method for assembly sequence generation.

Several researchers have discussed the development of the automatic method for assembly features extraction, and assembly codes determination: Li and Hwang [10] and Coma et al [11] used CAD drawing in B-Representation as input; while Ong et al [3] used CAD drawing in a 3D solid as input. The assembly codes are needed for assembly time estimation.

The development of automatic method for design improvement suggestion is still finding some difficulties in getting information of component function, because it usually needs non-geometric information. Mo et al [12] developed a graph model called FRM (*function relation model*) to analyze the function of a component. This model acquired DFA methodology into knowledge database and was intended to support the expert system. The component function was defined by the designer in the form of graph and matrix, and then the matrix values were used as an input to evaluate the product design in order to have a simpler structure. The model proposed by Mo et al [12] could be categorized as semi-automatic method, since there were still have some evaluation process that could not be done automatically.

Li and Hwang [8] proposed a framework based on the Boothroyd-Dewhurst DFA, while Dalgleish et al [7] proposed a framework based on the Lucas DFA called the Sandpit DFA. Lee et al [13] proposed an automatic DFA method based on DFA methodology in general, this framework was developed for INSPIRE (intelligent assembly planning integrated with redesign) which integrated the assembly planning and the design improvement. INSPIRE have interactive process with designer to meet the information needs in analysis. This interactive process is facilitated by an interface called RIC (revenge input constructor).

3. Proposed Framework for Automatic DFA Development

This article proposes a framework for the development of automatic DFA method that refer to product design evaluation steps in the Boothroyd-Dewhurst DFA methodology, namely [2]:

1. Minimize the number of component
2. Improve the assembly features of product design

This framework considers the assembly sequence problem in design improvement. The required data for product design evaluation are extracted automatically from the CAD system by using a 3D solid drawing as input. Design evaluation is performed by evaluating the basic function, assistant function [12] and assembly features of each component.

Product design evaluation is started by minimizing the number of components of a product. A component at the end position in assembly is usually supported by a number of other components that are assembled with it. If a component moves when the product is operated, then supporting components also move along with it. This set of mutually supportive components is defined as group in this paper. Grouping of product's components results the category of each component, i.e. main component or supporting component. Further, each component's group is examined to know the effect of removing one or a number of supporting component to the degree of freedom of the group.

Each supporting component is tried to be removed (reduction or elimination) from product structure by determining the function of components on its group first. The function of a component means the role of a component at the end position to support other components. The component function criteria proposed in this article refer to the criteria proposed by Mo et al [12]:

1. Cooperative is if a number of components support each other in performing the same function.
2. Independent is if a component has a function independently.
3. Compound is if a component performs more than one function simultaneously.
4. Transitional is if a number of components perform one function serially.
5. Redundant is if a number of components perform a same function independently.

Mo et al [12] recommended designing a component to have compound classification, the component which has the classification function of cooperative, transitional, or redundant become the candidates to be removed from the product structure. The framework in this article uses a technique to determine the classification of component function based on the degree of freedom of component and the group of components.

The output of component grouping analysis is the information of main component and supporting components, while the output of the component function analysis is recommendation for reduction or elimination of component. Reduction means transferring the function of several components into one component, so that the amount of components could be reduced. Elimination means removing a component from product structure without transferring it functions to other component. Elimination is suggested to be done only for supporting component.

The evaluation of product design, as an effort to provide some suggestions for design improvement, could be conducted through four steps, namely:

1. Grouping the components to indentify the main component and the supporting component.
2. Determine the value of degree of freedom in each group of components.
3. Determine the classifications of component function in the product.
4. Provide suggestion for reduction or elimination of component to minimize the number of components in the product.

BD-DFA has not yet discussed about assembly sequence and its effect to the product design, whereas DFA method requires assembly sequence to estimate assembly time. The framework proposed in this article considers the assembly sequence problem in identifying the component features that need to be improved, so that total assembly reorientation during assembly operation could be minimized. In DFA methodology, minimizing the number of assembly reorientations and minimizing the number of components are the efforts for assembly time reduction.

The BD-DFA provides some suggestions for reducing the assembly time of product. These suggestions are based on handling and insertion problem. The suggestions, for example, related to the symmetrical problem for easier handling, or the fillet for easier insertion. Assembly sequence analysis, handling and insertion analysis in product assembly evaluation processes could be done independently, so that it could be considered separately. This research will not discuss handling and insertion issues in features improvement, besides this problem could be considered independently, it has also been well defined in the BD-DFA.

The evaluation of product design involving the assembly sequence analysis is conducted by generating assembly sequences firstly. The feasible assembly sequences are generated by considering the precedence constraints and the component regions. The precedence constraints in this research, related to the geometric constraints, are generated based on the information of collision free path of assembly. The collision is a condition if a component blocked by other components when assembled. The region is a condition if the assembly sequence of product could be randomized without breaking the precedence constraints. Some assembly sequence alternatives could be generated by considering the region. The product design evaluation involving assembly sequence analysis could be done through the following steps:

1. Detect the collision free path of assembly component.
2. Generate the assembly precedence based on the information of collision free path of each assembly component.
3. Determined the region of each component of product.
4. Generate the feasible assembly sequence alternatives.

If some assembly sequence alternatives have been obtained, then select the best one. The best alternative is selected based on the minimum number of assembly reorientation criterion, so that the number of assembly reorientations of all assembly sequence alternatives must be calculated. If the minimum number of reorientations is more than once and successively, then it is suggested to improve the features of assembly component. These suggestions could be used by the designer but this research does not discuss how the feature improvement is done. This task really depends on the designer experiences. The assembly reorientation analysis could be conducted with the following steps:

1. Select the best assembly sequence based on the minimum number of assembly reorientation.
2. Check the assembly reorientation condition during assembly operations (more than once reorientation and successive or not).
3. Provide re-design suggestions for minimizing the number of assembly reorientations.

Design evaluation process could be done repeatedly until the designer feel confident that the product design already has a minimum number of components, a minimum number of assembly reorientations and the product could be functioned correspond to the design objective.

The proposed framework consists of three parts of model development, namely: the model for minimizing the number of product components, the model for analyzing assembly sequence, and the model for minimizing the number of assembly reorientations, see Figure 1. The model here is defined as criteria, propositions, rules and algorithms. The required information is extracted from CAD system database, and stored in Component Database.

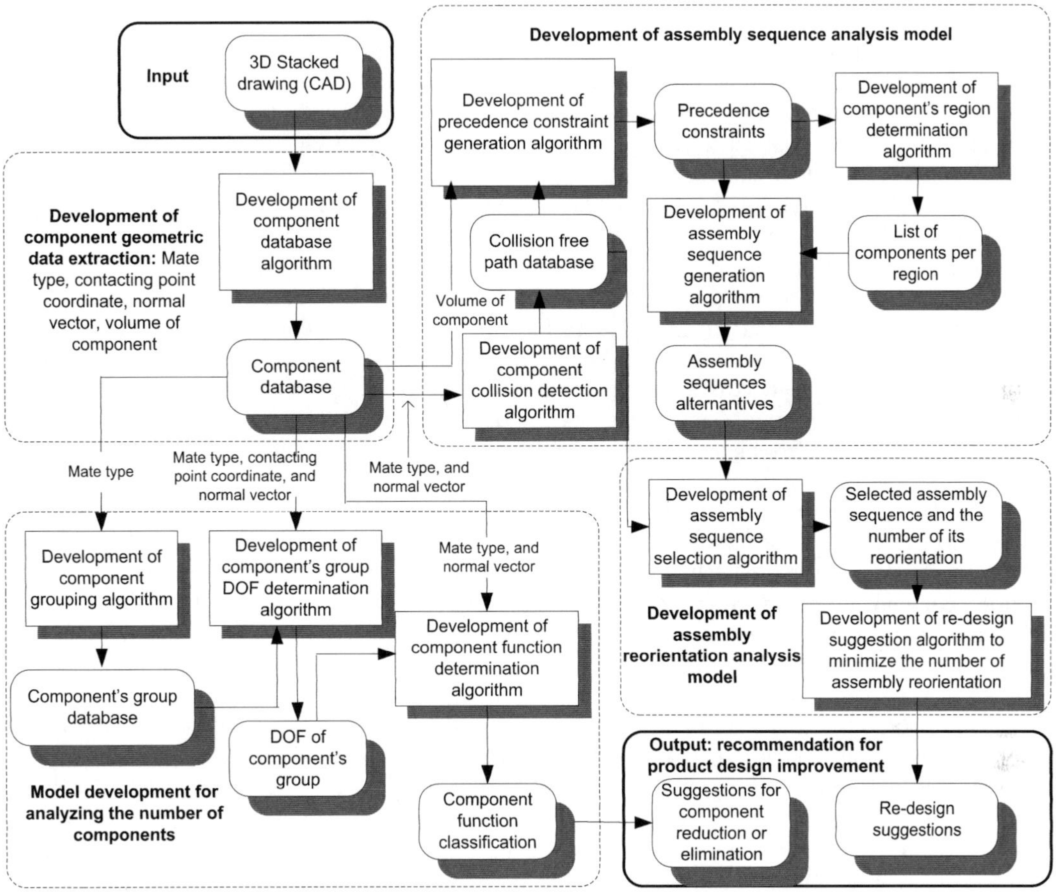

Figure 1. Framework for automatic DFA method development

3.1 Geometric data extraction model

The first stage in realizing automatic DFA method is the development of geometric data extraction model. Data are extracted from CAD database using an algorithm called Component Geometric Data Extraction Algorithm, this algorithm has been developed and published by Alfadhlani et al [14] as part of this research. The extracted data consist of: mate type, contacting point coordinate, normal vector, and volume of component. These data will be stored in Component Database

3.2 The number of components analysis model

The second stage in the framework is the development of a model for analyzing the number of components. Output of this analysis is suggestions for minimizing the number of components. Analysis task is performed through three steps, namely: grouping the component to identify the main components and supporting components, determining the degree of freedom (DOF) of component's group, and determining the function of supporting components.

Component grouping is determined by considering mating condition information, this information indicates the connections between components, components in a group will be connected each other. The category of components could be determined based on the number of connections, the main component is determined among components in a group with the largest connections. If there is more than one component with the largest number of connections, then all that components are defined as main components. The supporting components are all other components that are not defined as main component. It is necessary to develop an algorithm called Component Grouping Algorithm for analyzing the category of components. Input for this algorithm is mate type information of component, while it outputs are component's group, mating list of each group, and components category. These outputs are stored in Group Components Database.

The automatic DFA in this study is developed to have ability to identify the candidate of components that can be reduced or eliminated automatically. The candidates are identified by considering component function classifications. Component function is defined using the degree of freedom of components and group of components. To analyze the function of components, it is needed to develop the algorithm of group's degree of freedom determination. Inputs for this algorithm are component's group, mate type, contacting point coordinates, and normal vector of component's contacting face. While component function classification and identification of component candidate, for reduction or elimination, are determined using the algorithm of component function determination. Inputs for this algorithm are degree of freedom of component's group, mate type, and normal vector of component's contacting face.

3.3 Assembly sequence analysis model

The third stage in the proposed framework is the assembly sequence analysis model. This analysis is performed through four steps, namely: collision detection of assembly free path, precedence constraint generation, component region determination, and feasible assembly sequence generation. Assembly collision free path of component is specified based on information of mate type and normal vector direction. Mating indicate a connection, while normal vector indicate blocking path of a component by other components when assembled. An algorithm for detecting assembly collision free path need to be developed, this algorithm called Component Collision Detection. Inputs for this algorithm are mate type and normal vector information. Output of this algorithm is assembly collision free path of each component. This information is stored in Collision Free Path Database. The algorithm of Component Collision Free Path has been developed and published by Alfadhlani et al [14].

If information of assembly collision free path has been obtained, then precedence constraints of assembly could be generated by examining the information of assembly collision free path to disassemble each component. Disassembly approach is used because the information of assembly collision free path only shows the assembly path of final component that can be assembled in particular stage. Because of this approach, the direction of assembly collision free path should be changed in opposite way to obtain disassembly collision free path. After disassemble each component, update the information of assembly collision free path without taking into account the removed components.

The component candidate to be removed is determined based on the information of assembly collision free path. The priority of component to be removed, among candidates, is determined based on criteria: (1) the minimum number of connection among the candidates and the component that has not been selected; (2) component candidate with a minimum volume. Precedence constraints contain the information of component's predecessor, predecessor a component are all components that have been removed previously and connected with it. To generate precedence constraints, develop the algorithm of Precedence Constraint Generation. Inputs for this algorithm are information of assembly collision free path, and volume of component

Region information is used to find assembly sequence combination without breaking the precedence constraints. To determine region of components, develop the algorithm of Component's Region Determination, input for this algorithm is precedence constraints information. If precedence constraints and region information have been determined, then several alternatives of feasible assembly sequences could be generated. The assembly sequence are generated using the algorithm of Assembly Sequence Generation, input of this algorithm are precedence constraints and region of component. Precedence constraint information contains complete information of component's predecessor, but does not have the information of component's successor. Because of that, this precedence constraint could not be used directly to determine the component will be assembled next after a component is assembled. To get the explanation of this problem, then use the algorithm of Precedence Relation Determination.

To support Assembly Sequence Algorithm, it is required one step preparation to find all assembly sequence combinations from initial assembly sequence. This step is developed in the algorithm of Assembly Sequence Combination Determination. The algorithm of Precedence Relation Determination and the algorithm of Assembly Sequence Combination Determination are supporting algorithm, so both of these algorithms are not shown in Fig. 1.

3.4 Assembly reorientation analysis model

Re-design suggestions for the design improvement need to be provided in order to make the product could be assembled easier by minimizing the number of component and assembly reorientation. The number of assembly reorientation is analyzed by calculating the number of reorientation required in each alternative of assembly sequences generated, and then select the best alternative which has the minimum number of assembly reorientation. The number of assembly reorientation is attempted to be more minimum through improving the features of assembly component.

It needs to develop the algorithm of Assembly Sequence Selection for: analyzing and calculating the number of assembly reorientation on each assembly sequence alternatives; and selecting the best alternative. Inputs for this algorithm are assembly collision free path, and assembly sequence alternatives. While the Re-design Suggestion Algorithm needs to be developed to provide re-design suggestions, inputs for this algorithm are selected assembly sequence, and the number of its assembly reorientation.

4. Conclusion

This article discusses a framework proposed for developing automatic Design for Assembly (DFA) on CAD system base using a 3D solid drawing as input. This DFA method is addressed for design evaluation and re-designs suggestions, through: (1) reducing or eliminating or maintaining the current number of product components; and (2) minimizing the number of assembly reorientation. Re-design suggestion is provided for simplifying product structure and ease of assembly

This effort is performed through three general steps: first, defining components function; second, identifying components which could be reduced or eliminated based on geometric information; third, generating assembly sequences and selecting the best one. Some parts of this research that are: the algorithm of Component Geometric Data Extraction, and the algorithm of Collision Detection, have been developed and published, while other parts are being developed.

References

[1] Whitney DE 2004 *Mechanical Assemblies: Their Design, Manufacture, and Role in Product Development* Oxford University Press
[2] Boothroyd G, Dewhurst P, and Knight W 2011 Product Design for Manufacture and Assembly CRC Press Taylor & Francis Group
[3] Ong NS, Chua CK, and Lee EW 1997 Geometric analysis of part from a 3-D solid model for manual assembly time *Integrated Manufacturing Systems* **8**(3) 137 – 146

[4] De Fazio TL, Rhee SJ, and Whitney DE 1999 Design specific approach to design for assembly (DFA) for complex mechanical assemblies *IEEE Transactions on Robotics and Automation* **15**(5) 869-881

[5] Kretschmer R, Pfouga A, Rulhoff S, and Stjepandic J, 2017 Knowledge-based design for assembly in agile manufacturing by using Data Mining methods *Advanced Engineering Informatics* Vol 33 August 2017 Pages 285-299

[6] Barnes CJ, Dalgleish GF, and Jared GEM 1997 Assembly sequence structure in design for assembly *Proceedings of the 1997 IEEE, International Symposium on Assembly and Task Planning* 164-169

[7] Dalgleish GF, Jared, GEM, and Swift KG 2000 Design for assembly: influencing the design process *Journal of Engineering Design* **11**(1) 17-29

[8] Li RK, and Hwang CL 1992a A framework for automatic DFA system development *Computers Industrial Engineering* **22**(4) 403-413

[9] Soh SL, Ong SK, and Nee AYC 2016 Design for assembly and disassembly for remanufacturing *Assembly Automation* Vol. 36, Iss 1, pp. 12 – 24

[10] Li RK, and Hwang CL 1992b Assembly code generation from a feature based geometric model *International Journal of Production Research* **30** 627-646

[11] Coma O, Mascle C, and Veron P 2003 Geometric and form feature recognition tools applied to a design for assembly methodology *Computer Aided Design* **35** 1193-1210

[12] Mo J, Cai J, Zhang Z, and Lu Z 1999 DFA – oriented assembly relation modelling, *International Journal of Computer Integrated Manufacturing* **12**(3) 238-250

[13] Lee S, Kim GJ, and Bekey GA 1993 Combining assembly planning with redesign: An approach for more effective DFA *Proceedings of the 1993 IEEE Conference on Robotics and Automation* **3** 319-325

[14] Alfadhlani, Samadhi TMA, Ma'ruf A, and Toha IS, 2011 Automatic collision detection for assembly sequence planning using a three-dimensional solid model *Journal of Advanced Manufacturing Systems* **10**(2) 277–291

IOP Conf. Series: Materials Science and Engineering **319** (2018) 012084 doi:10.1088/1757-899X/319/1/012084

Determination of Economic Lot Size between Suppliers and Manufacturers for Imperfect Production System with Probabilistic Demand

S Yuniar[1], R Wangsaputra[1,2], A T Sinaga[1]

[1]Industrial Engineering and Management, Institut Teknologi Bandung, Ganeca Street 10, Bandung, West Java, Indonesia, 40132
[2]Manufacturing Systems Research Group, Institut Teknologi Bandung, Ganeca Street 10, Bandung, West Java, Indonesia , 40132

suciyuniar7@gmail.com; rwangsa@ti.itb.ac.id; astridtheresia27@gmail.com

Abstract. This study aims to develop a combined economical lot size model between supplier and manufacturer for imperfect production processes with probabilistic demand patterns and constant lead times. The supplier side produces the product within a certain time interval then sent to the manufacturer with a certain amount of lot size. Imperfect supplier production systems are characterized by the probability of defective product (γ). The model decision variables are the lot size of the manufacturer's ordering, supplier lot size, and the reorder point of the manufacturer. The optimal decision variables are obtained by minimizing the total expected cost of the combined costs between the suppliers and the manufacturers borne by both parties. The model is built compared to the transactional partnership model, in which the supplier does not participate in the efficiency of its inventory system. A numerical example is given as an illustration of the JELS model and the transactional partnership model. Sensitivity analysis of the model is done by changing the parameters aimed at analyzing the behavior of the developed model.

1. Introduction

In traditional management systems, inventory problems are viewed only from one aspect of the supplier or the manufacturer. Today, companies often manage inventory by involving outsiders directly related to the inventory system by determining the lot size of orders together. This process of integration of lot size determination can provide benefits to both parties known as Joint Economic Lot Sizing (JELS).

Some research has been done related to the development of this research. Goyal discusses the generic model of JELS on the problem of sub-division of product delivery from vendor to buyer[11]. The inventory of the entire system product that involves inventory on the vendor and inventory on the buyer will be more minimal when compared to the delayed equal-size shipment. Jonrinaldi and Suprayogi developed a combined economical lot size model between suppliers and buyers for the type of product that was deteriorated while in stock[16]. Jonrinaldi and Suprayogi conducted a model comparison considering the deterioration factor and a model that did not consider the deterioration factor[16]. Ben-Daya and Hariga developed a combined lot size model for one supplier and one buyer with deterministic demand with stochastic lead time[5]. Ouyang[19]developed research conducted by Ben-Daya and Hariga[5] where lead time can be controlled with normal distribution. The buyer orders a product of size Q to the supplier[19]. The supplier produces a product of nQ size with a one-time setup and sends the product to buyers of Q size. Saraswati examines the effect of fluctuating demand on the determination of the size of the production lot and the delivery schedule on an integrated inventory system with the total inventory cost involves the manufacturing and buyer inventory system simultaneously[21]. The search for a lot size production solution with fluctuating demand uses a forward dynamic programming approach. The integration model proposed considers two conditions, namely the condition of unlimited production capacity and limited production capacity. Jauhari developed a combined model of suppliers and buyers with probabilistic demand with equal shipping size. Every lot of orders will be shipped in multiple lot shipping and the supplier will produce the goods in batch production size which is an integer multiple of lot shipping[15]. Ekawati developed a combined lot size model for one supplier and one buyer with demand and lead time is probabilistic[10]. Buyers order goods of size Q and suppliers will produce goods in size nQ, but suppliers deliver goods to buyers in size Q. This research is essentially similar to the research conducted by Ouyang[19], the difference is Ekawati[10] develops in demand patterns and lead time buyers where deficiencies are permitted in research. In addition, Ekawati[10] comparing proposed models to 2 (two) partnership systems namely transactional partnerships and cooperative partnerships conducted by Cohen and Roussel[8].

In previous studies, the JELS model was built on the assumption that demand patterns are constant and are known. Few JELS studies are considering probabilistic requests. A probabilistic demand pattern will affect the form of inventory system management. Given the demand variance, companies must provide safety stocks outside the operating stock to anticipate fluctuations in emerging demand. The demand fluctuations depend heavily on lead time. The demand during lead time is influenced by two things: the demand pattern and the length of lead time. In addition, previous studies have assumed that all goods orders are always accepted into inventory, where the product conditions during the production process are considered perfect, meaning that there are no defective or defective products. Hsu and Hsu develop a combined lot size model with a deterministic demand pattern with constant lead times. Hsu and Hsu assume that the production process is not perfect, where the proportion of defective products on the lot of production is assumed to be uniformly distributed [13].

This study develops a combined lot size modeling for one supplier and one manufacturer for an imperfect production system with probabilistic demand patterns and constant lead times. This research uses 2 (two) reference paper that is research conducted by Ekawati[10] as the basis for JELS model development on the research that will be conducted and research conducted by Hsu and Hsu[13] which examines the determination of production lot size for the imperfect production system where the defect product produced is influenced by the limitations of the equipment and the operator.

2. Mathematical Model

The assumptions used in this research are:

- The rate of demand is probabilistic
- The supplier has a limited production rate and the production rate of the supplier is greater than the demand rate of the manufacturer (P>D)
- Suppliers deliver goods to manufacturers in Q size with unlimited delivery capacity
- The process is free from failure
- Shortage is permitted in the study
- Raw materials arrived at the supplier company in accordance with the agreed time, so that the fulfilment of production needs did not experience obstacles due to lack of raw materials
- Defective products cannot be reworked

Variables and parameters notation used in this research listed in Table 1.

Table 1. Notation

Notation	Definition	Unit of Measurement
D	Demand rate	Unit/ year

IOP Conf. Series: Materials Science and Engineering **319** (2018) 012084 doi:10.1088/1757-899X/319/1/012084

P	Production rate	Unit/ year
Q	Ordering lot size	Unit
nQ	Production lot size	Unit
N	Order frequency (in interger)	Unitless
r	Reorder point	Unit
L	Lead time	Year
γ	Probability of defective product	Unitless
$f(\gamma)$	Probabilistic density function	Unitless
σ	Standard deviation of demand	Unit/ year
k_b	Manufacturers' shortage cost	Rupiah/ unit
c_b	Manufacturers order cost	Rupiah/ order
h_v	Supplier inventory cost	Rupiah/ unit/ tahun
h_b	Manufacture inventory cost	Rupiah/ unit/ tahun
S	Setup cost	Rupiah/ setup
S_v	Supplier setup cost	Rupiah/ setup
Pi	Inspection cost	Rupiah/ unit
F_v	Supplier shipping cost	Rupiah/ shipping
$E(O_{kb})$	Expectation of manufacturing shortage cost	Rupiah/ unit/ tahun
$E(O_{cb})$	Expectation on manufacturer's order cost	Rupiah/ order/ tahun
$E(O_{hv})$	Expectation of supplier's inventory cost	Rupiah/ unit/ tahun
$E(O_{hb})$	Expectation of manufacturer's inventory cost	Rupiah/ unit/ tahun
$E(O_{sv})$	Expectation of supplier setup cost	Rupiah/ setup/ tahun
$E(O_{Pi})$	Expectation on inspection cost	Rupiah/ unit/ tahun
$E(O_{Fv})$	Expectation of supplier shipping cost	Rupiah/ shipping/ tahun
ETC_V	Total supplier cost	Rupiah/ tahun
ETC_b	Total manufacturer cost	Rupiah/ tahun
ETC_{Gab}	Total combined cost	Rupiah/ tahun

2.1. Manufacturer

When the inventory reaches the reorder point r, the manufacturer will place an order for the supplier in size Q. The ordered goods will arrive based on lead time L and added to the inventory system. SS is the safety stock level supplied by the manufacturer to overcome the demand fluctuation during the L time. The goods of Q will decrease at the rate of D and will run out within a cycle of ordering Q / D where D is the monthly demand rate coming to the manufacturer. Thus in one cycle there is D / Q order cycle.

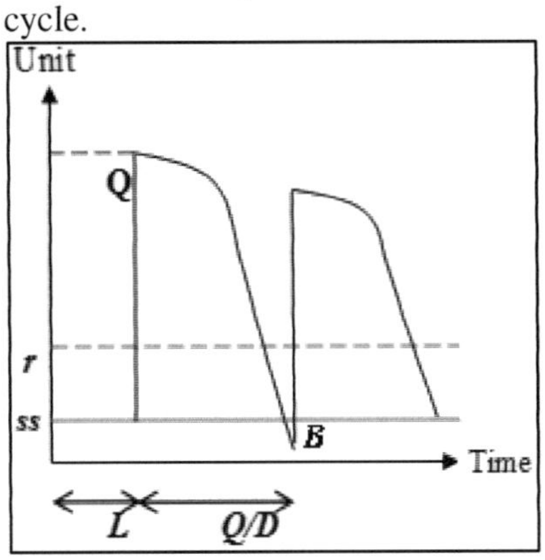

Figure 1. Manufacturer inventory cycle in JELS

Expectation of manufacturer inventory cost

$$= h_b \times \left(\frac{Q}{2} + r - DL\right) \quad (1)$$

Expectation of manufacturer order cost

$$= c_b \frac{D}{Q} \quad (2)$$

Expectation of manufacturer shortage cost

$$= \frac{D}{Q} k_b B(r) \quad (3)$$

Expectation of total manufacturer cost

$$= c_b \frac{D}{Q} + h_b \left(\frac{Q}{2} + r - D_L\right) + \frac{D}{Q} k_b B(r) \quad (4)$$

2.2 Supplier

To meet the demand of the manufacturer, the supplier produces the product within a certain time interval and then sent to the manufacturer with a certain amount of lot size. The imperfect supplier production process is characterized by the number of defective products with defective product probability (γ) uniformly distributed. The defective product cannot be fixed so it becomes scrap. However, in this study, manufacturers receive products from suppliers in a state free of defects.

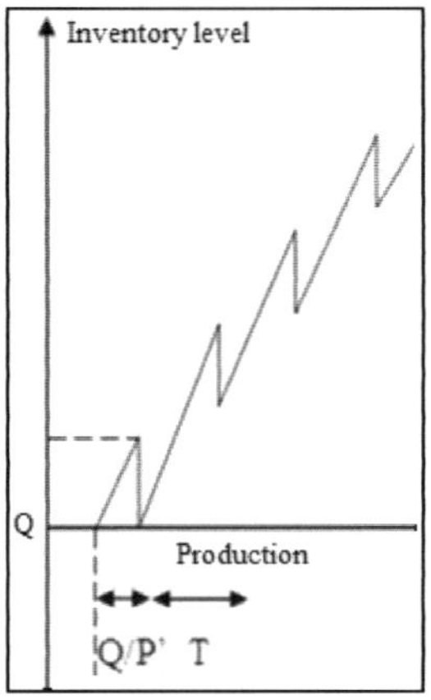

Figure 2. Supplier inventory cycle in JELS

Based on **Figure 2,** the supplier accepts the order from the manufacturer according to the manufacturer's order size Q. The order size Q will be produced in one setup with the production rate P so that the goods will be finished in Q/P time. With the defective product (γ) detected during inspection, the production rate becomes $P' = P(1-\gamma)$, so the goods will be finished in time $\frac{Q}{P(1-\gamma)}$.

The assumption of a constant and known lead time, the supplier sends a product of size Q. If there is a process of repeating the ordering cycle by the manufacturer, then the production process will recur every Q/D. To ensure that the supplier has sufficient production capacity to meet manufacturers' demand, it is assumed that the maximum value γ is less than $(1 - D/P)$. The probability of a defective product is uniformly distributed where the probability of the defective product being produced is likely to arise with the same value.

$$f(x,a,b) = \begin{cases} \frac{1}{b-a}, & a \le x \le b \\ 0, & other\ x \end{cases} \tag{5}$$

Expectation of x value is: (6)

$$E(x) = \int_a^b \frac{1}{b-a} x dx$$

Expectation of supplier inventory cost

$$= h_v\left(\left[\frac{Q}{2}\left((n-1)\left(1-\frac{D}{P(1-\gamma)}\right)+\frac{D}{P(1-\gamma)}\right)\right]+\left[\frac{nQ\gamma}{P(1-\gamma)}\right]\right) \tag{7}$$

Expectation of supplier inspection cost $= P_i nQ$ (8)

Expectation of supplier setup cost $= s_v \frac{D}{nQ}$ (9)

Expectation of supplier transportation cost $= F_v \frac{D}{Q}$ (10)

Expectation of supplier total cost

$$\begin{aligned} &= h_v\left(\left[\frac{Q}{2}\left((n-1)\left(1-\frac{D}{P(1-\gamma)}\right)+\frac{D}{P(1-\gamma)}\right)\right]\right. \\ &\quad + \left.\left[\frac{nQ\gamma}{P(1-\gamma)}\right]\right) h_v\left(\left[\frac{Q}{2}\left((n-1)\left(1-\frac{D}{P(1-\gamma)}\right)+\frac{D}{P(1-\gamma)}\right)\right]\right. \\ &\quad + \left.\left[\frac{nQ\gamma}{P(1-\gamma)}\right]\right) \end{aligned} \tag{11}$$

3. Model Analysis

The analysis is performed to determine the value of Q*, r*, n* for combined lot size between the manufacturer and the manufacturer by adjusting the total expected combined cost of (ETC_{gab}). The total expected combined cost is derived from the summation of the total expected cost of the manufacturer with the expected total cost of the supplier.

$$\begin{aligned} ETC_{Gab} &= c_b\frac{D}{Q} + h_b\left(\frac{Q}{2}+r-D_L\right)+\frac{D}{Q}k_b B(\mathrm{r}) + P_i nQ + s_v\frac{D}{nQ} + F_v\frac{D}{Q} \\ &\quad + h_v\left(\left[\frac{Q}{2}\left((n-1)\left(1-\frac{D}{P(1-\gamma)}\right)+\frac{D}{P(1-\gamma)}\right)\right]+\left[\frac{nQ\gamma}{P(1-\gamma)}\right]\right) \end{aligned} \tag{12}$$

To determine the value of Q^*, it must meet the following conditions:

$$\frac{\partial ETC_{Gab}(Q,r,n)}{\partial Q} = 0 \tag{13}$$

To determine the value of r^*, it must meet the following conditions:

$$\frac{\partial ETC_{Gab}(Q,r,n)}{\partial r} = 0 \tag{14}$$

And to meet the minimum criteria, must meet the following conditions:

$$\frac{\partial ETC_{Gab}{}^2(Q,r,n)}{\partial Q^2} > 0 \tag{15}$$

$$\frac{\partial ETC_{Gab}{}^2(Q,r,n)}{\partial r^2} > 0 \tag{16}$$

Iteration solution research model is as follows:

1) Apply n = 1, i = 1, dan $ETC_{Gab}(Q_{n-1}, r_{n-1}, 0) = \infty$
2) Calculate the value of Q_{n-1} by using equation (17) for $r_{n-1} = \infty$

$$Q_{n-1} = \sqrt{\frac{2D\left(c_b + F_v + \frac{S_v}{n}\right)}{h_b + h_v\left[\left((n-1)\left(1 - \frac{D}{P(1-\gamma)}\right) + \frac{D}{P(1-\gamma)}\right) + \left(\frac{n\gamma}{P(1-\gamma)}\right)\right] + 2P_i.n}} \tag{17}$$

3) Calculate r_{n-1} that completed

$$\alpha_{n,i} = \frac{h_b Q_{n-1}}{k_b\, D}$$

So that,

$$r_{n,i} = Z_{\alpha_{n,1}} \sigma\sqrt{L} + DL \tag{18}$$

$$ss_{n,i} = r_{n,i} - DL \tag{19}$$

4) Calculate $Q_{n,i}$ by using equation (20).

$$Q_{n,i} = \sqrt{\frac{2D\left(c_b + F_v + \frac{S_v}{n} + k_b B(r_{n,i})\right)}{h_b + h_v\left[\left((n-1)\left(1 - \frac{D}{P(1-\gamma)}\right) + \frac{D}{P(1-\gamma)}\right) + \left(\frac{n\gamma}{P(1-\gamma)}\right)\right] + 2P_i.n}} \tag{20}$$

5) Calculate $ETC_{Gab}(Q_{n,i}{}^*, r_{n,i}{}^*, n)$.
6) Add *i=i+1*, then calculate the value of $\alpha_{n,(i=i+1)}$ by using equation (21). Then calculate the value of $r_{n,(i=i+1)}$ and $ss_{n,(i=i+1)}$ by using equation (22) and (23).

$$\alpha_{n,(i=i+1)} = \frac{h_b Q_{n,(i=i+1)}{}^*}{k_b\, D} \tag{21}$$

So that,

$$r_{n,(i=i+1)} = Z_{\alpha_{n,(i=i+1)}} \sigma\sqrt{L} + DL \tag{22}$$

$$ss_{n,(i=i+1)} = r_{n,(i=i+1)}{}^* - DL \tag{23}$$

7) Calculate $ETC_{Gab}(Q_{n,(i=i+1)}{}^*, r_{n,(i=i+1)}{}^*, n)$. Ifs:
   - $ETC_{Gab}(Q_{n,(i=i+1)}{}^*, r_{n,(i=i+1)}{}^*, n) > ETC_{Gab}(Q_{n,i}{}^*, r_{n,i}{}^*, n)$, then the optimal solution is $n^* = n$, $Q^* = Q_{n,i}$, dan $r^* = r_{n,i}$.
   - $ETC_{Gab}(Q_{n,(i=i+1)}{}^*, r_{,(i=i+1)}{}^*, n) < ETC_{Gab}(Q_{n,i}{}^*, r_{n,i}{}^*, n)$, then apply $n^* = n + 1$ and back to step (3) by replacing the value of Q_{n-1} with $Q_{n,(i=i+1)}$ which has been calculated before.

Do each step to get the minimum ETC_{Gab} value.

4. Numerical Example

Table 2. Numerical Example

Notation	**Value**	**Measure**
D	500.000	Unit/year
P	900.000	Unit/year
s_v	100.000	Rupiah/*setup*
h_v	1500	Rupiah/unit/year
Pi	1000	Rupiah/unit
F_v	100.000	Rupiah/shipping
L	0,02	Year
c_b	15.000	Rupiah/order
h_b	2000	Rupiah/unit/year
σ	50.000	Unit/year
k_b	5.500	Rupiah/unit

The search results for Q *, r *, and n * values for model solutions using joint economic lot sizing (JELS) can be seen in Table 3.

Table 3. Result of Calculation of JELS Model Inventory Policy

Manufacturing Lot size (*Q*)	19.229,7 Unit
Manufacturing reorder point level (r)	26.820,6 Unit
N factor	2
Supplier's lot size production	38.459,4 Unit
Total cost (*ETC(Q, r, n)*	Rp.106.822.810,9/year

In comparison, this numerical example also presents an optimal solution for suppliers and manufacturers in the absence of labour in inventory management, a pattern of partnership called a transactional pattern. Mathematical model without cooperation in inventory system management can be seen in appendix. Full results can be seen on Table 4.

Table 4. Calculation Result of Inventory Policy without Cooperation

Manufacturing Lot size (*Q*)	23547,127 Unit
Manufacturing reorder point level (r)	25716,4488 Unit
Supplier's lot size production	23547,127 Unit
Total cost (*ETC(Q, r, n)*	Rp. 117.381.977,1/year

Based on the above calculation, it can be seen that the inventory policy with JELS model is smaller than the model without any cooperation. This is because in the decision-making obtained from the results of mutual decisions from both parties.

5. Sensitivity Analysis

Sensitivity analysis is done by changing parameters in joint economic lot sizing model developed. This parameter change aims to analyze the behavior of the developed model, both in terms of decision variables and performance generated by the model. Changes in parameters made are changes in the manufacturer's message-cost parameters, manufacturer's shipping cost, manufacturing shortage cost, supplier checking fee, supplier's store cost, supplier setup cost, supplier transportation cost, demand rate, probability of supplier defective product (γ), and length of lead time. The magnitude of the parameter value change is 10%, 25%, and 50% of the parameter value in the numerical example.

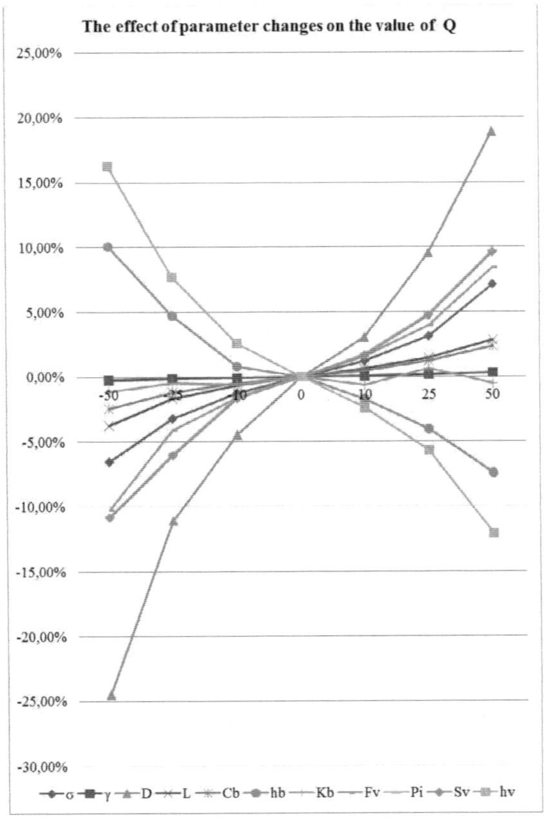

Figure 3. Result of Recapitulation Percentage of Parameter Change on Lot Size of Optimal Order (Q *)

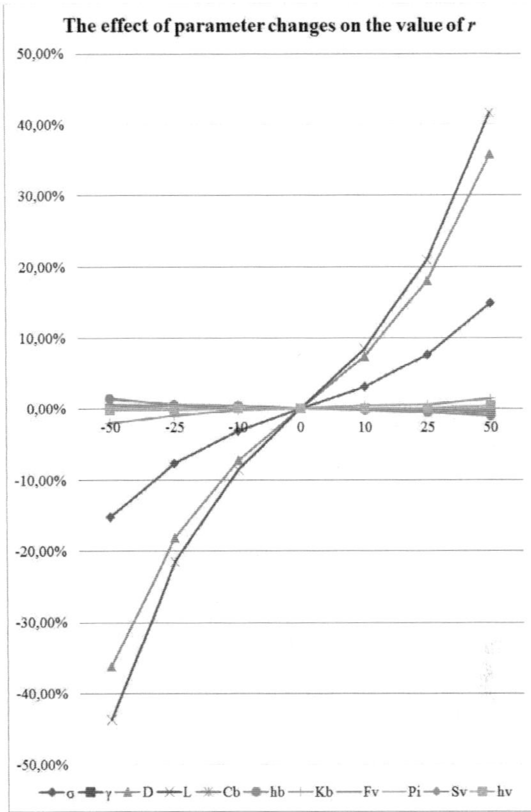

Figure 4. Result of Recapitulation Percentage of Parameter Change to Optimal Reorder Level (r *)

Based on Figure 3, it can be seen that the parameters that are sensitive to the optimal lot lot size (Q^*) are the demand rate (D), supplier's store cost (h_v), and the manufacturing shelf cost (h_b). According to Figure 3, the greater the change in the cost parameters of the supplier's store cost (h_v) and the manufacturing shelf cost (h_b) the smaller the lot size of order (Q). Unlike the change of demand rate parameter (D) to order lot size (Q), the greater the demand rate parameter (D) parameter, the larger the order lot size (Q).

Based on Figure 4, it can be seen that the parameters that are sensitive to the optimal reorder point (r*) are the delivery lead time (L), the demand (D) and the standard deviation (σ). The greater the parameter of the delivery lead time (L), the demand rate (D) and the standard deviation (σ), the greater the reorder point (r).

6. Conclusion

This study develops a combined economical lot size model between supplier and manufacturer for an imperfect production system with probabilistic demand patterns. The supplier side produces the product within a certain time interval then sent to the manufacturer with a certain amount of lot size. Incomplete supplier production systems are characterized by the probability of defective product (γ). The model decision variables are the lot size of the manufacturer's ordering, supplier lot size, and the reorder point of the manufacturer. The optimal decision variable is obtained by minimizing the total expected cost of the combined costs between the supplier and the manufacturer that is borne by both parties. The model is built compared to the transactional partnership model, in which the supplier does not participate in the efficiency of its inventory system. The comparison results show that JELS model gives more minimum cost compared with transactional model.

To see the effect of parameters that exist in the model to optimal lot size, then the sensitivity analysis is done. The parameters that are sensitive to optimal order lot size (Q^*) are demand rate (D), supplier's store cost (h_v), and manufacturer's shelf cost (h_b). The parameters that are sensitive to the optimal reorder point (r^*) are the delivery lead time parameter L, the demand rate (D) and the standard deviation (σ).

This research is conducted using a supply chain system consisting of one supplier and one manufacturer so that it can be developed by creating a model involving multiple suppliers and some manufacturers. In addition, in the study developed models with production systems that are not perfect on the supplier resulting in defective products. However, it is assumed that the defective product cannot be reworked and becomes scrap. So that can be developed for further research with the assumption that the defective product can be done rework.

Acknowledgement
We would like to express our gratitude to the ITB internal research grant and Manufacturing Systems Research Group-ITB for the financial support to the research.

Appendix
The mathematical model of economical lot size in the pattern of transactional partnership is as follows:
1. Manufacturer
The manufacturer will place the order of goods to the supplier in size Q when the inventory reaches the reorder point level. The ordered item will arrive in lead time (L) and added to the inventory system at once. *SS* is the level of safety stock supplied by the manufacturer to overcome the demand fluctuations over time L. The goods of Q will decrease with the demand rate per year (D) and will be exhausted within a cycle of ordering of Q/D. Thus within one year there is D/Q order cycle. Manufacturing inventory systems in transactional partnerships are similar to inventory systems on JELS models.

2. Supplier
In the transactional partnership pattern, the supplier receives an order from the manufacturer then manufactures it according to the manufacturer's ordering size Q. The order size Q is produced in one setup at the production rate P so that the product will be completed in Q / P time. Then the supplier sends the finished product to the manufacturer in Q size and if there is a repeat of the ordering cycle by the manufacturer, then the production process will repeat every Q / D. Supplier supply systems on transactional partnerships can be seen in Figure 5.

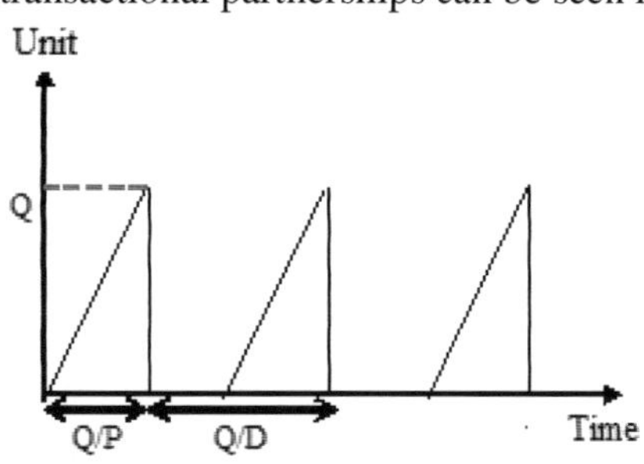

Figure 5 Supplier Inventory System In Transactional Partnership

With the defective product (γ) detected during inspection, the production rate becomes $P'=P(1-\gamma)$, so that the goods will be finished in $Q/(P(1-\gamma))$. To ensure that the supplier has sufficient production capacity to meet manufacturers' demand, it is assumed that the maximum value γ is less than (1 - D / P).

- Expectation of supplier inventory cost $=h_v\left(\left[\frac{DQ}{2P(1-\gamma)}\right]+\left[\frac{Q\gamma}{P(1-\gamma)}\right]\right)$ (24)
- Expectation of supplier setup cost $= s_v\ \frac{D}{Q}$ (25)
- Expectation of supplier transportation cost $= F_v\ \frac{D}{Q}$ (26)
- Expectation of supplier inspection cost $= P_iQ$ (27)

- Expectation of supplier total cost (28)

$$= h_v \left(\left[\frac{DQ}{2P(1-\gamma)} \right] + \left[\frac{Q\gamma}{P(1-\gamma)} \right] \right) + s_v \frac{D}{Q} + F_v \frac{D}{Q} + P_i Q$$

References

[1] Bahagia, S.N., (2006), *Sistem Inventori*, Penerbit ITB, Bandung.

[2] Banarjee, A., (1986), A joint economic-lot size for purchaser and vendor, *Decision Science* 185:292-311.

[3] Ben-Daya M., Darwish M., Ertogral K., (2008), The joint economic lot sizing problem: review and extensions, Eur J Oper Res 185 (2):726–742.

[4] Ben-Daya, M., dan Hariga, M., (2000), Economic lot scheduling problem with imperfect production processes. *Journal of the Operational Research Society, 51*, 875–881.

[5] Ben-Daya, M., and Hariga, M., (2004), Integrated Single Vendor Single Buyer Model with Stochastic Demand and Variable Lead Time, *International Journal of Production Economic,* Vol. 92, pp. 75-80.

[6] Buchan, Joseph, dan Koenigsberg, E, *Scientific Inventory Management,* Prentice Hall, Englewood Cliffs, NJ, 1963.

[7] Buffa, Elwood S. Dan Miller, J.F., *Production-Inventory System:Planning and Control,* 3nd, Richard D. Irwin Inc. 1979.

[8] Cohen, S., dan Roussel J., (2005), *Strategic Supply Chain Management: The Five Disciplines for Top Performance,* McGraw-Hill.

[9] Daellenbach, H.G., (1995), *System and Decision Making,* John Wiley & Sons, Inggris.

[10] Ekawati, I., Iskandar, B.P., Suprayogi., Cakravastia, A., (2010), Model penentuan ukuran lot gabungan ekonomis pemasok dan pemanufaktur dengan pola permintaan dan *lead time* probabilistik, *Tesis S-2, Teknik dan Manajemen Industri, ITB.*

[11] Goyal, SK., (1988), A joint economic-lot-size model for purchaser and vendor: a comment, Decis Sci 19(1): 236–241.

[12] Hadley, G., Whitin, T. M., (1963), *Analysis of Inventory System,* Prentice-Hall Inc, Englewood Cliffs, New Jersey.

[13] Hsu, J.T., dan Hsu, L.F., (2013), An integrated vendor-buyer cooperative inventory model for items with imperfect quality and shortage backordering. *Advances in Decision Sciences*, Article ID 679083, 19.

[14] Hsu, J.T., dan Hsu, L.F., (2015), Economic production quantity (EPQ) models under an imperfect production process with shortages backordered, *International Journal of Systems Science*, Department of Management, Zicklin School of Business, Baruch College, The City University of New York, New York, NY, USA.

[15] Jauhari, W.A., Pujawan, I.N., Wiratno, S.E., (2009), Model joint economic lot sizing kasus pemasok-pembeli dengan permintaan probabilistik, *Jurnal Teknik Industri,* Vol 11 no 1:1-14.

[16] Jonrinaldi dan Suprayogi, (2003), Model penentuan ukuran lot ekonomis gabungan antara pemasok dan pembeli untuk produk yang mengalami deteriorasi, *Seminar Nasional Perencanaan Sistem Industri 2013 (SNPSI 2003).*

[17] McLeavy, D.W., Narasimhan S.L., *Production Planning and Inventory Control,* Allyn and Bacon, Boston, MA, 1985.

[18] Monden, Yasuhiro, T*oyota Production System: Practica Approach to Production Management,* Institute of Industrial Engineers, Atlanta, GA, 1983.

[19] Ouyang, L.Y., Wu, K. S., dan Ho, C. H., 2004, Integrated Vendor-Buyer Cooperative Models with Stochastic Demand in Controllable Lead Time, *International Journal of Production Economics,* Vol. 92, pp. 255-266.

[20] Prasetyo, H., (2004), Model ukuran lot untuk proses produksi yang mengalami penurunan kinerja dengan pola permintaan berfluktuasi, *Tesis S-2, Teknik dan Manajemen Industri, ITB.*

[21] Saraswati, D., Halim, A.H., Iskandar, B.P., Cakravastia, A., (2009), Model kemitraan kooperasi antara pemasok dan pemanufaktur dalam penentuan ukuran lot gabungan. *Proceedings of Joint Seminar Japan-Indonesia Seminar on Technlogy Transfer (JITT) & National Seminar on Industrial System Planning 2008 (SNPSI).*

[22] Salameh, M.K., dan Jaber,M.Y., (2000), Economic production quantity model for items with imperfect quality, *International Journal of Production Economics*, *64*(1), 59–64.

[23] Shimci-Levi, D., Kamisnsky, P., dan Simchi-Levi, E., (2003), *Designing and Managing The Supply Chain: Concept, Strategy, and Case Studies,* Second Edition, McGraw-Hill, New Jersey.

[24] Tersine, R.J., (1994), *Principle of Inventory and Material Management*, Prentice Hall Int. Ed.4

[25] Walpole, R.E., Myers, R.H., Myers, S.L., Ye, K., (2002), *Probability and Statistics for Engineers and Scientists, Seventh Edition,* Prentice-Hall, Inc., New Jersey.

Bearing Procurement Analysis Method by Total Cost of Ownership Analysis and Reliability Prediction

Wildan Trusaji, Muhammad Akbar, Sukoyo and Dradjad Irianto

Faculty of Industrial Technology, Institut Teknologi Bandung, Bandung 40132, Indonesia

E-mail: wildan@mail.ti.itb.ac.id[1]

Abstract. In making bearing procurement analysis, price and its reliability must be considered as decision criteria, since price determines the direct cost as acquisition cost and reliability of bearing determine the indirect cost such as maintenance cost. Despite the indirect cost is hard to identify and measured, it has high contribution to overall cost that will be incurred. So, the indirect cost of reliability must be considered when making bearing procurement analysis. This paper tries to explain bearing evaluation method with the total cost of ownership analysis to consider price and maintenance cost as decision criteria. Furthermore, since there is a lack of failure data when bearing evaluation phase is conducted, reliability prediction method is used to predict bearing reliability from its dynamic load rating parameter. With this method, bearing with a higher price but has higher reliability is preferable for long-term planning. But for short-term planning the cheaper one but has lower reliability is preferable. This contextuality can give rise to conflict between stakeholders. Thus, the planning horizon needs to be agreed by all stakeholder before making a procurement decision.
Keywords: Total Cost of Ownership, Maintenance, Reliability Prediction, Bearing

1. Introduction
The total cost of ownership (TCO) is an application of cost accounting that can help decision maker to compare price between the value of a good to choose which one is better [1]. TCO analysis can quantify cost that incurred when transaction activity happens - for example; ordering, transportation, quality control – and cost that incurred because of low quality – for example; rejection, rework, and warranty [2]. TCO facilitates the firm to make the procurement to be more value oriented. Anderson, Glenn, & Sedatole (2000) see that it is important to uses value chain perspective that focuses on TCO then to use selling price to make a procurement decision. Empirical study shows that to get overall cost reduction and performance improvement from a supplier, a firm should not use a selling price criteria only [4]. Using TCO analysis is crucial to evaluate alternative procurement. Thus, in this paper, TCO analysis is used to evaluate bearing alternative that must be procured.

2. Modelling Assumptions
This research tries to identify what cost that can incur when a bearing is procured and model the TCO of bearing procurement. Identifying direct and indirect cost is a crucial step to do TCO analysis [5]. Thus direct and indirect cost is identified to model TCO of bearing. Maintenance activity is the source of direct and indirect cost. The direct cost of maintenance activity is based on cost of preventive and corrective maintenance cost. The indirect cost of maintenance activity is a downtime cost because of the incurrence of corrective maintenance.

A reliability prediction method by Naval Surface Warfare Center [6] is used the get the bearing reliability prediction. This method considers lubricant, water contaminant, operating temperature, service condition (smooth or rough and sudden movement), lubricant contamination, equivalent radial load, and dynamic load rating of the bearing. The latter is informed by bearing manufacturer, while the others are operating context specific. Predicted reliability is used to get the optimal maintenance planning. In this paper, block policy model is used to get the optimal solution [7]. This model considers preventive and corrective maintenance cost, and failure rate of the part to get the optimal maintenance period.

3. The Mathematical Model

3.1. Mathematical Notation

λ_{BE}	= Failure Rate of Bearing, failure/million hours	$ECC(T)$	= Expected Cost of One Cycle Maintenance \$
$\lambda_{BE,B}$	= Base Failure Rate of Bearing, failure/million hours	$ECC(Yr)$	= Expected Cost of Maintenance \$/yr
C_Y	= Multiplying Factor for Applied Load	$M(T)$	= Expectation of Corrective Maintenance Activity in One Cycle
L_S	= Dynamic Load Rating, lbf	C_f	= Corrective Maintenance Cost
L_A	= Equivalent Radial Load, lbf	C_p	= Preventive Maintenance Cost
C_R	= Life Adjustment factor for Reliability	T	= Block/Cycle Period hr
C_V	= Multiplying Factor for Lubricant	$J_\infty(T)$	= Expected Cost per Unit Time
C_{CW}	= Multiplying Factor for Water Contaminant Level	$f(T)$	= Probability Density Function of Failure
C_t	= Multiplying Factor for Operating Temperature	$F(T)$	= Cumulative Probability Function of Failure
C_{SF}	= Multiplying Factor for Service Conditions	C_{mf}	= Technician Rate for Corrective Maintenance \$/hr
C_C	= Multiplying Factor for Lubricant contamination	C_{mp}	= Technician Rate for Preventive Maintenance \$/hr
C_b	= Price of Bearing	D_f	= Downtime due Corrective Maintenance hr
C_d	= Downtime cost \$/hr	D_p	= Downtime due Preventive Maintenance hr

3.2. Reliability Prediction and Maintenance Block Policy Model

This paper uses reliability prediction model developed by Naval Surface Warfare Center. This model considers Dynamic Load Rating (L_S) as main reliability characteristic of bearing. In addition, several operating characteristics like radial load, lubricant, water, etc are considered. To predict failure rate of bearing, equation (1) is used.

$$\lambda_{BE} = \lambda_{BE,B}.C_Y.C_R.C_V.C_{CW}.C_t.C_{SF}.C_C \tag{1}$$

Equation (2) is used to find multiplying factor for applied load C_Y.

$$C_Y = (\frac{L_A}{L_s})^y \tag{2}$$

Where $y = 3$ for ball bearing and 3,3 for roller bearing.

Maintenance block policy model [7] is used to find the optimal maintenance period. In block policy, every preventive maintenance (PM) activity is a renewal point and interval between two PM activities is defined as one cycle. In that interval, failure(s) can occur and follow failure distribution function $f(T)$. That failure is repaired by corrective maintenance (CM) activity. Expected cost in one cycle $ECC(T)$ can calculated by equation (3).

$$ECC(T) = C_f.M(T) + C_p \tag{3}$$

To find corrective maintenance and preventive maintenance cost, bearing price, downtime cost, and technician cost is considered. Thus the equation to find C_f and C_p are

$$C_f = C_b + (C_{mf} + C_d).D_f \tag{4}$$

$$C_p = C_b + (C_{mp} + C_d).D_p \tag{5}$$

The optimal solution T^* is the period of a cycle that gives zero value for the derivative function of expected cost per unit time $ECC(T)/T$. Jiang & Murthy (2008) approximate the optimal solution T^* by equation (6).

$$Tf(T) - F(T) = C_p(C_f)^{-1} \tag{6}$$

Failure distribution is assumed follows the exponential distribution with mean time to failure $\boldsymbol{\beta} = \mathbf{1}/\boldsymbol{\lambda_{BE}}$. Thus $\boldsymbol{f(T)}$ and $\boldsymbol{F(T)}$ are probability density function and cumulative probability function of the exponential distribution with mean $\boldsymbol{\beta} = \boldsymbol{\lambda_{BE}}^{-1}$.

4. Numerical Example

Two alternatives of bearing (1&2) are considered to be chosen. The TCO of those bearing is calculated and bearing with lesser TCO is chosen. Consider the following bearing characteristic, operational characteristic, and maintenance characteristic parameter in Table 1.

Table 1. Bearing, Operational, and Maintenance Parameter

Bearing Characteristic	Operational Characteristic
L_{S1} = 267.523 lbf	$\lambda_{BE,B}$ = 1
L_{S2} = 229.305 lbf	C_R = 1
C_{b1} = 7.110 \$/unit	C_V = 0,947
C_{b2} = 4.444 \$/unit	C_{CW} = 1
Maintenance Characteristic	C_t = 1
C_{mf} = 137,5 \$/hr	C_{SF} = 1
C_{mp} = 68,75 \$/hr	C_C = 1,4
D_f = 30 hr	L_A = 184.047
D_p = 6 hr	C_d = 500 \$/hr

These characteristic parameters are calculated based on the equation (1) – (6). The output of the model are listed in table 2

Table 2. Model Output

Bearing 1	Bearing 2
β = 14.852 hr	β = 8.930 hr
C_p = 7.522 \$	C_p = 4.856 \$
C_f = 11.235 \$	C_f = 8.569 \$
T^* = 43.183 hr	T^* = 16.976 hr
$ECC(Yr)$ = 6.183 \$/yr	$ECC(Yr)$ = 8.846 \$/yr

From the result, it can be concluded that, although bearing 1 has higher bearing price, corrective and preventive maintenance cost, it has longer optimal maintenance period than bearing 2. Thus, bearing 1 has a lower expected cost per year than bearing 2. So bearing 1, the more expensive and reliable bearing, is preferable than bearing 2, the cheaper and the unreliable one.

To know whether the statement, more expensive and reliable bearing is preferable than the cheaper and unreliable bearing, holds true in every context, a sensitivity analysis is conducted. Downtime cost and maintenance period are considered as independent variables in sensitivity analysis. First, downtime cost is considered to be a variable and everything is fixed. The result is the higher the downtime the higher expected cost per year gap between bearing 1 and bearing 2, as can be seen in fig 1. So in every downtime cost, the statement holds true.

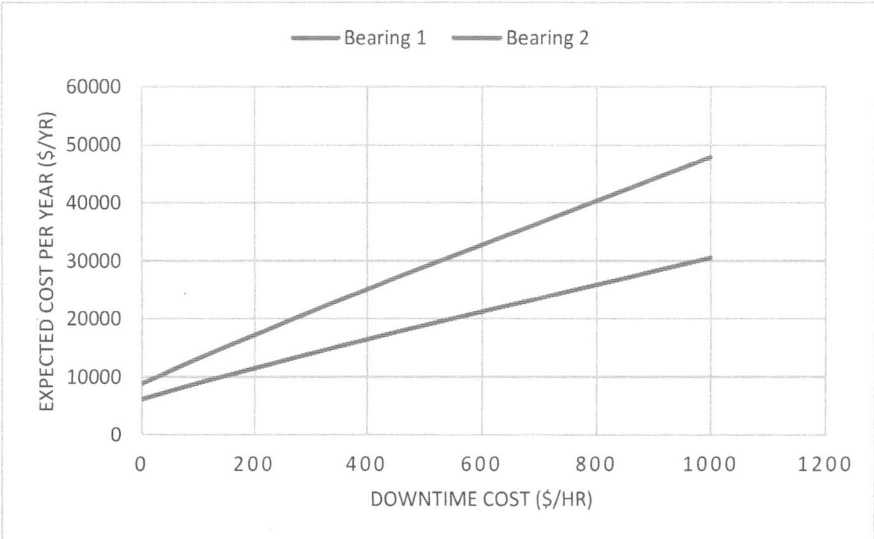

Figure 1. Sensitivity Analysis by Downtime Cost (C_d)

Second, maintenance period is assumed as a variable. Unlike before, the maintenance period policy for bearing 1 is assumed same with bearing 2 and it acts as an independent variable instead of a dependent variable. It is known, that the statement does not always hold true in every $T^* = T_1^* = T_2^*$. It is known that for $T^* > 6000$, the $ECC(Yr)_1 < ECC(Yr)_2$. Thus, when $T^* > 6000$, h the statement holds true. But it does not hold true for $T^* < 6000$ because when $T^* < 6000$ the $ECC(Yr)_2 < ECC(Yr)_1$. Thus, from sensitivity analysis, it can be concluded that more expensive and reliable bearing is preferable than the cheaper and unreliable bearing in the relatively long planning horizon.

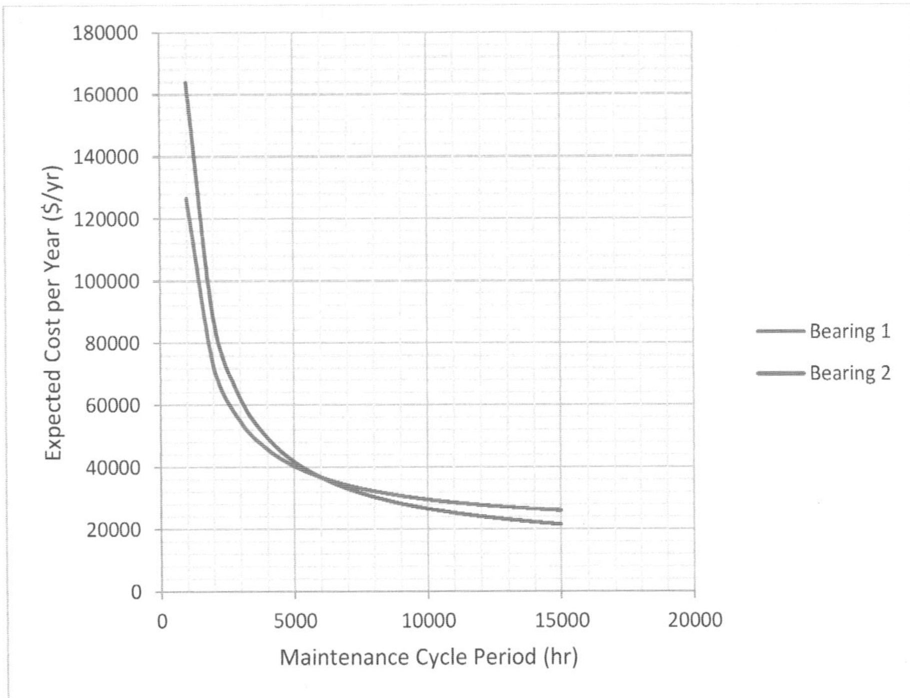

Figure 2. Sensitivity Analysis by Maintenance Cycle Period (T)

5. Conclusion

The TCO analysis can give insight to decision maker which one bearing is better to be procured. TCO analysis considers not only buying price, but also the direct and indirect cost of bearing's reliability. A block policy maintenance model is used to find optimal maintenance period and calculated the direct and indirect cost. To predict bearing's reliability from its dynamic rating, a reliability prediction model is used. For relatively long planning horizon, a more expensive and has higher dynamic rating is preferable than the other. But for relatively short-term planning, the cheaper and has lower dynamic rating is preferable.

This paper shows that the statement, the more expensive and reliable one is better, is contextual. It depends on horizon planning. This contextuality can give rise to conflict between an operational manager and a strategic manager. It happens because the operational manager has a short-term objective while the strategic manager has a long-term objective. Thus, the planning horizon needs to be agreed by all stakeholder before making a procurement decision.

Acknowledgement

This research was conducted at the Manufacturing System Research Group, funded by P3MI research grant of ITB

References

[1] Wouter M 2004 *The Adoption of Total Cost of Ownership for Sourcing Decision - A Structural Equations Analysis.* ISBM Report 4.

[2] Ellram L M 1995 Activity Based Costing and Total Cost of Ownership: a Critical Linkage *Journal of Cost Management*, pp. 22-30.

[3] Anderson S W, Glenn D and Sedatole K L 2000 Sourcing Parts of Complex Product: Evidence on Transaction Costs, High Powerd Incentives and Ex-post Opportunism *Accounting, Organization, and Society* pp. 723-749.

[4] Ittner C D et al 1999 Supplier Seletion, Monitoring Practices and Firm Performance *Journal of Accounting and Public Policy* pp. 253-281.

IOP Conf. Series: Materials Science and Engineering **319** (2018) 012085 doi:10.1088/1757-899X/319/1/012085

[5] West R and Dalgle S L 2004 *Total Cost of Ownership: A Strategic Tool for ERP Planning and Implementation.*, EDUCAUSE Center for Applied Research Bulletin.
[6] Naval Surface Warfare Center Carderock Division 2011 *Handbook of Reliability Prediction Procedures for Mechanical Equipment..*
[7] Jiang R and Murthy D 2008 *Maintenance: Decision Model for Management.* Beijing : Science Press

IOP Conf. Series: Materials Science and Engineering **319** (2018) 012086 doi:10.1088/1757-899X/319/1/012086

Design Of Measurements For Evaluating Readiness Of Technoware Components To Meet The Required Standard Of Products

Ilham Fauzi, Fariz Muharram Hasby, Dradjad Irianto

Institut Teknologi Bandung, Jalan Ganesha No. 10, Bandung 40151

Email: fariz@mail.ti.itb.ac.id, dradjad@mail.ti.itb.ac.id

Abstract. Although government is able to make mandatory standards that must be obeyed by the industry, the respective industries themselves often have difficulties to fulfil the requirements described in those standards. This is especially true in many small and medium sized enterprises that lack the required capital to invest in standard-compliant equipment and machineries. This study aims to develop a set of measurement tools for evaluating the level of readiness of production technology with respect to the requirements of a product standard based on the quality function deployment (QFD) method. By combining the QFD methodology, UNESCAP Technometric model [9] and Analytic Hierarchy Process (AHP), this model is used to measure a firm's capability to fulfill government standard in the toy making industry. Expert opinions from both the governmental officers responsible for setting and implementing standards and the industry practitioners responsible for managing manufacturing processes are collected and processed to find out the technological capabilities that should be improved by the firm to fulfill the existing standard. This study showed that the proposed model can be used successfully to measure the gap between the requirements of the standard and the readiness of technoware technological component in a particular firm.

1. Introduction

To support Indonesia's role in international trade, the government has set product standards expected to protect the national industry. This is in accordance with Government Rules (PERMEN 55/M-IND/PER/11/2013) that the goal of standards implementation is to support the productivity, production efficiency, and quality of goods, services, processes, systems, which were intended to improve the competitiveness, consumer protection, business, labor and community. However, most of the product standards designed by the governments nowadays are adopted from international standards. The requirements of those standards are relatively high, and thus it is difficult for small and medium-sized domestic businesses to meet those standards.One of the main obstacles to the implementation of product standard is the inability of domestic industry to fulfill the technological requirements of the standard, such as in terms of manufacturing technology like the equipment and machines.

Pudjiastuti [6] showed that the productivity of companies that use the National Standard (SNI) as a management tool is still not good enough, and the average contribution per worker is still low when compared to industries that use other standards. Furthermore, Puspita [7] has found that there areseveral factors, including the competence of human resources, raw materials and technology used, that can be the reasons why a product is unable to meet a standard [7]. This indicates that the application of product standards in Indonesia is still inefficient.

Other researchers have conducted several studies to design the method to help regulators in assessing the readiness of a company or to help a company to comply with standards. For example, Sinuhaji [8] examined in depth the readiness of humanware components (i.e.,the aspect of the competence of human resources (HR)) in the industry to meet the requirements of the standard. Puspita [7] also emphasized that the technology components should also be evaluated when examining a product of a company is in compliance with product standards.

2. Literature Study

In 2007, the Body of National Standardization of Indonesia conducted research that assessed the fulfillment for standards using the method of Standard Value Assessment Tool (SVAT). This study also maintained and assessed the feasibility of standards using voting process by stakeholders. Based on the results of voting, the committee form by the stakeholders determine whether the standard should be revised or kept as existing.

Gunawan [3] extended the prior study and pointed out that the prior research had not yet thoroughly considered the aspect of technology, which might be critical in determining the capability of a company in meeting government requirements or standards. To answer that problem, he designed a model to determine the contribution of various technology components in fulfilling standards by considering Technoware and Humanware components into consideration when assessing the technology components. These components are taken from the Technoware assessment model developed by UNESCAP.

Further development of this type of research is done by Islam [4], in which QFD approach is used to design a tool for measuring the readiness of Technoware and Humanware components in laboratory to meet the standard (SNI ISO/IEC 17025). Nurcahya et al. [5] have assessed the technology level of industrial estates with respect to environment standards using a similar QFD approach to identify the gap between the existing level and the expected level of technology. The same study pointed out this assessment can be applied in the existing business processes of an industrial estate and the standard requirements. Later Akbar et al. [1] also used the same approach as a tool for identifying targets for improvement of both Humanware and Technoware components in order to impose standards in Manufacturing Polytechnic Institutions.

3. Proposed Model

House of Quality (HOQ) is the primary planning tool used in QFD. HOQ is an important tool for QFD activities, containing information on "what, "how", relation between "what" and "how", and the relationship between the "how" factors themselves. HOQ describes the customer needs (as "what"), planning matrix (as "why"), product technical responses or characteristics (as "how"), relationship matrix, technical correlations, and target of technical matrix requirement (as "how much") [2]. The structure of HOQ can be thought of as a framework of a house, as shown in Figure 1.

In this study, we use the same approach to describe a relationship between implemented standards and the production process technology of a particular manufacturing firm. First, we define the standard requirements to take the place of customer needs & benefits in the HOQ matrix ("what"). Then we define the production processes to take the place of technical responses in the HOQ matrix ("how"). We can then define the relationship between the standards and production process using the classifications detailed in the HOQ approach.

The planning matrix ("why") in the traditional HOQ approach is then combined with the UNESCAP technometric model to assess the level of Technoware components needed to fulfill the prescribed standards and the level of Technoware components currently being used in the particular firm. Other information in this matrix includes the target value, improvement ratio, absolute weight and normalized absolute weight. These numbers are calculated to describe the gap between the current capability of the firm and the effort that they should concentrate to fulfill the required standards.

IOP Conf. Series: Materials Science and Engineering **319** (2018) 012086 doi:10.1088/1757-899X/319/1/012086

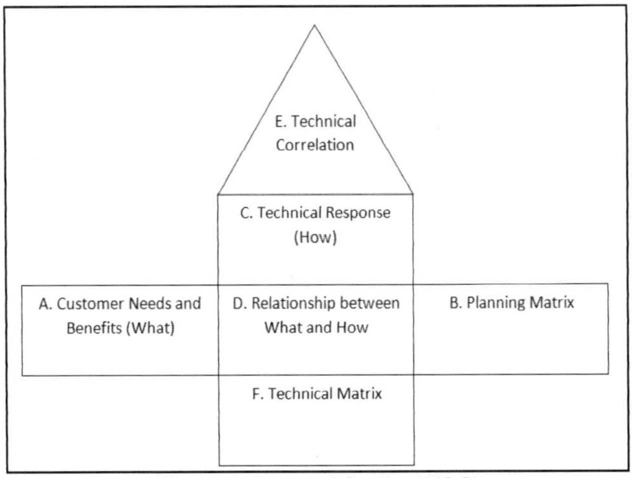

Figure 1. House of Quality (HOQ)

To fill in the planning matrix described before, data will be collected from experts to determine the level of Technoware components based on standard requirements (denoted as *TS*), the level of technoware of production processes in industry (denoted as *TI*) and the importance weight of standard requirements (denoted as WS) and production processes (denoted as *WP*). We can then proceed to calculate the Aggregate Technoware of Standards Requirements (*ATS*), Aggregate Technoware of Production Processes in industry (*ATI*), Improvement Ratio and Absolute Weight (*AW*) using the following equations:

$$ATS = \sum WP \times TS \quad (1)$$

$$ATI = \sum WP \times TI \quad (2)$$

$$Improvement\ Ratio = \frac{ATS}{ATI} \quad (3)$$

$$AW = WS \times Improvement\ Ratio \quad (4)$$

4. Data Collection

We use the proposed model to measure a manufacturing firm's Technoware component capability to meet the required standards in the toy industry (SNI ISO 8124-1:2010). To help in determining the variables *TS, TI, WP* and *WS*, 8 experts have been asked to provide their judgment. These experts are people with knowledge and experience in government standards (of the toy product) or people active in the industry for more than 10 years, including head officers of the testing laboratory, head officers of the body of certification and standardization, head of the toys laboratory, analysts of the laboratory, head officers of production department, head of R & D department, staff of production department, and staff of R & D department. Responses from the experts are collected using the Analytical Hierarchy Process (AHP) to produce one consistent value for the variables mentioned.

The standard mentioned before consists of 29 requirements of toy products in general and will be defined as customer needs and benefits in the House of Quality with the assumption that customers will want products that meet the standard. The experts then converted them into 20 requirements because not all requirements in the clauses stated in the original standards are related to the toy products analysed in this study (ride-on toys). On the other hand, discussion with the experts generated 9 main processes required to manufacture the toys. Figure 2 shows the clauses of the standard included in this research.

IOP Conf. Series: Materials Science and Engineering **319** (2018) 012086 doi:10.1088/1757-899X/319/1/012086

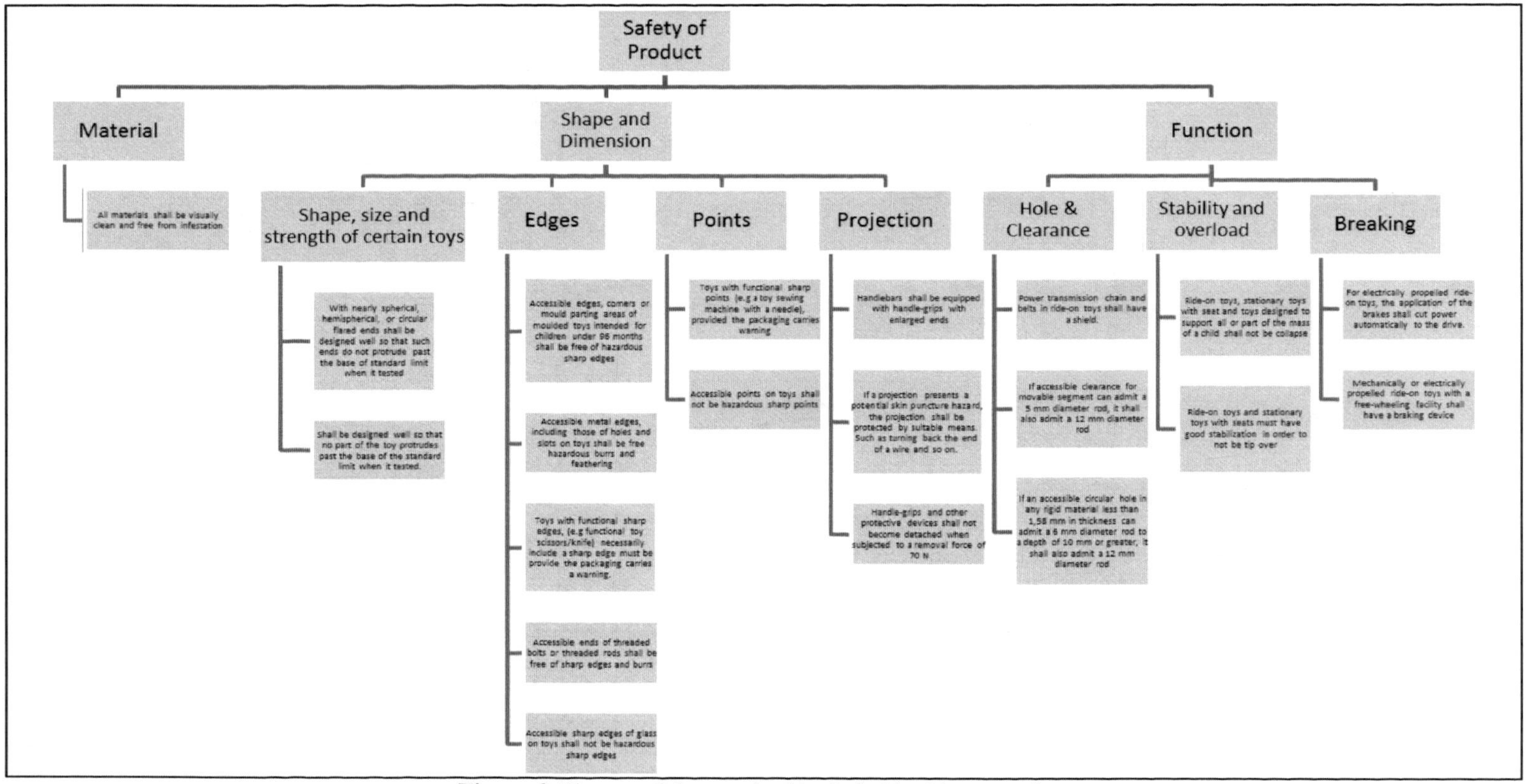

Figure 2 Clauses of the standard included in this research

Correlation between the standards and the technical processes can then be filled using the usual HOQ procedure. By using the weights and measurements of Technoware components, the planning matrix section of the modified HOQ matrix is then filled. The partial modified HOQ matrix is shown in Figure 3.

IOP Conf. Series: Materials Science and Engineering **319** (2018) 012086 doi:10.1088/1757-899X/319/1/012086

No	Function		Customer Requirements	Making of the image or design	Formulation of raw material	Making of the pattern	Making of the mold	Molding Process	Finishing	Assembling components	Packaging and labeling	Inspection	Importance Weight of standard	Goal (ATS)	Current Value (ATI)	Improvement Ratio	Absolute Weight	Normalized Weight
				Technical Responses														
3	Function	Holes and clearances	If an accessible circular hole in any rigid material less than 1,58 mm in thickness can admit a 6 mm diameter rod to a depth of 10 mm or greater, it shall also admit a 12 mm diameter rod			3						3	0.013	0.480	0.432	1.111	0.533	0.019
			If accessible clearance for movable segment can admit a 5 mm diameter rod, it shall also admit a 12 mm diameter rod			3						3	0.013	0.480	0.432	1.111	0.533	0.019
			Power transmission chain and belts in ride-on toys shall have a shield.			3			1	9		3	0.025	1.146	1.149	0.997	1.143	0.040
		Stability and overload	Ride-on toys and stationary toys with seats must have good stabilization in order to not be tip over	1		3			1	9		3	0.050	1.575	1.721	0.915	1.441	0.050
			Ride-on toys, stationary toys with seat and toys designed to support all or part of the mass of a child shall not be collapse	1		3			1	9		3	0.100	1.575	1.721	0.915	1.441	0.050
		Braking	Mechanically or electrically propelled ride-on toys with a free-wheeling facility shall have a braking device			1			1	9		3	0.075	1.146	1.149	0.997	1.143	0.040
			For electrically propelled ride-on toys, the application of the brakes shall cut power automatically to the drive.	0	0	1	0	0	1	9	0	3	0.075	1.146	1.149	0.997	1.143	0.040
			Absolute Weight Standard (AWS)	1.511	5.971	1.713	0.340	1.223	0.880	4.134	0.525	1.740						
			Absolute Weight Industry (AWI)	1.610	5.476	1.816	0.358	1.201	0.928	4.340	0.548	1.822						
			Relative Weight (Contribution Value)	1.419	6.511	1.618	0.323	1.252	0.836	3.946	0.502	1.666						
			GAP	-0.099	0.495	-0.103	-0.018	0.022	-0.048	-0.206	-0.024	-0.082						

Figure 3 Partial modified HOQ matrix

5. Results and Discussion

The modified HOQ of Figure 3 shows that the planning matrix allows us to prioritize customer needs indicated from the Normalized Absolute Weight value. The highest value is in the standard terms "toys with functional sharp edges" and "toys with functional sharp points" with a value of 0.074. Second order is on the standard of "Accessible edges, corners, mold parting area of molded toys" with a value of 0.066, while the lowest priority is in the "accessible clearance for moveable segment" with a value 0.019.

The goal value is set equal to the value of ATS because the fulfillment of the standard requirements of the toy product is the goal aimed by the industry. The effort required in achieving the goals is represented by Improvement Ratio, where a higher the number means that more effort is needed to improve that particular aspect. The highest value of the improvement ratio is in the clause of "hole and clearances" (standard clause regarding accessible circular hole in any rigid material and accessible clearance for moveable segment) which has a value of improvement ratio of 1.111. While the lowest value is found for the clause sub criteria of "edges" (standard clause regarding accessible edges, corners or mold parting areas of molded toys and accessible metal edges), which is equal to 0.900.

Filling in the modified HOQ matrix will also show which standard requirements are met by the technical response and which are not yet met. This is indicated by evaluating whether the ATS value is lower or higher than the ATI value. If ATS is lower than ATI, that means the process meets the standard requirements and vice versa. In this case the matrix shows that there are three standards that have not yet been met by the industry and these are as follows:

- All material shall be visually clean and free from infestation
- Accessible circular hole in any rigid material
- Accessible clearance for moveable segment

References

[1] Akbar, J., Muhammad, A., and Irianto, D. (2015).*Measurement and Development of Humanware and Technoware Competenciesin Order to Meet Pintle Chain Product Requirements in Bandung Manufacture Polytechnic*, Proceedings of the APCOMS Conference 2015

[2] Cohen L. (1995).Quality Function Deployment: How to make QFD Work for You, Addison Wesley Publishing Company

[3] Gunawan, H. (2010). *Perancangan Model kaji Ulang Standard dengan pendekatan Quality Function Deployment (QFD) dan Berbasis Teknologi (Dimensi Technoware dan Humanware untuk SNI IEC 60335-2-7:2009).* Bandung: Tesis TMI-ITB

[4] Islam, S. and Irianto, D. (2013). Design of Assessment Model for Technoware and Humanware Readiness of Laboratory in accordance to Requirements in SNI ISO/IEC 17025, Proceedings of 7th National Industrial Engineering Conference

[5] Nurcahya,D.F.D., Akbar, M., andIrianto, D. (2014). *Assessing Technology Level of Industrial Estate to Meet standard of Environment*, Proceedings the 15th APIEMS Conference. ISBN:978-979-18925-0-6

[6] Pudjiastuti, U. (2010). *Kajian PengaruhPenggunaan Standard Nasional Indonesia di Industri*. Prosiding Pertemuan dan Persentasi Ilmiah Standardisasi. BSN

[7] Puspita, I. (2009). *Analisis Pengendalian Mutu untuk Mencapai Standard Kualitas Produkpada PT Central Power Indonesia*. Jurusan Management, Fakultas Ekonomi, Universitas Gunadarma

[8] Sinuhaji, L.H. (2012). *Perancangan Model Pengukuran Kompetensi Sumber Daya Manusia (SDM) Industri dalam Memenuhi Standard Produk: Kasus Kajian pada Persyaratan Umum Rumah Meteran Air: SNI 05-0666-1997*, Tesis TMI-ITB

[9] UNESCAP, (1989)Technology Atlas Project: A Framework for Technology Based Development, United Nations-Economic and Social Commission for Asia and the Pacific (UNESCAP), Bangalore India